Prealgebra

An Integrated Approach

Margaret L. Lial
American River College

Diana L. Hestwood
Minneapolis Community and
Technical College

Boston • San Francisco • New York
London • Toronto • Sydney • Tokyo • Singapore • Madrid
Mexico City • Munich • Paris • Cape Town • Hong Kong • Montreal

Publisher	Greg Tobin
Editor in Chief	Maureen O'Connor
Senior Project Editor	Lauren Morse
Assistant Editor	Caroline Case
Senior Managing Editor	Karen Wernholm
Senior Production Supervisor	Kathleen A. Manley
Design Direction (text and cover)	Dennis Schaefer
Photo Researcher	Beth Anderson
Digital Assets Manager	Marianne Groth
Media Producer	Christine Stavrou
Software Development	Janet Szykowny and Ted Hartman
Marketing Manager	Michelle Renda
Marketing Assistant	Alexandra Waibel
Senior Prepress Supervisor	Caroline Fell
Manufacturing Manager	Evelyn Beaton
Production Services	Pre-Press Company, Inc.
Cover Photo	Quiet Path © Copyright Lorraine Cota Manley

Photo Credits All photos from PhotoDisc, except the following: Thinkstock RF, pp. 1 left, 316, 386, 503, 536 right, 557 left; PictureQuest/Image Farm RF, pp.1 right, 27, 429; Corbis RF, pp. 13, 169 left, 169 right, 181 right, 263 right, 272 left; 323 top and bottom, 361 bottom left, 398, 513 left and right, 601 top left, 603 bottom left, 604 bottom right, 604 left, 609 right, 653, 794 right; GE Healthcare, p. 14; Leo Mason/Corbis, p. 23; Brand X Pictures, pp. 38 left, 83 bottom right, 91, 117, 354 right, 585 bottom left, 616 right; Beth Anderson, pp. 38 right, 47 right, 79, 92 left, 137 bottom left, 151, 321, 396 left, 544 right, 545 left, 601 top right, 601 bottom right, 602 left, 602 right, 603 top left, 604 bottom left, 608, 626 top right, 626 bottom right; Ralph White/Corbis, p. 48 right; Comstock RF, pp. 83, 91 left, 94, 160, 352 left; Hattie Young/Photo Researchers (PAL), p. 137 top right, Digital Vision, pp. 149, 360 right, 361 bottom right, 421, 439, 448 right, 517 right, 586 right, 649, 724 right; Bohemian Nomad Picturemakers/Corbis, p. 169 right; Wally McNamee/Corbis, p. 186 left; Laurel Beager, p. 186 right; Getty Images/Taxi, pp. 201, 282; Reuters/Corbis, pp. 209 right, 351, 448 left, 451, 784 right; Bettmann/Corbis, p. 237; Phillip Gould/Corbis, p. 238 right; Diana Hestwood, p. 250; Public domain, pp. 266, 546; NationBill/Corbis Sygma, p. 352 right; Dorling Kindersley, pp. 354 center, 794 left; NASA, pp. 396 right, 547 right, 577 chapter opener, 587, 723 right; Tony Freeman/Photo Edit, p. 402; Bob Daemmrich/Image Works, p. 412 right; Robert Brenner/PhotoEdit, p. 544 left; Blend Images (RF), p. 566 right; Martin Harvey/Corbis, p. 584 left; Dwi Oblo/Stringer/Reuters/Corbis, p. 584 right; Richard Carson/Reuters/Corbis, p. 585 right; Custom Medical Stock Photography, p. 603 bottom right; Theirry Orban/Corbis Sygma, p. 613; Michael Newman/PhotoEdit, p. 626 top left; Reprinted courtesy of Caterpillar, Inc., p. 628; U.S. Geological Survey, p. 737; Robbie Jack/Corbis, p. 784 left.

Many of the designations used by manufacturers and sellers to distinguish their products are claimed as trademarks. Where those designations appear in this book, and Addison-Wesley was aware of a trademark claim, the designations have been printed in initial caps or all caps.

Library of Congress Cataloging-in-Publication Data

Lial, Margaret L.
    Prealgebra : an integrated approach / Margaret L. Lial, Diana L. Hestwood.—1st ed.
        p. cm.
    Includes index.
    ISBN 0-321-35639-X (student edition)—
    1. Mathematics. I. Hestwood, Diana. II. Title.
    QA39.3.L535 2007
    510—dc22
                                    2006048388

3 4 5 6 7 8 9 10—QWD—09

Contents

Index of Chapter Group Activities

Preface

Prealgebra: An Integrated Approach continues our ongoing commitment to provide the best possible text and supplements package that will help instructors teach and students succeed. To that end, we have addressed the diverse needs of today's students through an attractive design, up-to-date applications and graphs, helpful features, careful explanation of concepts, and a comprehensive package of supplements and study aids.

This text is designed for developmental students who are new to algebra, relearning what they studied in the past, or anxious about their ability to learn mathematics. Many students see algebra as mysterious and unintelligible. Why use letters instead of numbers? Why solve equations? When will they ever use it? Even students who have completed a high school algebra course may not "get the point" of the course. By bridging the gap between arithmetic and algebra, *Prealgebra: An Integrated Approach* will help students answer these questions.

The authors' main objectives are based on what has worked well for their students over many years of teaching. The emphasis is on:

- Building a solid understanding of the foundation of algebra, why we use it, and how it works;

- Explaining important terminology in everyday English that is mathematically correct but written in a friendly style;

- Showing why things work the way they do;

- Tying the content to the students' experiences and previous knowledge;

- Providing well-designed, carefully sequenced exercises that build skills and understanding but are not overwhelming.

HALLMARK FEATURES

▶ *Scope and Sequence of Topics* Basic algebraic concepts form the backbone of this text, with arithmetic review and geometry topics integrated at appropriate points.

- The book gets students into algebra topics early, in Section 1.2 Using Integers, and continues with integer computation and order of operations in the rest of Chapter 1. Then Chapters 2 and 3 introduce variables, expressions, and equations in such a way that students see the need for variables, the usefulness of algebraic expressions and formulas, and the power of writing and solving equations. These ideas are revisited and reinforced throughout Chapters 4–9 as students work with fractions, decimals, proportions, percents, measurements, and graphs. Thus, the review of arithmetic is accomplished within the framework of basic algebra.

- An entire chapter (Chapter 3) is devoted to the most difficult algebra topic, solving application problems, rather than the usual treatment in a single section. This allows for the development of a six-step problem solving strategy that students can actually use. The six-step strategy is then reviewed and reinforced throughout the rest of the text, providing the strong foundation that students need for later mathematics courses.

- Geometry topics are integrated at points where they help introduce or reinforce algebraic concepts, or provide interesting real-world applications. Geometry topics include perimeter, area, volume, surface area, Pythagorean Theorem, lines and angles, and congruent and similar triangles.

- Whole number computation is placed in a Review chapter at the end of the text, to emphasize that these computational skills are prerequisites for the course. A diagnostic test precedes the Review chapter; students can assess their whole number computation skills and then work only the sections of the Review chapter that are needed.

▶ *A friendly, understandable, mathematically correct writing style* All too often, text-books are written in a rigorous, mathematically correct style that is satisfying to instructors but confusing and frustrating to the students in developmental mathematics courses. In this text, the authors speak to students as capable adults but understand that they are also novices with regard to algebra. The explanations and definitions are clear, complete, and written in "plain English." Mathematical vocabulary and concepts are introduced carefully, in the context of students' experiences and ideas. Vocabulary is used consistently with frequent, brief reminders of its meaning. Occasional mention of historical connections adds depth. Overall, this style results in a book that students can truly read, relate to, and understand.

▶ *Chapter Openers* The chapter openers feature real-world applications of mathematics that are relevant to students and tied to specific material within the chapters. Examples of topics include finding the best buy on cell phone service, weather, home improvements, driving, discounts and sales tax, fishing, and work/career applications. (See pp. 83, 151, and 201.)

▶ *Real-Life Applications* The authors are always on the lookout for interesting data to use in real-life applications. As a result, they have included many examples and exercises throughout the text that focus on real-life applications of mathematics. Students are often asked to find data in a table, chart, graph, or advertisement. (See pp. 249, 323, and 549–550.) These applied problems provide an up-to-date flavor that will appeal to and motivate students.

▶ *Figures and Photos* Today's students are more visually oriented than ever. Thus, the authors have made a concerted effort to include mathematical figures, diagrams, tables, and graphs whenever possible. Many of the graphs use a style similar to that seen by students in today's print and electronic media. Photos are incorporated to enhance applications in examples and exercises. (See pp. 107–109, 169, and 238.)

▶ *Emphasis on Problem Solving* Chapter 3 introduces students to the authors' six-step process for solving application problems algebraically: *Read, Assign a Variable, Write an Equation, Solve, State the Answer,* and *Check.* By devoting an entire chapter to this process, students build a strong foundation for problem solving, which is then reinforced through specific problem-solving examples in Chapters 4, 5, 6, 7, and 8. (See pp. 175, 182, and 278.)

▶ *Learning Objectives* Each section begins with clearly stated numbered objectives, and the material within sections is keyed to these objectives so that students know exactly what concepts are covered. (See pp. 119, 161, and 239.)

▶ *Cautions and Notes* These color-coded boxed comments warn students about common errors and emphasize important ideas throughout the exposition. (See pp. 122, 173, and 251.) Cautions are highlighted in yellow and Notes are highlighted in purple.

▶ *Teaching Tips and Classroom Examples* Located in the margins of the Annotated Instructor's Edition, the Teaching Tips are based on the authors' extensive classroom experience with students in this course. They provide the new or adjunct instructor with methods for handling common trouble spots, ideas for simple manipulatives to illustrate key geometry and measurement concepts, and suggestions for points to emphasize during lecture. The Classroom Examples are correlated to the worked examples throughout the text and provide carefully designed problems that instructors can use as they explain each skill or concept. (See pp. 98, 174, 230.)

Calculator Tips These optional tips, marked with a calculator icon, offer basic information and instruction for students using calculators in the course. (See pp. 53, 87, and 220.) In addition, an Introduction to Scientific Calculators is included in the student material that precedes Chapter 1.

Ample and Varied Exercise Sets The text contains a wealth of exercises to provide students with opportunities to practice, apply, connect, and extend the skills they are learning. Numerous illustrations, tables, graphs, and photos are included in the exercise sets to help students visualize the problems they are solving. Problem types include skill building, writing, estimation, and calculator exercises, as well as applications and correct-the-error problems. In the Annotated Instructor's Edition of the text, the writing exercises are marked with an icon so that instructors may assign these problems at their discretion. Exercises suitable for calculator work are marked in both the student and instructor editions with a calculator icon. (See pp. 102, 178, and 209.)

Relating Concepts Exercises These sets of exercises help students tie concepts together and develop higher level problem-solving skills as they compare and contrast ideas, identify and describe patterns, and extend concepts to new situations. (See pp. 14, 170, and 271.) These exercises make great collaborative activities for pairs or small groups of students.

Summary Exercises There are eight sets of in-chapter summary exercises: operations with integers, variables and expressions, perimeter and area, fractions, decimals, ratio/rate/ proportion, percents, and exponents. These exercises provide students with the all-important *mixed* practice they need at these critical points in their skill development. (See pp. 58, 265, and 451.)

Study Skills Component Poor study skills are a major reason why students do not succeed in mathematics. A few generic tips sprinkled here and there are not enough to help students change their behavior. So, in this text, a desk-light icon at key points in the text directs students to one of 12 activities in **Appendix A: Study Skills.** These carefully designed activities, correlated directly to the text material, cover such essential skills as using the textbook, note taking, completing homework, making study cards, test preparation, test taking, and preparing for a final exam. This unique appendix explains *how* the brain actually learns and remembers so students understand *why* the study skills activities will help them succeed in the course. Students are introduced to the appendix in "To the Student" on page xxvi. They should also complete **Activity 1: How Your Brain Learns** in Appendix A before starting Chapter 1.

Chapter Group Activities Each one- or two-page activity presents a relevant and in-depth look at how mathematics is used in the real world. Designed to help instructors answer the often-asked question, "When will I ever use this stuff?," these activities ask students to read and interpret data from newspaper articles, the Internet, and other familiar, real-world sources. (See pp. 137–38, 187, and 292.)

Test Your Word Power This feature, incorporated into each chapter summary, helps students understand and master mathematical vocabulary. Key terms from the chapter are presented along with four possible definitions in a multiple-choice format. Answers and examples illustrating each term are provided. (See pp. 72, 139, and 294.)

Ample Opportunity for Review Each chapter ends with a Chapter Summary featuring: Key Terms with definitions and helpful graphics, New Symbols, New Formulas, Test Your Word Power, and a Quick Review of each section's content with additional examples. Also included is a comprehensive set of Chapter Review Exercises keyed to individual sections, a set of Mixed Review Exercises, and a Chapter Test. Beginning with Chapter 2, each chapter concludes with a set of Cumulative Review Exercises. (See pp. 139, 189, and 293.)

Diagnostic Pretest A diagnostic pretest is included on p. xxv of the text and covers all the material in the book, much like a sample final exam. This pretest can be used to facilitate student placement in the correct chapter according to skill level. The pretest also exposes students to the scope of the course content.

▶ *Use of raised red negative signs in Chapters 1–3* A red dash in the raised position is used to designate negative numbers in Chapters 1–3; this makes a negative number visually distinct from the dash used to indicate the operation of subtraction. Traditional notation, using parentheses around negative numbers when necessary, is introduced in Chapter 4 and used throughout the rest of the text. The raised red negative sign helps students develop a clear understanding of the difference between a negative number and the subtraction symbol. It also helps students avoid errors related to "forgetting" the negative sign. The authors and reviewers have found that students make the transition to traditional notation without difficulty.

Chapters 1–3	Chapters 4–10
$^-7 - {}^-3$	$-7 - (-3)$

WHAT SUPPLEMENTS ARE AVAILABLE

For a comprehensive list of the supplements and study aids that accompany *Prealgebra: An Integrated Approach*, see pages x and xi.

Student Supplements	**Instructor Supplements**

Student's Solutions Manual
- Provides detailed solutions to the odd-numbered section-level exercises and to all margin, Relating Concepts, Summary, Chapter Review, Chapter Test, and Cumulative Review Exercises
- ISBN: 0-321-44163-X

Digital Video Tutor
- Complete set of digitized videos on CD-ROM for students to use at home or on campus
- Ideal for distance learning or supplemental instruction
- ISBN: 0-321-47119-9

MathXL® Tutorials on CD
- Provides algorithmically generated practice exercises that correlate at the objective level to the content of the text
- Every exercise is accompanied by an example and a guided solution, and selected exercises may also include a video clip
- The software recognizes student errors and provides feedback; it also tracks student activity and scores and can generate printed summaries of students' progress
- ISBN 0-321-43355-6

Addison-Wesley Math Tutor Center
- Staffed by qualified mathematics instructors
- Provides tutoring on examples and odd-numbered exercises from the textbook through a registration number with a new textbook or purchased separately
- Accessible via toll-free telephone, toll-free fax, e-mail, or the Internet at www.aw-bc/tutorcenter

InterAct Math Tutorial Website
- Get practice and tutorial help online!
- Retry an exercise as many times as you like with new values each time for unlimited practice and mastery
- Every exercise is accompanied by an interactive guided solution that gives you helpful feedback when an incorrect answer is entered
- View the steps of a worked-out sample problem similar to the one you're working on
- Visit www.interactmath.com

Videotape Series
- Features an engaging team of lecturers
- Provides comprehensive coverage of each section and topic in the text
  ISBN: 0-321-27929-8

Annotated Instructor's Edition
- Provides answers to all text exercises in color next to the corresponding problems
- Icons identify writing ✍ and calculator ▦ exercises
- ISBN: 0-321-42781-5

Instructor's Solutions Manual
- Provides complete solutions to all even-numbered section-level exercises
- ISBN: 0-321-43666-0

Adjunct Support Manual
- Includes resources designed to help both new and adjunct faculty with course preparation and classroom management
- Offers helpful teaching tips correlated to the sections of the text
- ISBN: 0-321-27924-7

Printed Test Bank and Instructor's Resource Guide
- The test bank contains two diagnostic pretests, six free-response and two multiple-choice test forms per chapter, and two final exams
- The resource guide contains additional practice exercises for most objectives of every section and other helpful resources for instructors
- ISBN: 0-321-27932-8

TestGen
- Enables instructors to build, edit, print, and administer tests
- Features a computerized bank of questions developed to cover all text objectives
- Available on a dual-platform Windows/Macintosh CD-ROM
- ISBN: 0-321-38396-6

PowerPoint Lecture Slides
- Present key concepts and definitions from the text
- Provide additional exercise examples for use during lectures

Adjunct Support Center
The Math Adjunct Support Center is staffed by qualified mathematics instructors with over 50 years combined experience at both the community college and university level. Assistance is provided for faculty in the following areas:
- Suggested syllabus consultation
- Tips on using materials packaged with your book
- Book-specific content assistance
- Teaching suggestions including advice on classroom strategies

For more information visit
www.aw-bc.com/tutorcenter/math-adjunct.html

MathXL®: www.mathxl.com MathXL® is a powerful online homework, tutorial, and assessment system that accompanies your Addison-Wesley textbook in mathematics or statistics. With MathXL, instructors can create, edit, and assign online homework and tests using algorithmically generated exercises correlated at the objective level to the textbook. They can also create and assign their own online exercises and import TestGen tests for added flexibility. All student work is tracked in MathXL's online gradebook. Students can take chapter tests in MathXL and receive personalized study plans based on their test results. The study plan diagnoses weaknesses and links students directly to tutorial exercises for the objectives they need to study and retest. Students can also access supplemental video clips directly from selected exercises. MathXL is available to qualified adopters. For more information, visit our Web site at www.mathxl.com, or contact your Addison-Wesley sales representative.

MyMathLab MyMathLab is a series of text-specific, easily customizable online courses for Addison-Wesley textbooks in mathematics and statistics. MyMathLab is powered by CourseCompass™—Pearson Education's online teaching and learning environment—and by MathXL®—our online homework, tutorial, and assessment system. MyMathLab gives instructors the tools they need to deliver all or a portion of their course online, whether students are in a lab setting or working from home. MyMathLab provides a rich and flexible set of course materials, featuring free-response exercises that are algorithmically generated for unlimited practice and mastery. Students can also use online tools, such as video lectures, animations, and a multimedia textbook, to independently improve their understanding and performance. Instructors can use MyMathLab's homework and test managers to select and assign online exercises correlated directly to the textbook, and they can also create and assign their own online exercises and import TestGen tests for added flexibility. MyMathLab's online gradebook—designed specifically for mathematics and statistics—automatically tracks students' homework and test results and gives the instructor control over how to calculate final grades. Instructors can also add offline (paper-and-pencil) grades to the MathXL gradebook. MyMathLab is available to qualified adopters. For more information, visit our Web site at www.mymathlab.com or contact your Addison-Wesley sales representative.

ACKNOWLEDGMENTS

The comments, criticisms, and suggestions of instructors and students have positively shaped this textbook. We especially wish to thank these individuals who provided invaluable suggestions:

Carla Ainsworth, *Salt Lake Community College*
Barbara Brown, *Anoka-Ramsey Community College*
Kim Brown, *Tarrant County College—Northeast Campus*
Hien Bui, *Hillsborough Community College*
John Close, *Salt Lake Community College*
Jane Cuellar, *Taft College*
Ky Davis, *Muskingum Area Technical College*
Randy Gallaher, *Lewis and Clark Community College*
Rosemary Karr, *Collin County Community College*
Lou Ann Mahaney, *Tarrant County College—Northeast Campus*
Pam Miller, *Phoenix College*
Elizabeth Morrison, *Valencia Community College—West Campus*
Faith Peters, *Broward Community College*
Manoj Raghunandanan, *Temple University*
Janalyn Richards, *Idaho State University*
Julia Simms, *Southern Illinois University—Edwardsville*
Sounny Slitine, *Palo Alto College*
Sharon Testone, *Onondaga Community College*
Shae Thompson, *Montana State University*
Cheryl Wilcox, *Diablo Valley*
Kevin Yokoyama, *College of the Redwoods*
Karl Zilm, *Lewis and Clark Community College*

Our sincere thanks go to these dedicated individuals at Addison-Wesley who worked long and hard to make this book a success: Maureen O'Connor, Lauren Morse, Caroline Case, Kathy Manley, Dennis Schaefer, Michelle Renda, Alexandra Waibel, Christine Stavrou, and Beth Anderson. Sam Blake of Pre-Press Company, Inc. provided his customary excellent production work. We are most grateful to Peg Crider for assistance in researching and writing the Chapter Group Activities; Robie Grant for her accurate and useful index; Abby Tanenbaum for writing the Diagnostic Pretest; and Janis Cimperman, Sharon O'Donnell, Gary Williams, and Steve Ouellette for accuracy checking the manuscript.

Much of the material in Chapter 10, Exponents and Polynomials, was adapted from *Introductory Algebra*, Eighth Edition, by Lial, Hornsby, and McGinnis. Many thanks to Marge, John, and Terry for their contribution.

Special thanks also go to Barbara Brown, *Anoka-Ramsey Community College*, who reviewed all the material for this book in great detail. Her expertise in teaching developmental algebra students resulted in greater clarity, consistency, and creativity. Finally, it is the wonderful collaboration with Linda Russell, reading and study skills specialist at *Minneapolis Community and Technical College*, which resulted in the Study Skills Appendix.

The ultimate measure of this textbook's success is whether it helps students master algebra skills, develop problem-solving techniques, and increase their confidence in learning and using mathematics. Please let us know how we are doing by sending an e-mail to math@awl.com.

This book is dedicated to my husband, Earl Orf, who is the wind beneath my wings, and to my students at Minneapolis Community and Technical College, from whom I have learned so much.

Diana L. Hestwood

Feature
Walk-Through

Chapter Openers Chapter openers feature real-world applications of mathematics that are relevant to students and tied to specific material within the chapters (page 1).

The weather—we all talk about it and often complain about it. When reporting temperatures, we need both *positive* and *negative* numbers. If you live in Chicago, the highest temperature ever recorded is 104 °F and the lowest is −27 °F, a difference of 131 degrees! In Atlanta, the record high and low are 105 °F and −8 °F. And in Barrow, Alaska, the range of extreme temperatures is 79 °F to −56 °F, a difference of 135 degrees. To make things even more uncomfortable, humidity can make hot temperatures feel hotter and wind can make cold temperatures feel colder. Learn more about "windchill" as you work Exercises 47–48 in **Section 1.4**. (*Source:* National Climatic Data Center.)

1

---

22 Chapter 1 Introduction to Algebra: Integers

46. $^-5 + 2 + 5$	**47.** $^-7 + 6 + {}^-4$	**48.** $^-9 + 8 + {}^-2$
49. $^-3 + {}^-11 + 14$	**50.** $15 + {}^-7 + {}^-8$	**51.** $10 + {}^-6 + {}^-3 + 4$
52. $2 + {}^-1 + {}^-9 + 12$	**53.** $^-7 + 28 + {}^-56 + 3$	**54.** $4 + {}^-37 + 29 + {}^-5$

Write an addition problem for each situation and find the sum.

55. The football team gained 13 yards on the first play and lost 17 yards on the second play. How many yards did the team gain or lose in all?

56. At penguin breeding grounds on Antarctic islands, temperatures routinely drop to $^-15$ °C. Temperatures in the interior of the continent may drop another 60 °C below that. What is the temperature in the interior?

57. Nick's checking account was overdrawn by $62. He deposited $50 in his account. What is the balance in his account?

58. Cynthia had $100 in her checking account. She wrote a check for $83 and was charged $17 for overdrawing her account last month. What is her account balance?

59. $88 was stolen from Jay's car. He got $35 of it back. What was his net loss?

60. Marion lost 4 pounds in April, gained 2 pounds in May, and gained 3 pounds in June. How many pounds did she gain or lose in all?

61. Use the score sheet to find each player's point total after three rounds in a card game.

	Jeff	Terry
Round 1	Lost 20 pts	Won 42 pts
Round 2	Won 75 pts	Lost 15 pts
Round 3	Lost 55 pts	Won 20 pts

62. Use the information in the table on flood water depths to find the new flood level for each river.

	Red River	Mississippi
Monday	Rose 8 ft	Rose 4 ft
Tuesday	Fell 3 ft	Rose 7 ft
Wednesday	Fell 5 ft	Fell 13 ft

Figures and Photos Today's students are more visually oriented than ever. Thus, a concerted effort has been made to use mathematical figures, diagrams, tables, and graphs whenever possible. Photos are incorporated to enhance applications in examples and exercises (page 22).

Relating Concepts   These sets of exercises help students tie together topics and develop problem-solving skills as they compare and contrast ideas, identify and describe patterns, and extend concepts to new situations. These exercises make great collaborative activities for pairs or small groups of students (page 118).

118   Chapter 2 Understanding Variables and Solving Equations

65. During the summer months, Ernesto spends an average of only $45 per month on parking fees by riding his bike to work on nice days. This is $65 less per month than what he spends for parking in the winter. Solving the equation $p - 65 = 45$ will give you his monthly parking fees in the winter. Solve the equation.

66. By walking to work several times a week in the summer, Aimee spends an average of $56 less per month on parking fees. If she spends $98 per month on parking in the summer, solve the equation $p - 56 = 98$ to find her monthly parking fees in the winter.

Solve each equation. Show your work.

67. $^-17 - 1 + 26 - 38 = ^-3 - m - 8 + 2m$

68. $19 - 38 - 9 + 11 = ^-t - 6 + 2t - 6$

69. $^-6x + 2x + 6 + 5x = |0 - 9| - |^-6 + 5|$

70. $^-h - |^-9 - 9| + 8h - 6h = ^-12 - |^-5 + 0|$

RELATING CONCEPTS (EXERCISES 71–72) For Individual or Group Work

Use what you have learned about solving equations to **work Exercises 71 and 72 in order.**

71. (a) Write two *different* equations that have $^-2$ as the solution. Be sure that you have to use the *addition property of equality* to solve the equations. Show how to solve each equation. Use Exercises 7 to 22 as models.

  (b) Follow the directions in part (a), but this time write two equations that have 0 as the solution.

72. Not all equations have solutions that are integers. Try solving these equations.

  (a) $x + 1 = 1\frac{1}{2}$

  (b) $\frac{1}{4} = y - 1$

  (c) $\$2.50 + n = \$3.35$

  (d) Write two more equations that have fraction or decimal solutions.

Chapter 2 Group Activity   **137**

Chapter 2 Group Activity: REAL-WORLD EXPRESSIONS

Objective   Write, use, and interpret algebraic expressions that model real-world events.

A college with four campuses uses an auditorium for graduation that has 1500 seats. Each campus hosts its own graduation ceremony. Students are allocated a whole number of tickets. Round each result **down** to the nearest whole number.

1. How many tickets are allocated to each of 458 graduates at the **North** Campus graduation?
2. How many tickets are allocated to each of 297 graduates at the **East** Campus graduation?
3. How many tickets are allocated to each of 315 graduates at the **South** Campus graduation?
4. How many tickets are allocated to each of 186 graduates at the **West** Campus graduation?
5. Write an *expression* that represents the number of tickets that are allocated to each of *g* graduates.

6. What recommendations do you have for unallocated tickets?

Traffic engineers have to decide how long to have the red, yellow, and green lights showing on a traffic signal. To decide the number of seconds that a yellow light should be on, the engineers use the expression

$$\frac{5v}{100} + 1$$

where *v* is the speed limit in miles per hour (mph).

7. How many seconds should the yellow light be on if the speed limit is 20 mph? 40 mph? 60 mph?

8. Based on the answers you just calculated, how could you estimate the time for a yellow light if the speed limit is 30 mph? 50 mph?

9. Use the given expression to find the number of seconds that the yellow light should be on if the speed limit is 30 mph and 50 mph. Did you get the same result as in Problem 8?

To estimate the number of words in a child's vocabulary, a speech therapist uses the expression $60A - 900$ where *A* is the child's age in months.

10. Estimate the number of words that a child aged 20 months knows.

11. Estimate the number of words that a child aged 2 years knows. (*Hint:* How many months are in two years?)

12. How many words does a child learn between the ages of 20 months and 2 years?

13. Estimate the number of words in the vocabulary of a 3-year-old child.

14. How many words does a child learn between the ages of 2 years and 3 years?

15. Evaluate the expression for a child who is 15 months old. Do you think that the answer is reasonable? Explain why or why not.

16. Evaluate the expression for a child who is 12 months old. Do you think that the answer makes sense? Explain why or why not.

Group Activities   These one- or two-page activities, found near the end of each chapter, present even more relevant and in-depth looks at how mathematics is used in the real world. Designed to help instructors answer the often-asked question, "When will I ever use this stuff?," these activities ask students to read and interpret data from newspaper articles, the Internet, and other familiar, real sources. The activities are well suited to collaborative work and can also be completed by individuals or used for open-ended class discussions (page 137).

---

220   Chapter 4 Rational Numbers: Positive and Negative Fractions

Calculator Tip   You can use your calculator to find the prime factorization of a number. Here is an example that uses 539.

Try dividing 539 by the first prime number, 2.

539 ⊘ 2 ⊜ **269.5**   Does not divide evenly

Try dividing 539 by the next prime number, 3.

539 ⊘ 3 ⊜ **179.6666667**   Does not divide evenly

Keep trying the next prime numbers until you find one that divides evenly.

539 ⊘ 5 ⊜ **107.8**   Does not divide evenly

539 ⊘ 7 ⊜ **77**   Divides evenly

Once you have found that 7 works, try using it again on the quotient 77.

77 ⊘ 7 ⊜ **11**   Divides evenly

Because 11 is prime, you're finished. The prime factorization of 539 is 7 • 7 • 11.

Now try factoring 2431 using your calculator.

(The answer is eleven times thirteen times seventeen.)

OBJECTIVE 4 Write a fraction in lowest terms, using prime factorization.   Now you can use the second method for writing fractions in lowest terms: prime factorization. This is a good method to use when the numerator and denominator are larger numbers or include variables.

EXAMPLE 6   Using Prime Factorization to Write Fractions in Lowest Terms

(a) Write $\frac{20}{35}$ in lowest terms.

20 can be written as 2 • 2 • 5 ← Prime factors
35 can be written as 5 • 7 ← Prime factors

$$\frac{20}{35} = \frac{2 \cdot 2 \cdot 5}{5 \cdot 7}$$

The numerator and denominator have 5 as a common factor. Dividing both numerator and denominator by 5 will give an equivalent fraction.

Any number divided by itself is

Calculator Tips   These optional tips, marked with red calculator icons, offer basic information and instruction for students using calculators in the course (page 220).

End-of-Chapter Material One of the most admired features of the Lial textbooks is the extensive and well-thought-out end of chapter material. At the end of each chapter, students will find:

Key Terms Terms are listed, defined, and referenced back to the appropriate section number (page 139).

Test Your Word Power To help students understand and master mathematical vocabulary, Test Your Word Power has been incorporated in each Chapter Summary. Students are quizzed on Key Terms from the chapter in a multiple-choice format. Answers and examples illustrating each term are provided (page 139).

Quick Review These sections give students not only the main concepts from the chapter (referenced back to the appropriate section), but also an adjacent example of each concept (page 140).

Review Exercises These exercises are keyed to the appropriate sections so that students can refer to examples of that type of problem if they need help (page 145).

Chapter 2 — SUMMARY

KEY TERMS

2.1
- **variable** A variable is a letter that represents a number that varies or changes, depending on the situation.
- **constant** A constant is a number that is added or subtracted in an expression. It does not vary. For example, 5 is the constant in the expression $c + 5$.
- **expression** An expression expresses, or tells, the rule for doing something. It is a combination of operations on variables and numbers. An expression does *not* have an equal sign.
- **evaluate the expression** To evaluate an expression, replace each variable with specific values (numbers) and follow the order of operations.
- **coefficient** The number part in a multiplication expression is the coefficient. For example, 4 is the coefficient in the expression $4s$.

2.2
- **simplifying expressions** To simplify an expression, write it in a simpler way by combining all the like terms.
- **term** Each addend in an expression is a term.
- **variable term** A variable term has a number part (called the coefficient) multiplied by a variable part (a letter). An example is $4s$.
- **like terms** Like terms are terms with exactly the same variable parts (the same letters and exponents). The coefficients may be different.

2.3
- **equations** An equation has an equal sign. It shows the relationship between what is known about a problem and what isn't known.
- **solve the equation** To solve an equation, find a number that can replace the variable and make the equation balance.
- **solution** A solution of an equation is a number that can replace the variable and make the equation balance.
- **addition property of equality** The addition property of equality states that adding the same quantity to both sides of an equation will keep it balanced.
- **check the solution** To check the solution of an equation, go back to the *original* equation and replace the variable with the solution. If the equation balances, the solution is correct.

2.4
- **division property of equality** The division property of equality states that dividing both sides of an equation by the same nonzero number will keep it balanced.

TEST YOUR WORD POWER

See how well you have learned the vocabulary in this chapter. Answers follow the Quick Review.

1. A **variable**
   - A. can only be the letter x
   - B. is never an addend in an expression
   - C. is the solution of an equation
   - D. represents a number that varies.

2. Which expression has 2 as a **coefficient**?
   - A. x^2
   - B. $2x$
   - C. $x + 2$
   - D. $2 - x$

3. Which expression has 4 as a **constant** term?
   - A. $4y$
   - B. y^4
   - C. $4 + y$
   - D. $\dfrac{y}{4}$

4. Which expression has four **terms**?
   - A. $2 + 3x = {}^{-}6 + x$
   - B. $2 + 3x - 6 + x$
   - C. $(2)(3x)({}^{-}6)(x)$
   - D. $2(3x) = {}^{-}6(x)$

5. **Like terms**
   - A. can be multiplied but not added
   - B. have the same coefficients
   - C. have the same solutions
   - D. have the same variable parts.

6. To **simplify an expression,**
   - A. combine all the like terms
   - B. multiply the exponents
   - C. add all the numbers in the expression
   - D. add the same quantity to both sides.

---

140 Chapter 2 Understanding Variables and Solving Equations

QUICK REVIEW

Concepts	Examples
2.1 Evaluating Expressions Replace each variable with the specified value. Then follow the order of operations to simplify the expression.	The expression for ordering textbooks for two prealgebra classes is $2c + 10$, where c is the class limit. Evaluate the expression when the class limit is 24. $2c + 10$ Replace c with 24. $2 \cdot 24 + 10$ Multiply first. $48 + 10$ Add last. 58 Order 58 books.
Using Exponents with Variables The exponent next to a variable tells how many times to use the variable as a factor in multiplication.	Rewrite ${}^{-}6x^4$ without exponents. ${}^{-}6x^4$ can be written as ${}^{-}6 \cdot x \cdot x \cdot x \cdot x$ Coefficient is ${}^{-}6$ x is used as a factor 4 times.
Evaluating Expressions with Exponents To write the expression without exponents, replace each variable with the specified value, and multiply all the factors.	Evaluate x^3y when x is ${}^{-}4$ and y is 5. x^3y means $x \cdot x \cdot x \cdot y$ Replace x with ${}^{-}4$ and y with 5. ${}^{-}4 \cdot {}^{-}4 \cdot {}^{-}4 \cdot 5$ Multiply two factors at a time. $16 \cdot {}^{-}4 \cdot 5$ ${}^{-}64 \cdot 5$

Chapter 2 Review Exercises **145**

Chapter 2 — REVIEW EXERCISES

[2.1]

1. (a) Identify the variable, the coefficient, and the constant in this expression.
   $${}^{-}3 + 4k$$

   (b) Circle the expression that has 20 as the constant term and ${}^{-}9$ as the coefficient.
   $$20m - 9 \qquad {}^{-}9 + 20x$$
   $${}^{-}9y + 20 \qquad 20 + 9n$$

2. The expression for ordering test tubes for a chemistry lab is $4c + 10$, where c is the class limit. Evaluate the expression when
   (a) the class limit is 15.

   (b) the class limit is 24.

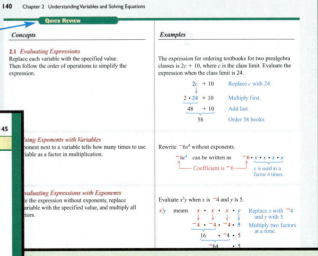

3. Rewrite each expression without exponents.
   (a) x^2y^4
   (b) $5ab^3$

4. Evaluate each expression when m is 2, n is ${}^{-}3$, and p is 4.
   (a) n^2
   (b) n^3
   (c) ${}^{-}4mp^2$
   (d) $5m^4n^2$

[2.2] Simplify.

5. $ab + ab^2 + 2ab$

6. ${}^{-}3x + 2y - x - 7$

7. ${}^{-}8({}^{-}2g^3)$

8. $4(3r^2t)$

9. $5(k + 2)$

10. ${}^{-}2(3b + 4)$

11. $3(2y - 4) + 12$

12. ${}^{-}4 + 6(4x + 1) - 4x$

13. Write an expression with four terms that *cannot* be simplified.

[2.3] *Solve each equation and check each solution.*

Mixed Review Exercises These exercises require students to solve problems without the help of section reference (page 146).

146   Chapter 2   Understanding Variables and Solving Equations

19. $-2p + 5p = 3 - 21$ **20.** $-30 = 3(-5r)$ **21.** $12 = -h$

[2.5] *Solve each equation. Show your work.*

22. $12w - 4 = 8w + 12$ **23.** $0 = -4(c + 2)$

24. Every Wednesday is "treat day" at the office. The person who brings the treats buys two treat items for each employee plus four extras. If Roger brings 34 doughnuts, solve the equation $34 = 2n + 4$ to find the number of employees.

MIXED REVIEW EXERCISES

Solve each equation. Show your work.

25. $12 + 7a = 4a - 3$ **26.** $-2(p - 3) = -14$ **27.** $10y = 6y + 20$

28. $2m - 7m = 5 - 20$ **29.** $20 = 3x - 7$ **30.** $b + 6 = 3b - 8$

31. $z + 3 = 0$ **32.** $3(2n - 1) = 3(n + 3)$ **33.** $-4 + 46 = 7(-3$

Study Skills Component A desk-light icon at key points in the text directs students to **Appendix A: Study Skills,** which contains activities correlated directly to the text. This material explains how the brain actually learns, so students understand *why* the study tips presented will help them succeed in the course.

Chapter Test This test helps students practice for the real thing (page 147).

Chapter 2 Test   **147**

Chapter 2

TEST

Appendix A: Study Skills
Activity 8: Test Preparation

1. Identify the parts of this expression: $-7w + 6$ Choose from these labels: variable, constant, coefficient.

2. The expression for buying hot dogs for the company picnic is $3a + 2c$, where a is the number of adults and c is the number of children. Evaluate the expression when there are 45 adults and 21 children.

Rewrite each expression without exponents.

3. $x^5 y^3$ **4.** $4ab^4$

5. Evaluate $-2s^2 t$ when s is -5 and t is 4.

Simplify each expression.

6. $3w^3 - 8w^3 + w^3$ **7.** $xy - xy$

8. $-6c - 5 + 7c + 5$ **9.** $3m^2 - 3m + 3mn$

10. $-10(4b^2)$ **11.** $-5(-3k)$

12. $7(3t + 4)$ **13.** $-4(a + 6)$

1. _____
2. _____
3. _____
4. _____
5. _____
6. _____
7. _____
8. _____
9. _____
10. _____
11. _____
12. _____
13.

Cumulative Review Exercises These exercises gather various types of exercises from preceding chapters to help students remember and retain what they are learning throughout the course (page 305).

Cumulative Review Exercises: Chapters 1–4   **305**

Cumulative Review Exercises

CHAPTERS 1–4

1. Write this number in words.
505,008,238

2. Write this number using digits.
thirty-five billion, six hundred million, nine hundred sixteen

3. (a) Round 60,719 to the nearest hundred.

(b) Round 99,505 to the nearest thousand.

(c) Round 3206 to the nearest ten.

4. Name the property illustrated by each example.
(a) $-3 \cdot 6 = 6 \cdot (-3)$

(b) $(7 + 18) + 2 = 7 + (18 + 2)$

(c) $5(-10 + 7) = 5 \cdot (-10) + 5 \cdot 7$

Simplify.

5. $9(-6)$ **6.** $-10 - 10$ **7.** $\dfrac{-14}{0}$ **8.** $6 + 3(2 - 7)^2$

9. $|-8| + |2|$ **10.** $-8 + 24 \div 2$ **11.** $(-4)^2 - 2^5$ **12.** $\dfrac{-45 \div (-5)(3)}{-5 - 4(0 - 8)}$

13. $-3 \div \dfrac{3}{8}$ **14.** $-\dfrac{5}{6}(-42)$ **15.** $\left(\dfrac{5a^2}{12}\right)\left(\dfrac{18}{a}\right)$ **16.** $\dfrac{7}{x^2} \div \dfrac{7y^2}{3x}$

17. $\dfrac{3}{10} - \dfrac{5}{6}$ **18.** $-\dfrac{3}{8} + \dfrac{11}{16}$ **19.** $\dfrac{2}{3} - \dfrac{b}{7}$ **20.** $\dfrac{8}{5} + \dfrac{3}{n}$

An Introduction to Scientific Calculators

Scientific Calculators

OBJECTIVES

1. Use the arithmetic computation keys.
2. Understand the "clear" and "clear entry" keys.
3. Understand the floating decimal point.
4. Use the %️ (percent) key.
5. Use the x^2️ (squaring) and $\sqrt{x}$️ (square root) keys.
6. Use the y^x️ (exponent or power) key.
7. Use the $a^{b/c}$️ fraction) key.
8. Solve problems with negative numbers.
9. Use the calculator memory function.
10. Solve chain calculations using the order of operations.
11. Use the parentheses keys.

Calculators are among the more popular inventions of the last four decades. Each year better calculators are developed and costs drop. One of the first consumer models available was the Texas Instruments SR-10, which sold for about $150 in 1973. It could perform the four operations of arithmetic and take square roots, but could do very little more. Today, a $5 calculator the size of a credit card can do all these operations, plus provide percent and memory functions.

Calculators now come in a large array of different types, sizes, and prices. **For the course for which this textbook is intended, the most appropriate type is the scientific calculator,** which costs $10–$20.

In this introduction, we explain some of the more common features of scientific calculators. However, remember that calculators vary among manufacturers and models, and that while the methods explained here apply to many of them, they may not apply to your specific calculator. For this reason, it is important to remember that *this introduction is only a guide and is not intended to take the place of your owner's manual.*

CAUTION
Be sure to keep the instruction booklet supplied with your calculator! Keys and functions vary from model to model, so always refer to the instruction booklet for details.

OBJECTIVE 1 Use the arithmetic computation keys. Most calculators use *algebraic logic* (such as models sold by Texas Instruments, Sharp, Casio, and Radio Shack). For example, enter 14 + 28 in the order it is written.

14 +️ 28 =️   Answer is 42

Enter 387 − 62 as follows.

387 −️ 62 =️   Answer is 325

If your calculator does *not* work problems in this way, check its instruction book to see how to proceed. (For example, Hewlett-Packard calculators use RPN, Reverse Polish Notation.)

OBJECTIVE 2 Understand the "clear" and "clear entry" keys. All calculators have a C️, ON/C️, or ON/AC️ key. Pressing this key erases everything in the calculator and prepares the calculator to begin a new problem. Some calculators also have a CE️ key. Pressing this key erases *only* the number displayed and allows you to correct a mistake without having to start the problem over.

Many calculators combine the Ⓒ key and the Ⓒᴱ key into an (ON/C) key. This key turns the calculator on and is also used to erase the calculator display. If the (ON/C) key is pressed after the ⊜ or one of the operation keys (⊕, ⊖, ⊗, ⊘), everything in the calculator is erased. If you happen to press the wrong operation key, immediately press the correct key to cancel the error. For example, 7 ⊕ ⊖ 3 ⊜ 4. Pressing the ⊖ key cancels out the previous ⊕ key entry.

> **CAUTION**
> Be sure to look at the directions that come with your calculator to see how to clear the memory.

OBJECTIVE 3 Understand the floating decimal point.   Most calculators have a *floating decimal* that locates the decimal point in the final result. For example, to buy 55.75 square yards of carpet at $18.99 per square yard, proceed as follows.

55.75 ⊗ 18.99 ⊜   Answer is 1058.6925

The decimal point is automatically placed in the answer. You usually *round* money answers to the nearest cent. Start by drawing a cut-off line after the hundredths place.

Look only at the first digit being cut off.
↓
1058.69│25
↑
Cent position (hundredths)

Because the first digit being cut off is *4 or less,* the part you are keeping stays the same. The answer is rounded to $1058.69. If the first digit being cut off had been *5 or more,* you would have rounded up by adding 1 to the cent position (see **Section 5.2**).

When using a calculator with a floating decimal, enter the decimal point as needed. For example, enter $47 by pressing 47 with no decimal point, but enter 95¢ as ⊙ 95 with a decimal point.

When you add $21.38 and $1.22, the answer is $22.60, but the calculator does *not* show the final 0.

21.38 ⊕ 1.22 ⊜   Answer is 22.6

You must remember that the problem dealt with money and write the final 0, making the answer $22.6**0**.

OBJECTIVE 4 Use the ⊗ (percent) key.   The ⊗ key moves the decimal point two places to the left when pressed following multiplication or division. Calculate 8% of $4205 as follows.

4205 ⊗ 8 ⊗ ⊜   Answer is 336.4

Because the problem involved money, write the answer as $336.4**0**.

OBJECTIVE 5 Use the (x²) (squaring) and the (√x) (square root) keys.   The squaring key, (x²), squares the number in the display (multiplies the number by itself). For example, find 7^2 (which means 7×7) as shown below. You do *not* need to press the ⊜ key.

7 (x²)   Answer is 49

The logic of the order of operations is built into most scientific calculators. To check your calculator, try entering $3 + 5 \times 2$ in the order it is written.

3 ⊕ 5 ⊗ 2 ⊜   Answer should be 13

Using Order of Operations

$3 + \underbrace{5 \times 2}$ Multiply first.

$\underbrace{3 + \quad 10}$ Then add.

13 ⟵——— Correct

Working from Left to Right

$3 + 5 \times 2$

$\underbrace{8 \quad} \times 2$

16 ⟵——— **Incorrect**

If your calculator uses the order of operations, it will automatically multiply 5×2 *before* adding 3. If your calculator gives the *incorrect* answer of 16, it does *not* follow the order of operations. You would have to know that multiplication is done ahead of addition and enter the problem as 5 ⊗ 2 ⊕ 3 ⊜ to get the correct result.

> **CAUTION**
> *Scientific* calculators keep track of the order of operations for us. However, the basic four-function calculator is *not* programmed to observe the order of operations and will calculate correctly *only* if you enter numbers in the proper order.

OBJECTIVE 11 Use the parentheses keys.   The parentheses keys allow you to group numbers in a chain calculation. For example, $\dfrac{24}{5 + 7}$ can be written with parentheses in the denominator as $\dfrac{24}{(5 + 7)}$ and entered as follows.

Left parentheses key

24 ÷ (5 + 7) =   Answer is 2

Right parentheses key

Without the parentheses the calculator would have automatically divided 24 by 5 *before* adding 7, giving an *incorrect* answer of 11.8.

Here is a more complicated example.

Solve $\dfrac{16 - 2.5}{55 - 29.4 \div 0.6}$ by writing it as $\dfrac{(16 - 2.5)}{(55 - 29.4 \div 0.6)}$

Use parentheses to keep the numerator separated from the denominator.

(16 − 2.5) ÷ (55 − 29.4 ÷ .6) =   Answer is 2.25

Numerator   Denominator

To the Student: Success in Mathematics

There are two main reasons why students have difficulty with mathematics:

- Students start in a course for which they do not have the necessary background knowledge.

- Students don't know how to study mathematics effectively.

Your instructor can help you decide whether this is the right course for you. We can give you some study tips.

Studying mathematics *is* different from studying subjects like English and history. The key to success is regular practice. This should not be surprising. After all, can you learn to play the piano or ski well without a lot of regular practice? The same is true for learning mathematics. Working problems nearly every day is the key to becoming successful. Here is a list of things that will help you succeed in studying mathematics.

1. *Attend class regularly.* Pay attention to what your instructor says and does in class, and take careful notes. In particular, note the problems the instructor works on the board and copy the complete solutions. Keep these notes separate from your homework to avoid confusion when you review them later.

2. Don't hesitate to *ask questions in class.* It is not a sign of weakness but of strength. There are always other students with the same question who are too shy to ask.

3. *Read your text carefully.* Many students read only enough to get by, usually only the examples. Reading the complete section will help you solve the homework problems. Most exercises are keyed to specific examples or objectives that will explain the procedures for working them.

4. Before you start on your homework assignment, *rework the problems the teacher worked in class.* This will reinforce what you have learned. Many students say, "I understand it perfectly when you do it in class, but I get stuck when I try to do the homework problems by myself."

5. Do your homework assignment only *after reading the text* and reviewing your notes from class. Check your work against the answers in the back of the book. If you get a problem wrong and are unable to understand why, mark that problem and ask your instructor about it. Then practice working additional problems of the same type to reinforce what you have learned.

6. *Work as neatly as you can.* Write your symbols clearly, and make sure the problems are clearly separated from each other. Working neatly will help you to think clearly and also make it easier to review the homework before a test.

7. After you complete a homework assignment, *look over the text again.* Try to identify the main ideas that are in the lesson. Often they are clearly highlighted or boxed in the text.

8. *Use the chapter test at the end of each chapter as a practice test.* Work through the problems under test conditions, without referring to the text or the answers until you are finished. You may want to time yourself to see how long it takes you. When you finish, check your answers against those in the back of the book, and rework the problems you missed.

9. *Keep all quizzes and tests that are returned to you,* and use them when you study for future tests and the final exam. These quizzes and tests indicate what concepts your instructor considers to be most important. Be sure to correct any problems that you missed on these tests, so you will have the corrected work to study.

10. *Don't worry if you do not understand a new topic right away.* As you read more about it and work through the problems, you will gain understanding. Each time you review a topic, you will understand it a little better. Few people understand each topic completely right from the start.

NOTE

Reading a list of study tips is a good start, but you may need some help actually *applying* the tips to your work in this math course.

 Watch for the desk-light icon as you work in this textbook, particularly in the first few chapters. It will direct you to one of 12 activities in **Appendix A: Study Skills** near the end of the book. Each activity helps you to actually *use* a study skills technique. These techniques will greatly improve your chances for success in this course.

- Find out *how your brain learns new material.* Then use that information to set up effective ways to learn math.

- Find out *why short-term memory is so short* and what you can do to help your brain remember new material weeks and months later.

- Find out *what happens when you "blank out" on a test* and simple ways to prevent it from happening.

All the activities in **Appendix A: Study Skills** are brain-friendly ways to enjoy and succeed at math. Whether you need help with note taking, managing home-work, taking tests, or preparing for a final exam, you'll find specific, clearly explained ideas that really work because they're based on research about how the brain learns and remembers.

CAUTION

Be sure to complete **Activity 1: Your Brain** in **Appendix A** before starting Chapter 1. You'll learn the basics about how your brain works and be ready to understand and use the techniques in the rest of the Appendix activities.

Diagnostic Pretest

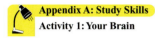

[Chapter 1]

1. Write this number using digits: twenty-five million, three thousand, seven hundred one

1. _____

2. Simplify. $-3(-4)^2 \div 12 - (-10)$

2. _____

3. Round 4,396,028 to the nearest ten-thousand.

3. _____

4. Max had $185 in his checking account. He deposited his $372 paycheck, and then wrote a $575 check for rent. What is the new balance in his account?

4. _____

[Chapter 2]

5. Evaluate $-5a^3b^2$ when a is 2 and b is -1.

5. _____

6. Simplify. $-5 + 4(r - 7)$

6. _____

7. Solve. $-18 = -2(t + 4)$

7. _____

8. Solve. $7n - 6 = -2n + 3$

8. _____

[Chapter 3]

9. Find the perimeter.

10. Find the area.

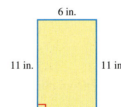

9. _____

10. _____

11. Translate this sentence into an equation and solve it. If 13 is added to four times a number, the result is 8 less than the number. What is the number?

11. _____

12. Write an equation and solve it to answer this problem. A rope is 238 ft long. Marc cut it into two pieces, with one piece 12 ft longer than the other. Find the lengths of both pieces.

12. _____

[Chapter 4]

13. _____

13. Write $\dfrac{39}{52}$ in lowest terms.

14. _____

14. Add: $-\dfrac{5}{6} + \dfrac{7}{12}$

15. _____

15. Solve. $\dfrac{4}{5}x + 9 = -7$

16. _____

16. Find the area.

[Chapter 5]

17. _____

17. Anita bought 2.6 pounds of salmon at $6.99 per pound. How much did she pay for the salmon, to the nearest cent?

18. _____

18. Solve. $-3r + 5.8 = -1.1$

19. _____

19. The hourly wages for six employees of a small company are $8.50, $9.40, $7.30, $13.75, $11, and $12. Find the median hourly wage for these employees.

20. _____

20. Find the circumference of this circle. Use 3.14 for π and round your answer to the nearest tenth.

[Chapter 6]

21. _____

21. Find the unknown number.

$$\dfrac{9}{8} = \dfrac{x}{96}$$

22. _____

22. On a road map, 1 inch represents 32 miles. If two towns are 3.8 inches apart on the map, what is the actual distance between them?

23. _____

23. Find the supplement of a 63° angle.

24. _____

24. Find the unknown lengths in these similar triangles.

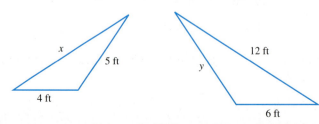

[Chapter 7]

25. (a) Write 250% as a decimal.

25. (a) _____

(b) Write 0.3 as a percent.

(b) _____

26. (a) Write 38% as a fraction in lowest terms.

26. (a) _____

(b) Write $\dfrac{3}{40}$ as a percent.

(b) _____

27. The price of a digital camera is $235 plus sales tax of 6%. Find the total cost of the camera including sales tax.

27. _____

28. Suppose that you go out to lunch with three friends and the bill comes to $48. You decide to add a 15% tip, and then, split the bill evenly. How much should each of you pay?

28. _____

[Chapter 8]

29. Convert each measurement.

29. (a) _____

(a) 432 cm to meters

(b) 0.08 kg to grams

(b) _____

30. The average annual precipitation over the period 1971–2000 is 64.16 inches in New Orleans, Louisiana, and 4.16 inches in Barrow, Alaska. What was the difference in precipitation between the two cities in feet?   (_Source: World Almanac and Book of Facts._)

30. _____

31. Each serving of punch at a graduation party is to be 160 mL. How many liters of punch are needed for 65 servings?

31. _____

32. Write the most reasonable metric unit in each blank. Choose from km, m, cm, mm, L, mL, kg, g, and mg.

32. (a) _____

(a) The length of Rosie's bedroom is about 4 _____.

(b) Satish takes a pill containing 500 _____ of vitamin C every morning.

(b) _____

[Chapter 9]

33. _____

33. This circle graph shows the budget for the Patel family. The total budget for one month is $3600. Find the monthly amount budgeted for savings.

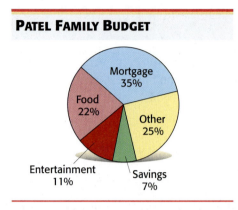

PATEL FAMILY BUDGET

34. **(a)** _____

   (b) _____

34. This graph shows the number of sections of Prealgebra and Introductory Algebra at River Bend Community College over a four-year period.

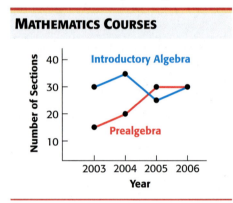

MATHEMATICS COURSES

(a) In what year(s) were there more sections of Introductory Algebra than of Prealgebra?

(b) In what year(s) were there the same number of sections of both courses? How many sections of each course were there in that year?

35. **(a)** _____

   (b) _____

   (c) _____

   (d) _____

35. Name the quadrant (if any) in which each point is located.

   (a) $(-3, 5)$

   (b) $(6, 0)$

   (c) $(-2, -6)$

   (d) $(10, 10)$

36.

36. Graph $y = 2x - 1$ on the grid at the left. Complete the table at the right using 0, 1, and 2 as the values of x.

x	y	(x, y)

[Chapter 10]

37. _____

37. Evaluate. $5^0 - 3^{-1}$

38. _____

38. Simplify. $-4(a^4 b^3)^2$

39. **(a)** _____

   (b) _____

39. **(a)** Write 54,500,000 in scientific notation.

   (b) Write 4.07×10^{-3} without exponents.

40. _____

40. Multiply. $(3x - 5)(2x + 1)$

Introduction to Algebra: Integers

The weather—we all talk about it and often complain about it. When reporting temperatures, we need both *positive* and *negative* numbers. If you live in Chicago, the highest temperature ever recorded is 104 °F and the lowest is −27 °F, a difference of 131 degrees! In Atlanta, the record high and low are 105 °F and −8 °F. And in Barrow, Alaska, the range of extreme temperatures is 79 °F to −56 °F, a difference of 135 degrees. To make things even more uncomfortable, humidity can make hot temperatures feel hotter and wind can make cold temperatures feel colder. Learn more about "windchill" as you work Exercises 47–48 in **Section 1.4.** (*Source:* National Climatic Data Center.)

1.1 Place Value

OBJECTIVES

1. Identify whole numbers.
2. Identify the place value of a digit through hundred-trillions.
3. Write a whole number in words or digits.

It would be nice to earn millions of dollars like some of our favorite entertainers or sports stars. But how much is a million? If you received $1 every second, 24 hours a day, day after day, how many days would it take for you to receive a million dollars? How long to receive a billion dollars? Or a trillion dollars? Make some guesses and write them here.

It would take _____ to receive a million dollars.

It would take _____ to receive a billion dollars.

It would take _____ to receive a trillion dollars.

The answers are at the bottom left of the page. Later, in the exercises for **Section 1.7,** you'll find out how to calculate the answers.

OBJECTIVE 1 Identify whole numbers. First we have to be able to write the number that represents *one million.* We can write *one* as 1. How do we make it 1 *million?* Our number system is a **place value system.** That means that the location, or place, in which a number is written gives it a different value. Using money as an example, you can see that

$1 is one dollar.

$10 is ten dollars.

$100 is one hundred dollars.

$1000 is one thousand dollars.

Each time the 1 moved to the left one place, it was worth *ten times* as much. Can you keep moving it to the left? Yes, as many times as you like.

The chart below shows the *value* of each *place.* In other words, you write the 1 in the correct place to represent the number you want to express. It is important to memorize the place value names shown on the chart.

Whole Number Place Value Chart

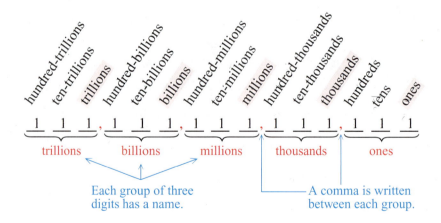

If there had been more room on this page, we could have continued to the left with quadrillions, quintillions, sextillions, septillions, octillions, and more.

Of course, we can use other *digits* besides 1. In our decimal system of writing numbers, we can use these ten **digits:** 0, 1, 2, 3, 4, 5, 6, 7, 8, and 9. In this section we will use the digits to write **whole numbers.**

ANSWERS

About $11\frac{1}{2}$ days to receive a million dollars; nearly 32 *years* to receive a billion dollars; about 31,710 *years* to receive a trillion dollars.

These are whole numbers.

0	8	37	100	24,014

These are *not* whole numbers.

-6 $\dfrac{3}{4}$ 7.528   0.3 $5\dfrac{2}{3}$

EXAMPLE 1 **Identifying Whole Numbers**

Circle the whole numbers in this list.

75   -4 0   1.5 $\dfrac{5}{8}$ 300   0.666 $7\dfrac{1}{2}$ 2

Whole numbers *do* include zero. If we started a list of *all* the whole numbers, it would look like this: 0, 1, 2, 3, 4, 5, . . . with the three dots indicating that the list goes on and on. So the whole numbers in this example are: 75, 0, 300, and 2.

Now Try Exercise 1. ▶▶▶

OBJECTIVE 2 Identify the place value of a digit through hundred-trillions.   The estimated world population in 2004 was 6,323,979,176 people. (*Source:* U.S. Census Bureau.) There are two 6s in this number but the value of each 6 is very different. The 6 on the right is in the ones place, so its value is simply 6. But the 6 on the left is worth a great deal more because of the *place* where it is written. Looking back at the place value chart, we see that this 6 is in the billions place, so its value is *6 billion*.

6 , 3 2 3 , 9 7 9 , 1 7 6 people in the world

↑ Value of 6 *billion*

↑ Value of 6 *ones*

EXAMPLE 2 **Identifying Place Value**

Identify the place value of each 7 in the number of people in the world.

6 , 3 2 3 , 9 7 9 , 1 7 6

Ten-thousands place ↑   ↑ Tens place

Now Try Exercise 13. ▶▶▶

OBJECTIVE 3 Write a whole number in words or digits.   To write a whole number in words, or to say it aloud, begin at the left. Write or say the number in each group of three, followed by the name for that group. When you get to the ones group, do *not* include the group name. Hyphens (dashes) are used whenever you write a number from 21 to 99, like twenty–one thousand, or thirty–seven million, or ninety–four billion.

EXAMPLE 3 **Writing Numbers in Words**

(a) Write 6,058,120 in words.
Start at the left.

6 , 0 5 8 , 1 2 0

six million, **fifty-eight thousand**, **one hundred twenty**

Group name   Group name   Do not use group name "ones."

Continued on Next Page

(b) Write 50,588,000,040,000 in words.
Start at the left.

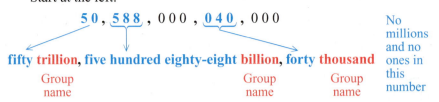

$$5\,0\,,\,5\,8\,8\,,\,0\,0\,0\,,\,0\,4\,0\,,\,0\,0\,0$$

fifty trillion, five hundred eighty-eight billion, forty thousand

Group name   Group name   Group name

No millions and no ones in this number

> **CAUTION**
> You often hear people say "and" when reading a group of three digits. For example, you may hear 120 as "one hundred *and* twenty," but this is **not** correct. The word *and* is used only when reading a *decimal point*, which we do not have here. The correct wording for 120 is "one hundred twenty."

◀◀◀ **Now Try Exercises 19 and 25.**

When you read or hear a number and want to write it in digits, look for the group names: **trillion**, **billion**, **million**, and **thousand**. Write the number in each group, followed by a comma. Do *not* put a comma at the end of the ones group.

EXAMPLE 4 **Writing Numbers in Digits**

Write each number using digits.

(a) Five hundred sixteen **thousand**, nine

The first group name is *thousand*, so you need to fill *two groups* of three digits: thousands and ones.

$$\underbrace{5\quad 1\quad 6}_{\text{thousands}}\,,\,\underbrace{0\quad 0\quad 9}_{\text{ones}}$$

comma between groups

The number is 516,009.

(b) Seventy-seven **billion**, thirty **thousand**, five hundred

The first group name is *billion*, so you need to fill *four groups* of three digits: billions, millions, thousands, and ones.

$$\underbrace{0\quad 7\quad 7}_{\text{billions}}\,,\,\underbrace{0\quad 0\quad 0}_{\text{millions}}\,,\,\underbrace{0\quad 3\quad 0}_{\text{thousands}}\,,\,\underbrace{5\quad 0\quad 0}_{\text{ones}}$$

There are no millions, so fill the millions group with zeros.
When writing the number, you can omit the leading **0** in the billions group.
The number is 77,000,030,500.

◀◀◀ **Now Try Exercises 29 and 35.**

1.1 Exercises

| **FOR EXTRA HELP** | 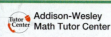 Addison-Wesley Math Tutor Center | 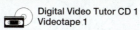 MathXL | Digital Video Tutor CD 1 Videotape 1 | Student's Solutions Manual | 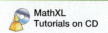 MyMathLab | MathXL Tutorials on CD |

Circle the whole numbers. See Example 1.

1. 15 $8\frac{3}{4}$ 0   3.781

83,001 -8 $\frac{7}{16}$ $\frac{9}{5}$

2. 33.7 -5 457 $\frac{8}{5}$

0   6 $1\frac{3}{4}$ -14.1

3. 5.8 -6 7 $\frac{5}{4}$

$\frac{1}{10}$ 362,049   0.1 $7\frac{7}{8}$

4. 75,039 $\frac{1}{3}$ -87 6.49

-0.5 $2\frac{7}{10}$ $\frac{15}{8}$ 4

Identify the place value of the digit 2 in each number. See Example 2.

5. 61,284

6. 82,110

7. 284,100

8. 823,415

9. 725,837,166

10. 442,653,199

11. 253,045,701,000

12. 823,000,419,567

13. From left to right, name the place value for each 0 in this number: 302,016,450,098,570.

14. From left to right, name the place value for each 0 in this number: 810,704,069,809,035.

In Exercises 15–26, write each number in words. See Example 3.

15. 8421

16. 1936

17. 46,205

18. 75,089

19. 3,064,801

20. 7,900,408

21. 840,111,003

22. 304,008,401

23. 51,006,888,321

24. 99,046,733,214

25. 3,000,712,000,000

26. 50,918,000,000,600

In Exercises 27–36, write each number using digits. See Example 4.

27. Forty-six thousand, eight hundred five

28. Seventy-nine thousand, forty-six

29. Five million, six hundred thousand, eighty-two

30. One million, thirty thousand, five

31. Two hundred seventy-one million, nine hundred thousand

32. Three hundred eleven million, four hundred

33. Twelve billion, four hundred seventeen million, six hundred twenty-five thousand, three hundred ten

34. Seventy-five billion, eight hundred sixty-nine million, four hundred eighty-eight thousand, five hundred six

35. Six hundred trillion, seventy-one million, four hundred

36. Four hundred forty trillion, thirty-six thousand, one hundred two

In Exercises 37–46, if the number is given in digits, write it in words. If the number is given in words, write it in digits. See Examples 3 and 4. (Source: The World Almanac, Statistical Abstract of the United States, Scholastic Book of World Records.)

37. In the United States, 6375 couples get married every day.

38. There are 3582 pairs of bowling shoes purchased each day.

39. One hundred one million, two hundred eighty thousand adults are on diets at any one time in the United States.

40. Two million, five thousand Americans suffer from heartburn on any given day.

41. The average amount spent each day by individuals in the United States on computers and software is $93,972,602.

42. Americans spend an average of $176,986,300 daily on toys.

43. People eat fifty-five million, eight hundred hot dogs each day.

44. Five hundred twenty-four million servings of cola drinks are consumed every day.

45. Dunkin' Donuts sells nearly 6,400,000 doughnuts every day. That's 2,336,000,000 doughnuts in one year!

46. The Hostess company bakeries are able to make 60,000 Twinkies every hour. That adds up to 525,600,000 Twinkies per year.

Use your knowledge of place value to **work Exercises 47–50 in order.**

47. Here is a group of digits: 6, 0, 9, 1, 5, 0, 7, 1. Using each digit exactly once, arrange them to make the largest possible whole number and the smallest possible whole number. Then write each number in words.

48. Write these numbers in digits and in words.

   (a) Your house number or apartment building number

   (b) Your phone number, including the area code

   (c) The approximate cost of your tuition and books for this quarter or semester

   (d) Your zip code

   Now tell how you *usually* say each of the numbers in parts (a)–(d) above. Why do you think we ignore the rules when saying these numbers in everyday situations?

49. Look again at the Whole Number Place Value Chart at the beginning of this section. As you move to the left, each place is worth *ten* times as much as the previous place. Computers work on a *binary* system where each place is worth *two* times as much as the previous place. Complete this place value chart based on 2s.

$$\underline{1}\ ,\ \underline{1}\ \ \underline{1}\ \ \underline{1}\ ,\ \underline{1}\ \ \underline{1}\ \ \underline{1}$$

Try writing some numbers using the 2s (binary) place value chart.

(a) 5 = _____   The *only* digits you may use in

(b) 10 = _____   the binary system are 0 and 1.

(c) 15 = _____   Here is an example.

$$6 = 110$$

one 4 + one 2 + zero 1s = 4 + 2 + 0 = 6

50. (a) Explain in your own words why our number system is called a *place value system*. Include an example as part of your explanation.

   (b) Find information on the Roman numeral system. Write these numbers using Roman numerals.

      8 = _____   38 = _____

    275 = _____   3322 = _____

   (c) Explain why the Roman system is *not* a place value system. What are the disadvantages of the Roman system?

1.2 Introduction to Signed Numbers

OBJECTIVE 1 Write positive and negative numbers used in everyday situations.   The whole numbers in **Section 1.1** were either 0 or greater than 0. Numbers *greater* than 0 are called *positive numbers*. But many everyday situations involve numbers that are *less* than 0, called *negative numbers*. Here are a few examples.

At midnight the temperature dropped to 4 degrees *below* zero.

−4 degrees

Jean had $30 in her checking account. She wrote a check for $40.75. We say she is now $10.75 "in the hole," or $10.75 "in the red," or *overdrawn* by $10.75.

−$10.75

The Packers football team *gained* 6 yards on the first play. On the second play they *lost* 9 yards. We can write the results using a positive and a negative number.

+6 yards   and   **−9 yards**

A plane took off from the airport and climbed to 20,000 feet *above* sea level. We can write this using a positive number.

+20,000 feet

A scuba diver swam down to $25\frac{1}{2}$ feet *below* the surface. We can write this using a negative number.

−$25\frac{1}{2}$ feet

NOTE
To write a negative number, put a negative sign (a dash) in front of it: −10. Notice that the negative sign looks exactly like the subtraction sign, as in 5 − 3 = 2. The negative sign and subtraction sign do *not* mean the same thing (more on that in the next section). To avoid confusion for now, we will write negative signs in **red** and put them up higher than subtraction signs.

⁻10   means   **negative** 10   14 − 10   means   14 **minus** 10
└── Raised dash

Starting in **Chapter 4**, we will write negative signs in the traditional way. However, if you use a graphing calculator, it may show negative signs in the raised position.

Positive numbers can be written two ways:

1. Write a positive sign in front of the number: ⁺2 is *positive* 2. We will write the sign in the raised position to avoid confusion with the sign for addition, as in 6 + 3 = 9.

2. Do not write any sign. For example, 16 is assumed to be *positive* 16.

EXAMPLE 1 **Writing Positive and Negative Numbers**

Write each negative number with a raised negative sign. Write each positive number in two ways.

(a) The river rose to 8 feet above flood stage.

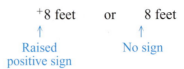

            $^{+}8$ feet   or   8 feet

              ↑   ↑

        Raised   No sign
    positive sign

(b) Michael lost $500 in the stock market.

                $^{-}\$500$

              ↑

           Raised negative sign

Now Try Exercises 5 and 7.

OBJECTIVE 2 **Graph signed numbers on a number line.** Mathematicians often use a **number line** to show how numbers relate to each other. A number line is like a thermometer turned sideways. *Zero* is the dividing point between the positive and negative numbers.

Negative numbers | Positive numbers

Zero is neither
positive nor negative

The number line could be shown with positive numbers on the left side of 0 instead of the right side. But it helps if everyone draws it the same way, as shown above. This method will also match what you do when graphing points and lines in **Chapter 9.**

EXAMPLE 2 **Graphing Numbers on a Number Line**

Graph each number on the number line.

(a) $^{-}5$ **(b)** 3   **(c)** $1\frac{1}{2}$ **(d)** 0   **(e)** $^{-}1$

Draw a dot at the correct location for each number.

Now Try Exercise 13.

OBJECTIVE 3 **Use the < and > symbols to compare integers.** In **Chapters 4 and 5** you will work with fractions and decimals. For the rest of this chapter, you will work only with *integers*. A list of **integers** can be written like this:

$$\ldots, {}^{-}6, {}^{-}5, {}^{-}4, {}^{-}3, {}^{-}2, {}^{-}1, 0, 1, 2, 3, 4, 5, 6, \ldots$$

The dots show that the list goes on forever in both directions.

We can use the number line to compare two integers.

When comparing 1 and 4,
1 is to the *left* of 4.
1 is *less than* 4.
Use < to mean "is less than."

$$1 \quad < \quad 4$$

1 is less than 4

When comparing 0 and ⁻3,
0 is to the *right* of ⁻3.
0 is *greater than* ⁻3.
Use > to mean "is greater than."

$$0 \quad > \quad ^-3$$

0 is greater than ⁻3

NOTE

One way to remember which symbol to use is that the "smaller end of the symbol" points to the "smaller number" (the number that is less).

$$1 < 4$$

Smaller number ⌐ ⌐ Smaller end
of symbol

$$0 > ^-3$$

Smaller end ⌐ ⌐ Smaller number
of symbol

EXAMPLE 3 **Comparing Integers, Using the < and > Symbols**

Write < or > between each pair of numbers to make a true statement.

(a) 0 ___ 2

 0 is to the *left* of 2 on the number line, so 0 is *less than* 2. Write 0 < 2.

(b) 1 ___ ⁻4

 1 is to the *right* of ⁻4, so 1 is *greater than* ⁻4. Write 1 > ⁻4.

(c) ⁻4 ___ ⁻2

 ⁻4 is to the *left* of ⁻2, so ⁻4 is *less than* ⁻2. Write ⁻4 < ⁻2.

Now Try Exercises 21 and 23.

OBJECTIVE 4 **Find the absolute value of integers.** In order to graph a number on the number line, you need to know two things:

1. Which *direction* is it from 0? It can be in a positive direction or a negative direction. You can tell the direction by looking for a positive sign (or no sign, which means positive), or a negative sign.

2. How *far* is it from 0? The distance from 0 is the *absolute value* of a number.

Absolute Value

The **absolute value** of a number is its distance from 0 on the number line. *Absolute value* is indicated by two vertical bars. For example,

 |6| is read "the **absolute value** of 6."

The absolute value of a number will *always* be positive (or 0), because it is the *distance* from 0. A distance is never negative. (You wouldn't say that your living room is ⁻16 feet long.) So absolute value concerns only *how far away* the number is from 0; we don't care which direction it is from 0.

EXAMPLE 4 **Finding Absolute Values**

Find each absolute value.

(a) $|4|$ The distance from 0 to 4 on the number line is 4 spaces. So, $|4| = 4$.

(b) $|{}^-4|$ The distance from 0 to ⁻4 on the number line is also 4 spaces. So, $|{}^-4| = 4$.

(c) $|0|$ $|0| = 0$ because the distance from 0 to 0 on the number line is 0 spaces.

Now Try Exercises 33 and 35.

1.2 Exercises

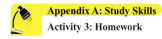
Write each negative number with a raised negative sign. Write each positive number in two ways. See Example 1.

1. Mount Everest, the tallest mountain in the world, rises 29,035 feet above sea level. (*Source: World Almanac.*)

2. The bottom of Lake Baikal in Siberia, Russia, is 5371 feet below the surface of the water. (*Source: Guinness Book of World Records.*)

Depth below surface

3. The coldest temperature ever recorded on Earth is 128.6 degrees below zero in Antarctica. (*Source: Fact Finder.*)

4. Normal body temperature is 98.6 degrees Fahrenheit, although it varies slightly for some people. (*Source:* Mayo Clinic *Health Letter.*)

5. During the first three plays of the football game, the Trojans lost a total of 18 yards.

6. The Jets gained 25 yards on a pass play.

7. Angelique won $100 in a prize drawing at the shopping mall.

8. Derice overdrew his checking account by $37.

9. Keith lost $6\frac{1}{2}$ pounds while he was sick with the flu.

10. The mice in the experiment gained $2\frac{1}{2}$ ounces.

Graph each set of numbers. See Example 2.

11. $^-3, 3, 0, ^-5$

12. $^-2, 2, 0, 5$

13. $^-1, 4, ^-2, 5$

14. $3, ^-4, 1, ^-5$

15. $^-4\frac{1}{2}, \frac{1}{2}, 0, ^-8$

16. $^-7, 1\frac{1}{2}, -\frac{1}{2}, ^-9$

Write < or > between each pair of numbers to make a true statement. See Example 3.

17. 10 ____ 2

18. 6 ____ 0

19. ⁻1 ____ 0

20. ⁻3 ____ ⁻1

21. ⁻10 ____ 2

22. ⁻9 ____ 7

23. ⁻3 ____ ⁻6

24. 0 ____ ⁻1

25. ⁻10 ____ ⁻2

26. ⁻1 ____ ⁻5

27. 0 ____ ⁻8

28. 6 ____ ⁻4

29. 10 ____ ⁻2

30. ⁻2 ____ 1

31. ⁻4 ____ 4

32. 9 ____ ⁻9

Find each absolute value. See Example 4.

33. |15|

34. |10|

35. |⁻3|

36. |⁻8|

37. |0|

38. |100|

39. |200|

40. |⁻99|

41. |⁻75|

42. |⁻6320|

43. |⁻8042|

44. |0|

RELATING CONCEPTS (EXERCISES 45–48) For Individual or Group Work

An ultrasound of a person's heel bone is a quick way to screen patients for osteoporosis (brittle bone disease). Use the information on the Patient Report Form to **work** *Exercises 45–48 in order.*

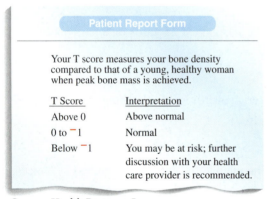

Patient Report Form

Your T score measures your bone density compared to that of a young, healthy woman when peak bone mass is achieved.

T Score	Interpretation
Above 0	Above normal
0 to ⁻1	Normal
Below ⁻1	You may be at risk; further discussion with your health care provider is recommended.

Source: Health Partners, Inc.

45. Here are the T scores for four patients. Draw a number line and graph the four scores.

Patient A: ⁻1.5   Patient C: ⁻1

Patient B: 0.5   Patient D: 0

46. List the patients' scores in order from lowest to highest.

47. What is the interpretation of each patient's score?

Taking an ultrasound of a patient's heel bone.

48. (a) What could happen if patient A did not understand the importance of a negative sign?

(b) For which patient does the sign of the score make no difference? Explain your answer.

1.3 Adding Integers

OBJECTIVES

1 Add integers.

2 Identify properties of addition.

OBJECTIVE 1 Add integers. Numbers that you are adding are called **addends,** and the result is called the **sum.** You can add integers while watching a football game. On each play, you can use a *positive* integer to stand for the yards *gained* by your team, and a *negative* integer for yards *lost.* Zero indicates no gain or loss. For example,

A drawing of a football field can help you add integers. Notice how similar the drawing is to a number line. Zero marks your team's starting point.

EXAMPLE 1 **Using a Number Line to Add Integers**

Use a number line to find ⁻5 + ⁻4.

Think of the number line as a football field. Your team starts at 0. On the first play it lost 5 yards. On the next play it lost 4 yards. The total loss is 9 yards.

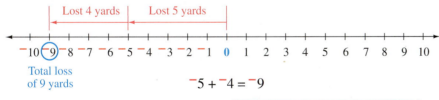

$$^-5 + {}^-4 = {}^-9$$

Now Try Exercise 3. ▶▶▶

Watch for a pattern as you find these sums using the number line.

$$^-2 + {}^-2 = \underline{\qquad} \qquad 2 + 2 = \underline{\qquad}$$

$$^-10 + {}^-1 = \underline{\qquad} \qquad 10 + 1 = \underline{\qquad}$$

$$^-3 + {}^-7 = \underline{\qquad} \qquad 3 + 7 = \underline{\qquad}$$

The first two sums are ⁻4 and 4. They are the same, *except for the sign.* The same is true for the next two sums, ⁻11 and 11, and for the last two sums, ⁻10 and 10. This pattern leads to a rule for adding two integers when the signs are the same.

Adding Two Integers with the Same Sign

Step 1 *Add* the absolute values of the numbers.

Step 2 Use the *common sign* as the sign of the sum. If both numbers are positive, the sum is positive. If both numbers are negative, the sum is negative.

EXAMPLE 2 **Adding Two Integers with the Same Sign**

Add.

(a) $^-8 + {}^-7$

> *Step 1* *Add* the absolute values.
>
> $$|{}^-8| = 8 \qquad \text{and} \qquad |{}^-7| = 7$$
>
> Add $8 + 7$ to get 15.
>
> *Step 2* Use the *common sign* as the sign of the sum. Both numbers are negative, so the sum is negative.
>
>
>
> $$^-8 + {}^-7 = {}^-15$$
>
> Both negative   Sum is negative.

(b) $3 + 6 = 9$

Both positive   Sum is positive.

In *Step 1*, when both numbers are positive, their absolute values are also positive, so we only need to show *Step 2*.

Now Try Exercise 9.

    You can also use a number line (or drawing of a football field) to add integers with *different* signs. For example, suppose that your team gained 2 yards on the first play and then lost 7 yards on the next play. We can represent this as $2 + {}^-7$.

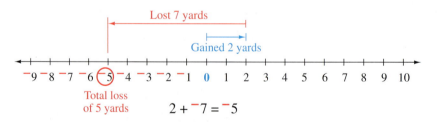

Total loss of 5 yards $2 + {}^-7 = {}^-5$

Or, try this one. On the first play your team gained 10 yards, but then it lost 4 yards on the next play. We can represent this as $10 + {}^-4$.

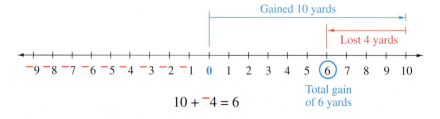

$$10 + {}^-4 = 6$$

Total gain of 6 yards

These examples illustrate the rule for adding two integers with unlike, or different, signs.

Adding Two Integers with Unlike Signs

Step 1 *Subtract* the smaller absolute value from the larger absolute value.

Step 2 Use the sign of the number with the *larger absolute value* as the sign of the sum.

EXAMPLE 3 **Adding Two Integers with Unlike Signs**

Add.

(a) $^-8 + 3$

Step 1 $|^-8| = 8$ and $|3| = 3$

Subtract $8 - 3$ to get 5.

Step 2 $^-8$ has the *larger absolute value* and is negative, so the sum is also negative.

$$^-8 + 3 = {}^-5$$

(b) $^-5 + 11$

Step 1 $|^-5| = 5$ and $|11| = 11$

Subtract $11 - 5$ to get 6.

Step 2 11 has the *larger absolute value* and is positive, so the sum is also positive.

$$^-5 + 11 = {}^+6 \quad \text{or} \quad 6$$

Now Try Exercise 19.

EXAMPLE 4 **Adding Several Integers**

A football team has to gain at least 10 yards during four plays in order to keep the ball. Suppose that your college team lost 6 yards on the first play, gained 8 yards on the second play, lost 2 yards on the third play, and gained 7 yards on the fourth play. Did the team gain enough to keep the ball?

When you're adding several integers, work from left to right.

No, the team didn't gain enough yards to keep the ball.

Now Try Exercise 61.

OBJECTIVE 2 Identify properties of addition. In our football model for adding integers, we said that 0 indicated no gain or loss, that is, no change in the position of the ball. This example illustrates one of the *properties of addition*. The properties of addition apply to all addition problems, regardless of the specific numbers you use.

Addition Property of 0
Adding 0 to any number leaves the number unchanged.

Some examples are shown below.

$$0 + 6 = 6 \qquad {}^-25 + 0 = {}^-25 \qquad 72{,}399 + 0 = 72{,}399$$
$$0 + {}^-100 = {}^-100$$

Another property of addition is that you can change the *order* of the addends and still get the same sum. For example,

Gaining 2 yards, then losing 7 yards, gives a result of ⁻5 yards.

Losing 7 yards, then gaining 2 yards, also gives a result of ⁻5 yards.

Commutative Property of Addition
Changing the **order** of two addends does *not* change the sum.

Here are some examples.

$$84 + 2 = 2 + 84 \qquad \text{Both sums are 86.}$$

$${}^-10 + 6 = 6 + {}^-10 \qquad \text{Both sums are } {}^-4.$$

EXAMPLE 5 **Using the Commutative Property of Addition**

Rewrite each sum, using the **commutative property of addition.** Check that the sum is unchanged.

(a) $65 + 35$

$$\underbrace{65 + 35}_{100} = \underbrace{35 + 65}_{100}$$
$$100 = 100$$

Both sums are 100, so the sum is unchanged.

(b) $^-20 + {}^-30$

$$\underbrace{{}^-20 + {}^-30}_{{}^-50} = \underbrace{{}^-30 + {}^-20}_{{}^-50}$$
$${}^-50 = {}^-50$$

Both sums are ⁻50, so the sum is unchanged.

Now Try Exercise 67.

When there are three addends, parentheses may be used to tell you which pair of numbers to add first, as shown below.

$$\underbrace{(3 + 4)}_{} + 2 \quad \text{First add } 3 + 4. \qquad\qquad 3 + \underbrace{(4 + 2)}_{} \quad \text{First add } 4 + 2.$$

$$\underbrace{7 + 2}_{} \quad \text{Then add } 7 + 2. \qquad\qquad \underbrace{3 + 6}_{} \quad \text{Then add } 3 + 6.$$

$$9 \qquad\qquad\qquad\qquad\qquad 9$$

Both sums are 9. This example illustrates another property of addition.

Associative Property of Addition

Changing the *grouping* of addends does *not* change the sum.

Some examples are shown below.

$$(^-5 + 5) + 8 = {}^-5 + (5 + 8)$$
$$\underbrace{0}\ \ + 8 = \underbrace{{}^-5 +\ \ 13}$$
$$\underbrace{8}\ \ =\ \ \ \ \ 8$$

Both sums are 8.

$$3 + (^-4 + {}^-6) = (3 + {}^-4) + {}^-6$$
$$3 +\ \ \underbrace{{}^-10}\ \ \ =\ \ \underbrace{{}^-1}\ \ + {}^-6$$
$$^-7\ \ \ \ \ \ \ \ =\ \ \ \ \ \ \ \ {}^-7$$

Both sums are $^-7$.

We can use the associative property to make addition problems easier. Notice in the first example in the box above that it is easier to group $^-5 + 5$ (which is 0) and then add 8. In the second example in the box, it is helpful to group $^-4 + {}^-6$ because the sum is $^-10$, and it is easy to work with multiples of 10. (See **Section R.3** for more on multiples of 10.)

EXAMPLE 6 **Using the Associative Property of Addition**

In each addition problem, pick out the two addends that would be easiest to add. Write parentheses around those addends. Then find the sum.

(a) $6 + 9 + {}^-9$

Group $9 + {}^-9$ because the sum is 0.

$$6 + \underbrace{(9 + {}^-9)}$$
$$\underbrace{6 +\ \ \ \ \ 0}$$
$$6$$

(b) $17 + 3 + {}^-25$

Group $17 + 3$ because the sum is 20, which is a multiple of 10.

$$\underbrace{(17 + 3)} + {}^-25$$
$$\underbrace{20\ \ \ + {}^-25}$$
$$^-5$$

NOTE

When using the associative property to make the addition of a group of numbers easier:

1. Look for two numbers whose sum is 0.

2. Look for two numbers whose sum is a multiple of 10 (the sum ends in 0, such as 10, 20, 30, or $^-100$, $^-200$, etc.).

If neither of these occurs, look for two numbers that are easier for you to add. For example, in $98 + 42 + 5$, you may find that adding $42 + 5$ is easier than adding $98 + 5$.

Now Try Exercise 71. ▶▶▶

1.3 Exercises

FOR EXTRA HELP
 Addison-Wesley Math Tutor Center MathXL Digital Video Tutor CD 1 Videotape 1 Student's Solutions Manual MyMathLab MathXL Tutorials on CD

Add by using the number line. See Example 1.

1. $^-2 + 5$

2. $^-3 + 4$

3. $^-5 + ^-2$

4. $^-2 + ^-2$

5. $3 + ^-4$

6. $5 + ^-1$

Add. See Example 2.

7. (a) $^-5 + ^-5$

(b) $5 + 5$

8. (a) $^-9 + ^-9$

(b) $9 + 9$

9. (a) $7 + 5$

(b) $^-7 + ^-5$

10. (a) $3 + 6$

(b) $^-3 + ^-6$

11. (a) $^-25 + ^-25$

(b) $25 + 25$

12. (a) $^-30 + ^-30$

(b) $30 + 30$

13. (a) $48 + 110$

(b) $^-48 + ^-110$

14. (a) $235 + 21$

(b) $^-235 + ^-21$

15. What pattern do you see in your answers to Exercises 7–14? Explain why this pattern occurs.

16. In your own words, explain how to add two integers that have the same sign.

Add. See Example 3.

17. (a) ⁻6 + 8

(b) 6 + ⁻8

18. (a) ⁻3 + 7

(b) 3 + ⁻7

19. (a) ⁻9 + 2

(b) 9 + ⁻2

20. (a) ⁻8 + 7

(b) 8 + ⁻7

21. (a) 20 + ⁻25

(b) ⁻20 + 25

22. (a) 30 + ⁻40

(b) ⁻30 + 40

23. (a) 200 + ⁻50

(b) ⁻200 + 50

24. (a) 150 + ⁻100

(b) ⁻150 + 100

25. What pattern do you see in your answers to Exercises 17–24? Explain why this pattern occurs.

26. In your own words, explain how to add two integers that have different signs.

Add. See Examples 2–4.

27. ⁻8 + 5

28. ⁻3 + 2

29. ⁻1 + 8

30. ⁻4 + 10

31. ⁻2 + ⁻5

32. ⁻7 + ⁻3

33. 6 + ⁻5

34. 11 + ⁻3

35. 4 + ⁻12

36. 9 + ⁻10

37. ⁻10 + ⁻10

38. ⁻5 + ⁻20

39. ⁻17 + 0

40. 0 + ⁻11

41. 1 + ⁻23

42. 13 + ⁻1

43. ⁻2 + ⁻12 + ⁻5

44. ⁻16 + ⁻1 + ⁻3

45. 8 + 6 + ⁻8

46. ⁻5 + 2 + 5

47. ⁻7 + 6 + ⁻4

48. ⁻9 + 8 + ⁻2

49. ⁻3 + ⁻11 + 14

50. 15 + ⁻7 + ⁻8

51. 10 + ⁻6 + ⁻3 + 4

52. 2 + ⁻1 + ⁻9 + 12

53. ⁻7 + 28 + ⁻56 + 3

54. 4 + ⁻37 + 29 + ⁻5

Write an addition problem for each situation and find the sum.

55. The football team gained 13 yards on the first play and lost 17 yards on the second play. How many yards did the team gain or lose in all?

56. At penguin breeding grounds on Antarctic islands, temperatures routinely drop to ⁻15 °C. Temperatures in the interior of the continent may drop another 60 °C below that. What is the temperature in the interior?

57. Nick's checking account was overdrawn by $62. He deposited $50 in his account. What is the balance in his account?

58. Cynthia had $100 in her checking account. She wrote a check for $83 and was charged $17 for overdrawing her account last month. What is her account balance?

59. $88 was stolen from Jay's car. He got $35 of it back. What was his net loss?

60. Marion lost 4 pounds in April, gained 2 pounds in May, and gained 3 pounds in June. How many pounds did she gain or lose in all?

61. Use the score sheet to find each player's point total after three rounds in a card game.

	Jeff	Terry
Round 1	Lost 20 pts	Won 42 pts
Round 2	Won 75 pts	Lost 15 pts
Round 3	Lost 55 pts	Won 20 pts

62. Use the information in the table on flood water depths to find the new flood level for each river.

	Red River	Mississippi
Monday	Rose 8 ft	Rose 4 ft
Tuesday	Fell 3 ft	Rose 7 ft
Wednesday	Fell 5 ft	Fell 13 ft

At the 2005 British Open golf tournament in St. Andrews, Scotland, 72 strokes was the "par" score for each round of play. A negative score indicates that the player had fewer than 72 strokes for that round, and a positive score indicates more than 72 strokes. Find the total score for each of these players in the tournament.

	Player	Round 1	Round 2	Round 3	Round 4	Total
63.	Tiger Woods	⁻6	⁻5	⁻1	⁻2	
64.	Colin Montgomerie	⁻1	⁻6	⁻2	0	
65.	Miguel Jimenez	⁻3	0	4	⁻1	
66.	Tom Lehman	3	⁻3	⁻2	⁻2	

Source: Associated Press.

Tiger Woods won the 2005 British Open, his 10th major golf title.

Rewrite each sum, using the commutative property of addition. Show that the sum is unchanged. See Example 5.

67. ⁻18 + ⁻5 = _____ + _____

Both sums are _____.

68. ⁻12 + 20 = _____ + _____

Both sums are _____.

69. ⁻4 + 15 = _____ + _____

Both sums are _____.

70. 17 + 1 = _____ + _____

Both sums are _____.

In each addition problem, write parentheses around the two addends that would be easiest to add. Then find the sum. See Example 6.

71. 6 + ⁻14 + 14

72. 9 + ⁻9 + ⁻8

73. ⁻14 + ⁻6 + ⁻7

74. ⁻18 + 3 + 7

75. Make up three of your own examples that illustrate the addition property of 0.

76. Make up three of your own examples that illustrate the associative property of addition. Show that the sum is unchanged.

 Find each sum.

77. ⁻7081 + 2965

78. ⁻1398 + 3802

79. ⁻179 + ⁻61 + 8926

80. 36 + ⁻6215 + 428

81. 86 + ⁻99,000 + 0 + 2837

82. ⁻16,719 + 0 + 8878 + ⁻14

1.4 Subtracting Integers

OBJECTIVE 1 Find the opposite of a signed number.   Look at how the integers match up on this number line.

Each integer is matched with its *opposite*. **Opposites** are the same *distance* from 0 on the number line but are on *opposite sides* of 0.

$^+2$ is the opposite of $^-2$ and $^-2$ is the opposite of $^+2$

When you add opposites, the sum is always 0. The opposite of a number is also called its *additive inverse*.

$$2 + {^-2} = \mathbf{0} \quad\text{and}\quad {^-2} + 2 = \mathbf{0}$$

EXAMPLE 1 Finding the Opposites of Signed Numbers

Find the opposite (additive inverse) of each number. Show that the sum of the number and its opposite is 0.

(a)   6   The opposite of 6 is $^-6$ and $6 + {^-6} = \mathbf{0}$

(b) $^-10$ The opposite of $^-10$ is 10   and $^-10 + 10 = \mathbf{0}$

(c)   0   The opposite of 0 is 0   and $0 + 0 = \mathbf{0}$

◀◀◀ **Now Try Exercise 3.**

OBJECTIVE 2 Subtract integers.   Now that you know how to add integers and how to find opposites, you can subtract integers. Every subtraction problem has the same answer as a related addition problem. The problems below illustrate how to change subtraction problems into addition problems.

$$6 - 2 = 4 \qquad\qquad 8 - 3 = 5$$
$$\downarrow \quad \downarrow \qquad\qquad \downarrow \quad \downarrow$$
Same answer   Same answer
$$6 + {^-2} = 4 \qquad\qquad 8 + {^-3} = 5$$

Subtracting Two Integers

To subtract two numbers, *add* the first number to the *opposite* of the second number. Remember to change *two* things:

Step 1   Make one pencil stroke to change the subtraction symbol to an addition symbol.

Step 2   Make a second pencil stroke to change the *second* number to its *opposite*. If the second number is positive, change it to negative. If the second number is negative, change it to positive.

CAUTION

When changing a subtraction problem to an addition problem, do ***not*** make any change in the *first* number. The pattern is

1st number − 2nd number = 1st number + opposite of 2nd number.

EXAMPLE 2 **Subtracting Two Integers**

Make *two* pencil strokes to change each subtraction problem into an addition problem. Then find the sum.

Change 10 to ⁻10.

(a) $4 - 10 = 4 + {}^-10 = {}^-6$

Change subtraction to addition.

Change ⁻6 to ⁺6.

(b) ${}^-9 - {}^-6 = {}^-9 + {}^+6 = {}^-3$

Change subtraction to addition.

Change ⁻5 to ⁺5.

(c) $3 - {}^-5 = 3 + {}^+5 = 8$ Make *two* pencil strokes.

Change subtraction to addition.

Change 9 to ⁻9.

(d) ${}^-2 - 9 = {}^-2 + {}^-9 = {}^-11$ Make *two* pencil strokes.

Change subtraction to addition.

Now Try Exercises 9, 13, and 17.

OBJECTIVE 3 Combine adding and subtracting of integers. When adding and subtracting more than two signed numbers, first change all subtractions to additions. Then add from left to right.

EXAMPLE 3 **Combining Addition and Subtraction**

Simplify by completing all the calculations.

${}^-5 - 10 - 12 + 1$ Change all subtractions to addition. Change
                          10 to ⁻10. Change 12 to ⁻12.

${}^-5 + {}^-10 + {}^-12 + 1$ Add from left to right. First add ⁻5 + ⁻10.

$\quad {}^-15 + {}^-12 + 1$ Then add ⁻15 + ⁻12.

$\qquad {}^-27 + 1$ Finally, add ⁻27 + 1.

$\qquad {}^-26$

Now Try Exercise 41.

Ⓗ **Calculator Tip**   You can use the *change of sign* key (+/−) or (+C−) on your *scientific* calculator to enter negative numbers. To enter ⁻5, press ⑤ (+/−). To enter ⁺5, just press ⑤. To enter Example 3 above, press the following keys.

$^-5$ Subtract.   The answer is ⁻26.

When using a calculator, you do *not* need to change subtraction to addition.

1.4 Exercises

Find the opposite (additive inverse) of each number. Show that the sum of the number and its opposite is 0. *See Example 1.*

1. 6

2. 10

3. $^-$13

4. $^-$3

5. 0

6. 1

Subtract by changing subtraction to addition. See Example 2.

7. $19 - 5$

8. $24 - 11$

9. $10 - 12$

10. $1 - 8$

11. $7 - 19$

12. $2 - 17$

13. $^-15 - 10$

14. $^-10 - 4$

15. $^-9 - 14$

16. $^-3 - 11$

17. $^-3 - ^-8$

18. $^-1 - ^-4$

19. $6 - ^-14$

20. $8 - ^-1$

21. $1 - ^-10$

22. $6 - ^-1$

23. $^-30 - 30$

24. $^-25 - 25$

25. $^-16 - ^-16$

26. $^-20 - ^-20$

27. $13 - 13$

28. $19 - 19$

29. $0 - 6$

30. $0 - 12$

31. (a) $3 - ^-5$

(b) $3 - 5$

(c) $^-3 - ^-5$

(d) $^-3 - 5$

32. (a) $9 - 6$

(b) $^-9 - 6$

(c) $9 - ^-6$

(d) $^-9 - ^-6$

33. (a) $4 - 7$

(b) $4 - ^-7$

(c) $^-4 - 7$

(d) $^-4 - ^-7$

34. (a) $8 - ^-2$

(b) $^-8 - ^-2$

(c) $8 - 2$

(d) $^-8 - 2$

Simplify. See Example 3.

35. $^-2 - 2 - 2$

36. $^-8 - 4 - 8$

37. $9 - 6 - 3 - 5$

38. $12 - 7 - 5 - 4$

39. $3 - ^-3 - 10 - ^-7$

40. $1 - 9 - ^-2 - ^-6$

41. $^-2 + ^-11 - ^-3$

42. $^-5 - ^-2 + ^-6$

43. $4 - ^-13 + ^-5$

44. $6 - ^-1 + ^-10$

45. $6 + 0 - 12 + 1$

46. $^-10 - 4 + 0 + 18$

WINDCHILL
Temperature (degrees Fahrenheit)

Calm	40	35	30	25	20	15	10	5	0	–5	–10	–15	–20	–25	–30
5	36	31	25	19	13	7	1	–6	–11	–16	–22	–28	–34	–40	–46
10	34	27	21	15	9	3	–4	–10	–16	–22	–28	–35	–41	–47	–53
15	32	25	19	13	6	0	–7	–13	–19	–26	–32	–39	–45	–51	–58
20	30	24	17	11	4	–2	–9	–15	–22	–29	–35	–42	–48	–55	–61
25	29	23	16	9	3	–4	–11	–17	–24	–31	–37	–44	–51	–58	–64
30	28	22	15	8	1	–5	–12	–19	–26	–33	–39	–46	–53	–60	–67
35	28	21	14	7	0	–7	–14	–21	–27	–34	–41	–48	–55	–62	–69
40	27	20	13	6	–1	–8	–15	–22	–29	–36	–43	–50	–57	–64	–71

Wind Speed (miles per hour) (left axis)

Shaded area: Frostbite occurs in 15 minutes or less.

Source: National Weather Service

This windchill table shows how wind increases a person's heat loss. For example, find the temperature of 15 °F along the top of the table. Then find a wind speed of 20 mph along the left side of the table. This column and row intersect at $^-2$ °F, the "windchill temperature." The actual temperature is 15 °F but the wind makes it feel like $^-2$ °F. The difference between the actual temperature and the windchill temperature is $15 - {}^-2 = 15 + {}^+2 = 17$ degrees difference.

Use the table to find the windchill temperature under each set of conditions in Exercises 47–48. Then write and solve a subtraction problem to calculate actual temperature minus windchill temperature.

47. (a) 30 °F; 10 mph wind

(b) 15 °F; 15 mph wind

(c) 5 °F; 25 mph wind

(d) $^-$10 °F; 35 mph wind

48. (a) 40 °F; 20 mph wind

(b) 20 °F; 35 mph wind

(c) 10 °F; 15 mph wind

(d) $^-$5 °F; 30 mph wind

49. Find, correct, and explain the mistake made in this subtraction.

$$\begin{array}{c} ^-6 - 6 \\ \downarrow \ \downarrow \\ ^-6 + 6 = 0 \end{array}$$

50. Find, correct, and explain the mistake made in this subtraction.

$$\begin{array}{c} ^-7 - \ \ 5 \\ \downarrow \ \ \ \downarrow \\ ^+7 + {}^-5 = 2 \end{array}$$

Simplify. Begin each exercise by working inside the absolute value bars or the parentheses.

51. $^-2 + {}^-11 + |{}^-2|$

52. $5 - |{}^-3| + 3$

53. $0 - |{}^-7 + 2|$

54. $|1 - 8| - |0|$

55. $^-3 - ({}^-2 + 4) + {}^-5$

56. $5 - 8 - (6 - 7) + 1$

RELATING CONCEPTS (EXERCISES 57–58) For Individual or Group Work

Use your knowledge of the properties of addition to **work Exercises 57 and 58 in order.**

57. Look for a pattern in these pairs of subtractions.

$$^-3 - 5 = \underline{\hspace{1cm}} \qquad ^-4 - {}^-3 = \underline{\hspace{1cm}}$$
$$5 - {}^-3 = \underline{\hspace{1cm}} \qquad ^-3 - {}^-4 = \underline{\hspace{1cm}}$$

Explain what happens when you try to apply the commutative property to subtraction.

58. Recall the addition property of 0. Can 0 be used in a subtraction problem without changing the other number? Explain what happens and give several examples. (*Hint:* Think about *order* in a subtraction problem.)

1.5 Problem Solving: Rounding and Estimating

1 Locate the place to which a number is to be rounded.

2 Round integers.

3 Use front end rounding to estimate answers in addition and subtraction.

One way to get a rough check on an answer is to *round* the numbers in the problem. **Rounding** a number means finding a number that is close to the original number, but easier to work with.

For example, a superintendent of schools in a large city might be discussing the need to build new schools. In making her point, it probably would not be necessary to say that the school district has 152,807 students—it probably would be sufficient to say that there are about 153,000 students, or even 150,000 students.

OBJECTIVE 1 Locate the place to which a number is to be rounded. The first step in rounding a number is to locate the *place to which the number is to be rounded*.

> **EXAMPLE 1 Finding the Place to Which a Number Is to Be Rounded**
>
> Locate and draw a line under the place to which each number is to be rounded. Then answer the question.
>
> **(a)** Round ⁻23 to the nearest ten. Is ⁻23 closer to ⁻20 or ⁻30?
>
>
>
> ⁻23 is closer to ⁻20
>
> Tens place
>
> **(b)** Round $381 to the nearest hundred. Is it closer to $300 or $400?
>
>
>
> $381 is closer to $400
>
> Hundreds place
>
> **(c)** Round ⁻54,702 to the nearest thousand. Is it closer to ⁻54,000 or ⁻55,000?
>
> ⁻54,702 is closer to ⁻55,000
>
> Thousands place ──────
>
> Now Try Exercise 3.

OBJECTIVE 2 Round integers. Use the following steps for rounding integers.

Rounding an Integer

Step 1 Locate the *place* to which the number is to be rounded. Draw a line under that place.

Step 2 Look only at the next digit to the right of the one you underlined. If the next digit is *5 or more, increase* the underlined digit by 1. If the next digit is *4 or less*, do *not* change the digit in the underlined place.

Step 3 *Change* all digits to the right of the underlined place to zeros.

CAUTION

If you are rounding a negative number, be careful to write the negative sign in front of the rounded number. For example, ⁻79 rounds to ⁻80.

EXAMPLE 2 **Using the Rounding Rule for 4 or Less**

Round 349 to the nearest hundred.

Step 1   Locate the place to which the number is being rounded. Draw a line under that place.

Step 2   Because the next digit to the right of the underlined place is 4, which is *4 or less*, do *not* change the digit in the underlined place.

Step 3   Change all digits to the right of the underlined place to zeros.

In other words, 349 is closer to 300 than to 400.

Now Try Exercise 25. ▶▶▶

EXAMPLE 3 **Using the Rounding Rule for 5 or More**

Round 36,833 to the nearest thousand.

Step 1   Find the place to which the number is to be rounded. Draw a line under that place.

Step 2   Because the next digit to the right of the underlined place is 8, which is *5 or more*, add 1 to the underlined place.

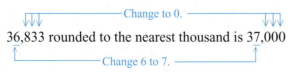

Step 3   Change all digits to the right of the underlined place to zeros.

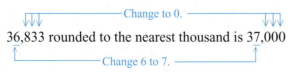

In other words, 36,833 is closer to 37,000 than to 36,000.

Now Try Exercise 19. ▶▶▶

EXAMPLE 4 **Using the Rules for Rounding**

(a) Round $^-$2382 to the nearest ten.

Step 1 $^-$238̠2
   └─ Tens place

Step 2 The next digit to the right is 2, which is *4 or less.*

Next digit is *4 or less.*
$^-$2382
Leave 8 as 8.

Change to 0.
Step 3 $^-$2382 rounds to $^-$2380
Leave 8 as 8.

$^-$2382 rounded to the nearest ten is $^-$2380.

(b) Round 13,961 to the nearest hundred.

Step 1 13,̠961
   └─ Hundreds place

Step 2 The next digit to the right is 6, which is *5 or more.*

Next digit is *5 or more.*
13,961
Change 9 to 10; write 0 and regroup 1 into thousands place.
3 + regrouped 1 = 4

Change to 0.
Step 3 13,961 rounds to 14,000
Change 39 to 40.

13,961 rounded to the nearest hundred is 14,000.

NOTE
In Step 2 above, when you added 1 to the hundreds place, notice that
the first three digits increased from 139 to 140.

13,961   rounded to   **14,0**00

Now Try Exercise 17.

EXAMPLE 5 Rounding Large Numbers

(a) Round ⁻37,892 to the nearest ten-thousand.

Step 1 ⁻37,892
└── Ten-thousands place

Step 2 The next digit to the right is 7, which is *5 or more*.

┌── Next digit is *5 or more*.
⁻37,892
└── Change 3 to 4.

Step 3 ⁻37,892 rounds to ⁻40,000
┌──Change to 0.──┐
└── Change 3 to 4.──┘

⁻37,892 rounded to the nearest ten-thousand is ⁻40,000.

(b) Round 528,498,675 to the nearest million.

Step 1 528,498,675
└── Millions place

Step 2 528,498,675
┌── Next digit is *4 or less*.
└── Leave 8 as 8.

Step 3 528,498,675 rounds to 528,000,000
┌──Change to 0.──┐
└── Leave 8 as 8.──┘

528,498,675 rounded to the nearest million is 528,000,000.

Now Try Exercises 29 and 33. ▶▶▶

OBJECTIVE 3 Use front end rounding to estimate answers in addition and subtraction.   In many everyday situations, we can round numbers and **estimate** the answer to a problem. For example, suppose that you're thinking about buying a sofa for $988 and a chair for $209. You can round the prices and estimate the total cost as $1000 + $200 ≈ $1200. The ≈ symbol means "approximately equal to." The estimated total of $1200 is close enough to help you decide whether you can afford both items. Of course, when it comes time to pay the bill, you'll want the *exact* total of $988 + $209 = $1197.

Sofa		**Chair**			
$988	+	$209	=	$1197	← Exact cost
Rounds to		Rounds to			
↓		↓			
$1000	+	$200	=	$1200	← Estimated cost

In front end rounding, each number is rounded to the highest possible place, so all the digits become 0 except the first digit. Once the numbers have lots of zeros, working with them is easy. Front end rounding is often used to estimate answers.

EXAMPLE 6 Using Front End Rounding

Use front end rounding to round each number.

(a) ⁻216

Round to the highest possible place, that is, the leftmost digit. In this case, the leftmost digit, 2, is in the hundreds place, so round to the nearest hundred.

```
        ┌─Next digit is 4 or less.            ┌─Change to 0.─┐
        ↓                                      ↓↓              ↓↓
    ⁻2 1 6                                  ⁻2 1 6 rounds to ⁻2 0 0
        ↑                                      ↑               ↑
        └─Leave 2 as 2.                        └─Leave 2 as 2.─┘
```

The rounded number is ⁻200. Notice that all the digits in the rounded number are 0, except the first digit. Also, remember to write the negative sign.

(b) 97,203

The leftmost digit, 9, is in the ten-thousands place, so round to the nearest ten-thousand.

```
        ┌─Next digit is 5 or more.            ┌─Change to 0.─┐
        ↓                                      ↓↓↓↓           ↓↓↓↓
    97,203                                  97,203 rounds to 100,000
        ↑                                      ↑               ↑
        └─Change 9 to 10.                      └─Change 9 to 10.─┘
                                               Regroup 1 into the hundred-thousands place.
```

The rounded number is 100,000. Notice that all the digits in the rounded number are 0, except the first digit.

◀◀◀ **Now Try Exercises 41 and 45.**

EXAMPLE 7 Using Front End Rounding to Estimate an Answer

Use front end rounding to estimate an answer. Then find the exact answer.

Meisha's paycheck showed gross pay of $823. It also listed deductions of $291. What is her net pay after deductions?

Estimate: Use front end rounding to round $823 and $291.

```
        ┌─Next digit is 4 or less.            ┌─Next digit is 5 or more.
        ↓                                      ↓
    $823 rounds to $800                    $291 rounds to $300
        ↑                                      ↑
        └─Leave 8 as 8.─┘                      └─Change 2 to 3.─┘
```

Use the rounded numbers and subtract to *estimate* Meisha's net pay.

$$\$800 - \$300 = \$500 \leftarrow \text{Estimate}$$

Exact: Use the original numbers and subtract to find the *exact* amount.

$$\$823 - \$291 = \$532 \leftarrow \text{Exact}$$

Meisha's paycheck will show the *exact* amount of $532. Because $532 is fairly close to the *estimate* of $500, Meisha can quickly see that the amount shown on her paycheck probably is correct. She might also use the estimate when talking to a friend, saying, "My net pay is about $500."

──── **Continued on Next Page**

CAUTION

Always *estimate* the answer first. Then, when you find the *exact* answer, check that it is close to the estimate. If your exact answer is very far off, rework the problem because you probably made an error.

Calculator Tip   It's easy to press the wrong key when using a calculator. If you use front end rounding and estimate the answer *before* entering the numbers, you can catch many such mistakes. For example, a student thought that he entered Example 7 correctly.

823 ⊖ 291 ⊜ | *1114*

Front end rounding gives an estimated answer of $800 - 300 = 500$, which is very different from 1114. Can you figure out which key the student pressed incorrectly? (Answer: The student pressed ⊕ instead of ⊖.)

Now Try Exercise 69. ▶▶▶

1.5 Exercises

Locate and draw a line under the place to which the number is to be rounded. Then answer the question. See Example 1.

1. 42 (nearest ten)
Is it closer to 40 or 50?

2. ⁻57 (nearest ten)
Is it closer to ⁻50 or ⁻60?

3. ⁻691 (nearest hundred)
Is it closer to ⁻600 or ⁻700?

4. 135 (nearest hundred)
Is it closer to 100 or 200?

5. 54,402 (nearest thousand)
Is it closer to 54,000 or 55,000?

6. 85,180 (nearest thousand)
Is it closer to 85,000 or 86,000?

Round each number to the indicated place. See Examples 2–5.

7. 625 to the nearest ten

8. 206 to the nearest ten

9. ⁻1083 to the nearest ten

10. ⁻2439 to the nearest ten

11. 7862 to the nearest hundred

12. 6746 to the nearest hundred

13. ⁻86,813 to the nearest hundred

14. ⁻17,211 to the nearest hundred

15. 42,495 to the nearest hundred

16. 18,273 to the nearest hundred

17. ⁻5996 to the nearest hundred

18. ⁻8451 to the nearest hundred

19. 15,758 to the nearest hundred

20. 28,065 to the nearest hundred

21. ⁻78,499 to the nearest thousand

22. ⁻14,314 to the nearest thousand

23. 5847 to the nearest thousand

24. 49,706 to the nearest thousand

25. 53,182 to the nearest thousand

26. 13,124 to the nearest thousand

27. 595,008 to the nearest ten-thousand

28. 725,182 to the nearest ten-thousand

29. 5,444,000 to the nearest hundred-thousand

30. 3,000,999 to the nearest hundred-thousand

31. ⁻8,906,422 to the nearest million

32. ⁻13,713,409 to the nearest million

33. 139,610,000 to the nearest million

34. 609,845,500 to the nearest million

35. 292,987,333 to the nearest ten-million

36. 54,001,775 to the nearest ten-million

37. 19,951,880,500 to the nearest hundred-million

38. 5,993,505,000 to the nearest hundred-million

39. 8,608,200,100 to the nearest billion

40. 703,750,678,005 to the nearest billion

Use front end rounding to round each number. See Example 6.

41. Tyrone's truck shows this number on the odometer.

42. Ezra bought a used car with this odometer reading.

43. From summer to winter the average temperature drops 56 degrees.

44. The flood waters fell 42 inches yesterday.

45. Jan earned $9942 working part time.

46. Carol deposited $285 in her checking account.

47. 60,950,000 Americans go to a video store each week. (*Source:* Video Software Dealer's Assoc.)

48. 99,375,000 U.S. households have at least one TV. (*Source:* Nielsen Media Research.)

49. The submarine will dive to 255 feet below the surface of the ocean.

50. DeAnne lost $1352 in the stock market.

51. The population of Alaska is 635,478 people, and the population of California is 34,501,994 people. (*Source:* U.S. Bureau of the Census.)

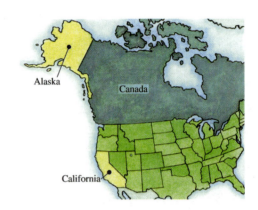

52. By 2010, it is estimated that the U.S population will be 300,118,000 people and Canada's population will be 34,253,000 people. (*Source:* U.S. Bureau of the Census.)

53. Explain in your own words how to do front end rounding. Also show two examples of numbers and how you round them.

54. Describe two situations in your own life when you might use rounded numbers. Describe two situations in which exact numbers are important.

First, use front end rounding to estimate each answer. Then find the exact answer.
In Exercises 63–68, change subtraction to addition. See Example 7.

55. ⁻42 + 89

Estimate: _____ + _____ = _____

Exact:

56. ⁻66 + 25

Estimate: _____ + _____ = _____

Exact:

57. 16 + ⁻97

Estimate: _____ + _____ = _____

Exact:

58. 58 + ⁻19

Estimate: _____ + _____ = _____

Exact:

59. $^-273 + {}^-399$

Estimate:

Exact:

60. $^-311 + {}^-582$

Estimate:

Exact:

61. $3081 + 6826$

Estimate:

Exact:

62. $4904 + 1181$

Estimate:

Exact:

63. $23 - 81$

Estimate:

Exact:

64. $72 - 84$

Estimate:

Exact:

65. $^-39 - 39$

Estimate:

Exact:

66. $^-91 - 91$

Estimate:

Exact:

67. $^-106 + 34 - {}^-72$

Estimate:

Exact:

68. $52 - {}^-87 - 139$

Estimate:

Exact:

First use front end rounding to estimate the answer to each application problem.
Then find the exact answer. See Example 7.

69. The community has raised $52,882 for the homeless shelter. If the goal is $78,650, how much more needs to be collected?

Estimate:

Exact:

70. A truck weighs 9250 pounds when empty. After being loaded with firewood, it weighs 21,375 pounds. What is the weight of the firewood?

Estimate:

Exact:

71. Dorene Cox decided to establish a monthly budget. She will spend $685 for rent, $325 for food, $320 for child care, $182 for transportation, $150 for other expenses, and put the remainder in savings. If her monthly take-home pay is $1920, find her monthly savings.

Estimate:

Exact:

72. Jared Ueda had $2874 in his checking account. He wrote checks for $308 for auto repairs, $580 for child support, and $778 for tuition. Find the amount remaining in his account.

Estimate:

Exact:

73. In a laboratory experiment, a mixture started at a temperature of ⁻102 degrees. First the temperature was raised 37 degrees and then raised 52 degrees. What was the final temperature?

Estimate:

Exact:

74. A scuba diver was photographing fish at 65 feet below the surface of the ocean. She swam up 24 feet and then swam down 49 feet. What was her final depth?

Estimate:

Exact:

75. The White House in Washington, D.C., has 132 rooms, 412 doors, and 147 windows. What is the total number of doors and windows? (*Source: Scholastic Book of World Records.*)

Estimate:

Exact:

76. There are 20,802 McDonald's restaurants throughout the world in 119 different countries. Burger King has 10,529 restaurants. How many restaurants do the two companies have together? (*Source: Scholastic Book of World Records.*)

Estimate:

Exact:

1.6 Multiplying Integers

OBJECTIVE 1 Use a raised dot or parentheses to express multiplication. In arithmetic we usually use "×" when writing multiplication problems. But in algebra, we use a raised dot or parentheses to show multiplication. The numbers being multiplied are called **factors** and the answer is called the **product.**

Arithmetic   ┌──────── **Algebra** ────────┐

$3 \times 5 = 15$ $3 \cdot 5 = 15$ or $3(5) = 15$ or $(3)(5) = 15$

Factors Product   Factors Product   Factors Product   Factors Product

OBJECTIVES

1 Use a raised dot or parentheses to express multiplication.

2 Multiply integers.

3 Identify properties of multiplication.

4 Estimate answers to application problems involving multiplication.

> **NOTE**
> Parentheses are used to show several different things in algebra. When we discussed the associative property of addition earlier in this chapter, we used parentheses as shown below.
>
> $$6 + \underbrace{(9 + {}^-9)} \quad \leftarrow \text{Parentheses show which numbers to add first.}$$
> $$\underbrace{6 + \quad 0}$$
> $$6$$
>
> Now we are using parentheses to indicate multiplication, as in $3(5)$ or $(3)(5)$.

OBJECTIVE 2 Multiply integers. Suppose that our football team gained 5 yards on the first play, gained 5 yards again on the second play, and gained 5 yards again on the third play. We can add to find the result.

$$5 \text{ yards} + 5 \text{ yards} + 5 \text{ yards} = 15 \text{ yards}$$

A quick way to add the same number several times is to multiply.

Our team made 3 plays and **gained 5 yards each time.**   **Our team gained a total of 15 yards.**

$$3 \qquad \cdot \qquad 5 \qquad = \qquad 15$$

Here are the rules for multiplying two integers.

Multiplying Two Integers

If two factors have *different signs*, the product is *negative*.
For example,

$$^-2 \cdot 6 = {}^-12 \qquad \text{and} \qquad 4 \cdot {}^-5 = {}^-20$$

If two factors have the *same sign*, the product is *positive*.
For example,

$$7 \cdot 3 = 21 \qquad \text{and} \qquad {}^-3 \cdot {}^-10 = 30$$

There are several ways to illustrate these rules. First we'll continue with football. Remember, you are interested in the results for *our* team. We will designate **our team** with a **positive sign** and **their team** with a **negative sign**.

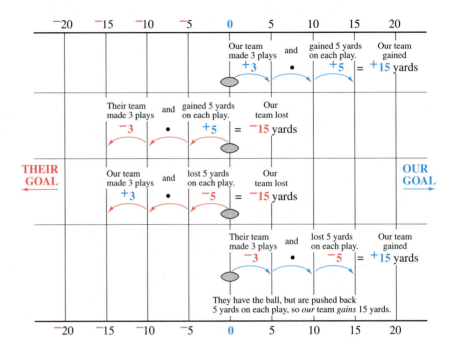

Here is a summary of the football examples.

When two factors have the *same* sign, the product is *positive*.

Both positive

$$3 \cdot 5 \ = 15$$ Product is positive.

$$^-3 \cdot {}^-5 = 15$$

Both negative

When two factors have *different* signs, the product is *negative*.

$$^-3 \cdot 5 \ = {}^-15$$ Product is negative.

$$3 \cdot {}^-5 = {}^-15$$

There is another way to look at these multiplication rules. In mathematics, the rules or patterns must always be consistent.

Look for a pattern in this list of products.

$$4 \cdot 2 = 8$$
$$3 \cdot 2 = 6$$
$$2 \cdot 2 = 4$$
$$1 \cdot 2 = 2$$
$$0 \cdot 2 = 0$$
$$^-1 \cdot 2 = ?$$

Blue numbers decrease by 1.

Red numbers *decrease* by 2.

To keep the red pattern going, replace the *?* with a number that is 2 *less than* 0, which is $^-2$.

So, $^-1 \cdot 2 = ^-2$. This pattern illustrates that the product of two numbers with *different* signs is *negative*.

Look for a pattern in this list of products.

$$4 \cdot {}^-2 = {}^-8$$
$$3 \cdot {}^-2 = {}^-6$$
$$2 \cdot {}^-2 = {}^-4$$
$$1 \cdot {}^-2 = {}^-2$$
$$0 \cdot {}^-2 = 0$$
$${}^-1 \cdot {}^-2 = ?$$

Blue numbers decrease by 1.

Red numbers *increase* by 2.

To keep the red pattern going, replace the *?* with a number that is 2 *more than* 0, which is $^+2$.

So, $^-1 \cdot {}^-2 = {}^+2$. This pattern illustrates that the product of two numbers with the *same* sign is *positive*.

EXAMPLE 1 **Multiplying Two Integers**

(a) $^-2 \cdot 8 = {}^-16$ The factors have *different signs,* so the product is *negative.*

Negative · Positive

(b) $^-10\,(^-6) = 60$ The factors have the *same sign,* so the product is *positive.*

Both negative

(c) $(9)(^-11) = {}^-99$ The factors have *different signs,* so the product is *negative.*

Positive · Negative

Now Try Exercise 3.

Sometimes there are more than two factors in a multiplication problem. If there are parentheses around two of the factors, multiply them first. If there aren't any parentheses, start at the left and work with two factors at a time.

EXAMPLE 2 **Multiplying Several Factors**

Multiply.

(a) $^-3 \cdot (4 \cdot {}^-5)$ Parentheses tell you to multiply $4 \cdot {}^-5$ first. The factors have *different* signs, so the product is *negative.*
$^-3 \cdot {}^-20$ Then multiply $^-3 \cdot {}^-20$. Both factors have the *same* sign, so the product is *positive.*
60

(b) $^-2 \cdot {}^-2 \cdot {}^-2$ There are no parentheses, so multiply $^-2 \cdot {}^-2$ first. The factors have the *same* sign, so the product is *positive.*
$4 \cdot {}^-2$ Then multiply $4 \cdot {}^-2$. The factors have *different* signs, so the product is *negative.*
$^-8$

CAUTION

In Example 2(b) you may be tempted to think that the final product will be *positive* because all the factors have the *same* sign. Be careful to work with just two factors at a time and keep track of the sign at each step.

Calculator Tip   You can use the *change of sign* key for multiplication and division, just as you did for adding and subtracting. To enter Example 2(b) on your scientific calculator, press the following keys.

$$\underbrace{2 \; +/-}_{-2} \; \times \; \underbrace{2 \; +/-}_{-2} \; \times \; \underbrace{2 \; +/-}_{-2} \; =$$ The answer is ⁻8.

Now Try Exercise 31.

OBJECTIVE 3 Identify properties of multiplication.   Addition involving 0 is unusual because adding 0 does *not* change the number. For example, $7 + 0$ is still 7. (See **Section 1.3**.) But what happens in multiplication? Let's use our football team as an example.

Our team made 3 plays	and	didn't gain or lose yards on any play.	Altogether our team didn't gain or lose any yards.
3	•	0	= 0

This example illustrates one of the properties of multiplication.

Multiplication Property of 0

Multiplying any number by 0 gives a product of 0.
Some examples are shown below.

$$^{-}16 \cdot 0 = 0 \qquad (0)(5) = 0 \qquad 32{,}977(0) = 0$$

So, can you multiply a number by something that will *not* change the number?

$$6 \cdot \mathbf{?} = 6 \qquad ^{-}12\,(\mathbf{?}) = ^{-}12 \qquad (\mathbf{?})(5876) = 5876$$

The number 1 can replace the **?** in each example. This illustrates another property of multiplication.

Multiplication Property of 1

Multiplying a number by 1 leaves the number unchanged.
Some examples are shown below.

$$6 \cdot 1 = 6 \qquad ^{-}12(1) = ^{-}12 \qquad (1)(5876) = 5876$$

EXAMPLE 3   **Using Properties of Multiplication**

Multiply. Then name the property illustrated by each example.

(a) $(0)(^{-}48) = 0$ Illustrates the **multiplication property of 0**.

(b) $615(1) = 615$ Illustrates the **multiplication property of 1**.

Now Try Exercises 15 and 21.

When adding, we said that changing the order of the addends did not change the sum (commutative property of addition). We also found that changing the *grouping* of addends did not change the sum (associative property of addition). These same ideas apply to multiplication.

Commutative Property of Multiplication

Changing the *order* of two factors does not change the product.
For example,

$$2 \cdot 5 = 5 \cdot 2 \quad \text{and} \quad {}^{-}4 \cdot 6 = 6 \cdot {}^{-}4$$

Associative Property of Multiplication

Changing the *grouping* of factors does not change the product.
For example,

$$9 \cdot (3 \cdot 2) = (9 \cdot 3) \cdot 2$$

EXAMPLE 4 **Using the Commutative and Associative Properties**

Show that the product is unchanged and name the property that is illustrated in each case.

(a) ${}^{-}7 \cdot {}^{-}4 = {}^{-}4 \cdot {}^{-}7$

        28   =   28   Both products are 28.

This example illustrates the **commutative property of multiplication.**

(b) $5 \cdot (10 \cdot 2) = (5 \cdot 10) \cdot 2$

     5 ·   20   =   50   · 2

       100   =   100   Both products are 100.

This example illustrates the **associative property of multiplication.**

Now Try Exercises 53 and 55.

Now that you are familiar with multiplication and addition, we can look at a property that involves both operations.

Distributive Property

Multiplication *distributes* over addition. An example is shown below.
$$3(6 + 2) = 3 \cdot 6 + 3 \cdot 2$$

What is the **distributive property** really saying? Notice that there is an understood multiplication symbol between the 3 and the parentheses. To "distribute" the 3 means to multiply 3 times each number inside the parentheses.

Understood to be $3(6 + 2)$
multiplying by 3   ↓
           $3 \cdot (6 + 2)$

Using the distributive property,

$3 \cdot (6 + 2)$ can be rewritten as $3 \cdot 6 + 3 \cdot 2$

EXAMPLE 5 **Using the Distributive Property**

Rewrite each product, using the distributive property. Show that the result is unchanged.

(a) $4(3 + 7)$

$$4(3 + 7) = 4 \cdot 3 + 4 \cdot 7$$
$$4(10) = 12 + 28$$
$$40 = 40 \qquad \text{Both results are 40.}$$

(b) $^-2(^-5 + 1)$

$$^-2(^-5 + 1) = {^-2} \cdot {^-5} + {^-2} \cdot 1$$
$$^-2(^-4) = 10 + {^-2}$$
$$8 = 8 \qquad \text{Both results are 8.}$$

Now Try Exercise 51.

OBJECTIVE 4 **Estimate answers to application problems involving multiplication.** Front end rounding can be used to estimate answers in multiplication, just as we did when adding and subtracting (see **Section 1.5**). Once the numbers have been rounded so that there are lots of zeros, we can use the multiplication shortcut described in the Review chapter (see **Section R.3**). As a brief review, look at the pattern in these examples.

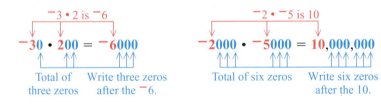

EXAMPLE 6 **Using Front End Rounding to Estimate an Answer**

Use front end rounding to estimate an answer. Then find the exact answer.

Last year the Video Land store had to replace 392 defective videos at a cost of $19 each. How much money did the store lose on defective videos? (*Hint:* Because it's a loss, use a negative number for the cost.)

Estimate: Use front end rounding: 392 rounds to 400 and $^-\$19$ rounds to $^-\$20$. Use the rounded numbers and multiply to estimate the total amount of money lost.

$$\overset{4 \cdot {^-2} \text{ is } {^-8}}{400 \cdot {^-\$20} = {^-\$8000}} \qquad \text{Estimate}$$

Total of three zeros   Write three zeros after the $^-8$.

Exact: $392 \cdot {^-\$19} = {^-\$7448}$

Because the exact answer of $^-\$7448$ is fairly close to the estimate of $^-\$8000$, you can see that $^-\$7448$ probably is correct. The store manager could also use the estimate to say, "We lost about $8000 on defective videos last year."

Now Try Exercise 61.

1.6 Exercises

FOR EXTRA HELP

 Addison-Wesley Math Tutor Center

 MathXL

 Digital Video Tutor CD 1 Videotape 1

Student's Solutions Manual

 MyMathLab

MathXL Tutorials on CD

Multiply. See Examples 1–3.

1. (a) $9 \cdot 7$
   (b) $^-9 \cdot ^-7$
   (c) $^-9 \cdot 7$
   (d) $9 \cdot ^-7$

2. (a) $^-6 \cdot 9$
   (b) $6 \cdot ^-9$
   (c) $^-6 \cdot ^-9$
   (d) $6 \cdot 9$

3. (a) $7(^-8)$
   (b) $^-7(8)$
   (c) $7(8)$
   (d) $^-7(^-8)$

4. (a) $8(6)$
   (b) $^-8(^-6)$
   (c) $^-8(6)$
   (d) $8(^-6)$

5. $^-5 \cdot 7$

6. $^-10 \cdot 2$

7. $(^-5)(9)$

8. $(^-9)(4)$

9. $3(^-6)$

10. $8(^-9)$

11. $10(^-5)$

12. $5(^-11)$

13. $(^-1)(40)$

14. $(75)(^-1)$

15. $^-56 \cdot 1$

16. $1 \cdot ^-87$

17. $^-8(^-4)$

18. $^-3(^-9)$

19. $11 \cdot 7$

20. $4 \cdot 25$

21. $25 \cdot 0$

22. $0 \cdot 30$

23. $^-19(^-7)$

24. $^-21(^-3)$

25. $^-13(^-1)$

26. $^-1(^-31)$

27. $(0)(^-25)$

28. $(^-50)(0)$

29. $^-4 \cdot ^-6 \cdot 2$

30. $^-9 \cdot 3 \cdot ^-3$

31. $(^-4)(^-2)(^-7)$

32. $(^-6)(^-2)(^-3)$

33. $5(^-8)(4)$

34. $5(4)(^-6)$

Write an integer in each blank to make a true statement.

35. $(^-3)(\_\_\_) = ^-15$

36. $6 \cdot \_\_\_ = ^-24$

37. $\_\_\_ \cdot 10 = ^-30$

38. $(\_\_\_)(^-4) = 16$

39. $^-17 = 17(\_\_\_)$

40. $29 = ^-29(\_\_\_)$

41. $(\_\_\_)(^-350) = 0$

42. $\_\_\_ \cdot 99 = 99$

43. $5 \cdot ^-4 \cdot \_\_\_ = ^-100$

44. $\_\_\_ \cdot 2 \cdot ^-2 = ^-24$

45. $(\_\_\_)(^-5)(^-2) = ^-40$

46. $(^-3)(\_\_\_)(^-3) = ^-27$

47. In your own words, explain the difference between the commutative and associative properties of multiplication. Show an example of each.

48. A student did this multiplication.

$$^-3 \cdot {}^-3 \cdot {}^-3 = 27$$

He knew that $3 \cdot 3 \cdot 3$ is 27. Since all the factors have the same sign, he made the product positive. Do you agree with his reasoning? Explain.

RELATING CONCEPTS (EXERCISES 49–50) For Individual or Group Work

Look for patterns as you **work Exercises 49 and 50 in order.**

49. Write three numerical examples for each of these situations:

(a) A positive number multiplied by $^-1$

(b) A negative number multiplied by $^-1$

Now write a rule that explains what happens when you multiply a signed number by $^-1$.

50. Do these multiplications.

$$^-2 \cdot {}^-2 = \underline{\quad}$$
$$^-2 \cdot {}^-2 \cdot {}^-2 = \underline{\quad}$$
$$^-2 \cdot {}^-2 \cdot {}^-2 \cdot {}^-2 = \underline{\quad}$$
$$^-2 \cdot {}^-2 \cdot {}^-2 \cdot {}^-2 \cdot {}^-2 = \underline{\quad}$$

Describe the pattern in the products. Then find the next three products without multiplying all the $^-2$s.

Rewrite each multiplication, using the stated property. Show that the result is unchanged. See Examples 4 and 5.

51. Distributive property
   $9(^-3 + 5)$

52. Distributive property
   $^-6(4 + {}^-5)$

53. Commutative property
   $25 \cdot 8$

54. Commutative property
   $^-7 \cdot {}^-11$

55. Associative property
   $^-3 \cdot (2 \cdot 5)$

56. Associative property
   $(5 \cdot 5) \cdot 10$

First use front end rounding to estimate the answer to each application problem.
Then find the exact answer. See Example 6.

57. Alliette receives $324 per week for doing child care in her home. How much income will she have for an entire year? There are 52 weeks in a year.

Estimate:

Exact:

58. Enrollment at our community college has increased by 875 students each of the last four semesters. What is the total increase?

Estimate:

Exact:

59. A new computer software store had losses of $9950 during each month of its first year. What was the total loss for the year?

Estimate:

Exact:

60. A long-distance phone company estimates that it is losing 95 customers each week. How many customers will it lose in a year?

Estimate:

Exact:

61. Tuition at the state university is $182 per credit for undergraduates. How much tuition will Wei Chen pay for 13 credits?

Estimate:

Exact:

62. Pat ate a dozen crackers as a snack. Each cracker had 17 calories. How many calories did Pat eat?

Estimate:

Exact:

63. There are 24 hours in one day. How many hours are in one year (365 days)?

Estimate:

Exact:

64. There are 5280 feet in one mile. How many feet are in 17 miles?

Estimate:

Exact:

Simplify.

65. $^-8 \cdot |^-8 \cdot 8|$

66. $^-7 \cdot |7| \cdot |^-7|$

67. $(^-37)(^-1)(85)(0)$

68. $^-1(9732)(^-1)(^-1)$

69. $|6 - 7| \cdot {}^-355{,}299$

70. $987 \cdot {}^-65{,}432 \cdot |9 - 9|$

Each of these application problems requires several steps and may involve addition and subtraction as well as multiplication.

71. Each of Maurice's four cats needed a $24 rabies shot and a $29 shot to prevent respiratory infections. There was also one $35 office visit charge. What was the total amount of Maurice's bill?

72. Chantele has three children. Her older daughter had a throat culture taken at the clinic today. Her baby received three immunization shots and her son received two shots. The co-pay amounts were $8 for each shot, an $18 office charge for each child, and $12 for the throat culture. How much did Chantele pay?

73. There is a 3-degree drop in temperature for every thousand feet that an airplane climbs into the sky. If the temperature on the ground is 50 degrees, what will be the temperature when the plane reaches an altitude of 24,000 feet? (*Source:* Lands' End.)

74. An unmanned research submarine descends to 150 feet below the surface of the ocean. Then it continues to go deeper, taking a water sample every 25 feet. What is its depth when it takes the 15th sample?

75. In Ms. Zubero's algebra class, there are six tests of 100 points each, eight quizzes of six points each, and 20 homework assignments of five points each. There are also four "bonus points" on each test. What is the total number of possible points?

76. In Mr. Jackson's prealgebra class, there are three group projects worth 25 points each, five 100-point tests, a 150-point final exam, and seven quizzes worth 12 points each. Find the total number of possible points.

1.7 Dividing Integers

OBJECTIVE 1 Divide integers.   In arithmetic, we usually use $\overline{)}$ to write division problems so that we can do them by hand. Calculator keys use the ÷ symbol for division. In algebra, we usually show division by using a fraction bar, a slash mark, or the ÷ symbol. The answer to a division problem is called the **quotient.**

OBJECTIVES

1. **Divide integers.**
2. **Identify properties of division.**
3. **Combine multiplying and dividing of integers.**
4. **Estimate answers to application problems involving division.**
5. **Interpret remainders in division application problems.**

Arithmetic

$$\text{Divisor} \rightarrow 2\overline{)16} \begin{array}{l} \leftarrow \text{Quotient} \\ \leftarrow \text{Dividend} \end{array}$$

$$\overset{8}{}$$

Calculator and Algebra

$$\underset{\text{Quotient}}{\underset{\uparrow}{16} \div \underset{\uparrow}{2} = \underset{\uparrow}{8}}$$

Dividend, Divisor

$$\text{Dividend} \rightarrow \frac{16}{2} = 8 \leftarrow \text{Quotient}$$
$$\text{Divisor} \rightarrow$$

For every division problem, we can write a related multiplication problem. (See **Section R.4.**) Because of this relationship, the rules for dividing integers are the same as the rules for multiplying integers. For example,

$$\frac{16}{8} = 2 \quad \text{because} \quad (2)(8) = 16$$

$$\frac{{}^-16}{{}^-8} = 2 \quad \text{because} \quad (2)({}^-8) = {}^-16$$

$$\frac{{}^-16}{8} = {}^-2 \quad \text{because} \quad ({}^-2)(8) = {}^-16$$

$$\frac{16}{{}^-8} = {}^-2 \quad \text{because} \quad ({}^-2)({}^-8) = 16$$

Dividing Two Integers

If two numbers have *different signs,* the quotient is *negative.*
Some examples are shown below.

$$\frac{{}^-18}{3} = {}^-6 \qquad \frac{40}{{}^-5} = {}^-8$$

If two numbers have the *same sign,* the quotient is *positive.*
Some examples are shown below.

$$\frac{{}^-30}{{}^-6} = 5 \qquad \frac{48}{8} = 6$$

EXAMPLE 1 Dividing Two Integers

Divide.

(a) $\dfrac{{}^-20}{5}$ ← Numbers have *different* signs, so the quotient is *negative.* $\dfrac{{}^-20}{5} = {}^-4$

(b) $\dfrac{{}^-24}{{}^-4}$ ← Numbers have the *same* sign, so the quotient is *positive.* $\dfrac{{}^-24}{{}^-4} = 6$

(c) $60 \div {}^-2$ Numbers have *different* signs, so the quotient is *negative.* $60 \div {}^-2 = {}^-30$

Now Try Exercise 1. ▶▶▶

OBJECTIVE 2 Identify properties of division. You have seen that 0 and 1 are used in special ways in addition and multiplication. This is also true in division.

Examples			Pattern (Division Property)
$\dfrac{5}{5} = 1$	$\dfrac{^-18}{^-18} = 1$	$\dfrac{^-793}{^-793} = 1$	When a nonzero number is divided by itself, the quotient is 1.
$\dfrac{5}{1} = 5$	$\dfrac{^-18}{1} = {^-18}$	$\dfrac{^-793}{1} = {^-793}$	When a number is divided by 1, the quotient is the number.
$\dfrac{0}{5} = 0$	$\dfrac{0}{^-18} = 0$	$\dfrac{0}{^-793} = 0$	When 0 is divided by any other number (except 0), the quotient is 0.
$\dfrac{5}{0}$ is undefined.	$\dfrac{^-18}{0}$ is undefined.		Division by 0 is *undefined*. There is no answer.

The most surprising property is that division by 0 *cannot be done*. Let's review the reason for that by rewriting this division problem as a related multiplication problem.

$$\frac{^-18}{0} = ? \quad \text{can be written as the multiplication} \quad ? \cdot 0 = {^-18}$$

If you thought the answer to $\frac{^-18}{0}$ should be 0, try replacing **?** with 0. It doesn't work in the related multiplication problem! Try replacing **?** with any number you like. The result in the related multiplication problem is always 0 instead of $^-18$. That is how we know that dividing by 0 cannot be done. Mathematicians say that it is *undefined* and have agreed never to divide by 0.

EXAMPLE 2 Using the Properties of Division

Divide. Then state the property illustrated by each example.

(a) $\dfrac{^-312}{^-312} = 1$ Any nonzero number divided by itself is 1.

(b) $\dfrac{75}{1} = 75$ Any number divided by 1 is the number.

(c) $\dfrac{0}{^-19} = 0$ Zero divided by any nonzero number is 0.

(d) $\dfrac{48}{0}$ is undefined.   Division by 0 is undefined.

Calculator Tip Try Examples 2(c) and 2(d) above on your calculator. Use the change of sign key to enter $^-19$ on a *scientific* calculator.

0 ÷ 19 +/- =   Answer is 0
$^-19$

48 ÷ 0 =   Calculator shows "Error" or "ERR" or "E" for error because it cannot divide by 0.

Now Try Exercises 5 and 7.

OBJECTIVE 3 **Combine multiplying and dividing of integers.** When a problem involves both multiplying and dividing, first check to see if there are any parentheses. Do what is inside parentheses first. Then start at the left and work toward the right, using two numbers at a time.

EXAMPLE 3 **Combining Multiplication and Division of Integers**

Simplify.

(a) $6\,(^-10) \div (^-3 \cdot 2)$ Do operations inside parentheses first: $^-3 \cdot 2$ is $^-6$. The signs are *different*, so the product is *negative*.

$6\,(^-10) \div \quad ^-6$ Start at the left: $6\,(^-10)$ is $^-60$. The signs are *different*, so the product is *negative*.

$^-60 \div \quad ^-6$ Finally, $^-60 \div\, ^-6$ is 10. The signs are the *same*, so the quotient is *positive*.

10

(b) $^-24 \div\, ^-2\,(4) \div\, ^-6$ No operations inside parentheses, so start at the left: $^-24 \div\, ^-2$ is 12. (*Same* signs, *positive* quotient.)

$\mathbf{12}\,(4) \div\, ^-6$ Next, $12\,(4)$ is 48. (*Same* signs, *positive* product.)

$48 \div\, ^-6$ Finally, $48 \div\, ^-6$ is $^-8$. (*Different* signs, *negative* quotient.)

$^-8$

(c) $^-50 \div\, ^-5 \div\, ^-2$ There are no parentheses, so start at the left: $^-50 \div\, ^-5$ is 10. (*Same* signs, *positive* quotient.)

$10 \div\, ^-2$ Now, $10 \div\, ^-2$ is $^-5$. (*Different* signs, *negative* quotient.)

$^-5$

Now Try Exercise 41. ▶▶▶

OBJECTIVE 4 **Estimate answers to application problems involving division.** Front end rounding can be used to estimate answers in division just as you did when multiplying (see **Section 1.6**). Once the numbers have been rounded so that there are lots of zeros, you can use the division short-cut described in the Review chapter (see **Section R.5**). As a brief review, look at the pattern in these examples.

$$4000 \div\, ^-50 = 400 \div\, ^-5 = ^-80$$

Drop one 0 from both
dividend and divisor.

$$^-6000 \div\, ^-3000 = ^-6 \div\, ^-3 = 2$$

Drop three zeros from both
dividend and divisor.

EXAMPLE 4 **Using Front End Rounding to Estimate an Answer in Division**

First use front end rounding to estimate an answer. Then find the exact answer.

During a 24-hour laboratory experiment, the temperature of a solution dropped 96 degrees. What was the average drop in temperature each hour?

Estimate: Use front end rounding: ⁻96 degrees rounds to ⁻100 degrees and 24 hours rounds to 20 hours. To estimate the average, divide the rounded number of degrees by the rounded number of hours.

$$\text{⁻}100 \text{ degrees} \div 20 \text{ hours} = \text{⁻}5 \text{ degrees each hour} \quad \leftarrow \text{Estimate}$$

Exact: $\quad$ ⁻96 degrees $\div$ 24 hours = ⁻4 degrees each hour $\quad \leftarrow$ Exact

Because the exact answer of ⁻4 degrees is close to the estimate of ⁻5 degrees, you can see that ⁻4 degrees probably is correct.

> ▦ **Calculator Tip** The answer in Example 4 above "came out even." In other words, the quotient was an integer. Suppose that the drop in temperature had been 97 degrees. Do the division on your calculator.
>
> $$97 \; \boxed{+/-} \; \boxed{\div} \; 24 \; \boxed{=} \quad \text{Calculator shows } \text{⁻}4.041666667$$
> $$\underbrace{\qquad\qquad}_{\text{⁻}97}$$
>
> The quotient is *not* an integer. We will work with numbers like these in **Chapter 5,** Positive and Negative Decimals.

◀◀◀ **Now Try Exercise 51.**

OBJECTIVE 5 **Interpret remainders in division application problems.** In arithmetic, division problems often have a remainder, as shown below.

$$\begin{array}{r} 14 \text{ } \mathbf{R10} \\ 25\overline{)360} \\ \underline{25} \\ 110 \\ \underline{100} \\ \mathbf{10} \leftarrow \text{Remainder} \end{array}$$

But what does **R10** really mean? Let's look at this same problem by using money amounts.

EXAMPLE 5 **Interpreting Remainders in Division Applications**

Divide; then interpret the remainder in each application.

(a) The math department at Lake Community College has $360 in its budget to buy scientific calculators for the math lab. If the calculators cost $25 each, how many can be purchased? How much money will be left over?

We can solve this problem by using the same division as shown above. But this time we can decide what the remainder really means.

$$\begin{array}{r} 14 \quad \leftarrow \text{Number of calculators purchased} \\ \text{Cost of one calculator} \rightarrow \$25\overline{)\$360} \quad \leftarrow \text{Budget} \\ \underline{25} \\ 110 \\ \underline{100} \\ \$10 \quad \leftarrow \text{Money left over} \end{array}$$

The department can buy 14 calculators. There will be $10 left over.

Continued on Next Page

▦ **Calculator Tip**   You can use your calculator to solve Example 5(a). Recall that digits on the *right* side of the decimal point show *part* of one whole. You cannot order *part* of one calculator, so ignore those digits and use only the *whole number part* of the quotient.

Now use your calculator to find how much money was spent to buy 14 calculators.

(ON/AC) 14 (×) 25 (=) 350 ← Cost of 14 calculators

↑
Clear 14.4 from calculator before multiplying by pressing (C), (ON/C), or (ON/AC).

(b) Luke's son is going on a Scout camping trip. There are 135 Scouts. Luke is renting tents that sleep 6 people each. How many tents should he rent?

We again use division to solve the problem. There is a remainder, but this time it must be interpreted differently than in the calculator example.

$$
\begin{array}{r}
22 \\
6\overline{)135} \\
12 \\
\hline
15 \\
12 \\
\hline
3
\end{array}
$$

Each tent holds → 6)135   ← Number of tents with 6 Scouts each
                            ← Total number of Scouts
                            3 ← Scouts left over

If Luke rents 22 tents, 3 Scouts will have to sleep out in the rain. He must rent 23 tents to accommodate all the Scouts. (One tent will have only 3 Scouts in it.)

▸ **Now Try Exercise 69.** ▸▸▸

1.7 Exercises

FOR EXTRA HELP   Addison-Wesley Math Tutor Center   MathXL   Digital Video Tutor CD 1 Videotape 1   Student's Solutions Manual   MyMathLab   MathXL Tutorials on CD

Divide. See Examples 1 and 2.

1. (a) $14 \div 2$

  (b) $^-14 \div\,^-2$

  (c) $14 \div\,^-2$

  (d) $^-14 \div 2$

2. (a) $^-18 \div\,^-3$

  (b) $18 \div 3$

  (c) $^-18 \div 3$

  (d) $18 \div\,^-3$

3. (a) $^-42 \div 6$

  (b) $^-42 \div\,^-6$

  (c) $42 \div\,^-6$

  (d) $42 \div 6$

4. (a) $45 \div 5$

  (b) $45 \div\,^-5$

  (c) $^-45 \div\,^-5$

  (d) $^-45 \div 5$

5. (a) $\dfrac{35}{35}$

  (b) $\dfrac{35}{1}$

  (c) $\dfrac{^-13}{1}$

  (d) $\dfrac{^-13}{^-13}$

6. (a) $\dfrac{^-23}{1}$

  (b) $\dfrac{^-23}{^-23}$

  (c) $\dfrac{17}{1}$

  (d) $\dfrac{17}{17}$

7. (a) $\dfrac{0}{50}$

  (b) $\dfrac{50}{0}$

  (c) $\dfrac{^-11}{0}$

  (d) $\dfrac{0}{^-11}$

8. (a) $\dfrac{^-85}{0}$

  (b) $\dfrac{0}{^-85}$

  (c) $\dfrac{6}{0}$

  (d) $\dfrac{0}{6}$

9. $\dfrac{^-8}{2}$

10. $\dfrac{^-14}{7}$

11. $\dfrac{21}{^-7}$

12. $\dfrac{30}{^-6}$

13. $\dfrac{^-54}{^-9}$

14. $\dfrac{^-48}{^-6}$

15. $\dfrac{55}{^-5}$

16. $\dfrac{70}{^-7}$

17. $\dfrac{^-28}{0}$

18. $\dfrac{^-40}{0}$

19. $\dfrac{14}{^-1}$

20. $\dfrac{25}{^-1}$

21. $\dfrac{^-20}{^-2}$

22. $\dfrac{^-80}{^-4}$

23. $\dfrac{^-48}{^-12}$

24. $\dfrac{^-30}{^-15}$

25. $\dfrac{^-18}{18}$

26. $\dfrac{50}{^-50}$

27. $\dfrac{0}{^-9}$

28. $\dfrac{0}{^-4}$

29. $\dfrac{^-573}{^-3}$

30. $\dfrac{^-580}{^-5}$

31. $\dfrac{163{,}672}{^-328}$

32. $\dfrac{^-69{,}496}{1022}$

Simplify. See Example 3.

33. $^-60 \div 10 \div {^-3}$

34. $36 \div {^-4} \div 3$

35. $^-64 \div {^-8} \div {^-2}$

36. $^-72 \div {^-9} \div {^-4}$

37. $100 \div {^-5}({^-2})$

38. $^-80 \div 4({^-5})$

39. $48 \div 3 \cdot (12 \div {^-4})$

40. $^-2 \cdot ({^-3} \cdot {^-7}) \div 7$

41. $^-5 \div {^-5}({^-10}) \div {^-2}$

42. $^-9(4) \div {^-36}(50)$

43. $64 \cdot 0 \div {^-8}(10)$

44. $^-88 \div {^-8} \div {^-11}(0)$

RELATING CONCEPTS (EXERCISES 45–50)   For Individual or Group Work

*Use your knowledge of the properties of multiplication as you **work Exercises 45–50 in order.***

45. Explain whether or not division is commutative like multiplication. Start by doing these two divisions on your calculator: $2 \div 1$ and $1 \div 2$.

46. Explain whether or not division is associative like multiplication. Start by doing these two divisions: $(12 \div 6) \div 2$ and $12 \div (6 \div 2)$.

47. Explain what is different and what is similar about multiplying and dividing two signed numbers.

48. In your own words, describe at least three division properties. Include examples to illustrate each property.

49. Write three numerical examples for each situation.
   (a) A negative number divided by $^-1$

   (b) A positive number divided by $^-1$

   Now write a rule that explains what happens when you divide a signed number by $^-1$.

50. Explain why $\frac{0}{^-3}$ and $\frac{^-3}{0}$ do not give the same result.

Solve these application problems by using addition, subtraction, multiplication, or division. First use front end rounding to estimate the answer. Then find the exact answer. See Example 4.

51. The greatest ocean depth is 35,836 feet below sea level. If an unmanned research sub dives to that depth in 17 equal steps, how far does it dive in each step? (*Source: The Top 10 of Everything.*)

Estimate:

Exact:

52. Our college enrollment dropped by 3245 students over the last 11 years. What was the average drop in enrollment each year?

Estimate:

Exact:

53. When Ashwini discovered that her checking account was overdrawn by $238, she quickly transferred $450 from her savings to her checking account. What is the new balance in her checking account?

Estimate:

Exact:

54. The Tigers offensive team lost a total of 48 yards during the first half of the football game. During the second half they gained 191 yards. How many yards did they gain or lose during the entire game?

Estimate:

Exact:

55. The foggiest place in the United States is Cape Disappointment, Washington. It is foggy there an average of 106 days each year. How many days is it not foggy each year? (*Source:* National Weather Service.)

Estimate:

Exact:

56. The number of cellular phone users worldwide in 1999 was 490,000,000. The number of users reached 1,341,000,000 in 2003. What was the increase in the number of users during this 5-year period? (*Source: The World Almanac.*)

Estimate:

Exact:

57. A plane descended an average of 730 feet each minute during a 37-minute landing. How far did the plane descend during the landing?

Estimate:

Exact:

58. A discount store found that 174 items were lost to shoplifting last month. The average value of each item was $24. What was the total loss due to shoplifting?

Estimate:

Exact:

59. Mr. and Mrs. Martinez drove on the Interstate for five hours and traveled 315 miles. What was the average number of miles they drove each hour?

Estimate:

Exact:

60. Rochelle has a 48-month car loan for $12,048. How much is her monthly payment?

Estimate:

Exact:

Find the exact answer in Exercises 61–66. Solving these problems requires more than one step.

61. Clarence bowled four games and had scores of 143, 190, 162, and 177. What was his average score? (*Hint:* To find the average, add all the scores and divide by the number of scores.)

62. Sheila kept track of her grocery expenses for six weeks. The amounts she spent were $84, $111, $82, $110, $98, and $79. What was the average weekly cost of her groceries?

63. On the back of an oatmeal box, it says that one serving weighs 40 grams and that there are 13 servings in the box. On the front of the box, it says that the weight of the contents is 510 grams. What is the difference in the total weight on the front and the back of the box? (*Source:* Quaker Oats.)

64. A 2000-calorie-per-day diet recommends that you eat no more than 65 grams of fat. If each gram of fat is 9 calories, how many calories can you consume in other types of food? (*Source:* U.S. Department of Agriculture.)

65. Stephanie had $302 in her checking account. She wrote a $116 check for day care and a $548 check for rent. She also deposited her $347 paycheck. What is the balance in her account?

66. Gary started a new checking account with a $500 deposit. The bank charged him $18 to print his checks. He also wrote a $193 check for car repairs and a $289 check to his credit card company. What is the balance in his account?

Divide; then interpret the remainder in each application. See Example 5.

67. A cellular phone company is offering 1000 free minutes of air time to new subscribers. How many hours of free time will a new subscriber receive?

68. Nikki is catering a large party. If one pie will serve eight guests, how many pies should she make for 100 guests?

69. Hurricane victims are being given temporary shelter in a hotel. Each room can hold five people. How many rooms are needed for 163 people?

70. A college has received a $250,000 donation to be used for scholarships. How many $3500 scholarships can be given to students?

Simplify:

71. $|{}^-8| \div {}^-4 \cdot |{}^-5| \cdot |1|$

72. ${}^-6 \cdot |{}^-3| \div |9| \cdot {}^-2$

73. ${}^-6({}^-8) \div ({}^-5 - {}^-5)$

74. ${}^-9 \div {}^-9({}^-9 \div 9) \div (12 - 13)$

75. Look back at page 2 of this chapter. You guessed how many days it would take to receive a million dollars if you got $1 each second. Here's how to use your calculator to get the answer. If you get $1 per second, it would take 1,000,000 seconds to receive $1,000,000. Press these keys.

1000000	÷ 60	= 16666.66667	÷ 60	= 277.7777778	÷ 24	= 11.57407407
	There are 60 seconds in one minute.	About 16,667 minutes	There are 60 minutes in one hour.	About 278 hours	There are 24 hours in one day.	About $11\frac{1}{2}$ days (11.5 is equivalent to $11\frac{1}{2}$.)

Notice that you do *not* have to re-enter the intermediate answers. When the answer 16666.66667 appears on your calculator display, just go ahead and enter ÷ 60.

Now use a *scientific* calculator to find how long it takes to receive a *billion* dollars. Start by entering 1000000000. Then follow the pattern shown above. You will need to do one more division step to get the number of years. (Assume that there are 365 days in one year.)

Summary Exercises on Operations with Integers

Simplify each expression.

1. $2 - 8$

2. $(^-16)(0)$

3. $^-14 - ^-7$

4. $\dfrac{^-42}{6}$

5. $^-9(^-7)$

6. $\dfrac{^-12}{12}$

7. $(1)(^-56)$

8. $1 + ^-23$

9. $5 - ^-7$

10. $^-88 \div ^-11$

11. $^-18 + 5$

12. $\dfrac{0}{^-10}$

13. $^-40 - ^-40$

14. $^-17 + 0$

15. $8(^-6)$

16. $^-1 - 9$

17. $^-5(10)$

18. $\dfrac{30}{0}$

19. $0 - 14$

20. $\dfrac{18}{^-3}$

21. $^-13 + 13$

22. $\dfrac{^-16}{^-1}$

23. $20 - 50$

24. $\dfrac{^-7}{0}$

25. $(^-4)(^-6)(2)$

26. $^-2 + ^-12 + ^-5$

27. $^-60 \div 10 \div ^-3$

28. $^-8 - 4 - 8$

29. $64(0) \div ^-8$

30. $2 - ^-5 + 9$

31. $^-9 + 8 + ^-2$

32. $(^-6)(^-2)(^-3)$

33. $8 + 6 + ^-8$

34. $9 - 0 - 16$

35. $^-25 \div ^-1 \div ^-5$

36. $1 - 32 + 0 \left(-31\right)$

37. $^-72 \div ^-9 \div ^-4$

38. $^-7 + 28 + ^-56 + 3 \left(-32\right)$

39. $9 - 6 - 3 - 5$

40. $^-6(^-8) \div (^-5 - 7)$

41. $^-1(9732)(^-1)(^-1) \left(-9732\right)$

42. $^-80 \div 4(^-5 \left(100\right)$

43. $^-10 - 4 + 0 + 18$

44. $^-7 \cdot |7| \cdot |^-7| \left(343\right)$

45. $5 - |^-3| + 3 \left(11\right)$

46. $^-2(^-3)(7) \div ^-7$

47. $^-3 - (^-2 + 4) - 5$

48. $0 - |^-7 + 2| \left(-5\right)$

49. Describe what happens in each situation.
  (a) Zero is divided by a nonzero number.

  (b) Any number is multiplied by 0.

  (c) A nonzero number is divided by itself.

50. Find, explain, and correct the errors in this student's work.

  (a) $\dfrac{15}{^-15} = 0$

  (b) $\dfrac{8}{0} = 8$

  (c) $^-10 \div ^-2 \div ^-5 = 1$

1.8 Exponents and Order of Operations

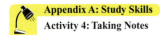

Appendix A: Study Skills

Activity 4: Taking Notes

OBJECTIVE 1 Use exponents to write repeated factors. An **exponent** is a quick way to write repeated multiplication. Here is an example.

$2 \cdot 2 \cdot 2 \cdot 2 \cdot 2$ can be written $2^5 \longleftarrow$ Exponent

$\uparrow$

Base

The *base* is the number being multiplied over and over, and the exponent tells how many times to use the number as a factor. This is called *exponential notation* or *exponential form*.

To simplify 2^5, actually do the multiplication.

$$2^5 = \underbrace{2 \cdot 2 \cdot 2 \cdot 2 \cdot 2}_{} = 32$$

$\qquad$ Exponential form $\qquad$ Factored form $\qquad$ Simplified form

Here are some more examples, using 2 as the base.

$2 = 2^1$ is read   "2 to the **first power**."

$2 \cdot 2 = 2^2$ is read   "2 to the **second power**" or, more commonly, "2 **squared**."

$2 \cdot 2 \cdot 2 = 2^3$ is read   "2 to the **third power**" or, more commonly, "2 **cubed**."

$2 \cdot 2 \cdot 2 \cdot 2 = 2^4$ is read   "2 to the **fourth power**."

$2 \cdot 2 \cdot 2 \cdot 2 \cdot 2 = 2^5$ is read   "2 to the **fifth power**."

and so on.

We usually don't write an exponent of 1, so if no exponent is shown, you can assume that it is 1. For example, 6 is actually 6^1, and 4 is actually 4^1.

> **NOTE**
> Exponents can also be negative numbers or 0, for example, 2^{-3} and 2^0. You will learn more about these exponents in **Chapter 10.**

EXAMPLE 1 Using Exponents

Complete this table.

	Exponential Form	Factored Form	Simplified	Read as
(a)		$5 \cdot 5 \cdot 5$		
(b)		$(4)(4)$		
(c)		7		

Now Try Exercises 1 and 3.

OBJECTIVE **2** **Simplify expressions containing exponents.** Exponents are also used with signed numbers, as shown below.

$(^-3)^2 = (^-3)(^-3) = 9$ The factors have the same sign, so the product is positive.

$(^-4)^3 = \underbrace{(^-4)(^-4)}(^-4)$ Multiply two numbers at a time.

$\underbrace{16 \quad (^-4)}$ First, $(^-4)(^-4)$ is positive 16.

$^-64$ Then, $16(^-4)$ is $^-64$.

Simplify exponents before you do other multiplications, as shown below. Notice that the exponent applies only to the *first* thing to its *left*.

Exponent applies only to the 2. $2^3(5)(4^2)$ Exponent applies only to the 4.

$(2)(2)(2)$ is 8. → $8\,(5)(16)$ ← $(4)(4)$ is 16.

$\underbrace{40\ (16)}$

640

EXAMPLE 2 **Using Exponents with Negative Numbers**

Simplify.

(a) $(^-5)^2 = (^-5)(^-5) = 25$

(b) $(^-5)^3 = \underbrace{(^-5)(^-5)}(^-5)$

$\qquad\qquad \rightarrow 25(^-5)$

$\qquad\qquad ^-125$

(c) $(^-2)^4 = (^-2)(^-2)(^-2)(^-2) = 16$

(d) $2^3(^-3)^2 = \underbrace{(2)(2)(2)}\,\underbrace{(^-3)(^-3)}$

$\qquad\qquad\quad \rightarrow (8)(9) \leftarrow$

$\qquad\qquad\qquad 72$

⊞ Calculator Tip On a *scientific* calculator, use the exponent key $\boxed{y^x}$ to enter exponents. To enter 5^8, press the following keys.

$5 \quad \boxed{y^x} \quad 8 \quad \boxed{=}$ Answer is 390,625.

$\qquad \uparrow \qquad\qquad \uparrow$

$\quad$ Base $\qquad$ Exponent

Be careful when using your calculator's exponent key with a negative number, such as $(^-5)^3$. Different calculators use different keystrokes, so check the instruction manual, or experiment to see how your calculator works.

Now Try Exercises 23 and 25. ▶▶▶

O B J E C T I V E 3 Use the order of operations.   In **Sections 1.4** and **1.7**, you worked examples that mixed either addition and subtraction or multiplication and division. In those situations, you worked from left to right. Example 3 below is a review.

EXAMPLE 3 **Working from Left to Right**

Simplify.

(a) $^-8 - {}^-6 + {}^-11$ Do additions and subtractions from left to right.

${}^-2 + {}^-11$

${}^-13$

(b) $^-15 \div {}^-3(6)$ Do multiplications and divisions from left to right.

$5(6)$

30

Now Try Exercise 41.

Now we're ready to do problems that use a mix of the four operations, parentheses, and exponents. Let's start with a simple example: $4 + 2 \cdot 3$.

If we work from left to right	If we multiply first
$4 + 2 \cdot 3$	$4 + 2 \cdot 3$
$6 \cdot 3$	$4 + 6$
18	10

To be sure that everyone gets the same answer to a problem like this, mathematicians have agreed to do things in a certain order. The following order of operations shows that multiplying is done ahead of adding, so the correct answer is 10.

Order of Operations

Step 1   Work inside **parentheses** or **other grouping symbols**.

Step 2   Simplify expressions with **exponents**.

Step 3   Do the remaining **multiplications and divisions** as they occur from left to right.

Step 4   Do the remaining **additions and subtractions** as they occur from left to right.

Calculator Tip   Enter the example above in your calculator.

$4 \; \boxed{+} \; 2 \; \boxed{\times} \; 3 \; \boxed{=}$

Which answer do you get? If you have a scientific calculator, it automatically uses the order of operations and multiplies first to get the correct answer of 10. Some standard, four-function calculators may *not* have the order of operations built into them and will give the *incorrect* answer of 18.

EXAMPLE 4   **Using the Order of Operations with Whole Numbers**

Simplify.

$9 + 3\,(20 - 4) \div 8$ Work inside parentheses first: $20 - 4$ is 16.
Bring down the other numbers and symbols that you haven't used.

$9 + \quad 3\,(16) \quad \div 8$ Look for exponents: none.
Move from left to right, looking for multiplying and dividing.

$9 + \quad 3\,(16) \quad \div 8$ Yes, here is multiplying: 3(16) is 48.

$9 + \quad 48 \quad \div 8$ Here is dividing: $48 \div 8$ is 6. There is no other multiplying or dividing, so look for adding and subtracting.

$9 + \quad\quad 6$ Add last: $9 + 6$ is 15.

15

Now Try Exercise 51. ▶▶▶

EXAMPLE 5   **Using the Order of Operations with Integers**

Simplify.

(a) $^-8 \div (7 - 5) - 9$ Work inside parentheses first: $7 - 5$ is 2.
Bring down the other numbers and symbols that you haven't used.
Look for exponents: none.
Move from left to right, looking for multiplying and dividing.

$^-8 \div \quad (2) \quad - 9$

$^-8 \div \quad (2) \quad - 9$ Here is dividing: $^-8 \div 2$ is $^-4$. No other multiplying or dividing, so look for adding and subtracting.

$^-4 \quad - \quad 9$ Change subtracting to adding. Change 9 to its opposite.

$^-4 \quad + \quad ^-9$ Add $^-4 + \,^-9$.

$^-13$

(b) $3 + 2(6 - 8) \cdot (15 \div 3)$ Work inside first set of parentheses.
Change $6 - 8$ to $6 + \,^-8$ to get $^-2$.

$3 + 2\,(^-2) \cdot (15 \div 3)$ Work inside second set of parentheses: $15 \div 3$ is 5.

$3 + 2\,(^-2) \cdot \quad 5$ Multiply and divide from left to right.
First multiply $2(^-2)$ to get $^-4$.

$3 + \quad ^-4 \quad \cdot \quad 5$ Then multiply $^-4 \cdot 5$ to get $^-20$.

$3 + \quad\quad ^-20$ Add last: $3 + \,^-20$ is $^-17$.

$^-17$

Now Try Exercise 53. ▶▶▶

EXAMPLE 6   **Using the Order of Operations with Exponents**

Simplify.

(a) $4^2 - (^-3)^2$ The only parentheses are around $^-3$, but there is no work to do inside these parentheses.

$\underbrace{4^2} - \underbrace{(^-3)^2}$ Simplify the exponents: $4^2 = (4)(4) = 16$, and $(^-3)^2 = (^-3)(^-3) = 9$.

$\underbrace{16 - 9}$ There is no multiplying or dividing, so add and subtract: $16 - 9$ is 7.

7

(b) $(^-4)^3 - \underbrace{(4 - 6)}^2 (^-3)$ Work inside parentheses: $4 - 6$ becomes $4 + {}^-6$, which is $^-2$.

$\underbrace{(^-4)^3} - \underbrace{(^-2)^2} (^-3)$ Simplify the exponents next: $(^-4)^3$ is $(^-4)(^-4)(^-4) = {}^-64$, and $(^-2)^2$ is $(^-2)(^-2) = 4$.

$^-64 - \underbrace{4(^-3)}$ Look for multiplying and dividing. Multiply $4(^-3)$ to get $^-12$.

$^-64 - {}^-\mathbf{12}$ Change subtraction to addition. Change $^-12$ to its opposite.

$\underbrace{^-64 + {}^+\mathbf{12}}$ Add: $^-64 + 12$ is $^-52$.

$^-52$

> **CAUTION**
>
> To help in remembering the order of operations, you may have memorized the letters **PEMDAS**, or the phrase "Please Excuse My Dear Aunt Sally."
>
> **Please   Excuse   My   Dear   Aunt   Sally**
>
> **P**arentheses; **E**xponents; **M**ultiply and **D**ivide; **A**dd and **S**ubtract
>
> Be careful! Do *not* automatically do all multiplication before division. Multiplying and dividing are done *from left to right* (after parentheses and exponents).

Now Try Exercise 69.

OBJECTIVE **4** **Simplify expressions with fraction bars.**   A fraction bar indicates division, as in $\dfrac{^-6}{2}$, which means $^-6 \div 2$. In an expression such as

$$\frac{^-5 + 3^2}{16 - 7(2)}$$

the fraction bar also acts as a grouping symbol, like parentheses. It tells us to do the work in the numerator (above the bar) and then the work in the denominator (below the bar). The last step is to divide the results.

$$\frac{^-5 + 3^2}{16 - 7(2)} \longrightarrow \frac{^-5 + 9}{16 - 14} \longrightarrow \frac{4}{2} \longrightarrow 4 \div 2 = 2$$

The final result is 2.

EXAMPLE 7   **Using the Order of Operations with Fraction Bars**

Simplify $\dfrac{^-8 + 5(4 - 6)}{4 - 4^2 \div 8}$.

First do the work in the numerator.

$$^-8 + 5(\underbrace{4 - 6})$$ Work inside the parentheses.

$$^-8 + \underbrace{5\,(^-2)}$$ Multiply.

$$\underbrace{^-8 + \ (^-10)}$$ Add.

Numerator $\longrightarrow$ $^-18$

Now do the work in the denominator.

$$4 - \underbrace{4^2} \div 8$$ There are no parentheses; simplify the exponent.

$$4 - \underbrace{16 \div 8}$$ Divide.

$$\underbrace{4 - \quad 2}$$ Subtract.

Denominator $\longrightarrow 2$

The last step is the division.

$$\begin{array}{c} \text{Numerator} \longrightarrow \\ \text{Denominator} \longrightarrow \end{array} \frac{^-18}{2} = {}^-9$$

Now Try Exercise 71.

1.8 Exercises

FOR EXTRA HELP   Addison-Wesley Math Tutor Center   MathXL   Digital Video Tutor CD 1 Videotape 1   Student's Solutions Manual   MyMathLab   MathXL Tutorials on CD

Complete this table. See Example 1.

	Exponential Form	Factored Form	Simplified	Read as
1.	4^3		64	
2.	10^2		100	
3.		2 • 2 • 2 • 2 • 2 • 2 • 2		
4.		3 • 3 • 3 • 3 • 3		
5.		5 • 5 • 5 • 5		
6.		2 • 2 • 2 • 2 • 2 • 2		
7.				7 squared
8.				6 cubed
9.				10 to the first power
10.				4 to the fourth power

Simplify. See Examples 1 and 2.

11. (a) 10^1
 (b) 10^2
 (c) 10^3
 (d) 10^4

12. (a) 5^1
 (b) 5^2
 (c) 5^3
 (d) 5^4

13. (a) 4^1
 (b) 4^2
 (c) 4^3
 (d) 4^4

14. (a) 3^1
 (b) 3^2
 (c) 3^3
 (d) 3^4

15. 5^{10}

16. 4^9

17. 2^{12}

18. 3^{10}

19. $(^-2)^2$

20. $(^-4)^2$

21. $(^-5)^2$

22. $(^-10)^2$

23. $(^-4)^3$

24. $(^-2)^3$

25. $(^-3)^4$

26. $(^-2)^4$

27. $(^-10)^3$

28. $(^-5)^3$

29. 1^4

30. 1^5

31. $3^3 \cdot 2^2$

32. $4^2 \cdot 5^2$

33. $2^3(^-5)^2$

34. $3^2(^-2)^2$

35. $6^1(^-5)^3$

36. $7^1(^-4)^3$

37. $(^-2)(^-2)^4$

38. $^-6(^-6)^2$

39. Simplify.

$(^-2)^2 = \underline{\hspace{1cm}}$ $(^-2)^6 = \underline{\hspace{1cm}}$

$(^-2)^3 = \underline{\hspace{1cm}}$ $(^-2)^7 = \underline{\hspace{1cm}}$

$(^-2)^4 = \underline{\hspace{1cm}}$ $(^-2)^8 = \underline{\hspace{1cm}}$

$(^-2)^5 = \underline{\hspace{1cm}}$ $(^-2)^9 = \underline{\hspace{1cm}}$

(a) Describe the pattern you see in the signs of the answers.

(b) What would be the sign of $(^-2)^{15}$ and the sign of $(^-2)^{24}$?

40. Explain why it is important to have rules for the order of operations. Why do you think our "natural instinct" is to just work from left to right?

Simplify. See Examples 3–7.

41. $12 \div 6(^-3)$

42. $10 - 30 \div 2$

43. $^-1 + 15 - 7 - 7$

44. $9 + {}^-5 + 2(^-2)$

45. $10 - 7^2$

46. $5 - 5^2$

47. $2 - {}^-5 + 3^2$

48. $6 - {}^-9 + 2^3$

49. $3 + 5(6 - 2)$

50. $4 + 3(8 - 3)$

51. $^-7 + 6(8 - 14)$

52. $^-3 + 5(9 - 12)$

53. $2(^-3 + 5) - (9 - 12)$

54. $3(2 - 7) - (^-5 + 1)$

55. $^-5(7 - 13) \div {}^-10$

56. $^-4(9 - 17) \div {}^-8$

57. $9 \div (^-3)^2 + {}^-1$

58. $^-48 \div (^-4)^2 + 3$

59. $2 - {}^-5(^-2)^3$

60. $1 - {}^-10(^-3)^3$

61. $^-2(^-7) + 3(9)$

62. $4(^-2) + {}^-3(^-5)$

63. $30 \div {}^-5 - 36 \div {}^-9$

64. $8 \div {}^-4 - 42 \div {}^-7$

65. $2(5) - {}^-3(4) + 5(3)$

66. $9(3) - 6(4) + 3(7)$

67. $4(3^2) + 7(3 + 9) - {}^-6$

68. $5(4^2) - 6(1 + 4) - {}^-3$

69. $(^-4)^2 \cdot (7 - 9)^2 \div 2^3$

70. $(^-5)^2 \cdot (9 - 17)^2 \div (^-10)^2$

71. $\dfrac{^-1 + 5^2 - {}^-3}{^-6 - 9 + 12}$

72. $\dfrac{^-6 + 3^2 - {}^-7}{7 - 9 - 3}$

73. $\dfrac{{}^-2(4^2) - 4(6 - 2)}{{}^-4\,(8 - 13) \div {}^-5}$ $\dfrac{-48}{-4}$

74. $\dfrac{3(3^2) - 5(9 - 2)}{8(6 - 9) \div {}^-3}$ ${}^-8$

75. $\dfrac{2^3 \cdot ({}^-2 - 5) + 4({}^-1)}{4 + 5({}^-6 \cdot 2) + (5 \cdot 11)}$

76. $\dfrac{3^3 + 4({}^-1 - 2) - 25}{{}^-4 + 4(3 \cdot 5) + ({}^-6 \cdot 9)}$

77. $5^2(9 - 11)({}^-3)({}^-3)^3$

78. $4^2(13 - 17)({}^-2)({}^-2)^3$

#71 $\dfrac{27}{-3} = -9$

79. $|{}^-12| \div 4 + 2 \cdot |({}^-2)^3| \div 4$

80. $6 - |2 - 3 \cdot 4| + ({}^-5)^2 \div 5^2$

81. $\dfrac{{}^-9 + 18 \div {}^-3({}^-6)}{32 - 4(12) \div 3(2)}$

82. $\dfrac{{}^-20 - 15({}^-4) - {}^-40}{14 + 27 \div 3({}^-2) - {}^-4}$

Chapter 1   Group Activity: 'TIL DEBT DO YOU PART! WEDDING EXPENSES

Objective   Solve application problems using all four operations; interpret remainders in division.

Bride's magazine reported that average wedding costs in the United States increased from $15,208 in 1990 to the grand total you will calculate in the table below (in 2002). The average number of wedding guests also increased, from 170 guests to 200 guests.

Category	Average Cost in 2002
Miscellaneous expenses (stationery, clergy, gifts, limousine)	$ 1260
Bouquets and other flowers	$ 855
Photography and videography	$ 3000
Music	$ 830
Engagement and wedding rings (bride and groom)	$ 4060
Rehearsal dinner	$ 698
Bride's wedding dress and headpiece	$ 989
Bridal attendants' dresses (average of 5 bridesmaids)	$ 790
Mother of the bride's apparel	$ 231
Men's formalwear (ushers, best man, groom)	$ 544
Wedding reception	$10,320
Grand Total	

Source: Bride's magazine.

Take turns being the person who sets up the problem and the person who finds the answer using a calculator. Show your work.

1. What is the grand total of expenses shown in the table?

2. How much more expensive was a wedding in 2002 compared to 1990?

3. If the groom pays for the bouquets and flowers, the rehearsal dinner, the bride's engagement and wedding rings ($3500), the clergy ($232), and his formalwear ($95), what is the total amount spent by him?

4. Using the information from Question 3 and the table, find the cost of the

   (a) groom's wedding ring;   **(b)** the total for stationery, gifts, and limousine.

5. If the bride and groom pay for the reception with six equal monthly payments of $1806, how much will they pay in all? How much could they have saved by paying the cash price listed in the table?

6. If you budgeted $48 per person for the wedding reception and you invited 200 guests to a wedding in 2005, how much money would you have spent compared to the 2002 wedding reception costs?

7. If you budget $5000 for the reception and the caterer charges $48 per person, how many guests can you invite? How much of your budget is left over?

8. If you budget $8500 for the reception and the caterer charges $48 per person, how many guests can you invite? How much of your budget is left over?

9. What type of arithmetic problem did you work to get the answers to Problems 7 and 8? What is the mathematical term for the "left over" budget?

10. What is the average cost of each bridesmaid's dress? When you solved this problem, which number was the dividend? the divisor? the quotient? Rewrite the division as a related multiplication.

Chapter 1

SUMMARY

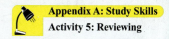

KEY TERMS

1.1	**place value system**	A place value system is a number system in which the location, or place, where a digit is written gives it a different value.
	digits	The 10 digits in our number system are 0, 1, 2, 3, 4, 5, 6, 7, 8, and 9.
	whole numbers	The whole numbers are 0, 1, 2, 3, and so on.
1.2	**number line**	A number line is like a thermometer turned sideways. It is used to show how numbers relate to each other.
	integers	Integers are the whole numbers and their opposites.
	absolute value	The absolute value of a number is its distance from 0 on the number line. Absolute value is indicated by two vertical bars and is always positive (or 0) but never negative.
1.3	**addends**	In an addition problem, the numbers being added are called addends.
	sum	The answer to an addition problem is called the sum.
	addition property of 0	Adding 0 to any number leaves the number unchanged.
	commutative property of addition	Changing the *order* of two addends does not change the sum.
	associative property of addition	Changing the *grouping* of addends does not change the sum.
1.4	**opposite**	The opposite of a number is the same distance from 0 on the number line but on the opposite side of 0. It is also called the *additive inverse* because a number plus its opposite equals 0.
1.5	**rounding**	Rounding a number means finding a number that is close to the original number but easier to work with.
	estimate	Use rounded numbers to get an approximate answer, or estimate.
	front end rounding	Front end rounding is rounding numbers to the highest possible place, so all the digits become 0 except the first digit.
1.6	**factors**	In a multiplication problem, the numbers being multiplied are called factors.
	product	The answer to a multiplication problem is called the product.
	multiplication property of 0	Multiplying any number by 0 gives a product of 0.
	multiplication property of 1	Multiplying a number by 1 leaves the number unchanged.
	commutative property of multiplication	Changing the *order* of two factors does not change the product.
	associative property of multiplication	Changing the *grouping* of factors does not change the product.
	distributive property	Multiplication distributes over addition. For example, $3(6 + 2) = 3 \cdot 6 + 3 \cdot 2$.
1.7	**quotient**	The answer to a division problem is called the quotient.
1.8	**exponent**	An exponent tells how many times a number is used as a factor in repeated multiplication.

NEW SYMBOLS

$<$	is less than
$>$	is greater than
$\lvert 6 \rvert$ and $\lvert {}^{-}2 \rvert$	absolute value of 6; absolute value of $^{-}2$

2^5 and 10^3 exponential form

See how well you have learned the vocabulary in this chapter. Answers follow the Quick Review.

1. In $(4)(^-6) = ^-24$, the 4 and $^-6$ are called
   A. products
   B. factors
   C. addends
   D. opposites.

2. A list of all the **whole numbers** is
   A. 1, 2, 3, 4, . . .
   B. . . . , $^-4$, $^-3$, $^-2$, $^-1$, 0, 1, 2, 3, . . .
   C. 0, 1, 2, 3, 4, 5, 6, 7, 8, 9
   D. 0, 1, 2, 3, 4, . . .

3. The **absolute value** of a number is
   A. its distance from 0 on the number line
   B. never positive
   C. used when multiplying negative factors
   D. less than 0.

4. An **exponent**
   A. tells how many times a number is added
   B. is the number being multiplied
   C. tells how many times a number is used as a factor
   D. applies only to the first thing to its right.

5. The **opposite** of a number is
   A. never negative
   B. called the additive inverse
   C. called the absolute value
   D. at the same point on the number line.

6. **Front end rounding** is rounding numbers
   A. to the nearest thousand
   B. so there are no zeros
   C. to the highest possible place
   D. so they become integers.

7. A list of all the **integers** is
   A. 1, 2, 3, 4, . . .
   B. . . . , $^-4$, $^-3$, $^-2$, $^-1$, 0, 1, 2, 3, . . .
   C. 0, 1, 2, 3, 4, 5, 6, 7, 8, 9
   D. 0, 1, 2, 3, 4, . . .

8. The **associative property of multiplication** says that
   A. changing the order of two factors does not change the product
   B. multiplication distributes over addition
   C. multiplying a number by 0 is undefined
   D. changing the grouping of factors does not change the product.

Concepts

1.1 *Reading and Writing Whole Numbers*
Do not use the word *and* when reading whole numbers. Commas separate groups of three digits. The first few group names are ones, thousands, millions, billions, trillions.

Examples

Write **3**, **008**, **160** in words.

three million, **eight thousand**, **one hundred sixty**

Write this number using digits:
twenty **billion**, sixty-five **thousand**, eighteen.

$$\underbrace{2\ \ 0}_{\text{billions}},\ \underbrace{0\ \ 0\ \ 0}_{\text{millions}},\ \underbrace{0\ \ 6\ \ 5}_{\text{thousands}},\ \underbrace{0\ \ 1\ \ 8}_{\text{ones}}$$

1.2 *Graphing Signed Numbers*
Place a dot at the correct location on the number line.

Graph: **(a)** $^-4$ **(b)** 0 **(c)** $^-1$ **(d)** $\dfrac{1}{2}$ **(e)** 2

1.2 *Comparing Integers*
When comparing two integers, the one that is farther to the left on the number line is less than the other.
Use the $<$ symbol for "is less than" and the $>$ symbol for "is greater than."

Write $<$ or $>$ between each pair of numbers to make a true statement.

$^-3 < ^-2$
$^-3$ **is less than** $^-2$
because $^-3$ is to the *left* of $^-2$ on the number line.

$0 > ^-4$
0 **is greater than** $^-4$
because 0 is to the *right* of $^-4$ on the number line.

Concepts	Examples

1.2 *Finding the Absolute Value of a Number*
Find the distance on the number line from 0 to the number. The absolute value is always positive (or 0) but never negative.

Find each absolute value.

$|^-5| = 5$ because $^-5$ is 5 steps away from 0 on the number line.

$|3| = 3$ because 3 is 3 steps away from 0 on the number line.

1.3 *Adding Two Integers*
When both integers have the *same sign*, add the absolute values and use the common sign as the sign of the sum.

When the integers have *different signs*, subtract the smaller absolute value from the larger absolute value. Use the sign of the number with the larger absolute value as the sign of the sum.

Add.

(a) $^-6 + {}^-7$

Add the absolute values.

$$|^-6| = 6 \quad \text{and} \quad |^-7| = 7$$

Add $6 + 7 = 13$ and use the common sign as the sign of the sum: $^-6 + {}^-7 = {}^-13$.

(b) $^-10 + 4$

Subtract the smaller absolute value from the larger.

$$|^-10| = 10 \quad \text{and} \quad |4| = 4$$

Subtract $10 - 4 = 6$; the number with the larger absolute value is negative, so the sum is negative: $^-10 + 4 = {}^-6$.

1.3 *Using Properties of Addition*
Addition Property of 0: Adding 0 to any number leaves the number unchanged.

Commutative Property of Addition: Changing the *order* of two addends does not change the sum.

Associative Property of Addition: Changing the *grouping* of addends does not change the sum.

Name the property illustrated by each case.

(a) $^-16 + 0 = {}^-16$

(b) $4 + 10 = 10 + 4$ Both sums are 14.

(c) $2 + (^-6 + 1) = (2 + {}^-6) + 1$ Both sums are $^-3$.

(a) Addition property of 0

(b) Commutative property of addition

(c) Associative property of addition

1.4 *Subtracting Two Integers*
To subtract two numbers, add the first number to the opposite of the second number.

Step 1 Make one pencil stroke to change the subtraction symbol to an addition symbol.

Step 2 Make a second pencil stroke to change the *second* number to its *opposite*.

Subtract.

(a) $7 - {}^-2$ Change subtraction to addition.
        ↓ ↓ ↓   Change $^-2$ to its opposite, $^+2$.
        $7 + {}^+2$

        9

(b) $^-9 - 12$ Change subtraction to addition.
        ↓ ↓ ↓   Change 12 to its opposite, $^-12$.
        $^-9 + {}^-12$

        $^-21$

Concepts	Examples

1.5 Rounding Integers

Step 1 Draw a line under the place to which the number is to be rounded.

Step 2 Look only at the next digit to the right of the underlined place. If the next digit is 5 or more, increase the underlined digit by 1. If the next digit is 4 or less, do not change the digit in the underlined place.

Step 3 Change all digits to the right of the underlined place to zeros.

Round 36,833 to the nearest thousand.

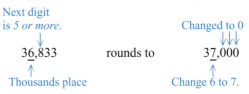

Round $^-$3582 to the nearest ten.

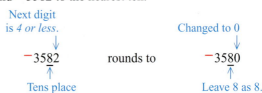

1.5 Front End Rounding

Round to the highest possible place so that all the digits become 0 except the first digit.

Use front end rounding.

Next digit
is *5 or more*.   Changed to 0

97,203   rounds to   100,000

Change 9 to 10. Regroup 1
into hundred-thousands place.

1.5 Estimating Answers in Addition and Subtraction

Use front end rounding to round the numbers in a problem. Then add or subtract the rounded numbers to estimate the answer.

First use front end rounding to estimate the answer. Then find the exact answer.

The temperature was 48 degrees below zero. During the morning it rose 21 degrees. What was the new temperature?

Estimate: $^-50 + 20 = {}^-30$ degrees

Exact: $^-48 + 21 = {}^-27$ degrees

1.6 Multiplying Two Integers

If two factors have *different signs*, the product is *negative*.

If two factors have the *same sign*, the product is *positive*.

Multiply.

(a) $^-5(6) = {}^-30$ The factors have *different* signs, so the product is *negative*.

(b) $(^-10)(^-2) = 20$ The factors have the *same* sign, so the product is *positive*.

1.6 Using Properties of Multiplication

Multiplication property of 0: Multiplying any number by 0 gives a product of 0.

Multiplication property of 1: Multiplying any number by 1 leaves the number unchanged.

Commutative property of multiplication: Changing the *order* of two factors does not change the product.

Associative property of multiplication: Changing the *grouping* of factors does not change the product.

Distributive property: Multiplication distributes over addition.

Name the property illustrated by each case.

(a) $^-49 \cdot 0 = 0$

(b) $1(675) = 675$

(c) $(^-8)(2) = (2)(^-8)$ Both products are $^-16$.

(d) $(^-3 \cdot {}^-2) \cdot 4 = {}^-3 \cdot (^-2 \cdot 4)$ Both products are 24.

(e) $5(2 + 4) = 5 \cdot 2 + 5 \cdot 4$ Both results are 30.

(a) Multiplication property of 0
(b) Multiplication property of 1
(c) Commutative property of multiplication
(d) Associative property of multiplication
(e) Distributive property

Concepts	**Examples**

1.6 *Estimating Answers in Multiplication*
First use front end rounding. Then multiply the rounded numbers using a shortcut: Multiply the nonzero digits in each factor; count the total number of zeros in the two factors and write that number of zeros in the product.

First use front end rounding to estimate the answer. Then find the exact answer.

At a PTA fund-raiser, Lionel sold 96 photo albums at $22 each. How much money did he take in?

Estimate: $100 \cdot \$20 = \$2000$

Exact: $96 \cdot \$22 = \$2112$

1.7 *Dividing Two Integers*
Use the same rules for dividing two integers as for multiplying two integers.

If two numbers have *different signs,* the quotient is *negative.*

If two numbers have the *same sign,* the quotient is *positive.*

Divide.

$\dfrac{^-24}{6} = {}^-4$ Numbers have *different* signs, so the quotient is *negative.*

$^-72 \div {}^-8 = 9$ Numbers have the *same* sign, so the quotient is *positive.*

$\dfrac{50}{^-5} = {}^-10$ Numbers have *different* signs, so the quotient is *negative.*

1.7 *Using Properties of Division*
(a) When a nonzero number is divided by itself, the quotient is 1.

(b) When a number is divided by 1, the quotient is the number.

(c) When 0 is divided by any other number (except 0), the quotient is 0.

(d) Division by 0 is *undefined.* There is no answer.

State the property illustrated by each case.

(a) $\dfrac{^-4}{^-4} = 1$ **(b)** $\dfrac{65}{1} = 65$

(c) $\dfrac{0}{9} = 0$ **(d)** $\dfrac{^-10}{0}$ is undefined.

The examples are in the same order as the properties listed at the left. Note that division is *not* commutative nor associative.

1.7 *Estimating Answers in Division*
First use front end rounding. Then divide the rounded numbers, using a shortcut: Drop the same number of zeros in both the divisor and the dividend.

First use front end rounding to estimate the answer. Then find the exact answer.

Joan has one year to pay off a $1020 loan. What is her monthly payment?

Estimate: $\$1000 \div 10 = \$100$

Exact: $\$1020 \div 12 = \$85$

1.7 *Interpreting Remainders in Division*
In some situations the remainder tells you how much is left over. In other situations, you must increase the quotient by 1 in order to accommodate the "left over."

Divide; then interpret the remainder.

Each chemistry student needs 35 milliliters of acid for an experiment. How many students can be served from a bottle holding 500 milliliters of acid?

$$\begin{array}{r} 14 \\ 35)\overline{500} \\ 35 \\ \hline 150 \\ 140 \\ \hline 10 \end{array}$$

14 → 14 students served

10 → 10 milliliters of acid left over

Concepts	*Examples*

1.8 *Using Exponents*

An exponent tells how many times a number is used as a factor in repeated multiplication. An exponent applies only to its base (the first thing to the left of the exponent).

Simplify.

Exponent
↓

(a) $2^5 = 2 \cdot 2 \cdot 2 \cdot 2 \cdot 2 = 32$

(b) $(^-3)^2 = (^-3)(^-3) = 9$

1.8 *Order of Operations*

Mathematicians have agreed to follow this order.

Step 1 Work inside parentheses or other grouping symbols.

Step 2 Simplify expressions with exponents.

Step 3 Do the remaining multiplications and divisions as they occur from left to right.

Step 4 Do the remaining additions and subtractions as they occur from left to right.

Simplify.

$(^-2)^4 + 3(^-4 - {}^-2)$ Work inside parentheses.

$(^-2)^4 + \quad 3(^-2)$ Simplify exponents.

$\quad 16 \;+ \quad 3(^-2)$ Multiply.

$\quad 16 \;+ \quad\quad {}^-6$ Add.

$\quad\quad\quad 10$

1.8 *Using the Order of Operations with Fraction Bars*

When there is a fraction bar, do all the work in the numerator. Then do all the work in the denominator. Finally, divide numerator by denominator.

Simplify.

$$\frac{^-10 + 4^2 - 6}{2 + 3(1 - 4)} = \frac{^-10 + 16 - 6}{2 + 3(^-3)} = \frac{0}{^-7} = 0$$

ANSWERS TO TEST YOUR WORD POWER

1. B; *Example:* In $7(10) = 70$, the factors are 7 and 10.
2. D; *Example:* 12, 0, 710, and 89,475 are all whole numbers.
3. A; *Example:* $|^-3| = 3$ and $|3| = 3$ because both $^-3$ and 3 are 3 steps away from 0 on the number line.
4. C; *Example:* In 2^5, the exponent is 5, which indicates that 2 is used as a factor 5 times, so $2^5 = 2 \cdot 2 \cdot 2 \cdot 2 \cdot 2 = 32$.
5. B; *Example:* The opposite of 5 is $^-5$; it is the additive inverse because $5 + (^-5) = 0$.
6. C; *Example:* Using front end rounding, round 48,299 to the highest possible place, which is ten-thousands. So, 48,299 rounds to 50,000. All digits are 0 except the first digit.
7. B; *Example:* $^-10$, 0, and 6 are all integers.
8. D; *Example:* $4 \cdot (^-2 \cdot 3)$ can be rewritten as $(4 \cdot {}^-2) \cdot 3$; both products are $^{--}24$.

Chapter **1**

REVIEW EXERCISES

If you need help with any of these Review Exercises, look in the section indicated in the red brackets.

[1.1] 1. Circle the whole numbers: 86   2.831 $^-4$ 0 $\dfrac{2}{3}$ 35,600

Write these numbers in words.

2. 806

3. 319,012

4. 60,003,200

5. 15,749,000,000,006

Write these numbers using digits.

6. Five hundred four thousand, one hundred

7. Six hundred twenty million, eighty thousand

8. Ninety-nine billion, seven million, three hundred fifty-six

[1.2] 9. Graph these numbers: $^-3\dfrac{1}{2}$, 2, $^-5$, 0.

Write < or > between each pair of numbers to make a true statement.

10. 0 __ $^-4$

11. $^-3$ __ $^-1$

12. 2 __ $^-2$

13. $^-2$ __ 1

Find each absolute value.

14. $|^-5|$

15. $|9|$

16. $|0|$

17. $|^-125|$

[1.3] *Add.*

18. $^-9 + 8$

19. $^-8 + {}^-5$

20. $16 + {}^-19$

21. $^-4 + 4$

22. $6 + {}^-5$

23. $^-12 + {}^-12$

24. $0 + {}^-7$

25. $^-16 + 19$

26. $9 + {}^-4 + {}^-8 + 3$

27. $^-11 + {}^-7 + 5 + {}^-4$

[1.4] *Find the opposite (additive inverse) of each number. Show that the sum of the number and its opposite is 0.*

28. $^-5$

29. 18

Subtract by changing subtraction to addition.

30. $5 - 12$

31. $24 - 7$

32. $^-12 - 4$

33. $4 - {}^-9$

34. $^-12 - {}^-30$

35. $^-8 - 14$

36. $^-6 - {}^-6$

37. $^-10 - 10$

38. $^-8 - {}^-7$

39. $0 - 3$

40. $1 - {}^-13$

41. $15 - 0$

Simplify.

42. $3 - 12 - 7$

43. $^-7 - {}^-3 + 7$

44. $4 + {}^-2 - 0 - 10$

45. $^-12 - 12 + 20 - {}^-4$

[1.5] *Round each number as indicated.*

46. 205 to the nearest ten

47. 59,499 to the nearest thousand

48. 85,066,000 to the nearest million

49. $^-2963$ to the nearest hundred

50. $^-7,063,885$ to the nearest ten-thousand

51. 399,712 to the nearest thousand

Use front end rounding to round each number.

52. The combined weight loss of 10 dieters was 197 pounds.

53. The land on the shore of the Dead Sea in the Middle East is 1312 feet below sea level. (*Source: Goode's World Atlas.*)

54. There are 362,000,000 Yellow Pages directories published in the United States each year. (*Source: Yellow Pages Publishers Association.*)

55. By 2050, the Census Bureau predicts a world population of 9,104,000,000 people. (*Source: U.S. Bureau of the Census.*)

[1.6] *Multiply.*

56. $^-6(9)$

57. $(^-7)(^-8)$

58. $10(^-10)$

59. $^-45 \cdot 0$

60. $^-1(^-24)$

61. $17 \cdot 1$

62. $4(^-12)$

63. $(^-5)(^-25)$

64. $^-3(^-4)(^-3)$

65. $^-5(2)(^-5)$

66. $(^-8)(^-1)(^-9)$

[1.7] *Simplify.*

67. $\dfrac{^-63}{^-7}$

68. $\dfrac{70}{^-10}$

69. $\dfrac{^-15}{0}$

70. $^-100 \div ^-20$

71. $18 \div ^-1$

72. $\dfrac{0}{12}$

73. $\dfrac{^-30}{^-2}$

74. $\dfrac{^-35}{35}$

75. $^-40 \div ^-4 \div ^-2$

76. $^-18 \div 3(^-3)$

77. $0 \div ^-10(5) \div 5$

78. Divide; then interpret the remainder. It took 1250 hours to build the set for a new play. How many working days of eight hours each did it take to build the set?

[1.8] *Simplify.*

79. 10^4

80. 2^5

81. 3^3

82. $(^-4)^2$

83. $(^-5)^3$

84. 8^1

85. $6^2 \cdot 3^2$

86. $5^2 (^-2)^3$

87. $^-30 \div 6 - 4(5)$

88. $6 + 8(2 - 3)$

89. $16 \div 4^2 + (^-6 + 9)^2$

90. $^-3(4) - 2(5) + 3(^-2)$

91. $\dfrac{^-10 + 3^2 - ^-9}{3 - 10 - 1}$

92. $\dfrac{^-1(1 - 3)^3 + 12 \div 4}{^-5 + 24 \div 8 \cdot 2(6 - 6) + 5}$

MIXED REVIEW EXERCISES

Name the property illustrated by each case.

93. $^-3 + (5 + 1) = (^-3 + 5) + 1$ **94.** $^-7(2) = 2(^-7)$ **95.** $0 + 19 = 19$

96. $^-42 \cdot 0 = 0$ **97.** $2(^-6 + 4) = 2 \cdot {}^-6 + 2 \cdot 4$ **98.** $(^-6 \cdot 3) \cdot {}^-1 = {}^-6 \cdot (3 \cdot {}^-1)$

First use front end rounding to estimate each answer. Then find the exact answer.

99. Last year, 192 Elvis jukeboxes were sold at a price of $11,900 each. What was the total value of the jukeboxes?

Estimate:

Exact:

100. Chad had $185 in his checking account. He deposited his $428 paycheck and then wrote a $706 check for car repairs. What is the balance in his account?

Estimate:

Exact:

101. Georgia's car used 24 gallons of gas on her 840-mile vacation trip. What was the average number of miles she drove on each gallon of gas?

Estimate:

Exact:

102. When inventory was taken at Mathtronic Company, 19 calculators and 12 computer modems were missing. Each calculator is worth $39 and each modem is worth $85. What is the total value of the missing items?

Estimate:

Exact:

Elena Sanchez opened a shop that does alterations and designs custom clothing. Use the table of her income and expenses to answer Exercises 103–106.

Month	Income	Expenses	Profit or Loss
Jan.	$2400	$3100	
Feb.	$1900	$2000	
Mar.	$2500	$1800	
Apr.	$2300	$1400	
May	$1600	$1600	
June	$1900	$1200	

103. Complete the table by finding Elena's profit or loss for each month.

104. Which month had the greatest loss? Which month had the greatest profit?

105. What was Elena's average monthly income?

106. What was the average monthly amount of expenses?

Chapter **1**

T E S T

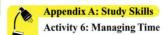
Appendix A: Study Skills
Activity 6: Managing Time

1. Write this number in words: 20,008,307

2. Write this number using digits:
   thirty billion, seven hundred thousand, five

3. Graph the numbers 3, $^-2$, 0, $-\dfrac{1}{2}$ on the number line at the right.

4. Write < or > between each pair of numbers to make a true statement.

   0 ___ $^-3$ $^-2$ ___ $^-1$

5. Find each absolute value: $|10|$ and $|^-14|$.

Add, subtract, multiply, or divide.

6. $3 - 9$

7. $^-12 + 7$

8. $\dfrac{-28}{-4}$

9. $^-1(40)$

10. $^-5 - {}^-15$

11. $(^-8)(^-8)$

12. $^-25 + {}^-25$

13. $\dfrac{17}{0}$

14. $^-30 - 30$

15. $\dfrac{50}{-10}$

16. $5(^-9)$

17. $0 - {}^-6$

Simplify.

18. $^-35 \div 7(^-5)$

19. $^-15 - {}^-8 + 7$

20. $3 - 7(^-2) - 8$

21. $(^-4)^2 \cdot 2^3$

22. $\dfrac{5^2 - 3^2}{(4)(^-2)}$

23. $^-2(^-4 + 10) + 5(4)$

24. $^-3 + (^-7 - {}^-10) + 4(6 - 10)$

25. Explain how an exponent is used. Include two examples, one using a positive number as the base and one using a negative number as the base.

1. _____

2. _____

3.
```
  -+--+--+--+--+--+--+-→
  ‾3 ‾2 ‾1  0  1  2  3
```

4. _____

5. _____

6. _____

7. _____

8. _____

9. _____

10. _____

11. _____

12. _____

13. _____

14. _____

15. _____

16. _____

17. _____

18. _____

19. _____

20. _____

21. _____

22. _____

23. _____

24. _____

25. _____

26. _____

26. Explain the commutative and associative properties of addition. Also give an example to illustrate each property.

Round each number as indicated.

27. _____

27. 851 to the nearest hundred

28. _____

28. 36,420,498,725 to the nearest million

29. _____

29. 349,812 to the nearest thousand

First use front end rounding to estimate the answer to each application problem. Then find the exact answer.

30. _Estimate:_ _____

Exact: _____

30. Lorene had $184 in her checking account. She deposited her $293 pay-check and then wrote a $506 check for tuition. What is the balance in her account?

31. _Estimate:_ _____

Exact: _____

31. The Cardinals football team had a bad year. It lost a total of 1140 yards in 12 games. What was the average loss in each game?

32. _Estimate:_ _____

Exact: _____

32. One kind of cereal has 220 calories in each serving. Another kind has 110 calories in each serving. During a month with 31 days, how many calories would you save by eating the second kind of cereal each morning for breakfast? (_Source:_ General Mills and Post Cereals.)

Write and solve a subtraction problem to answer this question.

33. _____

33. On the planet Mars the average high temperature is $^-10$ °F and the average low is $^-100$ °F. What is the difference between the high and low temperatures? (_Source:_ NASA.)

Divide; then interpret the remainder.

34. _____

34. Anthony has a part-time job as a shipping clerk. He is sending 1276 mathematics books to a college bookstore. Each shipping carton can hold 48 books. What is the minimum number of cartons he will need?

Understanding Variables and Solving Equations

2

A s many as 50,000 thunderstorms occur worldwide in a single day. Thunderstorms always include lightning, and in the United States an average of 300 people are struck by lightning each year. (*Source:* AccuWeather.com) But you don't need a weather forecaster to tell you how far away a thunderstorm is. You can use the algebraic expression in **Section 2.1,** Exercises 59–60, to estimate your distance from the storm and, if it is close, move to shelter.

There are other algebraic expressions related to weather. For example, in summertime, you can find the temperature by counting the chirps of a cricket and using the expression in **Section 2.1,** Exercise 14.

2.1 Introduction to Variables

OBJECTIVES

1. Identify variables, constants, and expressions.
2. Evaluate variable expressions for given replacement values.
3. Write properties of operations using variables.
4. Use exponents with variables.

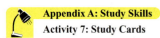

Appendix A: Study Skills
Activity 7: Study Cards

OBJECTIVE 1 Identify variables, constants, and expressions. You probably know that algebra uses letters, especially the letter x. But why use letters when numbers are easier to understand? Here is an example.

Suppose that you run your college bookstore. When deciding how many books to order for a certain class, you first find out the class limit, that is, the maximum number of students allowed in the class. You will need at least that many books. But you decide to order 5 extra copies for emergencies.

> Rule for ordering books: Order the class limit + 5 extra

How many books would you order for a prealgebra class with a limit of 25 students?

Class limit ↓   ↓ Extra
$$25 + 5$$ You would order 30 prealgebra books.

How many books would you order for a geometry class that allows 40 students to register?

Class limit ↓   ↓ Extra
$$40 + 5$$ You would order 45 geometry books.

You could set up a table to keep track of the number of books to order for various classes.

Class	Rule for Ordering Books: Class Limit + 5 Extra	Number of Books to Order
Prealgebra	25 + 5	30
Geometry	40 + 5	45
College algebra	35 + 5	40
Calculus 1	50 + 5	55

A shorthand way to write your rule is shown below.

$$c + 5$$
↑
The c stands for class limit.

You can't write your rule by using just numbers because the class limit varies, or changes, depending on which class you're talking about. So you use a letter, called a **variable,** to represent the part of the rule that varies. Notice the similarity in the words *varies* and *variable*. When part of a rule does *not* change, it is called a **constant.**

The variable, or the part of the rule that varies or changes
↓
$$c + 5$$
↑
The constant, or the part of the rule that does *not* change

$c + 5$ is called an **expression.** It expresses (tells) the rule for ordering books. You could use any letter you like for the variable part of the expression, such as $x + 5$, or $n + 5$, and so on. But one suggestion is to use a letter that reminds you of what it stands for. In this situation, the letter c reminds us of "class limit."

EXAMPLE 1 **Writing an Expression and Identifying the Variable and Constant**

Write an expression for this rule. Identify the variable and the constant.

Order the class limit minus 10 books because some students will buy used books.

Variable
↓
$c - 10$
↑
Constant

Now Try Exercise 3. ▶▶▶

OBJECTIVE 2 **Evaluate variable expressions for given replacement values.** When you need to figure out how many books to order for a particular class, you use a specific value for the class limit, like 25 students in prealgebra. Then you **evaluate the expression,** that is, you follow the rule.

Ordering Books for a Prealgebra Class

$c + 5$ Expression (rule for ordering books) is $c + 5$.
↓   Replace c with 25, the class limit for prealgebra.
$25 + 5$ Follow the rule. Add $25 + 5$.
30   Order 30 prealgebra books.

EXAMPLE 2 **Evaluating an Expression**

Use this rule for ordering books: Order the class limit minus 10. The expression is $c - 10$.

(a) Evaluate the expression when the class limit is 32.

$c - 10$ Replace c with 32.
↓
$32 - 10$ Follow the rule. Subtract to find $32 - 10$.
22   Order 22 books.

(b) Evaluate the expression when the class limit is 48.

$c - 10$ Replace c with 48.
↓
$48 - 10$ Follow the rule. Subtract to find $48 - 10$.
38   Order 38 books.

Now Try Exercise 13. ▶▶▶

In any career you choose, there will be many useful "rules" that need to be written using variables (letters) because part of the rule changes depending on the situation. This is one reason algebra is such a powerful tool. Here is another example.

Suppose that you work in a landscaping business. You are putting a fence around a square-shaped garden. Each side of the garden is 6 feet long. How much fencing material should you bring to finish the job? You could add the lengths of the four sides.

6 feet + 6 feet + 6 feet + 6 feet = 24 feet of fencing

Or, recall that multiplication is a quick way to do repeated addition. The square garden has 4 sides, so multiply by 4.

$$4 \cdot 6 \text{ feet} = 24 \text{ feet of fencing}$$

So the rule for calculating the amount of fencing for a square garden is

$$4 \cdot \text{length of one side.}$$

Other jobs may require fencing for larger or smaller square shapes. The following table shows how much fencing you will need.

Length of One Side of Square Shape	Expression (Rule) to Find Total Amount of Fencing: 4 • Length of One Side	Total Amount of Fencing Needed
6 feet	4 • 6 feet	24 feet
9 feet	4 • 9 feet	36 feet
10 feet	4 • 10 feet	40 feet
3 feet	4 • 3 feet	12 feet

The expression (rule) can be written in shorthand form as shown below.

The number part in a *multiplication* expression is called the numerical coefficient, or just the **coefficient.** We usually don't write multiplication dots in expressions, so we do the following.

$$4 \cdot s \quad \text{is written as} \quad 4s$$

You can use the expression $4s$ any time you need to know the *perimeter* of a square shape, that is, the total distance around all four sides of the square.

CAUTION
If an expression involves adding, subtracting, or dividing, then you **do** have to write $+$, $-$, or $\div$. It is *only* multiplication that is understood without writing an operation symbol.

$$4 + s \qquad 4 - s \qquad 4 \div s \qquad 4s$$

Add s.   Subtract s.   Divide by s.   Multiply by s.

EXAMPLE 3 **Evaluating an Expression with Multiplication**

The expression (rule) for finding the perimeter of a square shape is $4s$. Evaluate the expression when the length of one side of a square parking lot is 30 yards.

$4s$ Replace s with 30 yards.

$4 \cdot \textbf{30 yards}$ There is no operation symbol between the 4 and the s, so it is understood to be multiplication. 4 times 30 is 120.

120 yards ← Total distance around the lot (perimeter) is 120 yards.

Now Try Exercise 15.

Some expressions (rules) involve several different steps. An expression for finding the approximate systolic blood pressure of a person of a certain age is shown below.

$$100 + \frac{a}{2} \quad \longleftarrow \text{Age of person (the variable)}$$

Remember that a fraction bar means division, so $\frac{a}{2}$ is the person's age divided by 2. You also need to remember the order of operations, which means doing division before addition.

EXAMPLE 4 **Evaluating an Expression with Several Steps**

Evaluate the expression $100 + \frac{a}{2}$ when the age of the person is 24.

$$100 + \frac{a}{2} \qquad \text{Replace } a \text{ with 24, the age of the person.}$$

$$100 + \frac{24}{2} \qquad \begin{array}{l}\text{Follow the rule using the order of operations.}\\ \text{First divide: } 24 \div 2 \text{ is 12.}\end{array}$$

$$100 + 12 \qquad \text{Now add: } 100 + 12 \text{ is 112.}$$

$$112 \quad \longleftarrow \text{The approximate systolic blood pressure is 112.}$$

Now Try Exercise 17. ▶▶▶

Calculator Tip If you like to fish, you can use an expression (rule) like the one below to find the approximate weight (in pounds) of a fish you catch. Measure the length of the fish (in inches) and then use the correct expression for that type of fish. For a northern pike, the weight expression is shown below.

Variable (length of fish) ⟶

$$\frac{l^3}{3600}$$

where l is the length of the fish in inches. (*Source: InFisherman.*)

To evaluate this expression for a fish that is 43 inches long, follow the rule by calculating as follows.

$$\frac{43^3}{3600} \qquad \begin{array}{l}\text{Replace } l \text{ with 43, the length of the}\\ \text{fish in inches.}\end{array}$$

In the numerator, you can multiply $43 \cdot 43 \cdot 43$ or use the $\boxed{y^x}$ key on your calculator. Then divide by 3600.

Enter 43 $\boxed{y^x}$ 3 $\boxed{\div}$ 3600 $\boxed{=}$ Calculator shows 22.08527778

    ↑   ↑

  Base   Exponent

The fish weighs about 22 pounds.

Now use the expression to find the approximate weight of a northern pike that is 37 inches long. (Answer: about 14 pounds.)

(continued)

Notice that variables are used on your calculator keys. On the y^x key, y represents the base and x represents the exponent. You first evaluated y^x by entering 43 as the base and 3 as the exponent for the first fish. Then you evaluated y^x again by entering 37 as the base and 3 as the exponent for the second fish.

Some expressions (rules) involve several variables. For example, if you bowl three games and want to know your average score, you can use this expression.

$$\dfrac{t}{g}$$
  $t \leftarrow$ Total score for all games (variable)
  $g \leftarrow$ Number of games (variable)

EXAMPLE 5   **Evaluating Expressions with Two Variables**

(a) Find your average score if you bowl three games and your total score for all three games is 378.

Use the expression (rule) for finding your average score.

$$\dfrac{t}{g}$$ Replace t with your total score of 378.

Replace g with 3, the number of games.

$$\dfrac{378}{3}$$ Follow the rule. Divide 378 by 3.

126   Your average score is 126.

(b) Complete these tables to show how to evaluate each expression.

Value of x	Value of y	Expression (Rule) $x + y$
2	5	2 + 5 is 7
⁻6	4	__ + __ is __
0	16	__ + __ is __

Value of x	Value of y	Expression (Rule) xy
2	5	2 • 5 is 10
⁻6	4	__ • __ is __
0	16	__ • __ is __

The expression (rule) is to *add* the two variables. So the completed table is:

Value of x	Value of y	Expression (Rule) $x + y$
2	5	2 + 5 is 7
⁻6	4	⁻6 + 4 is ⁻2
0	16	0 + 16 is 16

The expression (rule) is to *multiply* the two variables. We know that it's multiplication because there is no operation symbol between the x and y. So the completed table is:

Value of x	Value of y	Expression (Rule) xy
2	5	2 • 5 is 10
⁻6	4	⁻6 • 4 is ⁻24
0	16	0 • 16 is 0

Now Try Exercise 23.

OBJECTIVE 3 **Write properties of operations using variables.**   Now you can use variables as a shorthand way to express the properties you learned about in **Sections 1.3, 1.6,** and **1.7.** We'll use the letters *a* and *b* to represent any two numbers.

Commutative Property of Addition	**Commutative Property of Multiplication**
$a + b = b + a$	$a \cdot b = b \cdot a$
	or, $ab = ba$

To get specific examples, you can pick values for *a* and *b*. For example, if *a* is ⁻3, replace every *a* with ⁻3. If *b is* **5**, replace every *b* with **5**.

$$ab = ba$$

$$a + b = b + a$$
$$\downarrow \quad \downarrow \quad \downarrow \quad \downarrow$$
$$\underbrace{^-3 + 5} = \underbrace{5 + {}^-3}$$
$$\quad 2 \quad = \quad 2$$

Both sums are 2

$$a \cdot b = b \cdot a$$
$$\downarrow \quad \downarrow \quad \downarrow \quad \downarrow$$
$$\underbrace{^-3 \cdot 5} = \underbrace{5 \cdot {}^-3}$$
$$\quad ^-15 \quad = \quad ^-15$$

Both products are ⁻15

Of course, you could pick many different values for *a* and *b*, because the commutative "rule" will always work for adding *any* two numbers or multiplying *any* two numbers.

EXAMPLE 6   **Writing Properties of Operations Using Variables**

Use the variable *b* to state this property: When any number is divided by 1, the quotient is the number.

Use the letter *b* to represent any number.

$$\frac{b}{1} = b$$

Now Try Exercise 27. ▶▶▶

OBJECTIVE 4 **Use exponents with variables.**   In **Section 1.8** we used an exponent as a quick way to write repeated multiplication. For example,

$$\underbrace{3 \cdot 3 \cdot 3 \cdot 3 \cdot 3}_{\text{3 is used as a factor 5 times.}} \quad \text{can be written} \quad 3^5 \leftarrow \text{Exponent}$$
$$\qquad\qquad\qquad\qquad\qquad\qquad \uparrow\!\!\!\text{——} \text{Base}$$

The meaning of an exponent remains the same when a variable (a letter) is the base.

$$\underbrace{c \cdot c \cdot c \cdot c \cdot c}_{c \text{ is used as a factor 5 times.}} \quad \text{can be written} \quad c^5 \leftarrow \text{Exponent}$$
$$\qquad\qquad\qquad\qquad\qquad\qquad \uparrow\!\!\!\text{——} \text{Base}$$

m^2 means $m \cdot m$ Here *m* is used as a factor 2 times.

$x^4 y^3$ means $x^4 \cdot y^3$ or

$7b^2$ means $7 \cdot b \cdot b$ The exponent applies *only* to *b*.

$^-4xy^2z$ means $^-4 \cdot x \cdot y \cdot y \cdot z$ The exponent applies *only* to *y*.

EXAMPLE 7 **Understanding Exponents Used with Variables**

Rewrite each expression without exponents.

(a) y^6 can be written as $y \cdot y \cdot y \cdot y \cdot y \cdot y$

y is used as a factor 6 times.

(b) $12bc^3$ can be written as $12 \cdot b \cdot c \cdot c \cdot c$ The exponent applies *only* to c.

Coefficient is 12. c^3

(c) $^-2m^2n^4$ can be written as $^-2 \cdot m \cdot m \cdot n \cdot n \cdot n \cdot n$

Coefficient is $^-2$. m^2 n^4

Now Try Exercise 35.

To evaluate an expression with exponents, multiply all the factors.

EXAMPLE 8 **Evaluating Expressions with Exponents**

Evaluate each expression.

(a) x^2 when x is $^-3$

x^2 means $x \cdot x$ Replace each x with $^-3$.

$^-3 \cdot {}^-3$ Multiply $^-3$ times $^-3$.

9

So x^2 becomes $(^-3)^2$, which is $(^-3)(^-3)$, or 9.

(b) x^3y when x is $^-4$ and y is $^-10$

x^3y means $x \cdot x \cdot x \cdot y$ Replace x with $^-4$, and replace y with $^-10$.

$^-4 \cdot {}^-4 \cdot {}^-4 \cdot {}^-10$ Multiply two factors at a time.

$16 \quad \cdot {}^-4 \cdot {}^-10$

$^-64 \quad \cdot {}^-10$

640

So x^3y becomes $(^-4)^3(^-10)$, which is $(^-4)(^-4)(^-4)(^-10)$, or 640.

(c) $^-5ab^2$ when a is 5 and b is 3

$^-5ab^2$ means $^-5 \cdot a \cdot b \cdot b$ Replace a with 5, and replace b with 3.

$^-5 \cdot 5 \cdot 3 \cdot 3$ Multiply two factors at a time.

$^-25 \cdot 3 \cdot 3$

$^-75 \quad \cdot 3$

$^-225$

So $^-5ab^2$ becomes $^-5(5)(3)^2$, which is $^-5(5)(3)(3)$, or $^-225$.

Coefficient is $^-5$.

Now Try Exercise 49.

2.1 Exercises

Identify the variable and the constant in each expression. See Example 1.

1. $c + 4$

2. $d + 6$

3. $^-3 + m$

4. $^-4 + n$

Identify the parts of each expression. Choose from these labels: variable, constant, *and* coefficient.

5. $5h$

6. $3s$

7. $2c - 10$

8. $6b - 1$

9. $x - y$

10. xy

11. $^-6g + 9$

12. $^-10k + 15$

Evaluate each expression. See Examples 2–5.

13. The expression (rule) for ordering robes for the graduation ceremony at West Community College is $g + 10$, where g is the number of graduates. Evaluate the expression when

(a) there are 654 graduates.

(b) there are 208 graduates.

(c) there are 95 graduates.

14. Crickets chirp faster as the weather gets hotter. The expression (rule) for finding the approximate temperature (in degrees Fahrenheit) is $c + 37$ where c is the number of chirps made in 15 seconds by a field cricket. (*Source:* AccuWeather.com) Evaluate the expression when the cricket

(a) chirps 45 times.

(b) chirps 33 times.

(c) chirps 58 times.

15. The expression for finding the perimeter of a triangle with sides of equal length is $3s$, where s is the length of one side. Evaluate the expression when

(a) the length of one side is 11 inches.

(b) the length of one side is 3 feet.

16. The expression for finding the perimeter of a pentagon with sides of equal length is $5s$, where s is the length of one side. Evaluate the expression when

(a) the length of one side is 25 meters.

(b) the length of one side is 8 inches.

17. The expression for ordering brushes for an art class is $3c - 5$, where c is the class limit. Evaluate the expression when

   (a) the class limit is 12.

   (b) the class limit is 16.

18. The expression for ordering doughnuts for the office staff is $2n - 4$, where n is the number of people at work. Evaluate the expression when

   (a) there are 13 people at work.

   (b) there are 18 people at work.

19. The expression for figuring a student's average test score is $\dfrac{p}{t}$, where p is the total points earned on all the tests and t is the number of tests. Evaluate the expression when

   (a) 332 points were earned on 4 tests.

   (b) there were 7 tests and 637 points were earned.

20. The expression for deciding how many buses are needed for a group trip is $\dfrac{p}{b}$, where p is the total number of people and b is the number of people that one bus will hold. Evaluate the expression when

   (a) 176 people are going on a trip and one bus holds 44 people.

   (b) A bus holds 36 people and 72 people are going on a trip.

Complete each table by evaluating the expressions. See Example 5.

21.

Value of x	Expression $x + x + x + x$	Expression $4x$
$^-2$	$^-2 + {}^-2 + {}^-2 + {}^-2$ is $^-8$	$4 \cdot {}^-2$ is $^-8$
12		
0		
$^-5$		

22.

Value of y	Expression $3y$	Expression $y + 2y$
$^-6$	$3(^-6)$ is $^-18$	$^-6 + 2(^-6)$ is $^-6 + {}^-12$, or $^-18$
10		
$^-3$		
0		

23.

Value of x	Value of y	Expression $^-2x + y$
3	7	$^-2(3) + 7$ is $^-6 + 7$, or 1
$^-4$	5	
$^-6$	$^-2$	
0	$^-8$	

24.

Value of x	Value of y	Expression ^-2xy
3	7	$^-2 \cdot 3 \cdot 7$ is $^-42$
$^-4$	5	
$^-6$	$^-2$	
0	$^-8$	

25. Explain the words *variable* and *expression*.

26. Explain the words *coefficient* and *constant*.

Use the variable b to express each of these properties. See Example 6.

27. Multiplying a number by 1 leaves the number unchanged.

28. Adding 0 to any number leaves the number unchanged.

29. Any number divided by 0 is undefined.

30. Multiplication distributes over addition. (Use *a*, *b*, and *c* as the variables.)

Rewrite each expression without exponents. See Example 7.

31. c^6

32. d^7

33. $x^4 y^3$

34. $c^2 d^5$

35. $^-3a^3 b$

36. $^-8m^2 n$

37. $9xy^2$

38. $5ab^4$

39. $^-2c^5 d$

40. $^-4x^3 y$

41. $a^3 bc^2$

42. $x^2 y z^6$

Evaluate each expression when r is $^-3$, s is 2, and t is $^-4$. See Example 8.

43. t^2

44. r^2

45. rs^3

46. $s^4 t$

47. $3rs$

48. $6st$

49. $^{-}2s^2t^2$

50. $^{-}4rs^4$

▦ **51.** $r^2s^5t^3$

▦ **52.** $r^3s^4t^2$

▦ **53.** $^{-}10r^5s^7$

▦ **54.** $^{-}5s^6t^5$

Evaluate each expression when x is 4, y is $^{-}2$, and z is $^{-}6$.

55. $|xy| + |xyz|$

56. $x + |y^2| + |xz|$

57. $\dfrac{z^2}{^{-}3y + z}$

58. $\dfrac{y^2}{x + 2y}$

RELATING CONCEPTS (EXERCISES 59–60) For Individual or Group Work

At the beginning of this chapter, you read that there is an expression that tells your approximate distance from a thunderstorm. First, count the number of seconds from the time you see a lightning flash until you start to hear the thunder. Then, to estimate the distance (in miles), use the expression $\dfrac{s}{5}$, where s is the number of seconds. Use this expression as you **work Exercises 59 and 60 in order.** *(This expression is based on the fact that the light from the lightning travels faster than the sound from the thunder.)*

59. Evaluate the thunderstorm expression for each number of seconds. How far away is the storm?

   (a) 15 seconds

   (b) 10 seconds

   (c) 5 seconds

60. Explain how you can use your answers from Exercise 59 to:

   (a) Estimate the distance when the time is $2\frac{1}{2}$ seconds.

   (b) Find the number of seconds when the distance is $1\frac{1}{2}$ miles.

   (c) Find the number of seconds when the distance is $2\frac{1}{2}$ miles.

2.2 Simplifying Expressions

OBJECTIVES

1 Combine like terms, using the distributive property.

2 Simplify expressions.

3 Use the distributive property to multiply.

OBJECTIVE 1 Combine like terms, using the distributive property.
In **Section 2.1,** the expression for ordering math textbooks was $c + 5$. This expression was simple and easy to use. Sometimes expressions are *not* written in the simplest possible way. For example:

Evaluate this expression when c is 20.	Evaluate this expression when c is 20.
$c + 5$ Replace c with 20.	$2c - 10 - c + 15$ Replace c with 20.
$\underbrace{20 + 5}$ Add 20 + 5.	$\underbrace{2 \cdot 20} - 10 - 20 + 15$ Multiply 2 · 20.
25	$40 - 10 - 20 + 15$ Change subtraction to adding the opposite.
	$\underbrace{40 + {}^-10} + {}^-20 + 15$ Add from left to right.
	$\underbrace{30 \qquad + {}^-20} + 15$
	$\underbrace{10 \qquad\qquad + 15}$
	25

These two expressions are actually equivalent. When you evaluate them, the final result is the same, but it takes a lot more work when you use the right-hand expression. To save a lot of work, you need to learn how to *simplify expressions.* Then you can rewrite $2c - 10 - c + 15$ in the simplest way possible, which is $c + 5$.

The basic idea in **simplifying expressions** is to *combine,* or *add,* like terms. Each addend in an expression is a **term.** Here are two examples.

In $6x^2 + {}^-2xy + 8$, the 8 is the *constant term.* There are also two *variable terms* in the expression: $6x^2$ is a variable term, and ^-2xy is a variable term. A **variable term** has a number part (coefficient) and a letter part (variable).

If no coefficient is shown, it is assumed to be 1. Remember from **Section 1.6** that multiplying any number by 1 does *not* change the number.

Also, ^-c can be written $^-1 \cdot c$. The coefficient of ^-c is understood to be $^-1$.

Like Terms

Like terms are terms with exactly the same variable parts (the same letters and exponents). The coefficients do *not* have to match.

Like Terms		Unlike Terms	
$5x$ and $3x$	Variable parts match; both are x.	$3x$ and $3x^2$	Variable parts do *not* match; exponents are different.
$^-6y^3$ and y^3	Variable parts match; both are y^3.	^-2x and ^-2y	Variable parts do *not* match; letters are different.
$4a^2b$ and $5a^2b$	Variable parts match; both are a^2b.	a^3b and a^2b	Variable parts do *not* match; exponents are different.
$^-8$ and 4	There are no variable parts; numbers are like terms.	^-8c and 4	Variable parts do *not* match; one term has a variable part, but the other term does not.

EXAMPLE 1   **Identifying Like Terms and Their Coefficients**

List the like terms in each expression. Then identify the coefficients of the like terms.

(a) $^-5x + ^-5x^2 + 3xy + x + ^-5$
The like terms are ^-5x and x.
The coefficient of ^-5x is $^-5$, and the coefficient of x is understood to be 1.

(b) $2yz^2 + 2y^2z + ^-3y^2z + 2 + ^-6yz$
The like terms are $2y^2z$ and $^-3y^2z$.
The coefficients are 2 and $^-3$.

(c) $10ab + 12 + ^-10a + 12b + ^-6$
The like terms are 12 and $^-6$.
The like terms are constants (there are no variable parts).

Now Try Exercise 3.

The distributive property (see **Section 1.6**) can be used "in reverse" to combine like terms. Here is an example.

$$3 \cdot x + 4 \cdot x \qquad \text{can be written as} \qquad (3 + 4) \cdot x$$
$$3x + 4x \qquad\qquad\qquad\qquad\qquad 7 \cdot x$$
$$7x$$

Thus, $3x + 4x$ can be written in *simplified form* as $7x$. To check that $3x + 4x$ is the same as $7x$, evaluate each expression when x is 2.

$$3x + 4x \qquad\qquad\qquad 7x$$
$$3 \cdot 2 + 4 \cdot 2 \qquad\qquad 7 \cdot 2$$
$$6 + 8 \qquad\qquad\qquad 14$$
$$14$$

Both results are 14, so the expressions are equivalent. But you can see how much easier it is to work with $7x$, the simplified expression.

CAUTION
Notice that $3x + 4x$ is simplified to $7x$, **not** to $7x^2$.

Variable part is unchanged.   Do *not* change x to x^2.

Combining Like Terms

Step 1   If there are any variable terms with no coefficient, write in the understood 1.

Step 2   If there are any subtractions, change each one to adding the opposite.

Step 3   Find *like* terms (the variable parts match).

Step 4   Add the coefficients (number parts) of like terms. ***The variable part stays the same.***

EXAMPLE 2   **Combining Like Terms**

Combine like terms.

(a) $2x + 4x + x$

$2x + 4x + x$ No coefficient; write understood 1.
               There are no subtractions to change.

$2x + 4x + 1x$ Find like terms: $2x$, $4x$, and $1x$ are like terms,
                  so add the coefficients, $2 + 4 + 1$.

$(2 + 4 + 1)x$ The variable part, x, stays the same.

$7x$

Therefore, $2x + 4x + x$ can be written as $7x$.

(b) $^{-}3y^2 - 8y^2$

$^{-}3y^2 - 8y^2$ Both coefficients are shown.
                     Change subtraction to adding the opposite.

$^{-}3y^2 + {^{-}8}y^2$ Find like terms: $^{-}3y^2$ and $^{-}8y^2$ are like terms,
                          so add the coefficients, $^{-}3 + {^{-}8}$.

$(^{-}3 + {^{-}8})y^2$ The variable part, y^2, stays the same.

$^{-}11y^2$

Therefore, $^{-}3y^2 - 8y^2$ can be written as $^{-}11y^2$.

Now Try Exercises 9 and 23.

OBJECTIVE 2 Simplify expressions.   When simplifying expressions, be careful to combine only *like* terms—those having variable parts that match. You *cannot* combine terms if the variable parts are different.

EXAMPLE 3   Simplifying Expressions

Simplify each expression by combining like terms.

(a) $6xy + 2y + 3xy$

The *like* terms are $6xy$ and $3xy$. We can use the commutative property to rewrite the expression so that the like terms are next to each other. This helps to organize our work.

$6xy + 3xy + 2y$ Combine *like* terms only.

$(6 + 3)xy + 2y$ Add the coefficients, $6 + 3$. The variable part, xy, stays the same.

$9xy \qquad + 2y$ Keep writing $2y$, the term that was *not* combined; it is still part of the expression.

The simplified expression is $9xy + 2y$.

(b) Here is the expression from the first page in this section.

$2c - 10 - c + 15$ Write the understood 1 as the coefficient of c.

$2c - 10 - 1c + 15$ Change subtractions to adding the opposite.

$2c + {}^-10 + {}^-1c + 15$ We can add in any order, so rewrite the expression so that like terms are next to each other.

$2c + {}^-1c + {}^-10 + 15$ Combine $2c + {}^-1c$. Also combine ${}^-10 + 15$.

$(2 + {}^-1)c + \qquad 5$

$1c \qquad + \quad 5$

The simplified expression is $1c + 5$ or just $c + 5$.

$1c$ is the same as c.

NOTE

In this book, when combining like terms we will usually write the variable terms in alphabetical order. A constant term (number only) will be written last. So, in Examples 3(a) and 3(b) above, the preferred and alternative ways of writing the expressions are as follows.

The simplified expression is $9xy + 2y$ (alphabetical order). However, by the commutative property of addition, $2y + 9xy$ is also correct.

The simplified expression is $c + 5$ (constant written last). However, by the commutative property of addition, $5 + c$ is also correct.

Now Try Exercises 29 and 33.

We can use the associative property of multiplication to simplify an expression such as $4(3x)$.

$4(3x)$ can be written as $4 \cdot (3 \cdot x)$

Understood multiplications

Using the associative property, we can regroup the factors.

$4 \cdot (3 \cdot x)$ can be written as $\underbrace{(4 \cdot 3)}_{} \cdot x$ To simplify, multiply $4 \cdot 3$.

$\underbrace{12 \quad \cdot x}_{}$ Write $12 \cdot x$ without the multiplication dot.

$12x$

The simplified expression is $12x$.

EXAMPLE 4 **Simplifying Multiplication Expressions**

Simplify.

(a) $5(10y)$

Use the associative property.

$5 \cdot (10 \cdot y)$ can be written as $\underbrace{(5 \cdot 10)}_{} \cdot y$ Multiply $5 \cdot 10$.

$\underbrace{50 \quad \cdot y}_{}$ Write $50 \cdot y$ without the multiplication dot.

$50y$

So, $5(10y)$ simplifies to $50y$.

(b) $^-6(3b)$

Use the associative property.

$^-6(3b)$ can be written as $\underbrace{(^-6 \cdot 3)}b$

^-18b

So, $^-6(3b)$ simplifies to ^-18b.

(c) $^-4(^-2x^2)$

Use the associative property.

$^-4(^-2x^2)$ can be written as $\underbrace{(^-4 \cdot ^-2)}x^2$

$8x^2$

So, $^-4(^-2x^2)$ simplifies to $8x^2$.

Now Try Exercise 41. ▶▶▶

OBJECTIVE 3 Use the distributive property to multiply.   The distributive property can also be used to simplify expressions such as $3(x + 5)$. You *cannot* add the terms inside the parentheses because x and 5 are *not* like terms. But notice the understood multiplication dot between the 3 and the parentheses.

$3(x + 5)$

$\downarrow$

$3 \cdot (x + 5)$

Thus you can *distribute* multiplication over addition, as you did in **Section 1.6.** That is, multiply 3 times each term inside the parentheses.

$3 \cdot (x + 5)$ can be written as $\underbrace{3 \cdot x}_{} + \underbrace{3 \cdot 5}_{}$

$3x \quad + \quad 15$

So, $3(x + 5)$ simplifies to $3x + 15$.

⬆ Stays as addition ⬆

Multiplication also distributes over subtraction.

$4(y - 2)$ means $4 \cdot (y - 2)$ and can be written as $4 \cdot y - 4 \cdot 2$

$$4y \quad - \quad 8$$

So, $4(y - 2)$ simplifies to $4y - 8$.   Notice that we did *not* need to change
└ Stays as subtraction ┘   subtraction to adding the opposite.

EXAMPLE 5 **Using the Distributive Property**

Simplify.

(a) $6(y - 4)$ can be written as $6 \cdot y - 6 \cdot 4$

$$6y \quad - \quad 24$$

————— Stays as subtraction —————

So, $6(y - 4)$ simplifies to $6y - 24$

(b) $5(3x + 2)$ can be written as $5 \cdot 3x + 5 \cdot 2$

$$5 \cdot 3 \cdot x + 10$$

$$15 \cdot x + 10$$

$$15x + 10$$

So, $5(3x + 2)$ simplifies to $15x + 10$

(c) $^-2(4a + 3)$ can be written as $^-2 \cdot 4a + {}^-2 \cdot 3$

$$^-2 \cdot 4 \cdot a + {}^-6$$

$$^-8 \cdot a + {}^-6$$

$$^-8a + {}^-6$$

Now we will use the definition of subtraction "in reverse" to rewrite $^-8a + {}^-6$.

Change $^-6$ to its
opposite, $^+6$.

Write $^-8a + {}^-6$ as $^-8a - 6$

Change addition
to subtraction.

Think back to the way we changed subtraction to adding the opposite. Here we are "working backward." From now on, whenever addition is followed by a negative number, we will change it to subtracting a positive number.

$$^-8a + {}^-6$$

Equivalent expressions

$$^-8a - 6$$

So, $^-2(4a + 3)$ simplifies to $^-8a - 6$.

◀◀◀ **Now Try Exercises 53 and 57.**

Sometimes you need to do several steps to simplify an expression.

> **EXAMPLE 6**   **Simplifying a More Complex Expression**

Simplify: $8 + 3(x - 2)$

$8 + 3(x - 2)$ Do *not* add $8 + 3$. Use the distributive property first because multiplying is done *before* adding.

$8 + 3 \cdot x - 3 \cdot 2$ Do the multiplications.

$8 + 3x - 6$ Rewrite so that like terms are next to each other.

$3x + 8 - 6$ Subtract to find $8 - 6$ or change to adding $8 + {}^-6$.

$3x + 2$

The simplified expression is $3x + 2$.

> **CAUTION**
>
> Do ***not*** add $8 + 3$ as the first step in Example 6 above. Remember that the order of operations tells you to do multiplying *before* adding.

Now Try Exercises 63 and 67. ▶▶▶

2.2 **Exercises**

 FOR EXTRA HELP   Addison-Wesley Math Tutor Center   MathXL   Digital Video Tutor CD 1 Videotape 2   📖 Student's Solutions Manual   *MyMathLab* MyMathLab   MathXL Tutorials on CD

Circle the like terms in each expression. Then identify the coefficients of the like terms. See Example 1.

1. $2b^2 + 2b + 2b^3 + b^2 + 6$

2. $3x + x^3 + 3x^2 + 3 + 2x^3$

3. $^-x^2y + {}^-xy + 2xy + {}^- 2xy^2$

4. $ab^2 + {}^-a^2b + 2ab + {}^-3a^2b$

5. $7 + 7c + 3 + 7c^3 + {}^-4$

6. $4d + {}^-5 + 1 + {}^-5d^2 + 4$

Simplify each expression. See Example 2.

7. $6r + 6r$

8. $4t + 10t$

9. $x^2 + 5x^2$

10. $9y^3 + y^3$

11. $p - 5p$

12. $n - 3n$

13. $^-2a^3 - a^3$

14. $^-10x^2 - x^2$

15. $c - c$

16. $b^2 - b^2$

17. $9xy + xy - 9xy$

18. $r^2s - 7r^2s + 7r^2s$

19. $5t^4 + 7t^4 - 6t^4$

20. $10mn - 9mn + 3mn$

21. $y^2 + y^2 + y^2 + y^2$

22. $a + a + a$

23. $^-x - 6x - x$

24. $^-y - y - 3y$

Simplify by combining like terms. Write each answer with the variables in alphabetical order and any constant term last. See Example 3.

25. $8a + 4b + 4a$

26. $6x + 5y + 4y$

27. $6 + 8 + 7rs$

28. $10 + 2c^2 + 15$

29. $a + ab^2 + ab^2$

30. $n + mn + n$

31. $6x + y - 8x + y$

32. $d + 3c - 7c + 3d$

33. $8b^2 - a^2 - b^2 + a^2$

34. $5ab - ab + 3a^2b - 4ab$

35. $^-x^3 + 3x - 3x^2 + 2$

36. $a^2b - 2ab - ab^3 + 3a^3b$

37. $^-9r + 6t - s - 5r + s + t - 6t + 5s - r$

38. $^-x - 3y + 4z + x - z + 5y - 8x - y$

Simplify by using the associative property of multiplication. See Example 4.

39. $3(10a)$

40. $8(4b)$

41. $^-4(2x^2)$

42. $^-7(3b^3)$

43. $5(^-4y^3)$

44. $2(^-6x)$

45. $^-9(^-2cd)$

46. $^-6(^-4rs)$

47. $7(3a^2bc)$

48. $4(2xy^2z^2)$

49. $^-12(^-w)$

50. $^-10(^-k)$

Use the distributive property to simplify each expression. See Example 5.

51. $6(b + 6)$

52. $5(a + 3)$

53. $7(x - 1)$

54. $4(y - 4)$

55. $3(7t + 1)$

56. $8(2c + 5)$

57. $^{-}2(5r + 3)$

58. $^{-}5(6z + 2)$

59. $^{-}9(k + 4)$

60. $^{-}3(p + 7)$

61. $50(m - 6)$

62. $25(n - 1)$

Simplify each expression. See Example 6.

63. $10 + 2(4y + 3)$

64. $4 + 7(x^2 + 3)$

65. $6(a^2 - 2) + 15$

66. $5(b - 4) + 25$

67. $2 + 9(m - 4)$

68. $6 + 3(n - 8)$

69. $^{-}5(k + 5) + 5k$

70. $^{-}7(p + 2) + 7p$

71. $4(6x - 3) + 12$

72. $6(3y - 3) + 18$

73. $5 + 2(3n + 4) - n$

74. $8 + 8(4z + 5) - z$

75. $^{-}p + 6(2p - 1) + 5$

76. $^{-}k + 3(4k - 1) + 2$

77. Explain the difference between *simplifying* an expression and *evaluating* an expression.

78. Simplify each expression. Are the answers equivalent? Explain why or why not.

$$5(3x + 2) \qquad 5(2 + 3x)$$

79. Explain what makes two terms *like* terms. Include several examples in your explanation.

80. Explain how to combine like terms. Include an example in your explanation.

81. Explain and correct the error made by a student who simplified this expression.

$$\underbrace{{}^-2x + 7x}_{5x^2} + 8$$

82. Explain and correct the error made by a student who simplified this expression.

$$\underbrace{{}^-10a + 6a}_{{}^-4a} - \underbrace{7 + 2}_{{}^-7 + 2}$$

$${}^-4a \quad + \quad {}^-7 + 2$$

$${}^-4a \quad + \quad {}^-5$$

$$4a - 5$$

Simplify.

83. $-4(3y) - 5 + 2(5y + 7)$

84. $6({}^-3x) - 9 + 3({}^-2x + 6)$

85. $-10 + 4({}^-3b + 3) + 2(6b - 1)$

86. $12 + 2(4a - 4) + 4({}^-2a - 1)$

87. $-5({}^-x + 2) + 8({}^-x) + 3({}^-2x - 2) + 16$

88. $-7({}^-y) + 6(y - 1) + 3({}^-2y) + 6 - y$

Summary Exercises on Variables and Expressions

1. The expression for finding the perimeter of an octagon with sides of equal length is $8s$. Evaluate the expression when

   (a) the length of one side is 4 yards.

   $\longleftrightarrow$
   s

   (b) the length of one side is 15 inches.

2. The expression for the total cost of a car is $d + mt$ where d is the down payment, m is the monthly payment, and t is the number of months you must make payments. Evaluate the expression when:

   (a) the down payment is \$3000 and you make 36 monthly payments of \$280.

   (b) the down payment is \$1750 and you make 48 monthly payments of \$429.

Evaluate each expression when w is 5, x is ⁻2, and y is ⁻6.

3. wxy

4. x^3

5. ^-4wy

6. $3xy^2$

7. w^2x^5

8. $^-7wx^4y^3$

Simplify each expression.

9. $10b + 4b + 10b$

10. $^-3x - 5 + 12 + 10x$

11. $^-8(c + 4)$

12. $^-9xy + 9xy$

13. $^-4(^-3c^2d)$

14. $3f - 5f - 4f$

15. $2(3w + 4)$

16. $^-a - 6b - a$

17. $^-10(^-5x^3y^2)$

18. $5r^2 + 2r - 2r^2 + 5r^3$

19. $21 + 7(h^2 - 3)$

20. $^-3(m + 3) + 3m$

21. $^-4(8y - 5) + 5$

22. $2 + 12(3x - 1)$

23. $^-n + 5(4n - 2) + 11$

24. $^-6 + 3(2h + 4) - 10h + 5(h^2 - 2)$

2.3 Solving Equations Using Addition

Now you are ready for a look at the "heart" of algebra, writing and solving **equations.** *Writing* an equation is a way to show the relationship between what you *know* about a problem and what you *don't* know. Then, *solving* the equation is a way to figure out the part that you didn't know and answer your question.

The questions you can answer by writing and solving equations are as varied as the careers people choose. A zookeeper can solve an equation that answers the question of how long to incubate the egg of a particular tropical bird. An aerobics instructor can solve an equation that answers the question of how hard a certain person should exercise for maximum benefit.

OBJECTIVES

1 Determine whether a given number is a solution of an equation.

2 Solve equations using the addition property of equality.

3 Simplify equations before using the addition property of equality.

OBJECTIVE **1** **Determine whether a given number is a solution of an equation.** Let's start with the example from the beginning of this chapter: ordering textbooks for math classes. The expression we used to order books was $c + 5$, where c was the class limit (the maximum number of students allowed in the class). Suppose that 30 prealgebra books were ordered. What is the class limit for prealgebra? To answer this question, write an equation showing the relationship between what you know and what you don't know.

You don't know the class limit. → ↓   ↓ You do know the total number of books ordered.

$$c + 5 = 30$$

↑——You do know that 5 extra books were ordered.

> **NOTE**
> An equation has an equal sign. Notice the similarity in the words **equa**tion and **equa**l. An expression does *not* have an equal sign.

The equal sign in an equation is like the balance point on a playground teeter-totter, or seesaw. To have a true equation, the two sides must balance.

These equations balance, so we can use the $=$ sign.

$6 + 8 = 14$

$10 = 5 \cdot 2$

$3 \cdot 2 = 5 + 1$

These equations do *not* balance, so we write $\neq$ to mean "not equal to."

$6 + 8 \neq 15$

$10 \neq 4 \cdot 2$

$4 + 5 \neq 5 \cdot 4$

When an equation has a variable, we **solve the equation** by finding a number that can replace the variable and make the equation balance. For the example about ordering prealgebra textbooks:

$$c + 5 = 30$$

What number can replace c so that the equation balances?

Try replacing c with **15**. $15 + 5 \neq 30$ Does *not* balance:
$15 + 5$ is only 20.

Try replacing c with **40**. $40 + 5 \neq 30$ Does *not* balance:
$40 + 5$ is more than 30.

Try replacing c with **25**. $25 + 5 = 30$ Balances: $25 + 5$ is 30.

The **solution** is **25** because **25** is the *only* number that makes the equation balance. By solving the equation, you have answered the question about the class limit for prealgebra. The class limit is 25.

NOTE

Most of the equations that you will solve in this book have only one solution, that is, one number that makes the equation balance. In **Chapter 9,** and in other algebra courses, you will solve equations that have two or more solutions.

EXAMPLE 1 **Identifying the Solution of an Equation**

Which of these numbers, 70, 40, or 60, is the solution of the equation $c - 10 = 50$?

Replace c with each of the numbers. The one that makes the equation balance is the solution.

$70 - 10 \neq 50$ $40 - 10 \neq 50$ $60 - 10 = 50$

Does *not* balance: Does *not* balance: Balances:
$70 - 10$ is more than 50. $40 - 10$ is only 30. $60 - 10$ is 50.

The solution is 60 because, when c is 60, the equation balances.

Now Try Exercise 3.

OBJECTIVE 2 **Solve equations using the addition property of equality.** When solving the book ordering equation, $c + 5 = 30$, you could just look at the equation and think, "What number, plus 5, would balance with 30?" You could easily see that c had to be 25. Not all equations can be solved this easily, so you'll need some tools for the harder ones. The first tool, called the **addition property of equality,** allows us to add the *same* number to *both* sides of an equation.

Addition Property of Equality

If $a = b$, then $a + c = b + c$.

In other words, you may add the same number to both sides of an equation and still keep it balanced.

Think of the teeter-totter. If there are 3 children of the same size on each side, it will balance. If 2 more children climb onto the left side, the only way to keep the balance is to have 2 more children of the same size climb onto the right side as well.

$$3 = 3 \qquad 3 + 2 = 3 + 2$$

All the tools you will learn to use with equations have one goal.

Goal in Solving an Equation

The goal is to end up with the variable (letter) on one side of the equal sign balancing a number on the other side.

We work on the original equation until we get:

variable = number   or   number = variable

Once we have arrived at that point, the number balancing the variable is the solution to the original equation.

EXAMPLE 2   Using the Addition Property of Equality

Solve each equation and check the solution.

(a) $c + 5 = 30$

We want to get the variable, c, by itself on the left side of the equal sign. To do that, we add the *opposite* of 5, which is $^-5$. Then $5 + ^-5$ will be 0.

$$c + 5 = 30$$

Add $^-5$ to the left side. $\longrightarrow$ $\underline{\quad ^-5 \quad ^-5 \quad}$ $\longleftarrow$ To keep the balance, add $^-5$ to the right side also.

$$c + 0 = 25 \quad \longleftarrow 30 + ^-5 \text{ is } 25.$$

$5 + ^-5$ is 0.

Recall that adding 0 to any number leaves the number unchanged, so $c + 0$ is c.

$$\underbrace{c + 0}_{c} = 25$$
$$c = 25$$

Because c *balances* with 25, the *solution* is 25.

Check the solution by replacing c with 25 in the *original equation*.

$c + 5 = 30$	Original equation
$25 + 5 = 30$	Replace c with 25.
$\underbrace{\qquad}$	
$30 = 30$	Balances

Because the equation balances when we use 25 to replace the variable, we know that **25 is the correct solution**. If it had *not* balanced, we would need to rework the problem, find our error, and correct it.

(b) $^-5 = x - 3$

We want the variable, x, by itself on the right side of the equal sign. (Remember, it doesn't matter which side of the equal sign the variable is on, just so it ends up by itself.) To see what number to add, we change the subtraction to adding the opposite.

$^-5 = x - 3$	Change subtraction to adding the opposite.
$^-5 = x + ^-3$	To get x by itself on the right side, add the opposite of $^-3$, which is 3. Then $^-3 + 3$ is 0.
$\underline{\quad 3 \qquad\qquad 3 \quad}$	
$^-2 = x + 0$	Adding 0 to x leaves x unchanged.
$^-2 = x$	x balances with $^-2$, so $^-2$ is the solution.

To keep the balance, add 3 to the left side also. $^-5 + 3$ is $^-2$.

We check the solution by replacing x with $^-2$ in the *original equation*. If the equation balances when we use $^-2$, we know that it is the correct solution. If the equation does *not* balance when we use $^-2$, we made an error and need to try solving the equation again.

Continued on Next Page

Check $^-5 = \ \ x - \ \ 3$ Original equation

  $^-5 = ^-2 - \ \ 3$ Replace x with $^-2$.

  $^-5 = \underbrace{^-2 + ^-3}$ Change subtraction to adding the opposite.

  $^-5 = \ \ \ \ ^-5$ Balances; this shows $^-2$ is the correct solution.

When x is replaced with $^-2$, the equation balances, so **$^-2$ is the correct solution**.

CAUTION
When checking the solution to Example 2(b) above, we ended up with $^-5 = ^-5$. Notice that $^-5$ is **not** the solution. The solution is $^-2$, the number used to replace x in the original equation.

Now Try Exercises 13 and 21. ▶▶▶

OBJECTIVE **3** **Simplify equations before using the addition property of equality.** Sometimes you can simplify the expression on one or both sides of the equal sign. Doing so will make it easier to solve the equation.

EXAMPLE 3 **Simplifying before Solving Equations**

Solve each equation and check each solution.

(a) $y + 8 = 3 - 7$

You cannot simplify the left side because y and 8 are *not* like terms.

$y + \ \ 8 = 3 - \ \ 7$ Simplify the right side by changing subtraction to adding the opposite.

$y + \ \ 8 = \underbrace{3 + \ ^-7}$ Add $3 + ^-7$.

To get y by itself on the left side, add the opposite of 8, which is $^-8$.

$y + \ \ 8 = \ \ \ ^-4$

$\underline{\quad ^-8 \quad\quad ^-8}$ To keep the balance, add $^-8$ to the right side also.

$8 + ^-8$ is 0. $\underbrace{y + \ \ 0} = \ ^-12$ $^-4 + ^-8$ is $^-12$.

$y \ \ \ \ \ = \ ^-12$

The solution is $^-12$. Now check the solution.

Check $y \ + 8 = 3 - \ \ 7$ Go back to the *original* equation and replace y with $^-12$.

Add $^-12 + 8$. $\underbrace{^-12 + 8} = 3 - \ \ 7$ Change $3 - 7$ to $3 + ^-7$.

  $^-4 \ \ \ = \underbrace{3 + ^-7}$ Add $3 + ^-7$.

  $^-4 \ \ \ = \ \ \ ^-4$ Balances; so $^-12$ is the correct solution.

When y is replaced with $^-12$, the equation balances, so **$^-12$ is the correct solution** (**not** $^-4$).

──── **Continued on Next Page**

(b) $^-2 + 2 = {}^-4b - 6 + 5b$

Simplify the left side by adding $^-2 + 2$.

$$^-2 + 2 = {}^-4b - 6 + 5b$$

Simplify the right side by changing subtraction to adding the opposite.

$$0 = {}^-4b + {}^-6 + 5b$$ Find like terms.

$$0 = \underbrace{{}^-4b + 5b} + {}^-6$$ Combine $^-4b + 5b$.

To keep the balance, add 6 to the left side also. ⟶

$$0 = 1b + {}^-6$$ To get $1b$ by itself, add the opposite of $^-6$, ⟵ which is 6.

$$\underline{6} \qquad\qquad \underline{6}$$

$$6 = \underbrace{1b + 0}$$

$$6 = 1b$$ $1b$ is equivalent to b.

$$6 = b$$

The solution is 6.

Check

$$^-2 + 2 = {}^-4\boldsymbol{b} - 6 + 5\boldsymbol{b}$$

Go back to the *original* equation and replace each b with 6.

Add $^-2 + 2$.

$$\underbrace{{}^-2 + 2} = \underbrace{{}^-4 \cdot \boldsymbol{6}} - 6 + \underbrace{5 \cdot \boldsymbol{6}}$$

On the right side, do multiplications first.

$$0 = {}^-24 + {}^-6 + 30$$

Change subtraction to adding the opposite.

$$0 = \underbrace{{}^-30 + 30}$$

Add from left to right.

$$0 = 0$$ Balances

When b is replaced with 6, the equation balances, so **6 is the correct solution** (**not** 0).

> **CAUTION**
> When checking a solution, always go back to the *original* equation. That way, you will catch any errors you made when simplifying each side of the equation.

◀◀◀ **Now Try Exercises 29 and 41.**

2.3 Exercises

In each list of numbers, find the one that is a solution of the given equation.
See Example 1.

1. $n - 50 = 8$

58, 42, 60

2. $r - 20 = 5$

15, 30, 25

3. $^-6 = y + 10$

$^-4, ^-16, 16$

4. $^-4 = x + 13$

$17, ^-17, ^-9$

5. $t + 12 = 0$

$0, ^-12, ^-24$

6. $b - 8 = 0$

$8, 0, ^-8$

Solve each equation and check each solution. See Example 2.

7. $p + 5 = 9$ *Check* $p + 5 = 9$

8. $a + 3 = 12$ *Check* $a + 3 = 12$

9. $8 = r - 2$ *Check* $8 = r - 2$

10. $3 = b - 5$ *Check* $3 = b - 5$

11. $^-5 = n + 3$ *Check*

12. $^-1 = a + 8$ *Check*

13. $^-4 + k = 14$ *Check*

14. $^-9 + y = 7$ *Check*

15. $y - 6 = 0$ *Check*

16. $k - 15 = 0$ *Check*

17. $7 = r + 13$ *Check*   **18.** $12 = z + 19$ *Check*

19. $x - 12 = {}^-1$ *Check*   **20.** $m - 3 = {}^-9$ *Check*

21. ${}^-5 = {}^-2 + t$ *Check*   **22.** ${}^-1 = {}^-10 + w$ *Check*

A solution is given for each equation. Show how to check the solution. If the solution is correct, leave it. If the solution is not correct, solve the equation and check your new solution. See Example 2.

23. $z - 5 = 3$ *Check* $z - 5 = 3$ **24.** $x - 9 = 4$ *Check* $x - 9 = 4$
The solution is ${}^-2$. $\downarrow$ The solution is 13. $\downarrow$

25. $7 + x = {}^-11$ *Check*   **26.** $2 + k = {}^-7$ *Check*
The solution is ${}^-18$.   The solution is ${}^-5$.

27. ${}^-10 = {}^-10 + b$ *Check*   **28.** $0 = {}^-14 + a$ *Check*
The solution is 10.   The solution is 0.

Simplify each side of the equation when possible. Then solve the equation and check the solution. Show your work. See Example 3.

29. $c - 4 = {}^{-}8 + 10$ *Check*

30. $b - 8 = 10 - 6$ *Check*

31. ${}^{-}1 + 4 = y - 2$ *Check*

32. $2 + 3 = k - 4$ *Check*

33. $10 + b = {}^{-}14 - 6$ *Check*

34. $1 + w = {}^{-}8 - 8$ *Check*

35. $t - 2 = 3 - 5$ *Check*

36. $p - 8 = {}^{-}10 + 2$ *Check*

37. $10z - 9z = {}^{-}15 + 8$ *Check*

38. $2r - r = 5 - 10$ *Check*

39. ${}^{-}5w + 2 + 6w = {}^{-}4 + 9$ *Check*

40. ${}^{-}2t + 4 + 3t = 6 - 7$ *Check*

Solve each equation. Show your work. See Examples 2 and 3.

41. $^-3 - 3 = 4 - 3x + 4x$

42. $^-5 - 5 = {}^-2 - 6b + 7b$

43. $^-3 + 7 - 4 = {}^-2a + 3a$

44. $6 - 11 + 5 = {}^-8c + 9c$

45. $y - 75 = {}^-100$

46. $a - 200 = {}^-100$

47. $^-x + 3 + 2x = 18$

48. $^-s + 2s - 4 = 13$

49. $82 = {}^-31 + k$

50. $^-5 = 72 + w$

51. $^-2 + 11 = 2b - 9 - b$

52. $^-6 + 7 = 2h - 1 - h$

53. $r - 6 = 7 - 10 - 8$

54. $m - 5 = 2 - 9 + 1$

55. $^-14 = n + 91$

56. $66 = x - 28$

57. $^-9 + 9 = 5 + h$

58. $18 - 18 = 6 + p$

59. A student did this work when solving an equation. Do you agree that the solution is $^-7$? Explain why or why not.

$$\underbrace{^-8 + 1} = x + 7$$
$$^-7 = x + 7$$
$$\underline{^-7 \qquad ^-7}$$
$$^-14 = \underbrace{x + 0}$$
$$^-14 = x$$

Check

$$^-8 + 1 = \quad x + 7$$
$$^-8 + 1 = \underbrace{^-14 + 7}$$
$$^-7 = \quad ^-7$$

Balances, so $^-7$ is the solution.

60. A student did this work when solving an equation. Show how to check the solution. If the solution does not check, find and correct the errors.

$$\underbrace{^-3 - 6} = n - 5$$
$$\underbrace{^-3 + 6} = n - 5$$
$$3 = n - 5$$
$$\underline{^-5 \qquad ^-5}$$
$$^-2 = \underbrace{n + 0}$$
$$^-2 = n$$

61. West Community College always orders 10 extra robes for the graduation ceremony. The college ordered 305 robes this year. Solving the equation $g + 10 = 305$ will give you the number of graduates (g) this year. Solve the equation.

62. Refer to Exercise 61. The college ordered 278 robes last year. Solve the equation $g + 10 = 278$ to find the number of graduates last year.

63. The warmer the temperature, the faster a field cricket chirps. Solving the equation $92 = c + 37$ will give you the number of chirps (in 15 seconds) when the temperature is 92 degrees. Solve the equation.

64. Refer to Exercise 63. Solve the equation $77 = c + 37$ to find the number of times a field cricket chirps (in 15 seconds) when the temperature is 77 degrees.

65. During the summer months, Ernesto spends an average of only $45 per month on parking fees by riding his bike to work on nice days. This is $65 less per month than what he spends for parking in the winter. Solving the equation $p - 65 = 45$ will give you his monthly parking fees in the winter. Solve the equation.

66. By walking to work several times a week in the summer, Aimee spends an average of $56 less per month on parking fees. If she spends $98 per month on parking in the summer, solve the equation $p - 56 = 98$ to find her monthly parking fees in the winter.

Solve each equation. Show your work.

67. $^-17 - 1 + 26 - 38 = {}^-3 - m - 8 + 2m$

68. $19 - 38 - 9 + 11 = {}^-t - 6 + 2t - 6$

69. $^-6x + 2x + 6 + 5x = |0 - 9| - |{}^-6 + 5|$

70. $^-h - |{}^-9 - 9| + 8h - 6h = {}^-12 - |{}^-5 + 0|$

RELATING CONCEPTS (EXERCISES 71–72) For Individual or Group Work

*Use what you have learned about solving equations to **work Exercises 71 and 72 in order.***

71. (a) Write two *different* equations that have $^-2$ as the solution. Be sure that you have to use the *addition property of equality* to solve the equations. Show how to solve each equation. Use Exercises 7 to 22 as models.

   (b) Follow the directions in part (a), but this time write two equations that have 0 as the solution.

72. Not all equations have solutions that are integers. Try solving these equations.

   (a) $x + 1 = 1\dfrac{1}{2}$

   (b) $\dfrac{1}{4} = y - 1$

   (c) $\$2.50 + n = \$3.35$

   (d) Write two more equations that have fraction or decimal solutions.

2.4 Solving Equations Using Division

OBJECTIVE 1 Solve equations, using the division property of equality. In **Section 2.1** you worked with the expression for finding the perimeter of a square-shaped garden, that is, finding the total distance around all four sides of the garden:

$$4s, \text{ where } s \text{ is the length of one side of the square.}$$

Suppose you know that 24 feet of fencing was used around a square-shaped garden. What was the length of one side of the garden? To answer this question, write an equation showing the relationship between what you know and what you don't know.

You don't know the length of one side.
You do know that there are 4 sides. — You do know the perimeter.

$$4s = 24$$

To solve the equation, what number can replace s so that the equation balances? You can see that s is 6 feet.

$$4 \cdot 6 = 24$$

Balances:
4 • 6 is exactly 24.

The **solution** is **6 feet** because 6 is the *only* number that makes the equation balance. You have answered the question about the length of one side: The length is 6 feet.

There is a tool that you can use to solve equations such as $4s = 24$. Called the **division property of equality,** it allows you to *divide* both sides of an equation by the *same* number. (The only exception is that you cannot divide by 0.)

Division Property of Equality

If $a = b$, then $\dfrac{a}{c} = \dfrac{b}{c}$ as long as c is not 0.

In other words, you may divide both sides of an equation by the same nonzero number and still keep it balanced.

In **Section 2.3,** you saw that *adding* the same number to both sides of an equation kept it balanced. We could also have *subtracted* the same number from both sides because subtraction is defined as adding the opposite. Now we're saying that you can *divide* both sides by the same number. In **Chapter 4** we'll *multiply* both sides by the same number.

Equality Principle for Solving an Equation

As long as you do the *same* thing to *both* sides of an equation, the balance is maintained. (The only exception is that you cannot divide by 0.)

OBJECTIVES

1 Solve equations using the division property of equality.

2 Simplify equations before using the division property of equality.

3 Solve equations such as $^-x = 5$.

EXAMPLE 1 **Using the Division Property of Equality**

Solve each equation and check each solution.

(a) $4s = 24$

As with any equation, the goal is to get the variable by itself on one side of the equal sign. On the left side we have $4s$, which means $4 \cdot s$. The variable is multiplied by 4. Division is the opposite of multiplication, so dividing by 4 can be used to "undo" multiplying by 4.

Divide $4s$ by 4. The fraction bar indicates division: $4s \div 4$ is s.
$$\frac{4s}{4} = \frac{24}{4}$$
To keep the balance, divide the right side by 4 also: $24 \div 4$ is 6.

$$s = 6$$

So, as we already knew, 6 is the solution. We check the solution by replacing s with 6 in the original equation.

Check

$$4s = 24 \quad \text{Original equation}$$
$$4 \cdot 6 = 24 \quad \text{Replace } s \text{ with 6.}$$
$$24 = 24 \quad \text{Balances}$$

When s is replaced with 6, the equation balances, so **6 is the correct solution** (**not** 24).

(b) $42 = {}^-6w$

On the right side of the equation, the variable is *multiplied* by $^-6$. To undo the multiplication, *divide* by $^-6$.

To keep the balance, divide by $^-6$ on the left side also.
$$\frac{42}{{}^-6} = \frac{{}^-6w}{{}^-6}$$
Use division to undo multiplication: $^-6w \div {}^-6$ is w.

$$^-7 = w$$

The solution is $^-7$.

Check

$$42 = {}^-6w \quad \text{Original equation}$$
$$42 = {}^-6 \cdot {}^-7 \quad \text{Replace } w \text{ with } ^-7.$$
$$42 = 42 \quad \text{Balances}$$

When w is replaced with $^-7$, the equation balances, so **$^-7$ is the correct solution** (**not** 42).

CAUTION
Be careful to divide both sides by the *same* number as the coefficient of the variable term. In Example 1(b) above, the coefficient of ^-6w is $^-6$, so divide both sides by $^-6$. (Do *not* divide by the *opposite* of $^-6$, which is 6. Use the opposite only when you're *adding* the same number to both sides.)

Now Try Exercises 1 and 11.

OBJECTIVE 2 Simplify equations before using the division property of equality. You can sometimes simplify the expression on one or both sides of the equal sign, as you did in **Section 2.3**.

EXAMPLE 2 **Simplifying before Solving Equations**

Solve each equation and check each solution.

(a) $4y - 7y = {}^-12$

Simplify the left side by combining like terms.	$4y - 7y = {}^-12$	The right side cannot be simplified.
Change subtraction to adding the opposite.	$4y + {}^-7y = {}^-12$	
Divide by the coefficient, which is $^-3$.	$\dfrac{{}^-3y}{{}^-3} = \dfrac{{}^-12}{{}^-3}$	To keep the balance, divide by $^-3$ on the right side.
	$y = 4$	$^-12 \div {}^-3$ is 4.

The solution is 4.

Check

	$4y - 7y = {}^-12$	Go back to the *original* equation and replace each y with 4.
Do multiplications first.	$4 \cdot 4 - 7 \cdot 4 = {}^-12$	
Change subtraction to adding the opposite.	$16 - 28 = {}^-12$	
	$16 + {}^-28 = {}^-12$	
	$^-12 = {}^-12$	Balances

When y is replaced with 4, the equation balances, so **4 is the correct solution**.

(b) $3 - 10 + 7 = h + 7h$

Change subtraction to adding the opposite.	$3 - 10 + 7 = h + 7h$	Write the understood 1 as the coefficient of h.
Add from left to right.	$3 + {}^-10 + 7 = 1h + 7h$	Combine like terms.
	$^-7 + 7 = 8h$	
To keep the balance, divide by 8 on the left side also.	$\dfrac{0}{8} = \dfrac{8h}{8}$	Divide by the coefficient, which is 8.
$0 \div 8$ is 0.	$0 = h$	

The solution is 0.

Check

$3 - 10 + 7 = h + 7h$	Go back to the *original* equation and replace each h with 0.
$3 + {}^-10 + 7 = 0 + 7 \cdot 0$	
$^-7 + 7 = 0 + 0$	
$0 = 0$	Balances

When h is replaced with 0, the equation balances, so **0 is the correct solution**.

Now Try Exercises 15 and 27.

OBJECTIVE **3** **Solve equations such as** $^-x = 5$**.** When solving equations, do *not* leave a negative sign in front of the variable.

EXAMPLE 3 **Solving an Equation of the Type** $^-x = 5$

Solve $^-x = 5$ and check the solution.

It may look as if there is nothing more we can do to the equation $^-x = 5$, but ^-x is *not* the same as x. To see this, we write in the understood $^-1$ as the coefficient of ^-x.

$$\text{Coefficient is understood to be } ^-1.$$
$$^-x = 5 \quad \text{can be written} \quad ^-1x = 5$$

We want the coefficient of x to be $^+1$, not $^-1$. To accomplish that, we can divide both sides by the coefficient of x, which is $^-1$.

$$\frac{^-1x}{^-1} = \frac{5}{^-1} \qquad \text{Divide both sides by } ^-1.$$

On the left side, $^-1 \div ^-1$ is 1. $\qquad 1x = ^-5 \qquad$ On the right side, $5 \div ^-1$ is $^-5$.

Now x is by itself on one side of the equal sign and has a coefficient of $^+1$. The solution is $^-5$.

Check $\qquad\qquad ^-x = 5 \qquad$ Go back to the *original* equation.

$\qquad\qquad\qquad ^-1\,x = 5 \qquad$ Write in the understood $^-1$ as the coefficient of ^-x.

$\qquad\qquad\qquad ^-1 \cdot ^-5 = 5 \qquad$ Replace x with $^-5$.

$\qquad\qquad\qquad\quad 5 \;\; = 5 \qquad$ Balances

When x is replaced with $^-5$, the equation balances, so $^-5$ **is the correct solution** (**not** 5).

CAUTION

As the last step in solving an equation, do *not* leave a negative sign in front of a variable. For example, do *not* leave $^-y = ^-8$. Write in the understood $^-1$ as the coefficient, so that

$$^-y = ^-8 \quad \text{is written as} \quad ^-1y = ^-8$$

Then divide both sides by $^-1$ to get $y = 8$. The solution is 8.

Now Try Exercise 41.

2.4 Exercises

Solve each equation and check each solution. See Example 1.

1. $6z = 12$ Check $6z = 12$ **2.** $8k = 24$ Check $8k = 24$

3. $48 = 12r$ Check   **4.** $99 = 11m$ Check

5. $3y = 0$ Check   **6.** $5a = 0$ Check

7. $^-7k = 70$ Check   **8.** $^-6y = 36$ Check

9. $^-54 = ^-9r$ Check   **10.** $^-36 = ^-4p$ Check

11. $^-25 = 5b$ Check   **12.** $^-70 = 10x$ Check

Simplify where possible. Then solve each equation and check each solution. See Example 2.

13. $2r = ^-7 + 13$ Check $2r = \underbrace{^-7 + 13}$ **14.** $6y = 28 - 4$ Check $6y = \underbrace{28 - 4}$

15. $^-12 = 5p - p$ Check   **16.** $20 = z - 11z$ Check

Solve each equation. Show your work. See Examples 1 and 2.

17. $3 - 28 = 5a$

18. $^-55 + 7 = 8n$

19. $x - 9x = 80$

20. $4c - c = {}^-27$

21. $13 - 13 = 2w - w$

22. $^-11 + 11 = 8t - 7t$

23. $3t + 9t = 20 - 10 + 26$

24. $6m + 6m = 40 + 20 - 12$

25. $0 = {}^-9t$

26. $^-10 = 10b$

27. $^-14m + 8m = 6 - 60$

28. $7w - 14w = 1 - 50$

29. $100 - 96 = 31y - 35y$

30. $150 - 139 = 20x - 9x$

Use multiplication to simplify the side of the equation with the variable. Then solve each equation.

31. $3(2z) = {}^-30$

32. $2(4k) = 16$

33. $50 = {}^-5(5p)$

34. $60 = 4({}^-3a)$

35. ${}^-2({}^-4k) = 56$

36. ${}^-5(4r) = {}^-80$

37. ${}^-90 = {}^-10({}^-3b)$

38. ${}^-90 = {}^-5({}^-2y)$

Solve each equation. See Example 3.

39. ${}^-x = 32$

40. ${}^-c = 23$

41. ${}^-2 = {}^-w$

42. ${}^-75 = {}^-t$

43. ${}^-n = {}^-50$

44. ${}^-x = {}^-1$

45. $10 = {}^-p$

46. $100 = {}^-k$

47. Look again at the solutions to Exercises 39–46. Describe the pattern you see. Then write a rule for solving equations with a negative sign in front of the variable, such as $^-x = 5$.

48. Explain the division property of equality in your own words.

49. Explain and correct the error made by a student who solved this equation.

$$3x = \underbrace{16 - 1}$$

$$\frac{3x}{^-3} = \frac{15}{^-3}$$

$$x = {}^-5$$

50. Write two *different* equations that have $^-4$ as the solution. Be sure that you have to use the division property of equality to solve the equations. Show how to solve each equation. Use Exercises 1–14 as models.

51. The perimeter of a triangle with sides of equal length is 3 times the length of one side (s). If the perimeter is 45 ft, solving the equation $3s = 45$ will give the length of one side. Solve the equation.

52. Refer to Exercise 51. If the perimeter of the triangle is 63 inches, solve the equation $3s = 63$ to find the length of one side.

53. The perimeter of a pentagon with sides of equal length is 5 times the length of one side (s). If the perimeter is 120 meters, solve the equation $120 = 5s$ to find the length of one side.

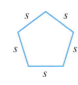

54. Refer to Exercise 53. If the pentagon has a perimeter of 335 yards, solving the equation $335 = 5s$ will give the length of one side. Solve the equation.

Solve each equation. Show your work.

55. $89 - 116 = {}^-4({}^-4y) - 9(2y) + y$

56. $58 - 208 = {}^-b + 8({}^-3b) + 5({}^-5b)$

57. $^-37(14x) + 28(21x) = |72 - 72| + |{}^-166 + 96|$

58. $6a - 10a - 3(2a) = |{}^-25 - 25| - 5(8)$

2.5 Solving Equations with Several Steps

OBJECTIVE 1 Solve equations using the addition and division properties of equality. To solve some equations, you need to use both the addition property of equality (see **Section 2.3**) and the division property of equality (see **Section 2.4**). Here are the steps.

Solving an Equation Using the Addition and Division Properties

Step 1 *Add* the same amount to both sides of the equation so that the variable term (the variable and its coefficient) ends up by itself on one side of the equal sign.

Step 2 *Divide* both sides by the coefficient of the variable term to find the solution.

Step 3 *Check* the solution by going back to the *original* equation.

EXAMPLE 1 Solving an Equation with Several Steps

Solve this equation and check the solution: $5m + 1 = 16$.

Step 1 Get the variable term by itself on one side of the equal sign. The variable term is $5m$. Adding $^-1$ to the left side of the equation will leave $5m$ by itself. To keep the balance, add $^-1$ to the right side also.

$$5m + 1 = 16$$
$$\underline{\quad {}^-1 \quad {}^-1}$$
$$\underbrace{5m + 0}_{} = 15$$
$$5m \quad = 15$$

Step 2 Divide both sides by the coefficient of the variable term. In $5m$, the coefficient is 5, so divide both sides by 5.

$$\frac{5m}{5} = \frac{15}{5}$$
$$m = 3$$

Step 3 Check the solution by going back to the *original* equation.

$$5\boldsymbol{m} + 1 = 16 \qquad \text{Use the original equation}$$
$$\downarrow \qquad\qquad\quad \text{and replace } m \text{ with 3.}$$
$$\underbrace{5(\mathbf{3})}_{} + 1 = 16$$
$$\underbrace{15 + 1}_{} = 16$$
$$16 \quad = 16 \qquad \text{Balances}$$

When m is replaced with 3, the equation balances, so **3 is the correct solution** (**not** 16).

Now Try Exercises 1 and 7. ▶▶▶

So far, variable terms have appeared on just one side of the equal sign. But some equations start with variable terms on both sides. In that case, you can use the addition property of equality to add the same *variable term* to both sides of the equation, just as you have added the same *number* to both sides.

First decide whether to keep the variable term on the left side, or to keep the variable term on the right side. It doesn't matter which one you keep; just pick one side or the other. Then use the addition property to "get rid of" the variable term on the *other* side by adding its opposite.

EXAMPLE 2 **Solving an Equation with Variable Terms on Both Sides**

Solve this equation and check the solution: $2x - 2 = 5x - 11$.

First let's keep $2x$, the variable term on the *left* side. That means we need to "get rid of" $5x$ on the *right* side. We can do that by adding the opposite of $5x$, which is ^-5x.

To keep the balance, add ^-5x to the left side also. Write ^-5x under $2x$, *not* under 2.

Write ^-5x under $5x$, *not* under 11.
$5x + ^-5x$ is $0x$, or 0.

$$2x - 2 = 5x - 11$$
$$\underline{^-5x \qquad\qquad ^-5x}$$
$$^-3x - 2 = 0 - 11$$

Change subtractions to adding the opposite.

$$^-3x + ^-2 = 0 + ^-11$$

To get ^-3x by itself, add 2 to both sides.

$$\underline{\qquad 2 \qquad\qquad\qquad 2}$$
$$^-3x + 0 = 0 + ^-9$$

Divide both sides by $^-3$, the coefficient of the variable term.

$$\frac{^-3x}{^-3} = \frac{^-9}{^-3}$$
$$x = 3$$

Suppose that, in the first step, we decided to keep $5x$ on the *right* side and "get rid of" $2x$ on the *left* side. Let's see what happens.

$$2x - 2 = 5x - 11$$

Add ^-2x to both sides.

$$\underline{^-2x \qquad\qquad ^-2x}$$
$$0 - 2 = 3x - 11$$

Change subtractions to adding the opposite. To get $3x$ by itself, add 11 to both sides.

$$0 + ^-2 = 3x + ^-11$$
$$\underline{\qquad 11 \qquad\qquad\qquad 11}$$
$$0 + 9 = 3x + 0$$

Divide both sides by 3.

$$\frac{9}{3} = \frac{3x}{3}$$
$$3 = x$$

The two solutions are the same. In both cases, x balances with 3.

Notice that we used the addition principle *twice:* once to "get rid of" the variable term $2x$ and once to "get rid of" the number $^-11$. We could have done those steps in the reverse order without changing the result.

> **NOTE**
> More than one sequence of steps will work to solve complicated equations. The basic approach is the following:
>
> • Simplify each side of the equation, if possible.
> • Get the variable term by itself on one side of the equal sign and a number by itself on the other side.
> • Divide both sides by the coefficient of the variable term.

Now Try Exercise 13.

OBJECTIVE 2 Solve equations using the distributive, addition, and division properties.   If an equation contains parentheses, check to see whether you can use the distributive property to remove them.

> **EXAMPLE 3** Solving an Equation Using the Distributive Property

Solve this equation and check the solution: $^-6 = 3(y - 2)$.

We can use the distributive property to simplify the right side of the equation. Recall from **Section 2.2** that

$$3(y - 2) \qquad \text{can be written as} \qquad 3 \bullet y - 3 \bullet 2$$
$$3y \quad - \quad 6$$

So the original equation $^-6 = 3(y - 2)$ becomes $^-6 = 3y - 6$.

$$^-6 = 3y - 6 \qquad \text{Change subtraction to adding the opposite.}$$

$$^-6 = 3y + {}^-6$$

$$\underline{6 \qquad\qquad 6} \qquad \text{To get } 3y \text{ by itself, add 6 to both sides.}$$

$$\frac{0}{3} = \frac{3y}{3} \qquad \text{Divide both sides by 3, the coefficient of } 3y.$$

$$0 = y$$

The solution is 0.

Check
$$^-6 = 3(y - 2) \qquad \text{Go back to the } \textit{original} \text{ equation and replace } y \text{ with 0.}$$

$$^-6 = 3(0 - 2) \qquad \text{Follow the order of operations: work inside parentheses first.}$$

$$^-6 = 3(0 + {}^-2) \qquad \text{Change subtraction to addition.}$$

$$^-6 = \quad 3({}^-2)$$

$$^-6 = \quad {}^-6 \qquad \text{Balances}$$

When y is replaced with 0, the equation balances, so **0 is the correct solution**.

> **Now Try Exercises 19 and 23.**))

Here is a summary of all the steps you can use to solve an equation. Sometimes you will use only two or three steps, and sometimes you will need all five steps, as in the next example.

Solving an Equation

Step 1   If possible, use the ***distributive property*** to remove parentheses.

Step 2   ***Combine*** any like terms on the left side of the equation. Combine any like terms on the right side of the equation.

Step 3   ***Add*** the same amount to both sides of the equation so that the variable term ends up by itself on one side of the equal sign and a number is by itself on the other side. You may have to do this step more than once.

Step 4   ***Divide*** both sides by the coefficient of the variable term to find the solution.

Step 5   ***Check*** your solution by going back to the *original* equation. Replace the variable with your solution. Follow the order of operations to complete the calculations. If the two sides of the equation balance, your solution is correct.

EXAMPLE 4 Solving an Equation

Solve this equation and check the solution: $8 + 5(m + 2) = 6 + 2m$.

Step 1 Use the distributive property on the left side.

$$8 + 5(m + 2) = 6 + 2m$$

Step 2 Combine like terms on the left side.

$$8 + 5m + 10 = 6 + 2m$$

No like terms on the right side.

$$5m + 18 = 6 + 2m$$

Step 3 Add ^-2m to both sides.

$$\underline{^-2m^-2m}$$
$$3m + 18 = 6 + 0$$
$$3m + 18 = 6$$

Step 3 To get $3m$ by itself, add $^-18$ to both sides.

$$\underline{^-18^-18}$$
$$3m + 0 = {^-12}$$

Step 4 Divide both sides by 3, the coefficient of the variable term $3m$.

$$\frac{3m}{3} = \frac{^-12}{3}$$
$$m = {^-4}$$

The solution is $^-4$.

Step 5 Check

$$8 + 5(m + 2) = 6 + 2m$$

Replace each m with $^-4$.

$$8 + 5(^-4 + 2) = 6 + 2(^-4)$$
$$8 + 5(^-2) = 6 + {^-8}$$
$$8 + {^-10} = {^-2}$$
$$^-2 = {^-2}$$

Balances

When m is replaced with $^-4$, the equation balances, so **$^-4$ is the correct solution**.

Now Try Exercise 43.

2.5 Exercises

Solve each equation and check each solution. See Example 1.

1. $7p + 5 = 12$ *Check* $7p + 5 = 12$
                                              $\downarrow$

2. $6k + 3 = 15$ *Check* $6k + 3 = 15$
                                              $\downarrow$

3. $2 = 8y - 6$ *Check*

4. $10 = 11p - 12$ *Check*

5. $^{-}3m + 1 = 1$ *Check*

6. $^{-}4k + 5 = 5$ *Check*

7. $28 = {}^{-}9a + 10$ *Check*

8. $75 = {}^{-}10w + 25$ *Check*

9. $^{-}5x - 4 = 16$ *Check*

10. $^{-}12b - 3 = 21$ *Check*

*In Exercises 11–16, solve each equation **two** ways. First keep the variable term on the left side when you solve it. Then solve it again, keeping the variable term on the right side. Finally, check your solution. See Example 2.*

11. $6p - 2 = 4p + 6$ $6p - 2 = 4p + 6$ *Check* $6p - 2 = 4p + 6$
                                                                                                            $\downarrow$ $\downarrow$

12. $5y - 5 = 2y + 10$ $5y - 5 = 2y + 10$ *Check* $5y - 5 = 2y + 10$
                                                                                                            $\downarrow$ $\downarrow$

13. $^-2k - 6 = 6k + 10$ $^-2k - 6 = 6k + 10$ *Check*

14. $5x + 4 = {}^-3x - 4$ $5x + 4 = {}^-3x - 4$ *Check*

15. $^-18 + 7a = 2a + 7$ $^-18 + 7a = 2a + 7$ *Check*

16. $^-9 + 2z = 9z + 12$ $^-9 + 2z = 9z + 12$ *Check*

Use the distributive property to help you solve each equation. Show your work.
See Example 3.

17. $8(w - 2) = 32$ **18.** $9(b - 4) = 27$ **19.** $^-10 = 2(y + 4)$

20. $^-3 = 3(x + 6)$ **21.** $^-4(t + 2) = 12$ **22.** $^-5(k + 3) = 25$

23. $6(x - 5) = ^-30$ **24.** $7(r - 7) = ^-49$ **25.** $^-12 = 12(h - 2)$

26. $^-11 = 11(c - 3)$ **27.** $0 = ^-2(y + 2)$ **28.** $0 = ^-9(b + 1)$

Solve each equation. Show your work. See Example 4.

29. $6m + 18 = 0$

30. $8p - 40 = 0$

31. $6 = 9w - 12$

32. $8 = 8h + 24$

33. $5x = 3x + 10$

34. $7n = {}^-2n - 36$

35. $2a + 11 = 8a - 7$

36. $r - 10 = 10r + 8$

37. $7 - 5b = 28 + 2b$

38. $1 - 8t = {}^-9 - 3t$

39. ${}^-20 + 2k = k - 4k$

40. $6y - y = {}^-16 + y$

41. $10(c - 6) + 4 = 2 + c - 58$

42. $8(z + 7) - 6 = z + 60 - 10$

43. ${}^-18 + 13y + 3 = 3(5y - 1) - 2$

44. $3 + 5h - 9 = 4(3h + 4) - 1$

45. $6 - 4n + 3n = 20 - 35$

46. $^-19 + 8 = 6p - 7p - 5$

47. $6(c - 2) = 7(c - 6)$

48. $^-3(5 + x) = 4(x - 2)$

49. $^-5(2p + 2) - 7 = 3(2p + 5)$

50. $4(3m - 6) = 72 + 3(m - 8)$

51. $^-6b - 4b + 7b = 10 - b + 3b$

52. $w + 8 - 5w = ^-w - 15w + 11w$

53. Solve $^-2t - 10 = 3t + 5$. Show each step you take while solving it. Next to each step, write a sentence that explains what you did in that step. Be sure to tell when you used the addition property of equality and when you used the division property of equality.

54. Explain the distributive property in your own words. Show two examples of using the distributive property to remove parentheses in an expression.

55. Here is one student's solution to an equation. Show how to check the solution. If the solution doesn't check, explain the error and correct it.

$$
\begin{aligned}
^-8 + 4a &= 2a + 2 \\
\underline{^-2a} &\quad \underline{^-2a} \\
^-10 + 4a &= 0 + 2 \\
^-10 + 4a &= 2 \\
\underline{10} &\quad \underline{10} \\
0 + 4a &= 12 \\
\frac{4a}{4} &= \frac{12}{4} \\
a &= 3
\end{aligned}
$$

56. Here is one student's solution to an equation. Show how to check the solution. If the solution doesn't check, explain the error and correct it.

$$
\begin{aligned}
2(x + 4) &= {}^-16 \\
2x + 4 &= {}^-16 \\
\underline{^-4} &\quad \underline{^-4} \\
2x + 0 &= {}^-20 \\
\frac{2x}{2} &= \frac{^-20}{2} \\
x &= {}^-10
\end{aligned}
$$

RELATING CONCEPTS (EXERCISES 57–60) For Individual or Group Work

Work Exercises 57–60 in order.

57. **(a)** Suppose that the sum of two numbers is negative and you know that one of the numbers is positive. What can you conclude about the other number?

(b) How can you tell, just by looking, that the solution to $x + 5 = {}^-7$ must be a negative number? Recall your answer from part (a).

58. **(a)** Suppose that the sum of two numbers is positive, and you know that one of the numbers is negative. What can you conclude about the other number?

(b) How can you tell, just by looking, that the solution to $^-8 + d = 2$ must be a positive number? Recall your answer from part (a).

59. **(a)** Suppose the product of two numbers is negative and you know that one of the numbers is negative. What can you conclude about the other number?

(b) How can you tell, just by looking, that the solution to $^-15n = {}^-255$ must be positive?

60. **(a)** Suppose the product of two numbers is positive and you know that one of the numbers is negative. What can you conclude about the other number?

(b) How can you tell, just by looking, that the solution to $437 = {}^-23y$ must be negative?

Objective **Write, use, and interpret algebraic expressions that model real-world events.**

A college with four campuses uses an auditorium for graduation that has 1500 seats. Each campus hosts its own graduation ceremony. Students are allocated a whole number of tickets. Round each result **down** to the nearest whole number.

1. How many tickets are allocated to each of 458 graduates at the **North** Campus graduation?

2. How many tickets are allocated to each of 297 graduates at the **East** Campus graduation?

3. How many tickets are allocated to each of 315 graduates at the **South** Campus graduation?

4. How many tickets are allocated to each of 186 graduates at the **West** Campus graduation?

5. Write an *expression* that represents the number of tickets that are allocated to each of g graduates.

6. What recommendations do you have for unallocated tickets?

Traffic engineers have to decide how long to have the red, yellow, and green lights showing on a traffic signal. To decide the number of seconds that a yellow light should be on, the engineers use the expression

$$\frac{5v}{100} + 1$$

where v is the speed limit in miles per hour (mph).

7. How many seconds should the yellow light be on if the speed limit is 20 mph? 40 mph? 60 mph?

8. Based on the answers you just calculated, how could you estimate the time for a yellow light if the speed limit is 30 mph? 50 mph?

9. Use the given expression to find the number of seconds that the yellow light should be on if the speed limit is 30 mph and 50 mph. Did you get the same result as in Problem 8?

To estimate the number of words in a child's vocabulary, a speech therapist uses the expression $60A - 900$ where A is the child's age in months.

10. Estimate the number of words that a child aged 20 months knows.

11. Estimate the number of words that a child aged 2 years knows. (*Hint:* How many months are in two years?)

12. How many words does a child learn between the ages of 20 months and 2 years?

13. Estimate the number of words in the vocabulary of a 3-year-old child.

14. How many words does a child learn between the ages of 2 years and 3 years?

15. Evaluate the expression for a child who is 15 months old. Do you think that the answer is reasonable? Explain why or why not.

16. Evaluate the expression for a child who is 12 months old. Do you think that the answer makes sense? Explain why or why not.

Tuition and fees for resident students in 2004 at North Harris Montgomery Community College District (NHMCCD) in Texas are described below.

Fees Required at NHMCCD

[*Residents of the district pay*] tuition at the rate of $32 per credit hour, a $6 per credit hour technology fee, a $2 per credit hour student activity fee, and a registration fee of $12. (*Source:* www.nhmccd.edu)

17. Calculate the tuition and fees for a resident student who enrolls at NHMCCD for the number of credit hours listed in parts **(a), (b),** and **(c)** below. Then, in part **(d),** write the algebraic expression for *x* credit hours.

    (a) 3 credit hours: _____

    (b) 9 credit hours: _____

    (c) 12 credit hours: _____

    (d) *x* credit hours: _____ dollars

Write the algebraic expression that represents the tuition and fees for each institution for one semester. Let *x* represent the number of credit hours. If you have difficulty, first calculate the costs for 3 or 9 credit hours and focus on the process that you used to get the answer.

18. American River College, California (nonresident student) (*Source:* www.arc.losrios.edu)
    Enrollment: $18 per credit hour; parking: $30 per semester; an additional nonresident enrollment: $149 per credit hour; other fees: $6

19. Austin Community College, Texas (out-of-district student) (*Source:* www.austin.edu)
    Tuition: $34 per credit hour; building fee: $13 per credit hour; parking: $10 per semester; an additional out-of-district tuition: $55 per credit hour; student service fee: $3

20. Your college tuition and fees

KEY TERMS

2.1	**variable**	A variable is a letter that represents a number that varies or changes, depending on the situation.
	constant	A constant is a number that is added or subtracted in an expression. It does not vary. For example, 5 is the constant in the expression $c + 5$.
	expression	An expression expresses, or tells, the rule for doing something. It is a combination of operations on variables and numbers. An expression does *not* have an equal sign.
	evaluate the expression	To evaluate an expression, replace each variable with specific values (numbers) and follow the order of operations.
	coefficient	The number part in a multiplication expression is the coefficient. For example, 4 is the coefficient in the expression $4s$.
2.2	**simplifying expressions**	To simplify an expression, write it in a simpler way by combining all the like terms.
	term	Each addend in an expression is a term.
	variable term	A variable term has a number part (called the coefficient) multiplied by a variable part (a letter). An example is $4s$.
	like terms	Like terms are terms with exactly the same variable parts (the same letters and exponents). The coefficients may be different.
2.3	**equations**	An equation has an equal sign. It shows the relationship between what is known about a problem and what isn't known.
	solve the equation	To solve an equation, find a number that can replace the variable and make the equation balance.
	solution	A solution of an equation is a number that can replace the variable and make the equation balance.
	addition property of equality	The addition property of equality states that adding the same quantity to both sides of an equation will keep it balanced.
	check the solution	To check the solution of an equation, go back to the *original* equation and replace the variable with the solution. If the equation balances, the solution is correct.
2.4	**division property of equality**	The division property of equality states that dividing both sides of an equation by the same nonzero number will keep it balanced.

TEST YOUR WORD POWER

See how well you have learned the vocabulary in this chapter. Answers follow the Quick Review.

1. **A variable**
   - **A.** can only be the letter x
   - **B.** is never an addend in an expression
   - **C.** is the solution of an equation
   - **D.** represents a number that varies.

2. Which expression has 2 as a **coefficient**?
   - **A.** x^2
   - **B.** $2x$
   - **C.** $x + 2$
   - **D.** $2 - x$

3. Which expression has 4 as a **constant** term?
   - **A.** $4y$
   - **B.** y^4
   - **C.** $4 + y$
   - **D.** $\dfrac{y}{4}$

4. Which expression has four **terms**?
   - **A.** $2 + 3x = {}^{-}6 + x$
   - **B.** $2 + 3x - 6 + x$
   - **C.** $(2)(3x)({}^{-}6)(x)$
   - **D.** $2(3x) = {}^{-}6(x)$

5. **Like terms**
   - **A.** can be multiplied but not added
   - **B.** have the same coefficients
   - **C.** have the same solutions
   - **D.** have the same variable parts.

6. To **simplify an expression,**
   - **A.** combine all the like terms
   - **B.** multiply the exponents
   - **C.** add all the numbers in the expression
   - **D.** add the same quantity to both sides.

QUICK REVIEW

Concepts	Examples

2.1 Evaluating Expressions

Replace each variable with the specified value. Then follow the order of operations to simplify the expression.

The expression for ordering textbooks for two prealgebra classes is $2c + 10$, where c is the class limit. Evaluate the expression when the class limit is 24.

$$2c + 10 \qquad \text{Replace } c \text{ with 24.}$$
$$2 \cdot 24 + 10 \qquad \text{Multiply first.}$$
$$48 \quad + 10 \qquad \text{Add last.}$$
$$58 \qquad \text{Order 58 books.}$$

2.1 Using Exponents with Variables

An exponent next to a variable tells how many times to use the variable as a factor in multiplication.

Rewrite $^-6x^4$ without exponents.

$^-6x^4$ can be written as $^-6 \cdot x \cdot x \cdot x \cdot x$

Coefficient is $^-6$ — x is used as a factor 4 times.

2.1 Evaluating Expressions with Exponents

Rewrite the expression without exponents, replace each variable with the specified value, and multiply all the factors.

Evaluate x^3y when x is $^-4$ and y is 5.

x^3y means $x \cdot x \cdot x \cdot y$ Replace x with $^-4$ and y with 5.

$^-4 \cdot {}^-4 \cdot {}^-4 \cdot 5$ Multiply two factors at a time.

$16 \cdot {}^-4 \cdot 5$

$^-64 \cdot 5$

$^-320$

2.2 Identifying Like Terms

Like terms have *exactly* the same letters and exponents. The coefficients may be different.

List the like terms in this expression. Then identify the coefficients of the like terms.

$$^-3b + {}^-3b^2 + 3ab + b + 3$$

The like terms are ^-3b and b. The coefficient of ^-3b is $^-3$, and the coefficient of b is understood to be 1.

2.2 Combining Like Terms

Step 1 If there are any variable terms with no coefficient, write in the understood 1.

Step 2 Change any subtractions to adding the opposite.

Step 3 Find like terms.

Step 4 Add the coefficients of like terms, keeping the variable part the same.

Simplify $4x^2 - 10 + x^2 + 15$.

Write understood 1.

$4x^2 - 10 + 1x^2 + 15$ Change subtraction to adding the opposite.

$4x^2 + {}^-10 + 1x^2 + 15$

$4x^2 + 1x^2 + {}^-10 + 15$ Combine $4x^2 + 1x^2$. The variable part stays the same.

$(4 + 1)x^2 + 5$ Also combine $^-10 + 15$.

$5x^2 + 5$

The simplified expression is $5x^2 + 5$.

Concepts	**Examples**

2.2 *Simplifying Multiplication Expressions*

Use the associative property to rewrite the expression so that the two number parts can be multiplied. The variable part stays the same.

Simplify: $^-7(5k)$.

Use the associative property of multiplication.

$$^-7 \cdot (5 \cdot k) \text{ can be written as } (^-7 \cdot 5) \cdot k$$

$$\underbrace{^-35} \cdot k$$

$$^-35k$$

The simplified expression is ^-35k.

2.2 *Using the Distributive Property*

Multiplication distributes over addition and over subtraction. Be careful to multiply *every* term inside the parentheses by the number outside the parentheses.

Simplify.

(a) $6(w - 4)$ can be written as $\underbrace{6 \cdot w} - \underbrace{6 \cdot 4}$

$$6w - 24$$

The simplified expression is $6w - 24$.

(b) $^-3(2b + 5)$ can be written as $\underbrace{^-3 \cdot 2b} + \underbrace{^-3 \cdot 5}$

$$^-6b + ^-15$$

Use the definition of subtraction "in reverse" to write $^-6b + ^-15$ as $^-6b - 15$.

The simplified expression is $^-6b - 15$.

2.3 *Solving and Checking Equations Using the Addition Property of Equality*

If possible, *simplify* the expression on one or both sides of the equal sign.

Next, to get the variable by itself on one side of the equal sign, *add* the same number to both sides.

Finally, *check* the solution by going back to the original equation and replacing the variable with the solution. If the equation balances, the solution is correct.

Solve this equation and check the solution.

$$\underbrace{^-5 + 8} = 9 + r \qquad \text{Simplify the left side by}$$
$$\qquad\qquad\qquad\qquad \text{adding } ^-5 + 8.$$
$$3 = 9 + r \qquad \text{To get } r \text{ by itself, add}$$
$$\qquad\qquad\qquad\qquad \text{the opposite of } 9,$$
$$\underline{^-9 \qquad ^-9} \qquad \text{which is } ^-9, \text{ to both sides.}$$
$$^-6 = \underbrace{0 + r}$$
$$^-6 = r$$

The solution is $^-6$.

Check $^-5 + 8 = 9 + r$ Use the original equation and replace r with $^-6$.

$$^-5 + 8 = 9 + {}^-6$$
$$\underbrace{\qquad} \quad \underbrace{\qquad}$$
$$3 = 3 \qquad \text{Balances}$$

When r is replaced with $^-6$, the equation balances, so $^-6$ is the correct solution.

Concepts	Examples

2.4 *Solving and Checking Equations Using the Division Property of Equality*

If possible, *simplify* the expression on one or both sides of the equal sign.

Next, to get the variable by itself on one side of the equal sign, *divide* both sides by the coefficient of the variable term.

Finally, *check* the solution by going back to the original equation and replacing the variable with the solution. If the equation balances, the solution is correct.

Solve this equation and check the solution.

Simplify the left side. Change subtraction to adding the opposite.

$$2h - 6h = 18 + 22$$ Simplify the right side; add $18 + 22$.

$$2h + {}^-6h = 40$$

Divide by $^-4$, the coefficient of ^-4h.

$$\frac{^-4h}{^-4} = \frac{40}{^-4}$$ Also divide 40 by $^-4$ to keep the balance.

$$h = {}^-10$$

The solution is $^-10$.

Check $$2h - 6h = 18 + 22$$ Original equation; replace h with $^-10$.

$$2(^-10) - 6(^-10) = 40$$

$$^-20 - {}^-60 = 40$$

$$^-20 + {}^+60$$

$$40 = 40$$ Balances

When h is replaced with $^-10$, the equation balances, so $^-10$ is the correct solution.

2.4 *Solving Equations Such as* $^-x = 5$

As the last step in solving an equation, do **not** leave a negative sign in front of the variable, such as $^-x = 5$, because ^-x is **not** the same as x. Divide both sides by $^-1$, the understood coefficient of ^-x.

Solve this equation and check the solution.

$$9 = {}^-n$$

Write the understood $^-1$ as the coefficient of n.

$$9 = {}^-n \quad \text{can be written as} \quad 9 = {}^-1n$$

Now divide both sides by $^-1$.

$$\frac{9}{^-1} = \frac{^-1n}{^-1}$$

$$^-9 = n$$

The solution is $^-9$.

Check $$9 = {}^-n$$ Original equation

$$9 = {}^-1n$$ Write understood $^-1$.

$$9 = {}^-1\,({}^-9)$$ Replace n with $^-9$.

$$9 = 9$$ Balances

When n is replaced with $^-9$, the equation balances, so $^-9$ is the correct solution.

Concepts	Examples
2.5 *Solving Equations with Several Steps*	Solve this equation and check the solution.

Step 1 If possible, use the distributive property to remove parentheses.

Step 1
$$3 + 2(y + 8) = 5y + 4$$

Step 2 Combine any like terms on the left side of the equal sign. Combine any like terms on the right side of the equal sign.

Step 2
$$\mathbf{3 + 2y + 16} = 5y + 4$$
$$2y + \mathbf{19} = 5y + 4$$

Step 3 Add the same amount to both sides of the equation so that the variable term ends up by itself on one side of the equal sign, and a number is by itself on the other side. You may have to do this step more than once.

Step 3
$$\begin{array}{c} \underline{{}^-2y \qquad\qquad {}^-2y} \\ 0 + 19 = 3y + 4 \end{array}$$

Step 3
$$\begin{array}{c} \underline{\qquad {}^-4 \qquad\qquad {}^-4} \\ 0 + 15 = 3y + 0 \end{array}$$

Step 4 Divide both sides by the coefficient of the variable term to find the solution.

Step 4
$$\frac{15}{3} = \frac{3y}{3}$$
$$5 = y$$

The solution is 5.

Step 5 Check the solution by going back to the original equation. Replace the variable with the solution. If the equation balances, the solution is correct.

Check
Step 5

$$3 + 2(\mathbf{y} + 8) = 5\mathbf{y} + 4 \quad \text{Original equation}$$
$$3 + 2(\mathbf{5} + 8) = 5(\mathbf{5}) + 4$$
$$3 + 2(13) = 25 + 4$$
$$3 + 26 = 29$$
$$29 = 29 \quad \text{Balances}$$

When y is replaced with 5, the equation balances, so 5 is the correct solution.

ANSWERS TO TEST YOUR WORD POWER

1. D; *Example:* In $c + 5$, the variable is c.
2. B; *Example:* $2y^3$ and $2n$ also have 2 as a coefficient.
3. C; *Example:* $^-5a^2 + 4$ also has 4 as a constant term.
4. B; *Example:* $3y^2 - 6y + 2y - 5$ also has four terms. Choices (A) and (D) are equations, not expressions; choice (C) has four *factors*.
5. D; *Example:* $7n^2$ and $^-3n^2$ are like terms.
6. A; *Example:* To simplify $4a - 9 + 6a$, combine $4a$ and $6a$ by adding the coefficients. The simplified expression is $10a - 9$.

Chapter **2**
REVIEW EXERCISES

[2.1]

1. (a) Identify the variable, the coefficient, and the constant in this expression.
$$^-3 + 4k$$

(b) Circle the expression that has 20 as the constant term and $^-9$ as the coefficient.

$20m - 9$ $\qquad$ $^-9 + 20x$

$^-9y + 20$ $\qquad$ $20 + 9n$

2. The expression for ordering test tubes for a chemistry lab is $4c + 10$, where c is the class limit. Evaluate the expression when

(a) the class limit is 15.

(b) the class limit is 24.

3. Rewrite each expression without exponents.
(a) x^2y^4
(b) $5ab^3$

4. Evaluate each expression when m is 2, n is $^-3$, and p is 4.
(a) n^2 $\qquad$ **(b)** n^3
(c) $^-4mp^2$ $\qquad$ **(d)** $5m^4n^2$

[2.2] *Simplify.*

5. $ab + ab^2 + 2ab$

6. $^-3x + 2y - x - 7$

7. $^-8(^-2g^3)$

8. $4(3r^2t)$

9. $5(k + 2)$

10. $^-2(3b + 4)$

11. $3(2y - 4) + 12$

12. $^-4 + 6(4x + 1) - 4x$

13. Write an expression with four terms that *cannot* be simplified.

[2.3] *Solve each equation and check each solution.*

14. $16 + n = 5$ $\qquad$ *Check*

15. $^-4 + 2 = 2a - 6 - a$ $\qquad$ *Check*

[2.4] *Solve each equation. Show your work.*

16. $48 = ^-6m$

17. $k - 5k = ^-40$

18. $^-17 + 11 + 6 = 7t$

19. $^-2p + 5p = 3 - 21$

20. $^-30 = 3(^-5r)$

21. $12 = ^-h$

[2.5] *Solve each equation. Show your work.*

22. $12w - 4 = 8w + 12$

23. $0 = ^-4(c + 2)$

24. Every Wednesday is "treat day" at the office. The person who brings the treats buys two treat items for each employee plus four extras. If Roger brings 34 doughnuts, solve the equation $34 = 2n + 4$ to find the number of employees.

MIXED REVIEW EXERCISES

Solve each equation. Show your work.

25. $12 + 7a = 4a - 3$

26. $^-2(p - 3) = ^-14$

27. $10y = 6y + 20$

28. $2m - 7m = 5 - 20$

29. $20 = 3x - 7$

30. $b + 6 = 3b - 8$

31. $z + 3 = 0$

32. $3(2n - 1) = 3(n + 3)$

33. $^-4 + 46 = 7(^-3t + 6)$

34. $6 + 10d - 19 = 2(3d + 4) - 1$

35. $^-4(3b + 9) = 24 + 3(2b - 8)$

Chapter **2**

T E S T

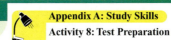

Appendix A: Study Skills
Activity 8: Test Preparation

1. Identify the parts of this expression: $^-7w + 6$
   Choose from these labels: variable, constant, coefficient.

   1. _____

2. The expression for buying hot dogs for the company picnic is $3a + 2c$,
   where a is the number of adults and c is the number of children.
   Evaluate the expression when there are 45 adults and 21 children.

   2. _____

Rewrite each expression without exponents.

3. x^5y^3 4. $4ab^4$

   3. _____

   4. _____

5. Evaluate $^-2s^2t$ when s is $^-5$ and t is 4.

   5. _____

Simplify each expression.

6. $3w^3 - 8w^3 + w^3$ 7. $xy - xy$

   6. _____

   7. _____

8. $^-6c - 5 + 7c + 5$ 9. $3m^2 - 3m + 3mn$

   8. _____

   9. _____

10. $^-10(4b^2)$ 11. $^-5(^-3k)$

    10. _____

    11. _____

12. $7(3t + 4)$ 13. $^-4(a + 6)$

    12. _____

    13. _____

14. $^-8 + 6(x - 2) + 5$ 15. $^-9b - c - 3 + 9 + 2c$

    14. _____

    15. _____

Solve each equation and check each solution.

16. $^-4 = x - 9$ *Check*   **17.** $^-7w = 77$ *Check*

16. _____

17. _____

18. $^-p = 14$ *Check*   **19.** $^-15 = ^-3(a + 2)$ *Check*

18. _____

19. _____

Solve each equation. Show your work.

20. $6n + 8 - 5n = ^-4 + 4$ **21.** $5 - 20 = 2m - 3m$

20. _____

21. _____

22. $^-2x + 2 = 5x + 9$ **23.** $3m - 5 = 7m - 13$

22. _____

23. _____

24. $2 + 7b - 44 = ^-3b + 12 + 9b$ **25.** $3c - 24 = 6(c - 4)$

24. _____

25. _____

26. _____

26. Write an equation that requires the *addition* property of equality to solve it and has $^-4$ as its solution. Then write a different equation that requires the *division* property of equality to solve it and has $^-4$ as its solution. Show how to solve each equation.

Cumulative Review Exercises

CHAPTERS 1–2

1. Write this number in words.
306,000,004,210

2. Write this number, using digits.
Eight hundred million, sixty-six thousand

3. Write $<$ or $>$ between each pair of numbers to make a true statement.

   (a) $^-3$ ___ $^-10$ **(b)** $^-1$ ___ 0

4. Name the property illustrated by each example.

   (a) $^-6 + 2 = 2 + \,^-6$

   (b) $0 \cdot 25 = 0$

   (c) $5(^-6 + 4) = 5 \cdot \,^-6 + 5 \cdot 4$

5. **(a)** Round 9047 to the nearest hundred.

   (b) Round 289,610 to the nearest thousand.

Simplify

6. $0 - 8$

7. $|^-6| + |4|$

8. $^-3(^-10)$

9. $(^-5)^2$

10. $\dfrac{^-42}{^-6}$

11. $^-19 + 19$

12. $(^-4)^3$

13. $\dfrac{^-14}{0}$

14. $^-5 \cdot 12$

15. $^-20 - 20$

16. $\dfrac{45}{^-5}$

17. $^-50 + 25$

18. $^-10 + 6(4 - 7)$

19. $\dfrac{^-20 - 3(^-5) + 16}{(^-4)^2 - 3^3}$

First use front end rounding to estimate the answer to each application problem. Then find the exact answer.

20. One Siberian tiger was tracked for 22 days while it was searching for food. It traveled 616 miles. What was the average distance it traveled each day?

    Estimate:

    Exact:

21. The temperature in Siberia got down to $^-48$ degrees one night. The next day the temperature rose 23 degrees. What was the daytime temperature?

    Estimate:

    Exact:

22. Doug owned 52 shares of Mathtronic stock that had a total value of $2132. Yesterday the value of each share dropped $8. What are his shares worth now?

    Estimate:

    Exact:

23. Ikuko's monthly rent is $552 plus $35 for parking. How much will she spend for rent and parking in one year?

    Estimate:

    Exact:

24. Rewrite $^-4ab^3c^2$ without exponents.

25. Evaluate $3xy^3$ when x is $^-5$ and y is $^-2$.

Simplify.

26. $3h - 7h + 5h$

27. $c^2d - c^2d$

28. $4n^2 - 4n + 6 - 8 + n^2$

29. $^-10(3b^2)$

30. $7(4p - 4)$

31. $3 + 5(^-2w^2 - 3) + w^2$

Solve each equation and check each solution.

32. $3x = x - 8$ ***Check***

33. $^-44 = ^-2 + 7y$ ***Check***

34. $2k - 5k = ^-21$ ***Check***

35. $m - 6 = ^-2m + 6$ ***Check***

Solve each equation. Show your work.

36. $4 - 4x = 18 + 10x$

37. $18 = ^-r$

38. $^-8b - 11 + 7b = b - 1$

39. $^-2(t + 1) = 4(1 - 2t)$

40. $5 + 6y - 23 = 5(2y + 8) - 10$

Solving Application Problems

3

A century ago there were only 8000 cars and 144 miles of paved roads in the United States. But now there are 136,000,000 registered cars and 4,000,000 miles of paved roads. One 12-lane segment of Interstate Highway 5 in California carries over 365,000 vehicles every day! (*Source:* Federal Highway Administration.)

A handy formula for drivers to use is the distance formula, $d = rt$. In **Section 3.1,** Exercises 51–54, you'll see how to use the formula to find your driving time, rate (speed), or distance traveled on long trips.

3.1 Problem Solving: Perimeter

OBJECTIVE 1 Use the formula for perimeter of a square to find the perimeter or the length of one side. If you have ever studied geometry, you probably used several different formulas such as $P = 2l + 2w$ and $A = lw$. A **formula** is just a shorthand way of writing a rule for solving a particular type of problem. A formula uses variables (letters) and it has an equal sign, so it is an equation. That means you can use the equation-solving techniques you learned in **Chapter 2** to work with formulas.

But let's start at the beginning. Geometry was developed centuries ago when people needed a way to measure land. The name *geometry* comes from the Greek words *ge,* meaning earth, and *metron,* meaning measure. Today we still use geometry to measure land. It is also important in architecture, construction, navigation, art and design, physics, chemistry, and astronomy. You can use geometry at home when you buy carpet or wallpaper, hang a picture, or do home repairs. In this chapter you'll learn about two basic ideas, perimeter and area. Other geometry concepts will appear in later chapters.

In **Section 2.1,** you found the *perimeter* of a square garden.

> **Perimeter**
>
> The distance around the outside edges of any flat shape is called the **perimeter** of the shape.

To review, a **square** has four sides that are all the same length. Also, the sides meet to form *right angles,* which measure 90° (90 degrees). This means that the sides form "square corners." (For more information on angles, see **Section 6.5.**) Two examples of squares are shown below.

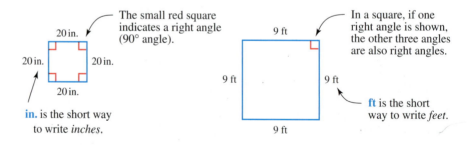

To find the *perimeter* of a square, we can "unfold" the shape so the four sides lie end-to-end, as shown below.

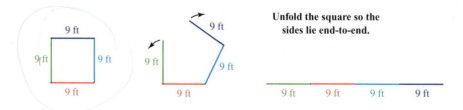

Now we can see the total length of the four sides. The total length is the *perimeter* of the square. We can find the perimeter by adding.

$$\text{Perimeter} = \mathbf{9\ ft} + \mathbf{9\ ft} + \mathbf{9\ ft} + \mathbf{9\ ft} = 36\ ft$$

A shorter way is to multiply the length of one side times 4, because all 4 sides are the same length.

Finding the Perimeter of a Square

$$\text{Perimeter of a square} = \text{side} + \text{side} + \text{side} + \text{side}$$
$$\text{or} \quad P = 4 \cdot \text{side}$$
$$P = 4s$$

EXAMPLE 1 Finding the Perimeter of a Square

Find the perimeter of the square on the previous page that measures 9 ft on each side.

Use the formula for perimeter of a square, $P = 4s$. You know that for this particular square, the value of s is 9 ft.

$P = 4s$ Formula for perimeter of a square

$P = 4 \cdot 9 \text{ ft}$ Replace s with 9 ft. Multiply 4 times 9 ft.

$P = 36 \text{ ft}$ Write 36 ft; ft is the unit of measure.

The perimeter of the square is 36 ft. Notice that this answer matches the result obtained from adding the four sides.

Now Try Exercise 3.

EXAMPLE 2 Finding the Length of One Side of a Square

If the perimeter of a square is 40 cm, find the length of one side.
(Note: **cm** is the short way to write *centimeters*.)

Use the formula for perimeter of a square, $P = 4s$. This time you know that the value of P (the perimeter) is 40 cm.

$P = 4s$ Formula for perimeter of a square

$40 \text{ cm} = 4s$ Replace P with 40 cm

$\dfrac{40 \text{ cm}}{4} = \dfrac{4s}{4}$ To get the variable by itself on the right side, divide both sides by 4.

$10 \text{ cm} = s$

The length of one side of the square is 10 cm.

Check Check the solution by drawing a square and labeling the length of each side as 10 cm. The perimeter is 10 cm + 10 cm + 10 cm + 10 cm = 40 cm. This result matches the perimeter given in the problem.

Now Try Exercise 13.

OBJECTIVE 2 Use the formula for perimeter of a rectangle to find the perimeter, the length, or the width. A **rectangle** is a figure with four sides that meet to form right angles, which measure 90°. Each set of opposite sides is parallel and congruent (has the same length). Three examples of rectangles are shown below.

In a rectangle, if one right angle is shown, the other three angles are also right angles.

Each longer side of a rectangle is called the length (l) and each shorter side is called the width (w).

Look at the rectangle above with the lengths of the sides labeled. To find the perimeter (distance around), you could unfold the shape so the sides lie end-to-end. Then add the lengths of the sides.

$$P = \textbf{12 cm} + \textbf{7 cm} + \textbf{12 cm} + \textbf{7 cm} = 38 \text{ cm}$$

or $\quad P = \textbf{12 cm} + \textbf{12 cm} + \textbf{7 cm} + \textbf{7 cm} = 38 \text{ cm} \qquad$ Commutative property

Because the two long sides are both 12 cm, and the two short sides are both 7 cm, you can also use the formula below.

Finding the Perimeter of a Rectangle

Perimeter of a rectangle = length + length + width + width

$$P = \quad (2 \cdot \text{length}) \quad + \quad (2 \cdot \text{width})$$
$$P = \qquad\quad 2l \qquad\quad + \qquad 2w$$

EXAMPLE 3 Finding the Perimeter of a Rectangle

Find the perimeter of this rectangle.

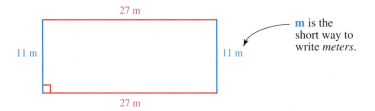

m is the short way to write *meters.*

The length is **27 m**, and the width is **11 m**.

$$P = \quad 2\textbf{\textit{l}} \quad + \quad 2\,\textbf{\textit{w}} \qquad \text{Replace } l \text{ with 27 m and } w \text{ with 11 m.}$$
$$P = 2 \cdot \textbf{27 m} + 2 \cdot \textbf{11 m} \qquad \text{Do the multiplications first.}$$
$$P = \quad 54 \text{ m} \quad + \quad 22 \text{ m} \qquad \text{Add last.}$$
$$P = 76 \text{ m}$$

The perimeter of the rectangle (the distance you would walk around the outside edges of the rectangle) is 76 m.

Check To check the solution, add the lengths of the four sides.

$$P = \textbf{27 m} + \textbf{27 m} + \textbf{11 m} + \textbf{11 m}$$
$$P = 76 \text{ m} \qquad \text{Matches the solution above}$$

◀◀◀ **Now Try Exercise 17.**

EXAMPLE 4 **Finding the Length or Width of a Rectangle**

If the perimeter of a rectangle is 20 ft and the width is 3 ft, find the length.

First draw a sketch of the rectangle and label the widths as 3 ft.

Then use the formula for perimeter of a rectangle, $P = 2l + 2w$. The value of P is 20 ft and the value of w is 3 ft.

$P = 2l + 2w$ Formula for perimeter of a rectangle

$20 \text{ ft} = 2l + 2 \cdot 3 \text{ ft}$ Replace P with 20 ft and w with 3 ft.
                                 Simplify the right side by multiplying 2 • 3 ft.

$$20 \text{ ft} = 2l + 6 \text{ ft}$$
$$\underline{\phantom{20 \text{ ft}} {}^{-}6 \text{ ft} \qquad {}^{-}6 \text{ ft}}$$ To get $2l$ by itself, add $^{-}6$ ft to both sides:
$$14 \text{ ft} = 2l + \quad 0 \qquad\qquad 6 + {}^{-}6 \text{ is } 0.$$

$$\frac{14 \text{ft}}{2} = \frac{2l}{2}$$ To get l by itself, divide both sides by 2.

$$7 \text{ ft} = l$$

The length is 7 ft.

Check   To check the solution, put the length measurements on your sketch. Then add the four measurements.

$$P = 7 \text{ ft} + 7 \text{ ft} + 3 \text{ ft} + 3 \text{ ft}$$

$$P = 20 \text{ ft}$$

A perimeter of 20 ft matches the information in the original problem, so 7 ft is the correct length of the rectangle.

Now Try Exercise 25. ▶▶▶

OBJECTIVE 3 **Find the perimeter of parallelograms, triangles, and irregular shapes.** A **parallelogram** is a four-sided figure in which opposite sides are both parallel and equal in length. Some examples are shown below. Notice that opposite sides have the same length.

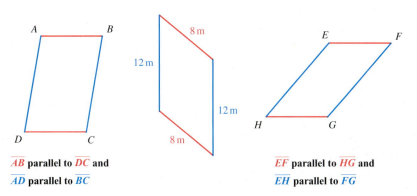

$\overline{AB}$ parallel to $\overline{DC}$ and
$\overline{AD}$ parallel to $\overline{BC}$

$\overline{EF}$ parallel to $\overline{HG}$ and
$\overline{EH}$ parallel to $\overline{FG}$

Perimeter is the distance around a flat shape, so the easiest way to find the perimeter of a parallelogram is to add the lengths of the four sides.

EXAMPLE 5 Finding the Perimeter of a Parallelogram

Find the perimeter of the middle parallelogram on the previous page.

$$P = \textcolor{blue}{12 \text{ m}} + \textcolor{blue}{12 \text{ m}} + \textcolor{red}{8 \text{ m}} + \textcolor{red}{8 \text{ m}}$$

$$P = 40 \text{ m}$$

Now Try Exercise 35.

A **triangle** is a figure with exactly three sides. Some examples are shown below.

mm is the short way to write *millimeters*.

To find the perimeter of a triangle (the distance around the edges), add the lengths of the three sides.

EXAMPLE 6 Finding the Perimeter of a Triangle

Find the perimeter of the triangle above on the right.

To find the perimeter, add the lengths of the sides.

$$P = 16 \text{ mm} + 30 \text{ mm} + 34 \text{ mm}$$

$$P = 80 \text{ mm}$$

Now Try Exercise 37.

As with any other shape, you can find the perimeter of (distance around) an irregular shape by adding the lengths of the sides.

EXAMPLE 7 Finding the Perimeter of an Irregular Shape

The floor of a room has the shape shown below.

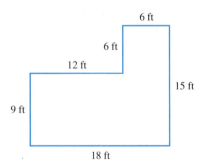

Suppose you want to put a new wallpaper border along the top of all the walls. How much material do you need?

Find the perimeter of the room by adding the lengths of the sides.

$$P = 9 \text{ ft} + 12 \text{ ft} + 6 \text{ ft} + 6 \text{ ft} + 15 \text{ ft} + 18 \text{ ft}$$

$$P = 66 \text{ ft}$$

You need 66 ft of wallpaper border.

Now Try Exercise 39.

3.1 Exercises

| FOR EXTRA HELP | Addison-Wesley Math Tutor Center | MathXL | Digital Video Tutor CD 2 Videotape 3 | Student's Solutions Manual | MyMathLab | MathXL Tutorials on CD |

Find the perimeter of each square, using the appropriate formula. See Example 1.

1.

2.

3.

4.

Draw a sketch of each square and label the lengths of the sides. Then find the perimeter. (Sketches may vary; show your sketches to your instructor.)

5. A square park measuring 1 mile on each side

6. A square garden measuring 4 meters on each side

7. A 22 mm square postage stamp

8. A 10 in. square piece of cardboard

For the given perimeter of each square, find the length of one side using the appropriate formula. See Example 2.

9. The perimeter is 120 ft.

10. The perimeter is 52 cm.

11. The perimeter is 4 mm.

12. The perimeter is 20 miles.

13. A square parking lot with a perimeter of 92 yards

14. A square building with a perimeter of 144 meters

15. A square closet with a perimeter of 8 ft

16. A square bedroom with a perimeter of 44 ft

Find the perimeter of each rectangle, using the appropriate formula. Check your solutions by adding the lengths of the four sides. See Example 3.

17.

18.

19.

20.

Draw a sketch of each rectangle and label the lengths of the sides. Then find the perimeter by using the appropriate formula. (Sketches may vary; show your sketches to your instructor.)

21. A rectangular living room 20 ft long by 16 ft wide

22. A rectangular placemat 45 cm long by 30 cm wide

23. An 8 in. by 5 in. rectangular piece of paper

24. A 2 ft by 3 ft rectangular window

For each rectangle, you are given the perimeter and either the length or width. Find the unknown measurement using the appropriate formula. Draw a sketch of each rectangle and use it to check your solution. See Example 4. (Show your sketches to your instructor.)

25. The perimeter is 30 cm and the width is 6 cm.

26. The perimeter is 48 yards and the length is 14 yards.

27. The length is 4 miles and the perimeter is 10 miles.

28. The width is 8 meters and the perimeter is 34 meters.

29. A 6 ft long rectangular table has a perimeter of 16 ft.

30. A 13 in. wide rectangular picture frame has a perimeter of 56 in.

31. A rectangular door 1 meter wide has a perimeter of 6 meters.

32. A rectangular house 33 ft long has a perimeter of 118 ft.

In exercises 33–44, find the perimeter of each shape. The figures in Exercises 33–36 are parallelograms. See Examples 5–7.

33.

34.

35.

36.
1000 ft
1240 ft

37.
26 mm
12 mm   16 mm

38.
9 yd   7 yd
11 yd

39.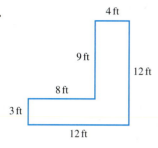
4 ft
9 ft
12 ft
8 ft
3 ft
12 ft

40.
7 m
3 m
5 m
12 m
9 m
2 m

41.
8 in.
18 in.
13 in.
13 in.
8 in.
18 in.

42.
13 ft
13 ft
12 ft
22 ft

43.
22 m
34 m
27 m
20 m
22 m

44.
87 cm
48 cm   41 cm
32 cm   44 cm
92 cm

For each shape you are given the perimeter and the lengths of all sides except one. Find the length of the unlabeled side.

45. The perimeter is 115 cm.

30 cm
10 cm
25 cm
10 cm
?

46. The perimeter is 63 in.

?   9 in.
20 in.   21 in.

47. The perimeter is 78 in.

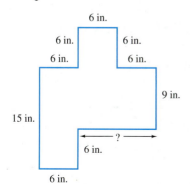
6 in.
6 in.   6 in.
6 in.   6 in.
9 in.
15 in.
?
6 in.
6 in.

48. The perimeter is 196 ft.

12 ft
18 ft   20 ft
16 ft   10 ft
40 ft
32 ft
?

49. In an *equilateral* triangle, all sides have the same length.

   (a) Draw sketches of four different equilateral triangles, label the lengths of the sides, and find the perimeters.

   (b) Write a "shortcut" rule (a formula) for finding the perimeter of an equilateral triangle.

   (c) Will your formula work for other kinds of triangles that are not equilateral? Explain why or why not.

50. Be sure that you have done Exercise 49 first.

   (a) Draw a sketch of a figure with five sides of equal length. Write a "shortcut" rule (a formula) for finding the perimeter of this shape.

   (b) Draw a sketch of a figure with six sides of equal length. Write a formula for finding the perimeter of the shape.

   (c) Write a formula for finding the perimeter of a shape with 10 sides of equal length.

   (d) Write a formula for finding the perimeter of a shape with n sides of equal length.

RELATING CONCEPTS (EXERCISES 51–54) For Individual or Group Work

A formula that has many uses for drivers is $d = rt$, called the distance formula. If you are driving a car, then

   d is the *distance* you travel (how many miles)
   r is the *rate* (how fast you are driving in miles per hour)
   t is the *time* (how many hours you drive).

Use the distance formula as you **work Exercises 51–54 in order.**

51. Suppose you are driving on Interstate highways at an average rate of 70 miles per hour. Use the distance formula to find out how far you will travel in **(a)** 2 hours; **(b)** 5 hours; **(c)** 8 hours.

52. If an ice storm slows your driving rate to 35 miles per hour, how far will you travel in **(a)** 2 hours, **(b)** 5 hours, **(c)** 8 hours? Show how to find each answer using the formula. **(d)** Explain how to find each answer using the results from Exercise 51 instead of the formula.

53. Use the distance formula to find out how many hours you would have to drive to travel the 3000 miles from Boston to San Francisco if your average rate is **(a)** 60 miles per hour; **(b)** 50 miles per hour; **(c)** 20 miles per hour (which was the speed limit 100 years ago).

54. Use the distance formula to find the average driving rate (speed) on each of these trips. (Distances are from *World Almanac*.)

   (a) It took 11 hours to drive 671 miles from Atlanta to Chicago.

   (b) Sam drove 1539 miles from New York City to Dallas in 27 hours.

   (c) Carlita drove 16 hours to travel 1040 miles from Memphis to Denver.

3.2 Problem Solving: Area

OBJECTIVE 1 Use the formula for area of a rectangle to find the area, the length, or the width.

Knowing the Difference between Perimeter and Area
Perimeter is the *distance around the outside edges* of a flat shape.
Area is the amount of *surface inside* a flat shape.

The *perimeter* of a rectangle is the distance around the *outside edges*. Recall that we unfolded a shape and laid the sides end-to-end so we could see the total distance. The *area* of a rectangle is the amount of surface *inside* the rectangle. We measure area by finding the number of squares of a certain size needed to cover the surface inside the rectangle. Think of covering the floor of a living room with carpet. Carpet is measured in square yards, that is, square pieces that measure 1 yard along each side. Here is a drawing of a rectangular living room floor.

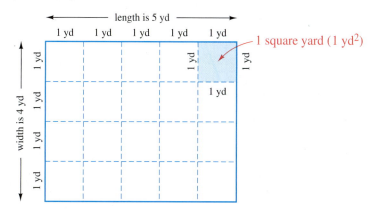

You can see from the drawing that it takes 20 squares to cover the floor. We say that the area of the floor is 20 *square yards*. A short way to write square yards is yd².

20 **square yards**   can be written as   20 **yd²**

To find the number of squares, you can count them, or you can multiply the number of squares in the length (5) times the number of squares in the width (4) to get 20. The formula is given below.

Finding the Area of a Rectangle
Area of a rectangle = length • width
$$A = lw$$
Remember to use *square units* when measuring area.

Squares of many sizes can be used to measure area. For smaller areas, you might use the ones shown at the right.

Actual-size drawings

Other sizes of squares that are often used to measure area are listed here, but they are too large to draw on this page.

1 square meter (1 m²)   1 square foot (1 ft²)
1 square kilometer (1 km²)   1 square yard (1 yd²)
                                1 square mile (1 mi²)

> **CAUTION**
> The raised 2 in 4^2 means that you multiply $4 \cdot 4$ to get 16. The raised 2 in cm² or yd² is a short way to write the word "square." It means that you multiplied cm times cm to get cm², or yd times yd to get yd². Recall that a short way to write $x \cdot x$ is x^2. Similarly, cm $\cdot$ cm is cm². When you see 5 cm², say "five square centimeters." Do *not* multiply $5 \cdot 5$, because the exponent applies only to the *first* thing to its left. The exponent applies to cm, *not* to the number.

EXAMPLE 1 Finding the Area of Rectangles

Find the area of each rectangle.

(a)

8 m

13 m

The length of this rectangle is 13 m and the width is 8 m. Use the formula $A = lw$, which means $A = l \cdot w$.

$A = \quad l \quad \cdot \quad w$ Replace l with 13 m and w with 8 m.

$A = \mathbf{13\ m} \cdot \mathbf{8\ m}$ Multiply 13 times 8 to get 104.

$A = 104\ m^2$ Multiply m times m to get m².

The area of the rectangle is 104 m². If you count the number of squares in the sketch, you will also get 104 m². Each square in the sketch represents 1 m by 1 m, which is 1 square meter (1 m²).

(b) A rectangle measuring 7 cm by 21 cm

First make a sketch of the rectangle. The length is 21 cm (the longer measurement) and the width is 7 cm. Then use the formula for area of a rectangle, $A = lw$.

7 cm

21 cm

$A = \quad l \quad \cdot \quad w$ Replace l with 21 cm and w with 7 cm.

$A = \mathbf{21\ cm} \cdot \mathbf{7\ cm}$ Multiply $21 \cdot 7$ to get 147.

$A = 147\ cm^2$ Multiply cm $\cdot$ cm to get cm².

The area of the rectangle is 147 cm².

Now Try Exercise 9.

> **CAUTION**
> The units for *area* will always be *square* units (cm², m², yd², mi², and so on). The units for *perimeter* will always be *linear* units (cm, m, yd, mi, and so on), *not* square units.

EXAMPLE 2 Finding the Length or Width of a Rectangle

If the area of a rectangular rug is 12 yd² and the length is 4 yd, find the width.

First draw a sketch of the rug and label the length as 4 yd.

Use the formula for area of a rectangle, $A = lw$.

The value of A is 12 yd² and the value of l is 4 yd.

4 yd

$$A = l \cdot w$$ Replace A with 12 yd² and replace l with 4 yd.

$$12 \text{ yd}^2 = 4 \text{ yd} \cdot w$$ To get w by itself, divide both sides by 4 yd.

$$\frac{12 \text{ yd} \cdot \text{yd}}{4 \text{ yd}} = \frac{4 \text{ yd} \cdot w}{4 \text{ yd}}$$ On the left side, rewrite yd² as yd • yd. Then $\frac{\text{yd}}{\text{yd}}$ is 1, so they "cancel out."

$$3 \text{ yd} = w$$ On the left side, 12 yd ÷ 4 is 3 yd.

The width of the rug is 3 yd.

Check To check the solution, put the width measurement on your sketch. Then use the area formula.

$$A = l \cdot w$$

$$A = 4 \text{ yd} \cdot 3 \text{ yd}$$

$$A = 12 \text{ yd}^2$$

3 yd

4 yd

An area of 12 yd² matches the information in the original problem. So 3 yd is the correct width of the rug.

Now Try Exercise 17. ▶▶▶

OBJECTIVE 2 Use the formula for area of a square to find the area or the length of one side. As with a rectangle, you can multiply length times width to find the area (surface inside) of a square. Because the length and the width are the same in a square, the formula is written as shown below.

Finding the Area of a Square

Area of a square = side • side

$$A = s \cdot s$$

$$A = s^2$$

Remember to use *square units* when measuring area.

EXAMPLE 3 Finding the Area of a Square

Find the area of a square highway sign that is 4 ft on each side.

Use the formula for area of a square, $A = s^2$.

$$A = s^2$$ Remember that s^2 means $s \cdot s$.

$$A = s \cdot s$$ Replace s with 4 ft.

$$A = 4 \text{ ft} \cdot 4 \text{ ft}$$ Multiply 4 • 4 to get 16.

$$A = 16 \text{ ft}^2$$ Multiply ft • ft to get ft².

The area of the sign is 16 ft².

Continued on Next Page

> **CAUTION**
> Be careful! s^2 means $s \cdot s$. It does *not* mean $2 \cdot s$. In this example s is 4 ft, so $(4 \text{ ft})^2$ is 4 ft $\cdot$ 4 ft $= 16 \text{ ft}^2$. It is *not* $2 \cdot 4$ ft $= 8$ ft.

Check   Check the solution by drawing a square and labeling each side as 4 ft. You can multiply length (4 ft) times width (4 ft), as you did for a rectangle. So the area is 4 ft $\cdot$ 4 ft, or 16 ft^2. This result matches the solution you got by using the formula $A = s^2$.

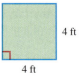

4 ft

4 ft

Now Try Exercise 13.

EXAMPLE 4   **Finding the Length of One Side of a Square**

If the area of a square township is 49 mi^2, what is the length of one side of the township?

Use the formula for area of a square, $A = s^2$. The value of A is 49 mi^2.

$A \;=\; s^2$		Replace A with 49 mi^2.
$49 \text{ mi}^2 \;=\; s^2$		To get s by itself, we have to "undo" the squaring of s. This is called *finding the square root* (more on square roots in **Chapter 5**).
$49 \text{ mi}^2 \;=\; s \cdot s$		For now, solve by inspection. Ask, what number times itself gives 49?
$49 \text{ mi}^2 = 7 \text{ mi} \cdot 7 \text{ mi}$		$7 \cdot 7$ is 49, so 7 mi $\cdot$ 7 mi is 49 mi^2.

The value of s is 7 mi, so the length of one side of the township is 7 mi.

Now Try Exercise 23.

OBJECTIVE 3 **Use the formula for area of a parallelogram to find the area, the base, or the height.**   To find the area of a parallelogram, first draw a dashed line inside the figure, as shown here.

height

base

As an experiment, try this yourself by tracing this parallelogram onto a piece of paper.

The length of the dashed line is the *height* of the parallelogram. It forms a 90° angle (a right angle) with the base. The height is the shortest distance between the base and the opposite side.

Now cut off the triangle created on the left side of the parallelogram and move it to the right side, as shown below.

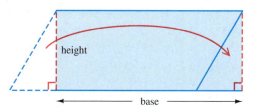

You have made the parallelogram into a rectangle. You can see that the area of the parallelogram and the rectangle are the same.

Equal areas

Area of the rectangle = length • width

Area of the parallelogram = base • height

Finding the Area of a Parallelogram

Area of a parallelogram = base • height

$$A = bh$$

Remember to use *square units* when measuring area.

EXAMPLE 5   **Finding the Area of Parallelograms**

Find the area of each parallelogram.

(a)

19 cm

24 cm

(b)

47 m

30 m   24 m   30 m

47 m

(a) The base is 24 cm and the height is 19 cm. The formula for the area of a parallelogram is $A = bh$.

$A = \quad b \quad • \quad h$ Replace b with 24 cm and h with 19 cm.

$A = $ **24 cm** • **19 cm**   Multiply 24 • 19 to get 456.

$A = 456 \text{ cm}^2$ Multiply cm • cm to get cm^2.

The area of the parallelogram is 456 cm^2.

(b) Use the formula for area of a parallelogram, $A = bh$.

$A = \quad b \quad • \quad h$ Replace b with 47 m and h with 24 m.

$A = $ **47 m** • **24 m**   Multiply 47 • 24 to get 1128.

$A = 1128 \text{ m}^2$ Multiply m • m to get m^2.

Notice that the 30 m sides are *not* used in finding the area. But you would use them when finding the *perimeter* of the parallelogram.

Now Try Exercise 5. ▶▶▶

EXAMPLE 6   **Finding the Base or Height of a Parallelogram**

The area of a parallelogram is 24 ft^2 and the base is 6 ft. Find the height.

First draw a sketch of the parallelogram and label the base as 6 ft.

?

6 ft

Continued on Next Page

Use the formula for the area of a parallelogram, $A = bh$. The value of A is 24 ft^2, and the value of b is 6 ft.

$$A \;=\; b \cdot h$$ Replace A with 24 ft^2 and b with 6 ft.

$$24 \text{ ft}^2 \;=\; 6 \text{ ft} \cdot h$$ To get h by itself, divide both sides by 6 ft.

$$\frac{24 \text{ ft} \cdot \text{ft}}{6 \text{ ft}} = \frac{6 \text{ ft} \cdot h}{6 \text{ ft}}$$ On the left side, rewrite ft^2 as ft $\cdot$ ft. Then $\frac{\text{ft}}{\text{ft}}$ is 1, so they "cancel out."

$$4 \text{ ft} = h$$ On the left side, 24 ft $\div$ 6 is 4 ft.

The height of the parallelogram is 4 ft.

Check   To check the solution, put the height measurement on your sketch. Then use the area formula.

$$A = b \cdot h$$
$$A = 6 \text{ ft} \cdot 4 \text{ ft}$$
$$A = 24 \text{ ft}^2$$

An area of 24 ft^2 matches the information in the original problem. So 4 ft is the correct height of the parallelogram.

Now Try Exercise 29.

OBJECTIVE 4 Solve application problems involving perimeter and area of rectangles, squares, or parallelograms. When you are solving problems, first decide whether you need to find the perimeter or the area.

EXAMPLE 7 **Solving an Application Problem Involving Perimeter or Area**

A group of neighbors is fixing up a playground for their children. The rectangular lot is 22 yd by 16 yd. If chain-link fencing costs \$6 per yard, how much will they spend to put a fence around the lot?

First draw a sketch of the rectangular lot and label the lengths of the sides. The fence will go around the edges of the lot, so you need to find the *perimeter* of the lot.

$$P = 2l + 2w$$ Formula for perimeter of a rectangle

$$P = 2 \cdot 22 \text{ yd} + 2 \cdot 16 \text{ yd}$$ Replace l with 22 yd and w with 16 yd.

$$P = 44 \text{ yd} + 32 \text{ yd}$$

$$P = 76 \text{ yd}$$

The perimeter of the lot is 76 yd, so the neighbors need to buy 76 yd of fencing. The cost of the fencing is \$6 *per yard*, which means \$6 *for 1 yard*. To find the cost for 76 yd, multiply \$6 $\cdot$ 76. The neighbors will spend \$456 on the fence.

Now Try Exercise 37.

3.2 Exercises

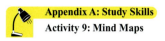
Appendix A: Study Skills
Activity 9: Mind Maps

Find the area of each rectangle, square, or parallelogram using the appropriate formula. See Examples 1, 3, and 5.

1.
7 ft
11 ft   11 ft
7 ft

2.
15 yd
5 yd   5 yd
15 yd

3.
10 m
10 m   10 m
10 m

4.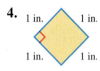
1 in.   1 in.
1 in.   1 in.

5.
31 mm
31 mm
25 mm   31 mm
31 mm

6.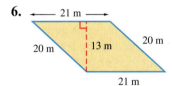
21 m
20 m   13 m   20 m
21 m

7.
6 in.

8.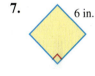
20 cm

In Exercises 9–16, first draw a sketch of the shape and label the lengths of the sides or base and height. Then find the area. (Sketches may vary; show your sketches to your instructor.)

9. A rectangular calculator that measures 15 cm by 7 cm

10. A rectangular piece of plywood that is 8 ft long and 2 ft wide

11. A parallelogram with height of 9 ft and base of 8 ft

12. A parallelogram measuring 18 mm on the base and 3 mm on the height

13. A fire burned a square-shaped forest 25 mi on a side.

14. An 11 in. square pillow

15. A piece of window glass 1 m on each side

16. A table 12 ft long by 3 ft wide

In Exercises 17–22, use the area of each rectangle and either its length or width, and the appropriate formula, to find the other measurement. Draw a sketch of each rectangle and use it to check your solution. See Example 2. (Sketches may vary; show your sketches to your instructor.)

17. The area of a desk is 18 ft², and the width is 3 ft. Find its length.

18. The area of a classroom is 630 ft², and the length is 30 ft. Find its width.

19. A parking lot is 90 yd long and has an area of 7200 yd². Find its width.

20. A playground is 60 yd wide and has an area of 6000 yd². Find its length.

21. A 154 in.² photo has a width of 11 in. Find its length.

22. A 15 in.² note card has a width of 3 in. Find its length.

Given the area of each square, find the length of one side by inspection. See Example 4.

23. A square floor has an area of 36 m².

24. A square stamp has an area of 9 cm².

25. The area of a square sign is 4 ft².

26. The area of a square piece of metal is 64 in.².

Use the area of each parallelogram and either its base or height, and the appropriate formula, to find the other measurement. Draw a sketch of each parallelogram and use it to check your solution. See Example 6. (Sketches may vary; show your sketches to your instructor.)

27. The area is 500 cm², and the base is 25 cm. Find the height.

28. The area is 1500 m², and the height is 30 m. Find the base.

29. The height is 13 in. and the area is 221 in.². Find the base.

30. The base is 19 cm, and the area is 114 cm². Find the height.

31. The base is 9 m, and the area is 9 m². Find the height.

32. The area is 25 mm², and the height is 5 mm. Find the base.

*Explain and correct the **two** errors made by students in Exercises 33 and 34.*

33.

25 cm   25 cm
24 cm
25 cm   25 cm

$P = 25 \text{ cm} + 24 \text{ cm} + 25 \text{ cm} + 25 \text{ cm} + 25 \text{ cm}$

$P = 124 \text{ cm}^2$

34.

7 ft

$A = s^2$

$A = 2 \cdot 7 \text{ ft}$

$A = 14 \text{ ft}$

Solve each application problem. You may need to find the perimeter, the area, or one of the side measurements. In Exercises 35–40 draw a sketch for each problem and label the sketch with the appropriate measurements. See Example 7. (Sketches may vary; show your sketches to your instructor.)

35. Gymnastic floor exercises are performed on a square mat that is 12 meters on a side. Find the perimeter and area of a mat. (*Source:* www.nist.gov)

36. A regulation volleyball court is 18 meters by 9 meters. Find the perimeter and area of a regulation court. (*Source:* www.nist.gov)

37. Tyra's kitchen is 4 m wide and 5 m long. She is pasting a decorative strip that costs $6 per meter around the top edge of all the walls. How much will she spend?

38. The Wang's family room measures 20 ft by 25 ft. They are covering the floor with square tiles that measure 1 ft on a side and cost $1 each. How much will they spend on tile?

39. Mr. and Mrs. Gomez are buying carpet for their square-shaped bedroom, which measures 5 yd along each wall. The carpet is $23 per square yard and padding and installation is another $6 per square yard. How much will they spend in all?

40. A page in this book measures about 27 cm from top to bottom and 21 cm from side to side. Find the perimeter and the area of the page.

41. A regulation football field is 100 yd long (excluding end zones) and has an area of 5300 yd^2. Find the width of the field. (*Source:* NFL.)

42. There are 14,790 ft^2 of ice in the rectangular playing area for a major league hockey game (excluding the area behind the goal lines). If the playing area is 85 ft wide, how long is it? (*Source:* NHL.)

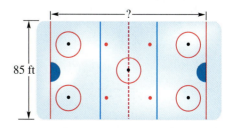

43. Advanced Photo System (APS) cameras allow you to choose from three different print sizes each time you snap a photo. The choices are shown below. Find the perimeter and area of each size print. (*Source:* Kodak.)

Panoramic 4 in. × 14 in.

4 in. × 7 in.

4 in. × 6 in.

44. The Monterey Bay Aquarium in California lets visitors look into a million-gallon tank through an acrylic panel that is 13 in. thick. The panel is 54 ft long and 15 ft high. What is the perimeter and the area of the panel? (*Source: AAA California Tour Book.*)

RELATING CONCEPTS (EXERCISES 45–48) For Individual or Group Work

Use your knowledge of perimeter and area to **work Exercises 45–48 in order.**

45. Suppose you have 12 ft of fencing to make a square or rectangular garden plot. Draw sketches of *all* the possible plots that use exactly 12 ft of fencing and label the lengths of the sides. Use only *whole number* lengths. (*Hint:* There are three possibilities.)

46. (a) Find the area of each plot in Exercise 45.

(b) Which plot has the greatest area?

47. Repeat Exercise 45 using 16 ft of fencing. Be sure to draw *all* possible square or rectangular plots that have whole number lengths for the sides.

48. (a) Find the area of each plot in Exercise 47.

(b) Compare your results to those from Exercise 46. What do you notice about the plots with the greatest area?

Summary Exercises on Perimeter and Area

Name each figure and find its perimeter and area using the appropriate formulas.

1.

3m

13 m   13 m

3m

2.

26 ft

26 ft   26 ft

26 ft

3.

←7 yd→

10 yd   10 yd   8 yd

7 yd

4.

8 cm

10 cm

5.

9 in.

9 in.

6.

9 m

6 m   5 m   6 m

← 9 m →

7.

4 ft

9 ft

9 ft

4 ft

8.

44 ft

50 ft   42 ft

50 ft

44 ft

In Exercises 9–14, use the appropriate formula to find the unknown measurement. Draw a sketch of the figure and use it to check your solution.

9. A rectangle with a perimeter of 14 ft has a length of 6 ft. Find the width.

10. A parallelogram has an area of 4 yd². The height is 1 yd. Find the base.

11. A square clock face has an area of 81 in.² Find the length of one side.

12. A hallway is 31 m long and has an area of 93 m². Find the width of the hall.

13. The perimeter of a square map is 64 cm. What is the length of one side?

14. A rectangular patio is 9 ft wide and has a perimeter of 48 ft. What is the patio's length?

Solve each application problem using the appropriate formula. Show your work.

15. How much fencing is needed to enclose a triangular building site that measures 168 meters on each side?

16. Regulation soccer fields can measure 50 to 100 yards wide and 100 to 130 yards long, depending upon the age and skill level of the players. Find the difference in playing room between the smallest and largest fields. (*Source:* Womenssportsnet.com)

17. A Toshiba laptop computer has a rectangular screen with a perimeter of 42 inches. The width of the screen is 12 inches. Find the height of the screen. (*Source:* Toshiba.)

18. Kari is decorating the cover of a square-shaped photo album that is 22 cm along each side. She is gluing braid along the edges of the front and back covers. How much braid will she need?

19. A school flag measures 3 feet by 5 feet. Find the amount of fabric needed to make seven flags, and the amount of binding needed to go around all the edges of all the flags.

20. The lobby floor of a new skyscraper is in the shape of a parallelogram, with an area of 3024 ft². If the base of the parallelogram measures 63 ft, what is its height?

3.3 Solving Application Problems with One Unknown Quantity

OBJECTIVE 1 **Translate word phrases into algebraic expressions.** In Sections **3.1** and **3.2** you worked with applications involving perimeter and area. You were able to use well-known rules (formulas) to set up equations that could be solved. However, you will encounter many problems for which no formula is available. Then you need to analyze the problem and translate the words into an equation that fits the particular situation. We'll start by translating word phrases into algebraic expressions.

OBJECTIVES

1 Translate word phrases into algebraic expressions.

2 Translate sentences into equations.

3 Solve application problems with one unknown quantity.

EXAMPLE 1 Translating Word Phrases into Algebraic Expressions

Write each phrase as an algebraic expression. Use x as the variable.

Words	Algebraic Expression	
A number **plus** 2	$x + 2$ or $2 + x$	
The **sum** of 8 and a number	$8 + x$ or $x + 8$	Two correct ways to write each addition expression
5 **more than** a number	$x + 5$ or $5 + x$	
$^-35$ **added to** a number	$^-35 + x$ or $x + {}^-35$	
A number **increased by** 6	$x + 6$ or $6 + x$	
9 **less than** a number	$x - 9$	
A number **subtracted from** 3	$3 - x$	Only one correct way to write each subtraction expression
3 **subtracted from** a number	$x - 3$	
A number **decreased by** 4	$x - 4$	
10 **minus** a number	$10 - x$	

CAUTION

Recall that addition can be done in any order, so $x + 2$ gives the same result as $2 + x$. This is *not* true in subtraction, so be careful. $10 - x$ does *not* give the same result as $x - 10$.

Now Try Exercises 3 and 7. ▶▶▶

EXAMPLE 2 Translating Word Phrases into Algebraic Expressions

Write each phrase as an algebraic expression. Use x as the variable.

Words	Algebraic Expression
8 **times** a number	$8x$
The **product** of 12 and a number	$12x$
Double a number (meaning "2 times")	$2x$
The **quotient** of $^-6$ and a number	$\dfrac{^-6}{x}$
A number **divided by** 10	$\dfrac{x}{10}$
15 **subtracted from** 4 **times** a number	$4x - 15$
The result **is**	$=$

Now Try Exercises 15 and 19. ▶▶▶

OBJECTIVE 2 Translate sentences into equations. The next example shows you how to translate a sentence into an equation that you can solve.

EXAMPLE 3 Translating a Sentence into an Equation

If 5 times a number is added to 11, the result is 26. Find the number.

Let x represent the unknown number. Use the information in the problem to write an equation.

$$\underbrace{5 \text{ times a number}}_{5x} \quad \underbrace{\text{added to } 11}_{+ \quad 11} \quad \underset{=}{\overset{\text{is}}{\downarrow}} \quad \underset{26}{\overset{26}{\downarrow}}$$

Next, solve the equation.

$$
\begin{array}{rcl}
5x + 11 &=& 26 \\
\underline{-11} && \underline{-11} \\
\underbrace{5x + 0} &=& 15 \\
\dfrac{5x}{5} &=& \dfrac{15}{5} \\
x &=& 3
\end{array}
$$

To get $5x$ by itself, add $^-11$ to both sides.

To get x by itself, divide both sides by 5.

The number is 3.

Check Go back to the words of the original problem.

$$\overset{\downarrow}{5} \quad \overset{\downarrow}{\bullet} \quad \underbrace{\text{times a number}}_{3} \quad \underbrace{\text{is added to}}_{+} \quad \underbrace{11,}_{11} \quad \underbrace{\text{the result is}}_{=} \quad \overset{\downarrow}{26.}$$

Does $5 \bullet 3 + 11$ really equal 26? Yes, $5 \bullet 3 + 11 = 15 + 11 = 26$.

So 3 is the correct solution because it "works" when you put it back into the original problem.

Now Try Exercise 23.

OBJECTIVE 3 Solve application problems with one unknown quantity. Now you are ready to tackle application problems. The steps we will use are summarized below.

Solving an Application Problem

Step 1 **Read** the problem once to see what it is about. Read it carefully a second time. As you read, make a sketch or write word phrases that identify the known and the unknown parts of the problem.

Step 2(a) If there is one unknown quantity, **assign a variable** to represent it. Write down what your variable represents.

Step 2(b) If there is more than one unknown quantity, **assign a variable** to represent "the thing you know the least about." Then write variable expression(s), using the same variable, to show the relationship of the other unknown quantities to the first one.

Step 3 **Write an equation,** using your sketch or word phrases as the guide.

Step 4 **Solve** the equation.

Step 5 **State the answer** to the question in the problem and label your answer.

Step 6 **Check** whether your answer fits all the facts given in the *original* statement of the problem. If it does, you are done. If it doesn't, start again at Step 1.

EXAMPLE 4 **Solving an Application Problem with One Unknown Quantity**

Heather had put some money aside in an envelope for household expenses. Yesterday she took out $20 for groceries. Today a friend paid back a loan and Heather put the $34 in the envelope. Now she has $43 in the envelope. How much was in the envelope at the start?

Step 1   **Read** the problem once. It is about money in an envelope. Read it a second time and write word phrases.

Unknown: amount of money in the envelope at the start
Known: took out $20; put in $34; ended up with $43

Step 2(a)   There is only one unknown quantity, so **assign a variable** to represent it. Let m represent the money at the start.

Step 3   **Write an equation,** using the phrases you wrote as a guide.

Money at the start	Took out $20	Put in $34	Ended up with $43
m	$- \$20$	$+ \$34$	$= \$43$

Step 4   **Solve** the equation.

$$m - 20 + 34 = 43 \qquad \text{Change subtraction to adding the opposite.}$$
$$m + {}^{-}20 + 34 = 43 \qquad \text{Simplify the left side.}$$
$$m + 14 = 43$$
$$\qquad \qquad \text{To get } m \text{ by itself,}$$
$$\qquad \qquad \text{add } {}^{-}14 \text{ to both sides.}$$
$$\underline{\quad {}^{-}14 \qquad \quad {}^{-}14 \quad}$$
$$m + 0 = 29$$
$$m = 29$$

Step 5   **State the answer** to the question, "How much was in the envelope at the start?" There was $29 in the envelope.

Step 6   **Check** the solution by going back to the *original* problem and inserting the solution.

Started with $29 in the envelope

Took out $20, so $29 − $20 = $9 in the envelope

Put in $34, so $9 + $34 = $43

Now has $43 ⟵ Matches ⟶

Because $29 "works" when you put it back into the original problem, you know it is the correct solution.

Now Try Exercise 31. ▶▶▶

EXAMPLE 5 Solving an Application Problem with One Unknown Quantity

Three friends each put in the same amount of money to buy a gift. After they spent $2 for a card and $31 for the gift, they had $6 left. How much money had each friend put in originally?

Step 1   **Read** the problem. It is about 3 friends buying a gift.

Unknown: amount of money each friend contributed
Known: 3 friends put in money; spent $2 and $31; had $6 left

Step 2(a)   There is only one unknown quantity. **Assign a variable,** m, to represent the amount of money each friend contributed.

Step 3   **Write an equation.**

Number of friends	Amount each friend put in	Spent on card	Spent on gift	Left over
3	$\cdot$ m	$-$ $2	$-$ $31	$=$ $6

To see why this is multiplication, think of an example. If each friend put in $10, how much money would there be? $3 \cdot \$10$, or $30

Step 4   **Solve.**

$$3m - 2 - 31 = 6$$

Changes subtractions to adding the opposite.

$$3m + {}^-2 + {}^-31 = 6$$

Simplify the left side.

$$3m + {}^-33 = 6$$

To get $3m$ by itself, add 33 to both sides.

$$ {}^+33 \qquad {}^+33$$

$$3m + 0 = 39$$

$$\frac{3m}{3} = \frac{39}{3}$$

To get m by itself, divide both sides by 3.

$$m = 13$$

Step 5   **State the answer.** Each friend put in $13.

Step 6   **Check** the solution by putting it back into the *original* problem.

3 friends each put in $13, so $3 \cdot \$13 = \$39$.

Spent $2, spent $31, so $39 - \$2 - \$31 = \$6$.

Had $6 left ⟵———— Matches ————⟶

$13 is the correct solution because it "works."

Now Try Exercise 35.

EXAMPLE 6 **Solving a More Complex Application Problem with One Unknown Quantity**

Michael has completed 5 less than three times as many lab experiments as David. If Michael has completed 13 experiments, how many experiments has David completed?

Step 1 **Read** the problem. It is about the number of experiments done by two students.

> **Unknown:** number of experiments David did
> **Known:** Michael did 5 less than 3 times the number David did; Michael did 13.

Step 2(a) **Assign a variable.** Let n represent the number of experiments David did.

Step 3 **Write an equation.**

The number Michael did	is	5 less than 3 times David's number.
13	=	$3n - 5$

Step 4 **Solve.**

$$13 = 3n - 5 \qquad \text{Change subtraction to adding the opposite.}$$

$$13 = 3n + {}^-5$$

$$\underline{+5 \qquad\qquad +5} \qquad \text{To get } 3n \text{ by itself, add 5 to both sides.}$$

$$18 = \underbrace{3n + \quad 0}$$

$$\frac{18}{3} = \frac{3n}{3} \qquad \text{To get } n \text{ by itself, divide both sides by 3.}$$

$$6 = n$$

Step 5 **State the answer.** David did 6 experiments.

Step 6 **Check** the solution by putting it back into the *original* problem.

3 times David's number $3 \cdot 6 = 18$

Less 5 $18 - 5 = 13$

Michael did 13. ⟵——— Matches ———

The correct solution is: David did 6 experiments.

Now Try Exercise 41. ▶▶▶

3.3 Exercises

FOR EXTRA HELP

  Addison-Wesley Math Tutor Center

 MathXL

 Digital Video Tutor CD 2 Videotape 3

Student's Solutions Manual

 MyMathLab

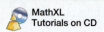 MathXL Tutorials on CD

Write an algebraic expression, using x as the variable. See Examples 1 and 2.

1. 14 plus a number

2. The sum of a number and ⁻8

3. ⁻5 added to a number

4. 16 more than a number

5. 20 minus a number

6. A number decreased by 25

7. 9 less than a number

8. A number subtracted from ⁻7

9. Subtract 4 from a number.

10. 3 fewer than a number

11. ⁻6 times a number

12. The product of ⁻3 and a number

13. Double a number

14. A number times 10

15. A number divided by 2

16. 4 divided by a number

17. Twice a number added to 8

18. Five times a number plus 5

19. 10 fewer than seven times a number

20. 12 less than six times a number

21. The sum of twice a number and the number

22. Triple a number subtracted from the number

Translate each sentence into an equation and solve it. Check your solution by going back to the words in the original problem. See Example 3.

23. If four times a number is decreased by 2, the result is 26. Find the number.

24. The sum of 8 and five times a number is 53. Find the number.

25. If a number is added to twice the number, the result is ⁻15. What is the number?

26. If a number is subtracted from three times the number, the result is ⁻8. What is the number?

27. If the product of some number and 5 is increased by 12, the result is seven times the number. Find the number.

28. If eight times a number is subtracted from eleven times the number, the result is ⁻9. Find the number.

29. When three times a number is subtracted from 30, the result is 2 plus the number. What is the number?

30. When twice a number is decreased by 8, the result is the number increased by 7. Find the number.

Solve each application problem. Use the six problem-solving steps you learned in this section. See Examples 4–6.

31. Ricardo gained 15 pounds over the winter. He went on a diet and lost 28 pounds. Then he regained 5 pounds and weighed 177 pounds. How much did he weigh originally?

32. Mr. Chee deposited $80 into his checking account. Then, after writing a $23 check for gas and a $90 check for his child's day care, the balance in his account was $67. How much was in his account before he made the deposit?

33. There were 18 cookies in Magan's cookie jar. While she was busy in another room, her children ate some of the cookies. Magan bought three dozen cookies and added them to the jar. At that point she had 49 cookies in the jar. How many cookies did her children eat?

34. The Greens had a 20-pound bag of bird seed in their garage. Mice got into the bag and ate some of it. The Greens then bought an 8-pound bag of seed and put all the seed in a metal container. They now have 24 pounds of seed. How much did the mice eat?

35. A college bookstore ordered six boxes of red pens. The store sold 32 red pens last week and 35 red pens this week. Five pens were left on the shelf. How many pens were in each box?

36. A local charity received a donation of eight cartons filled with cans of soup. The charity gave out 100 cans of soup yesterday and 92 cans today before running out. How many cans were in each carton?

37. The 14 music club members each paid the same amount for dues. The club also earned $340 selling magazine subscriptions. They spent $575 to organize a jazz festival. Now their bank account is overdrawn by $25. How much did each member pay in dues?

38. The manager of an apartment complex had 11 packages of lightbulbs on hand. He replaced 29 burned out bulbs in hallway lights and 7 bulbs in the party room. Eight bulbs were left. How many bulbs were in each package?

39. When 75 is subtracted from four times Tamu's age, the result is Tamu's age. How old is Tamu?

40. If three times Linda's age is decreased by 36, the result is twice Linda's age. How old is Linda?

41. While shopping for clothes, Consuelo spent $3 less than twice what Brenda spent. Consuelo spent $81. How much did Brenda spend?

42. Dennis weighs 184 pounds. His weight is 2 pounds less than six times his child's weight. How much does his child weigh?

43. Paige bought five bags of candy for Halloween. Forty-eight children visited her home and she gave each child three pieces of candy. At the end of the night she still had one bag of candy. How many pieces of candy were in each bag?

44. A restaurant ordered four packages of paper napkins. Yesterday they used up one package, and today they used up 140 napkins. Two packages plus 60 napkins remain. How many napkins are in each package?

45. The recommended daily intake of iron for an adult female is 3 mg more than twice the recommended amount for a newborn infant. The amount for an adult female is 15 mg. How much should the infant receive? (*Source:* Food and Nutrition Board.)

46. A cheetah's sprinting speed is 61 miles per hour less than three times a zebra's running speed. A cheetah can sprint 68 miles per hour. Find the zebra's running speed. (*Source: Grolier Multimedia Encyclopedia.*)

3.4 Solving Application Problems with Two Unknown Quantities

OBJECTIVES

1. Solve application problems with two unknown quantities.

OBJECTIVE 1 Solve application problems with two unknown quantities. In the preceding section, the problems had only one unknown quantity. As a result, we used Step 2(a) rather than Step 2(b) in the problem-solving steps. For easy reference, we repeat the steps here.

Solving an Application Problem

Step 1   **Read** the problem once to see what it is about. Read it carefully a second time. As you read, make a sketch or write word phrases that identify the known and the unknown parts of the problem.

Step 2(a)   If there is one unknown quantity, **assign a variable** to represent it. Write down what your variable represents.

Step 2(b)   If there is more than one unknown quantity, **assign a variable** to represent "the thing you know the least about." Then write variable expression(s), using the same variable, to show the relationship of the other unknown quantities to the first one.

Step 3   **Write an equation**, using your sketch or word phrases as the guide.

Step 4   **Solve** the equation.

Step 5   **State the answer** to the question in the problem and label your answer.

Step 6   **Check** whether your answer fits all the facts given in the *original* statement of the problem. If it does, you are done. If it doesn't, start again at Step 1.

Now you are ready to solve problems with two unknown quantities.

EXAMPLE 1 Solving an Application Problem with Two Unknown Quantities

Last month, Sheila worked 72 hours more than Russell. Together they worked a total of 232 hours. Find the number of hours each person worked last month.

Step 1   **Read** the problem. It is about the number of hours worked by Sheila and by Russell.

   Unknowns: hours worked by Sheila;
                      hours worked by Russell
   Known: Sheila worked 72 hours more than Russell;
                   232 hours total for Sheila and Russell

Step 2(b)   There are *two* unknowns so **assign a variable** to represent "the thing you know the least about." You know the *least* about the hours worked by Russell, so let h represent Russell's hours.

   Sheila worked 72 hours more than Russell, so her hours are $h + 72$, that is, Russell's hours (h) plus 72 more.

Step 3   **Write an equation.**

Hours worked by Russell		Hours worked by Sheila		Total hours worked
h	$+$	$h + 72$	$=$	232

Continued on Next Page

Step 4   **Solve.**

$$h + h + 72 = 232$$

Simplify the left side by combining like terms.

$$2h + 72 = 232$$

$$\underline{\quad\quad -72 \quad -72 \quad}$$

To get $2h$ by itself, add $^-72$ to both sides.

$$2h + 0 = 160$$

$$\frac{2h}{2} = \frac{160}{2}$$

To get h by itself, divide both sides by 2.

$$h = 80$$

Step 5   **State the answer.**

Because h represents Russell's hours, and the solution of the equation is $h = 80$, Russell worked 80 hours.

$h + 72$ represents Sheila's hours. Replace h with 80.

$80 + 72 = 152$, so Sheila worked 152 hours.

The final answer is: Russell worked 80 hours and Sheila worked 152 hours.

Step 6   **Check** the solution by putting both numbers back into the *original* problem.

"Sheila worked 72 hours more than Russell."
Sheila's 152 hours are 72 more than Russell's 80 hours, so the solution checks.

$$\begin{array}{r} 152 \\ -\ 72 \\ \hline 80 \end{array}$$

"Together they worked a total of 232 hours."
Sheila's 152 hours + Russell's 80 hours = 232 hours so the solution checks.

$$\begin{array}{r} 152 \\ +\ 80 \\ \hline 232 \end{array}$$

You've answered the question correctly because 80 hours and 152 hours fit all the facts given in the problem.

> **CAUTION**
> Check the solution to an application problem by putting the numbers back in the *original* problem. If they do *not* work, recheck your work or try solving the problem in a different way.

Now Try Exercise 1. ▶▶▶

EXAMPLE 2   **Solving a Geometry Application with Two Unknown Quantities**

The length of a rectangle is 2 cm more than the width. The perimeter is 68 cm. Find the length and width.

Step 1   **Read** the problem. It is about a rectangle. Make a sketch of a rectangle.

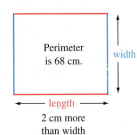

Perimeter is 68 cm.

width

length

2 cm more than width

Unknowns: length of the rectangle; width of the rectangle

Known: The length is 2 cm more than the width; the perimeter is 68 cm.

Continued on Next Page

Step 2(b) There are *two* unknowns so **assign a variable** to represent "the thing you know the least about." You know the *least* about the width, so let **w** represent the **width**.

The length is 2 cm more than the width, so the **length** is **w + 2**.

Step 3 **Write an equation.**
Use the formula for perimeter of a rectangle, $P = 2l + 2w$, to help you write the equation.

$$P = 2 \quad l \quad + 2 \quad w \qquad \text{Replace } P \text{ with 68.}$$
$$\downarrow \quad \downarrow \quad \downarrow \quad \quad \downarrow \quad \downarrow \qquad \text{Replace } l \text{ with } (w + 2).$$
$$68 = 2(w + 2) + 2 \cdot w$$

Step 4 **Solve.**

$$68 = 2(w + 2) + 2w \qquad \text{Use the distributive property.}$$
$$68 = 2w + 4 + 2w \qquad \text{Combine like terms.}$$

$$68 = 4w + 4$$

$$\underline{-4 \qquad \qquad -4} \qquad \text{To get } 4w \text{ by itself, add } ^-4 \text{ to both sides.}$$

$$64 = 4w + 0$$

$$\frac{64}{4} = \frac{4w}{4} \qquad \text{To get } w \text{ by itself, divide both sides by 4.}$$

$$16 = w$$

Step 5 **State the answer.**

w represents the width, and $w = 16$, so the width is 16 cm.

$w + 2$ represents the length. Replace w with 16.

$16 + 2 = 18$, so the length is 18 cm. ⟵ The label for both answers is cm.

The final answer is: The width is 16 cm and the length is 18 cm.

Step 6 **Check** the solution by putting the measurements on your sketch and going back to the original problem.

"The length of a rectangle is 2 cm more than the width."

18 cm is 2 cm more than 16 cm, so the solution checks.

"The perimeter is 68 cm."

$$P = \underbrace{2 \cdot 18 \text{ cm}} + \underbrace{2 \cdot 16 \text{ cm}}$$
$$P = \quad 36 \text{ cm} \quad + \quad 32 \text{ cm}$$

$$P = 68 \text{ cm} \quad ← \text{ This matches the perimeter given in the original problem, so the solution checks.}$$

Now Try Exercise 17.

3.4 Exercises

Solve each application problem using the six problem-solving steps you learned in this section. See Example 1.

1. My sister is 9 years older than I am. The sum of our ages is 51. Find our ages.

2. Ed and Marge were candidates for city council. Marge won, with 93 more votes than Ed. The total number of votes cast in the election was 587. Find the number of votes received by each candidate.

3. Last year, Lien earned $1500 more than her husband. Together they earned $37,500. How much did each of them earn?

4. A $149,000 estate is to be divided between two charities so that one charity receives $18,000 less than the other. How much will each charity receive?

5. Jason paid five times as much for his computer as he did for his printer. He paid a total of $1320 for both items. What did each item cost?

6. The attendance at the Saturday night baseball game was three times the attendance at Sunday's game. In all, 56,000 fans attended the games. How many fans were at each game?

7. A board is 78 cm long. Rosa cut the board into two pieces, with one piece 10 cm longer than the other. Find the length of both pieces. (*Hint:* Make a sketch of the board. Which piece do you know the least about? Let *x* represent the length of that piece.)

Longer piece   Shorter piece

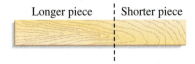

8. A rope is 21 yd long. Marcos cut it into two pieces so that one piece is 3 yd longer than the other. Find the length of each piece.

Longer piece   Shorter piece

9. A wire is cut into two pieces, with one piece 7 ft shorter than the other. The wire was 31 ft long before it was cut. How long was each piece?

10. A 90 cm pipe is cut into two pieces so that one piece is 6 cm shorter than the other. Find the length of each piece.

11. In the U.S. Congress, the number of Representatives is 65 less than five times the number of Senators. There are a total of 535 members of Congress. Find the number of Senators and the number of Representatives. (*Source: World Almanac.*)

12. Florida's record low temperature is 68 degrees higher than Montana's record low. The sum of the two record lows is ⁻72 degrees. What is the record low for each state? (*Source:* National Climatic Data Center.)

13. A fence is 706 m long. It is to be cut into three parts. Two parts are the same length, and the third part is 25 m longer than either of the other two. Find the length of each part.

14. A wooden railing is 82 m long. It is to be divided into four pieces. Three pieces will be the same length, and the fourth piece will be 2 m longer than each of the other three. Find the length of each piece.

In Exercises 15–20, use the formula for the perimeter of a rectangle, P = 2l + 2w. Make a sketch to help you solve each problem. See Example 2. (Sketches may vary; show your sketches to your instructor.)

15. The perimeter of a rectangle is 48 yd. The width is 5 yd. Find the length.

16. The length of a rectangle is 27 cm, and the perimeter is 74 cm. Find the width of the rectangle.

17. A rectangular dog pen is twice as long as it is wide. The perimeter of the pen is 36 ft. Find the length and the width of the pen.

18. A new city park is a rectangular shape. The length is triple the width. It will take 240 meters of fencing to go around the park. Find the length and width of the park.

19. The length of a rectangular jewelry box is 3 in. more than twice the width. The perimeter is 36 in. Find the length and the width.

20. The perimeter of a rectangular house is 122 ft. The width is 5 ft less than the length. Find the length and the width.

21. A photo measures 8 in. by 10 in. Earl put it in a frame that added 2 in. to each side. Find the outside perimeter and total area of the photo and frame.

22. Barb had a 16 in. by 20 in. photograph. She cropped 3 in. off each side of the photo. What are the perimeter and the area of the cropped photo?

Chapter **3**   **Group Activity: USING FORMULAS**

Objective: Solve equations and interpret the solutions.

In **Section 3.1**, you found the perimeter of a triangle by adding the lengths of the three sides. A formula for this is shown here.

$$P = a + b + c$$

where P is the perimeter and a, b, and c are the lengths of the sides.

1. Two sides of a triangle measure 12 ft and 7 ft, and the perimeter is 33 ft. Use the formula to find the length of the third side in this triangle.

Perimeter is 33 ft.

Use the formula to find the length of the third side in each of these triangles.

2.

Perimeter is 24 cm.

3.

Perimeter is 64 in.

4. Describe *in words* what you were doing as you found the length of the third side. What things were being added? What were you doing with the perimeter?

5. How can you simplify the perimeter formula when *two* sides of the triangle are the *same* length?

Use the simplified formula to find the unknown lengths in these triangles.

6.

Perimeter is 33 m.

7.

Perimeter is 24 cm.

In the **Chapter 2 Group Activity**, you worked with the expression $60A - 900$ for the vocabulary of a child. Now we can use this formula.

$$V = 60A - 900$$

where V is the number of vocabulary words and A is the age of the child in months.

Use the formula to find the *age* of each child.

8. The child's vocabulary is 180 words.

9. The child's vocabulary is 1440 words.

10. The child's vocabulary is 60 words.

11. Describe *in words* what you were doing as you found each child's age.

12. Find the child's age if the child's vocabulary is 0 words.

13. Refer to Problem 12 above. Without solving an equation, are there any other ages at which children would have a vocabulary of 0 words?

14. Is the formula useful for all ages of children? Explain your answer.

Chapter 3

SUMMARY

KEY TERMS

3.1 **formula**
Formulas are well-known rules for solving common types of problems. They are written in a shorthand form that uses variables.

perimeter
Perimeter is the distance around the outside edges of a flat shape. It is measured in linear units such as in., ft, yd, mm, cm, m, and so on.

square
A square is a figure with four sides that are all the same length and meet to form right angles, which measure 90°. See example at the right.

rectangle
A rectangle is a four-sided figure in which all sides meet to form right angles, which measure 90°. The opposite sides are the same length. See example at the right.

parallelogram
A parallelogram is a four-sided figure in which opposite sides are both parallel and equal in length. See example at the right.

triangle
A triangle is a figure with exactly three sides. See example at the right.

3.2 **area**
Area is the surface inside a two-dimensional (flat) shape. It is measured by determining the number of squares of a certain size needed to cover the surface inside the shape. Some of the commonly used units for measuring area are square inches (in.²), square feet (ft²), square yards (yd²), square centimeters (cm²), and square meters (m²).

NEW SYMBOLS

Right angle:
(90° angle)

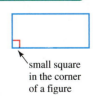

small square in the corner of a figure

Square units: in.²   ft²   yd²   mi²
(for measuring area)   mm²   cm²   m²   km²

NEW FORMULAS

Perimeter of a square: $P = 4s$

Perimeter of a rectangle: $P = 2l + 2w$

Area of a square: $A = s^2$

Area of a rectangle: $A = lw$

Area of a parallelogram: $A = bh$

TEST YOUR WORD POWER

See how well you have learned the vocabulary in this chapter. Answers follow the Quick Review.

1. The **perimeter** of a flat shape is
   A. measured in square units
   B. found by adding the lengths of the sides
   C. measured in cubic units
   D. found by multiplying length times width.

2. The **area** of a flat shape is
   A. found by adding length plus width
   B. measured in linear units
   C. found by squaring the height
   D. measured in square units.

3. When working with a **square shape,**
   A. the perimeter formula is $P = s^2$
   B. the area formula is $A = \frac{1}{2}bh$
   C. all sides have the same length
   D. all angles measure 180°.

4. When working with a **rectangular shape,**
   A. the area formula is $A = lw$
   B. opposite sides have different lengths
   C. the perimeter formula is $P = bh$
   D. the width is the longer measurement.

5. In *all* **triangles**
   A. the sides meet at 90° angles
   B. the sides have the same length
   C. there are exactly three sides
   D. the area formula is $A = s^2$.

6. In *all* **parallelograms**
   A. there are exactly six sides
   B. the height is parallel to the base
   C. all sides have the same length
   D. opposite sides are parallel and equal in length.

QUICK REVIEW

| *Concepts* | *Examples* |

3.1 *Finding Perimeter*

To find the perimeter of *any* shape, add the lengths of the sides. Perimeter is measured in linear units (cm, m, ft, yd, and so on).

Or, for squares and rectangles, you can use the formulas shown below.

Find the perimeter of each figure.

$P = 9 \text{ cm} + 11 \text{ cm} + 9 \text{ cm} + 11 \text{ cm}$

$P = 40 \text{ cm}$

Perimeter of a square: $P = 4s$

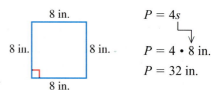

$P = 4s$

$P = 4 \cdot 8 \text{ in.}$

$P = 32 \text{ in.}$

Perimeter of a rectangle: $P = 2l + 2w$

$P = 2 \cdot l + 2 \cdot w$

$P = 2 \cdot 3 \text{ yd} + 2 \cdot 2 \text{ yd}$

$P = 6 \text{ yd} + 4 \text{ yd}$

$P = 10 \text{ yd}$

Concepts	**Examples**

3.1 *Finding the Length of One Side of a Square*

If you know the perimeter of a square, use the formula $P = 4s$. Replace P with the value for the perimeter and solve the equation for s.

If the perimeter of a square room is 44 ft, find the length of one side.

$$P = 4s \qquad \text{Replace } P \text{ with 44 ft.}$$

$$44\text{ ft} = 4s$$

$$\frac{44\text{ ft}}{4} = \frac{4s}{4} \qquad \text{Divide both sides by 4.}$$

$$11\text{ ft} = s$$

The length of one side is 11 ft.

Check Check the solution by drawing a sketch of the room. The perimeter is
11 ft + 11 ft + 11 ft + 11 ft = 44 ft. The result matches the perimeter given in the problem.

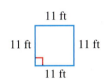

3.1 *Finding the Length or Width of a Rectangle*

If you know the perimeter of a rectangle and either its width or length, use the formula $P = 2l + 2w$. Replace P and either l or w with the values that you know. Then solve the equation.

The width of a rectangular rug is 8 ft. The perimeter is 36 ft. Find the length.

$$P = 2l + 2w \qquad \text{Replace } P \text{ with 36 ft and}$$
$$\qquad\qquad\qquad\qquad w \text{ with 8 ft.}$$

$$36\text{ ft} = 2l + 2 \cdot 8\text{ ft}$$

$$36\text{ ft} = 2l + 16\text{ ft}$$

$$\underline{{}^{-}16\text{ ft} \qquad\qquad {}^{-}16\text{ ft}} \qquad \text{Add } {}^{-}16 \text{ ft to both sides.}$$

$$20\text{ ft} = 2l + 0$$

$$\frac{20\text{ ft}}{2} = \frac{2l}{2} \qquad \text{Divide both sides by 2.}$$

$$10\text{ ft} = l \qquad \text{The length is 10 ft.}$$

Check To check the solution, draw a sketch of the rectangle and label the lengths of the sides. Then add the four measurements.

10 ft + 10 ft + 8 ft + 8 ft = 36 ft

The result matches the perimeter given in the problem.

Concepts	Examples

3.2 Finding Area

Use the appropriate formula. Remember to measure area in *square* units (cm², m², ft², yd², and so on).

Area of a rectangle: $A = lw$,
   where l is the length and w is the width.

Area of a square: $A = s^2$, which means $s \cdot s$,
   where s is the length of one side.

Area of a parallelogram: $A = bh$,
   where b is the base and h is the height.

Find the area of each figure.

$A = lw$

$A = 3\ \text{m} \cdot 2\ \text{m}$

$A = 6\ \text{m}^2$

$A = s^2$

$A = s \cdot s$

$A = 8\ \text{in.} \cdot 8\ \text{in.}$

$A = 64\ \text{in.}^2$

$A = bh$

$A = 4\ \text{cm} \cdot 3\ \text{cm}$

$A = 12\ \text{cm}^2$

3.2 Finding the Unknown Length in a Rectangle or Parallelogram

If you know the area of a rectangle or parallelogram and one of the other measurements, use the appropriate area formula (see above). Replace A and one of the other variables with the values that you know. Then solve the equation.

The area of a parallelogram is 72 yd², and its height is 9 yd. Find the base.

$$A = b\textbf{\textit{h}}$$
Replace A with 72 yd² and h with 9 yd.

$$\textbf{72 yd}^2 = b \cdot \textbf{9 yd}$$

$$\frac{72\ \text{yd} \cdot \text{yd}}{\textbf{9 yd}} = \frac{b \cdot 9\ \text{yd}}{\textbf{9 yd}}$$
Divide both sides by 9 yd.

$$8\ \text{yd} = b$$

The base of the parallelogram is 8 yd.

3.2 Finding the Length of One Side of a Square

If you know the area of a square, use the formula $A = s^2$. Replace A with the value that you know. Then solve the equation by asking, "What number, times itself, gives the value of A?"

A square ceiling has an area of 100 ft². What is the length of each side of the ceiling?

$$A = s^2$$
Replace A with 100 ft².

$$\textbf{100 ft}^2 = s^2$$
Rewrite s^2 as $s \cdot s$.

$$100\ \text{ft}^2 = s \cdot s$$
Ask, "What number times itself gives 100?"

$$100\ \text{ft}^2 = 10\ \text{ft} \cdot 10\ \text{ft}$$

The value of s is 10 ft, so each side of the ceiling is 10 ft long.

Concepts

3.3 *Translating Sentences into Equations*

Translate word phrases into symbols using x (or any other letter) as the variable. Then solve the equation. Check the solution by putting it back in the original problem.

Examples

If 10 is subtracted from three times a number, the result is 14. Find the number.

Let x represent the unknown number.

$$
\begin{array}{rcll}
3x - 10 &=& 14 & \text{Change subtraction} \\
3x + {}^{-}10 &=& 14 & \quad\text{to adding the opposite.} \\
\underline{{}^{+}10 \quad {}^{+}10} & & & \text{Add 10 to both sides.} \\
3x + 0 &=& 24 & \\
\dfrac{3x}{3} &=& \dfrac{24}{3} & \text{Divide both sides by 3.} \\
x &=& 8 &
\end{array}
$$

The number is 8.

Check If 10 is subtracted from three times 8, do you get 14?   Yes, $3 \cdot 8 - 10 = 24 - 10 = 14$.

3.3 *Solving Application Problems with One Unknown Quantity*

Use the six problem-solving steps outlined in **Section 3.3.** They are listed below in abbreviated form.

Step 1 **Read** the problem and identify what is known and what is unknown.

Step 2 **Assign a variable** to represent the unknown quantity.

Step 3 **Write an equation.**

Step 4 **Solve** the equation.

Step 5 **State the answer.**

Step 6 **Check** whether your answer fits all the facts given in the *original* statement of the problem.

Denise had some money in her purse this morning. She gave $15 to her daughter and paid $4 to park in the lot at work. At that point she still had $27. How much was in her purse this morning?

Step 1 **Unknown:** money in purse this morning
Known: took out $15; took out $4; still had $27.

Step 2 Let m represent the money in her purse this morning.

Step 3 $m - \$15 - \$4 = \$27$

Step 4
$$
\begin{array}{rcll}
m - 15 - 4 &=& 27 & \\
m + {}^{-}15 + {}^{-}4 &=& 27 & \text{Combine like terms.} \\
m + {}^{-}19 &=& 27 & \\
\underline{{}^{+}19 \qquad {}^{+}19} & & & \text{Add 19 to both side.} \\
m + 0 &=& 46 & \\
m &=& 46 &
\end{array}
$$

Step 5 She had $46 in her purse this morning.

Step 6 Started with $46.
Took out $15, so $46 − $15 = $31
Took out $4, so $31 − $4 = $27
Had $27 left ⟵— Matches —⟶

Concepts

3.4 Solving Application Problems with Two Unknown Quantities

Use the six problem-solving steps outlined in **Section 3.3.**

In *Step 2*, there are *two* unknown quantities, so assign a variable to represent "the thing you know the least about." Then write a variable expression, using the same variable, to show the relationship of the other unknown quantity to the first one.

Examples

Last week, Brian earned $50 more than twice what Dan earned. How much did each person earn if the total for both of them was $254?

Step 1 **Unknowns:** Brian's earnings;
Dan's earnings

Known: Brian earned $50 more than twice what Dan earned; the sum of their earnings was $254.

Step 2 You know the least about Dan's earnings, so let m represent Dan's earnings.

Brian earned $50 more than twice what Dan earned, so Brian's earnings are $2m + \$50$.

Step 3 $\underline{m + 2m} + 50 = 254$

Step 4 $3m \quad + 50 = 254$

$\qquad \dfrac{^-50 \qquad ^-50}{3m \;+\; 0 = 204}$ Add $^-50$ to both sides.

$\qquad \dfrac{3m}{3} = \dfrac{204}{3}$ Divide both sides by 3.

$\qquad m \;=\; 68$

Step 5 m represents Dan's earnings, so Dan earned $68.

$2m + \$50$ represents Brian's earnings,

and $2 \cdot \$68 + \$50$ is $\$136 + \$50 = \$186$.

Dan earned $68; Brian earned $186.

Step 6 Is $186 actually $50 more than twice $68?
Yes, the solution checks.

Does $68 + \$186 = \$254$?
Yes, the solution checks.

ANSWERS TO TEST YOUR WORD POWER

1. B; *Example:* If a triangle has sides measuring 4 ft, 10 ft, and 8 ft, then $P = 4$ ft $+ 10$ ft $+ 8$ ft $= 22$ ft.
2. D; *Example:* Area is measured in square units such as in.^{2}, ft^2, yd^2, cm^2, and km^2.
3. C; *Example:* If one side of a square is 5 in. long, all the other sides will also be 5 in. long.
4. A; *Example:* If the length of a rectangle is 12 cm and the width is 8 cm, then $A = 12$ cm $\cdot$ 8 cm $= 96$ cm^2.
5. C; *Examples:*

6. D; *Example*: In parallelogram *ABCD*, sides *AB* and *DC* are parallel and equal in length. Also, sides *AD* and *BC* are parallel and equal in length.

Chapter **3**
REVIEW EXERCISES

[3.1] *In Exercises 1–4, find the perimeter of each figure. Also, name each figure in Exercises 1–3.*

1.

28 cm

28 cm

2.

8 mi

3 mi

3.

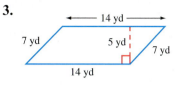

14 yd

7 yd   5 yd   7 yd

14 yd

4.

44 m

14 m

26 m

20 m

13 m

24 m

In Exercises 5–7, use the appropriate formula to find the unknown measurement.

5. A square card table has a perimeter of 12 ft. Find the length of one side of the table.

6. A rectangular playground has a perimeter of 128 yd. Find its length if it is 31 yd wide.

7. A rectangular watercolor painting that is 21 in. long has a perimeter of 72 in. What is the width of the painting?

[3.2] *In Exercises 8–10, draw a sketch of each shape and label the lengths of the sides or the base and height. Then find the area, using the appropriate formula. Sketches may vary; show your sketches to your instructor.*

8. A tablecloth that measures 5 ft by 8 ft

9. A 25 m square dance floor

10. A parallelogram-shaped lot with a base of 16 yd and a height of 13 yd

In Exercises 11–13, use the appropriate formula to find the unknown measurement.

11. A rectangular patio that is 14 ft long has an area of 126 ft². Find its width.

12. A parallelogram has an area of 88 cm². If the base is 11 cm, what is the height?

13. The area of a square piece of land is 100 mi². What is the length of one side?

[3.3] *Write each phrase as an algebraic expression. Use x as the variable.*

14. A number subtracted from 57

15. The sum of 15 and twice a number

16. The product of ⁻9 and a number

Translate each sentence into an equation and solve it. Show your work.

17. The sum of four times a number and 6 is ⁻30. What is the number?

18. When twice a number is subtracted from 10, the result is 4 plus the number. Find the number.

[3.3–3.4]   *Use the six problem-solving steps to solve each problem. Show your work.*

19. Grace wrote a $600 check for her rent. Then she deposited her $750 paycheck and a $75 tax refund into her account. The new balance was $309. How much was in her account before she wrote the rent check?

20. Yoku ordered four boxes of candles for his restaurant. One candle was put on each of the 25 tables. There were 23 candles left. How many candles were originally in each box?

21. $1000 in prize money in an essay contest is being split between Reggie and Donald. Donald should get $300 more than Reggie. How much will each man receive?

22. A rectangular photograph is twice as long as it is wide. The perimeter of the photograph is 84 cm. Find the length and the width of the photograph.

MIXED REVIEW EXERCISES

Use the information in the advertisement and the appropriate formulas to work Exercises 23–26.

NOW ON SALE

Build Your Own Dog Pen

Kit #1 includes 20 feet of fencing.

Kit #2 includes 36 feet of fencing.

23. Anthony made a square dog pen using Kit #2.

   (a) What was the length of each side of the pen?

   (b) What was the area of the pen?

24. First draw sketches of two *different* rectangular dog pens that you could build using all the fencing in Kit #1. Label the lengths of the sides. Then find the area of each pen.

25. Timotha bought Kit #2. But she used some of the fencing around her garden, so she went back and bought Kit #1. Now she has 41 ft of fencing for a dog pen. How much fencing did she use around her garden? Use the six problem-solving steps.

26. Diana bought Kit #2. The pen she built had a length that was 2 ft more than the width. Find the length and width of the pen. Use the six problem-solving steps.

Chapter 3

TEST

Appendix A: Study Skills
Activity 10: Taking Tests

Find the perimeter of each shape.

1.

72 m
59 m   46 m   59 m
72 m

2.

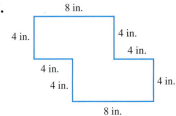

8 in.
4 in.   4 in.
             4 in.
4 in.
4 in.   4 in.
        8 in.

3. A square wetland 3 miles on a side

4. A rectangular mirror that measures 2 ft by 4 ft

5.

50 cm
                15 cm
45 cm

Find the area of each shape.

6.

27 mm
18 mm   18 mm
27 mm

7.

11 cm
        10 cm
14 cm
                14 cm
11 cm

8. A rectangular animal preserve is 55 mi wide and 68 mi long.

9. A square room measures 6 m on each side.

Solve Exercises 10–14 using the appropriate formulas. Show your work.

10. A square table has a perimeter of 12 ft. Find the length of one side.

11. The Mercado family has 34 ft of fencing to put around a garden plot. The plot is rectangular in shape. If it is 6 ft wide, find the length of the plot.

12. The area of a parallelogram is 65 in.^{2}, and the base is 13 in. What is the height of the parallelogram?

1. _____

2. _____

3. _____

4. _____

5. _____

6. _____

7. _____

8. _____

9. _____

10. _____

11. _____

12. _____

13. A rectangular postage stamp has a length of 4 cm and an area of 12 cm². Find its width.

13. _____

14. A square bulletin board has an area of 16 ft². How long is each side of the bulletin board?

14. _____

15. Explain the difference between ft and ft². For which types of problems might you use each of these units?

15. _____

Translate each sentence into an equation and solve it. Show your work.

16. _____

16. If 40 is added to four times a number, the result is zero. Find the number.

17. _____

17. When 7 times a number is decreased by 23, the result is the number plus 7. What is the number?

Solve each application problem, using the six problem-solving steps.

18. _____

18. Josephine had $43 in her wallet. Her son used some of it when he bought groceries. Josephine found $16 in her desk drawer and put it in her wallet. She counted $44 in the wallet. How much money did her son spend on groceries?

19. _____

19. Ray is 39 years old. His age is 4 years more than five times his daughter's age. How old is his daughter?

20. _____

20. A board is 118 cm long. Karin cut it into two pieces, with one piece 4 cm longer than the other. Find the length of both pieces.

21. _____

21. The perimeter of a rectangular building is 420 ft. The length is four times as long as the width. Find the length and the width. Draw a sketch to help you solve this problem.

22. _____

22. Marcella and her husband, Tim, spent a total of 19 hours redecorating their living room. Tim spent 3 hours less time than Marcella. How long did each person work on the room?

Cumulative Review Exercises

CHAPTERS 1–3

1. Write this number in words.
   4,000,206,300

2. Write this number, using digits.
   seventy million, five thousand, four hundred eighty-nine

3. Write $<$ or $>$ between each pair of numbers to make a true statement.

   $^-7$ ___ $^-1$ 0 ___ $^-5$

4. Name the property illustrated by each equation.
   (a) $1(97) = 97$

   (b) $^-10 + 0 = ^-10$

5. **(a)** Round 3795 to the nearest ten.
   (b) Round 493,662 to the nearest ten-thousand.

   (c) $(3 \cdot {}^-7) \cdot 6 = 3 \cdot ({}^-7 \cdot 6)$

Simplify.

6. $^-12 - 12$

7. $^-3({}^-9)$

8. $|7| - |{}^-10|$

9. $^-40 \div 2 \cdot 5$

10. $3 - 8 + 10$

11. $\dfrac{0}{^-6}$

12. $^-8 + 5(2 - 3)$

13. $({}^-3)^2 + 4^2$

14. $4 - 3({}^-6 \div 3) + 7(0 - 6)$

15. $\dfrac{4 - 2^3 + 5^2 - 3({}^-2)}{^-1(3) - 6({}^-2) - 9}$

16. Rewrite $10w^2xy^4$ without exponents.

17. Evaluate $^-6cd^3$ when c is 5 and d is $^-2$.

Simplify.

18. $^-4k + k + 5k$

19. $m^2 + 2m + 2m^2$

20. $xy^3 - xy^3$

21. $5({}^-4a)$

22. $^-8 + x + 5 - 2x^2 - x$

23. $^-3(4n + 3) + 10$

Solve each equation and check each solution.

24. $6 - 20 = 2x - 9x$ *Check*

25. $^-5y = y + 6$ *Check*

Solve each equation. Show your work.

26. $3b - 9 = 19 - 4b$

27. $^-16 - h + 2 = h - 10$

28. $^-5(2x + 4) = 3x - 20$

29. $6 + 4(a + 8) = ^-8a + 5 + a$

Find the perimeter and area of each shape. The figure in Exercise 30 is a parallelogram.

30.

11 in.

18 in.

12 in.

31. A square plaza is 15 m on a side.

32. A rectangular piece of plywood is 4 ft wide and 8 ft long.

Translate each sentence into an equation and solve it. Show your work.

33. When $^-50$ is added to five times a number, the result is 0. What is the number?

34. When three times a number is subtracted from 10, the result is two times the number. Find the number.

Solve each application problem, using the six problem-solving steps.

35. Some people were waiting in line at the bank. Three people made deposits and left. Six more people got in line. Then two more people were helped, but five people were still in line. How many were in the line originally?

36. Twelve soccer players each paid the same amount toward a team trip. Expenses for the trip were $2200, and now the team bank account is overdrawn by $40. How much did each player originally pay?

37. A group of 192 students was divided into two smaller groups. One smaller group was three times the size of the other. How many students were in each smaller group?

38. A rectangular swimming pool is 14 ft longer than it is wide. If the perimeter of the pool is 92 ft, find the length and the width.

Rational Numbers: Positive and Negative Fractions

4

Americans spend over $270 billion each year buying materials for their home repair, construction, or remodeling projects. Both men and women take on home improvement projects in about equal numbers. One of the small but essential items for many projects is nails. There are many kinds and sizes of nails, and you'll need to buy the right size nail for each project. Nails are sized according to the "penny" system, which is a number followed by the abbreviation "d." (The "d" is the traditional British abbreviation for "penny.") An algebraic expression, and the ability to solve equations with fractions, will help you find the relationship between a nail's "penny size" and its length in inches. See **Section 4.7,** Exercises 33–36. (*Source:* U.S. Bureau of the Census; Opinion Research Corp.; *Season by Season Home Maintenance.*)

4.1 Introduction to Signed Fractions

OBJECTIVE **1** **Use a fraction to name the part of a whole that is shaded.** In **Chapters 1–3** you worked with integers. Recall that a list of the integers can be written as follows.

$$\ldots, {}^-6, {}^-5, {}^-4, {}^-3, {}^-2, {}^-1, 0, 1, 2, 3, 4, 5, 6, \ldots$$

The dots show that the list goes on forever in both directions.

Now we will work with *fractions*.

Fractions

A **fraction** is a number of the form $\dfrac{a}{b}$ where a and b are integers and b is not 0.

One use for fractions is situations in which we need a number that is between two integers. Here is an example.

$\frac{2}{3}$ cup

A recipe uses $\frac{2}{3}$ cup of milk.

$\frac{2}{3}$ is between 0 and 1.

$\frac{2}{3}$ is a fraction because it is of the form $\dfrac{a}{b}$ and 2 and 3 are integers.

The number $\frac{2}{3}$ is a fraction that represents 2 of 3 equal parts. In this example, the cup is divided into 3 equal parts and we use enough milk to fill 2 of the parts.

We read $\frac{2}{3}$ as "two-thirds."

EXAMPLE 1 **Using Fractions to Represent Part of One Whole**

Use fractions to represent the shaded portion and the unshaded portion of each figure.

(a)

The figure has 3 equal parts. The 2 shaded parts are represented by the fraction $\frac{2}{3}$. The *un*shaded part is $\frac{1}{3}$ of the figure.

(b)

The figure has 7 equal parts. The 4 shaded parts are represented by the fraction $\frac{4}{7}$. The *un*shaded part is $\frac{3}{7}$ of the figure.

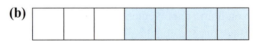

Now Try Exercise 1.

Fractions can also be used to represent more than one whole object.

EXAMPLE 2 **Using Fractions to Represent More Than One Whole**

Use a fraction to represent the shaded parts.

(a)
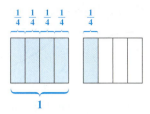

An area equal to 5 of the $\frac{1}{4}$ parts is shaded, so $\frac{5}{4}$ is shaded.

(b)
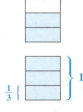

An area equal to 5 of the $\frac{1}{3}$ parts is shaded, so $\frac{5}{3}$ is shaded.

Now Try Exercise 7. ▶▶▶

OBJECTIVE 2 Identify numerators, denominators, proper fractions, and improper fractions. In the fraction $\frac{2}{3}$, the number 2 is the *numerator* and 3 is the *denominator*. The bar between the numerator and the denominator is the *fraction bar*.

$$\text{Fraction bar} \rightarrow \frac{2}{3} \begin{array}{l} \leftarrow \text{Numerator} \\ \leftarrow \text{Denominator} \end{array}$$

Numerator and Denominator

The **denominator** of a fraction shows the number of equal parts in the whole, and the **numerator** shows how many parts are being considered.

NOTE
Recall that a fraction bar, —, is a symbol for division and division by 0 is undefined. Therefore a fraction with a denominator of 0 is also undefined.

EXAMPLE 3 **Identifying Numerators and Denominators**

Identify the numerator and denominator in each fraction. Then state the number of equal parts in the whole.

(a) $\frac{5}{9}$

$$\frac{5}{9} \begin{array}{l} \leftarrow \text{Numerator} \\ \leftarrow \text{Denominator} \end{array}$$

9 equal parts in the whole

(b) $\frac{11}{7}$

$$\frac{11}{7} \begin{array}{l} \leftarrow \text{Numerator} \\ \leftarrow \text{Denominator} \end{array}$$

7 equal parts in the whole

Now Try Exercise 19. ▶▶▶

Fractions are sometimes called *proper* or *improper* fractions.

Proper and Improper Fractions

If the numerator of a fraction is *smaller* than the denominator, the fraction is a **proper fraction.** A proper fraction is less than 1.

If the numerator is *greater than or equal to* the denominator, the fraction is an **improper fraction.** An improper fraction is greater than or equal to 1.

Proper Fractions			Improper Fractions		
$\dfrac{1}{2}$	$\dfrac{5}{11}$	$\dfrac{35}{36}$	$\dfrac{9}{7}$	$\dfrac{126}{125}$	$\dfrac{7}{7}$

EXAMPLE 4 Classifying Types of Fractions

(a) Identify all proper fractions in this list.

$$\frac{3}{4} \qquad \frac{5}{9} \qquad \frac{17}{5} \qquad \frac{9}{7} \qquad \frac{3}{3} \qquad \frac{12}{25} \qquad \frac{1}{9} \qquad \frac{5}{3}$$

Proper fractions have a numerator that is *smaller* than the denominator. The proper fractions in the list are shown below.

$$\frac{3}{4} \leftarrow \text{3 is smaller than 4.} \qquad\qquad \frac{5}{9} \qquad \frac{12}{25} \qquad \frac{1}{9}$$

(b) Identify all improper fractions in the list in part (a).

Improper fractions have a numerator that is *equal to or greater* than the denominator. The improper fractions in the list are shown below.

$$\frac{17}{5} \leftarrow \text{17 is greater than 5.} \qquad\qquad \frac{9}{7} \qquad \frac{3}{3} \qquad \frac{5}{3}$$

Now Try Exercise 23.

CAUTION

In **Chapters 1–3** we used a raised negative sign to help you avoid confusion between negative numbers and subtraction. Now you are ready to start writing the negative sign in the more traditional way. In this chapter, the negative sign will still be red, but it will be centered on the number instead of raised: for example, -2 instead of $^-2$. When the negative sign might be confused with the sign for subtraction, we will write parentheses around the negative number. Here is an example.

$$3 - (-2) \qquad \text{means} \qquad 3 \text{ minus } (\text{negative } 2)$$

For fractions, the negative sign will be written in front of the fraction bar: for example, $-\frac{3}{4}$. As with integers, the negative sign tells you that a fraction is *less than 0;* it is to the *left* of 0 on the number line. When there is *no* sign in front of a fraction, the fraction is assumed to be positive. For example, $\frac{3}{4}$ is assumed to be $+\frac{3}{4}$. It is to the *right* of 0 on the number line.

OBJECTIVE 3 Graph positive and negative fractions on a number line. Sometimes we need *negative* numbers that are between two integers. For example, $-\frac{3}{4}$ is between 0 and -1. Graphing numbers on a number line helps us see the difference between $\frac{3}{4}$ and $-\frac{3}{4}$. Both represent 3 out of 4 equal parts, but they are in opposite directions from 0 on the number line. For $\frac{3}{4}$, divide the distance from 0 to 1 into 4 equal parts. Then start at 0, count over 3 parts, and make a dot. For $-\frac{3}{4}$, repeat the same process between 0 and -1.

EXAMPLE 5   **Graphing Positive and Negative Fractions**

Graph each fraction on the number line.

(a) $\dfrac{2}{5}$

There is *no* sign in front of $\frac{2}{5}$, so it is *positive*. Because $\frac{2}{5}$ is between 0 and 1, we divide that space into 5 equal parts. Then we start at 0 and count to the right 2 parts.

(b) $-\dfrac{4}{5}$

The fraction is *negative,* so it is between 0 and -1. We divide that space into 5 equal parts. Then we start at 0 and count to the left 4 parts.

Now Try Exercise 29. ▶▶▶

O B J E C T I V E **4** **Find the absolute value of a fraction.** In **Section 1.2** we said that the *absolute value* of a number was its distance from 0 on the number line. Two vertical bars indicate absolute value, as shown below.

$$\left| -\dfrac{3}{4} \right| \text{ is read "the absolute value of negative three-fourths."}$$

As with integers, the absolute value of fractions will *always* be positive (or 0) because it is the *distance* from 0 on the number line.

EXAMPLE 6   **Finding the Absolute Value of Fractions**

Find each absolute value: $\left| \dfrac{1}{2} \right|$ and $\left| -\dfrac{1}{2} \right|$.

The distance from 0 to $\frac{1}{2}$ on the number line is $\frac{1}{2}$ space, so $|\frac{1}{2}| = \frac{1}{2}$.

The distance from 0 to $-\frac{1}{2}$ is also $\frac{1}{2}$ space, so $|-\frac{1}{2}| = \frac{1}{2}$.

Now Try Exercises 41 and 43. ▶▶▶

OBJECTIVE 5 **Write equivalent fractions.** You may already know that $\frac{2}{4}$ and $\frac{1}{2}$ are at the same point on the number line. Both of them are halfway between 0 and 1. There are actually *many* different names for this point. We illustrate some of them below.

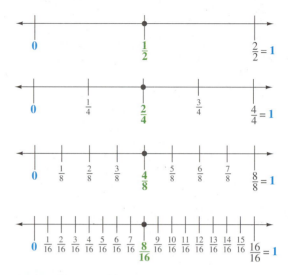

That is, $\frac{8}{16} = \frac{4}{8} = \frac{2}{4} = \frac{1}{2}$. If you have used a standard ruler with inches divided into sixteenths, you probably already noticed that these distances are the same. Although the fractions look different, they all name the same point that is halfway between 0 and 1. In other words, they have the same value. We say that they are *equivalent fractions*.

> **Equivalent Fractions**
>
> Fractions that represent the same number (the same point on a number line) are **equivalent fractions.**

Drawing number lines is tedious, so we usually find equivalent fractions by multiplying or dividing both the numerator and denominator by the same number. We can use some of the fractions that we just graphed to illustrate this method.

$$\frac{1}{2} = \frac{1 \cdot \mathbf{2}}{2 \cdot \mathbf{2}} = \frac{2}{4} \qquad\qquad \frac{8}{16} = \frac{8 \div \mathbf{4}}{16 \div \mathbf{4}} = \frac{2}{4}$$

Multiply both numerator and denominator by 2.   Divide both numerator and denominator by 4.

> **Writing Equivalent Fractions**
>
> If a, b, and c are numbers (and b and c are not 0), then
>
> $$\frac{a}{b} = \frac{a \cdot c}{b \cdot c} \qquad \text{or} \qquad \frac{a}{b} = \frac{a \div c}{b \div c}.$$

In other words, if the numerator and denominator of a fraction are multiplied or divided by the *same* nonzero number, the result is an *equivalent* fraction.

EXAMPLE 7   **Writing Equivalent Fractions**

(a) Write $-\dfrac{1}{2}$ as an equivalent fraction with a denominator of 16.

In other words, $-\dfrac{1}{2} = -\dfrac{?}{16}$.

The original denominator is 2. *Multiplying* 2 times 8 gives 16, the new denominator. To write an equivalent fraction, multiply *both* the numerator and denominator by 8.

$$-\frac{1}{2} = -\frac{1 \cdot 8}{2 \cdot 8} = -\frac{8}{16}$$

Keep the negative sign.

So, $-\dfrac{1}{2}$ is equivalent to $-\dfrac{8}{16}$.

(b) Write $\dfrac{12}{15}$ as an equivalent fraction with a denominator of 5.

In other words, $\dfrac{12}{15} = \dfrac{?}{5}$.

The original denominator is 15. *Dividing* 15 by 3 gives 5, the new denominator. To write an equivalent fraction, divide *both* the numerator and denominator by 3.

$$\frac{12}{15} = \frac{12 \div 3}{15 \div 3} = \frac{4}{5}$$

So, $\dfrac{12}{15}$ is equivalent to $\dfrac{4}{5}$.

Now Try Exercises 47(c) and 49(c). ▶▶▶

Look back at the set of four number lines on the previous page. Notice that there are many different names for 1.

$$\frac{2}{2} = 1 \qquad \frac{4}{4} = 1 \qquad \frac{8}{8} = 1 \qquad \frac{16}{16} = 1$$

Because a fraction bar is a symbol for division, you can think of $\frac{2}{2}$ as $2 \div 2$, which equals 1. Similarly, $\frac{4}{4}$ is $4 \div 4$, which also is 1, and so on. These examples illustrate one of the division properties from **Section 1.7.**

Division Properties

If a is any number (except 0), then $\dfrac{a}{a} = 1$. In other words, when a nonzero number is divided by itself, the result is 1.

For example, $\dfrac{6}{6} = 1$ and $\dfrac{-4}{-4} = 1$.

Also recall that when any number is divided by 1, the result is the number. That is, $\dfrac{a}{1} = a$.

For example, $\dfrac{6}{1} = 6$ and $-\dfrac{12}{1} = -12$.

EXAMPLE 8 **Using Division to Simplify Fractions**

Simplify each fraction by dividing the numerator by the denominator.

(a) $\dfrac{5}{5}$ Think of $\dfrac{5}{5}$ as $5 \div 5$. The result is 1, so $\dfrac{5}{5} = 1$.

(b) $-\dfrac{12}{4}$ Think of $-\dfrac{12}{4}$ as $-12 \div 4$. The result is -3, so $-\dfrac{12}{4} = -3$.

Keep the negative sign.

(c) $\dfrac{6}{1}$ Think of $\dfrac{6}{1}$ as $6 \div 1$. The result is 6, so $\dfrac{6}{1} = 6$.

Now Try Exercises 63, 65, and 67.

NOTE

The title of this chapter is "Rational Numbers: Positive and Negative Fractions." *Rational numbers* are numbers that can be written in the form $\dfrac{a}{b}$, where a and b are integers and b is not 0. In Example 8(c) above, you saw that an integer can be written in the form $\dfrac{a}{b}$ (6 can be written as $\frac{6}{1}$). So rational numbers include all the integers and all the fractions. In **Chapter 5** you'll work with rational numbers that are in decimal form.

4.1 Exercises

FOR EXTRA HELP

 Addison-Wesley Math Tutor Center

 MathXL

 Digital Video Tutor CD 2 Videotape 4

Student's Solutions Manual

MyMathLab MyMathLab

 MathXL Tutorials on CD

Write the fractions that represent the shaded and unshaded portions of each figure. See Examples 1 and 2.

1.

2.

3.

4.

5.

6.

7.

8.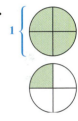

9. What fraction of these 11 coins are dimes? What fraction are pennies? What fraction are nickels?

10. What fraction of these six recording artists are men? What fraction are women? What fraction are wearing something orange?

11. In an American Sign Language (A.S.L.) class of 25 students, eight are deaf. What fraction of the students are deaf? What fraction are not deaf?

12. Of 35 motorcycles in the parking lot, 17 are Harley-Davidsons. What fraction of the motorcycles are *not* Harley-Davidsons? What fraction are Harley-Davidsons?

13. Of the 71 computers in the lab, 58 are laptops. What fraction of the computers are *not* laptops? What fraction are laptops?

14. A community college basketball team has 12 members. If five of the players are sophomores and the rest are freshmen, find the fraction of the members that are sophomores and the fraction that are freshmen.

The circle graph shows the results of a survey on where women would like to have flowers delivered on Valentine's Day. Use the graph to answer Exercises 15–18.

DELIVERING FLOWERS ON VALENTINE'S DAY

13 At home
6 At work
1 — Other

Out of every 20 women surveyed, the number who would like flowers delivered at home, at work, or elsewhere.

Source: FTD, Inc.

15. What fraction of the women would like flowers delivered at work?

16. What fraction of the women would like flowers delivered at home?

17. What fraction would like flowers delivered either at home or at work?

18. What fraction picked a location other than home or work?

Identify the numerator and denominator in each fraction. Then state the number of equal parts in the whole. See Example 3.

19. $\dfrac{3}{4}$

20. $\dfrac{5}{8}$

21. $\dfrac{12}{7}$

22. $\dfrac{8}{3}$

List the proper and improper fractions in each group of numbers. See Example 4.

	Proper	Improper
23. $\dfrac{8}{5}, \dfrac{1}{3}, \dfrac{5}{8}, \dfrac{6}{6}, \dfrac{12}{2}, \dfrac{7}{16}$	_____	_____
24. $\dfrac{1}{6}, \dfrac{5}{8}, \dfrac{15}{14}, \dfrac{11}{9}, \dfrac{7}{7}, \dfrac{3}{4}$	_____	_____
25. $\dfrac{3}{4}, \dfrac{3}{2}, \dfrac{5}{5}, \dfrac{9}{11}, \dfrac{7}{15}, \dfrac{19}{18}$	_____	_____
26. $\dfrac{12}{12}, \dfrac{15}{11}, \dfrac{13}{12}, \dfrac{11}{8}, \dfrac{17}{17}, \dfrac{19}{12}$	_____	_____

27. Write a fraction of your own choice. Label the *three* parts of the fraction and write a sentence describing what each part represents. Draw a figure with shaded parts to illustrate your fraction.

28. Give one example of a proper fraction and one example of an improper fraction. What determines whether a fraction is proper or improper? Draw figures with shaded parts to illustrate your fractions.

Graph each pair of fractions on the number line. See Example 5.

29. $\dfrac{1}{4}, -\dfrac{1}{4}$

30. $-\dfrac{1}{3}, \dfrac{1}{3}$

31. $-\dfrac{3}{5}, \dfrac{3}{5}$

32. $\dfrac{5}{6}, -\dfrac{5}{6}$

33. $\dfrac{7}{8}, -\dfrac{7}{8}$

34. $-\dfrac{3}{4}, \dfrac{3}{4}$

Write a positive or negative fraction to describe each situation.

35. The baby lost $\frac{3}{4}$ pound in weight while she was sick.

36. Greta needed $\frac{1}{3}$ cup of brown sugar for the cookie recipe.

37. The oil level in my car is $\frac{1}{2}$ quart below normal.

38. The mice who were on an experimental diet lost an average of $\frac{1}{4}$ ounce.

39. Barb Brown's driveway is $\frac{3}{10}$ mile long.

40. Marcel cut $\frac{5}{8}$ in. from the bottom of the door so that it wouldn't scrape the floor.

Find each absolute value. See Example 6.

41. $\left| -\dfrac{2}{5} \right|$

42. $\left| -\dfrac{2}{3} \right|$

43. $\left| \dfrac{9}{10} \right|$

44. $|0|$

45. $\left| -\dfrac{13}{6} \right|$

46. $\left| -\dfrac{4}{4} \right|$

47. Rewrite each fraction as an equivalent fraction with a denominator of 24. See Example 7.

(a) $\dfrac{1}{2} = \dfrac{\;\;\;}{24}$ **(b)** $\dfrac{1}{3} = \text{---}$ **(c)** $\dfrac{2}{3} = \text{---}$ **(d)** $\dfrac{1}{4} = \text{---}$ **(e)** $\dfrac{3}{4} = \text{---}$

(f) $\dfrac{1}{6} = \text{---}$ **(g)** $\dfrac{5}{6} = \text{---}$ **(h)** $\dfrac{1}{8} = \text{---}$ **(i)** $\dfrac{3}{8} = \text{---}$ **(j)** $\dfrac{5}{8} = \text{---}$

48. Rewrite each fraction as an equivalent fraction with a denominator of 36. See Example 7.

(a) $\dfrac{1}{2} = \dfrac{\;\;\;}{36}$ **(b)** $\dfrac{1}{3} = \text{---}$ **(c)** $\dfrac{2}{3} = \text{---}$ **(d)** $\dfrac{1}{4} = \text{---}$ **(e)** $\dfrac{3}{4} = \text{---}$

(f) $\dfrac{1}{6} = \text{---}$ **(g)** $\dfrac{5}{6} = \text{---}$ **(h)** $\dfrac{1}{9} = \text{---}$ **(i)** $\dfrac{4}{9} = \text{---}$ **(j)** $\dfrac{8}{9} = \text{---}$

49. Rewrite each fraction as an equivalent fraction with a denominator of 3. See Example 7.

(a) $-\dfrac{2}{6} = -\dfrac{\;\;\;}{3}$ **(b)** $-\dfrac{4}{6} = \text{---}$ **(c)** $-\dfrac{12}{18} = \text{---}$ **(d)** $-\dfrac{6}{18} = \text{---}$ **(e)** $-\dfrac{200}{300} = \text{---}$

(f) Write two more fractions that are equivalent to $-\frac{1}{3}$ and two more fractions equivalent to $-\frac{2}{3}$.

50. Rewrite each fraction as an equivalent fraction with a denominator of 4. See Example 7.

(a) $-\dfrac{2}{8} = \dfrac{\;\;\;}{4}$ **(b)** $-\dfrac{6}{8} = \text{---}$ **(c)** $-\dfrac{15}{20} = \text{---}$ **(d)** $-\dfrac{50}{200} = \text{---}$ **(e)** $-\dfrac{150}{200} = \text{---}$

(f) Write two more fractions that are equivalent to $-\frac{1}{4}$ and two more fractions equivalent to $-\frac{3}{4}$.

RELATING CONCEPTS (EXERCISES 51–58) For Individual or Group Work

Use your calculator as you **work Exercises 51–58 in order.**

51. (a) Write $\frac{3}{8}$ as an equivalent fraction with a denominator of 3912.

　　(b) Explain how you solved part (a).

52. (a) Write $\frac{7}{9}$ as an equivalent fraction with a denominator of 5472.

　　(b) Explain how you solved part (a).

53. (a) Is $-\dfrac{697}{3485}$ equivalent to $-\dfrac{1}{2}$, $-\dfrac{1}{3}$, or $-\dfrac{1}{5}$?

　　(b) Explain how you solved part (a).

54. (a) Is $-\dfrac{817}{4902}$ equivalent to $-\dfrac{1}{4}$, $-\dfrac{1}{6}$, or $-\dfrac{1}{8}$?

　　(b) Explain how you solved part (a).

(continued)

Find a number to replace the ? that will make the two fractions equivalent.

55. $\dfrac{1183}{2028} = \dfrac{?}{12}$

56. $\dfrac{2775}{6105} = \dfrac{?}{11}$

57. $\dfrac{13}{?} = \dfrac{1157}{1335}$

58. $\dfrac{9}{?} = \dfrac{891}{1584}$

59. Explain how to write equivalent fractions. Show one example in which the new denominator is larger than the original denominator, and one example in which the new denominator is smaller.

60. Explain how to find the absolute value of a fraction. Draw a number line to illustrate your explanation and include a positive fraction and a negative fraction as examples.

61. Can you write $\frac{3}{5}$ as an equivalent fraction with a denominator of 18? Explain why or why not. If not, what denominators could you use instead of 18?

62. Can you write $\frac{3}{4}$ as an equivalent fraction with a denominator of 0? Explain why or why not.

Simplify each fraction by dividing the numerator by the denominator. See Example 8.

63. $\dfrac{10}{1}$

64. $\dfrac{9}{9}$

65. $-\dfrac{16}{16}$

66. $-\dfrac{7}{1}$

67. $-\dfrac{18}{3}$

68. $-\dfrac{40}{4}$

69. $\dfrac{24}{8}$

70. $\dfrac{42}{6}$

71. $\dfrac{12}{12}$

72. $-\dfrac{5}{5}$

73. $\dfrac{14}{7}$

74. $\dfrac{8}{2}$

75. $-\dfrac{5}{1}$

76. $-\dfrac{90}{10}$

77. $-\dfrac{45}{9}$

78. $\dfrac{16}{1}$

79. $\dfrac{150}{150}$

80. $\dfrac{55}{5}$

81. $-\dfrac{32}{4}$

82. $-\dfrac{200}{200}$

There are many correct ways to draw the answers for Exercises 83–100, so ask your instructor to check your work.

83. Shade $\frac{3}{5}$ of this figure. What fraction is unshaded?

84. Shade $\frac{5}{6}$ of this figure. What fraction is unshaded?

85. Shade $\frac{3}{8}$ of this figure. What fraction is unshaded?

86. Shade $\frac{1}{3}$ of this figure. What fraction is unshaded?

87. Shade $\frac{7}{4}$ of this figure. What fraction is unshaded?

88. Shade $\frac{6}{5}$ of this figure. What fraction is unshaded?

89. Shade $\frac{4}{3}$ of this figure. What fraction is unshaded?

90. Shade $\frac{11}{8}$ of this figure. What fraction is unshaded?

91. Shade $\frac{6}{6}$ of this figure. What fraction is unshaded?

92. Shade $\frac{10}{10}$ of this figure. What fraction is unshaded?

93. Shade $\frac{10}{5}$ of this figure.

94. Shade $\frac{8}{4}$ of this figure.

95. Draw a group of seven faces. Draw some of the faces smiling and some sad. What fraction of your faces are smiling? What fraction are sad?

96. Draw a group of eight apples. Draw some of the apples with a stem and some of them with a bite taken out. What fraction of your apples have stems? What fraction have a bite taken out?

97. Draw a group of figures. Make $\frac{1}{10}$ of the figures circles, $\frac{6}{10}$ of the figures squares, and $\frac{3}{10}$ of the figures triangles. Then shade $\frac{1}{6}$ of the squares and $\frac{2}{3}$ of the triangles.

98. Write a group of capital letters. Make $\frac{4}{9}$ of the letters A's, $\frac{2}{9}$ of the letters B's, and $\frac{3}{9}$ of the letters C's. Then draw a line under $\frac{3}{4}$ of the A's and $\frac{1}{2}$ of the B's.

99. Draw a group of punctuation marks. Make $\frac{5}{12}$ of them exclamation points, $\frac{1}{12}$ of them commas, $\frac{3}{12}$ of them periods, and add enough question marks to make a full $\frac{12}{12}$ in all. Then circle $\frac{2}{5}$ of the exclamation points.

100. Draw a group of symbols. Make $\frac{2}{15}$ of them addition signs, $\frac{4}{15}$ of them subtraction signs, $\frac{2}{15}$ of them division signs, and add enough equal signs to make a full $\frac{15}{15}$ in all. Then circle $\frac{3}{4}$ of the subtraction signs.

4.2 Writing Fractions in Lowest Terms

OBJECTIVE **1** **Identify fractions written in lowest terms.**   You can see from these drawings that $\frac{1}{2}$ and $\frac{4}{8}$ are different names for the same amount of pizza.

$\frac{1}{2}$ of the pizza has pepperoni on it.

$\frac{4}{8}$ of the pizza has pepperoni on it.

OBJECTIVES

1 Identify fractions written in lowest terms.

2 Write a fraction in lowest terms using common factors.

3 Write a number as a product of prime factors.

4 Write a fraction in lowest terms using prime factorization.

5 Write a fraction with variables in lowest terms.

You saw in the last section that $\frac{1}{2}$ and $\frac{4}{8}$ are equivalent fractions. But we say that the fraction $\frac{1}{2}$ is in *lowest terms* because the numerator and denominator have no *common factor* other than 1. That means that 1 is the only number that divides evenly into both 1 and 2. However, the fraction $\frac{4}{8}$ is *not* in lowest terms because its numerator and denominator have a common factor of 4. That means 4 will divide evenly into both 4 and 8.

> **NOTE**
> Recall that *factors* are numbers being multiplied to give a product. For example,
>
> $1 \cdot 4 = 4$, so 1 and 4 are factors of 4.
>
> $2 \cdot 4 = 8$, so 2 and 4 are factors of 8.
>
> 4 is a factor of both 4 and 8, so 4 is a *common factor* of those numbers.

Writing a Fraction in Lowest Terms

A fraction is written in **lowest terms** when the numerator and denominator have no common factor other than 1. Examples are $\frac{1}{3}$, $\frac{3}{4}$, $\frac{2}{5}$, and $\frac{7}{10}$.

When you work with fractions, always write the final answer in lowest terms.

EXAMPLE 1 **Identifying Fractions Written in Lowest Terms**

Are the following fractions in lowest terms? If not, find a common factor of the numerator and denominator (other than 1.)

(a) $\dfrac{3}{8}$

The numerator and denominator have no common factor other than 1, so the fraction is in lowest terms.

(b) $\dfrac{21}{36}$

The numerator and denominator have a common factor of 3, so the fraction is *not* in lowest terms.

Now Try Exercises 1(a) and 1(e). ▶▶▶

OBJECTIVE **2** **Write a fraction in lowest terms using common factors.** We will show you two methods for writing a fraction in lowest terms. The first method, dividing by a common factor, works best when the numerator and denominator are small numbers.

EXAMPLE 2 **Using Common Factors to Write Fractions in Lowest Terms**

Write each fraction in lowest terms. Use the method of dividing by a common factor.

(a) $\dfrac{20}{24}$

The *largest* common factor of 20 and 24 is 4. Divide both numerator and denominator by **4**.

$$\frac{20}{24} = \frac{20 \div 4}{24 \div 4} = \frac{5}{6}$$

(b) $\dfrac{30}{50} = \dfrac{30 \div 10}{50 \div 10} = \dfrac{3}{5}$ Divide both numerator and denominator by 10.

(c) $-\dfrac{24}{42} = -\dfrac{24 \div 6}{42 \div 6} = -\dfrac{4}{7}$ Divide both numerator and denominator by 6. Keep the negative sign.

(d) $\dfrac{60}{72}$

Suppose we made an error and thought that 4 was the largest common factor of 60 and 72. Dividing by 4 gives the following.

$$\frac{60}{72} = \frac{60 \div 4}{72 \div 4} = \frac{15}{18}$$

But $\frac{15}{18}$ is *not* in lowest terms because 15 and 18 have a common factor of 3. Therefore, divide the numerator and denominator by 3.

$$\frac{15}{18} = \frac{15 \div 3}{18 \div 3} = \frac{5}{6} \quad \leftarrow \text{Lowest terms}$$

The fraction $\frac{60}{72}$ could have been written in lowest terms in one step by dividing by 12, the *largest* common factor of 60 and 72.

$$\frac{60}{72} = \frac{60 \div 12}{72 \div 12} = \frac{5}{6} \quad \left\{ \begin{array}{l} \text{Same answer} \\ \text{as above} \end{array} \right.$$

Either way works. Just keep dividing until the fraction is in lowest terms.

The method of dividing by a common factor is summarized below.

Dividing by a Common Factor to Write a Fraction in Lowest Terms

Step 1   Find the *largest* number that will divide evenly into both the numerator and denominator. This number is a **common factor.**

Step 2   **Divide** both numerator and denominator by the common factor.

Step 3   **Check** to see if the new numerator and denominator have any common factors (besides 1). If they do, repeat Steps 2 and 3. If the only common factor is 1, the fraction is in lowest terms.

◀◀◀ Now Try Exercises 3(a) and 3(f).

OBJECTIVE 3 **Write a number as a product of prime factors.**   In Example 2(d) on the previous page, the largest common factor of 60 and 72 was difficult to see quickly. You can handle a problem like that by writing the numerator and denominator as a product of *prime numbers*.

Prime Numbers

A **prime number** is a whole number that has exactly *two different* factors, itself and 1.

The number 3 is a prime number because it can be divided evenly only by itself and 1. The number 8 is *not* a prime number. The number 8 is a *composite number* because it can be divided evenly by 2 and 4, as well as by itself and 1.

Composite Numbers

A number with a factor other than itself or 1 is called a **composite number.**

CAUTION

A prime number has *only two* different factors, itself and 1. The number 1 is *not* a prime number because it does not have *two different* factors; the only factor of 1 is 1. Also, 0 is *not* a prime number. Therefore, 0 and 1 are *neither* prime nor composite numbers.

EXAMPLE 3 **Finding Prime Numbers**

Label each number as *prime* or *composite* or *neither*.

$$0 \quad 2 \quad 5 \quad 10 \quad 11 \quad 15$$

First, 0 is *neither* prime nor composite. Next, 2, 5, and 11, are *prime*. Each of these numbers is divisible only by itself and 1. The number 10 can be divided by 5 and 2, so it is *composite*. Also, 15 is a *composite* number because 15 can be divided by 5 and 3.

Now Try Exercise 5. ▶▶▶

For reference, here are the prime numbers smaller than 100.

2	3	5	7	11
13	17	19	23	29
31	37	41	43	47
53	59	61	67	71
73	79	83	89	97

CAUTION

All prime numbers are odd numbers except the number 2. Be careful though, because *not all odd numbers are prime numbers*. For example, 9, 15, and 21 are odd numbers but they are *not* prime numbers.

The *prime factorization* of a number can be especially useful when working with fractions.

Prime Factorization

A **prime factorization** of a number is a factorization in which every factor is a prime number.

EXAMPLE 4 Factoring Using the Division Method

(a) Find the prime factorization of 48.

$2\overline{)48}$ ⟵ Divide 48 by 2 (the first prime number); quotient is 24

$2\overline{)24}$ ⟵ Divide 24 by 2; quotient is 12

All the divisors are prime factors. $2\overline{)12}$ ⟵ Divide 12 by 2; quotient is 6

$2\overline{)6}$ ⟵ Divide 6 by 2; quotient is 3

$3\overline{)3}$ ⟵ Divide 3 by 3; quotient is 1

1 ⟵ Continue to divide until the quotient is 1

Because all the factors (divisors) are prime, the prime factorization of 48 is

$$2 \cdot 2 \cdot 2 \cdot 2 \cdot 3$$

Check by multiplying the factors to see if the product is 48.
Yes, 2 • 2 • 2 • 2 • 3 does equal 48.

> **NOTE**
> You may write the factors in any order because multiplication is commutative. So you could write the factorization of 48 as 3 • 2 • 2 • 2 • 2. We will show the factors from smallest to largest in our examples.

(b) Find the prime factorization of 225.

$3\overline{)225}$ ⟵ 225 is not divisible by 2 (first prime) so use 3 (next prime)

All the divisors are prime factors. $3\overline{)75}$ ⟵ Divide 75 by 3

$5\overline{)25}$ ⟵ 25 is not divisible by 3; use 5

$5\overline{)5}$ ⟵ Divide 5 by 5

1 ⟵ Quotient is 1

So, $225 = 3 \cdot 3 \cdot 5 \cdot 5$.

> **CAUTION**
> When you're using the division method of factoring, the last quotient is 1. Do **not** list 1 as a prime factor because 1 is not a prime number.

◀◀◀ **Now Try Exercises 9 and 13.**

Another method of factoring uses what is called a *factor tree*.

EXAMPLE 5 Factoring Using a Factor Tree

Find the prime factorization of each number.

(a) 60
Try to divide 60 by the first prime number, 2. Write the factors under the 60. Circle the 2, because it is a prime.

$$60$$
$$②\quad 30$$

Continued on Next Page

Try dividing 30 by 2. Write the factors under the 30.

Because 15 cannot be divided evenly by 2, try dividing 15 by the next prime number, 3.

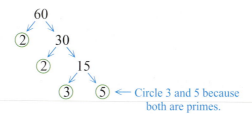

No uncircled factors remain, so you have found the prime factorization (the circled factors).

$$60 = \mathbf{2 \cdot 2 \cdot 3 \cdot 5}$$

(b) 72

Divide 72 by 2, the first prime number.

So, $72 = \mathbf{2 \cdot 2 \cdot 2 \cdot 3 \cdot 3}$.

(c) 45

Because 45 cannot be divided evenly by 2, try dividing by the next prime, 3.

$$\begin{array}{c} 45 \\ \swarrow \quad \searrow \\ ③ \quad 15 \quad \leftarrow \text{Divide by 3 again.} \\ \swarrow \quad \searrow \\ ③ \quad ⑤ \end{array}$$

So, $45 = \mathbf{3 \cdot 3 \cdot 5}$.

NOTE

Here is a reminder about the quick way to see whether a number is *divisible* by 2, 3, or 5; in other words, there is no remainder when you do the division.

A number is divisible by 2 if the ones digit is 0, 2, 4, 6, or 8. For example, 30, 512, 76, and 3018 are all divisible by 2.

A number is divisible by 3 if the *sum* of the digits is divisible by 3. For example, 129 is divisible by 3 because $1 + 2 + 9 = 12$ and 12 is divisible by 3.

A number is divisible by 5 if it has 0 or 5 in the ones place. For example, 85, 610, and 1725 are all divisible by 5.

See **Section R.4** for more information.

Now Try Exercise 17. ▶▶▶

⊞ **Calculator Tip**   You can use your calculator to find the prime factorization of a number. Here is an example that uses 539.

Try dividing 539 by the first prime number, 2.

$$539 \; ÷ \; 2 \; = \; \mathbf{269.5} \qquad \text{Does not divide evenly}$$

Try dividing 539 by the next prime number, 3.

$$539 \; ÷ \; 3 \; = \; \mathbf{179.6666667} \qquad \text{Does not divide evenly}$$

Keep trying the next prime numbers until you find one that divides evenly.

$$539 \; ÷ \; 5 \; = \; \mathbf{107.8} \qquad \text{Does not divide evenly}$$

$$539 \; ÷ \; 7 \; = \; \mathbf{77} \qquad \text{Divides evenly}$$

Once you have found that 7 works, try using it again on the quotient 77.

$$77 \; ÷ \; 7 \; = \; \mathbf{11} \qquad \text{Divides evenly}$$

Because 11 is prime, you're finished. The prime factorization of 539 is $7 \cdot 7 \cdot 11$.

Now try factoring 2431 using your calculator.

(The answer is eleven times thirteen times seventeen.)

OBJECTIVE 4 **Write a fraction in lowest terms, using prime factorization.**   Now you can use the second method for writing fractions in lowest terms: prime factorization. This is a good method to use when the numerator and denominator are larger numbers or include variables.

EXAMPLE 6 Using Prime Factorization to Write Fractions in Lowest Terms

(a) Write $\frac{20}{35}$ in lowest terms.

20 can be written as $2 \cdot 2 \cdot 5 \leftarrow$ Prime factors
35 can be written as $5 \cdot 7 \leftarrow$ Prime factors

$$\frac{20}{35} = \frac{2 \cdot 2 \cdot 5}{5 \cdot 7}$$

The numerator and denominator have 5 as a common factor. Dividing both numerator and denominator by 5 will give an equivalent fraction.

$$\frac{20}{35} = \frac{2 \cdot 2 \cdot 5}{5 \cdot 7} = \frac{2 \cdot 2 \cdot 5}{7 \cdot 5} = \frac{2 \cdot 2 \cdot \boxed{5 ÷ 5}}{7 \cdot \boxed{5 ÷ 5}} = \frac{2 \cdot 2 \cdot \mathbf{1}}{7 \cdot \mathbf{1}} = \frac{4}{7}$$

Any number divided by itself is 1

Multiplication is commutative.

$\frac{20}{35}$ is written in lowest terms as $\frac{4}{7}$.

To shorten the work, you may use slashes to indicate the divisions. For example, the work on $\frac{20}{35}$ can be shown as follows.

$$\frac{20}{35} = \frac{2 \cdot 2 \cdot \overset{1}{\cancel{5}}}{\underset{1}{\cancel{5}} \cdot 7} \qquad \begin{array}{l}\text{Slashes indicate}\\ 5 ÷ 5, \text{ and}\\ \text{the result is 1}\end{array}$$

Continued on Next Page

(b) Write $\frac{60}{72}$ in lowest terms.

Use the prime factorizations of 60 and 72 from Examples 5(a) and 5(b).

$$\frac{60}{72} = \frac{2 \cdot 2 \cdot 3 \cdot 5}{2 \cdot 2 \cdot 2 \cdot 3 \cdot 3}$$

This time there are three common factors. Use slashes to show the three divisions.

$$\frac{60}{72} = \frac{\overset{1}{2} \cdot \overset{1}{2} \cdot \overset{1}{3} \cdot 5}{\underset{1}{2} \cdot \underset{1}{2} \cdot 2 \cdot \underset{1}{3} \cdot 3} = \frac{5}{6}$$

2 ÷ 2 is 1
2 ÷ 2 is 1 ⟶   ⟵ 3 ÷ 3 is 1

← Multiply 1 • 1 • 1 • 5 to get 5
← Multiply 1 • 1 • 2 • 1 • 3 to get 6

(c) $\dfrac{18}{90}$

$$\frac{18}{90} = \frac{\overset{1}{2} \cdot \overset{1}{3} \cdot \overset{1}{3}}{\underset{1}{2} \cdot \underset{1}{3} \cdot \underset{1}{3} \cdot 5} = \frac{1}{5}$$

← Multiply 1 • 1 • 1 to get 1
← Multiply 1 • 1 • 1 • 5 to get 5

> **CAUTION**
> In Example 6(c) above, all factors of the numerator divided out. But
> 1 • 1 • 1 is still 1, so the final answer is $\frac{1}{5}$ (**not** 5).

The method of using prime factorization to write a fraction in lowest terms is summarized below.

Using Prime Factorization to Write a Fraction in Lowest Terms

Step 1 Write the **prime factorization** of both numerator and denominator.

Step 2 Use slashes to show where you are **dividing** the numerator and denominator by any common factors.

Step 3 **Multiply** the remaining factors in the numerator and in the denominator.

$\dfrac{3}{4}$

Now Try Exercises 25 and 35. ▶▶▶

OBJECTIVE **5** **Write a fraction with variables in lowest terms.**
Fractions may have variables in the numerator or denominator. Examples are shown below.

$$\frac{6}{2x} \qquad \frac{3xy}{9xy} \qquad \frac{4b^3}{8ab} \qquad \frac{7ab^2}{n^2}$$

You can use prime factorization to write these fractions in lowest terms.

EXAMPLE 7 **Writing Fractions with Variables in Lowest Terms**

Write each fraction in lowest terms.

(a) $\dfrac{6}{2x}$ ← Prime factors of 6 are 2 • 3
   ← $2x$ means 2 • x.

$$\frac{6}{2x} = \frac{\overset{1}{\cancel{2}} \cdot 3}{\underset{1}{\cancel{2}} \cdot x} = \frac{3}{x} \begin{array}{l} \leftarrow 1 \cdot 3 \text{ is } 3 \\ \leftarrow 1 \cdot x \text{ is } x \end{array}$$

$3xy$ means
$3 \cdot x \cdot y$

(b) $\dfrac{3xy}{9xy} = \dfrac{\overbrace{3 \cdot x \cdot y}}{\underbrace{3 \cdot 3} \cdot x \cdot y} = \dfrac{\overset{1}{\cancel{3}} \cdot \overset{1}{\cancel{x}} \cdot \overset{1}{\cancel{y}}}{\underset{1}{\cancel{3}} \cdot 3 \cdot \underset{1}{\cancel{x}} \cdot \underset{1}{\cancel{y}}} = \dfrac{1}{3}$

The prime
factors of
9 are 3 • 3

b^3 means $b \cdot b \cdot b$

(c) $\dfrac{4b^3}{8ab} = \dfrac{2 \cdot 2 \cdot \overbrace{b \cdot b \cdot b}}{\underbrace{2 \cdot 2 \cdot 2} \cdot a \cdot b} = \dfrac{\overset{1}{\cancel{2}} \cdot \overset{1}{\cancel{2}} \cdot \overset{1}{\cancel{b}} \cdot b \cdot b}{\underset{1}{\cancel{2}} \cdot \underset{1}{\cancel{2}} \cdot 2 \cdot a \cdot \underset{1}{\cancel{b}}} = \dfrac{b^2}{2a} \begin{array}{l} \leftarrow b \cdot b \text{ is } b^2 \\ \leftarrow 2 \cdot a \text{ is } 2a \end{array}$

The prime
factors of 8
are 2 • 2 • 2

(d) $\dfrac{7ab^2}{n^2} = \dfrac{7 \cdot a \cdot b \cdot b}{n \cdot n}$ There are no
        common factors.

$\dfrac{7ab^2}{n^2}$ is already in lowest terms.

Now Try Exercises 53 and 57.

4.2 Exercises

Are the following fractions in lowest terms? If not, find a common factor of the numerator and denominator (other than 1). See Example 1.

1. (a) $-\dfrac{3}{10}$　(b) $\dfrac{10}{15}$　(c) $\dfrac{9}{16}$　(d) $-\dfrac{4}{21}$　(e) $\dfrac{6}{9}$　(f) $-\dfrac{7}{28}$

2. (a) $\dfrac{10}{12}$　(b) $\dfrac{3}{18}$　(c) $-\dfrac{8}{15}$　(d) $-\dfrac{22}{33}$　(e) $\dfrac{2}{25}$　(f) $-\dfrac{14}{15}$

Write each fraction in lowest terms. Use the method of dividing by a common factor. See Example 2.

3. (a) $\dfrac{10}{15}$　(b) $\dfrac{6}{9}$　(c) $-\dfrac{7}{28}$　(d) $-\dfrac{25}{50}$　(e) $\dfrac{16}{18}$　(f) $-\dfrac{8}{20}$

4. (a) $\dfrac{10}{12}$　(b) $\dfrac{3}{18}$　(c) $-\dfrac{22}{33}$　(d) $\dfrac{12}{16}$　(e) $-\dfrac{15}{20}$　(f) $-\dfrac{9}{15}$

Label each number as prime *or* composite *or* neither. *See Example 3.*

5.　9　　2　　8　　1　　5　　11　　10　　21

6. 12　　3　　7　　6　　0　　15　　13　　25

Find the prime factorization of each number. See Examples 4 and 5.

7. 6　　　　　**8.** 12　　　　　**9.** 20　　　　　**10.** 30

11. 25　　　　**12.** 18　　　　**13.** 36　　　　**14.** 56

15. (a) 44　　**16.** (a) 45　　**17.** (a) 75　　**18.** (a) 80

(b) 88　　　　　(b) 135　　　　(b) 68　　　　　(b) 64

　　　　　　　　　　　　　　　　(c) 189　　　　(c) 385

Write each numerator and denominator as a product of prime factors. Then use the prime factorization to write the fraction in lowest terms. See Example 6.

19. $\dfrac{8}{16}$ **20.** $\dfrac{6}{8}$ **21.** $\dfrac{32}{48}$

22. $\dfrac{9}{27}$ **23.** $\dfrac{14}{21}$ **24.** $\dfrac{20}{32}$

25. $\dfrac{36}{42}$ **26.** $\dfrac{22}{33}$ **27.** $\dfrac{50}{63}$

28. $\dfrac{72}{80}$ **29.** $\dfrac{27}{45}$ **30.** $\dfrac{36}{63}$

31. $\dfrac{12}{18}$ **32.** $\dfrac{63}{90}$ **33.** $\dfrac{35}{40}$

34. $\dfrac{36}{48}$ **35.** $\dfrac{90}{180}$ **36.** $\dfrac{16}{64}$

37. $\dfrac{210}{315}$ **38.** $\dfrac{96}{192}$

39. $\dfrac{429}{495}$ **40.** $\dfrac{135}{182}$

Write your answers to Exercises 41–46 in lowest terms.

41. There are 60 minutes in an hour. What fraction of an hour is

    (a) 15 minutes?   **(b)** 30 minutes?

    (c) 6 minutes?   **(d)** 60 minutes?

42. There are 24 hours in a day. What fraction of a day is

    (a) 8 hours?   **(b)** 18 hours?

    (c) 12 hours?   **(d)** 3 hours?

43. SueLynn's monthly income is $1500.

    (a) She spends $500 on rent. What fraction of her income is spent on rent?

    (b) She spends $300 on food. What fraction of her income is spent on food?

    (c) What fraction of her income is left for other expenses?

44. There are 10,000 students at Minneapolis Community and Technical College.

    (a) 7500 of the students receive some form of financial aid. What fraction of the students receive financial aid?

    (b) 6000 of the students are women. What fraction are women?

    (c) What fraction of the students are men?

45. What fraction of the time spent on household chores is done by **(a)** husbands, **(b)** wives, **(c)** children?

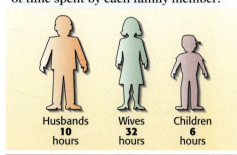

Of the typical 48 hours spent weekly on household chores, here is the amount of time spent by each family member.

Husbands	Wives	Children
10 hours	**32** hours	**6** hours

Source: *Journal of Marriage and the Family.*

46. A survey asked people whether certain types of advertising were believable. Out of every 100 people in the survey, here is the number who said the advertising was believable.

Type of Advertising	Number Who Said Advertising Was Believable
Computer software	35
Pharmaceutical companies	28
Auto manufacturers	18
Insurance companies	15

Source: Porter Novella.

What fraction of the people said each type of advertising was believable?

47. Explain the error in each of these problems and correct it.

    (a) $\dfrac{9}{36} = \dfrac{\cancel{3} \cdot \cancel{3}}{2 \cdot 2 \cdot \cancel{3} \cdot \cancel{3}} = 4$

    (b) $\dfrac{9}{16} = \dfrac{9 \div 3}{16 \div 4} = \dfrac{3}{4}$

48. (a) Explain how you could use your calculator to find the prime factorization of 437. Then find the prime factorization.

    (b) The text lists all the prime numbers less than 100. Use the divisibility rules and your calculator to find at least five prime numbers between 100 and 150.

Write each fraction in lowest terms. See Example 7.

49. $\dfrac{16c}{40}$

50. $\dfrac{36}{54a}$

51. $\dfrac{20x}{35x}$

52. $\dfrac{21n}{28n}$

53. $\dfrac{18r^2}{15rs}$

54. $\dfrac{18ab}{48b^2}$

55. $\dfrac{6m}{42mn^2}$

56. $\dfrac{10g^2}{90g^2h}$

57. $\dfrac{9x^2}{16y^2}$

58. $\dfrac{5rst}{8st}$

59. $\dfrac{7xz}{9xyz}$

60. $\dfrac{6a^3}{23b^3}$

61. $\dfrac{21k^3}{6k^2}$

62. $\dfrac{16x^3}{12x^4}$

63. $\dfrac{13a^2bc^3}{39a^2bc^3}$

64. $\dfrac{22m^3n^4}{55m^3n^4}$

65. $\dfrac{14c^2d}{14cd^2}$

66. $\dfrac{19rs}{19s^3}$

67. $\dfrac{210ab^3c}{35b^2c^2}$

68. $\dfrac{81w^4xy^2}{300wy^4}$

69. $\dfrac{25m^3rt^2}{36n^2s^3w^2}$

70. $\dfrac{42a^5b^4c^3}{7a^4b^3c^2}$

71. $\dfrac{33e^2fg^3}{11efg}$

72. $\dfrac{21xy^2z^3}{17ab^2c^3}$

4.3 Multiplying and Dividing Signed Fractions

OBJECTIVE 1   Multiply signed fractions.   Suppose that you give $\frac{1}{3}$ of your candy bar to your friend Ann. Then Ann gives $\frac{1}{2}$ of her share to Tim. How much of the bar does Tim get to eat?

$\frac{1}{3}$ $\frac{1}{3}$ $\frac{1}{3}$ to Ann

A sketch of the candy bar shows that Tim will get $\frac{1}{6}$ of the bar.

$\frac{1}{2}$ of $\frac{1}{3}$ to Tim

Tim's share is $\frac{1}{2}$ **of** $\frac{1}{3}$ candy bar. When used with fractions, the word **of** indicates multiplication.

$$\frac{1}{2} \text{ of } \frac{1}{3} \quad \text{means} \quad \frac{1}{2} \cdot \frac{1}{3}$$

Tim's share is $\frac{1}{6}$ bar, so $\frac{1}{2} \cdot \frac{1}{3} = \frac{1}{6}$.

This example illustrates the rule for multiplying fractions.

> **Multiplying Fractions**
>
> If a, b, c, and d are numbers (but b and d are not 0), then
>
> $$\frac{a}{b} \cdot \frac{c}{d} = \frac{a \cdot c}{b \cdot d}$$
>
> In other words, multiply the numerators and multiply the denominators.

When we apply this rule to find Tim's part of the candy bar, we get

$$\frac{1}{2} \cdot \frac{1}{3} = \frac{1 \cdot 1}{2 \cdot 3} = \frac{1}{6} \quad \leftarrow \text{Multiply numerators.} \atop \leftarrow \text{Multiply denominators.}$$

EXAMPLE 1   Multiplying Signed Fractions

Multiply.

(a) $-\dfrac{5}{8} \cdot -\dfrac{3}{4}$ Recall that the product of two negative numbers is a positive number.

Multiply the numerators and multiply the denominators.

$$-\frac{5}{8} \cdot -\frac{3}{4} = \frac{5 \cdot 3}{8 \cdot 4} = \frac{15}{32} \quad \leftarrow \text{Lowest terms}$$

The product of two negative numbers is positive.

The answer is in lowest terms because 15 and 32 have no common factor other than 1.

Continued on Next Page

(b) $\left(\dfrac{4}{7}\right)\left(-\dfrac{2}{5}\right) = -\dfrac{4 \cdot 2}{7 \cdot 5} = -\dfrac{8}{35}$ Recall that the product of a negative number and a positive number is negative.

Now Try Exercise 1.

Sometimes the result won't be in lowest terms. For example, find $\frac{3}{10}$ of $\frac{5}{6}$.

$$\dfrac{3}{10} \textbf{ of } \dfrac{5}{6} \quad \text{means} \quad \dfrac{3}{10} \cdot \dfrac{5}{6} = \dfrac{3 \cdot 5}{10 \cdot 6} = \dfrac{15}{60} \quad \left\{ \begin{array}{l} \text{Not in lowest} \\ \text{terms} \end{array} \right.$$

Now write $\frac{15}{60}$ in lowest terms.

$$\dfrac{15}{60} = \dfrac{\overset{1}{\cancel{3}} \cdot \overset{1}{\cancel{5}}}{2 \cdot 2 \cdot \underset{1}{\cancel{3}} \cdot \underset{1}{\cancel{5}}} = \dfrac{1}{4} \quad \leftarrow \text{Lowest terms}$$

You used prime factorization in **Section 4.2** to write fractions in lowest terms. You can also use it when multiplying fractions. Writing the prime factors of the original fractions and dividing out common factors *before* multiplying usually saves time. If you divide out *all* the common factors, the result will automatically be in lowest terms. Let's see how that works when finding $\frac{3}{10}$ of $\frac{5}{6}$.

3 and 5 are already prime.

Write 10 as 2 · 5

Write 6 as 2 · 3

Divide out the common factors.

$$\dfrac{3}{10} \cdot \dfrac{5}{6} = \dfrac{3 \cdot 5}{2 \cdot 5 \cdot 2 \cdot 3} = \dfrac{\overset{1}{\cancel{3}} \cdot \overset{1}{\cancel{5}}}{2 \cdot \underset{1}{\cancel{5}} \cdot 2 \cdot \underset{1}{\cancel{3}}} = \dfrac{1}{4} \quad \left\{ \begin{array}{l} \text{Same result} \\ \text{as above} \end{array} \right.$$

CAUTION

When you are working with fractions, always write the final result in lowest terms. Visualizing $\frac{15}{60}$ is hard to do. But when $\frac{15}{60}$ is written as $\frac{1}{4}$, working with it is much easier. A fraction is *simplified* when it is written in lowest terms.

EXAMPLE 2   **Using Prime Factorization to Multiply Fractions**

(a) $-\dfrac{8}{5}\left(\dfrac{5}{12}\right)$

Multiplying a negative number times a positive number gives a negative product.

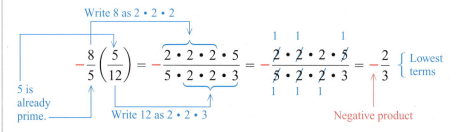

Write 8 as 2 · 2 · 2

5 is already prime.

Write 12 as 2 · 2 · 3

$$-\dfrac{8}{5}\left(\dfrac{5}{12}\right) = -\dfrac{2 \cdot 2 \cdot 2 \cdot 5}{5 \cdot 2 \cdot 2 \cdot 3} = -\dfrac{\overset{1}{\cancel{2}} \cdot \overset{1}{\cancel{2}} \cdot 2 \cdot \overset{1}{\cancel{5}}}{\underset{1}{\cancel{5}} \cdot \underset{1}{\cancel{2}} \cdot \underset{1}{\cancel{2}} \cdot 3} = -\dfrac{2}{3} \quad \left\{ \begin{array}{l} \text{Lowest} \\ \text{terms} \end{array} \right.$$

Negative product

Continued on Next Page

(b) Find $\dfrac{2}{9}$ of $\dfrac{15}{16}$.

Recall that, when used with fractions, *of* indicates multiplication.

2 is already prime.   Write 15 as 3 • 5

$$\dfrac{2}{9} \cdot \dfrac{15}{16} = \dfrac{2 \cdot 3 \cdot 5}{3 \cdot 3 \cdot 2 \cdot 2 \cdot 2 \cdot 2} = \dfrac{\overset{1}{\cancel{2}} \cdot \overset{1}{\cancel{3}} \cdot 5}{\cancel{3} \cdot 3 \cdot \cancel{2} \cdot 2 \cdot 2 \cdot 2} = \dfrac{5}{24} \quad \begin{cases}\text{Lowest}\\\text{terms}\end{cases}$$

Write 9 as 3 • 3

Write 16 as 2 • 2 • 2 • 2

> **Now Try Exercise 5.** ▶▶▶

▦ **Calculator Tip**   If your *scientific* calculator has a fraction key ⬭$^{a^{b/c}}$, you can do calculations with fractions. You'll also need the change of sign key ⬭$^{+⌀-}$ to enter negative fractions, as you did to enter negative integers in **Section 1.4.**

Start by entering several different fractions. Clear your calculator after each one.

To enter $\frac{3}{4}$, press 3 ⬭$^{a^{b/c}}$ 4.   The display will show **3 ⌐ 4**

↑
Fraction bar

To enter $-\frac{9}{10}$, press 9 ⬭$^{a^{b/c}}$ 10 ⬭$^{+⌀-}$.   The display will show **−9 ⌐ 10** .

Try entering a fraction that is *not* in lowest terms. As soon as you press an operation key, such as ⊗ or ⊘, most calculators will automatically show the fraction in lowest terms. Suppose that you start to enter the multiplication problem $\frac{4}{16} \cdot \frac{2}{3}$. Press 4 ⬭$^{a^{b/c}}$ 16 ⊗. The display shows **1 ⌐ 4** , or $\frac{1}{4}$, which is $\frac{4}{16}$ in lowest terms. The calculator will always show fractions in lowest terms.

Let's check the result of Example 2(a): Multiply $-\frac{8}{5}\left(\frac{5}{12}\right)$ by pressing

8 ⬭$^{a^{b/c}}$ 5 ⬭$^{+⌀-}$ ⊗ 5 ⬭$^{a^{b/c}}$ 12 ⊜.   The display shows **−2 ⌐ 3** .

$-\dfrac{8}{5}$ $\dfrac{5}{12}$ $-\dfrac{2}{3}$

Now try Example 2(b) above: Find $\frac{2}{9}$ of $\frac{15}{16}$. (Did you get $\frac{5}{24}$?)

There are some limitations to the calculations that you can do using the fraction key.

Try entering the fraction $\frac{9}{1000}$. What happens?

(You can't enter denominators >999.)

Try doing this multiplication: $\frac{7}{10} \cdot \frac{3}{100}$. The result should be $\frac{21}{1000}$. What happens?

(The answer is given in decimal form because the denominator is >999.)

CAUTION

The fraction key on a calculator is useful for *checking* your work. But knowing the rules for fraction computation is important because you'll need them when fractions involve variables. You *cannot* enter fractions such as $\dfrac{3x}{5}$ or $\dfrac{9}{m^2}$ on your calculator (see Example 4 below).

EXAMPLE 3 Multiplying a Fraction and an Integer

Find $\frac{2}{3}$ of 6.

We can write 6 in fraction form as $\frac{6}{1}$. Recall that $\frac{6}{1}$ means $6 \div 1$, which is 6. So we can write any integer a as $\frac{a}{1}$.

$$\frac{2}{3}\ \textbf{of}\ 6 \quad \text{means} \quad \frac{2}{3} \cdot \frac{6}{1} = \frac{2 \cdot 2 \cdot \overset{1}{\cancel{3}}}{\underset{1}{\cancel{3}} \cdot 1} = \frac{4}{1} = 4$$

$4 \div 1$ is 4

Now Try Exercise 9.

OBJECTIVE 2 Multiply fractions that involve variables. The multiplication method that uses prime factors also works when there are variables in the numerators and/or denominators of the fractions.

EXAMPLE 4 Multiplying Fractions with Variables

Multiply.

(a) $\dfrac{3x}{5} \cdot \dfrac{2}{9x}$

$3x$ means $3 \cdot x$.

$$\frac{3x}{5} \cdot \frac{2}{9x} = \frac{3 \cdot x \cdot 2}{5 \cdot \underbrace{3 \cdot 3 \cdot x}} = \frac{\overset{1}{\cancel{3}} \cdot \overset{1}{\cancel{x}} \cdot 2}{5 \cdot \underset{1}{\cancel{3}} \cdot 3 \cdot \underset{1}{\cancel{x}}} = \frac{2}{15}$$

$\dfrac{x}{x}$ is 1

The prime factors of 9 are $3 \cdot 3$, so $9x$ is $3 \cdot 3 \cdot x$.

(b) $\left(\dfrac{3y}{4x}\right)\left(\dfrac{2x^2}{y}\right)$

$2x^2$ means $2 \cdot x \cdot x$.

$$\left(\frac{3y}{4x}\right)\left(\frac{2x^2}{y}\right) = \frac{3 \cdot y \cdot 2 \cdot x \cdot x}{\underbrace{2 \cdot 2 \cdot x} \cdot y} = \frac{3 \cdot \overset{1}{\cancel{y}} \cdot \overset{1}{\cancel{2}} \cdot \overset{1}{\cancel{x}} \cdot x}{2 \cdot \underset{1}{\cancel{2}} \cdot \underset{1}{\cancel{x}} \cdot \underset{1}{\cancel{y}}} = \frac{3x}{2}$$

The prime factors of 4 are $2 \cdot 2$, so $4x$ is $2 \cdot 2 \cdot x$.

Now Try Exercise 11.

OBJECTIVE 3 Divide signed fractions. To divide fractions, you will rewrite division problems as multiplication problems. For division, you will leave the first number (the dividend) as it is, but change the second number (the divisor) to its *reciprocal*.

Reciprocal of a Fraction

Two numbers are **reciprocals** of each other if their product is 1.

The reciprocal of the fraction $\dfrac{a}{b}$ is $\dfrac{b}{a}$ because

$$\frac{a}{b} \cdot \frac{b}{a} = \frac{\overset{1}{\cancel{a}} \cdot \overset{1}{\cancel{b}}}{\underset{1}{\cancel{b}} \cdot \underset{1}{\cancel{a}}} = \frac{1}{1} = 1$$

Notice that you "flip" or "invert" a fraction to find its reciprocal. Here are some examples.

Number	Reciprocal	Reason
$\dfrac{1}{6}$	$\dfrac{6}{1}$	Because $\dfrac{1}{6} \cdot \dfrac{6}{1} = \dfrac{6}{6} = 1$
$-\dfrac{2}{5}$	$-\dfrac{5}{2}$	Because $\left(-\dfrac{2}{5}\right)\left(-\dfrac{5}{2}\right) = \dfrac{10}{10} = 1$
4 Think of 4 as $\frac{4}{1}$	$\dfrac{1}{4}$	Because $4 \cdot \dfrac{1}{4} = \dfrac{4}{1} \cdot \dfrac{1}{4} = \dfrac{4}{4} = 1$

NOTE

Every number has a reciprocal except 0. Why not 0? Recall that a number times its reciprocal equals 1. But that doesn't work for 0.

$$0 \cdot (\text{reciprocal}) \neq 1$$

⤷ Put any number here. When you multiply it by 0, you get 0, never 1

Dividing Fractions

If a, b, c, and d are numbers (but b, c, and d are not 0), then we have the following.

$$\frac{a}{b} \div \frac{c}{d} = \frac{a}{b} \cdot \frac{d}{c}$$

Reciprocals

In other words, change division to multiplying by the reciprocal of the divisor.

Use this method to find the quotient for $\frac{2}{3} \div \frac{1}{6}$. Rewrite it as a multiplication problem and then use the steps for multiplying fractions.

Change division to multiplication.

$$\frac{2}{3} \div \frac{1}{6} = \frac{2}{3} \cdot \frac{6}{1} = \frac{2 \cdot 2 \cdot \overset{1}{\cancel{3}}}{\underset{1}{\cancel{3}} \cdot 1} = \frac{4}{1} = 4$$

The reciprocal of $\frac{1}{6}$ is $\frac{6}{1}$

Does it make sense that $\frac{2}{3} \div \frac{1}{6} = 4$? Let's compare dividing fractions to dividing whole numbers.

$15 \div 3$ is asking, "How many 3s are in 15?"

$\frac{2}{3} \div \frac{1}{6}$ is asking, "How many $\frac{1}{6}$s are in $\frac{2}{3}$?"

The figure below illustrates $\frac{2}{3} \div \frac{1}{6}$.

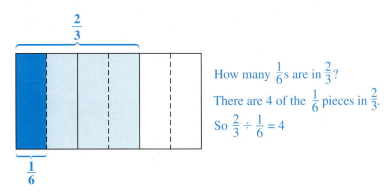

How many $\frac{1}{6}$s are in $\frac{2}{3}$?

There are 4 of the $\frac{1}{6}$ pieces in $\frac{2}{3}$.

So $\frac{2}{3} \div \frac{1}{6} = 4$

As a final check on this method of dividing, try changing $15 \div 3$ into a multiplication problem. You know that the quotient should be 5.

$$15 \div 3 = 15 \cdot \frac{1}{3} = \frac{15}{1} \cdot \frac{1}{3} = \frac{\overset{1}{\cancel{3}} \cdot 5 \cdot 1}{1 \cdot \underset{1}{\cancel{3}}} = \frac{5}{1} = 5 \quad \left\{ \begin{array}{l} \text{The quotient} \\ \text{we expected} \end{array} \right.$$

Reciprocals

So, you can see that *dividing* by a fraction is the same as *multiplying* by the *reciprocal* of the fraction.

EXAMPLE 5 Dividing Signed Fractions

Rewrite each division problem as a multiplication problem. Then multiply.

(a) $\dfrac{3}{10} \div \dfrac{4}{5} = \dfrac{3}{10} \cdot \dfrac{5}{4} = \dfrac{3 \cdot \overset{1}{\cancel{5}}}{2 \cdot \underset{1}{\cancel{5}} \cdot 2 \cdot 2} = \dfrac{3}{8}$

Reciprocals

> **NOTE**
> When multiplying fractions, you don't always have to factor the numerator and denominator completely into prime numbers. In part (a) above, if you notice that 5 is a common factor of the numerator and denominator, you can write
>
> $$\frac{3}{10} \cdot \frac{5}{4} = \frac{3 \cdot \overset{1}{\cancel{5}}}{2 \cdot \underset{1}{\cancel{5}} \cdot 4} = \frac{3}{8}$$
>
> Factor 10 into $2 \cdot 5$ Leave 4 as it is.
>
> If no common factors are obvious to you, then write out the complete prime factorization to help find the common factors.

Continued on Next Page

(b) $2 \div \left(-\dfrac{1}{3} \right)$

First notice that the numbers have different signs. In a division problem, different signs mean that the quotient is negative. Then write 2 in fraction form as $\frac{2}{1}$.

$$2 \div \left(-\frac{1}{3} \right) = \frac{2}{1} \cdot \left(-\frac{3}{1} \right) = -\frac{2 \cdot 3}{1 \cdot 1} = -\frac{6}{1} = -6$$

Reciprocals   Negative product

(c) $-\dfrac{3}{4} \div (-8) = -\dfrac{3}{4} \cdot \left(-\dfrac{1}{8} \right) = \dfrac{3 \cdot 1}{4 \cdot 8} = \dfrac{3}{32}$

Reciprocals   No common factor to divide out

The quotient is positive because both numbers in the problem were negative. When signs match, the quotient is positive.

(d) $\dfrac{9}{16} \div 0$

This is *undefined*, just as dividing by 0 was undefined for integers (see **Section 1.7**). Recall that 0 does *not* have a reciprocal, so you can't change the division to multiplying by the reciprocal of the divisor.

(e) $0 \div \dfrac{9}{16} = 0 \cdot \dfrac{16}{9} = 0$ Recall that 0 divided by any nonzero number gives a result of 0.

Reciprocals

Now Try Exercises 15 and 19. ▶▶▶

OBJECTIVE 4 Divide fractions that involve variables.   The method for dividing fractions also works when there are variables in the numerators and/or denominators of the fractions.

EXAMPLE 6 Dividing Fractions with Variables

Divide. (Assume that none of the variables represents zero.)

(a) $\dfrac{x^2}{y} \div \dfrac{x}{3y} = \dfrac{x^2}{y} \cdot \dfrac{3y}{x} = \dfrac{x \cdot x \cdot 3 \cdot y}{y \cdot x} = \dfrac{3x}{1} = 3x$

Reciprocals

(b) $\dfrac{8b}{5} \div b^2 = \dfrac{8b}{5} \cdot \dfrac{1}{b^2} = \dfrac{8 \cdot b \cdot 1}{5 \cdot b \cdot b} = \dfrac{8}{5b}$

Think of b^2 as $\dfrac{b^2}{1}$ so the reciprocal is $\dfrac{1}{b^2}$.

Now Try Exercise 21. ▶▶▶

OBJECTIVE **5** **Solve application problems involving multiplying and dividing fractions.** When you're solving application problems, some indicator words are used to suggest multiplication and some are used to suggest division.

INDICATOR WORDS FOR MULTIPLICATION	INDICATOR WORDS FOR DIVISION
product	per
double	each
triple	goes into
times	divided by
twice	divided into
of (when *of* follows a fraction)	divided equally

Look for these indicator words in the following examples. However, *you won't always find an indicator word.* Then, you need to think through the problem to decide what to do. Sometimes, drawing a sketch of the situation described in the problem will help you decide which operation to use.

EXAMPLE 7 **Using Indicator Words and Sketches to Solve Application Problems**

(a) Lois gives $\frac{1}{10}$ **of** her income to her church. Last month she earned $1980. How much of that did she give to her church?

Notice the word **of**. Because the word **of** *follows the fraction* $\frac{1}{10}$, it indicates multiplication.

$$\frac{1}{10} \quad \textbf{of} \quad 1980 = \frac{1}{10} \cdot \frac{1980}{1} = \frac{1 \cdot \overset{1}{\cancel{10}} \cdot 198}{\underset{1}{\cancel{10}} \cdot 1} = \frac{198}{1} = 198$$

Lois gave $198 to her church.

(b) The apparel design class is making infant snowsuits to give to a local shelter. A fabric store donated 12 yd of fabric for the project. If one snowsuit needs $\frac{2}{3}$ yd of fabric, how many suits can the class make?

The word **of** appears in the second sentence: "A fabric store donated 12 yd **of** fabric." But the word **of** does *not follow a fraction,* so it is *not* an indicator to multiply. Let's try a sketch. There is a 12 yd piece of fabric. One snowsuit will use $\frac{2}{3}$ yd. The question is, how many $\frac{2}{3}$ yd pieces can be cut from the 12 yards?

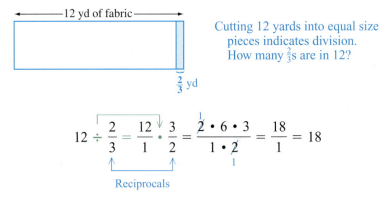

Cutting 12 yards into equal size pieces indicates division. How many $\frac{2}{3}$s are in 12?

$$12 \div \frac{2}{3} = \frac{12}{1} \cdot \frac{3}{2} = \frac{\overset{1}{\cancel{2}} \cdot 6 \cdot 3}{1 \cdot \underset{1}{\cancel{2}}} = \frac{18}{1} = 18$$

Reciprocals

The class can make 18 snowsuits.

Now Try Exercise 53.

4.3 Exercises

FOR EXTRA HELP Addison-Wesley Math Tutor Center   MathXL   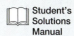 Digital Video Tutor CD 2 Videotape 4   📖 Student's Solutions Manual   MyMathLab   MathXL Tutorials on CD

Multiply. Write the products in lowest terms. See Example 1–4.

1. $-\dfrac{3}{8} \cdot \dfrac{1}{2}$

2. $\left(\dfrac{2}{3}\right)\left(-\dfrac{5}{7}\right)$

3. $\left(-\dfrac{3}{8}\right)\left(-\dfrac{12}{5}\right)$

4. $\dfrac{4}{9} \cdot \dfrac{12}{7}$

5. $\dfrac{21}{30}\left(\dfrac{5}{7}\right)$

6. $\left(-\dfrac{6}{11}\right)\left(-\dfrac{22}{15}\right)$

7. $10\left(-\dfrac{3}{5}\right)$

8. $-20\left(\dfrac{3}{4}\right)$

9. $\dfrac{4}{9}$ of 81

10. $\dfrac{2}{3}$ of 48

11. $\left(\dfrac{3x}{4}\right)\left(\dfrac{5}{xy}\right)$

12. $\left(\dfrac{2}{5a^2}\right)\left(\dfrac{a}{8}\right)$

Divide. Write the quotients in lowest terms. See Examples 5 and 6.

13. $\dfrac{1}{6} \div \dfrac{1}{3}$

14. $-\dfrac{1}{2} \div \dfrac{2}{3}$

15. $-\dfrac{3}{4} \div \left(-\dfrac{5}{8}\right)$

16. $\dfrac{7}{10} \div \dfrac{2}{5}$

17. $6 \div \left(-\dfrac{2}{3}\right)$

18. $-7 \div \left(-\dfrac{1}{4}\right)$

19. $-\dfrac{2}{3} \div 4$

20. $\dfrac{5}{6} \div (-15)$

21. $\dfrac{11c}{5d} \div 3c$

22. $8x^2 \div \dfrac{4x}{7}$

23. $\dfrac{ab^2}{c} \div \dfrac{ab}{c}$

24. $\dfrac{mn}{6} \div \dfrac{n}{3m}$

25. Explain and correct the error in each of these calculations.

(a) $\dfrac{3}{14} \cdot \dfrac{7}{9} = \dfrac{\cancel{3} \cdot \cancel{7}}{2 \cdot \cancel{7} \cdot \cancel{3} \cdot 3} = 6$

(b) $8 \cdot \dfrac{2}{3} = \dfrac{8}{1} \cdot \dfrac{3}{2} = \dfrac{\overset{1}{\cancel{2}} \cdot 4 \cdot 3}{1 \cdot \underset{1}{\cancel{2}}} = \dfrac{12}{1} = 12$

26. Explain and correct the error in each of these calculations.

(a) $\dfrac{3}{4} \cdot \dfrac{8}{9} = \dfrac{3 \cdot 8}{4 \cdot 9} = \dfrac{24}{36}$

(b) $\dfrac{2}{5} \cdot \dfrac{3}{8} = \dfrac{\cancel{2}}{5} \cdot \dfrac{3}{\underset{2}{\cancel{8}}}\overset{1}{} = \dfrac{3}{10}$

27. Explain and correct the error in each of these calculations.

(a) $\dfrac{2}{3} \div 4 = \dfrac{2}{3} \cdot \dfrac{4}{1} = \dfrac{2 \cdot 4}{3 \cdot 1} = \dfrac{8}{3}$

(b) $\dfrac{5}{6} \div \dfrac{10}{9} = \dfrac{6}{5} \cdot \dfrac{10}{9} = \dfrac{2 \cdot \cancel{3}^{1} \cdot 2 \cdot \cancel{5}^{1}}{\cancel{5}_{1} \cdot \cancel{3}_{1} \cdot 3} = \dfrac{4}{3}$

28. Explain and correct the error in each of these calculations.

(a) $\dfrac{1}{2} \div 0 = 0$

(b) $\dfrac{\cancel{3}^{1}}{10} \div \dfrac{1}{\cancel{3}_{1}} = \dfrac{1}{10}$

29. Your friend missed class and is confused about how to divide fractions. Write a short explanation for your friend.

30. Mary spilled coffee on her math homework, and part of one problem is covered up.

$$\blacksquare \div \dfrac{4}{5}$$

She knows the answer given in the back of the book is $\frac{3}{4}$. Describe how to find the missing number.

Find each product or quotient. Write all answers in lowest terms. See Examples 1–6.

31. $\dfrac{4}{5} \div 3$

32. $\left(-\dfrac{20}{21}\right)\left(-\dfrac{14}{15}\right)$

33. $-\dfrac{3}{8}\left(\dfrac{3}{4}\right)$

34. $-\dfrac{8}{17} \div \dfrac{4}{5}$

35. $\dfrac{3}{5}$ of 35

36. $\dfrac{2}{3} \div (-6)$

37. $-9 \div \left(-\dfrac{3}{5}\right)$

38. $\dfrac{7}{8} \cdot \dfrac{25}{21}$

39. $\dfrac{12}{7} \div 0$

40. $\dfrac{5}{8}$ of (-48)

41. $\left(\dfrac{11}{2}\right)\left(-\dfrac{5}{6}\right)$

42. $\dfrac{3}{4} \div \dfrac{3}{16}$

43. $\dfrac{4}{7}$ of $14b$

44. $\dfrac{ab}{6} \div \dfrac{b}{9}$

45. $\dfrac{12}{5} \div 4d$

46. $\dfrac{18}{7} \div 2t$

47. $\dfrac{x^2}{y} \div \dfrac{w}{2y}$

48. $\dfrac{5}{6}$ of $18w$

Solve each application problem. See Example 7.

49. Al is helping Tim make a mahogany lamp table for Jill's birthday. Find the area of the rectangular top of the table if it is $\frac{4}{5}$ yd long by $\frac{3}{8}$ yd wide.

50. A rectangular dog bed is $\frac{7}{8}$ yd by $\frac{10}{9}$ yd. Find its area.

51. How many $\frac{1}{8}$-ounce eyedrop dispensers can be filled from a container that holds 10 ounces of eyedrops?

52. Ms. Shaffer has a piece of property with an area that is $\frac{9}{10}$ acre. She wishes to divide it into three equal parts for her children. How many acres of land will each child get?

53. Todd estimates that it will cost him $12,400 to attend a community college for one year. He thinks he can earn $\frac{3}{4}$ of the cost and borrow the balance. Find the amount he must earn and the amount he must borrow.

54. Joyce Chen wants to make vests to sell at a craft fair. Each vest requires $\frac{3}{4}$ yd of material. She has 36 yd of material. Find the number of vests she can make.

55. Pam Trizlia has a small pickup truck that can carry $\frac{2}{3}$ cord of firewood. Find the number of trips needed to deliver 6 cords of wood.

56. At the Garlic Festival Fun Run, $\frac{5}{12}$ of the runners are women. If there are 780 runners, how many are women? How many are men?

57. There are about 210 baseball players in the Baseball Hall of Fame. One-third of these men played at infield positions (1st base, 2nd base, short stop, or 3rd base). How many infield players are in the Hall of Fame? (*Source:* National Baseball Hall of Fame.)

58. Parking lot A is $\frac{1}{4}$ mile long and $\frac{3}{16}$ mile wide, and parking lot B is $\frac{3}{8}$ mile long and $\frac{1}{8}$ mile wide. Which parking lot has the larger area?

59. An adult male alligator may weigh 400 pounds. A newly hatched alligator weighs only $\frac{1}{8}$ pound. The adult weighs how many times the hatchling? (*Source:* St. Marks NWR.)

60. A female alligator lays about 35 eggs in a nest of marsh grass and mud. But $\frac{4}{5}$ of the eggs or hatchlings will fall prey to raccoons, wading birds, or larger alligators. How many of the 35 eggs will hatch and survive? (*Source:* St. Marks NWR.)

The table below shows the income for the Gomez family last year and the circle graph shows how they spent their income. Use this information to work Exercises 61–64.

Month	Income	Month	Income
January	$4575	July	$5540
February	$4312	August	$3732
March	$4988	September	$4170
April	$4530	October	$5512
May	$4320	November	$4965
June	$4898	December	$6458

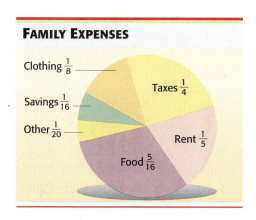

FAMILY EXPENSES

Clothing $\frac{1}{8}$

Savings $\frac{1}{16}$

Other $\frac{1}{20}$

Taxes $\frac{1}{4}$

Rent $\frac{1}{5}$

Food $\frac{5}{16}$

61. (a) What was the family's total income for the year?

   (b) Find the amount of the family's rent for the year.

62. (a) How much did the family pay in taxes during the year?

   (b) How much more did the family spend on taxes than on rent?

63. How much did the family spend for food and clothing last year?

64. Find the amount the family saved during the year.

There are a total of about 150 million companion pets in the United States. This table shows the fraction of pets that are dogs, cats, birds, and horses. Use the table to answer Exercises 65–68.

COMPANION PETS IN THE UNITED STATES

Type of Pet	Fraction of All Pets
Dog	$\frac{2}{5}$
Cat	$\frac{23}{50}$
Bird	$\frac{1}{10}$
Horse	$\frac{3}{100}$

Source: American Veterinary Medical Association.

65. How many U.S. pets are birds?

66. How many U.S. pets are dogs?

67. How many dogs and cats are pets?

68. How many birds and cats are pets?

4.4 Adding and Subtracting Signed Fractions

OBJECTIVE **1** **Add and subtract like fractions.** You probably remember learning something about "common denominators" in other math classes. When fractions have the *same* denominator, we say that they have a *common* denominator, which makes them **like fractions.** When fractions have different denominators, they are called **unlike fractions.** Here are some examples.

Like Fractions	**Unlike Fractions**
Common denominator	Different denominators
Common denominator	Different denominators

You can add or subtract fractions *only* when they have a common denominator. To see why, let's look at more pizzas.

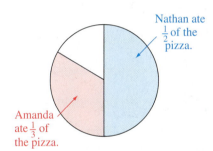

Nathan ate $\frac{1}{2}$ of the pizza.

Amanda ate $\frac{1}{3}$ of the pizza.

What fraction of the pizza has been eaten? We can't write a fraction until the pizza is cut into pieces of *equal* size. That's what the denominator of a fraction tell us: the number of *equal* size pieces in the pizza.

Now the pizza is cut into 6 *equal* pieces, and we can find out how much was eaten.

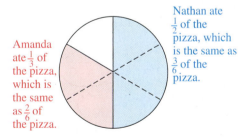

Amanda ate $\frac{1}{3}$ of the pizza, which is the same as $\frac{2}{6}$ of the pizza.

Nathan ate $\frac{1}{2}$ of the pizza, which is the same as $\frac{3}{6}$ of the pizza.

$$\frac{1}{2} + \frac{1}{3}$$
$$\downarrow \qquad \downarrow$$
$$\frac{3}{6} + \frac{2}{6}$$
$$= \frac{5}{6} \; \leftarrow \text{Number of pieces eaten}$$
$$\qquad\; \leftarrow \text{Number of equal pieces in the pizza}$$

Adding and Subtracting Like Fractions

You can add or subtract fractions *only* when they have a common denominator. If a, b, and c are numbers (and b is not 0), then

$$\frac{a}{b} + \frac{c}{b} = \frac{a + c}{b} \qquad \text{and} \qquad \frac{a}{b} - \frac{c}{b} = \frac{a - c}{b}$$

In other words, add or subtract the numerators and write the result over the common denominator. Then check to be sure that the answer is in lowest terms.

> **EXAMPLE 1** Adding and Subtracting Like Fractions

Find each sum or difference.

(a) $\dfrac{1}{8} + \dfrac{3}{8}$

These are *like* fractions because they have a *common denominator of 8*. So they are ready to be added. Add the numerators and write the sum over the common denominator.

$$\frac{1}{8} + \frac{3}{8} = \frac{1+3}{8} = \frac{4}{8}$$

Common denominator

Now write $\frac{4}{8}$ in lowest terms.

$$\frac{4}{8} = \frac{\overset{1}{\cancel{2}} \cdot \overset{1}{\cancel{2}}}{\underset{1}{\cancel{2}} \cdot \underset{1}{\cancel{2}} \cdot 2} = \frac{1}{2}$$

The sum, in lowest terms, is $\frac{1}{2}$.

> **CAUTION**
> Add *only* the numerators. ***Do not add the denominators.*** In part (a) above we ***kept the common denominator.***
>
> $$\frac{1}{8} + \frac{3}{8} = \frac{1+3}{8} \qquad \textbf{not} \qquad \frac{1}{8} + \frac{3}{8} = \frac{1+3}{8+8} = \frac{4}{16}$$
>
> Incorrect
>
> To help you understand why we add *only* the numerators, think of $\frac{1}{8}$ as $1\left(\frac{1}{8}\right)$ and $\frac{3}{8}$ as $3\left(\frac{1}{8}\right)$. Then we use the distributive property.
>
> $$\frac{1}{8} + \frac{3}{8} = 1\left(\frac{1}{8}\right) + 3\left(\frac{1}{8}\right) = (1+3)\left(\frac{1}{8}\right) = 4\left(\frac{1}{8}\right) = \frac{4}{8} = \frac{1}{2}$$
>
> Use the distributive property.
>
> Or think about a pie cut into 8 equal pieces. If you eat 1 piece and your friend eats 3 pieces, together you've eaten 4 of the 8 pieces or $\frac{4}{8}$ of the pie (***not*** $\frac{4}{16}$ of the pie, which would be 4 out of 16 pieces).

(b) $-\dfrac{3}{5} + \dfrac{4}{5} = \dfrac{-3+4}{5} = \dfrac{1}{5}$ ← Lowest terms

Common denominator

Rewrite subtraction as adding the opposite.

(c) $\dfrac{3}{10} - \dfrac{7}{10} = \dfrac{3-7}{10} = \dfrac{3+(-7)}{10} = \dfrac{-4}{10}$ or $-\dfrac{4}{10}$

Common denominator

Now write $-\frac{4}{10}$ in lowest terms.

$$-\frac{4}{10} = -\frac{\overset{1}{\cancel{2}} \cdot 2}{\underset{1}{\cancel{2}} \cdot 5} = -\frac{2}{5} \quad ← \text{Lowest terms}$$

(d) $\dfrac{5}{x^2} - \dfrac{2}{x^2} = \dfrac{5-2}{x^2} = \dfrac{3}{x^2}$

Common denominator

◀◀◀ **Now Try Exercises 11 and 23.**

OBJECTIVE 2 Find the lowest common denominator for unlike fractions.   When we first tried to add the pizza eaten by Nathan and Amanda, we could *not* do so because the pizza was *not cut into pieces of the same size*. So we rewrote $\frac{1}{2}$ and $\frac{1}{3}$ as equivalent fractions that both had 6 as the common denominator.

$$\frac{1}{2} = \frac{1 \cdot 3}{2 \cdot 3} = \frac{3}{6}$$
$$\frac{1}{3} = \frac{1 \cdot 2}{3 \cdot 2} = \frac{2}{6}$$

Common denominator of 6

Then we could add the fractions, because the pizza was cut into pieces of *equal* size.

$$\frac{1}{2} + \frac{1}{3} = \frac{3}{6} + \frac{2}{6} = \frac{3 + 2}{6} = \frac{5}{6} \leftarrow \text{Lowest terms}$$

In other words, when you want to add or subtract unlike fractions, the first thing you must do is rewrite them so that they have a common denominator.

> **A Common Denominator for Unlike Fractions**
>
> To find a common denominator for two unlike fractions, find a number that is divisible by *both* of the original denominators.
>   For example, a common denominator for $\frac{1}{2}$ and $\frac{1}{3}$ is 6 because 2 goes into 6 evenly and 3 goes into 6 evenly.

Notice that 12 is also a common denominator for $\frac{1}{2}$ and $\frac{1}{3}$ because 2 and 3 both go into 12 evenly.

$$\frac{1}{2} = \frac{1 \cdot 6}{2 \cdot 6} = \frac{6}{12}$$
$$\frac{1}{3} = \frac{1 \cdot 4}{3 \cdot 4} = \frac{4}{12}$$

Common denominator of 12

Now that the fractions have a common denominator, we can add them.

$$\frac{1}{2} + \frac{1}{3} = \frac{6}{12} + \frac{4}{12} = \frac{6 + 4}{12} = \frac{10}{12} \; \left\{ \begin{array}{l} \text{Not in} \\ \text{lowest} \\ \text{terms} \end{array} \right. \text{but} \quad \frac{10}{12} = \frac{\overset{1}{\cancel{2}} \cdot 5}{\underset{1}{\cancel{2}} \cdot 6} = \frac{5}{6} \; \left\{ \begin{array}{l} \text{Same} \\ \text{result as} \\ \text{above} \end{array} \right.$$

Both 6 and 12 worked as common denominators for adding $\frac{1}{2}$ and $\frac{1}{3}$, but using the smaller number saved some work. You should always try to find the smallest common denominator. If you don't for some reason, you can still work the problem—but it may take you longer. You'll have to divide out some common factors at the end in order to write the answer in lowest terms.

> **Least Common Denominator (LCD)**
>
> The **least common denominator** (LCD) for two fractions is the *smallest* positive number divisible by both denominators of the original fractions. For example, both 6 and 12 are common denominators for $\frac{1}{2}$ and $\frac{1}{3}$, but 6 is smaller, so it is the LCD.

There are several ways to find the LCD. When the original denominators are small numbers, you can often find the LCD by inspection. *Hint:* Always check to see if the larger denominator will work as the LCD.

EXAMPLE 2 Finding the LCD by Inspection

(a) Find the LCD for $\frac{2}{3}$ and $\frac{1}{9}$.

Check to see if 9 (the larger denominator) will work as the LCD. Is 9 divisible by 3 (the other denominator)? Yes, so 9 is the LCD for $\frac{2}{3}$ and $\frac{1}{9}$.

(b) Find the LCD for $\frac{5}{8}$ and $\frac{5}{6}$.

Check to see if 8 (the larger denominator) will work. No, 8 is not divisible by 6. So start checking numbers that are multiples of 8, that is, 16, 24, and 32. Notice that 24 will work because it is divisible by 8 and by 6.

The LCD for $\frac{5}{8}$ and $\frac{5}{6}$ is 24.

Now Try Exercises 1(a) and 1(b).

For larger denominators, you can use prime factorization to find the LCD. Factor each denominator completely into prime numbers. Then use the factors to build the LCD.

EXAMPLE 3 Using Prime Factors to Find the LCD

(a) What is the LCD for $\frac{7}{12}$ and $\frac{13}{18}$?

Write 12 and 18 as the product of prime factors. Then use prime factors in the LCD that "cover" both 12 and 18.

$$12 = 2 \cdot 2 \cdot 3$$
$$18 = 2 \cdot 3 \cdot 3$$

Factors of 12

$$\text{LCD} = 2 \cdot 2 \cdot 3 \cdot 3 = 36$$

Factors of 18

Check whether 36 is divisible by 12 (yes) and by 18 (yes). So 36 is the LCD for $\frac{7}{12}$ and $\frac{13}{18}$.

> **CAUTION**
> When finding the LCD, notice that we did *not* have to repeat the factors that 12 and 18 have in common. If we had used *all* the 2s and 3s, we would get a common denominator, but not the *smallest* one.

(b) What is the LCD for $\frac{11}{15}$ and $\frac{9}{70}$?

$$15 = 3 \cdot 5$$
$$70 = 2 \cdot 5 \cdot 7$$

Factors of 15

$$\text{LCD} = 3 \cdot 5 \cdot 2 \cdot 7 = 210$$

Factors of 70

Check whether 210 is divisible by 15 (yes) and divisible by 70 (yes). So 210 is the LCD for $\frac{11}{15}$ and $\frac{9}{70}$.

Now Try Exercise 3(a).

OBJECTIVE 3 **Add and subtract unlike fractions.**   Here are the steps for adding or subtracting unlike fractions. The key idea is that you must rewrite the fractions so that they have a common denominator before you can add or subtract them.

Adding and Subtracting Unlike Fractions

Step 1 Find the LCD, the smallest number divisible by both denominators in the problem.

Step 2 Rewrite each original fraction as an equivalent fraction whose denominator is the LCD.

Step 3 Add or subtract the numerators of the like fractions. Keep the common denominator.

Step 4 Write the sum or difference in lowest terms.

EXAMPLE 4 Adding and Subtracting Unlike Fractions

Find each sum or difference.

(a) $\dfrac{1}{5} + \dfrac{3}{10}$

Step 1 The larger denominator (10) is the LCD.

Step 2 $\dfrac{1}{5} = \dfrac{1 \cdot 2}{5 \cdot 2} = \dfrac{2}{10} \leftarrow \text{LCD}$ and $\dfrac{3}{10}$ already has the LCD.

Step 3 Add the numerators. Write the sum over the common denominator.

$$\frac{1}{5} + \frac{3}{10} = \frac{2}{10} + \frac{3}{10} = \frac{2 + 3}{10} = \frac{5}{10}$$

Step 4 Write $\frac{5}{10}$ in lowest terms.

$$\frac{5}{10} = \frac{\cancel{5}^{1}}{2 \cdot \cancel{5}_{1}} = \frac{1}{2} \leftarrow \text{Lowest terms}$$

(b) $\dfrac{3}{4} - \dfrac{5}{6}$

Step 1 The LCD is 12.

Step 2 $\dfrac{3}{4} = \dfrac{3 \cdot 3}{4 \cdot 3} = \dfrac{9}{12} \leftarrow \text{LCD}$ and $\dfrac{5}{6} = \dfrac{5 \cdot 2}{6 \cdot 2} = \dfrac{10}{12} \leftarrow \text{LCD}$

Step 3 Subtract the numerators. Write the difference over the common denominator.

$$9 + (-10) \text{ is } -1$$

$$\frac{3}{4} - \frac{5}{6} = \frac{9}{12} - \frac{10}{12} = \frac{9 - 10}{12} = \frac{-1}{12} \quad \text{or} \quad -\frac{1}{12}$$

Step 4 $-\frac{1}{12}$ is in lowest terms.

Continued on Next Page

(c) $-\dfrac{5}{12} + \dfrac{5}{9}$

Step 1 Use prime factorization to find the LCD.

$$12 = 2 \cdot 2 \cdot 3$$
$$9 = 3 \cdot 3$$

Factors of 12

$$\text{LCD} = 2 \cdot 2 \cdot 3 \cdot 3 = 36$$

Factors of 9

Step 2 $-\dfrac{5}{12} = -\dfrac{5 \cdot 3}{12 \cdot 3} = -\dfrac{15}{36}$ and $\dfrac{5}{9} = \dfrac{5 \cdot 4}{9 \cdot 4} = \dfrac{20}{36}$

Step 3 Add the numerators. Keep the common denominator.

$$-\dfrac{5}{12} + \dfrac{5}{9} = -\dfrac{15}{36} + \dfrac{20}{36} = \dfrac{-15 + 20}{36} = \dfrac{5}{36}$$

Step 4 $\dfrac{5}{36}$ is in lowest terms.

(d) $4 - \dfrac{2}{3}$

Step 1 Think of 4 as $\dfrac{4}{1}$. The LCD for $\dfrac{4}{1}$ and $\dfrac{2}{3}$ is 3, the larger denominator.

Step 2 $\dfrac{4}{1} = \dfrac{4 \cdot 3}{1 \cdot 3} = \dfrac{12}{3}$ and $\dfrac{2}{3}$ already has the LCD.

Step 3 Subtract the numerators. Keep the common denominator.

$$\dfrac{4}{1} - \dfrac{2}{3} = \dfrac{12}{3} - \dfrac{2}{3} = \dfrac{12 - 2}{3} = \dfrac{10}{3}$$

Step 4 $\dfrac{10}{3}$ is in lowest terms.

Now Try Exercises 15 and 17.

OBJECTIVE 4 Add and subtract unlike fractions that contain variables.
We use the same steps to add or subtract unlike fractions with variables in the numerators or denominators.

EXAMPLE 5 Adding and Subtracting Unlike Fractions with Variables

Find each sum or difference.

(a) $\dfrac{1}{4} + \dfrac{b}{5}$

Step 1 The LCD is 20.

Step 2 $\dfrac{1}{4} = \dfrac{1 \cdot 5}{4 \cdot 5} = \dfrac{5}{20}$ and $\dfrac{b}{5} = \dfrac{b \cdot 4}{5 \cdot 4} = \dfrac{4b}{20}$

Step 3 $\dfrac{1}{4} + \dfrac{b}{5} = \dfrac{5}{20} + \dfrac{4b}{20} = \dfrac{5 + 4b}{20}$ ← Add the numerators.
← Keep the common denominator.

Step 4 $\dfrac{5 + 4b}{20}$ is in lowest terms.

Continued on Next Page

CAUTION
In part (a), we could **not** add $5 + 4b$ in the numerator of the answer because 5 and $4b$ are **not** like terms. We *could* add $5b + 4b$ but **not** $5 + 4b$.

Variable parts match.

(b) $\dfrac{2}{3} - \dfrac{6}{x}$

Step 1 The LCD is $3 \cdot x$, or $3x$.

Step 2 $\dfrac{2}{3} = \dfrac{2 \cdot x}{3 \cdot x} = \dfrac{2x}{3x}$ and $\dfrac{6}{x} = \dfrac{6 \cdot 3}{x \cdot 3} = \dfrac{18}{3x}$

Step 3 $\dfrac{2}{3} - \dfrac{6}{x} = \dfrac{2x}{3x} - \dfrac{18}{3x} = \dfrac{2x - 18}{3x}$ ← Subtract the numerators.
← Keep the common denominator.

Step 4 $\dfrac{2x - 18}{3x}$ is in lowest terms.

NOTE
Notice in part (b) above that we found the LCD for $\frac{2}{3} - \frac{6}{x}$ by multiplying the two denominators. The LCD is $3 \cdot x$ or $3x$.

Multiplying the two denominators will *always* give you a common denominator, but it may not be the *smallest* common denominator. Here are more examples.

$\dfrac{1}{3} - \dfrac{2}{5}$ If you multiply the denominators, $3 \cdot 5 = 15$ and 15 is the LCD.

$\dfrac{5}{6} + \dfrac{3}{4}$ If you multiply the denominators, $6 \cdot 4 = 24$ and 24 will work. But you'll save some time by using the smallest common denominator, which is 12.

$\dfrac{7}{y} + \dfrac{a}{4}$ If you multiply the denominators, $y \cdot 4 = 4y$ and $4y$ is the LCD.

Now Try Exercises 19 and 21. ▶▶▶

4.4 Exercises

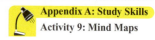
Find the LCD for each pair of fractions by inspection. See Example 2.

1. (a) $\dfrac{2}{5}$ and $\dfrac{1}{2}$ **(b)** $\dfrac{5}{6}$ and $\dfrac{5}{18}$ **(c)** $\dfrac{3}{10}$ and $\dfrac{1}{4}$ **(d)** $\dfrac{1}{3}$ and $\dfrac{6}{7}$

2. (a) $\dfrac{2}{3}$ and $\dfrac{2}{15}$ **(b)** $\dfrac{3}{4}$ and $\dfrac{1}{3}$ **(c)** $\dfrac{1}{8}$ and $\dfrac{1}{10}$ **(d)** $\dfrac{4}{9}$ and $\dfrac{5}{6}$

Use prime factorization to find the LCD for each pair of fractions. See Example 3.

3. (a) $\dfrac{5}{6}$ and $\dfrac{7}{16}$ **(b)** $\dfrac{11}{15}$ and $\dfrac{11}{12}$ **(c)** $\dfrac{1}{24}$ and $\dfrac{10}{21}$

4. (a) $\dfrac{5}{12}$ and $\dfrac{5}{9}$ **(b)** $\dfrac{1}{15}$ and $\dfrac{1}{18}$ **(c)** $\dfrac{17}{22}$ and $\dfrac{3}{20}$

Find each sum or difference. Write all answers in lowest terms. See Examples 1, 4, and 5.

5. $\dfrac{2}{3} - \dfrac{1}{6}$ **6.** $\dfrac{5}{12} - \dfrac{1}{4}$ **7.** $\dfrac{3}{4} + \dfrac{1}{8}$

8. $\dfrac{1}{3} + \dfrac{1}{2}$ **9.** $\dfrac{3}{8} - \dfrac{3}{5}$ **10.** $\dfrac{1}{3} - \dfrac{3}{5}$

11. $-\dfrac{7}{20} - \dfrac{5}{20}$ **12.** $-\dfrac{7}{9} - \dfrac{5}{6}$ **13.** $-\dfrac{3}{10} + 1$

14. $-\dfrac{7}{8} + 3$ **15.** $2 - \dfrac{6}{7}$ **16.** $5 - \dfrac{2}{5}$

17. $-\dfrac{1}{2} + \dfrac{3}{24}$ **18.** $\dfrac{7}{10} + \dfrac{7}{15}$ **19.** $\dfrac{1}{5} + \dfrac{c}{3}$

20. $\dfrac{x}{4} + \dfrac{2}{3}$

21. $\dfrac{5}{m} - \dfrac{1}{2}$

22. $\dfrac{2}{9} - \dfrac{4}{y}$

23. $\dfrac{3}{b^2} + \dfrac{5}{b^2}$

24. $\dfrac{10}{xy} - \dfrac{7}{xy}$

25. $\dfrac{c}{7} + \dfrac{3}{b}$

26. $\dfrac{2}{x} - \dfrac{y}{5}$

27. $-\dfrac{4}{c^2} - \dfrac{d}{c}$

28. $-\dfrac{1}{n} + \dfrac{m}{n^2}$

29. $-\dfrac{11}{42} - \dfrac{11}{70}$

30. $\dfrac{7}{45} - \dfrac{7}{20}$

31. A key step in adding or subtracting unlike fractions is to rewrite the fractions so that they have a common denominator. Explain why this step is necessary.

32. Explain how to write a fraction with an indicated denominator. As part of your explanation, show how to rewrite $\frac{3}{4}$ as an equivalent fraction with a denominator of 12.

33. Explain the error in each calculation and correct it.

(a) $\dfrac{3}{4} + \dfrac{2}{5} = \dfrac{3+2}{4+5} = \dfrac{5}{9}$

(b) $\dfrac{5}{6} - \dfrac{4}{9} = \dfrac{5}{18} - \dfrac{4}{18} = \dfrac{5-4}{18} = \dfrac{1}{18}$

34. Explain the error in each calculation and correct it.

(a) $-\dfrac{1}{4} + \dfrac{7}{12} = -\dfrac{3}{12} + \dfrac{7}{12} = \dfrac{-3+7}{12} = \dfrac{4}{12}$

(b) $\dfrac{3}{10} - \dfrac{1}{4} = \dfrac{3-1}{10-4} = \dfrac{2}{6} = \dfrac{1}{3}$

RELATING CONCEPTS (EXERCISES 35–36) For Individual or Group Work

*As you **work Exercises 35 and 36 in order,** think about the properties you learned when working with integers. Then explain what each pair of problems illustrates. (For a review of the properties, see the **Chapter 1 Summary.**)*

35. (a) $-\dfrac{2}{3} + \dfrac{3}{4} =$ _____ $\dfrac{3}{4} + \left(-\dfrac{2}{3}\right) =$ _____

   (b) $\dfrac{5}{6} - \dfrac{1}{2} =$ _____ $\dfrac{1}{2} - \dfrac{5}{6} =$ _____

   (c) $\left(-\dfrac{2}{3}\right)\left(\dfrac{9}{10}\right) =$ _____ $\left(\dfrac{9}{10}\right)\left(-\dfrac{2}{3}\right) =$ _____

   (d) $\dfrac{2}{5} \div \dfrac{1}{15} =$ _____ $\dfrac{1}{15} \div \dfrac{2}{5} =$ _____

36. (a) $-\dfrac{7}{12} + \dfrac{7}{12} =$ _____ $\dfrac{3}{5} + \left(-\dfrac{3}{5}\right) =$ _____

   (b) $-\dfrac{13}{16} \div \left(-\dfrac{13}{16}\right) =$ _____ $\dfrac{1}{8} \div \dfrac{1}{8} =$ _____

   (c) $\dfrac{5}{6} \cdot 1 =$ _____ $1\left(-\dfrac{17}{20}\right) =$ _____

   (d) $\left(-\dfrac{4}{5}\right)\left(-\dfrac{5}{4}\right) =$ _____ $7 \cdot \dfrac{1}{7} =$ _____

Solve each application problem. Write all answers in lowest terms.

37. When assembling shelving units, Pam Phelps must use the proper type and size of hardware. Find the total length of the bolt shown.

$\frac{1}{5}$ in. $\frac{1}{3}$ in. $\frac{1}{4}$ in.

38. How much fencing will be needed to enclose this rectangular wildflower preserve?

$\frac{1}{6}$ mile

$\frac{1}{3}$ mile

39. The owner of Racy's Feed Store ordered $\frac{1}{3}$ cubic yard of corn, $\frac{3}{8}$ cubic yard of oats, and $\frac{1}{4}$ cubic yard of washed medium mesh gravel. How many cubic yards of material were ordered?

40. A flower grower purchased $\frac{9}{10}$ acre of land one year and $\frac{3}{10}$ acre the next year. She then sold $\frac{7}{10}$ acre of land. How much land does she now have?

41. A forester planted $\frac{5}{12}$ acre of seedlings in the morning and $\frac{11}{12}$ acre in the afternoon. The next day, $\frac{7}{12}$ acre of seedlings were destroyed by a brush fire. How many acres of seedlings remained?

42. Adrian Ortega drives a tanker for the British Petroleum Company. He leaves the refinery with his tanker filled to $\frac{7}{8}$ of capacity. If he delivers $\frac{1}{4}$ of the tanker's capacity at the first stop and $\frac{1}{3}$ of the tanker's capacity at the second stop, find the fraction of the tanker's capacity remaining.

The circle graph shows the fraction of U.S. workers who used various methods to learn about computers. Use the graph to answer Exercises 43–46.

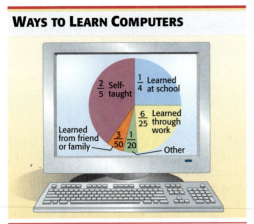

WAYS TO LEARN COMPUTERS

$\frac{2}{5}$ Self-taught

$\frac{1}{4}$ Learned at school

$\frac{6}{25}$ Learned through work

Learned from friend or family — $\frac{3}{50}$ $\frac{1}{20}$ — Other

Source: John J. Heldrich Center for Workforce Development.

43. What fraction of workers are self-taught or learned from friends or family?

44. What fraction of workers learned at school or through work?

45. What is the difference in the fraction of workers who are self-taught and those who learned at work?

46. What is the difference in the fraction of workers who are self-taught and those who learned at school?

Refer to the circle graph at the right to answer Exercises 47–50.

47. What fraction of the day was spent in class and study? How many hours is that?

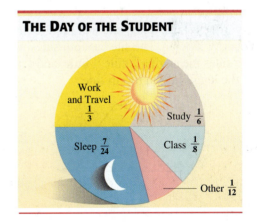

THE DAY OF THE STUDENT

Work and Travel $\frac{1}{3}$

Study $\frac{1}{6}$

Sleep $\frac{7}{24}$

Class $\frac{1}{8}$

Other $\frac{1}{12}$

48. What fraction of the day was spent in work and travel and sleep? How many hours is that?

49. How much more of the day was spent sleeping than studying? Write your answer as a fraction of the day.

50. How much more of the day was spent working and traveling than in class? Write your answer as a fraction of the day.

Use the photo to answer Exercises 51–54. A nut driver is like a screwdriver but is used to tighten nuts instead of screws. The ends of the drivers are sized from $\frac{3}{16}$ inch to $\frac{1}{2}$ inch to fit smaller or larger nuts. (The symbol " is for inches.)

$$\frac{3}{16}" \quad \frac{1}{4}" \quad \frac{5}{16}" \quad \frac{11}{32}" \quad \frac{3}{8}" \quad \frac{7}{16}" \quad \frac{1}{2}"$$

Source: Author's tool collection.

51. The rightmost driver fits a nut that is how much larger than the nut for the leftmost driver?

52. The nut size for the yellow-handled driver is how much less than the nut size for the blue-handled driver?

53. Find the difference in nut sizes for the middle driver compared to the blue-handled driver.

54. What is the increase in nut size from the black-handled driver to the orange-handled driver?

55. A hazardous waste dump site will require $\frac{7}{8}$ mile of security fencing. The site has four sides with three of the sides measuring $\frac{1}{4}$ mile, $\frac{1}{6}$ mile, and $\frac{3}{8}$ mile. Find the length of the fourth side.

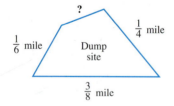

56. Chakotay is fitting a turquoise stone into a bear claw pendant. Find the diameter of the hole in the pendant. (The diameter is the distance across the center of the hole.)

4.5 Problem Solving: Mixed Numbers and Estimating

OBJECTIVE 1   Identify mixed numbers and graph them on a number line.   When a fraction and a whole number are written together, the result is a **mixed number.** For example, the mixed number

$$3\frac{1}{2} \qquad \text{represents} \qquad 3 + \frac{1}{2}$$

or 3 wholes and $\frac{1}{2}$ of a whole. Read $3\frac{1}{2}$ as "three and one half."

One common use of mixed numbers is to measure things. Examples are shown below.

Juan worked $5\frac{1}{2}$ hours.   The box weighs $2\frac{3}{4}$ pounds.

The park is $1\frac{7}{10}$ miles long.   Add $1\frac{2}{3}$ cups of flour.

OBJECTIVES

1   Identify mixed numbers and graph them on a number line.

2   Rewrite mixed numbers as improper fractions, or the reverse.

3   Estimate the answer and multiply or divide mixed numbers.

4   Estimate the answer and add or subtract mixed numbers.

5   Solve application problems containing mixed numbers.

EXAMPLE 1   Illustrating a Mixed Number with a Diagram and a Number Line

As this diagram shows, the mixed number $3\frac{1}{2}$ is equivalent to the improper fraction $\frac{7}{2}$.

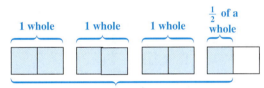

$\dfrac{7}{2}$ ←—Number of shaded parts

$\phantom{\dfrac{7}{2}}$ ←—Number of equal parts in each whole

We can also use a number line to show mixed numbers, as in this graph of $3\frac{1}{2}$ and $-3\frac{1}{2}$.

The number line shows the following.

$$3\frac{1}{2} \quad \text{is equivalent to} \quad \frac{7}{2}$$

$$-3\frac{1}{2} \quad \text{is equivalent to} \quad -\frac{7}{2}$$

NOTE

$3\frac{1}{2}$ represents $3 + \frac{1}{2}$.

$-3\frac{1}{2}$ represents $-3 + (-\frac{1}{2})$, which can also be written as $-3 - \frac{1}{2}$.

In algebra we usually work with the improper fraction form of mixed numbers, especially for negative mixed numbers. However, positive mixed numbers are frequently used in daily life, so it's important to know how to work with them. For example, we usually say $3\frac{1}{2}$ inches rather than $\frac{7}{2}$ inches.

Now Try Exercise 3. ▶▶▶

OBJECTIVE **2** **Rewrite mixed numbers as improper fractions, or the reverse.**   You can use the following steps to write $3\frac{1}{2}$ as an improper fraction without drawing a diagram or a number line.

Step 1   Multiply 2 times 3 and add 1 to the product.

$$3\frac{1}{2} \qquad 2 \cdot 3 = 6 \qquad \text{Then } 6 + 1 = 7$$

Step 2   Use 7 (from Step 1) as the numerator and 2 as the denominator.

$$3\frac{1}{2} = \frac{7}{2} \leftarrow (2 \cdot 3) + 1$$

Same denominator

To see why this method works, recall that $3\frac{1}{2}$ represents $3 + \frac{1}{2}$. Let's add $3 + \frac{1}{2}$.

$$3 + \frac{1}{2} = \frac{3}{1} + \frac{1}{2} = \frac{6}{2} + \frac{1}{2} = \frac{6 + 1}{2} = \frac{7}{2} \begin{cases} \text{Same result} \\ \text{as above} \end{cases}$$

Common denominator

In summary, use the following steps to *write a mixed number as an improper fraction.*

> **Writing a Mixed Number as an Improper Fraction**
>
> *Step 1*   ***Multiply*** the denominator of the fraction times the whole number and ***add*** the numerator of the fraction to the product.
>
> *Step 2*   Write the result of Step 1 as the ***numerator*** and keep the original ***denominator.***

EXAMPLE 2 **Writing a Mixed Number as an Improper Fraction**

Write $7\frac{2}{3}$ as an improper fraction (numerator greater than denominator).

Step 1 $7\frac{2}{3}$ $3 \cdot 7 = 21$ Then $21 + 2 = 23$

Step 2 $7\frac{2}{3} = \frac{23}{3} \leftarrow (3 \cdot 7) + 2$

Same denominator

Now Try Exercises 5 and 11.

We used *multiplication* for the first step in writing a mixed number as an improper fraction. To work in *reverse*, writing an improper fraction as a mixed number, we will use *division*. Recall that the fraction bar is a symbol for division.

Writing an Improper Fraction as a Mixed Number

To write an *improper fraction* as a mixed number, divide the numerator by the denominator. The quotient is the whole number part (of the mixed number), the remainder is the numerator of the fraction part, and the denominator remains the same.

Always check to be sure that the fraction part of the mixed number is in lowest terms. Then the mixed number is in *simplest form*.

EXAMPLE 3 **Writing Improper Fractions as Mixed Numbers**

Write each improper fraction as an equivalent mixed number in simplest form.

(a) $\dfrac{17}{5}$

Divide 17 by 5.

$$\begin{array}{r} \textbf{3} \quad \leftarrow \text{Whole number part} \\ 5\overline{)17} \\ \underline{15} \\ \textbf{2} \quad \leftarrow \text{Remainder} \end{array}$$

The quotient **3** is the whole number part of the mixed number. The remainder **2** is the numerator of the fraction, and the denominator stays as **5**.

$$\frac{17}{5} = 3\frac{2}{5} \quad \leftarrow \text{Remainder}$$

$\qquad\qquad\qquad$ Same denominator

Let's look at a drawing of $\frac{17}{5}$ to check our work.

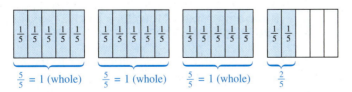

$\frac{5}{5} = 1 \text{ (whole)} \qquad \frac{5}{5} = 1 \text{ (whole)} \qquad \frac{5}{5} = 1 \text{ (whole)} \qquad \frac{2}{5}$

Continued on Next Page

In other words,

$$\frac{17}{5} = \frac{5}{5} + \frac{5}{5} + \frac{5}{5} + \frac{2}{5}$$

$$= \underbrace{1 + 1 + 1}_{} + \frac{2}{5}$$

$$= \quad 3 \quad + \frac{2}{5}$$

$$= 3\frac{2}{5}$$

(b) $\dfrac{26}{4}$

Divide 26 by 4.

$$4\overline{)26} \quad \begin{array}{c} 6 \\ 24 \\ \hline 2 \end{array} \quad \text{so} \quad \frac{26}{4} = 6\frac{2}{4} = 6\frac{1}{2} \quad \left\{ \begin{array}{l} \text{Simplest} \\ \text{form} \end{array} \right.$$

Write $\frac{2}{4}$ in lowest terms.

You could write $\frac{26}{4}$ in lowest terms first.

$$\frac{26}{4} = \frac{\overset{1}{\cancel{2}} \cdot 13}{\underset{1}{\cancel{2}} \cdot 2} = \frac{13}{2} \qquad \text{Then} \qquad 2\overline{)13} \quad \begin{array}{c} 6 \\ 12 \\ \hline 1 \end{array} \quad \text{so} \quad \frac{13}{2} = 6\frac{1}{2} \quad \left\{ \begin{array}{l} \text{Same result} \\ \text{as above} \end{array} \right.$$

Now Try Exercises 15 and 21.

OBJECTIVE 3 **Estimate the answer and multiply or divide mixed numbers.** Once you have rewritten mixed numbers as improper fractions, you can use the steps you learned in **Section 4.3** to multiply and divide. However, it's a good idea to estimate the answer before you start any other work.

EXAMPLE 4 **Rounding Mixed Numbers to the Nearest Whole Number**

To estimate answers, first round each mixed number to the *nearest whole number*. If the numerator is *half* of the denominator or *more,* round up the whole number part. If the numerator is *less* than half the denominator, leave the whole number as it is.

(a) Round $1\dfrac{5}{8}$ ← 5 is more than 4
← Half of 8 is 4
$\qquad 1\dfrac{5}{8}$ rounds up to 2

(b) Round $3\dfrac{2}{5}$ ← 2 is less than $2\frac{1}{2}$
← Half of 5 is $2\frac{1}{2}$
$\qquad 3\dfrac{2}{5}$ rounds to 3

Now Try Exercise 23(c).

Multiplying and Dividing Mixed Numbers

Step 1 **Rewrite** each mixed number as an improper fraction.

Step 2 **Multiply** or **divide** the improper fractions.

Step 3 Write the answer in lowest terms and change it to a mixed number or whole number where possible. This step gives you an answer that is in **simplest form.**

EXAMPLE 5 Estimating the Answer and Multiplying Mixed Numbers

First, round the numbers and estimate each answer. Then find the exact answer. Write exact answers in simplest form.

(a) $2\frac{1}{2} \cdot 3\frac{1}{5}$

Estimate the answer by rounding the mixed numbers.

$$2\frac{1}{2} \text{ rounds to } 3 \quad \text{and} \quad 3\frac{1}{5} \text{ rounds to } 3$$

$$3 \cdot 3 = 9 \leftarrow \text{Estimated answer}$$

To find the exact answer, first rewrite each mixed number as an improper fraction.

$$\text{Step 1} \quad 2\frac{1}{2} = \frac{5}{2} \quad \text{and} \quad 3\frac{1}{5} = \frac{16}{5}$$

Next, multiply.

$$2\frac{1}{2} \cdot 3\frac{1}{5} = \frac{5}{2} \cdot \frac{16}{5} = \frac{\overset{1}{\cancel{5}} \cdot \overset{1}{\cancel{2}} \cdot 8}{\underset{1}{\cancel{2}} \cdot \underset{1}{\cancel{5}}} = \frac{8}{1} = 8 \quad \left\{\begin{array}{l}\text{Simplest} \\ \text{form}\end{array}\right.$$

The estimate was 9, so an exact answer of 8 is reasonable.

(b) $\left(3\frac{5}{8}\right)\left(4\frac{4}{5}\right)$

First, round each mixed number and estimate the answer.

$$3\frac{5}{8} \text{ rounds to } 4 \quad \text{and} \quad 4\frac{4}{5} \text{ rounds to } 5$$

$$4 \cdot 5 = 20 \leftarrow \text{Estimated answer}$$

Now find the exact answer.

$$\left(3\frac{5}{8}\right)\left(4\frac{4}{5}\right) = \left(\frac{29}{8}\right)\left(\frac{24}{5}\right) = \frac{29 \cdot 3 \cdot \overset{1}{\cancel{8}}}{\underset{1}{\cancel{8}} \cdot 5} = \frac{87}{5} = 17\frac{2}{5} \quad \left\{\begin{array}{l}\text{Simplest} \\ \text{form}\end{array}\right.$$

The estimate was 20, so an exact answer of $17\frac{2}{5}$ is reasonable.

Now Try Exercise 25. ▶▶▶

EXAMPLE 6 Estimating the Answer and Dividing Mixed Numbers

First, round the numbers and estimate each answer. Then find the exact answer. Write exact answers in simplest form.

(a) $3\frac{3}{5} \div 1\frac{1}{2}$

To estimate the answer, round each mixed number to the nearest whole number.

$$
\begin{array}{ccc}
3\frac{3}{5} & \div & 1\frac{1}{2} \\
\downarrow \quad \text{Rounded} \quad \downarrow & \\
4 & \div \quad 2 & = \quad 2 \leftarrow \text{Estimate}
\end{array}
$$

To find the exact answer, first rewrite each mixed number as an improper fraction.

$$3\frac{3}{5} \div 1\frac{1}{2} = \frac{18}{5} \div \frac{3}{2}$$

Now rewrite the problem as multiplying by the reciprocal of $\frac{3}{2}$.

$$\frac{18}{5} \div \frac{3}{2} = \frac{18}{5} \cdot \frac{2}{3} = \frac{\overset{1}{\cancel{3}} \cdot 6 \cdot 2}{5 \cdot \cancel{3}} = \frac{12}{5} = 2\frac{2}{5} \quad \left\{\begin{array}{l}\text{Simplest} \\ \text{form}\end{array}\right.$$

Reciprocals

The estimate was 2, so an exact answer of $2\frac{2}{5}$ is reasonable.

(b) $4\frac{3}{8} \div 5$

First, round the numbers and estimate the answer.

$$
\begin{array}{ccc}
4\frac{3}{8} & \div & 5 \\
\downarrow \quad \text{Rounded} \quad \downarrow & \\
4 & \div \quad 5 & \quad \begin{array}{l}\text{Write } 4 \div 5 \text{ using} \\ \text{a fraction bar.}\end{array} \left.\begin{array}{l}\\ \frac{4}{5}\end{array}\right\} \leftarrow \text{Estimate}
\end{array}
$$

Now find the exact answer.

Write 5 as $\frac{5}{1}$

$$4\frac{3}{8} \div 5 = \frac{35}{8} \div \frac{5}{1} = \frac{35}{8} \cdot \frac{1}{5} = \frac{\overset{1}{\cancel{5}} \cdot 7 \cdot 1}{8 \cdot \cancel{5}} = \frac{7}{8} \quad \left\{\begin{array}{l}\text{Simplest} \\ \text{form}\end{array}\right.$$

Reciprocals

The estimate was $\frac{4}{5}$, so an exact answer of $\frac{7}{8}$ is reasonable. They are both less than 1.

Now Try Exercise 27.

OBJECTIVE 4 **Estimate the answer and add or subtract mixed numbers.** The steps you learned for adding and subtracting fractions in Section 4.4 will also work for mixed numbers: Just rewrite the mixed numbers as equivalent improper fractions. Again, it is a good idea to estimate the answer before you start any other work.

EXAMPLE 7 **Estimating the Answer and Adding or Subtracting Mixed Numbers**

First, estimate each answer. Then add or subtract to find the exact answer. Write exact answers in simplest form.

(a) $2\dfrac{3}{8} + 3\dfrac{3}{4}$

To estimate the answer, round each mixed number to the nearest whole number.

$$2\frac{3}{8} + 3\frac{3}{4}$$
$$\downarrow \qquad \downarrow$$
$$2 \;+\; 4 = 6 \;\leftarrow\; \text{Estimate}$$

To find the exact answer, first rewrite each mixed number as an equivalent improper fraction.

$$2\frac{3}{8} + 3\frac{3}{4} = \frac{19}{8} + \frac{15}{4}$$

You can't add fractions until they have a common denominator. The LCD for $\frac{19}{8}$ and $\frac{15}{4}$ is 8. Rewrite $\frac{15}{4}$ as an equivalent fraction with a denominator of 8.

$$\frac{19}{8} + \frac{15}{4} = \frac{19}{8} + \frac{30}{8} = \frac{19 + 30}{8} = \frac{49}{8} = 6\frac{1}{8} \;\left\{\begin{array}{l}\text{Simplest}\\\text{form}\end{array}\right.$$

Common denominator

The estimate was 6, so an exact answer of $6\frac{1}{8}$ is reasonable.

(b) $4\dfrac{2}{3} - 2\dfrac{4}{5}$

Round each number and estimate the answer.

$$4\frac{2}{3} - 2\frac{4}{5}$$
$$\downarrow \qquad \downarrow$$
$$5 \;-\; 3 = 2 \;\leftarrow\; \text{Estimate}$$

To find the exact answer, rewrite the mixed numbers as improper fractions and subtract.

$$4\frac{2}{3} - 2\frac{4}{5} = \frac{14}{3} - \frac{14}{5} = \frac{70}{15} - \frac{42}{15} = \frac{70 - 42}{15} = \frac{28}{15} = 1\frac{13}{15} \;\left\{\begin{array}{l}\text{Simplest}\\\text{form}\end{array}\right.$$

LCD is 15

The estimate was 2, so an exact answer of $1\frac{13}{15}$ is reasonable.

Continued on Next Page

(c) $5 - 1\dfrac{3}{8}$

$$5 - 1\dfrac{3}{8}$$
$$\downarrow \qquad \downarrow$$
$$5 - 1 = 4 \leftarrow \text{Estimate}$$

Write 5 as $\dfrac{5}{1}$

$$5 - 1\dfrac{3}{8} = \dfrac{5}{1} - \dfrac{11}{8} = \dfrac{40}{8} - \dfrac{11}{8} = \dfrac{29}{8} = 3\dfrac{5}{8} \begin{cases} \text{Simplest} \\ \text{form} \end{cases}$$

LCD is 8

The estimate was 4, so an exact answer of $3\dfrac{5}{8}$ is reasonable.

Now Try Exercises 29 and 31.

NOTE
In some situations the method of rewriting mixed numbers as improper fractions may result in very large numerators. Consider this example.

Last year Hue's child was $48\dfrac{3}{8}$ in. tall. This year the child is $51\dfrac{1}{4}$ in. tall. How much has the child grown?

First, estimate the answer by rounding each mixed number to the nearest whole number.

$$51\dfrac{1}{4} - 48\dfrac{3}{8}$$
$$\downarrow \qquad \downarrow$$
$$51 - 48 = 3 \text{ in.} \leftarrow \text{Estimate}$$

To find the exact answer, rewrite the mixed numbers as improper fractions.

Rewrite $\dfrac{205}{4}$ as $\dfrac{410}{8}$

$$51\dfrac{1}{4} - 48\dfrac{3}{8} = \dfrac{205}{4} - \dfrac{387}{8} = \dfrac{410}{8} - \dfrac{387}{8} = \dfrac{410 - 387}{8} = \dfrac{23}{8} = 2\dfrac{7}{8} \text{ in.}$$

LCD is 8

You can also use the fraction key $\boxed{a^{b/c}}$ on your *scientific* calculator to solve this problem.

To enter $51\dfrac{1}{4}$, press

51 $\boxed{a^{b/c}}$ 1 $\boxed{a^{b/c}}$ 4 $\boxed{a^{b/c}}$. The display shows $\boxed{51 \lrcorner 1 \lrcorner 4}$.

Whole   Numerator Denominator
number

Then press $\boxed{-}$ 48 $\boxed{a^{b/c}}$ 3 $\boxed{a^{b/c}}$ 8 $\boxed{=}$. The display shows $\boxed{2 \lrcorner 7 \lrcorner 8}$.

Subtract. $48\dfrac{3}{8}$ $2\dfrac{7}{8}$

Either way the exact answer is $2\dfrac{7}{8}$ in., which is close to the estimate of 3 in.

Another efficient method for handling large mixed numbers is to rewrite them in decimal form. You will learn how to do that in **Chapter 5**.

OBJECTIVE 5 Solve application problems containing mixed numbers.
Rounding mixed numbers to the nearest whole number can also help you decide whether to solve an application problem by adding, subtracting, multiplying, or dividing.

EXAMPLE 8 Solving Application Problems with Mixed Numbers

First, estimate the answer to each application problem. Then find the exact answer.

(a) Gary needs to haul $15\frac{3}{4}$ **tons** of sand to a construction site. His truck can carry $2\frac{1}{4}$ **tons**. How many trips will he need to make?
First, round each mixed number to the nearest whole number.

$$15\frac{3}{4} \quad \text{rounds to} \quad 16 \quad \text{and} \quad 2\frac{1}{4} \quad \text{rounds to} \quad 2$$

Now read the problem again, *using the rounded numbers.*
    Gary needs to haul **16 tons** of sand to a construction site. His truck can carry **2 tons**. How many trips will he need to make?
    Using the rounded numbers in the problem makes it easier to see that you need to *divide.*

$$16 \div 2 = 8 \text{ trips} \quad \leftarrow \text{Estimate}$$

To find the exact answer, use the original mixed numbers and divide.

$$15\frac{3}{4} \div 2\frac{1}{4} = \frac{63}{4} \div \frac{9}{4} = \frac{63}{4} \cdot \frac{4}{9} = \frac{7 \cdot \overset{1}{\cancel{9}} \cdot \overset{1}{\cancel{4}}}{\cancel{4} \cdot \cancel{9}} = \frac{7}{1} = 7 \quad \begin{cases} \text{Simplest} \\ \text{form} \end{cases}$$

Reciprocals

Gary needs to make 7 trips to haul all the sand. This result is close to the estimate of 8 trips.

(b) Zenitia worked $3\frac{5}{6}$ **hours** on Monday and $6\frac{1}{2}$ **hours** on Tuesday. How much longer did she work on Tuesday than on Monday?
First, round each mixed number to the nearest whole number.

$$3\frac{5}{6} \quad \text{rounds to} \quad 4 \quad \text{and} \quad 6\frac{1}{2} \quad \text{rounds to} \quad 7$$

Now read the problem again, *using the rounded numbers.*
    Zenitia worked **4 hours** on Monday and **7 hours** on Tuesday. How much longer did she work on Tuesday than on Monday?
    Using the rounded numbers in the problem makes it easier to see that you need to *subtract.*

$$7 - 4 = 3 \text{ hours} \quad \leftarrow \text{Estimate}$$

To find the exact answer, use the original mixed numbers and subtract.

Write answer in simplest form.

$$6\frac{1}{2} - 3\frac{5}{6} = \frac{13}{2} - \frac{23}{6} = \frac{39}{6} - \frac{23}{6} = \frac{39 - 23}{6} = \frac{16}{6} = 2\frac{4}{6} = 2\frac{2}{3}$$

LCD is 6

Zenitia worked $2\frac{2}{3}$ hours longer on Tuesday. This result is close to the estimate of 3 hours.

Now Try Exercise 43. ▶▶▶

4.5 Exercises

Graph the mixed numbers or improper fractions on the number line. See Example 1.

1. Graph $2\frac{1}{3}$ and $-2\frac{1}{3}$.

2. Graph $1\frac{3}{4}$ and $-1\frac{3}{4}$.

3. Graph $\frac{3}{2}$ and $-\frac{3}{2}$.

4. Graph $\frac{11}{3}$ and $-\frac{11}{3}$.

Write each mixed number as an improper fraction. See Example 2.

5. $4\frac{1}{2}$ **6.** $2\frac{1}{4}$ **7.** $-1\frac{3}{5}$ **8.** $-1\frac{5}{6}$ **9.** $2\frac{3}{8}$

10. $3\frac{4}{9}$ **11.** $-5\frac{7}{10}$ **12.** $-4\frac{5}{7}$ **13.** $10\frac{11}{15}$ **14.** $12\frac{9}{11}$

Write each improper fraction as a mixed number in simplest form. See Example 3.

15. $\frac{13}{3}$ **16.** $\frac{11}{2}$ **17.** $-\frac{10}{4}$ **18.** $-\frac{14}{5}$

19. $\frac{22}{6}$ **20.** $\frac{28}{8}$ **21.** $-\frac{51}{9}$ **22.** $-\frac{44}{10}$

Round each mixed number to the nearest whole number.

23. (a) $10\frac{1}{6}$ **(b)** $1\frac{2}{3}$ **(c)** $7\frac{7}{16}$ **(d)** $4\frac{1}{2}$ **(e)** $3\frac{5}{9}$ **(f)** $9\frac{2}{5}$

24. (a) $2\frac{1}{3}$ **(b)** $5\frac{3}{8}$ **(c)** $8\frac{3}{5}$ **(d)** $1\frac{7}{15}$ **(e)** $4\frac{7}{12}$ **(f)** $6\frac{1}{2}$

First, round the mixed numbers to the nearest whole number and estimate each answer.
Then find the exact answer. Write exact answers in simplest form. See Examples 4–7.

25. *Exact:*

$2\dfrac{1}{4} \cdot 3\dfrac{1}{2}$

Estimate:

___ • ___ = ___

26. *Exact:*

$\left(1\dfrac{1}{2}\right)\left(3\dfrac{3}{4}\right)$

Estimate:

___ • ___ = ___

27. *Exact:*

$3\dfrac{1}{4} \div 2\dfrac{5}{8}$

Estimate:

___ ÷ ___ = ___

28. *Exact:*

$2\dfrac{1}{4} \div 1\dfrac{1}{8}$

Estimate:

___ ÷ ___ = ___

29. *Exact:*

$3\dfrac{2}{3} + 1\dfrac{5}{6}$

Estimate:

___ + ___ = ___

30. *Exact:*

$4\dfrac{4}{5} + 2\dfrac{1}{3}$

Estimate:

___ + ___ = ___

31. *Exact:*

$4\dfrac{1}{4} - \dfrac{7}{12}$

Estimate:

___ − ___ = ___

32. *Exact:*

$10\dfrac{1}{3} - 6\dfrac{5}{6}$

Estimate:

___ − ___ = ___

33. *Exact:*

$5\dfrac{2}{3} \div 6$

Estimate:

___ ÷ ___ = ___

34. *Exact:*

$1\dfrac{7}{8} \div 6\dfrac{1}{4}$

Estimate:

___ ÷ ___ = ___

35. *Exact:*

$8 - 1\dfrac{4}{5}$

Estimate:

___ − ___ = ___

36. *Exact:*

$7 - 3\dfrac{3}{10}$

Estimate:

___ − ___ = ___

Find the perimeter and the area of each square or rectangle. Write all answers in simplest form.

37.

$1\frac{3}{4}$ in.

$1\frac{3}{4}$ in.

$1\frac{3}{4}$ in.

38.

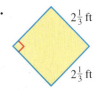

$2\frac{1}{3}$ ft

$2\frac{1}{3}$ ft

39.

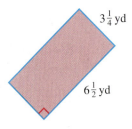

$3\frac{1}{4}$ yd

$6\frac{1}{2}$ yd

40.

$2\frac{3}{10}$ mi

$\frac{2}{5}$ mi $\frac{2}{5}$ mi

$2\frac{3}{10}$ mi

First, estimate the answer to each application problem. Then find the exact answer
Write all answers in simplest form. See Example 8.

41. A carpenter has two pieces of oak trim. One piece of trim is $12\frac{1}{2}$ ft long and the other is $8\frac{2}{3}$ ft long. How many feet of oak trim does he have in all?

Estimate:

Exact:

42. On Monday, $5\frac{3}{4}$ tons of cans were recycled, and, on Tuesday, $9\frac{3}{5}$ tons were recycled. How many tons were recycled on these two days?

Estimate:

Exact:

43. The directions for mixing an insect spray say to use $1\frac{3}{4}$ ounces of chemical in each gallon of water. How many ounces of chemical should be mixed with $5\frac{1}{2}$ gallons of water?

Estimate:

Exact:

44. Shirley Cicero wants to make 16 holiday wreaths to sell at the craft fair. Each wreath requires $2\frac{1}{4}$ yd of ribbon. How many yards does she need?

Estimate:

Exact:

45. The Boy Scout troop has volunteered to pick up trash along a 4-mile stretch of highway. So far they have done $1\frac{7}{10}$ miles. How much do they have left to do?

Estimate:

Exact:

46. The gas tank on a Jeep Cherokee has a capacity of $21\frac{3}{8}$ gallons. Scott started with a full tank and then used $8\frac{1}{2}$ gallons of gasoline. Find the number of gallons that remain.

Estimate:

Exact:

47. Suppose that a bridesmaid's floor-length dress requires $3\frac{3}{4}$ yd of material. How much material would be needed to make dresses for 5 bridesmaids?

Estimate:

Exact:

48. A cookie recipe uses $\frac{2}{3}$ cup brown sugar. How much brown sugar is needed to make $2\frac{1}{2}$ times the original recipe?

Estimate:

Exact:

49. A landscaper has $9\frac{5}{8}$ cubic yards of peat moss in a truck. If she unloads $1\frac{1}{2}$ cubic yards at the first stop and 3 cubic yards at the second stop, how much peat moss remains in the truck?

Estimate:

Exact:

50. Marv bought 10 yd of Italian silk fabric. He used $3\frac{7}{8}$ yd to make a jacket. How much fabric is left for other sewing projects?

Estimate:

Exact:

51. Melissa worked $18\frac{3}{4}$ hours over the last five days. If she worked the same amount each day, how long was she at work each day?

Estimate:

Exact:

52. Michael is cutting a $10\frac{1}{2}$ ft board into shelves for a bookcase. Each shelf will be $1\frac{3}{4}$ ft long. How many shelves can he cut?

Estimate:

Exact:

53. Find the length of the arrow shaft.

Estimate:

Exact:

54. Find the length of the indented section on this board.

Estimate:

Exact:

First, estimate the answer to each application problem. Then use your calculator to find the exact answer.

55. A craftsperson must attach a lead strip around all four sides of a rectangular stained glass window before it is installed. Find the length of lead stripping needed for the window shown.

$23\frac{3}{4}$ in.

$34\frac{1}{2}$ in.

Estimate:

Exact:

56. To complete a custom order, Zak Morten of Home Depot must find the number of inches of brass trim needed to go around the four sides of the lamp base plate shown. Find the length of brass trim needed.

$5\frac{1}{8}$ in.

$9\frac{7}{8}$ in.

Estimate:

Exact:

57. A fishing boat anchor requires $10\frac{3}{8}$ pounds of steel. Find the number of anchors that can be manufactured with 25,730 pounds of steel.

Estimate:

Exact:

58. Each apartment requires $62\frac{1}{2}$ square yards of carpet. Find the number of apartments that can be carpeted with 6750 square yards of carpet.

Estimate:

Exact:

59. Claire and Deb create custom hat bands that people can put around the crowns of their hats. The finished bands are the lengths shown in the table. The strip of fabric for each band must include the finished length plus an extra $\frac{3}{4}$ in. for the seam.

Band Size	Finished Length
Small	$21\frac{7}{8}$ in.
Medium	$22\frac{5}{8}$ in.
Large	$23\frac{1}{2}$ in.

What length of fabric strip is needed to make 4 small bands, 5 medium bands, and 3 large bands, including the seam allowance?

Estimate:

Exact:

60. Three sides of a parking lot are $108\frac{1}{4}$ ft, $162\frac{3}{8}$ ft, and $143\frac{1}{2}$ ft. If the distance around the lot is $518\frac{3}{4}$ ft, find the length of the fourth side.

$108\frac{1}{4}$ ft

$162\frac{3}{8}$ ft

? ft

$143\frac{1}{2}$ ft

Estimate:

Exact:

Summary Exercises on Fractions

1. Write fractions that represent the shaded and unshaded portion of each figure.

(a)

(b)

2. Graph $\frac{2}{3}$ and $-\frac{2}{3}$ on the number line.

3. Rewrite each fraction with the indicated denominator.

(a) $-\frac{4}{5} = -\frac{}{30}$ **(b)** $\frac{2}{7} = \frac{}{14}$

4. Simplify.

(a) $\frac{15}{15}$ **(b)** $-\frac{24}{6}$ **(c)** $\frac{9}{1}$

5. Write the prime factorization of each number.

(a) 72   **(b)** 105

6. Write each fraction in lowest terms.

(a) $\frac{24}{30}$ **(b)** $\frac{175}{200}$

Simplify.

7. $\left(-\frac{3}{4}\right)\left(-\frac{2}{3}\right)$

8. $-\frac{7}{8} + \frac{2}{3}$

9. $\frac{7}{16} + \frac{5}{8}$

10. $\frac{5}{8} \div \frac{3}{4}$

11. $\frac{2}{3} - \frac{4}{5}$

12. $\frac{7}{12}\left(-\frac{9}{14}\right)$

13. $-21 \div \left(-\frac{3}{8}\right)$

14. $\frac{7}{8} - \frac{5}{12}$

15. $-\frac{35}{45} \div \frac{10}{15}$

16. $-\frac{5}{6} - \frac{3}{4}$

17. $\frac{7}{12} + \frac{5}{6} + \frac{2}{3}$

18. $\frac{5}{8}$ of 56

First round the numbers and estimate each answer. Then find the exact answer.

19. *Exact:*

$4\frac{3}{4} + 2\frac{5}{6}$

Estimate:

___ + ___ = ___

20. *Exact:*

$2\frac{2}{9} \cdot 5\frac{1}{7}$

Estimate:

___ • ___ = ___

21. *Exact:*

$6 - 2\frac{7}{10}$

Estimate:

___ − ___ = ___

22. *Exact:*

$1\frac{3}{5} \div 3\frac{1}{2}$

Estimate:

___ ÷ ___ = ___

23. *Exact:*

$4\frac{2}{3} \div 1\frac{1}{6}$

Estimate:

___ ÷ ___ = ___

24. *Exact:*

$3\frac{5}{12} - \frac{3}{4}$

Estimate:

___ − ___ = ___

Solve each application problem. Write all answers in simplest form.

25. When installing cabinets, Cecil Feathers must be certain that the proper type and size of mounting screw is used.

(a) Find the total length of the screw shown.

(b) If the screw is put into a board that is $1\frac{3}{4}$ in. thick, how much of the screw will stick out the back of the board?

26. Find the perimeter and the area of this postage stamp.

27. A batch of cookies requires $\frac{3}{4}$ pound of chocolate chips. If you have nine pounds of chocolate chips, how many batches of cookies can you make?

28. The Municipal Utility District says that the cost of operating a hair dryer is $\frac{1}{5}$¢ per minute. Find the cost of operating the hair dryer for a half hour.

29. A survey asked 1500 adults if UFOs (unidentified flying objects) are real. The circle graph shows the fraction of the adults who gave each answer. How many adults gave each answer?

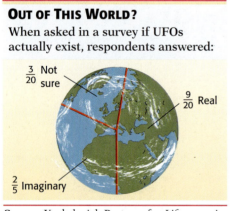

OUT OF THIS WORLD?
When asked in a survey if UFOs actually exist, respondents answered:

$\frac{3}{20}$ Not sure

$\frac{9}{20}$ Real

$\frac{2}{5}$ Imaginary

Source: Yankelovich Partners for *Life* magazine.

30. Find the diameter of the hole in the rectangular mounting bracket shown. (The diameter is the distance across the center of the hole.) Then find the perimeter of the bracket.

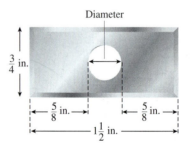

31. A bottle of contact lens daily cleaning solution holds $\frac{2}{3}$ fluid ounce. How many bottles can be filled with $15\frac{1}{3}$ fluid ounces? (*Source:* Alcon Laboratories, Inc.)

32. A home builder bought two parcels of land that were $5\frac{7}{8}$ acres and $10\frac{3}{4}$ acres. She is setting aside $2\frac{7}{8}$ acres for a park and using the rest for $1\frac{1}{4}$ acre home lots. How many lots will be in the development?

4.6 Exponents, Order of Operations, and Complex Fractions

OBJECTIVES

1 Simplify fractions with exponents.

2 Use the order of operations with fractions.

3 Simplify complex fractions.

OBJECTIVE 1 Simplify fractions with exponents. We have used exponents as a quick way to write repeated multiplication of integers and variables. Here are two examples as a review.

$$(-3)^2 = (-3)(-3) = 9 \quad \text{and} \quad x^3 = x \cdot x \cdot x$$

Base — Two factors of -3 Base Three factors of x

The meaning of an exponent remains the same when a fraction is the base.

EXAMPLE 1 Simplifying Fractions with Exponents

Simplify.

(a) $\left(-\dfrac{1}{2}\right)^3$

The base is $-\frac{1}{2}$. The exponent indicates that there are three factors of $-\frac{1}{2}$.

$$\left(-\frac{1}{2}\right)^3 = \left(-\frac{1}{2}\right)\left(-\frac{1}{2}\right)\left(-\frac{1}{2}\right)$$

Three factors of –
Watch the signs carefully.
Multiply –   –   to get

$$= \frac{1}{4}\left(-\frac{1}{2}\right)$$

Now multiply   –   to get –

$$= -\frac{1}{8}$$

The product is negative.

(b) $\left(\dfrac{3}{4}\right)^2 \left(\dfrac{2}{3}\right)^3$

Two factors of
Three factors of

$$\left(\frac{3}{4}\right)^2 \left(\frac{2}{3}\right)^3 = \left(\frac{3}{4}\right)\left(\frac{3}{4}\right)\left(\frac{2}{3}\right)\left(\frac{2}{3}\right)\left(\frac{2}{3}\right)$$

$$= \frac{\overset{1}{\cancel{3}} \cdot \overset{1}{\cancel{3}} \cdot \overset{1}{\cancel{2}} \cdot \overset{1}{\cancel{2}} \cdot \overset{1}{\cancel{2}}}{\underset{1}{\cancel{2}} \cdot \underset{1}{\cancel{2}} \cdot \underset{1}{\cancel{2}} \cdot 2 \cdot \underset{1}{\cancel{3}} \cdot \underset{1}{\cancel{3}} \cdot 3}$$

Divide out all common factors.

$$= \frac{1}{6}$$

Now Try Exercises 5 and 15.

OBJECTIVE 2 Use the order of operations with fractions. The order of operations that you used in **Section 1.8** for integers also applies to fractions.

Order of Operations

Step 1 Work inside **parentheses** or **other grouping symbols.**

Step 2 Simplify expressions with **exponents.**

Step 3 Do the remaining **multiplications and divisions** as they occur from left to right.

Step 4 Do the remaining **additions and subtractions** as they occur from left to right.

EXAMPLE 2 **Using the Order of Operations with Fractions**

Simplify.

(a) $-\dfrac{1}{3} + \dfrac{1}{2}\left(\dfrac{4}{5}\right)$

There is no work to be done inside the parentheses. There are no exponents, so start with Step 3, multiplying and dividing.

$-\dfrac{1}{3} + \dfrac{1 \cdot 4}{2 \cdot 5}$

Multiply.

$-\dfrac{1}{3} + \dfrac{4}{10}$

Now add. The LCD is 30.

$-\dfrac{1}{3} = -\dfrac{10}{30}$ and $\dfrac{4}{10} = \dfrac{12}{30}$

$\dfrac{-10 + 12}{30}$

Add the numerators; keep the common denominator.

$\dfrac{2}{30}$

Write $\dfrac{2}{30}$ in lowest terms: $\dfrac{\overset{1}{\cancel{2}}}{\underset{1}{\cancel{2}} \cdot 15} = \dfrac{1}{15}$

$\dfrac{1}{15}$

The answer is in lowest terms.

(b) $-2 + \left(\dfrac{1}{4} - \dfrac{3}{2}\right)^2$

Work inside parentheses. The LCD for $\dfrac{1}{4}$ and $\dfrac{3}{2}$ is 4

Rewrite $\dfrac{3}{2}$ as $\dfrac{6}{4}$ and subtract.

$-2 + \left(\dfrac{1 - 6}{4}\right)^2$

$-2 + \left(\dfrac{-5}{4}\right)^2$

Simplify the term with the exponent. Multiply $\left(-\dfrac{5}{4}\right)\left(-\dfrac{5}{4}\right)$. Signs match, so the product is positive.

$-2 + \left(\dfrac{25}{16}\right)$

Add last. Write -2 as $-\dfrac{2}{1}$

$-\dfrac{2}{1} + \dfrac{25}{16}$

The LCD is 16. Rewrite $-\dfrac{2}{1}$ as $-\dfrac{32}{16}$

$\dfrac{-32 + 25}{16}$

Add the numerators; keep the common denominator.

$-\dfrac{7}{16}$

The answer is in lowest terms.

Now Try Exercise 29.

OBJECTIVE 3 Simplify complex fractions. We have used both the symbol $\div$ and a fraction bar to indicate division. For example,

Indicates division $\rightarrow \dfrac{6}{2}$ can be written as $6 \div 2$

Indicates division

That means we could write $-\frac{4}{5} \div \left(-\frac{3}{10}\right)$ using a fraction bar instead of $\div$.

$-\dfrac{4}{5} \div \left(-\dfrac{3}{10}\right)$ can be written as $\dfrac{-\dfrac{4}{5}}{-\dfrac{3}{10}}$ $\leftarrow$ Indicates division

Indicates division

The result looks a bit complicated, and its name reflects that fact. We call it a *complex fraction*.

> ## Complex Fractions
>
> A **complex fraction** is a fraction in which the numerator and/or denominator contain one or more fractions.

To simplify a complex fraction, rewrite it in horizontal format using the $\div$ symbol for division.

EXAMPLE 3 **Simplifying a Complex Fraction**

Simplify: **(a)** $\dfrac{-\dfrac{4}{5}}{-\dfrac{3}{10}}$ **(b)** $\dfrac{\left(\dfrac{2}{3}\right)^2}{6}$

(a) Rewrite the complex fraction using the $\div$ symbol for division. Then follow the steps for dividing fractions.

$$\dfrac{-\dfrac{4}{5}}{-\dfrac{3}{10}} = -\dfrac{4}{5} \div -\dfrac{3}{10} = -\dfrac{4}{5} \cdot -\dfrac{10}{3} = \dfrac{4 \cdot 2 \cdot \cancel{5}^{1}}{\cancel{5}_{1} \cdot 3} = \dfrac{8}{3} \quad \text{or} \quad 2\dfrac{2}{3}$$

Reciprocals

The quotient is positive because the numbers in the problem had matching signs (both were negative).

(b) Rewrite the complex fraction using the $\div$ symbol for division. Then follow the order of operations.

$$\dfrac{\left(\dfrac{2}{3}\right)^2}{6} = \left(\dfrac{2}{3}\right)^2 \div 6 = \dfrac{4}{9} \div 6 = \dfrac{4}{9} \cdot \dfrac{1}{6} = \dfrac{2 \cdot 2 \cdot 1}{9 \cdot 2 \cdot 3} = \dfrac{2}{27}$$

Reciprocals

Now Try Exercise 47. ⟫⟫

4.6 Exercises

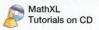
Simplify. See Example 1.

1. $\left(-\dfrac{3}{4}\right)^2$

2. $\left(-\dfrac{4}{5}\right)^2$

3. $\left(\dfrac{2}{5}\right)^3$

4. $\left(\dfrac{1}{4}\right)^3$

5. $\left(-\dfrac{1}{3}\right)^3$

6. $\left(-\dfrac{3}{5}\right)^3$

7. $\left(\dfrac{1}{2}\right)^5$

8. $\left(\dfrac{1}{3}\right)^4$

9. $\left(\dfrac{7}{10}\right)^2$

10. $\left(\dfrac{8}{9}\right)^2$

11. $\left(-\dfrac{6}{5}\right)^2$

12. $\left(-\dfrac{8}{7}\right)^2$

13. $\dfrac{15}{16}\left(\dfrac{4}{5}\right)^3$

14. $-8\left(-\dfrac{3}{8}\right)^2$

15. $\left(\dfrac{1}{3}\right)^4\left(\dfrac{9}{10}\right)^2$

16. $\left(\dfrac{4}{5}\right)^2\left(\dfrac{1}{2}\right)^6$

17. $\left(-\dfrac{3}{2}\right)^3\left(-\dfrac{2}{3}\right)^2$

18. $\left(\dfrac{5}{6}\right)^2\left(-\dfrac{2}{5}\right)^3$

RELATING CONCEPTS (EXERCISES 19–20) For Individual or Group Work

*Use your knowledge of exponents as you **work Exercises 19 and 20 in order.***

19. (a) Evaluate this series of examples.

$$\left(-\frac{1}{2}\right)^2 = \underline{\hspace{1cm}} \qquad \left(-\frac{1}{2}\right)^6 = \underline{\hspace{1cm}}$$

$$\left(-\frac{1}{2}\right)^3 = \underline{\hspace{1cm}} \qquad \left(-\frac{1}{2}\right)^7 = \underline{\hspace{1cm}}$$

$$\left(-\frac{1}{2}\right)^4 = \underline{\hspace{1cm}} \qquad \left(-\frac{1}{2}\right)^8 = \underline{\hspace{1cm}}$$

$$\left(-\frac{1}{2}\right)^5 = \underline{\hspace{1cm}} \qquad \left(-\frac{1}{2}\right)^9 = \underline{\hspace{1cm}}$$

20. Several drops of ketchup fell on Ron's homework. Explain how he can figure out what real number is covered by each drop. Be careful. More than one number may work, or there may not be any real number that works.

(a) $\left(\blacksquare\right)^2 = \frac{4}{9}$ **(b)** $\left(\blacksquare\right)^3 = -\frac{1}{27}$

(c) $\left(\blacksquare\right)^4 = \frac{1}{16}$ **(d)** $\left(\blacksquare\right)^2 = -\frac{9}{16}$

(e) $\left(\blacksquare\right)^2 \left(\blacksquare\right)^2 = \frac{1}{36}$

(b) Explain the pattern in the sign of the answers.

Simplify. See Example 2.

21. $\dfrac{1}{5} - 6\left(\dfrac{7}{10}\right)$

22. $\dfrac{2}{9} - 4\left(\dfrac{5}{6}\right)$

23. $\left(\dfrac{4}{3} \div \dfrac{8}{3}\right) + \left(-\dfrac{3}{4} \cdot \dfrac{1}{4}\right)$

24. $\left(-\dfrac{1}{3} \cdot \dfrac{3}{5}\right) + \left(\dfrac{3}{4} \div \dfrac{1}{4}\right)$

25. $-\dfrac{3}{10} \div \dfrac{3}{5}\left(-\dfrac{2}{3}\right)$

26. $5 \div \left(-\dfrac{10}{3}\right)\left(-\dfrac{4}{9}\right)$

27. $\dfrac{8}{3}\left(\dfrac{1}{4}-\dfrac{1}{2}\right)^2$

28. $\dfrac{1}{3}\left(\dfrac{4}{5}-\dfrac{3}{10}\right)^3$

29. $-\dfrac{3}{8}+\dfrac{2}{3}\left(-\dfrac{2}{3}+\dfrac{1}{6}\right)$

30. $\dfrac{1}{6}+4\left(\dfrac{2}{5}-\dfrac{7}{10}\right)$

31. $2\left(\dfrac{1}{3}\right)^3-\dfrac{2}{9}$

32. $8\left(-\dfrac{3}{4}\right)^2+\dfrac{3}{2}$

33. $\left(-\dfrac{2}{3}\right)^3\left(\dfrac{1}{8}-\dfrac{1}{2}\right)-\dfrac{2}{3}\left(\dfrac{1}{8}\right)$

34. $\left(\dfrac{3}{5}\right)^2\left(\dfrac{5}{9}-\dfrac{2}{3}\right)\div\left(-\dfrac{1}{5}\right)^2$

35. A square operation key on a calculator is $\frac{3}{8}$ inch on each side. What is the area of the key? Use the formula $A = s^2$. (*Source:* Texas Instruments.)

$\frac{3}{8}$ in.

36. A square lot for sale in the country is $\frac{3}{10}$ mile on a side. Find the area of the lot by using the formula $A = s^2$.

$\frac{3}{10}$ mi

37. A rectangular parking lot at the megamall is $\frac{7}{10}$ mile long and $\frac{1}{4}$ mile wide. How much fencing is needed to enclose the lot? Use the formula $P = 2l + 2w$.

38. A computer chip in a rectangular shape is $\frac{7}{8}$ inch long and $\frac{5}{16}$ inch wide. An insulating strip must be put around all sides of the chip. Find the length of the strip. Use the formula $P = 2l + 2w$.

$\frac{5}{16}$ in.

$\frac{7}{8}$ in.

Simplify. See Example 3.

39. $\dfrac{-\dfrac{7}{9}}{-\dfrac{7}{36}}$

40. $\dfrac{\dfrac{15}{32}}{-\dfrac{5}{64}}$

41. $\dfrac{-15}{\dfrac{6}{5}}$

42. $\dfrac{-6}{-\dfrac{5}{8}}$

43. $\dfrac{\dfrac{4}{7}}{8}$

44. $\dfrac{-\dfrac{11}{5}}{3}$

45. $\dfrac{-\dfrac{2}{3}}{-2\dfrac{2}{5}}$

46. $\dfrac{\dfrac{1}{2}}{3\dfrac{1}{3}}$ $\circ\ \dfrac{6}{1}$

47. $\dfrac{-4\dfrac{1}{2}}{\left(\dfrac{3}{4}\right)^2}$

48. $\dfrac{1\dfrac{2}{3}}{\left(-\dfrac{1}{6}\right)^2}$ $\cdot\ \dfrac{36}{2}\ =\ \dfrac{73}{6}$

$\dfrac{2}{36}$

49. $\dfrac{\left(\dfrac{2}{5}\right)^2}{\left(-\dfrac{4}{3}\right)^2}$

50. $\dfrac{\left(-\dfrac{5}{6}\right)^2}{\left(\dfrac{1}{2}\right)^3}$

$\dfrac{23}{27}$

$\dfrac{92}{108}$ $\dfrac{200}{108}$

$\dfrac{46}{54}$

$\dfrac{25}{36}\cdot\dfrac{8}{3}$

$\dfrac{3}{0}$

4.7 Problem Solving: Equations Containing Fractions

OBJECTIVE 1 Use the multiplication property of equality to solve equations containing fractions. In **Section 2.4** you used the division property of equality to solve an equation such as $4s = 24$. The division property says that you may divide both sides of an equation by the same nonzero number and the equation will still be balanced. Now that you have some experience with fractions, let's look again at how the division property works.

$4s$ means $4 \cdot s$
$$4s = 24$$
$$\frac{4 \cdot s}{4} = \frac{24}{4} \quad \text{Divide both sides by 4}$$

Divide out the common factor of 4
$$\frac{\overset{1}{\cancel{4}} \cdot s}{\underset{1}{\cancel{4}}} = 6$$

$$s = 6$$

Because multiplication and division are related to each other, we can also *multiply* both sides of an equation by the same nonzero number and keep it balanced.

> **Multiplication and Division Properties of Equality**
>
> The **multiplication property of equality** says, if $a = b$, then $a \cdot c = b \cdot c$.
>
> Also, if $a = b$, then $\dfrac{a}{c} = \dfrac{b}{c}$ as long as c is not 0.
>
> This is the **division property of equality.**
>
> In other words, you may multiply or divide both sides of an equation by the same nonzero number and it will still be balanced.

EXAMPLE 1 Using the Multiplication Property of Equality

Solve each equation and check each solution.

(a) $\dfrac{1}{2}b = 5$

As in **Chapter 2**, you want the variable by itself on one side of the equal sign. In this example, you want $1b$, not $\frac{1}{2}b$, on the left side. (Recall that $1b$ is equivalent to b.)

In **Section 4.3** you learned that the product of a number and its reciprocal is 1. Thus, multiplying $\frac{1}{2}$ by $\frac{2}{1}$ will give the desired coefficient of 1.

$$\frac{1}{2}b = 5$$

$$\frac{2}{1}\left(\frac{1}{2}b\right) = \frac{2}{1}(5) \quad \text{Multiply both sides by } \tfrac{2}{1}$$
$$\text{(the reciprocal of } \tfrac{1}{2}\text{).}$$

On the left side, use the associative property to regroup the factors.
$$\left(\frac{2}{1} \cdot \frac{1}{2}\right)b = \frac{2}{1}\left(\frac{5}{1}\right) \quad \text{On the right side, 5 is equivalent to } \tfrac{5}{1}$$

$$\left(\frac{\overset{1}{\cancel{2}}}{1} \cdot \frac{1}{\underset{1}{\cancel{2}}}\right)b = \frac{10}{1}$$

$1b$ is equivalent to b.
$$1b = 10$$
$$b = 10$$

Continued on Next Page

Once you understand the process, you don't have to show every step. Here is a shorthand solution of the same problem.

$$\frac{1}{2}b = 5$$

$$\frac{\overset{1}{\cancel{2}}}{\underset{1}{1}}\left(\frac{1}{\underset{1}{\cancel{2}}}b\right) = \frac{2}{1}(5)$$

$$b = 10$$

The solution is 10. Check the solution by going back to the *original* equation.

Check $\dfrac{1}{2}\boldsymbol{b} = 5$ Replace *b* with 10 in the original equation.

$\dfrac{1}{2}(\mathbf{10}) = 5$ Multiply on the left side: $\frac{1}{2}(10)$ is $\frac{1}{2}\cdot\frac{10}{1}$, or $\frac{1}{2}\cdot\frac{2\cdot 5}{1}$

$\dfrac{1\cdot\overset{1}{\cancel{2}}\cdot 5}{\underset{1}{\cancel{2}}\cdot 1} = 5$

$5 = 5$ Balances

When *b* is 10, the equation balances, so 10 is the correct solution (**not** 5).

(b) $12 = -\dfrac{3}{4}x$

$12 = -\dfrac{3}{4}x$ Multiply both sides by $-\frac{4}{3}$ (the reciprocal of $-\frac{3}{4}$). The reciprocal of a negative number is also negative.

$-\dfrac{4}{3}(12) = -\dfrac{\overset{1}{\cancel{4}}}{\cancel{3}}\left(-\dfrac{\cancel{3}}{\cancel{4}}x\right)$ On the left side $-\frac{4}{3}(12)$ is $-\frac{4}{3}\cdot\frac{12}{1}$ or $-\dfrac{4\cdot\overset{1}{\cancel{3}}\cdot 4}{\underset{1}{\cancel{3}}\cdot 1} = -16$

$-16 = x$

The solution is -16. Check the solution by going back to the *original* equation.

Check $12 = -\dfrac{3}{4}\boldsymbol{x}$ Replace *x* with -16 in the original equation.

$12 = -\dfrac{3}{4}(\mathbf{-16})$ The product of two negative numbers is positive.

$12 = \dfrac{3\cdot\overset{1}{\cancel{4}}\cdot 4}{\underset{1}{\cancel{4}}\cdot 1}$ $-\frac{3}{4}(-16)$ is $\frac{3}{4}\cdot\frac{16}{1}$, or $\frac{3}{4}\cdot\frac{4\cdot 4}{1}$

$12 = 12$ Balances

When *x* is -16, the equation balances, so -16 is the correct solution (*not* 12).

Continued on Next Page

(c) $-\dfrac{2}{5}n = -\dfrac{1}{3}$

$$-\dfrac{\overset{1}{\cancel{5}}}{\underset{1}{\cancel{2}}}\left(-\dfrac{\overset{1}{\cancel{2}}}{\underset{1}{\cancel{5}}}n\right) = \left(-\dfrac{5}{2}\right)\left(-\dfrac{1}{3}\right)$$ Multiply both sides by $-\dfrac{5}{2}$ (the reciprocal of $-\dfrac{2}{5}$).

$$n = \dfrac{5 \cdot 1}{2 \cdot 3}$$ The product of two negative numbers is positive.

$$n = \dfrac{5}{6}$$

Check $\qquad -\dfrac{2}{5}\boldsymbol{n} = -\dfrac{1}{3}$ $\qquad$ Original equation

$$\left(-\dfrac{2}{5}\right)\left(\dfrac{5}{6}\right) = -\dfrac{1}{3}$$ $\qquad$ Replace n with $\frac{5}{6}$

$$-\dfrac{\overset{1}{\cancel{2}} \cdot \overset{1}{\cancel{5}}}{\underset{1}{\cancel{5}} \cdot \underset{1}{\cancel{2}} \cdot 3} = -\dfrac{1}{3}$$ $\qquad$ Multiply on the left side.

$$-\dfrac{1}{3} = -\dfrac{1}{3}$$ $\qquad$ Balances

When n is $\frac{5}{6}$, the equation balances, so $\frac{5}{6}$ is the correct solution (***not*** $-\frac{1}{3}$).

>>> **Now Try Exercises 1 and 9.**

OBJECTIVE 2 **Use both the addition and multiplication properties of equality to solve equations containing fractions.** In **Section 2.5** you used both the addition and *division* properties of equality to solve equations. Now you can use both the addition and *multiplication* properties.

EXAMPLE 2 **Using the Addition and Multiplication Properties of Equality**

Solve each equation and check each solution.

(a) $\dfrac{1}{3}c + 5 = 7$

The first step is to get the variable term $\frac{1}{3}c$ by itself on the left side of the equal sign. Recall that to "get rid of" the 5 on the left side, add the opposite of 5, which is -5, to both sides.

$$\dfrac{1}{3}c + 5 = 7$$

$$\underline{\phantom{\dfrac{1}{3}c} -5 \quad -5}$$ Add -5 to both sides.

$$\underbrace{\dfrac{1}{3}c + 0}= 2$$

$$\dfrac{1}{3}c = 2$$

$$\dfrac{\overset{1}{\cancel{3}}}{1}\left(\dfrac{1}{\cancel{3}}c\right) = \dfrac{3}{1}\left(\dfrac{2}{1}\right)$$ Multiply both sides by $\dfrac{3}{1}$ (the reciprocal of $\frac{1}{3}$).

$$c = 6$$

The solution is 6. Check the solution by going back to the *original* equation.

—— **Continued on Next Page**

Check $\dfrac{1}{3}c + 5 = 7$ Original equation

$\dfrac{1}{3}(6) + 5 = 7$ Replace c with 6.

$2 + 5 = 7$

$7 = 7$ Balances

When c is 6, the equation balances, so 6 is the correct solution (*not* 7).

(b) $-3 = \dfrac{2}{3}y + 7$

To get the variable term $\frac{2}{3}y$ by itself on the right side, add -7 to both sides.

$$-3 = \dfrac{2}{3}y + 7$$

$$\underline{-7 \qquad\qquad -7}\qquad \text{Add } -7 \text{ to both sides.}$$

$$-10 = \dfrac{2}{3}y + 0$$

$$\dfrac{3}{2}(-10) = \dfrac{\overset{1}{\cancel{3}}}{\underset{1}{\cancel{2}}}\left(\dfrac{\overset{1}{\cancel{2}}}{\underset{1}{\cancel{3}}}y\right) \qquad \begin{array}{l}\text{Multiply both sides by } \frac{3}{2} \\ \text{(the reciprocal of } \frac{2}{3}\text{).}\end{array}$$

$$-15 = y$$

Check $-3 = \dfrac{2}{3}y + 7$ Original equation

$-3 = \dfrac{2}{3}(-15) + 7$ Replace y with -15

$-3 = -10 + 7$

$-3 = -3$ Balances

When y is -15, the equation balances, so -15 is the correct solution (*not* -3).

Now Try Exercise 11. ▶▶▶

OBJECTIVE 3 Solve application problems using equations containing fractions.   Use the six problem-solving steps from **Section 3.3** to solve application problems.

EXAMPLE 3 Solving an Application Problem Using an Equation with Fractions

The expression for finding a person's approximate systolic blood pressure is $100 + \dfrac{\text{age}}{2}$. Suppose your friend's systolic blood pressure is 116 (and he has normal blood pressure). Find his age.

Step 1 **Read** the problem. It is about blood pressure and age.

Unknown: friend's age

Known: Blood pressure expression is $100 + \dfrac{\text{age}}{2}$; friend's pressure is 116.

Step 2 **Assign a variable.** Let *a* represent the friend's age.

Step 3 **Write an equation.**

$$100 + \frac{\textbf{age}}{2} \text{ is blood pressure}$$

$$100 + \frac{a}{2} = 116$$

Step 4 **Solve.**

$$100 + \frac{a}{2} = 116$$

$$\underline{-100 \qquad\qquad -100} \qquad \text{Add } -100 \text{ to both sides.}$$

$$0 + \frac{a}{2} = 16$$

$$\frac{a}{2} = 16$$

$\frac{1}{2}a$ is equivalent to $\frac{a}{2}$ because $\frac{1}{2}a$ is $\frac{1}{2} \cdot \frac{a}{1}$

$$\frac{1}{2}\,a = 16$$

$$\frac{\overset{1}{\cancel{2}}}{1}\left(\frac{1}{\underset{1}{\cancel{2}}}\,a\right) = \frac{2}{1}(16) \qquad \text{Multiply both sides by } \frac{2}{1} \text{ (the reciprocal of } \frac{1}{2}\text{).}$$

$$a = 32$$

Step 5 **State the answer.** Your friend is 32 years old.

Step 6 **Check** the solution by putting it back into the *original* problem.

Approximate systolic blood pressure is $100 + \dfrac{\text{age}}{2}$.

If the age is 32, then $100 + \frac{32}{2} = 100 + 16 = 116$.

Friend's blood pressure is 116. ⟵ Matches

Age 32 is the correct solution because it "works" when you put it back into the original problem.

Now Try Exercise 29.

4.7 Exercises

FOR EXTRA HELP   Addison-Wesley Math Tutor Center   MathXL   Digital Video Tutor CD 2 Videotape 4   📖 Student's Solutions Manual   MyMathLab   💿 MathXL Tutorials on CD

Solve each equation and check each solution. See Examples 1 and 2.

1. $\dfrac{1}{3}a = 10$ *Check*

2. $7 = \dfrac{1}{5}y$ *Check*

3. $-20 = \dfrac{5}{6}b$ *Check*

4. $-\dfrac{4}{9}w = 16$ *Check*

5. $\dfrac{1}{?} - \dfrac{7}{2}c = -21$ *Check*

6. $-25 = \dfrac{5}{3}x$ *Check*

7. $\dfrac{9}{16} = \dfrac{3}{4}m$ *Check*

8. $\dfrac{5}{12}k = \dfrac{15}{16}$ *Check*

9. $\dfrac{3}{10} = -\dfrac{1}{4}d$ *Check*

10. $-\dfrac{7}{8}h = -\dfrac{1}{6}$ *Check*

11. $\dfrac{1}{6}n + 7 = 9$ *Check*

12. $3 + \dfrac{1}{4}p = 5$ *Check*

13. $-10 = \dfrac{5}{3}r + 5$ *Check*

14. $0 = 6 + \dfrac{3}{2}t$ *Check*

15. $\dfrac{3}{8}x - 9 = 0$ *Check*

16. $\dfrac{1}{3}s - 10 = -5$ *Check*

Solve each equation. Show your work.

17. $7 - 2 = \dfrac{1}{5}y - 4$

18. $0 - 8 = \dfrac{1}{10}k - 3$

19. $4 + \dfrac{2}{3}n = -10 + 2$

20. $-\dfrac{2}{5}m - 3 = -9 + 0$

21. $3x + \dfrac{1}{2} = \dfrac{3}{4}$

22. $4y + \dfrac{1}{3} = \dfrac{7}{9}$

23. $\dfrac{3}{10} = -4b - \dfrac{1}{5}$

24. $\dfrac{5}{6} = -3c - \dfrac{2}{3}$

25. Check the solution given for each equation. If a solution doesn't check, show how to find the correct solution.

(a) $\dfrac{1}{6}x + 1 = -2$

$x = 18$

(b) $-\dfrac{3}{2} = \dfrac{9}{4}k$

$k = -\dfrac{2}{3}$

26. Check the solution given for each equation. If a solution doesn't check, show how to find the correct solution.

(a) $-\dfrac{3}{4}y = -\dfrac{5}{8}$

$y = \dfrac{5}{6}$

(b) $16 = -\dfrac{7}{3}w + 2$

$w = 6$

27. Write two different equations that have 8 as a solution. Write your equations with a fraction as the coefficient of the variable term. Use Exercises 1–10 as models.

28. Write two different equations that have -12 as the solution. Write your equations with a fraction as the coefficient of the variable term. Use Exercises 1–10 as models.

In Exercises 29–32, find each person's age using the six problem-solving steps and this expression for approximate systolic blood pressure: $100 + \dfrac{age}{2}$. *Assume that all the people have normal blood pressure. See Example 3.*

29. A man has systolic blood pressure of 109. How old is he?

30. A man has systolic blood pressure of 118. How old is he?

31. A woman has systolic blood pressure of 122. How old is she?

32. A woman has systolic blood pressure of 113. How old is she?

As you read at the start of this chapter, nails for home repair projects are classi-fied by the "penny" system, which is a number that indicates the nail's length. The expression for finding the length of a nail, in inches, is $\dfrac{\text{penny size}}{4} + \dfrac{1}{2}$ inch. In Exercises 33–36, find the penny size for each nail using this expression and the six problem-solving steps. (Source: Season by Season Home Maintenance.)

33. The length of a common nail is 3 inches. What is its penny size?

3 in.

34. The length of a drywall nail is 2 inches. Find the nail's penny size.

2 in.

35. The length of a box nail is $2\frac{1}{2}$ inches. What penny size would you ask for when buying these nails?

$2\frac{1}{2}$ in.

36. The length of a finishing nail is $1\frac{1}{2}$ inches. What is its penny size?

$1\frac{1}{2}$ in.

An expression for the recommended weight of an adult is $\frac{11}{2}$ (height in inches) − 220. In Exercises 37–40, find each person's height using this expression and the six problem-solving steps. Assume that all the people are at their recommended weight.

37. A man weighs 209 pounds. What is his height in inches?

38. A woman weighs 165 pounds. What is her height in inches?

39. A woman weighs 132 pounds. What is her height in inches?

40. A man weighs 176 pounds. What is his height in inches?

4.8 Geometry Applications: Area and Volume

OBJECTIVE 1 Find the area of a triangle. In **Section 3.1** you worked with triangles, which are flat shapes that have exactly three sides. You found the perimeter of a triangle by adding the lengths of the three sides. Now you are ready to find the area of a triangle (the amount of surface inside the triangle).

You can find the *height* of a triangle by measuring the distance from one vertex of the triangle to the opposite side (the base). The height must be *perpendicular* to the base, that is, it must form a right angle with the base. Sometimes you have to extend the base before you can draw the height perpendicular to it, as shown on the right-hand figure below.

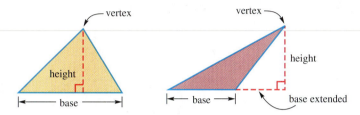

If you cut out two identical triangles and turn one upside down, you can fit them together to form a parallelogram, as shown below.

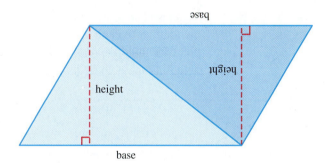

Recall from **Section 3.2** that the area of the parallelogram is *base* times *height*. Because each triangle is *half* of the parallelogram, the area of one triangle is

$$\frac{1}{2} \text{ of base times height.}$$

Use the following formula to find the area of a triangle.

> **Finding the Area of a Triangle**
>
> $$\text{Area of triangle} = \frac{1}{2} \cdot \text{base} \cdot \text{height}$$
>
> $$A = \frac{1}{2}bh$$
>
> Remember to use *square units* when measuring area.

OBJECTIVES

1 Find the area of a triangle.
2 Find the volume of a rectangular solid.
3 Find the volume of a pyramid.

EXAMPLE 1 Finding the Area of Triangles

Find the area of each triangle.

(a)

26 ft   22 ft $39\frac{3}{4}$ ft

47 ft

The base is 47 ft and the height is 22 ft. You do *not* need the 26 ft or $39\frac{3}{4}$ ft sides to find the area.

$$A = \frac{1}{2} \cdot b \cdot h$$

Replace *b* with 47 ft and *h* with 22 ft.

$$A = \frac{1}{2} \cdot 47 \text{ ft} \cdot 22 \text{ ft}$$

$$A = \frac{1}{2} \cdot \frac{47 \text{ ft}}{1} \cdot \frac{22 \text{ ft}}{1}$$

Divide out the common factor of 2

$$A = \frac{1 \cdot 47 \text{ ft} \cdot \overset{1}{\cancel{2}} \cdot 11 \text{ ft}}{\underset{1}{\cancel{2}} \cdot 1 \cdot 1}$$

Multiply 47 · 11 to get 517

$$A = 517 \text{ ft}^2 \quad \text{Multiply ft} \cdot \text{ft to get ft}^2.$$

(b)

$11\frac{1}{10}$ in. $6\frac{1}{2}$ in.

9 in.

Because two sides of the triangle are perpendicular to each other, use those sides as the base and the height. (Remember that the height must be perpendicular to the base.)

$$A = \frac{1}{2} bh$$

Formula for area of a triangle

$$A = \frac{1}{2} \cdot 9 \text{ in.} \cdot 6\frac{1}{2} \text{ in.}$$

Replace *b* with 9 in. and *h* with $6\frac{1}{2}$ in.

$$A = \frac{1}{2} \cdot \frac{9 \text{ in.}}{1} \cdot \frac{13 \text{ in.}}{2}$$

Write 9 in. and $6\frac{1}{2}$ in. as improper fractions.

$$A = \frac{1 \cdot 9 \text{ in.} \cdot 13 \text{ in.}}{2 \cdot 1 \cdot 2}$$

Multiply 9 · 13 to get 117 and in. · in. to get in.^{2}

$$A = \frac{117}{4} \text{ in.}^2 \quad \text{or} \quad 29\frac{1}{4} \text{ in.}^2$$

> **Now Try Exercise 5.**

EXAMPLE 2 Using the Concept of Area

Find the area of the shaded part in this figure.

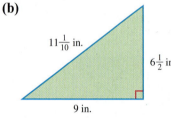

32 cm   40 cm

30 cm

The *entire* figure is a rectangle.

$$A = lw$$
$$A = 30 \text{ cm} \cdot 40 \text{ cm} = 1200 \text{ cm}^2$$

Continued on Next Page

The *un*shaded part
is a triangle.

$$A = \frac{1}{2}bh$$

$$A = \frac{1}{2} \cdot \frac{30 \text{ cm}}{1} \cdot \frac{32 \text{ cm}}{1}$$

$$A = \frac{1 \cdot \overset{1}{2} \cdot 15 \text{ cm} \cdot 32 \text{ cm}}{\underset{1}{2} \cdot 1 \cdot 1}$$

$$A = 480 \text{ cm}^2$$

Subtract to find the area of the shaded part.

$$\overset{\text{Entire area}}{A = 1200 \text{ cm}^2} - \overset{\text{Unshaded part}}{480 \text{ cm}^2} = \overset{\text{Shaded part}}{720 \text{ cm}^2}$$

> **Now Try Exercise 11.**

OBJECTIVE 2 Find the volume of a rectangular solid.   A shoe box
and a cereal box are examples of three-dimensional (or solid) figures. The
three dimensions are length, width, and height. (A rectangle or square is
a two-dimensional figure. The two dimensions are length and width.) If we
want to know how much a shoe box will hold, we find its *volume*. We measure volume by seeing how many cubes of a certain size will fill the space
inside the box. Three sizes of *cubic units* are shown here. Notice that all the
edges of a cube have the same length and all sides meet at right angles.

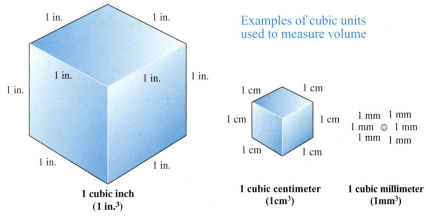

Examples of cubic units
used to measure volume

1 cubic inch
(1 in.³)

1 cubic centimeter
(1cm³)

1 cubic millimeter
(1mm³)

Some other sizes of cubes that are used to measure volume are 1 cubic foot
(1 ft³), 1 cubic yard (1 yd³), and 1 cubic meter (1 m³).

> **CAUTION**
> The raised 3 in 4^3 means that you multiply $4 \cdot 4 \cdot 4$ to get 64. The raised
> 3 in cm³ or ft³ is a short way to write the word "cubic." It means that you
> multiplied cm times cm times cm to get cm³, or ft times ft times ft to get
> ft³. Recall that a short way to write $x \cdot x \cdot x$ is x^3. Similarly, cm $\cdot$ cm $\cdot$ cm
> is cm³. When you see 5 cm³, say "five cubic centimeters." Do *not* multi-
> ply $5 \cdot 5 \cdot 5$ because the exponent applies only to the *first* thing to its left.
> The exponent applies to cm, *not* to the 5.

Volume

Volume is a measure of the space inside a solid shape. The volume of a
solid is the number of cubic units it takes to fill the solid.

Use the formula below to find the volume of *rectangular solids* (box-like shapes).

> **Finding the Volume of Rectangular Solids**
>
> Volume of a rectangular solid = length • width • height
> $$V = lwh$$
>
> Remember to use *cubic units* when measuring volume.

EXAMPLE 3 **Finding the Volume of Rectangular Solids**

Find the volume of each box.

(a)

Each cube that fits in the box is 1 cubic centimeter (1 cm^3). To find the volume, you can count the number of cubes.

Bottom layer has 12 cubes.

Top layer has 12 cubes.

Total of 24 cubes (24 cm^3)

Or you can use the formula for rectangular solids.

$$V = \quad l \quad • \quad w \quad • \quad h$$

$$V = \textbf{4 cm} • \textbf{3 cm} • \textbf{2 cm} \qquad \text{Multiply } 4 • 3 • 2 \text{ to get } 24$$

$$V = 24 \text{ cm}^3 \qquad \text{Multiply cm • cm • cm to get cm}^3.$$

(b)

10 in.

7 in.

$2\frac{1}{2}$ in.

Use the formula $V = lwh$.

$$V = \textbf{7 in.} • \textbf{2}\frac{1}{2}\textbf{ in.} • \textbf{10 in.} \qquad \text{Write each measurement as an improper fraction.}$$

$$V = \frac{7 \text{ in.}}{1} • \frac{5 \text{ in.}}{2} • \frac{10 \text{ in.}}{1}$$

$$V = \frac{7 \text{ in.} • 5 \text{ in.} • \overset{1}{\cancel{2}} • 5 \text{ in.}}{1 • \underset{1}{\cancel{2}} • 1} \qquad \text{Divide out the common factor of 2}$$

$$V = 175 \text{ in.}^3 \qquad \text{Cubic units for volume}$$

Now Try Exercise 21.

OBJECTIVE 3 Find the volume of a pyramid. A pyramid is a solid shape like the one shown below. We will study pyramids with square or rectangular bases.

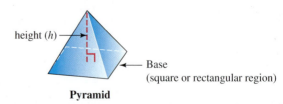

height (*h*)

Base (square or rectangular region)

Pyramid

The height is the distance from the base to the highest point of the pyramid. The height must be perpendicular to the base.

> **NOTE**
> In this book we will work only with pyramids that have a base with four sides that is square or rectangular. In later math courses you may work with pyramids that have a base with three sides (triangle), five sides (pentagon), six sides (hexagon), and so on.

Use this formula to find the volume of a pyramid.

Finding the Volume of a Pyramid

$$\text{Volume of a pyramid } = \frac{1}{3} \bullet B \bullet h$$

$$V = \frac{1}{3} Bh$$

where B is the area of the base of the pyramid and h is the height of the pyramid.

Remember to use *cubic units* when measuring volume.

EXAMPLE 4   **Finding the Volume of a Pyramid**

Find the volume of this pyramid with a rectangular base.

11 cm

4 cm

5 cm

First find the value of B in the formula, which is the *area of the rectangular base*. Recall that the area of a rectangle is found by multiplying length times width.

$$B = 5 \text{ cm} \bullet 4 \text{ cm}$$

$$\mathbf{B = 20 \text{ cm}^2}$$

Next, find the volume.

$$V = \frac{1}{3} \boldsymbol{Bh}$$ Formula for volume of pyramid

$$V = \frac{1}{3} \bullet \mathbf{20 \text{ cm}^2} \bullet \mathbf{11 \text{ cm}}$$ Replace B with 20 cm² and h with 11 cm.

$$V = \frac{1}{3} \bullet \frac{20 \text{ cm}^2}{1} \bullet \frac{11 \text{ cm}}{1}$$

$$V = \frac{1 \bullet 20 \text{ cm}^2 \bullet 11 \text{ cm}}{3 \bullet 1 \bullet 1}$$ There are no common factors to divide out.

$$V = \frac{220}{3} \text{ cm}^3 \quad \text{or} \quad 73\frac{1}{3} \text{ cm}^3$$ Cubic units for volume

Now Try Exercise 25. ▶▶▶

4.8 Exercises

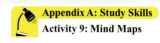
Find the perimeter and area of each triangle. See Example 1.

1.
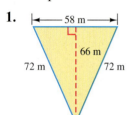
58 m
66 m
72 m   72 m

2.
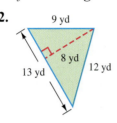
9 yd
8 yd
13 yd   12 yd

3.
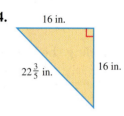
$2\frac{1}{4}$ ft
$1\frac{1}{4}$ ft $1\frac{1}{2}$ ft
$\frac{3}{4}$ ft

4.
16 in.
$22\frac{3}{5}$ in.   16 in.

5.
9 yd   6 yd   7 yd
$10\frac{1}{4}$ yd

6.
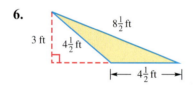
$8\frac{1}{2}$ ft
3 ft $4\frac{1}{2}$ ft
$4\frac{1}{2}$ ft

7.
$12\frac{3}{5}$ yd $7\frac{2}{3}$ yd
10 yd

8.
26 cm   26 cm
24 cm
20 cm

Find the shaded area in each figure. See Example 2.

9.

9 m
12 m
12 m   12 m
12 m

10.

22 m
20 m   19 m
27 m   20 m
22 m

11.
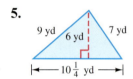
52 m
8 m
37 m   37 m
52 m

12.
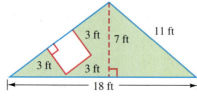
3 ft   7 ft   11 ft
3 ft
3 ft   3 ft
18 ft

13. Explain the difference between perimeter, area, and volume.

14. Explain where the $\frac{1}{2}$ comes from in the formula for area of a triangle. Draw a figure to illustrate your explanation.

Solve each application problem.

15. A triangular tent flap measures $3\frac{1}{2}$ ft along the base and has a height of $4\frac{1}{2}$ ft. How much canvas is needed to make the flap?

16. A wooden sign in the shape of a right triangle has perpendicular sides measuring $1\frac{1}{2}$ yd and 1 yd. How much surface area does the sign have?

17. A triangular space between three streets has the measurements shown below. How much new curbing will be needed to go around the space? How much sod will be needed to cover the space?

18. Each gable end of a new house has a span of 36 ft and a rise of $9\frac{1}{2}$ ft. What is the total area of both gable ends of the house?

19. A city lot with an unusual shape is shown below.

   (a) How much frontage (distance along streets) does the lot have?

   (b) What is the area of the lot?

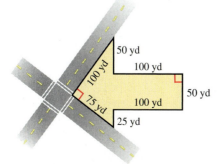

20. The sketch below shows the plan for an office building. The shaded part will be the parking lot. Find the area of the parking lot.

In Exercises 21–26, name each solid and find its volume. See Examples 3 and 4.

21.

12 cm

11 cm

4 cm

22.

10 in.

10 in.

10 in.

23.

$2\frac{1}{2}$ in.

$2\frac{1}{2}$ in.

$2\frac{1}{2}$ in.

24.

$3\frac{1}{2}$ ft

8 ft

2 ft

25.

20 cm

15 cm

8 cm

26.

15 m

5 m

7 m

27. Find the volume of a pyramid with a height of 5 ft and a square base that is 8 ft on each side.

28. A pyramid has a height of 4 yd. The rectangular base measures 10 yd by 4 yd. What is the volume of the pyramid?

29. A box to hold pencils measures 3 in. by 8 in. by $\frac{3}{4}$ in. high. Find the volume of the box. (*Source:* Faber Castell.)

$\frac{3}{4}$ in.

8 in.

3 in.

30. A train is being loaded with shipping crates. Each crate is 12 ft long, 8 ft wide, and $2\frac{1}{4}$ ft high. How much space will each crate take?

$2\frac{1}{4}$ ft

8 ft

12 ft

31. One of the ancient stone pyramids in Egypt has a square base that measures 145 m on each side. The height is 93 m. What is the volume of the pyramid? (*Source: Columbia Encyclopedia.*)

32. A cardboard model of an ancient stone pyramid has a square base that is $10\frac{3}{8}$ in. on each side. The height is $6\frac{1}{2}$ in. Find the volume of the model.

height $6\frac{1}{2}$ in. $10\frac{3}{8}$ in. $10\frac{3}{8}$ in.

33. Find the volume.

2 cm   12 cm   3 cm   9 cm   11 cm

34. Find the volume. (*Hint:* Notice the square hole that goes through the center of the rectangular solid.)

14 cm   8 cm   8 cm   14 cm   15 cm

The following are some answers from a student's test. The number part of each answer is correct, but some of the units are not. Find the errors, explain what is wrong, and correct the errors.

35. There are *two* errors in this list of answers.

$$A = 135^2 \text{ ft}$$
$$V = 76 \text{ yd}^3$$
$$P = 5\frac{1}{2} \text{ in.}^2$$

36. There are *two* errors in this list of answers.

$$P = 215 \text{ cm}$$
$$A = 8\frac{1}{2} \text{ m}$$
$$V = 98^3 \text{ in.}$$

Chapter **4** Group Activity: **MUSIC**

Objective   Add and subtract unlike fractions.

The time signature at the beginning of a piece of music looks like a fraction. Commonly used time signatures are $\frac{2}{4}$, $\frac{3}{4}$, $\frac{4}{4}$, and $\frac{6}{8}$. Musicians use the time signature to tell how long to hold each note. The values of different notes can be written as fractions:

$$\mathbf{o} = 1 \qquad = \frac{1}{2} \qquad = \frac{1}{4} \qquad = \frac{1}{8} \qquad = \frac{1}{16}$$

Music is divided into measures. In $\frac{4}{4}$ time, each measure contains notes that add up to $\frac{4}{4}$ (or 1).

In $\frac{2}{4}$ time the notes in each measure add up to $\frac{2}{4}\left(\text{or } \frac{1}{2}\right)$, and so on for $\frac{3}{4}$ time and $\frac{6}{8}$ time.

Write one or more notes in each measure to make it add up to its time signature. Use as many different kinds of notes as possible. Write the addition problem underneath the notes that shows how the notes add up.

$\left.\begin{array}{l}\text{Addition} \\ \text{problems}\end{array}\right\}$ $\frac{1}{4} + \frac{1}{4} + \frac{1}{4} +$

$\left.\begin{array}{l}\text{Addition} \\ \text{problems}\end{array}\right\}$ $\frac{1}{16} + \frac{1}{16} +$

$\left.\begin{array}{l}\text{Addition} \\ \text{problems}\end{array}\right\}$ $\frac{1}{2} +$

Below are excerpts from "Jingle Bells" and "The Star-Spangled Banner." Divide each line of music into measures based on the time signature. Write an addition problem underneath each measure to show how the notes add up.

4.1	**fraction**	A fraction is a number of the form $\dfrac{a}{b}$, where a and b are integers and b is not 0.
	numerator	The top number in a fraction is the numerator. It shows how many of the equal parts are being considered.
	denominator	The bottom number in a fraction is the denominator. It shows how many equal parts are in the whole.
	proper fraction	In a proper fraction, the numerator is smaller than the denominator. The fraction is less than 1.
	improper fraction	In an improper fraction, the numerator is greater than or equal to the denominator. The fraction is greater than or equal to 1.
	equivalent fractions	Equivalent fractions have the same value even though they look different. When graphed on a number line, they are names for the same point.
4.2	**lowest terms**	A fraction is written in lowest terms when its numerator and denominator have no common factor other than 1.
	prime number	A prime number is a whole number that has exactly two different factors, itself and 1. The first few prime numbers are 2, 3, 5, 7, 11, 13, and 17.
	composite number	A composite number has at least one factor other than itself and 1. Examples are 4, 6, 9, and 10. The numbers 0 and 1 are neither prime nor composite.
	prime factorization	In a prime factorization, every factor is a prime number. For example, the prime factorization of 24 is $2 \cdot 2 \cdot 2 \cdot 3$.
4.3	**reciprocal**	Two numbers are reciprocals of each other if their product is 1. The reciprocal of $\dfrac{a}{b}$ is $\dfrac{b}{a}$ because $\dfrac{a}{b} \cdot \dfrac{b}{a} = 1$.
4.4	**like fractions**	Like fractions have the same denominator.
	unlike fractions	Unlike fractions have different denominators.
	least common denominator	The least common denominator (LCD) for two fractions is the smallest positive number that can be divided evenly by both denominators.
4.5	**mixed number**	A mixed number is a fraction and a whole number written together. It represents the sum of the whole number and the fraction. For example, $5\frac{1}{3}$ represents $5 + \frac{1}{3}$.
4.6	**complex fraction**	A complex fraction is a fraction in which the numerator and/or denominator contain one or more fractions.
4.7	**multiplication property of equality**	The multiplication property of equality states that you may multiply both sides of an equation by the same nonzero number and it will still be balanced.
	division property of equality	The division property of equality states that you may divide both sides of an equation by the same nonzero number and it will still be balanced.
4.8	**volume**	Volume is a measure of the space inside a solid shape. Volume is measured in cubic units, such as in.^{3}, ft^3, yd^3, mm^3, cm^3, and so on.

Cubic units (for measuring volume)   in.^{3} ft^3 yd^3 mm^3 cm^3 m^3

NEW FORMULAS

Area of a triangle: $A = \frac{1}{2}bh$

Volume of a rectangular solid: $V = lwh$

Volume of a pyramid: $V = \frac{1}{3}Bh$

TEST YOUR WORD POWER

See how well you have learned the vocabulary in this chapter. Answers follow the Quick Review.

1. A **fraction** is in lowest terms if
   A. it has a value less than 1
   B. its numerator and denominator have no common factor other than 1
   C. its numerator and denominator are composite numbers
   D. it is rewritten as a mixed number.

2. The **denominator** of a fraction
   A. is written above the fraction bar
   B. is a prime number
   C. shows how many equal parts are in the whole
   D. is the smallest number divisible by the numerator.

3. The **LCD** of two fractions is the
   A. smallest number divisible by both denominators
   B. largest factor common to both denominators
   C. smallest number divisible by both numerators
   D. smallest prime number that divides evenly into both denominators.

4. Two numbers are **reciprocals** of each other if
   A. they have the same prime factorizations
   B. their sum is 0
   C. they are written in lowest terms
   D. their product is 1.

5. **Volume** is
   A. measured in square units
   B. the space inside a solid shape
   C. the sum of the lengths of the sides of a shape
   D. found by multiplying base times height.

6. A **mixed number**
   A. has a value equal to 1
   B. is the reciprocal of an improper fraction
   C. is the sum of a whole number and a fraction
   D. has a value less than 1.

7. A whole number is **prime** if
   A. it is divisible by itself and 1
   B. it has only composite factors
   C. it cannot be divided
   D. it has exactly two factors, itself and 1.

8. **Equivalent fractions**
   A. have the same denominators
   B. are written in lowest terms
   C. name the same point on a number line
   D. are reciprocals of each other.

QUICK REVIEW

Concepts	Examples
4.1 Understanding Fraction Terminology The *numerator* is the top number. The *denominator* is the bottom number. In a *proper fraction,* the numerator is smaller than the denominator, so the fraction is less than 1. In an *improper fraction,* the numerator is greater than or equal to the denominator, so the fraction is greater than or equal to 1.	**Proper fractions** $\dfrac{2}{3}, \dfrac{3}{4}, \dfrac{15}{16}, \dfrac{1}{8}$ ← Numerator ← Denominator **Improper fractions** $\dfrac{17}{8}, \dfrac{19}{12}, \dfrac{11}{2}, \dfrac{5}{3}, \dfrac{7}{7}$
4.1 Writing Equivalent Fractions Multiply or divide the numerator and denominator by the same nonzero number. The result is an equivalent fraction.	$\dfrac{1}{2} = \dfrac{1 \cdot 8}{2 \cdot 8} = \dfrac{8}{16}$ ← Equivalent to $\frac{1}{2}$ $-\dfrac{12}{15} = -\dfrac{12 \div 3}{15 \div 3} = -\dfrac{4}{5}$ ← Equivalent to $-\frac{12}{15}$

Concepts	**Examples**

4.2 *Finding Prime Factorizations*

A prime factorization of a number shows the number as the product of prime numbers. The first few prime numbers are 2, 3, 5, 7, 11, 13, and 17. You can use a division method or a factor tree to find the prime factorization.

Find the prime factorization of 24.

Division Method:

$2\overline{)24}$ ←Divide 24 by 2, the first prime; quotient is 12

$2\overline{)12}$ ←Divide 12 by 2; quotient is 6

$2\overline{)6}$ ←Divide 6 by 2; quotient is 3

$3\overline{)3}$ ← Divide 3 by 3; quotient is 1

1 ← Continue to divide until the quotient is 1

$24 = 2 \cdot 2 \cdot 2 \cdot 3$

Factor Tree Method:

Circle each prime number.

```
        24
       /  \
      2    12
          /  \
         2    6
             / \
            2   3
```

$24 = 2 \cdot 2 \cdot 2 \cdot 3$

4.2 *Writing Fractions in Lowest Terms*

Write the prime factorization of both numerator and denominator. Divide out all common factors, using slashes to show the division. Multiply any remaining factors in the numerator and in the denominator.

$$\frac{18}{90} = \frac{2 \cdot 3 \cdot 3}{2 \cdot 3 \cdot 3 \cdot 5} = \frac{1}{5}$$

$$\frac{2b^3}{8ab} = \frac{2 \cdot b \cdot b \cdot b}{2 \cdot 2 \cdot 2 \cdot a \cdot b} = \frac{b^2}{4a}$$

4.3 *Multiplying Fractions*

Multiply the numerators and multiply the denominators. The product must be written in lowest terms. One way to do this is to write each original numerator and each original denominator as the product of primes and divide out any common factors before multiplying.

$$\left(-\frac{7}{10}\right)\left(\frac{5}{6}\right) = -\frac{7 \cdot 5}{2 \cdot 5 \cdot 2 \cdot 3} = -\frac{7}{12}$$

$$\frac{3x^2}{5} \cdot \frac{2}{9x} = \frac{3 \cdot x \cdot x \cdot 2}{5 \cdot 3 \cdot 3 \cdot x} = \frac{2x}{15}$$

4.3 *Dividing Fractions*

Rewrite the division problem as multiplying by the reciprocal of the divisor. In other words, the first number (dividend) stays the same and the second number (divisor) is changed to its reciprocal. Then use the steps for multiplying fractions. The quotient must be in lowest terms.

Division by 0 is undefined.

$$2 \div \left(-\frac{1}{3}\right) = \frac{2}{1} \cdot \left(-\frac{3}{1}\right) = -\frac{2 \cdot 3}{1 \cdot 1} = -6$$

Reciprocals

$$\frac{x^2}{y^2} \div \frac{x}{3y} = \frac{x^2}{y^2} \cdot \frac{3y}{x} = \frac{x \cdot x \cdot 3 \cdot y}{y \cdot y \cdot x} = \frac{3x}{y}$$

Reciprocals

Concepts

4.4 *Adding and Subtracting Like Fractions*

You can add or subtract fractions *only* when they have the *same* denominator. Add or subtract the numerators and write the result over the common denominator. Be sure that the final result is in lowest terms.

4.4 *Finding the Lowest Common Denominator (LCD)*

Write the prime factorization of each denominator. Then use enough prime factors in the LCD to "cover" both denominators.

4.4 *Adding and Subtracting Unlike Fractions*

Find the LCD. Rewrite each original fraction as an equivalent fraction whose denominator is the LCD. Then add or subtract the numerators and keep the common denominator. Be sure that the final result is in lowest terms.

Examples

$$\frac{3}{10} - \frac{7}{10} = \frac{3-7}{10} = \frac{-4}{10} \quad \text{or} \quad -\frac{4}{10}$$

Write $-\frac{4}{10}$ in lowest terms.
$$-\frac{4}{10} = -\frac{\overset{1}{\cancel{2}} \cdot 2}{\underset{1}{\cancel{2}} \cdot 5} = -\frac{2}{5} \left\{ \begin{array}{l} \text{Lowest} \\ \text{terms} \end{array} \right.$$

$$\frac{5}{a} + \frac{7}{a} = \frac{5+7}{a} = \frac{12}{a} \left\{ \begin{array}{l} \text{Lowest} \\ \text{terms} \end{array} \right.$$

What is the LCD for $\frac{5}{12}$ and $\frac{5}{18}$?

$$\begin{array}{l} 12 = 2 \cdot 2 \cdot 3 \\ 18 = 2 \cdot 3 \cdot 3 \end{array} \qquad \text{LCD} = 2 \cdot 2 \cdot 3 \cdot 3 = 36$$

Factors of 12 ↗↖ ↙↑↗ Factors of 18

The LCD for $\frac{5}{12}$ and $\frac{5}{18}$ is 36.

$$-\frac{5}{12} + \frac{7}{9}$$

The LCD is 36.

Rewrite: $\quad -\dfrac{5}{12} = -\dfrac{5 \cdot \mathbf{3}}{12 \cdot \mathbf{3}} = -\dfrac{15}{36}$

Rewrite: $\quad \dfrac{7}{9} = \dfrac{7 \cdot \mathbf{4}}{9 \cdot \mathbf{4}} = \dfrac{28}{36}$

Add: $\quad -\dfrac{15}{36} + \dfrac{28}{36} = \dfrac{-15 + 28}{36} = \dfrac{13}{36} \left\{ \begin{array}{l} \text{Lowest} \\ \text{terms} \end{array} \right.$

$$\frac{2}{3} - \frac{6}{x}$$

The LCD is $3 \cdot x$ or $3x$.

Rewrite: $\quad \dfrac{2}{3} = \dfrac{2 \cdot \mathbf{x}}{3 \cdot \mathbf{x}} = \dfrac{2x}{3x}$

Rewrite: $\quad \dfrac{6}{x} = \dfrac{6 \cdot \mathbf{3}}{x \cdot \mathbf{3}} = \dfrac{18}{3x}$

Subtract: $\quad \dfrac{2x}{3x} - \dfrac{18}{3x} = \dfrac{2x - 18}{3x} \left\{ \begin{array}{l} \text{Lowest} \\ \text{terms} \end{array} \right.$

Concepts	**Examples**

4.5 *Mixed Numbers and Improper Fractions*

Changing Mixed Numbers to Improper Fractions
Multiply the denominator by the whole number, add the numerator, and place the result over the original denominator.

Changing Improper Fractions to Mixed Numbers
Divide the numerator by the denominator and place the remainder over the original denominator.

Mixed to improper $7\frac{2}{3} = \frac{23}{3} \leftarrow (3 \cdot 7) + 2$

Same denominator

Improper to mixed $\frac{17}{5} = 3\frac{2}{5}$

Same denominator

4.5 *Multiplying Mixed Numbers*

First, round the numbers and estimate the answer. Then follow these steps to find the exact answer.

Step 1   Rewrite each mixed number as an improper fraction.
Step 2   Multiply.
Step 3   Write the answer in lowest terms and change the answer to a mixed number if desired. Then the answer is in simplest form.

Estimate:

$$1\frac{3}{5} \cdot 3\frac{1}{3}$$

Rounded

$$2 \cdot 3 = 6$$

Exact:

$$1\frac{3}{5} \cdot 3\frac{1}{3} = \frac{8}{5} \cdot \frac{10}{3}$$

$$= \frac{8 \cdot 2 \cdot \overset{1}{\cancel{5}}}{\cancel{5} \cdot 3}$$

$$= \frac{16}{3} = 5\frac{1}{3} \leftarrow$$

Close to estimate

4.5 *Dividing Mixed Numbers*

First, round the numbers and estimate the answer. Then follow these steps to find the exact answer.

Step 1   Rewrite each mixed number as an improper fraction.
Step 2   Divide. (Rewrite as multiplication using the reciprocal of the divisor.)
Step 3   Write the answer in lowest terms and change the answer to a mixed number if desired. Then the answer is in simplest form.

Estimate:

$$3\frac{3}{4} \div 2\frac{2}{5}$$

Rounded

$$4 \div 2 = 2$$

Exact:

$$3\frac{3}{4} \div 2\frac{2}{5} = \frac{15}{4} \div \frac{12}{5}$$

Reciprocal of $\frac{12}{5}$ is $\frac{5}{12}$

$$= \frac{15}{4} \cdot \frac{5}{12}$$

$$= \frac{\overset{1}{\cancel{3}} \cdot 5 \cdot 5}{4 \cdot \underset{1}{\cancel{3}} \cdot 4}$$

$$= \frac{25}{16} = 1\frac{9}{16} \leftarrow$$

Close to estimate

4.5 *Adding and Subtracting Mixed Numbers*

First round the numbers and estimate the answer. Then rewrite the mixed numbers as improper fractions and follow the steps for adding and subtracting fractions. Write the answer in simplest form.

Estimate:

$$2\frac{3}{8} + 3\frac{3}{4}$$

$$2 + 4 = 6 \leftarrow \text{Estimate}$$

Exact:

$$2\frac{3}{8} + 3\frac{3}{4} = \frac{19}{8} + \frac{15}{4} = \frac{19}{8} + \frac{30}{8} = \frac{19 + 30}{8}$$

$$= \frac{49}{8} = 6\frac{1}{8} \leftarrow \text{Close to estimate}$$

Concepts	**Examples**

4.6 *Simplifying Fractions with Exponents*

The meaning of an exponent is the same for fractions as it is for integers. An exponent is a way to write repeated multiplication.

$$\left(-\frac{2}{3}\right)^2 \quad \text{means} \quad \left(-\frac{2}{3}\right)\left(-\frac{2}{3}\right) = \frac{2 \cdot 2}{3 \cdot 3} = \frac{4}{9}$$

The product of two negative numbers is positive.

4.6 *Order of Operations*

The order of operations is the same for fractions as for integers.

1. Work inside *parentheses* or *other grouping symbols*.
2. Simplify expressions with *exponents*.
3. Do the remaining *multiplications and divisions* as they occur from left to right.
4. Do the remaining *additions and subtractions* as they occur from left to right.

Simplify.

$$-\frac{2}{3} + 3\left(\frac{1}{4}\right)^2 \qquad$$ Cannot work inside parentheses. Apply the exponent: $\frac{1}{4} \cdot \frac{1}{4}$ is $\frac{1}{16}$

$$-\frac{2}{3} + 3\left(\frac{1}{16}\right) \qquad$$ Multiply next: $3\left(\frac{1}{16}\right)$ is $\frac{3}{1} \cdot \frac{1}{16} = \frac{3}{16}$

$$-\frac{2}{3} + \frac{3}{16} \qquad$$ Add last. The LCD is 48

$$-\frac{32}{48} + \frac{9}{48} \qquad$$ Rewrite $-\frac{2}{3}$ as $-\frac{32}{48}$ Rewrite $\frac{3}{16}$ as $\frac{9}{48}$

$$\frac{-32 + 9}{48} \qquad$$ Add the numerators. Keep the common denominator.

$$-\frac{23}{48} \qquad$$ The answer is in lowest terms.

4.6 *Simplifying Complex Fractions*

Recall that the fraction bar indicates division. Rewrite the complex fraction using the $\div$ symbol for division. Then follow the steps for dividing fractions.

Simplify. $\quad \dfrac{-\dfrac{4}{5}}{10}$

Rewrite as $-\dfrac{4}{5} \div 10$.

$$-\frac{4}{5} \div 10 = -\frac{4}{5} \cdot \frac{1}{10} = -\frac{\overset{1}{\cancel{2}} \cdot 2 \cdot 1}{5 \cdot \underset{1}{\cancel{2}} \cdot 5} = -\frac{2}{25}$$

Reciprocals

Concepts	Examples
4.7 *Solving Equations Containing Fractions*	Solve the equation. Check the solution.

$$\frac{1}{3}b + 6 = 10$$

Step 1 If necessary, add the same number to both sides of the equation so that the variable term is by itself on one side of the equal sign.

$$\frac{\underline{-6} \quad \underline{-6}}{\frac{1}{3}b + 0 = 4} \qquad \text{Add } -6 \text{ to both sides.}$$

$$\frac{1}{3}b = 4$$

Step 2 Multiply both sides by the reciprocal of the coefficient of the variable term.

$$\frac{\overset{1}{\cancel{3}}}{1}\left(\frac{1}{\cancel{3}}b\right) = \frac{3}{1}(4) \qquad \begin{array}{l}\text{Multiply both sides by } \frac{3}{1}\\ \text{(the reciprocal of } \frac{1}{3}\text{).}\end{array}$$

$$b = 12$$

The solution is 12.

Step 3 To check your solution, go back to the *original* equation and replace the variable with your solution. If the equation balances, your solution is correct. If it does not balance, rework the problem.

Check $\quad \dfrac{1}{3}\,b\; + 6 = 10 \qquad$ Original equation

$$\frac{1}{3}(12) + 6 = 10 \qquad \text{Replace } b \text{ with 12}$$

$$\underbrace{4}\; + 6 = 10$$

$$10\; = 10 \qquad \text{Balances}$$

When b is 12, the equation balances, so 12 is the correct solution (**not** 10).

4.8 *Finding the Area of a Triangle*

Use this formula to find the area of a triangle.

$$\text{Area} = \frac{1}{2} \cdot \text{base} \cdot \text{height}$$

$$A = \frac{1}{2}bh$$

Remember that area is measured in **square units**.

Find the area of this triangle.

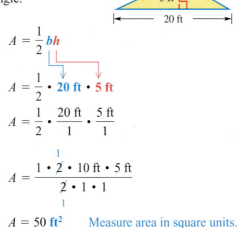

$$A = \frac{1}{2}\,\boldsymbol{bh}$$

$$A = \frac{1}{2} \cdot \textbf{20 ft} \cdot \textbf{5 ft}$$

$$A = \frac{1}{2} \cdot \frac{20 \text{ ft}}{1} \cdot \frac{5 \text{ ft}}{1}$$

$$A = \frac{1 \cdot \overset{1}{\cancel{2}} \cdot 10 \text{ ft} \cdot 5 \text{ ft}}{\underset{1}{\cancel{2}} \cdot 1 \cdot 1}$$

$$A = 50 \textbf{ ft}^2 \qquad \text{Measure area in square units.}$$

4.8 *Finding the Volume of a Rectangular Solid*

Use this formula to find the volume of box-like solids.

$$\text{Volume} = \text{length} \cdot \text{width} \cdot \text{height}$$

$$V = lwh$$

Volume is measured in **cubic units**.

Find the volume of this box.

$$V = \boldsymbol{l} \cdot \boldsymbol{w} \cdot \boldsymbol{h}$$

$$V = \textbf{5 cm} \cdot \textbf{3 cm} \cdot \textbf{6 cm}$$

$$V = 90 \textbf{ cm}^3$$

Measure volume in cubic units.

Concepts	Examples

4.8 *Finding the Volume of a Pyramid*

Use this formula to find the volume of a pyramid.

$$\text{Volume} = \frac{1}{3} \cdot B \cdot h$$

$$V = \frac{1}{3} Bh$$

where B is the area of the base and h is the height of the pyramid.

Volume is measured in **cubic units**.

Find the volume of a pyramid with a square base 2 cm by 2 cm and a height of 6 cm.

$$\text{Area of square base} = 2 \text{ cm} \cdot 2 \text{ cm}$$
$$B = \textbf{4 cm}^2$$

$$V = \frac{1}{3} \cdot \boldsymbol{B} \cdot \boldsymbol{h}$$

$$V = \frac{1}{3} \cdot \textbf{4 cm}^2 \cdot \textbf{6 cm}$$

$$V = \frac{1}{3} \cdot \frac{4 \text{ cm}^2}{1} \cdot \frac{6 \text{ cm}}{1}$$

$$V = \frac{1 \cdot 4 \text{ cm}^2 \cdot \overset{1}{\cancel{3}} \cdot 2 \text{ cm}}{\underset{1}{\cancel{3}} \cdot 1 \cdot 1}$$

$$V = 8 \textbf{ cm}^3 \qquad \text{Measure volume in cubic units.}$$

Chapter 4

REVIEW EXERCISES

[4.1] **1.** What fraction of these figures are squares? What fraction are circles?

2. Write fractions to represent the shaded and unshaded portions of this figure.

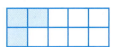

3. Graph $-\frac{1}{2}$ and $1\frac{1}{2}$ on the number line.

4. Simplify each fraction.

(a) $-\dfrac{20}{5}$ (b) $\dfrac{8}{1}$ (c) $-\dfrac{3}{3}$

[4.2] *Write each fraction in lowest terms.*

5. $\dfrac{28}{32}$

6. $\dfrac{54}{90}$

7. $\dfrac{16}{25}$

8. $\dfrac{15x^2}{40x}$

9. $\dfrac{7a^3}{35a^3b}$

10. $\dfrac{12mn^2}{21m^3n}$

[4.3] *Multiply or divide. Write all answers in lowest terms.*

11. $-\dfrac{3}{8} \div (-6)$

12. $\dfrac{2}{5}$ of (-30)

13. $\dfrac{4}{9}\left(\dfrac{2}{3}\right)$

14. $\left(\dfrac{7}{3x^3}\right)\left(\dfrac{x^2}{14}\right)$

15. $\dfrac{ab}{5} \div \dfrac{b}{10a}$

16. $\dfrac{18}{7} \div 3k$

[4.4] *Add or subtract. Write all answers in lowest terms.*

17. $-\dfrac{5}{12} + \dfrac{5}{8}$

18. $\dfrac{2}{3} - \dfrac{4}{5}$

19. $4 - \dfrac{5}{6}$

20. $\dfrac{7}{9} + \dfrac{13}{18}$

21. $\dfrac{n}{5} + \dfrac{3}{4}$

22. $\dfrac{3}{10} - \dfrac{7}{y}$

[4.5] *First, round the mixed numbers to the nearest whole number and estimate each answer. Then find the exact answer.*

23. *Exact:*

$$2\frac{1}{4} \div 1\frac{5}{8}$$

Estimate:

____ ÷ ____ = ____

24. *Exact:*

$$7\frac{1}{3} - 4\frac{5}{6}$$

Estimate:

____ − ____ = ____

25. *Exact:*

$$1\frac{3}{4} + 2\frac{3}{10}$$

Estimate:

____ + ____ = ____

[4.6] *Simplify.*

26. $\left(-\frac{3}{4}\right)^3$

27. $\left(\frac{2}{3}\right)^2\left(-\frac{1}{2}\right)^4$

28. $\frac{2}{5} + \frac{3}{10}(-4)$

29. $-\frac{5}{8} \div \left(-\frac{1}{2}\right)\left(\frac{14}{15}\right)$

30. $\dfrac{\frac{5}{8}}{\frac{1}{16}}$

31. $\dfrac{\frac{8}{9}}{-6}$

[4.7] *Solve each equation. Show your work.*

32. $-12 = -\frac{3}{5}w$

33. $18 + \frac{6}{5}r = 0$

34. $3x - \frac{2}{3} = \frac{5}{6}$

[4.8] *Find the area of the triangle. Name each solid and find its volume.*

35.

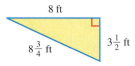

8 ft

$8\frac{3}{4}$ ft $3\frac{1}{2}$ ft

36.

$2\frac{1}{2}$ in.

4 in. $3\frac{1}{4}$ in.

37.

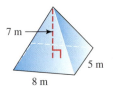

7 m

5 m

8 m

MIXED REVIEW EXERCISES

Solve each application problem.

38. A chili recipe that makes 10 servings uses $2\frac{1}{2}$ pounds of meat. How much meat will be in each serving? How much meat would be needed to make 30 servings?

39. Yanli worked as a math tutor for $4\frac{1}{2}$ hours on Monday, $2\frac{3}{4}$ hours on Tuesday, and $3\frac{2}{3}$ hours on Friday. How much longer did she work on Monday than on Friday? How many hours did she work in all?

40. There are 60 children in the day care center. If $\frac{1}{5}$ of the children are preschoolers, $\frac{2}{3}$ of the children are toddlers, and the rest are infants, find the number of children in each age group.

41. A rectangular city park is $\frac{3}{4}$ mile long and $\frac{3}{10}$ mile wide. Find the perimeter and area of the park.

Chapter **4**

TEST

Appendix A: Study Skills
Activity 11: Analyzing Test Results

1. Write fractions to represent the shaded and unshaded portions of the figure.

2. Graph $-\frac{2}{3}$ and $2\frac{1}{3}$ on the number line below.

Write each fraction in lowest terms.

3. $\dfrac{21}{84}$

4. $\dfrac{25}{54}$

5. $\dfrac{6a^2b}{9b^2}$

Add, subtract, multiply, or divide, as indicated. Write all answers in lowest terms.

6. $\dfrac{1}{6} + \dfrac{7}{10}$

7. $-\dfrac{3}{4} \div \dfrac{3}{8}$

8. $\dfrac{5}{8} - \dfrac{4}{5}$

9. $(-20)\left(-\dfrac{7}{10}\right)$

10. $\dfrac{\frac{4}{9}}{-6}$

11. $4 - \dfrac{7}{8}$

12. $-\dfrac{2}{9} + \dfrac{2}{3}$

13. $\dfrac{21}{24}\left(\dfrac{9}{14}\right)$

14. $\dfrac{12x}{7y} \div 3x$

15. $\dfrac{6}{n} - \dfrac{1}{4}$

16. $\dfrac{2}{3} + \dfrac{a}{5}$

17. $\left(\dfrac{5}{9b^2}\right)\left(\dfrac{b}{10}\right)$

18. Simplify.

$$\left(-\dfrac{1}{2}\right)^3\left(\dfrac{2}{3}\right)^2$$

19. Simplify.

$$\dfrac{1}{6} + 4\left(\dfrac{2}{5} - \dfrac{7}{10}\right)$$

1. _____

2. _____

3. _____

4. _____

5. _____

6. _____

7. _____

8. _____

9. _____

10. _____

11. _____

12. _____

13. _____

14. _____

15. _____

16. _____

17. _____

18. _____

19. _____

20. *Estimate:* _____

   Exact: _____

21. *Estimate:* _____

   Exact: _____

22. _____

23. _____

24. _____

25. _____

26. _____

27. _____

28. _____

29. _____

30. _____

31. _____

32. _____

First, round the numbers and estimate each answer. Then find the exact answer. Write exact answers in simplest form.

20. $4\dfrac{4}{5} \div 1\dfrac{1}{8}$

21. $3\dfrac{2}{5} - 1\dfrac{9}{10}$

Solve each equation. Show your work.

22. $7 = \dfrac{1}{5}d$

23. $-\dfrac{3}{10}t = \dfrac{9}{14}$

24. $0 = \dfrac{1}{4}b - 2$

25. $\dfrac{4}{3}x + 7 = -13$

Find the area of each triangle.

26.

12 m   8 m   9 m   13 m

27.

9 yd   13 yd $15\dfrac{4}{5}$ yd

Name each solid and find its volume.

28.

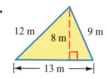

12 m   30 m   18 m

29.

4 yd   3 yd   4 yd

Solve each application problem.

30. Ann-Marie Sargent is training for an upcoming wheelchair race. She rides $4\dfrac{5}{6}$ hours on Monday, $6\dfrac{2}{3}$ hours on Tuesday, and $3\dfrac{1}{4}$ hours on Wednesday. How many hours did she spend in all? How many more hours did she train on Tuesday than on Monday?

31. A new vaccine is synthesized at the rate of $2\dfrac{1}{2}$ ounces per day. How long will it take to synthesize $8\dfrac{3}{4}$ ounces?

32. There are 8448 students at the Metro Community College campus. If $\dfrac{7}{8}$ of the students work either full time or part time, find the total number of students who work.

Cumulative Review Exercises

CHAPTERS 1–4

1. Write this number in words.
505,008,238

2. Write this number using digits.
thirty-five billion, six hundred million,
nine hundred sixteen

3. (a) Round 60,719 to the nearest hundred.

(b) Round 99,505 to the nearest thousand.

(c) Round 3206 to the nearest ten.

4. Name the property illustrated by each example.

(a) $-3 \cdot 6 = 6 \cdot (-3)$

(b) $(7 + 18) + 2 = 7 + (18 + 2)$

(c) $5(-10 + 7) = 5 \cdot (-10) + 5 \cdot 7$

Simplify.

5. $9(-6)$

6. $-10 - 10$

7. $\dfrac{-14}{0}$

8. $6 + 3(2 - 7)^2$

9. $|-8| + |2|$

10. $-8 + 24 \div 2$

11. $(-4)^2 - 2^5$

12. $\dfrac{-45 \div (-5)(3)}{-5 - 4(0 - 8)}$

13. $-3 \div \dfrac{3}{8}$

14. $-\dfrac{5}{6}(-42)$

15. $\left(\dfrac{5a^2}{12}\right)\left(\dfrac{18}{a}\right)$

16. $\dfrac{7}{x^2} \div \dfrac{7y^2}{3x}$

17. $\dfrac{3}{10} - \dfrac{5}{6}$

18. $-\dfrac{3}{8} + \dfrac{11}{16}$

19. $\dfrac{2}{3} - \dfrac{b}{7}$

20. $\dfrac{8}{5} + \dfrac{3}{n}$

21. $3\dfrac{1}{4} \div 2\dfrac{1}{4}$

22. $2\dfrac{2}{5} - 1\dfrac{3}{4}$

23. $\left(\dfrac{1}{2}\right)^3 (-2)^3$

24. $\dfrac{\frac{7}{12}}{-\frac{14}{15}}$

Solve each equation. Show your work.

25. $-5a + 2a = 12 - 15$

26. $y = -7y - 40$

27. $-10 = 6 + \dfrac{4}{9}k$

28. $20 + 5x = -2x - 8$

29. $3(-2m + 5) = -4m + 9 + m$

Name each figure and find its perimeter and area.

30.

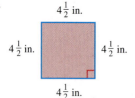

$4\frac{1}{2}$ in.

$4\frac{1}{2}$ in. $4\frac{1}{2}$ in.

$4\frac{1}{2}$ in.

31.

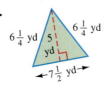

$6\frac{1}{4}$ yd   5 yd $6\frac{1}{4}$ yd

$7\frac{1}{2}$ yd

32.

24 mm

15 mm   12 mm   15 mm

24 mm

Solve each application problem by using the six problem-solving steps.

33. Three new bags of birdseed each weighed the same amount. Sixteen pounds of seed were used from one bag and 25 pounds from another. There were still 79 pounds of seed left. How much did each bag weigh originally?

34. A rectangular parking lot is twice as long as it is wide. If the perimeter of the lot is 102 yd, find the length and width of the parking lot.

Rational Numbers: Positive and Negative Decimals

5

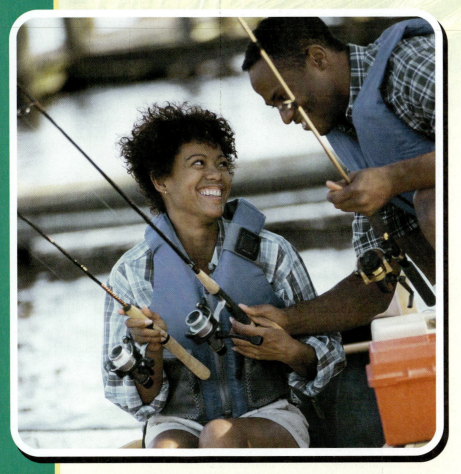

Nearly 49 million Americans go fishing at least once a year, making it America's fourth most popular recreational activity. (*Source:* National Sporting Goods Association.) In **Section 5.3,** Exercises 57–60, these friends will use decimal numbers when paying for new equipment. But will decimals help them catch their limit? (See **Section 5.1,** Exercises 59–62, and **Section 5.6,** Exercises 67–68.)

5.1   Reading and Writing Decimal Numbers

In **Chapter 4,** you worked with rational numbers written in fraction form to represent parts of a whole. In this chapter, we will use rational numbers written as **decimals** to show parts of a whole. For example, our money system is based on decimals. One dollar is divided into 100 equivalent parts. One cent ($0.01) is one of the parts, and a dime ($0.10) is 10 of the parts. Metric measurement (see **Chapter 8**) is also based on decimals.

OBJECTIVE 1 **Write parts of a whole using decimals.** Decimals are used when a whole is divided into 10 equivalent parts or into 100 or 1000 or 10,000 equivalent parts, and so on. In other words, decimals are fractions with denominators that are a power of 10. For example, the square at the right is cut into 10 equivalent parts. Written as a fraction, each part is $\frac{1}{10}$ of the whole. Written as a decimal, each part is **0.1**. Both $\frac{1}{10}$ and 0.1 are read as "*one tenth.*"

One tenth of the square is shaded.

The dot in 0.1 is called the **decimal point**.

$$0.1$$
↑
Decimal point

The square at the right has **7** of its 10 parts shaded.
Written as a *fraction*, $\frac{7}{10}$ of the square is shaded.
Written as a *decimal*, **0.7** of the square is shaded.
Both $\frac{7}{10}$ and 0.7 are read as "*seven tenths.*"

Seven tenths of the square is shaded.

Each square below is cut into 100 equivalent parts.
Written as a fraction, each part is $\frac{1}{100}$ of the whole.
Written as a decimal, each part is **0.01** of the whole.
Both $\frac{1}{100}$ and 0.01 are read as "*one hundredth.*"

Eighty-seven hundredths of the square is shaded.

The square above on the right has 87 of its 100 parts shaded.
Written as a fraction, $\frac{87}{100}$ of the total area is shaded.
Written as a decimal, **0.87** of the total area is shaded.
Both $\frac{87}{100}$ and 0.87 are read as "*eighty-seven hundredths.*"

The example below shows several numbers written as fractions, as decimals, and in words.

EXAMPLE 1 **Using the Decimal Forms of Fractions**

Fraction	Decimal	Read As
(a) $\dfrac{4}{10}$	0.4	four tenths
(b) $-\dfrac{9}{100}$	-0.09	negative nine hundredths
(c) $\dfrac{71}{100}$	0.71	seventy-one hundredths
(d) $\dfrac{8}{1000}$	0.008	eight thousandths
(e) $-\dfrac{45}{1000}$	-0.045	negative forty-five thousandths
(f) $\dfrac{832}{1000}$	0.832	eight hundred thirty-two thousandths

Now Try Exercise 1(b). ▶▶▶

OBJECTIVE 2 **Identify the place value of a digit.** The decimal point separates the *whole number part* from the *fractional part* in a decimal number. In the chart below, you see that the **place value** names for fractional parts are similar to those on the whole number side but end in "*ths*."

Decimal Place Value Chart

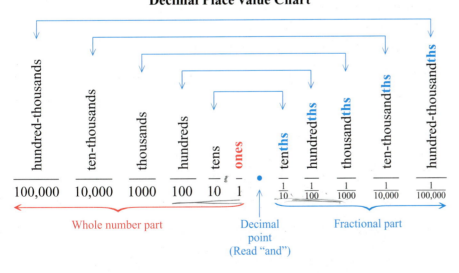

hundred-thousands	ten-thousands	thousands	hundreds	tens	ones		tenths	hundredths	thousandths	ten-thousandths	hundred-thousandths
100,000	10,000	1000	100	10	1	•	$\frac{1}{10}$	$\frac{1}{100}$	$\frac{1}{1000}$	$\frac{1}{10,000}$	$\frac{1}{100,000}$

Whole number part   Decimal point (Read "and")   Fractional part

NOTE

Notice that the **ones** place is at the center of the place value chart above. There is no "oneths" place.

Also notice that each place is 10 times the value of the place to its right.

Finally, be sure to write a hyphen (dash) in ten-thousand**ths** and hundred-thousand**ths**.

> **CAUTION**
> If a number does *not* have a decimal point, it is an *integer*. An integer has no fractional part. If you want to show the decimal point in an integer, it is just to the ***right*** of the digit in the ones place. Here are three examples.
>
> $$8 = 8. \qquad\qquad 306 = 306. \qquad -42 = -42.$$
>
> Decimal point   Decimal point   Decimal point

EXAMPLE 2 Identifying the Place Value of a Digit

Identify the place value of each digit.

(a) 178.36

(b) 0.00935

hundreds	tens	ones	tenths	hundredths
1	7	8 .	3	6

	ones	tenths	hundredths	thousandths	ten-thousandths	hundred-thousandths
	0 .	0	0	9	3	5

Notice in Example 2(b) that we do *not* use commas on the right side of the decimal point.

◀◀◀ **Now Try Exercise 5.**

OBJECTIVE 3 Read decimal numbers. A decimal number is read according to its form as a fraction.

ones
tenths
0.**9**

We read 0.9 as "nine tenths" because 0.9 is the same as $\frac{9}{10}$. Notice that 0.9 ends in the tenths place.

ones
tenths
hundredths
0.0 **2**

We read 0.02 as "two hundredths" because 0.02 is the same as $\frac{2}{100}$. Notice that 0.02 ends in the hundredths place.

EXAMPLE 3 Reading Decimal Numbers

Tell how to read each decimal in words.

(a) 0.3

Because $0.3 = \frac{3}{10}$, read the decimal as: three ten**ths**.

(b) 0.49   Read it as: forty-nine hundred**ths**.

(c) 0.08   Read it as: eight hundred**ths**.

(d) 0.918   Read it as: nine hundred eighteen thousand**ths**.

(e) 0.0106   Read it as: one hundred six ten-thousand**ths**.

◀◀◀ **Now Try Exercise 41.**

Reading a Decimal Number

Step 1 Read any whole number part to the *left* of the decimal point as you normally would.

Step 2 Read the decimal point as "*and*."

Step 3 Read the part of the number to the *right* of the decimal point as if it were an ordinary whole number.

Step 4 Finish with the place value name of the rightmost digit; these names all end in "*ths*."

NOTE
If there is *no whole number part,* you will use only Steps 3 and 4.

EXAMPLE 4 **Reading Decimal Numbers**

Read each decimal.

(a)
→ 9 is in tenths place.

16.9

sixteen **and** nine **tenths** ←

16.9 is read "sixteen and nine tenths."

(b)
→ 5 is in hundredths place.

482.35

four hundred eighty-two **and** thirty-five **hundredths** ←

482.35 is read "four hundred eighty-two and thirty-five hundredths."

→ 3 is in thousandths place.

(c) 0.063 is "sixty-three **thousandths**." (No whole number part)

(d) 11.1085 is "eleven **and** one thousand eighty-five **ten-thousandths**."

CAUTION
Use "and" *only* when reading a decimal point. A common mistake is to read the whole number 405 as "four hundred *and* five." But there is *no decimal point* shown in 405, so it is read "four hundred five."

Now Try Exercise 43. ▶▶▶

OBJECTIVE 4 Write decimals as fractions or mixed numbers.
Knowing how to read decimals will help you when writing decimals as fractions or mixed numbers.

Writing a Decimal as a Fraction or Mixed Number

Step 1 The digits to the right of the decimal point are the numerator of the fraction.

Step 2 The denominator is 10 for tenths, 100 for hundredths, 1000 for thousandths, 10,000 for ten-thousandths, and so on.

Step 3 If the decimal has a whole number part, the fraction will be a mixed number with the same whole number part.

EXAMPLE 5 **Writing Decimals as Fractions or Mixed Numbers**

Write each decimal as a fraction or mixed number.

(a) 0.19

The digits to the right of the decimal point, 19, are the numerator of the fraction. The denominator is 100 for hundredths because the rightmost digit is in the hundredths place.

$$0.19 = \frac{19}{100} \leftarrow \text{100 for hundredths}$$

↑ Hundredths place

(b) 0.863

$$0.863 = \frac{863}{1000} \leftarrow \text{1000 for thousandths}$$

↑ Thousandths place

(c) 4.0099

The whole number part stays the same.

$$4.0099 = 4\frac{99}{10,000} \leftarrow \text{10,000 for ten-thousandths}$$

↑ Ten-thousandths place

Now Try Exercise 27.

EXAMPLE 6 **Writing Decimals as Fractions or Mixed Numbers**

Write each decimal as a fraction or mixed number in lowest terms.

(a) $0.4 = \frac{4}{10} \leftarrow \text{10 for tenths}$

Write $\frac{4}{10}$ in lowest terms. $\frac{4}{10} = \frac{4 \div 2}{10 \div 2} = \frac{2}{5} \leftarrow$ Lowest terms

(b) $0.75 = \frac{75}{100} = \frac{75 \div 25}{100 \div 25} = \frac{3}{4} \leftarrow$ Lowest terms

(c) $18.105 = 18\frac{105}{1000} = 18\frac{105 \div 5}{1000 \div 5} = 18\frac{21}{200} \leftarrow$ Lowest terms

(d) $42.8085 = 42\frac{8085}{10,000} = 42\frac{8085 \div 5}{10,000 \div 5} = 42\frac{1617}{2000} \leftarrow$ Lowest terms

CAUTION
Always check that your fraction answers are in lowest terms.

Now Try Exercise 29.

 Calculator Tip In this book, we will write **0**.45 instead of just .45, to emphasize that there is no whole number. Your *scientific* calculator shows these zeros also. Enter [·] [4] [5] and notice that the display automatically shows 0.45 even though you did not press 0. For comparison, enter the whole number 45 by pressing [4] [5] [+] and notice that the decimal point automatically appears to the *right* of the 5. (*Graphing* calculators may not automatically show a 0 in the ones place.)

5.1 Exercises

For help with Exercises 1 and 2, see the first page of this section and Example 1.

1. There are 10 dimes in one dollar. Each dime is $\frac{1}{10}$ of a dollar. Write a fraction, a decimal, and the words that name the yellow shaded portion of each dollar.

   (a)   (b)   (c)

2. There are 100 equal parts in each square. Write a fraction, a decimal, and the words that name the shaded portion of each square.

   (a)   (b)   (c)

Identify the digit that has the given place value. See Example 2.

3. 70.489
  tens
  ones
  tenths

4. 135.296
  ones
  tenths
  tens

5. 0.2518
  hundredths
  thousandths
  ten-thousandths

6. 0.9347
  hundredths
  thousandths
  ten-thousandths

7. 93.01472
  thousandths
  ten-thousandths
  tenths

8. 0.51968
  tenths
  ten-thousandths
  hundredths

9. 314.658
  tens
  tenths
  hundreds

10. 51.325
  tens
  tenths
  hundredths

11. 149.0832
  hundreds
  hundredths
  ones

12. 3458.712
  hundreds
  hundredths
  tenths

Write the decimal number that has the specified place values. See Example 2.

13. 0 ones, 5 hundredths, 1 ten, 4 hundreds, 2 tenths

14. 7 tens, 9 tenths, 3 ones, 6 hundredths, 8 hundreds

15. 3 thousandths, 4 hundredths, 6 ones, 2 ten-thousandths, 5 tenths

16. 8 ten-thousandths, 4 hundredths, 0 ones, 2 tenths, 6 thousandths

17. 4 hundredths, 4 hundreds, 0 tens, 0 tenths, 5 thousandths, 5 thousands, 6 ones

18. 7 tens, 7 tenths, 6 thousands, 6 thousandths, 3 hundreds, 3 hundredths, 2 ones

Write each decimal as a fraction or mixed number in lowest terms. See Examples 5 and 6.

19. 0.7 **20.** 0.1 **21.** 13.4 **22.** 9.8 **23.** 0.35

24. 0.85 **25.** 0.66 **26.** 0.33 **27.** 10.17 **28.** 31.99

29. 0.06 **30.** 0.08 **31.** 0.205 **32.** 0.805

33. 5.002 **34.** 4.008 **35.** 0.686 **36.** 0.492

Tell how to read each decimal in words. See Examples 1, 3, and 4.

37. 0.5 **38.** 0.2

39. 0.78 **40.** 0.55

41. 0.105 **42.** 0.609

43. 12.04

44. 86.09

45. 1.075

46. 4.025

Write each decimal in numbers. See Examples 1, 3 and 4.

47. Six and seven tenths

48. Eight and twelve hundredths

49. Thirty-two hundredths

50. One hundred eleven thousandths

51. Four hundred twenty and eight thousandths

52. Two hundred and twenty-four thousandths

53. Seven hundred three ten-thousandths

54. Eight hundred and six hundredths

55. Seventy-five and thirty thousandths

56. Sixty and fifty hundredths

57. Anne read the number 4302 as "four thousand three hundred and two." Explain what is wrong with the way Anne read the number.

58. Jerry read the number 9.0106 as "nine and one hundred and six ten-thousandths." Explain the error he made.

The friends on the first page of this chapter need to select the correct fishing line for their reels. Fishing line is sold according to how many pounds of "pull" the line can withstand before breaking. Use the table to answer Exercises 59–62. Write all fractions in lowest terms. (Note: The diameter of the fishing line is its thickness.)

FISHING LINE

Test Strength (pounds)	Average Diameter (inches)
4	0.008
8	0.010
12	0.013
14	0.014
17	0.015
20	0.016

Source: Berkley Outdoor Technologies Group.

The diameter is the distance across the end of the line (or its thickness).

59. Write the diameter of 8-pound test line in words and as a fraction.

60. Write the diameter of 17-pound test line in words and as a fraction.

61. What is the test strength of the line with a diameter of $\frac{13}{1000}$ inch?

62. What is the test strength of the line with a diameter of sixteen thousandths inch?

Suppose your job is to take phone orders for precision parts. Use the table below. In Exercises 63–68, write the correct part number that matches what you hear the customer say over the phone. In Exercises 67–68, write the words you would say to the customer.

Part Number	Size in Centimeters
3-A	0.06
3-B	0.26
3-C	0.6
3-D	0.86
4-A	1.006
4-B	1.026
4-C	1.06
4-D	1.6
4-E	1.602

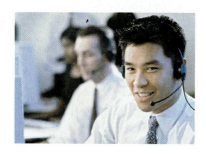

63. "Please send the six tenths centimeter bolt."

Part number _____

64. "The part missing from our order was the one and six hundredths size."

Part number _____

65. "The size we need is one and six thousandths centimeters."

Part number _____

66. "Do you still stock the twenty-six hundredths centimeter bolt?"

Part number _____

67. "What size is part number 4-E?" Write your answer in words.

68. "What size is part number 4-B?" Write your answer in words.

RELATING CONCEPTS (EXERCISES 69–76) For Individual or Group Work

*Use your knowledge of place value to **work Exercises 69–76 in order.***

69. Look back at the decimal place value chart on page 309 of this section. What do you think would be the names of the next four places to the *right* of hundred-thousandths? What information did you use to come up with these names?

70. A common mistake is to think that the first place to the right of the decimal point is "oneths" and the second place is "tenths." Why might someone make that mistake? How would you explain why there is no "oneths" place?

71. Use your answer from Exercise 69 to write 0.72436955 in words.

72. Use your answer from Exercise 69 to write 0.000678554 in words.

73. Write 8006.500001 in words.

74. Write 20,060.000505 in words.

75. Write this decimal using digits.
three hundred two thousand forty ten-millionths

76. Write this decimal using digits.
nine billion, eight hundred seventy-six million, five hundred forty-three thousand, two hundred ten and one hundred million two hundred thousand three hundred billionths

5.2 Rounding Decimal Numbers

Section 1.5 showed how to round integers. For example, 89 rounded to the nearest ten is 90, and 8512 rounded to the nearest hundred is 8500.

OBJECTIVE 1 Learn the rules for rounding decimals. It is also important to be able to **round** decimals. For example, a store is selling 2 candy mints for $0.75 but you want only one mint. The price of each mint is $0.75 ÷ 2, which is $0.375, but you cannot pay part of a cent. Is $0.375 closer to $0.37 or to $0.38? Actually, it's exactly halfway between. When this happens in everyday situations, the rule is to round *up*. The store will charge you $0.38 for the mint.

OBJECTIVES

1. Learn the rules for rounding decimals.
2. Round decimals to any given place.
3. Round money amounts to the nearest cent or nearest dollar.

Rounding a Decimal Number

Step 1   Find the place to which the rounding is being done. Draw a "cut-off" line *after* that place to show that you are cutting off and dropping the rest of the digits.

Step 2   Look *only* at the *first* digit you are cutting off.

Step 3A   If this digit is *4 or less,* the part of the number you are keeping *stays the same.*

Step 3B   If this digit is *5 or more,* you must *round up* the part of the number you are keeping.

Step 4   You can use the ≈ symbol or the ≐ symbol to indicate that the rounded number is now an approximation (close, but *not exact*). Both symbols mean "is approximately equal to." (In this book we will use the ≈ symbol.)

CAUTION
Do *not* move the decimal point when rounding.

OBJECTIVE 2 Round decimals to any given place. These examples show you how to round decimals.

EXAMPLE 1 Rounding a Decimal Number

Round 14.39652 to the nearest thousandth. (Is it closer to 14.396 or to 14.397?)

Step 1   Draw a "cut-off" line after the thousandths place.

$$1\ 4\ .\ 3\ 9\ 6\ \big|\ 5\ 2$$

You are cutting off the 5 and 2
They will be dropped.

Thousandths

Step 2   Look *only* at the *first* digit you are cutting off. Ignore the other digits you are cutting off.

$$1\ 4\ .\ 3\ 9\ 6\ \big|\ 5\ 2$$

Look *only* at the 5
Ignore the 2

Continued on Next Page

Step 3 If the first digit you are cutting off is *5 or more*, round up the part of the number you are keeping.

$$
\begin{array}{r}
14.396 \quad 52 \\
+\ 0.001 \\
\hline
14.397
\end{array}
$$

First digit cut is *5 or more,* so round up by adding 1 thousandth to the part you are keeping.

So, 14.39652 rounded to the nearest thousandth is 14.397.
You can write 14.39652 $\approx$ 14.397.

> **CAUTION**
> When rounding integers in **Section 1.5**, you kept all the digits but changed some to zeros. With decimals, you cut off and *drop the extra digits*. In the example above, 14.39652 rounds to 14.397 *not* 14.39700.

Now Try Exercise 7.

In Example 1 above, the rounded number 14.397 had *three decimal places*. **Decimal places** are the number of digits to the *right* of the decimal point. The first decimal place is tenths, the second is hundredths, the third is thousandths, and so on.

EXAMPLE 2 **Rounding Decimals to Different Places**

Round to the place indicated.

(a) Round 5.3496 to the nearest tenth. (Is it closer to 5.3 or to 5.4?)

Step 1 Draw a cut-off line after the tenths place.

$$5.3 \quad 496$$

You are cutting off the 4, 9, and 6

Tenths

Step 2 $\quad 5.3 \quad 4\ 9\ 6$

Look *only* at the 4

Ignore these digits.

Step 3 $\quad 5.3 \quad 4\ 9\ 6$

First digit cut is *4 or less,* so the part you are keeping stays the same

$5.3 \leftarrow$ Stays the same

So, 5.3496 rounded to the nearest tenth is 5.3 (one decimal place for tenths). You can write 5.3496 $\approx$ 5.3. Notice that it does *not* round to 5.3000 (which would be ten-thousandths instead of tenths).

(b) Round 0.69738 to the nearest hundredth. (Is it closer to 0.69 or to 0.70?)

Step 1 $\quad 0.69|738$

Draw a cut-off line after the hundredths place.

Hundredths

Look *only* at the 7.

Step 2 $\quad 0.69|738$

Continued on Next Page

First digit cut is *5 or more,* so round up by adding 1 hundredth to the part you are keeping.

Step 3   0 . 6 9 | 7 3 8

                 1
      0 . 6 9 ← Keep this part.
    + 0 . 0 1 ← To round up, add 1 hundredth.
    ───────────
      0 . 7 0 ← 9 + 1 is 10; write 0 and regroup 1 to the tenths place.

So, 0.69738 rounded to the nearest hundredth is 0.70. Hundredths is *two* decimal places so you *must* write the 0 in the hundredths place.
    You can write 0.69738 ≈ 0.70.

> **CAUTION**
> If a *rounded* number has a 0 in the rightmost place, you *must* keep the 0. As shown above, 0.69738 rounded to the nearest hundredth is 0.7**0**. Do ***not*** write 0.7, which is rounded to tenths instead of hundredths.

(c) Round 0.01806 to the nearest thousandth. (Is it closer to 0.018 or to 0.019?)

First digit cut is *4 or less,* so the part you are keeping stays the same.

        0 . 0 1 8 | 0 6

        0 . 0 1 8 ← Stays the same

So, 0.01806 rounded to the nearest thousandth is 0.018 (three decimal places for thousandths).
    You can write 0.01806 ≈ 0.018.

(d) Round 57.976 to the nearest tenth. (Is it closer to 57.9 or to 58.0?)

First digit cut is *5 or more,* so round up by adding 1 tenth to the part you are keeping.

        57.9 | 76

          1
        57.9
      + 0.1
      ─────────
        58.0 ← 9 + 1 is 10; write the 0 and regroup the 1 to the ones place.

So, 57.976 rounded to the nearest tenth is 58.0. You can write 57.976 ≈ 58.0. You *must* write the 0 in the tenths place to show that the number was rounded to the nearest tenth.

> **CAUTION**
> Check that your rounded answer shows *exactly* the number of decimal places asked for in the problem. Be sure your answer shows *one decimal place* if you rounded to *tenths,* two decimal places for *hundredths,* three *decimal places* for *thousandths,* and so on.

Now Try Exercises 5 and 15. ▶▶▶

OBJECTIVE **3** **Round money amounts to the nearest cent or nearest dollar.** When you are shopping in a store, money amounts are usually rounded to the nearest cent. There are 100 cents in a dollar.

$$\text{Each cent is } \frac{1}{100} \text{ of a dollar.}$$

Another way to write $\frac{1}{100}$ is 0.01. So rounding to the *nearest cent* is the same as rounding to the *nearest hundredth of a dollar.*

EXAMPLE 3 Rounding to the Nearest Cent

Round each money amount to the nearest cent.

(a) $2.4238 (Is it closer to $2.42 or to $2.43?)

$$\underbrace{\$2.42}|38$$

First digit cut is *4 or less,* so the part you are keeping stays the same.

$2.42 ⟵ You pay $2.42

$2.4238 rounded to the nearest cent is $2.42.

(b) $0.695 (Is it closer to $0.69 or to $0.70?)

$$\underbrace{\$0.69}|5$$

5 or more; round up.

$$\begin{array}{r} \$0.69 \\ +\ \$0.01 \\ \hline \$0.70 \end{array}$$

To round up, add 1 hundredth (1 cent).

You pay $0.70

$0.695 rounded to the nearest cent is $0.70.

◀◀◀ **Now Try Exercises 21 and 23.**

NOTE

Some stores round *all* money amounts up to the next higher cent, even if the next digit is *4 or less*. In Example 3(a) above, some stores would round $2.4238 *up* to $2.43, even though it is closer to $2.42.

It is also common to round money amounts to the nearest dollar. For example, you can do that on your federal and state income tax returns to make the calculations easier.

EXAMPLE 4 Rounding to the Nearest Dollar

Round to the nearest dollar.

(a) $48.69 (Is it closer to $48 or to $49?)

$$\underbrace{\$48.}|69$$

First digit cut is *5 or more,* so round up by adding $1

$$\begin{array}{r} \$48 \\ +\ \ 1 \\ \hline \$49 \end{array}$$

$48.69 rounded to the nearest dollar is $49.

CAUTION

$48.69 rounded to the nearest dollar is $49. Be careful to write the answer as $49 to show that the rounding is to the *nearest dollar*. Writing $49.00 would show rounding to the *nearest cent*.

Continued on Next Page

(b) $594.36 (Is it closer to $594 or to $595?)

$$\underbrace{\$594.}\,\big|\,36$$
$$\$594$$

First digit cut is *4 or less,*
so the part you are keeping
stays the same.

$594.36 rounded to the nearest dollar is $594.

(c) $399.88 (Is it closer to $399 or to $400?)

$$\underbrace{\$399.}\,\big|\,88$$

5 or more so round up
by adding $1

$$\begin{array}{r} \$399 \\ +\quad 1 \\ \hline \$400 \end{array}$$

$399.88 rounded to the nearest dollar is $400.

(d) $2689.50 (Is it closer to $2689 or to $2690?)

$$\underbrace{\$2689.}\,\big|\,50$$

5 or more, so round up
by adding $1

$$\begin{array}{r} \$2689 \\ +\quad 1 \\ \hline \$2690 \end{array}$$

$2689.50 rounded to the nearest dollar is $2690.

> **NOTE**
> When rounding $2689.50 to the nearest dollar, above, notice that it
> is exactly halfway between $2689 and $2690. When this happens
> in everyday situations, the rule is to round *up*. (Scientists working
> with technical data may use a more complicated rule when rounding
> numbers that are exactly in the middle.)

(e) $0.61 (Is it closer to $0 or to $1?)

$$\$0.\,\big|\,61$$

5 or more, so round up.

$0.61 rounded to the nearest dollar is $1.

⊞ **Calculator Tip** Accountants and other people who work with money
amounts often set their calculators to automatically round to two decimal
places (nearest cent) or to round to zero decimal places (nearest dollar).
Your calculator may have this feature.

Now Try Exercises 25 and 27. ▶▶▶

5.2 Exercises

Round each number to the place indicated. See Examples 1 and 2.

1. 16.8974 to the nearest tenth

2. 193.845 to the nearest hundredth

3. 0.95647 to the nearest thousandth

4. 96.81584 to the nearest ten-thousandth

5. 0.799 to the nearest hundredth

6. 0.952 to the nearest tenth

7. 3.66062 to the nearest thousandth

8. 1.5074 to the nearest hundredth

9. 793.988 to the nearest tenth

10. 476.1196 to the nearest thousandth

11. 0.09804 to the nearest ten-thousandth

12. 176.004 to the nearest tenth

13. 48.512 to the nearest one

14. 3.385 to the nearest one

15. 9.0906 to the nearest hundredth

16. 30.1290 to the nearest thousandth

17. 82.000151 to the nearest ten-thousandth

18. 0.400594 to the nearest ten-thousandth

Nardos is grocery shopping. The store will round the amount she pays for each item to the nearest cent. Write the rounded amounts. See Example 3.

19. Soup is three cans for $2.45, so one can is $0.81666. Nardos pays ___.82___.

20. Orange juice is two cartons for $3.89, so one carton is $1.945. Nardos pays _____.

21. Facial tissue is four boxes for $4.89, so one box is $1.2225. Nardos pays _____.

22. Muffin mix is three packages for $1.75, so one package is $0.58333. Nardos pays _____.

23. Candy bars are six for $2.99, so one bar is $0.4983. Nardos pays _____.

24. Boxes of spaghetti are four for $4.39, so one box is $1.0975. Nardos pays _____.

As she gets ready to do her income tax return, Ms. Chen rounds each amount to the nearest dollar. Write the rounded amounts. See Example 4.

25. Income from job, $48,649.60

26. Income from interest on bank account, $69.58

27. Union dues, $310.08

28. Federal withholding, $6064.49

29. Donations to charity, $848.91

30. Medical expenses, $609.38

Round each money amount as indicated.

31. $499.98 to the nearest dollar

32. $9899.59 to the nearest dollar

33. $0.996 to the nearest cent

34. $0.09929 to the nearest cent

35. $999.73 to the nearest dollar

36. $9999.80 to the nearest dollar

The table lists speed records for various types of transportation. Use the table to answer Exercises 37–40.

Record	Speed (miles per hour)
Land speed record (specially built car)	763.04
Motorcycle speed record (specially adapted motorcycle)	322.15
Fastest rollercoaster	106.9
Fastest military jet	2193.167
Boeing 737-300 airplane (regular passenger service)	495
Indianapolis 500 auto race (fastest average winning speed)	185.981
Daytona 500 auto race (fastest average winning speed)	177.602

Source: The Top 10 of Everything and The World Almanac.

37. Round these speed records to the nearest whole number.

(a) Motorcycle

(b) Rollercoaster

38. Round these speed records to the nearest hundredth.

(a) Daytona 500 average winning speed

(b) Indianapolis 500 average winning speed

39. Round these speed records to the nearest tenth.

(a) Indianapolis 500 average winning speed

(b) Land speed record

40. Round these speed records to the nearest hundred.

(a) military jet

(b) Boeing 737-300 airplane

RELATING CONCEPTS (EXERCISES 41–44) For Individual or Group Work

*Use your knowledge about rounding money amounts to **work Exercises 41–44 in order.***

41. Explain what happens when you round $0.499 to the nearest dollar. Why does this happen?

42. Look again at Exercise 41. How else could you round $0.499 that would be more helpful? What kind of guideline does this suggest about rounding to the nearest dollar?

43. Explain what happens when you round $0.0015 to the nearest cent. Why does this happen?

44. Suppose you want to know which of these amounts is less, so you round them both to the nearest cent.

$0.5968 $0.6014

Explain what happens. Describe what you could do instead of rounding to the nearest cent.

5.3 Adding and Subtracting Signed Decimal Numbers

OBJECTIVES

1. Add and subtract positive decimals.
2. Add and subtract negative decimals.
3. Estimate the answer when adding or subtracting decimals.

OBJECTIVE 1 Add and subtract positive decimals. When adding or subtracting *whole* numbers, you line up the numbers in columns so that you are adding ones to ones, tens to tens, and so on. A similar idea applies to adding or subtracting *decimal* numbers. With decimals, you line up the decimal points to be sure that you are adding tenths to tenths, hundredths to hundredths, and so on.

Adding and Subtracting Decimal Numbers

Step 1 Write the numbers in columns with the decimal points lined up.

Step 2 If necessary, write in zeros so both numbers have the same number of decimal places. Then add or subtract as if they were whole numbers.

Step 3 Line up the decimal point in the answer directly below the decimal points in the problem.

EXAMPLE 1 Adding Decimal Numbers

Find each sum.

(a) 16.92 and 48.34

Step 1 Write the numbers in columns with the decimal points lined up.

$$
\begin{array}{r}
\text{\small tens\ ones . tenths\ hundredths} \\
1\,6.9\,2 \\
+\,4\,8.3\,4 \\
\end{array}
$$

Decimal points are lined up.

Step 2 Add as if these were whole numbers.

$$
\begin{array}{r}
1\,1 \\
16.92 \\
+\,48.34 \\
\hline
65.26 \\
\end{array}
$$

Step 3

Decimal point in answer is lined up under decimal points in problem.

(b) 5.897 + 4.632 + 12.174

Write the numbers in columns with the decimal points lined up. Then add.

$$
\begin{array}{r}
11\ \ 21 \\
5.897 \\
4.632 \\
+\,12.174 \\
\hline
22.703 \\
\end{array}
$$

Decimal points are lined up.

◀◀◀ Now Try Exercise 1.

In Example 1(a) above, both numbers had *two* decimal places (two digits to the right of the decimal point). In Example 1(b), all the numbers had *three decimal places* (three digits to the right of the decimal point). That made it easy to add tenths to tenths, hundredths to hundredths, and so on.

If the numbers of decimal places do *not* match, you can write in zeros as placeholders to make them match. This is shown in Example 2 below.

EXAMPLE 2 **Writing Zeros as Placeholders before Adding**

Find each sum.

(a) 7.3 + 0.85

There are two decimal places in 0.85 (tenths and hundredths), so write a 0 in the hundredths place in 7.3 so that it has two decimal places also.

$$\begin{array}{r} 7.3\mathbf{0} \\ +\ 0.85 \\ \hline 8.15 \end{array}$$ ← One 0 is written in.   7.30   is equivalent to   7.3   because

$$7\frac{30}{100}$$ in lowest terms is $$7\frac{3}{10}$$

(b) 6.42 + 9 + 2.576

Write in zeros so that all the addends have three decimal places. Notice how the whole number 9 is written with the decimal point on the *right* side. (If you put the decimal point on the *left* side of the 9, you would turn it into the decimal fraction 0.9.)

$$\begin{array}{r} 6.42\mathbf{0} \\ 9.\mathbf{000} \\ +\ 2.576 \\ \hline 17.996 \end{array}$$

6.42**0** ← One 0 is written in.
9.**000** ← 9 is a whole number; decimal point and three zeros are written in.
+ 2.576 ← No zeros are needed.

NOTE

Writing zeros to the right of a *decimal* number does *not* change the value of the number, as shown in Example 2(a) above.

Now Try Exercise 5. ▶▶▶

EXAMPLE 3 **Subtracting Decimal Numbers**

Find each difference.

(a) 15.82 from 28.93

Step 1
$$\begin{array}{r} 28.93 \\ -\ 15.82 \end{array}$$
Line up decimal points. Then you will be subtracting hundredths from hundredths and tenths from tenths.

Step 2
$$\begin{array}{r} 28.93 \\ -\ 15.82 \\ \hline 13\ 11 \end{array}$$
Both numbers have two decimal places; no need to write in zeros.
← Subtract as if they were whole numbers.

Step 3
$$\begin{array}{r} 28.93 \\ -\ 15.82 \\ \hline 13.11 \end{array}$$
← Decimal point in answer is lined up.

Continued on Next Page

(b) 146.35 minus 58.98

Regrouping is needed here.

Line up decimal points.

$$
\begin{array}{r}
{\scriptstyle 0\ 13\ 15\ \ 12\ 15} \\
1\,4\,6\,.\,3\,5 \\
-\ \ 5\,8\,.\,9\,8 \\
\hline
8\,7\,.\,3\,7
\end{array}
$$

Now Try Exercise 13.

EXAMPLE 4 **Writing Zeros as Placeholders before Subtracting**

Find each difference.

(a) 16.5 from 28.362

Use the same steps as in Example 3 above. Remember to write in zeros so both numbers have three decimal places.

Line up decimal points.

$$
\begin{array}{r}
28.362 \quad \leftarrow \text{Write two zeros.}\\
-\ 16.5\mathbf{00} \quad \leftarrow \text{Subtract as usual.}\\
\hline
11.862
\end{array}
$$

(b) 59.7 − 38.914

$$
\begin{array}{r}
59.7\mathbf{00} \quad \leftarrow \text{Write two zeros.}\\
-\ 38.914 \\
\hline
20.786 \quad \leftarrow \text{Subtract as usual.}
\end{array}
$$

(c) 12 less 5.83

$$
\begin{array}{r}
12.\mathbf{00} \quad \leftarrow \text{Write a decimal point and two zeros.}\\
-\ 5.83 \\
\hline
6.17 \quad \leftarrow \text{Subtract as usual.}
\end{array}
$$

Now Try Exercises 15 and 19.

▦ **Calculator Tip** If you are *adding* decimal numbers, you can enter them in any order on your calculator. Try these; jot down the answers.

9.82 ⊕ 1.86 ⊜ _____   1.86 ⊕ 9.82 ⊜ _____

The answers are the same because addition is *commutative* (see **Section 1.3**). But subtraction is *not* commutative. It *does* matter which number you enter first. Try these:

9.82 ⊖ 1.86 ⊜ _____   1.86 ⊖ 9.82 ⊜ _____

The answers are 7.96 and −7.96. As you know, positive numbers are *greater* than 0, but negative numbers are *less* than 0. So it is important to do subtraction in the correct order, particularly if it is in your checkbook!

OBJECTIVE 2 Add and subtract negative decimals. The rules that you used to add integers in **Section 1.3** will also work for positive and negative decimal numbers.

Adding Signed Numbers

To add two numbers with the *same* sign, add the absolute values of the numbers. Use the common sign as the sign of the sum.

To add two numbers with *unlike* signs, subtract the smaller absolute value from the larger absolute value. Use the sign of the number with the larger absolute value as the sign of the sum.

EXAMPLE 5 **Adding Positive and Negative Decimal Numbers**

Find each sum.

(a) $-3.7 + (-16)$

Both addends are negative, so the sum will be negative. To begin, $|-3.7|$ is 3.7 and $|-16|$ is 16. Then add the absolute values.

$$
\begin{array}{r}
3.7 \\
+\ 16.\mathbf{0} \\
\hline
19.7
\end{array}
$$
$\leftarrow$ Write a decimal point and one 0

$$-3.7 + (-16) = -19.7$$

Both negative   Negative sum

NOTE

In **Chapters 1–4**, the negative sign was **red** to help you distinguish it from the subtraction symbol. From now on it will be **black**. We will continue to write parentheses around negative numbers when the negative sign might be confused with other symbols. Thus in part (a) above,

$$-3.7 + (-16) \quad \text{means} \quad \textbf{negative } 3.7 \textbf{ plus negative } 16$$

(b) $-5.23 + 0.792$

The addends have different signs. To begin, $|-5.23|$ is 5.23 and $|0.792|$ is 0.792. Then subtract the smaller absolute value from the larger.

$$
\begin{array}{r}
5.23\mathbf{0} \\
-\ 0.792 \\
\hline
4.438
\end{array}
$$
$\leftarrow$ Write one 0

$$-5.23 + 0.792 = -4.438$$

Number with larger   Answer is
absolute value   negative.
is negative.

Now Try Exercise 35. ▶▶▶

In **Section 1.4** you rewrote subtraction of integers as addition of the first number to the opposite of the second number. This same strategy works with positive and negative decimal numbers.

EXAMPLE 6 **Subtracting Positive and Negative Decimal Numbers**

Find each difference.

(a) $4.3 - 12.73$

Rewrite subtraction as adding the opposite.

$$4.3 - \mathbf{12.73} \qquad \text{The opposite of 12.73 is}$$
$$\downarrow \quad \downarrow \qquad\qquad -12.73$$
$$4.3 + (-\mathbf{12.73})$$

-12.73 has the larger absolute value and is negative, so the answer will be negative.

$$4.3 + (-12.73) = -8.43 \qquad \text{Subtract the absolute values:}$$
$$\uparrow \qquad\qquad\qquad 12.73$$
$$\text{Answer is} \qquad\qquad -\ 4.30$$
$$\text{negative.} \qquad\qquad \overline{8.43}$$

(b) $-3.65 - (-4.8)$

Rewrite subtraction as adding the opposite.

$$-3.65 - (-\mathbf{4.8}) \qquad \text{The opposite of } -4.8$$
$$\downarrow \quad \downarrow \qquad\qquad \text{is 4.8}$$
$$-3.65 + \mathbf{4.8}$$

4.8 has the larger absolute value and is positive, so the answer will be positive.

$$-3.65 + 4.8 = 1.15 \qquad \text{Subtract the absolute values:}$$
$$\uparrow \qquad\qquad\qquad 4.80$$
$$\text{Answer is} \qquad\qquad -3.65$$
$$\text{positive.} \qquad\qquad \overline{1.15}$$

(c) $14.2 - (1.69 + 0.48)$ Work inside parentheses first.

$14.2 - \quad (\mathbf{2.17})$ Change subtraction to adding the opposite.

$14.2 + (-\mathbf{2.17})$ 14.2 has the larger absolute value and is positive,
                            so the answer will be positive.

12.03

Now Try Exercises 31 and 39.

OBJECTIVE **3** **Estimate the answer when adding or subtracting decimals.** A common error when working decimal problems by hand is to misplace the decimal point in the answer. Or, when using a calculator, you may accidentally press the wrong key. Using *front end rounding* to estimate the answer will help you avoid these mistakes. Start by rounding each number to the highest possible place (as you did in **Section 1.5**). Here are several examples. Notice that in the rounded numbers, only the leftmost digit is something other than 0.

3.25	rounds to	3	6.812	rounds to	7
532.6	rounds to	500	26.397	rounds to	30
7094.2	rounds to	7000	351.24	rounds to	400

EXAMPLE 7   **Estimating Decimal Answers**

First, use front end rounding to round each number. Then add or subtract the rounded numbers to get an estimated answer. Finally, find the exact answer.

(a) Add 194.2 and 6.825

$$
\begin{array}{ll}
\textit{Estimate:} & \textit{Exact:} \\
\end{array}
$$

$$
\begin{array}{cc}
\begin{array}{r} \mathbf{200} \\ +\ \ \ \mathbf{7} \\ \hline 207 \end{array}
& \begin{array}{r} 194.200 \\ +\ \ \ 6.825 \\ \hline 201.025 \end{array}
\end{array}
$$

Rounds to ← 200
Rounds to ← 7

The estimate goes out to the hundreds place (three places to the *left* of the decimal point), and so does the exact answer. Therefore, the decimal point is probably in the correct place in the exact answer.

(b) Subtract $13.78 from $69.42

$$
\begin{array}{ll}
\textit{Estimate:} & \textit{Exact:} \\
\end{array}
$$

$$
\begin{array}{cc}
\begin{array}{r} \$70 \\ -\ \ 10 \\ \hline \$60 \end{array}
& \begin{array}{r} \$69.42 \\ -\ 13.78 \\ \hline \$55.64 \end{array}
\end{array}
$$

Rounds to
Rounds to

$55.64 ← Exact answer is close to estimate, so it is probably correct.

(c) $-1.861 - 7.3$

Rewrite subtraction as adding the opposite. Then use front end rounding to get an estimated answer.

$$
\begin{array}{ccccc}
-1.861 & - & & 7.3 & \\
\downarrow & & & \downarrow & \\
-1.861 & + & & (-7.3) & \\
\downarrow & & & \downarrow & \\
\text{Rounded}\quad -2 & + & & (-7) & = -9 \leftarrow \text{Estimate}
\end{array}
$$

To find the exact answer, add the absolute values.

$$
\begin{array}{r}
1.861 \\
+\ 7.300 \\
\hline
9.161
\end{array}
$$

← Absolute value of -1.861
← Absolute value of -7.3 with two zeros written in

The answer will be negative because both numbers are negative.

$$
-1.861 + (-7.3) = -9.161 \leftarrow \text{Exact}
$$

The exact answer of -9.161 is reasonable because it is close to the estimated answer of -9.

Now Try Exercise 45. ▶▶▶

5.3 Exercises

Find each sum. See Examples 1 and 2.

1.
```
  5.69
  0.24
+ 11.79
```
17.72

2.
```
  372.1
   33.7
+  42.3
```
448.1

3.
```
  0.38
  7
+ 4.6
```
11.98

4.
```
  3.7
  0.812
+ 55
```
— 59.512

5. $14.23 + 8 + 74.63 + 18.715 + 0.286$

6. $197.4 + 0.72 + 17.43 + 25 + 1.4$

7. $27.65 + 18.714 + 9.749 + 3.21$

8. $58.546 + 19.2 + 8.735 + 14.58$

9. Explain and correct the error that a student made when he added $0.72 + 6 + 39.5$ this way:
```
  0.72
  6
+ 39.50
  40.28
```

10. Explain and correct the error that a student made when she added $7.21 + 65 + 13.15$ this way:
```
  7.21
   .65
+ 13.15
  21.01
```
21.01

11. Show why 0.3 is equivalent to 0.3000.

12. Explain why 7 may be written as 7.0 but not as 0.7.

Find each difference. See Examples 3 and 4.

13. $90.5 - 0.8$

14. $303.72 - 0.68$

15. 0.4 less 0.291

16. 0.35 less 0.088
```
  -0.35
   1.38
```

17. 6 minus 5.09

18. 80 minus 16.3

19. Subtract 8.339 from 15

20. Subtract 0.08 from 44

21. Explain and correct the error that a student made when he subtracted 7.45 from 15.32 this way:
```
   7.45
- 15.32
  12.13
```

22. Explain the difference between saying "subtract 2.9 from 8" and saying "2.9 minus 8."

This drawing of a human skeleton shows the average length of the longest bones, in inches.
Use the drawing to answer Exercises 23–26.

7th rib
9.45 in.

Humerus 14.35 in.

8th rib
9.06 in.

Radius 10.4 in.

Ulna 11.1 in.

Femur 19.88 in.

16.94
36.82   8.06
        88

Tibia 16.94 in.   Fibula 15.94 in.

Source: *The Top 10 of Everything.*

23. (a) What is the combined length of the humerus and radius bones?

(b) What is the difference in the lengths of these two bones?

24. (a) What is the total length of the femur and tibia bones?

(b) How much longer is the femur than the tibia?

25. (a) Find the sum of the lengths of the humerus, ulna, femur, and tibia.

(b) How much shorter is the 8th rib than the 7th rib?

26. (a) What is the difference in the lengths of the two bones in the lower arm?

(b) What is the difference in the lengths of the two bones in the lower leg?

Find each sum or difference. See Examples 5 and 6.

27. $24.008 + (-0.995)$

28. $0.77 - 3.06$

29. $-6.05 + (-39.7)$

30. $-6.409 + 8.224$

31. $0.9 - 7.59$

32. $-489.7 - 38$

33. $-2 - 4.99$

34. $2.068 - (-32.7)$

35. $-5.009 + 0.73$

36. $-0.33 - 65$

37. $-1.7035 - (5 - 6.7)$

38. $60 + (-0.9345 + 1.4)$

39. $8000 - (8002.63 - 8)$

40. $-210 - (-0.7306 + 0.5)$

Use front end rounding to estimate each answer. Then use your estimate to select the correct exact answer. Circle your choice. See Example 7.

41. $18 - 11.725$

Estimate:

Exact: 29.725   6.275 -11.545

42. $20 - 1.37$

Estimate:

Exact: -21.37 -1.863 18.63

43. $-6.5 + 0.7$

Estimate:

Exact: -5.8 7.2 -0.58

44. $-9.67 + 3.09$

Estimate:

Exact: -12.76 -6.58 6.58

45. $-42.671 - 194.9$

Estimate:

Exact: -152.229 -23.7571
-237.571

46. $-803.25 - 0.6$

Estimate:

Exact: -803.85 80.385
803.85

47. $8.4 - (-50.83)$

Estimate:

Exact: -42.43 -59.23 59.23

48. $14.98 - (-6.506)$

Estimate:

Exact: -8.474 21.486   8.004

Use front end rounding to estimate each sum or difference. Then find the exact answer to each application problem. Use the information in the table for Exercises 49–52.

INTERNET USERS IN SELECTED COUNTRIES

Country	Number of Users
United States	134.6 million
Japan	33.9 million
China	22.5 million
South Korea	19 million
Canada	15.4 million
France	9 million
World total	**413.7 million**

Source: Computer Industry Almanac.

49. How many fewer Internet users are there in Canada compared to South Korea?

Estimate:

Exact:

50. How many more users are there in Japan compared to France?

Estimate:

Exact:

51. How many Internet users are there in all the countries listed in the table?

Estimate:

Exact:

52. Using the answer from Exercise 51, calculate the number of worldwide Internet users in countries other than the ones in the table.

Estimate:

Exact:

53. The tallest known land mammal is a prehistoric ancestor of the rhino, measuring 6.4 m. Compare the rhino's height to the combined heights of these NBA basketball stars: Kevin Garnett at 2.1 m, Allen Iverson at 1.83 m, and Shaquille O'Neal at 2.16 m. Is their combined height greater or less than the prehistoric rhino's? By how much? (*Source: www.NBA.com/players*)

6.4 m

Estimate:

Exact:

54. Sammy works in a veterinarian's office. He weighed two kittens. One was 3.9 ounces and the other was 4.05 ounces. What was the difference in the weight of the two kittens?

Estimate:

Exact:

55. Steven One Feather gave the cashier a $20 bill to pay for $9.12 worth of groceries. How much change did he get?

Estimate:

Exact:

56. The cost of Julie's tennis racket, with tax, was $41.09. She gave the clerk two $20 bills and a $10 bill. What amount of change did Julie receive?

Estimate:

Exact:

The friends buying fishing equipment on the first page of this chapter brought along the store's sale insert from the Sunday paper. Use the information below on sale prices to answer Exercises 57–60. When estimating, round to the nearest whole number.

Fishing Opener Sale
Catch your limit of savings!

Bobbers 3 for 87¢

Environmentally safe tin split shot
$2.07

Leaded split shot
94¢

8-pound test fishing line
regular $4.84
invisible $7.47
fluorescent $5.14

No-See Line

Tackle boxes

Two trays $7.96   Three trays $9.96

Spinning reels: $9.88, $12.54, $18.84, $24.96
Spinning rods: $9.97, $18.97, $22.96, $28.94

Source: Wal-Mart.

57. What is the difference in price between the fluorescent and regular fishing line?

Estimate:

Exact:

58. How much more does the least expensive spinning rod cost than the least expensive reel?

Estimate:

Exact:

59. Find the total cost of the second-highest-priced spinning reel, two packages of tin split shot, and a three-tray tackle box. Sales tax for all the items is $2.31.

Estimate:

Exact:

60. One friend bought three bobbers on sale. He also bought some SPF15 sunscreen for $7.53 and a flotation vest for $44.96. Sales tax was $3.74. How much did he spend in all?

Estimate:

Exact:

Olivia Sanchez kept track of her expenses for one month. Use her list to answer Exercises 61–64.

61. What were Olivia's total expenses for the month?

Monthly Expenses	
Rent	$994
Car payment	$190.78
Car repairs, gas	$205
Cable TV	$39.95
Internet access	$19.95
Electricity	$40.80
Telephone	$57.32
Groceries	$186.81
Entertainment	$97.75
Clothing, laundry	$107

62. How much did Olivia pay for telephone, cable TV, and Internet access?

63. What was the difference in the amounts spent for groceries and for the car payment?

64. How much more did Olivia spend on rent than on all her car expenses?

Find the length of the dashed line in each rectangle or circle.

65.

0.91 cm   0.7 cm   *b*

3 cm

66.

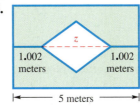

1.002 meters   *z*   1.002 meters

5 meters

67.

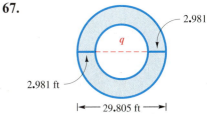

2.981

q

2.981 ft

29.805 ft

5.4 Multiplying Signed Decimal Numbers

OBJECTIVES

1 Multiply positive and negative decimals.

2 Estimate the answer when multiplying decimals.

OBJECTIVE 1 Multiply positive and negative decimals. The decimals 0.3 and 0.07 can be multiplied by writing them as fractions.

$$0.3 \times 0.07 = \frac{3}{10} \times \frac{7}{100} = \frac{3 \times 7}{10 \times 100} = \frac{21}{1000} = 0.021$$

1 decimal place + 2 decimal places → 3 decimal places

Can you see a way to multiply decimals without writing them as fractions? Try these steps. Remember that each number in a multiplication problem is called a *factor*, and the answer is called the *product*.

Multiplying Two Decimal Numbers

Step 1 Multiply the factors (the numbers being multiplied) as if they were whole numbers.

Step 2 Find the *total* number of decimal places in *both* factors.

Step 3 Write the decimal point in the product (the answer) so it has the same number of decimal places as the total from Step 2. You may need to write in extra zeros on the left side of the product in order to get the correct number of decimal places.

Step 4 If two factors have the *same sign*, the product is *positive*. If two factors have *different signs*, the product is *negative*.

> **NOTE**
> When multiplying decimals, you do **not** need to line up decimal points. (You **do** need to line up decimal points when adding or subtracting.)

EXAMPLE 1 Multiplying Decimal Numbers

Find the product of 8.34 and -4.2

Step 1 Multiply the numbers as if they were whole numbers.

$$
\begin{array}{r}
8.34 \\
\times\ 4.2 \\
\hline
1668 \\
3336 \\
\hline
35028 \\
\end{array}
$$

Step 2 Count the total number of decimal places in both factors.

$$
\begin{array}{r}
8.34 \quad \leftarrow \text{2 decimal places}\\
\times\ 4.2 \quad \leftarrow \text{1 decimal place}\\
\hline
1668 \qquad \text{3 total decimal places}\\
3336 \\
\hline
35028 \\
\end{array}
$$

Continued on Next Page

Step 3   Count over 3 places in the product and write the decimal point.
Count from *right to left*.

$$
\begin{array}{r}
8.3\,4 \quad \leftarrow \text{2 decimal places} \\
\times \quad 4.2 \quad \leftarrow \text{1 decimal place} \\
\hline
1\,6\,6\,8 \quad \text{3 total decimal places} \\
3\,3\,3\,6 \\
\hline
3\,5.0\,2\,8 \quad \leftarrow \text{3 decimal places in product}
\end{array}
$$

Count over 3 places from right to left
to position the decimal point.

Step 4   The factors have *different* signs, so the product is *negative:*
8.34 times (-4.2) is -35.028.

Now Try Exercise 5. ▶▶▶

EXAMPLE 2   **Writing Zeros as Placeholders in the Product**

Find the product: $(-0.042)(-0.03)$

Start by multiplying, then count decimal places.

$$
\begin{array}{r}
0.0\,4\,2 \quad \leftarrow \text{3 decimal places} \\
\times \quad 0.0\,3 \quad \leftarrow \text{2 decimal places} \\
\hline
1\,2\,6 \quad \leftarrow \text{5 decimal places needed in product}
\end{array}
$$

After multiplying, the answer has only three decimal places, but five are
needed. So write two zeros on the *left* side of the answer.

$$
\begin{array}{r}
0.0\,4\,2 \\
\times \quad 0.0\,3 \\
\hline
0\,0\,1\,2\,6 \\
\uparrow\,\uparrow
\end{array}
\qquad
\begin{array}{r}
0.0\,4\,2 \quad \leftarrow \text{3 decimal places} \\
\times \quad 0.0\,3 \quad \leftarrow \text{2 decimal places} \\
\hline
.0\,0\,1\,2\,6 \quad \leftarrow \text{5 decimal places}
\end{array}
$$

Write two zeros
on *left* side of answer.

Now count over 5 places
and write in the decimal point.

The final product is 0.00126, which has five decimal places. The product is
positive because the factors have the *same* sign (both factors are negative).

Now Try Exercise 17. ▶▶▶

📟 **Calculator Tip**   When working with money amounts, you may need
to write a 0 in your answer. For example, try multiplying $\$3.54 \times 5$ on
your calculator. Write down the result.

$$3.54 \ \times\ 5 \ = \ \underline{\qquad}$$

Notice that the result is 17.7, which is *not* the way to write a money
amount. You have to write the 0 in the hundredths place: $\$17.70$ is correct.
The calculator does not show the "extra" 0 because:

$$17.70 \text{ or } 17\frac{70}{100} \quad \text{simplifies to} \quad 17\frac{7}{10} \text{ or } 17.7$$

So keep an eye on your calculator—it doesn't know when you're working
with money amounts.

OBJECTIVE **2** **Estimate the answer when multiplying decimals.**   If you are doing multiplication problems by hand, estimating the answer helps you check that the decimal point is in the right place. When you are using a calculator, estimating helps you catch an error like pressing the ⊘ key instead of the ⊗ key.

EXAMPLE 3   **Estimating before Multiplying**

First, use front end rounding to estimate $(76.34)(12.5)$. Then find the exact answer.

$$
\begin{array}{rl}
\textit{Estimate:} & \\
80 & \xleftarrow{\text{Rounds to}} \\
\times\ 10 & \xleftarrow{\text{Rounds to}} \\
\hline
800 &
\end{array}
\qquad
\begin{array}{rl}
\textit{Exact:} & \\
7\,6.3\,4 & \leftarrow \text{2 decimal places} \\
\times\quad 1\,2.5 & \leftarrow \text{1 decimal place} \\
\hline
3\,8\,1\,7\,0 & \quad \text{3 decimal places are} \\
1\,5\,2\,6\,8 & \qquad \text{in the product.} \\
7\,6\,3\,4 & \\
\hline
9\,5\,4.2\,5\,0 &
\end{array}
$$

Both the estimate and the exact answer go out to the hundreds place, so the decimal point in 954.250 is probably in the correct place.

Now Try Exercise 25.

5.4 Exercises

Find each product. See Example 1.

1. 0.042
   ×   3.2

2. 0.571
   ×   2.9

3. $-21.5(7.4)$

4. $-85.4(-3.5)$

5. $(-23.4)(-0.66)$

6. $0.896(-0.7)$

📱 **7.** $51.88
   ×   665

📱 **8.** $736.75
   ×   118

Use the fact that $(72)(6) = 432$ *to solve Exercises 9–16 by simply counting decimal places and writing the decimal point in the correct location. Be sure to indicate the sign of the product.*

9. $72(-0.6) = $ 4 3 2

10. $7.2(-6) = $ 4 3 2

11. $(7.2)(0.06) = $ 4 3 2

12. $(0.72)(0.6) = $ 4 3 2

13. $-0.72(-0.06) = $ 4 3 2

14. $-72(-0.0006) = $ 4 3 2

15. $(0.0072)(0.6) = $ 4 3 2

16. $(0.072)(0.006) = $ 4 3 2

Find each product. See Example 2.

17. $(0.006)(0.0052)$

18. $(0.0052)(0.009)$

19. $(-0.003)^2$

20. $(0.0004)^2$

RELATING CONCEPTS (EXERCISES 21–22)   For Individual or Group Work

Look for patterns as you **work Exercises 21 and 22 in order**.

21. Do these multiplications:

$(5.96)(10) = $ _____ $(3.2)(10) = $ _____

$(0.476)(10) = $ _____ $(80.35)(10) = $ _____

$(722.6)(10) = $ _____ $(0.9)(10) = $ _____

What pattern do you see? Write a "rule" for multiplying by 10. What do you think the rule is for multiplying by 100? by 1000? Write the rules and try them out on the numbers above.

22. Do these multiplications:

$(59.6)(0.1) = $ _____ $(3.2)(0.1) = $ _____

$(0.476)(0.1) = $ _____ $(80.35)(0.1) = $ _____

$(65)(0.1) = $ _____ $(523)(0.1) = $ _____

What pattern do you see? Write a "rule" for multiplying by 0.1. What do you think the rule is for multiplying by 0.01? by 0.001? Write the rules and try them out on the numbers above.

First, use front end rounding to estimate the answer. Then find the exact answer. See Example 3.

23. *Estimate:*   *Exact:*

←── Rounds to ──   39.6

× ____   ←── Rounds to ──   × 4.8

24. *Estimate:*   *Exact:*

18.7

× ____   × 2.3

25. *Estimate:*   *Exact:*

37.1

× ____   × 42

26. *Estimate:*   *Exact:*

5.08

× ____   × 71

27. *Estimate:*   *Exact:*

6.53

× ____   × 4.6

28. *Estimate:*   *Exact:*

7.51

× ____   × 8.2

29. *Estimate:*   *Exact:*

2.809

× ____   × 6.85

30. *Estimate:*   *Exact:*

73.52

× ____   × 22.34

Even with most of the problem missing, you can tell whether or not these answers are reasonable. Circle reasonable *or* unreasonable. *If the answer is unreasonable, move the decimal point or insert a decimal point to make the answer reasonable.*

31. How much was his car payment? $28.90

reasonable

unreasonable, should be _____

32. How many hours did she work today? 25 hours

reasonable

unreasonable, should be _____

33. How tall is her son? 60.5 inches

reasonable

unreasonable, should be _____

34. How much does he pay for rent now? $6.92

reasonable

unreasonable, should be _____

35. What is the price of one gallon of milk? $349

reasonable

unreasonable, should be _____

36. How long is the living room? 16.8 feet

reasonable

unreasonable, should be _____

37. How much did the baby weigh? 0.095 pounds

reasonable

unreasonable, should be _____

38. What was the sale price of the jacket? $1.49

reasonable

unreasonable, should be _____

Solve each application problem. Round money answers to the nearest cent when necessary.

39. LaTasha worked 50.5 hours over the last two weeks. She earns $18.73 per hour. How much did she make?

40. Michael's time card shows 42.2 hours at $10.03 per hour. What are his earnings?

41. Sid needs 0.6 meter of canvas material to make a carry-all bag that fits on his wheelchair. If canvas is $4.09 per meter, how much will Sid spend? (*Note:* $4.09 *per* meter means $4.09 for *one* meter.)

42. How much will Mrs. Nguyen pay for 3.5 yards of lace trim that costs $0.87 per yard?

43. Michelle filled the tank of her SUV with regular unleaded gas. Use the information shown on the pump to find how much she paid for gas.

Source: Holiday.

44. Ground beef and chicken legs are on sale. Juma bought 1.7 pounds of legs. Use the information in the ad to find the amount she paid.

45. Ms. Rolack is a real estate broker who helps people sell their homes. Her fee is 0.07 times the price of the home. What was her fee for selling a $289,500 home?

46. Manny Ramirez of the Boston Red Sox had a batting average of 0.308 in the 2004 season. He went to bat 568 times. How many hits did he make? (*Hint:* Multiply the number of times at bat by his batting average.) Round to the nearest whole number. (*Source: World Almanac.*)

Paper money in the United States has not always been the same size. Shown below are the measurements of bills printed before 1929 and the measurements from 1929 on. Use this information to answer Exercises 47–50. Recall that perimeter is the total distance around the edges of a figure (see **Section 3.1**). *The area of a rectangle is found by multiplying the length by the width (see* **Section 3.2**). *(Source: www.moneyfactory.com)*

Before 1929

3.125 in.

7.4218 in.

From 1929 on

2.61 in.

6.14 in.

47. **(a)** Find the area of each bill, rounded to the nearest tenth.

(b) What is the difference in the rounded areas?

48. **(a)** Find the perimeter of each bill, to the nearest hundredth.

(b) How much less is the perimeter of today's bills than the bills printed before 1929?

49. The thickness of one piece of today's paper money is 0.0043 inch.

(a) If you had a pile of 100 bills, how high would it be?

(b) How high would a pile of 1000 bills be?

50. **(a)** Use your answers from Exercise 49 to find the number of bills in a pile that is 43 inches high.

(b) How much money would you have if the pile is all $20 bills?

51. Judy Lewis pays $38.96 per month for basic cable TV. The one-time installation fee was $49. How much will she pay for cable over two years? How much would she pay in two years for the deluxe cable package that costs $89.95 per month?

52. Chuck's SUV payment is $420.27 per month for four years. He also made a down payment of $5000 at the time he bought the SUV. How much will he pay altogether?

53. Paper for the copy machine at the library costs $0.015 per sheet. How much will the library pay for 5100 sheets?

54. A student group collected 2200 pounds of plastic as a fund-raiser. How much will they make if the recycling center pays $0.142 per pound?

55. Barry bought 16.5 meters of rope at $0.47 per meter and three meters of wire at $1.05 per meter. How much change did he get from three $5 bills?

56. Susan bought a 32-inch flat-tube TV that cost $549.99. She paid $55 down and $98.83 per month for six months. How much could she have saved by paying cash?

Use the information below from the Look Smart mail order catalog to answer Exercises 57–60.

Knit Shirt Ordering Information				Total Price of Items (excluding monograms and gift boxes)	Shipping, Packing, and Handling
43–2A	short sleeved, solid colors	$14.75 each		$0–25.00	$3.50
43–2B	short sleeved, stripes	$16.75 each		$25.01–75.00	$5.95
43–3A	long sleeved, solid colors	$18.95 each		$75.01–125.00	$7.95
43–3B	long sleeved, stripes	$21.95 each		$125.01+	$9.95
XXL size, add $2 per shirt.					
Monogram, $4.95 each. Gift box, $5 each.				Shipping to each additional address, add $4.25.	

57. Find the total cost of ordering four long-sleeved, solid-color shirts and two short-sleeved, striped shirts, all in the XXL size, and all shipped to your home.

58. What is the total cost of eight long-sleeved shirts, five in solid colors and three striped? Include the cost of shipping the solid shirts to your home and the striped shirts to your brother's home.

59. (a) What is the total cost, including shipping, of sending three short-sleeved, solid-color shirts, with monograms, in a gift box to your aunt for her birthday?

   (b) How much did the monograms, gift box, and shipping add to the cost of your gift?

60. (a) Suppose you order one of each type of shirt for yourself, adding a monogram on each of the solid-color shirts. At the same time, you order three long-sleeved striped shirts, in the XXL size, shipped to your dad in a gift box. Find the total cost of your order.

   (b) What is the difference in total cost (excluding shipping) between the shirts for yourself and the gift for your dad?

5.5 Dividing Signed Decimal Numbers

There are two kinds of decimal division problems: those in which a decimal is divided by an integer, and those in which a number is divided by a decimal. First recall the parts of a division problem.

Divisor → 2)$\overline{16}$ ← Dividend, 8 ← Quotient

$16 \div 2 = 8$ (Dividend, Divisor, Quotient)

Dividend → $\dfrac{16}{2} = 8$ ← Quotient (Divisor)

OBJECTIVES

1. Divide a decimal by an integer.
2. Divide a number by a decimal.
3. Estimate the answer when dividing decimals.
4. Use the order of operations with decimals.

OBJECTIVE 1 Divide a decimal by an integer.   When the divisor is an integer, use these steps.

Dividing a Decimal Number by an Integer

Step 1   Write the decimal point in the quotient (answer) directly above the decimal point in the dividend.

Step 2   Divide as if both numbers were whole numbers.

Step 3   If both numbers have the *same sign*, the quotient is *positive*. If they have *different signs*, the quotient is *negative*.

EXAMPLE 1 Dividing Decimals by Integers

Find each quotient.

(a) $21.93 \div (-3)$

Dividend, Divisor

First consider $21.93 \div 3$.   3)$\overline{21.93}$

Step 1   Write the decimal point in the quotient directly above the decimal point in the dividend.

Decimal points lined up

3)$\overline{21.93}$

Step 2   Divide as if the numbers were whole numbers.

Check by multiplying the quotient times the divisor.

$\dfrac{7.31}{3)\overline{21.93}}$

Matches, so 7.31 is correct.

Check

$\begin{array}{r} 7.31 \\ \times\ \ \ 3 \\ \hline 21.93 \end{array}$

Step 3   The quotient is -7.31 because the numbers have *different* signs.

$21.93 \div (-3) = -7.31$

Different signs — Negative quotient

Continued on Next Page

(b)

$$9\overline{)470.7}$$

Divisor ⟶ Dividend

Write the decimal point in the quotient directly above the decimal point in the dividend. Then divide as if they were whole numbers.

Decimal points lined up

Check

$$
\begin{array}{r}
52.3 \\
9\overline{)470.7} \\
\underline{45} \\
20 \\
\underline{18} \\
27 \\
\underline{27} \\
0
\end{array}
$$

Matches

$$
\begin{array}{r}
52.3 \\
\times\quad 9 \\
\hline
470.7
\end{array}
$$

The quotient is 52.3 and is *positive* because both numbers have the *same* sign.

◀◀◀ **Now Try Exercise 1.**

EXAMPLE 2   **Writing Extra Zeros to Complete a Division**

Divide 1.5 by 8.

Keep dividing until the remainder is 0, or until the digits in the quotient begin to repeat in a pattern. In Example 1(b) above, you ended up with a remainder of 0. But sometimes you run out of digits in the dividend before that happens. If so, write extra zeros on the right side of the dividend so you can continue dividing.

$$
\begin{array}{r}
0.1 \\
8\overline{)1.5} \\
\underline{8} \\
7
\end{array}
$$

← All digits have been used.

← Remainder is not yet 0.

Write a 0 after the 5 in the dividend so you can continue dividing. Keep writing more zeros in the dividend if needed. Recall that writing zeros to the *right* of a decimal number does **not** change its value.

$$
\begin{array}{r}
0.1\,8\,7\,5 \\
8\overline{)1.5\,0\,0\,0} \\
\underline{8} \\
7\,0 \\
\underline{6\,4} \\
6\,0 \\
\underline{5\,6} \\
4\,0 \\
\underline{4\,0} \\
0
\end{array}
$$

← Three zeros needed to complete the division

← Stop dividing when the remainder is 0.

Check

$$
\begin{array}{r}
0.1875 \\
\times\quad\ \ 8 \\
\hline
1.5000
\end{array}
$$

└ Matches dividend, so 0.1875 is correct

CAUTION

Notice that in decimals the dividend may *not* be the larger number. In Example 2 above, the dividend is 1.5, which is *smaller* than 8.

Continued on Next Page

⊞ **Calculator Tip**   When *multiplying* numbers, you can enter them in any order because multiplication is commutative (see **Section 1.6**). But division is *not* commutative. It *does* matter which number you enter first. Try Example 2 both ways; jot down your answers.

1.5 ⊕ 8 ⊜ _____   8 ⊕ 1.5 ⊜ _____

Notice that the first answer, 0.1875, matches the result from Example 2. But the second answer is much different: 5.333333333. Be careful to enter the dividend first.

Now Try Exercise 3(a). ▶▶▶

In the next example the remainder is never 0, even if we keep dividing.

EXAMPLE 3 **Rounding a Decimal Quotient**

Divide 4.7 by 3. Round the quotient to the nearest thousandth.

Write extra zeros in the dividend so that you can continue dividing.

```
      1.5 6 6 6
  3)4.7 0 0 0   ← Three zeros added so far
    3
    ──
    1 7
    1 5
    ───
      2 0
      1 8
      ───
        2 0
        1 8
        ───
          2 0
          1 8
          ───
            2   ← Remainder is still not 0.
```

Notice that the digit 6 in the quotient is repeating. It will continue to do so. The remainder will *never be 0*. There are two ways to show that an answer is a **repeating decimal** that goes on forever. You can write three dots after the answer, or you can write a bar above the digits that repeat (in this case, the 6).

$$1.5\underbrace{666}\ldots \quad \text{or} \quad 1.5\overline{6}$$

← Bar above repeating digit

Three dots

CAUTION
Do not use *both* the dots *and* the bar at the same time. Use three dots *or* the bar.

When repeating decimals occur, round the quotient according to the directions in the problem. In this example, to round to thousandths, divide out one *more* place, to ten-thousandths.

$$4.7 \div 3 = 1.5666\ldots \quad \text{rounds to} \quad 1.567$$

Check the answer by multiplying 1.567 by 3. Because 1.567 is a rounded answer, the check will *not* give exactly 4.7, but it should be very close.

$$(1.567)(3) = 4.701 \quad \text{← Does not equal exactly 4.7}$$
because 1.567 was rounded

Continued on Next Page

> **CAUTION**
> When checking quotients that you've rounded, the check will *not* match the dividend exactly, but it should be very close.

 Now Try Exercise 21.

OBJECTIVE 2 Divide a number by a decimal. To divide by a *decimal* divisor, first change the divisor to a whole number. Then divide as before. To see how this is done, write the problem in fraction form. Here is an example.

$$1.2\overline{)6.36} \quad \text{can be written} \quad \frac{6.36}{1.2}$$

In **Section 4.1** you learned that multiplying the numerator and denominator by the same number gives an equivalent fraction. We want the divisor (1.2) to be a whole number. Multiplying by 10 will accomplish that.

$$\text{Decimal divisor} \rightarrow \frac{6.36}{1.2} = \frac{(6.36)\,(10)}{(1.2)\,(10)} = \frac{63.6}{12} \leftarrow \text{Whole number divisor}$$

The short way to multiply by 10 is to move the decimal point *one place* to the *right* in both the divisor and the dividend.

$$1.2\overline{)6.36} \quad \text{is equivalent to} \quad 12\overline{)63.6}$$

> **NOTE**
> Moving the decimal points the **same** number of places in **both** the divisor and dividend will **not** change the answer.

Dividing by a Decimal Number

Step 1   Count the number of decimal places in the divisor and move the decimal point that many places to the *right*. (This changes the divisor to a whole number.)

Step 2   Move the decimal point in the dividend the *same* number of places to the *right*. (Write in extra zeros if needed.)

Step 3   Write the decimal point in the quotient directly above the decimal point in the dividend. Then divide as usual.

Step 4   If both numbers have the *same sign,* the quotient is *positive.* If they have *different signs,* the quotient is *negative.*

EXAMPLE 4 **Dividing by Decimal Numbers**

(a) $\dfrac{27.69}{0.003}$

Move the decimal point in the divisor *three* places to the *right* so that 0.003 becomes the whole number 3. In order to move the decimal point in the dividend the same number of places, write in an extra 0.

Moving decimal point three places is the same as multiplying by 1000

$0.003\overline{)27.690}$ Move decimal points in divisor and dividend. Then line up decimal point in the quotient.

$\dfrac{9230.}{3\overline{)27690.}}$ Divide as usual.

(b) Divide -5 by -4.2 and round the quotient to the nearest hundredth.

First consider $5 \div 4.2$. Move the decimal point in the divisor one place to the right so that 4.2 becomes the whole number 42. The decimal point in the dividend starts on the right side of 5 and is also moved one place to the right.

$$
\begin{array}{r}
1.190 \\
4.2\overline{)5.0000} \\
42 \\
\hline
80 \\
42 \\
\hline
380 \\
378 \\
\hline
20
\end{array}
$$

← In order to round to hundredths, divide out one *more* place, to thousandths.

Rounding the quotient to the nearest hundredth gives 1.19. The quotient is *positive* because both the divisor and dividend have the *same* sign (both are negative).

$$-5 \div (-4.2) \approx 1.19$$

Same sign   Positive quotient

Now Try Exercise 19.

OBJECTIVE 3 Estimate the answer when dividing decimals. Estimating the answer to a division problem helps you catch errors. Compare the estimate to your exact answer. If they are very different, do the division again.

EXAMPLE 5 Estimating before Dividing

First, use front end rounding to estimate the answer. Then divide to find the exact answer.

$$580.44 \div 2.8$$

Here is how one student solved this problem. She rounded 580.44 to 600 and 2.8 to 3 to estimate the answer.

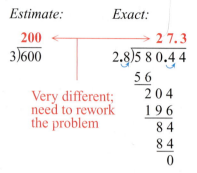

Estimate:

$$
\begin{array}{r}
\mathbf{200} \\
3\overline{)600}
\end{array}
$$

Very different; need to rework the problem

Exact:

$$
\begin{array}{r}
2\,7.3 \\
2.8\overline{)5\,8\,0.4\,4} \\
5\,6 \\
\hline
2\,0\,4 \\
1\,9\,6 \\
\hline
8\,4 \\
8\,4 \\
\hline
0
\end{array}
$$

Notice that the estimate, which is in the hundreds, is very different from the exact answer, which is only in the tens. This tells the student that she needs to rework the problem. Can you find the error? (The exact answer should be 207.3, which fits with the estimate of 200.)

Now Try Exercise 33.

OBJECTIVE 4 Use the order of operations with decimals. Use the order of operations from **Section 1.8** when a decimal problem involves more than one operation.

Order of Operations

Step 1 Work inside **parentheses** or **other grouping symbols.**

Step 2 Simplify expressions with **exponents.**

Step 3 Do the remaining **multiplications and divisions** as they occur from left to right.

Step 4 Do the remaining **additions and subtractions** as they occur from left to right.

EXAMPLE 6 Using the Order of Operations

Simplify by using the order of operations

(a) $2.5 + (-6.3)^2 + 9.62$ Apply the exponent: $(-6.3)(-6.3)$ is 39.69

$\quad\;\; 2.5 + 39.69 \;\; + 9.62$ Add from left to right.

$\qquad\;\; 42.19 \qquad + 9.62$

$\qquad\qquad\; 51.81$

(b) $1.82 + (5.2 - 6.7)(5.8)$ Work inside parentheses.

$\quad\; 1.82 + \quad (-1.5) \;\; (5.8)$ Multiply next.

$\quad\; 1.82 + \qquad (-8.7)$ Add last.

$\qquad\quad -6.88$

Continued on Next Page

(c) $3.7^2 - 1.8 \div 5(1.5)$ Apply the exponent.

$13.69 - 1.8 \div 5(1.5)$ Multiply and divide from *left to right,* so first divide 1.8 by 5

$13.69 - 0.36\,(1.5)$ Multiply 0.36 by 1.5

$13.69 - 0.54$ Subtract last.

13.15

Now Try Exercise 53. ▶▶▶

Calculator Tip   Most calculators that have parentheses keys ⦅ ⦆ can handle calculations like those in Example 6 just by entering the numbers in the order given. For example, the keystrokes for Example 6(b) are:

Parentheses

1.82 ⊕ ⦅ 5.2 ⊖ 6.7 ⦆ ⊗ 5.8 ⊜   Answer is −6.88

Standard, four-function calculators generally do *not* give the correct answer if you enter the numbers in the order given. Check the instruction manual that came with your calculator for information on "order of calculations" to see if your model has the rules for order of operations built into it. For a quick check, try entering this problem:

2 ⊕ 2 ⊗ 2 ⊜

If the result is 6, the calculator follows the order of operations. If the result is 8, it does *not* have the rules built into it. Use the space below to explain how this test works.

Answer: The test works because a calculator that follows the order of operations will automatically do the multiplication first. If the calculator does *not* have the rules built into it, it will work from left to right.

Following Order of Operations   **Working from Left to Right**

$2 + 2 \times 2$ Multiply before adding. $2 + 2 \times 2$

$2 + \quad 4$ $4 \quad \times 2$

6 ← Correct 8 ← Incorrect

5.5 Exercises

Find each quotient. See Examples 1, 2, and 4.

1. (a) $27.3 \div (-7)$ (b) $\dfrac{4.23}{9}$ **2.** (a) $-50.4 \div 8$ (b) $\dfrac{1.62}{6}$

3. (a) $0.65 \div 4$ (b) $\dfrac{138.5}{-50}$ **4.** (a) $\dfrac{-2.25}{6}$ (b) $-0.5 \div (-8)$

5. $-20.01 \div (-0.05)$ **6.** $-16.04 \div (-0.08)$ **7.** $1.5\overline{)54}$ **8.** $2.4\overline{)132}$

Use the fact that $108 \div 18 = 6$ to work Exercises 9–16 simply by moving decimal points. See Examples 1, 2, and 4.

9. $1.8\overline{)0.108}$ **10.** $18\overline{)10.8}$ **11.** $0.018\overline{)108}$ **12.** $0.18\overline{)1.08}$

13. $0.18\overline{)10.8}$ **14.** $0.18\overline{)108}$ **15.** $18\overline{)0.0108}$ **16.** $1.8\overline{)0.0108}$

Divide. Round quotients to the nearest hundredth when necessary. See Examples 3 and 4.

17. $4.6\overline{)116.38}$ **18.** $2.6\overline{)4.992}$ **19.** $\dfrac{-3.1}{-0.006}$ **20.** $\dfrac{-1.7}{0.09}$

Divide. Round quotients to the nearest thousandth. See Examples 3 and 4.

21. $-240.8 \div 9$ **22.** $-76.43 \div (-7)$ **23.** $0.034\overline{)342.81}$ **24.** $0.043\overline{)1748.4}$

RELATING CONCEPTS (EXERCISES 25–26) For Individual or Group Work

*First, look back at your work in **Section 5.4**, Exercises 21 and 22. Then look for patterns as you **work Exercises 25 and 26 in order.***

25. Do these division problems.

$3.77 \div 10 =$ _____ $9.1 \div 10 =$ _____ $0.886 \div 10 =$ _____

$30.19 \div 10 =$ _____ $406.5 \div 10 =$ _____ $6625.7 \div 10 =$ _____

(a) What do you see? Write a "rule" for dividing by 10. What do you think the rule is for dividing by 100? by 1000? Write the rules and try them out on the numbers above.

(b) Compare your rules to the ones you wrote in **Section 5.4**, Exercise 21.

(continued)

26. Do these division problems.

$40.2 \div 0.1 =$ _____ $7.1 \div 0.1 =$ _____ $0.339 \div 0.1 =$ _____

$15.77 \div 0.1 =$ _____ $46 \div 0.1 =$ _____ $873 \div 0.1 =$ _____

(a) What pattern do you see? Write a "rule" for dividing by 0.1. What do you think the rule is for dividing by 0.01? by 0.001? Write the rules and try them out on the numbers above.

(b) Compare your rules to the ones you wrote in **Section 5.4**, Exercise 22.

Decide whether each answer is reasonable *or* unreasonable *by using front end rounding to estimate the answer. If the exact answer is not reasonable, find and correct the error. See Example 5.*

27. $37.8 \div 8 = 47.25$

Estimate:

28. $345.6 \div 3 = 11.52$

Estimate:

29. $54.6 \div 48.1 \approx 1.135$

Estimate:

30. $2428.8 \div 4.8 = 50.6$

Estimate:

31. $307.02 \div 5.1 = 6.2$

Estimate:

32. $395.415 \div 5.05 = 78.3$

Estimate:

33. $9.3 \div 1.25 = 0.744$

Estimate:

34. $78 \div 14.2 = 0.182$

Estimate:

Solve each application problem. Round money answers to the nearest cent when necessary.

35. Rob has discovered that his daughter's favorite brand of tights are on sale. He decided to buy one pair as a surprise for her. How much did he pay?

36. The bookstore has a special price on notepads. How much did Randall pay for one notepad?

37. It will take 21 months for Aimee to pay off her credit card balance of $1408.66. How much is she paying each month?

38. Marcella Anderson bought 2.6 meters of microfiber woven suede fabric for $33.77. How much did she pay per meter?

39. Adrian Webb bought 619 bricks to build a barbecue pit, paying $185.70. Find the cost per brick. (*Hint:* Cost *per* brick means the cost for *one* brick.)

40. Lupe Wilson is a newspaper distributor. Last week she paid the newspaper $130.51 for 842 copies. Find the cost per copy.

41. Darren Jackson earned $476.80 for 40 hours of work. Find his earnings per hour.

42. At a CD manufacturing company, 400 CDs cost $289. Find the cost per CD.

43. It took 16.35 gallons of gas to fill the gas tank of Kim's car. She had driven 346.2 miles since her last fill-up. How many miles per gallon did she get? Round to the nearest tenth.

44. Mr. Rodriquez pays $53.19 each month to Household Finance. How many months will it take him to pay off $1436.13?

Use the table of women's longest long jumps (through June 2005) to answer Exercises 45–50. To find an average, add up the values you are interested in and then divide the sum by the number of values. Round your answers to the nearest hundredth.

Athlete	Country	Year	Length (meters)
Galina Christyakova	USSR	1988	7.52
Jackie Joyner-Kersee	U.S.	1994	7.49
Heike Drechsler	Germany	1992	7.48
Jackie Joyner-Kersee	U.S.	1987	7.45
Jackie Joyner-Kersee	U.S.	1988	7.40
Jackie Joyner-Kersee	U.S.	1991	7.32
Jackie Joyner-Kersee	U.S.	1996	7.20
Chioma Ajunwa	Nigeria	1996	7.12
Fiona May	Italy	2000	7.09
Tatyana Lebedeva	Russia	2004	7.07

Source: CNNSI.com and IAFF.org

45. Find the average length of the long jumps made by Jackie Joyner-Kersee.

46. Find the average length of all the long jumps listed in the table.

47. How much longer was the fifth-longest jump than the sixth-longest jump?

48. If the first-place athlete made five jumps of the same length, what would be the total distance jumped?

49. What was the total length jumped by the top three athletes in the table?

50. How much less was the last-place jump than the next-to-last-place jump?

Simplify by using the order of operations. See Example 6.

51. $7.2 - 5.2 + 3.5^2$

52. $6.2 + 4.3^2 - 9.72$

53. $38.6 + 11.6(10.4 - 13.4)$

54. $2.25 - 1.06(0.85 - 3.95)$

55. $-8.68 - 4.6(10.4) \div 6.4$

56. $25.1 + 11.4 \div 7.5(-3.75)$

57. $33 - 3.2(0.68 + 9) + (-1.3)^2$

58. $0.6 + (-1.89 + 0.11) \div 0.004(0.5)$

Solve each application problem.

59. Soup is on sale at six cans for $3.25, or you can purchase individual cans for $0.57. How much will you save per can if you buy six cans? Round to the nearest cent.

60. Nadia's diet says she can eat 3.5 ounces of chicken nuggets. The package weighs 10.5 ounces and contains 15 nuggets. How many nuggets can Nadia eat?

61. The U.S. Treasury prints 37,000,000 pieces of paper money each day. The printing presses run 24 hours a day. How many pieces of money are printed, to the nearest whole number:

(a) each hour

37,000,000 pieces of paper money are printed each day.

(b) each minute

(c) each second

(*Source:* www.moneyfactory.com.)

62. Mach 1 is the speed of sound. Dividing a vehicle's speed by the speed of sound gives its speed on the Mach scale. In 1997, a specially built car with two 110,000-horsepower engines broke the world land speed record by traveling 763.035 miles per hour. The speed of sound changes slightly with the weather. That day it was 748.11 miles per hour. What was the car's Mach speed, to the nearest hundredth? (*Source:* Associated Press.)

General Mills will give a school 10¢ for each box top logo from its cereals and other products. A school can earn up to $10,000 per year. Use this information to answer Exercises 63–66. Round answers to the nearest whole number. (Source: General Mills.)

63. How many box tops would a school need to collect in one year to earn the maximum amount?

64. (Complete Exercise 63 first.) If a school has 550 children, how many box tops would each child need to collect to reach the maximum?

65. How many box tops would need to be collected during each of the 38 weeks in the school year to reach the maximum amount?

66. How many box tops would each of the 550 children need to collect during each of the 38 weeks of school to reach the maximum amount?

Summary Exercises on Decimals

Write each decimal as a fraction or mixed number in lowest terms.

1. 0.8

2. 6.004

3. 0.35

Write each decimal in words.

4. 94.5

5. 2.0003

6. 0.706

Write each decimal in numbers.

7. five hundredths

8. three hundred nine ten-thousandths

9. ten and seven tenths

Round to the place indicated.

10. 6.1873 to the nearest hundredth

11. 0.95 to the nearest tenth

12. 0.42025 to the nearest thousandth

13. $0.893 to the nearest cent

14. $3.0017 to the nearest cent

15. $99.64 to the nearest dollar

Simplify.

16. $-0.27(3.5)$

17. $50 - 0.3801$

18. $0.35 \div (-0.007)$

19. $\dfrac{-90.18}{-6}$

20. $(0.004)(1.22)$

21. $1.55 - 3.7$

22. $-0.95 + 10.005$

23. $3.6 + 0.718 + 9 + 5.0829$

24. $32.305 - 40 + 0.7$

25. $-8.9 + 4^2 \div (-0.02)$

26. $(-0.18 + 2.5) + 4(-0.05)$

27. $0.64 \div 16.3$ Round your answer to the nearest hundredth.

Find the perimeter of each figure.

28.

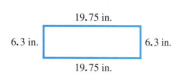

19.75 in.

6.3 in.   6.3 in.

19.75 in.

29.

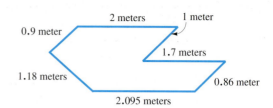

2 meters   1 meter

0.9 meter

1.7 meters

1.18 meters

0.86 meter

2.095 meters

Solve each application problem.

30. At a bakery, Sue Chee bought $7.42 worth of muffins and $10.09 worth of croissants for a staff party and a $0.69 cookie for herself. How much change did she receive from two $10 bills?

31. The Bell family rented a motor home for $375 per week plus $0.35 per mile. What was the rental cost for their three-week vacation trip of 2650 miles?

The table below shows information on two tents for camping. Use the table to answer Exercises 32 and 33.

Tents	Coleman family dome	Eddie Bauer dome tent
Dimensions	13 ft × 13 ft	12 ft × 12 ft
Sleeps	8 campers	6 campers
Sale price	$127	$99

(*Source:* target.com.)

32. For the Coleman tent, find:

(a) the area of the tent floor

(b) the cost per square foot of floor space, to the nearest cent.

33. Find the same information for the Eddie Bauer tent as you did for the Coleman tent.

Use the information in the table to answer Exercises 34–36.

Animal	Average Weight of Animal (ounces)	Average Weight of Food Eaten Each Day (ounces)
Hamster	3.5	0.4
Queen bee	0.004	?
Hummingbird	?	0.07

Source: NCTM News Bulletin.

34. (a) In how many days will a hamster eat enough food to equal its body weight? Round to the nearest whole number of days.

(b) If a 140-pound woman ate her body weight of food in the same number of days as the hamster, how much would she eat each day? Round to the nearest tenth.

35. While laying eggs, a queen bee eats eighty times her weight each day. Use this information to fill in one of the missing values in the table.

36. A hummingbird's body weight is about 1.6 times the weight of its daily food intake. Find its body weight.

5.6 Fractions and Decimals

Writing fractions as equivalent decimals can help you do calculations or compare the size of two numbers more easily.

OBJECTIVE 1 Write fractions as equivalent decimals. Recall that a fraction is one way to show division (see **Section 1.7**). For example, $\frac{3}{4}$ means $3 \div 4$. If you are doing the division by hand, write it as $4\overline{)3}$. When you do the division, the result is 0.75, the decimal equivalent of $\frac{3}{4}$.

> **Writing a Fraction as an Equivalent Decimal**
>
> *Step 1*   Divide the numerator of the fraction by the denominator.
>
> *Step 2*   If necessary, round the answer to the place indicated.

EXAMPLE 1 Writing Fractions or Mixed Numbers as Decimals

(a) Write $\frac{1}{8}$ as a decimal.

$\frac{1}{8}$ means $1 \div 8$. Write it as $8\overline{)1}$. The decimal point in the dividend is on the *right* side of the 1. Write extra zeros in the dividend so you can continue dividing until the remainder is 0.

$$\frac{1}{8} \implies 1 \div 8 \implies 8\overline{)1} \implies \begin{array}{r} 0.125 \\ 8\overline{)1.000} \\ \underline{8} \\ 20 \\ \underline{16} \\ 40 \\ \underline{40} \\ 0 \end{array}$$

Decimal points lined up

Three extra zeros needed

← Remainder is 0

Therefore, $\frac{1}{8} = 0.125$

To check, write 0.125 as a fraction, then write it in lowest terms.

$$0.125 = \frac{125}{1000} \quad \text{In lowest terms:} \quad \frac{125 \div 125}{1000 \div 125} = \frac{1}{8} \quad \left\{\begin{array}{l} \text{Original} \\ \text{fraction} \end{array}\right.$$

Calculator Tip When using your calculator to write fractions as decimals, enter the numbers from the top down. Remember that the *order* in which you enter the numbers *does* matter in division. Example 1(a) above works like this:

$\frac{1}{8}$ ↓ Top down   Enter 1 ÷ 8 =   Answer is 0.125

What happens if you enter 8 ÷ 1 =? Do you see why that cannot possibly be correct? (Answer: $8 \div 1 = 8$. A proper fraction like $\frac{1}{8}$ *cannot* be equivalent to a whole number.)

Continued on Next Page

(b) Write $2\frac{3}{4}$ as a decimal.

One method is to divide 3 by 4 to get 0.75 for the fraction part. Then add the whole number part to 0.75.

$$\frac{3}{4} \;\longrightarrow\; \begin{array}{r} 0.75 \\ 4\overline{)3.00} \\ \underline{2\,8} \\ 20 \\ \underline{20} \\ 0 \end{array}$$

Fraction part $\longrightarrow$

$$\begin{array}{r} 2.00 \quad \leftarrow \text{Whole number part} \\ +\,0.75 \\ \hline 2.75 \end{array}$$

So, $2\frac{3}{4} = 2.75$

Whole number parts match.

Check $2.75 = 2\frac{75}{100} = 2\frac{3}{4} \;\leftarrow$ Lowest terms

A second method is to write $2\frac{3}{4}$ as an improper fraction before dividing numerator by denominator.

$$2\frac{3}{4} = \frac{11}{4}$$

$$\frac{11}{4} \;\longrightarrow\; 11 \div 4 \;\longrightarrow\; 4\overline{)11} \;\longrightarrow\; \begin{array}{r} 2.7\,5 \\ 4\overline{)11.0\,0} \\ \underline{8} \\ 3\,0 \\ \underline{2\,8} \\ 2\,0 \\ \underline{2\,0} \\ 0 \end{array} \leftarrow \text{Two extra zeros needed}$$

Whole number parts match.

So, $2\frac{3}{4} = 2.75$

$\frac{3}{4}$ is equivalent to $\frac{75}{100}$ or 0.75

Now Try Exercises 11 and 13.

EXAMPLE 2 **Writing a Fraction as a Decimal with Rounding**

Write $\frac{2}{3}$ as a decimal and round to the nearest thousandth.

$\frac{2}{3}$ means $2 \div 3$. To round to thousandths, divide out one *more* place, to ten-thousandths.

$$\frac{2}{3} \;\longrightarrow\; 2 \div 3 \;\longrightarrow\; 3\overline{)2} \;\longrightarrow\; \begin{array}{r} 0.6666 \\ 3\overline{)2.0000} \\ \underline{1\,8} \\ 20 \\ \underline{18} \\ 20 \\ \underline{18} \\ 20 \\ \underline{18} \\ 2 \end{array} \leftarrow \begin{array}{l}\text{Four zeros needed} \\ \text{for ten-thousandths}\end{array}$$

Written as a repeating decimal, $\frac{2}{3} = 0.\overline{6} \;\leftarrow$ Bar above repeating digit

Rounded to the nearest thousandth, $\frac{2}{3} \approx 0.667$

Continued on Next Page

Calculator Tip Try Example 2 on your calculator. Enter 2 ÷ 3. Which answer do you get?

| 0.6666666667 | or | 0.666666666 |

Most *scientific* and *graphing* calculators will show a 7 as the last digit. Because the 6s keep on repeating forever, the calculator automatically rounds in the last decimal place it has room to show. If you have a 10-digit display space, the calculator is rounding as shown below.

0.6666666666 (11 digits)   rounds to   0.666666667

Next digit is *5 or more,* so 6 rounds to 7

Other calculators, especially standard, four-function ones, may *not* round. They just cut off, or *truncate,* the extra digits. Such a calculator would show 0.6666666 in the display.

Would this difference in calculators show up when changing $\frac{1}{3}$ to a decimal? Why not? (Answer: The repeating digit is a 3, which is *4 or less,* so it stays as a 3 whether it's rounded or not.)

Now Try Exercise 23. ▶▶▶

OBJECTIVE 2 Compare the size of fractions and decimals. You can use a number line to compare fractions and decimals. For example, the number line below shows the space between 0 and 1. The locations of some commonly used fractions are marked, along with their decimal equivalents.

The next number line shows the locations of some commonly used fractions between 0 and 1 that are equivalent to repeating decimals. The decimal equivalents use a bar above repeating digits.

EXAMPLE 3 Using a Number Line to Compare Numbers

Use the number lines above to decide whether to write >, <, or = in the blank between each pair of numbers.

(a) 0.6875 _____ 0.625

You learned in **Section 1.2** that the number farther to the right on the number line is the greater number. On the first number line, 0.6875 is to the *right* of 0.625, so use the > symbol.

0.6875   **is greater than**   0.625   0.6875 > 0.625

—— **Continued on Next Page**

(b) $\frac{3}{4}$ _____ 0.75

On the first number line, $\frac{3}{4}$ and 0.75 are at the same point on the number line. They are equivalent, so use the $=$ symbol.

$$\frac{3}{4} = 0.75$$

(c) 0.5 _____ $0.\overline{5}$

On the second number line, 0.5 is to the *left* of $0.\overline{5}$ (which is actually $0.555\ldots$), so use the $<$ symbol.

0.5   **is less than** $0.\overline{5}$ $0.5 < 0.\overline{5}$

(d) $\frac{2}{6}$ _____ $0.\overline{3}$

Write $\frac{2}{6}$ in lowest terms as $\frac{1}{3}$.

On the second number line you can see that $\frac{1}{3}$ and $0.\overline{3}$ are equivalent.

$$\frac{1}{3} = 0.\overline{3}$$

> **Now Try Exercise 53(a).**

You can also compare fractions by first writing each one as a decimal. You can then compare the decimals by writing each one with the same number of decimal places.

EXAMPLE 4 **Arranging Numbers in Order**

Write each group of numbers in order, from smallest to largest.

(a) 0.49   0.487   0.4903

It is easier to compare decimals if they are all tenths, or all hundredths, and so on. Because 0.4903 has four decimal places (ten-thousandths), write zeros to the right of 0.49 and 0.487 so they also have four decimal places. Writing zeros to the right of a decimal number does *not* change its value (see **Section 5.3**). Then find the smallest and largest number of ten-thousandths.

0.49 = 0.4900 = **4900** ten-thousandths ← 4900 is in the middle.

0.487 = 0.4870 = **4870** ten-thousandths ← 4870 is the smallest.

0.4903 = **4903** ten-thousandths ← 4903 is the largest.

From smallest to largest, the correct order is shown below.

0.487   0.49   0.4903

(b) $2\frac{5}{8}$ 2.63   2.6

Write $2\frac{5}{8}$ as $\frac{21}{8}$ and divide $8\overline{)21}$ to get the decimal form, 2.625. Then, because 2.625 has three decimal places, write zeros so all the numbers have three decimal places.

$2\frac{5}{8} = 2.625 = 2$ and **625** thousandths ← 625 is in the middle.

$2.63 = 2.630 = 2$ and **630** thousandths ← 630 is the largest.

$2.6 = 2.600 = 2$ and **600** thousandths ← 600 is the smallest.

From smallest to largest, the correct order is shown below.

2.6 $2\frac{5}{8}$ 2.63

> **Now Try Exercises 57 and 65.**

5.6 Exercises

Write each fraction or mixed number as a decimal. Round to the nearest thousandth when necessary. See Examples 1 and 2.

1. $\dfrac{1}{2}$ **2.** $\dfrac{1}{4}$ **3.** $\dfrac{3}{4}$ **4.** $\dfrac{1}{10}$ **5.** $\dfrac{3}{10}$

6. $\dfrac{7}{10}$ **7.** $\dfrac{9}{10}$ **8.** $\dfrac{4}{5}$ **9.** $\dfrac{3}{5}$ **10.** $\dfrac{2}{5}$

11. $\dfrac{7}{8}$ **12.** $\dfrac{3}{8}$ **13.** $2\dfrac{1}{4}$ **14.** $1\dfrac{1}{2}$ **15.** $14\dfrac{7}{10}$

16. $23\dfrac{3}{5}$ **17.** $3\dfrac{5}{8}$ **18.** $2\dfrac{7}{8}$ **19.** $6\dfrac{1}{3}$ **20.** $5\dfrac{2}{3}$

21. $\dfrac{5}{6}$ **22.** $\dfrac{1}{6}$ **23.** $1\dfrac{8}{9}$ **24.** $5\dfrac{4}{7}$

RELATING CONCEPTS (EXERCISES 25–28) For Individual or Group Work

Use your knowledge of fractions and decimals as you **work Exercises 25–28 in order.**

25. (a) Explain how you can tell that Keith made an error *just by looking at his final answer*. Here is his work.

$$\frac{5}{9} = 5)\overline{9.0}^{\,1.8} \quad \text{so} \quad \frac{5}{9} = 1.8$$

(b) Show the correct way to change $\frac{5}{9}$ to a decimal. Explain why your answer makes sense.

26. (a) How can you prove to Sandra that $2\frac{7}{20}$ is *not* equivalent to 2.035? Here is her work.

$$2\frac{7}{20} = 20)\overline{7.00}^{\,0.35} \quad \text{so} \quad 2\frac{7}{20} = 2.035$$

(b) What is the correct answer? Show how to prove that it is correct.

27. Ving knows that $\frac{3}{8} = 0.375$. How can he write $1\frac{3}{8}$ as a decimal *without* having to do a division? How can he write $3\frac{3}{8}$ as a decimal? $295\frac{3}{8}$? Explain your answer.

28. Iris has found a shortcut for writing mixed numbers as decimals.

$$2\frac{7}{10} = 2.7 \qquad 1\frac{13}{100} = 1.13$$

Does her shortcut work for all mixed numbers? Explain when it works and why it works.

Find the decimal or fraction equivalent for each number. Write fractions in lowest terms.

Fraction	Decimal	Fraction	Decimal
29. _____	0.4	**30.** _____	0.75
31. _____	0.625	**32.** _____	0.111
33. _____	0.35	**34.** _____	0.9
35. $\dfrac{7}{20}$	_____	**36.** $\dfrac{1}{40}$	_____
37. _____	0.04	**38.** _____	0.52
39. _____	0.15	**40.** _____	0.85
41. $\dfrac{1}{5}$	_____	**42.** $\dfrac{1}{8}$	_____
43. _____	0.09	**44.** _____	0.02

Solve each application problem.

45. The average length of a newborn baby is 20.8 inches. Charlene's baby is 20.08 inches long. Is her baby longer or shorter than the average? By how much?

46. The patient in room 830 is supposed to get 8.3 milligrams of medicine. She was actually given 8.03 milligrams. Did she get too much or too little medicine? What was the difference?

47. The label on the bottle of vitamins says that each capsule contains 0.5 gram of calcium. When checked, each capsule had 0.505 gram of calcium. Was there too much or too little calcium? What was the difference?

48. The glass mirror of the Hubble telescope had to be repaired in space because it would not focus properly. The problem was that the mirror's outer edge had a thickness of 0.6248 cm when it was supposed to be 0.625 cm. Was the edge too thick or too thin? By how much? (*Source:* NASA.)

49. Precision Medical Parts makes an artificial heart valve that must measure between 0.998 centimeter and 1.002 centimeters. Circle the lengths that are acceptable.

1.01 cm   0.9991 cm   1.0007 cm   0.99 cm

50. The white rats in a medical experiment must start out weighing between 2.95 ounces and 3.05 ounces. Circle the weights that can be used.

3.0 ounces   2.995 ounces   3.055 ounces

3.005 ounces

51. Ginny Brown had hoped her crops would get $3\frac{3}{4}$ inches of rain this month. The newspaper said the area received 3.8 inches of rain. Was that more or less than Ginny had hoped for? By how much?

52. The rats in the experiment in Exercise 50 gained $\frac{3}{8}$ ounce. They were expected to gain 0.3 ounce. Was their actual gain more or less than expected? By how much?

Use the number lines near Example 3 in the text to decide whether to write >, <, or = between each pair of numbers. See Example 3.

53. (a) 0.3125 ____ 0.375   **(b)** $\frac{6}{8}$ ____ 0.75   **(c)** $0.\overline{8}$ ____ $0.8\overline{3}$ **(d)** 0.5 ____ $\frac{5}{9}$

54. (a) 0.125 ____ 0.0625   **(b)** $\frac{4}{9}$ ____ $\frac{3}{6}$ **(c)** 1.0 ____ $\frac{4}{4}$ **(d)** $\frac{1}{6}$ ____ $0.\overline{1}$

Arrange each group of numbers in order from smallest to largest. See Example 4.

55. 0.54, 0.5455, 0.5399

56. 0.76, 0.7, 0.7006

57. 5.8, 5.79, 5.0079, 5.804

58. 12.99, 12.5, 13.0001, 12.77

59. 0.628, 0.62812, 0.609, 0.6009

60. 0.27, 0.281, 0.296, 0.3

61. 5.8751, 4.876, 2.8902, 3.88

62. 0.98, 0.89, 0.904, 0.9

63. 0.043, 0.051, 0.006, $\dfrac{1}{20}$

64. 0.629, $\dfrac{5}{8}$, 0.65, $\dfrac{7}{10}$

0.625   0.7

65. $\dfrac{3}{8}$, $\dfrac{2}{5}$, 0.37, 0.4001

66. 0.1501, 0.25, $\dfrac{1}{10}$, $\dfrac{1}{5}$

0.1   0.2

*The friends on the first page of this chapter found four boxes of fishing line in a sale bin.
They know that the thicker the line, the stronger it is. The diameter of the fishing line is its
thickness. Use the information on the boxes to answer Exercises 67–68.*

Fishing Line 0.018 in. diameter — Fishing Line 0.01 in. diameter — Fishing Line 0.008 in. diameter — Fishing Line 0.010 in. diameter

67. (a) Which color box has the strongest line? The weakest line?

(b) What is the difference in line diameter between the weakest and strongest lines?

68. (a) Which color box has the line that is $\dfrac{1}{125}$ inch in diameter?

(b) What is the difference in line diameter between the blue and purple boxes?

*Some rulers for technical drawings show each inch divided into tenths. Use this scale
drawing for Exercises 69–74. Change the measurements on the drawing to decimals and
round them to the nearest tenth of an inch.*

69. Length **(a)** is _____

70. Length **(b)** is _____

71. Length **(c)** is _____

72. Length **(d)** is _____

73. Length **(e)** is _____

74. Length **(f)** is _____

(a) $1\frac{7}{16}$ in.   (d) $\frac{1}{2}$ in.   (b) $1\frac{1}{8}$ in.   (e) $\frac{3}{8}$ in.   (f) $\frac{11}{16}$ in.   (c) $\frac{1}{4}$ in.

5.7 Problem Solving with Statistics: Mean, Median, Mode, and Variability

The word *statistics* originally came from words that mean *state numbers.* State numbers refer to numerical information, or *data,* gathered by the government such as the number of births, deaths, or marriages in a population. Today the word *statistics* has a much broader meaning; data from the fields of economics, social science, science, and business can all be organized and studied under the branch of mathematics called *statistics.*

OBJECTIVES

1. Find the mean of a list of numbers.
2. Find a weighted mean.
3. Find the median.
4. Find the mode.
5. Evaluate the variability of a set of data by finding the range of values.

OBJECTIVE 1   Find the mean of a list of numbers.   Making sense of a long list of numbers can be difficult. So when you analyze data, one of the first things to look for is a *measure of central tendency*—a single number that you can use to represent the entire list of numbers. One such measure is the *average* or **mean.** The mean can be found with the following formula.

Finding the Mean (Average)

$$\text{mean} = \frac{\text{sum of all values}}{\text{number of values}}$$

EXAMPLE 1   Finding the Mean (Average)

David had test scores of 84, 90, 95, 98, and 88. Find his mean (average) score.

Use the formula for finding the mean. Add up all the test scores and then divide the sum by the number of tests.

$$\text{mean} = \frac{84 + 90 + 95 + 98 + 88}{5} \quad \begin{array}{l} \leftarrow \text{Sum of test scores} \\ \leftarrow \text{Number of tests} \end{array}$$

$$\text{mean} = \frac{455}{5} \quad \text{Divide.}$$

$$\text{mean} = 91$$

David has a mean (average) score of 91.

Now Try Exercise 1. ▶▶▶

EXAMPLE 2   Applying the Mean (Average)

The sales of photo albums at Sarah's Card Shop for each day last week were $86, $103, $118, $117, $126, $158, and $149. Find the mean daily sales of photo albums, to the nearest cent.

To find the mean, add all the daily sales amounts and then divide the sum by the number of days (7).

$$\text{mean} = \frac{\$86 + \$103 + \$118 + \$117 + \$126 + \$158 + \$149}{7} \quad \begin{array}{l} \leftarrow \text{Sum of sales} \\ \leftarrow \text{Number of days} \end{array}$$

$$\text{mean} = \frac{\$857}{7}$$

$$\text{mean} \approx \$122.43 \quad \text{Rounded to nearest cent}$$

Now Try Exercise 7. ▶▶▶

OBJECTIVE **2** **Find a weighted mean.** Some items in a list of data might appear more than once. In this case, we find a **weighted mean,** in which each value is "weighted" by multiplying it by the number of times it occurs.

EXAMPLE 3 Finding a Weighted Mean

The table below shows the amount of contribution and the number of times the amount was given (frequency) to a food pantry. Find the weighted mean.

Contribution Value	Frequency	
$ 3	4	← 4 people each contributed $3
$ 5	2	
$ 7	1	
$ 8	5	
$ 9	3	
$10	2	
$12	1	
$13	2	

The same amount was given by more than one person: for example, $3 was given by four people, and $8 was given by five people. Other amounts, such as $12, were given by only one person.

To find the mean, multiply each contribution value by its frequency. Then add the products. Next, add the numbers in the *frequency* column to find the total number of values, that is, the total number of people who contributed money.

Value	Frequency	Product
$ 3	4	($3 • 4) = $12
$ 5	2	($5 • 2) = $10
$ 7	1	($7 • 1) = $ 7
$ 8	5	($8 • 5) = $40
$ 9	3	($9 • 3) = $27
$10	2	($10 • 2) = $20
$12	1	($12 • 1) = $12
$13	2	($13 • 2) = $26
Totals	**20**	**$154**

Finally, divide the totals.

$$\text{mean} = \frac{\$154}{20} = \$7.70$$

The mean contribution to the food pantry was $7.70.

Now Try Exercise 11.

A common use of the weighted mean is to find a student's *grade point average (GPA),* as shown in the next example.

EXAMPLE 4 Applying the Weighted Mean

Find the GPA (grade point average) for a student who earned the following grades last semester. Assume A = 4, B = 3, C = 2, D = 1, and F = 0. The number of credits determines how many times the grade is counted (the frequency).

Course	Credits	Grade	Credits · Grade
Mathematics	4	A (= 4)	4 · 4 = 16
Speech	3	C (= 2)	3 · 2 = 6
English	3	B (= 3)	3 · 3 = 9
Computer science	2	A (− 4)	2 · 4 = 8
Theater	2	D (= 1)	2 · 1 = 2
Totals	**14**		**41**

It is common to round grade point averages to the nearest hundredth. So the grade point average for this student is rounded to 2.93.

$$\text{GPA} = \frac{41}{14} \approx 2.93$$

Now Try Exercise 15. ▶▶▶

OBJECTIVE 3 Find the median. Because it can be affected by extremely high or low numbers, the mean is often a poor indicator of central tendency for a list of numbers. In cases like this, another measure of central tendency, called the *median,* can be used. The **median** divides a group of numbers in half; half the numbers lie above the median, and half lie below the median.

Find the median by listing the numbers *in order* from *smallest* to *largest.* If the list contains an *odd* number of items, the median is the *middle number.*

EXAMPLE 5 Finding the Median (Odd Number of Items)

Find the median for this list of prices.

$$\$7, \$23, \$15, \$6, \$18, \$12, \$24$$

First arrange the numbers in numerical order from smallest to largest.

Smallest ⟶ 6, 7, 12, 15, 18, 23, 24 ⟵ Largest

Next, find the middle number in the list.

6, 7, 12, **15**, 18, 23, 24

Three are below. ↓ Three are above.
Middle number

The median price is $15.

Now Try Exercise 19. ▶▶▶

If a list contains an *even* number of items, there is no single middle number. In this case, the median is defined as the mean (average) of the *middle two* numbers.

EXAMPLE 6 Finding the Median (Even Number of Items)

Find the median for this list of ages, in years.

$$74, 7, 15, 13, 25, 28, 47, 59, 33, 68$$

First, arrange the numbers in numerical order from smallest to largest. Then, because the list has an even number of ages, find the middle *two* numbers.

Smallest ⟶ 7, 13, 15, 25, **28, 33**, 47, 59, 68, 74 ⟵ Largest

Middle two numbers

The median age is the mean of the two middle numbers.

$$\text{median} = \frac{\mathbf{28 + 33}}{2} = \frac{61}{2} = 30.5 \text{ years}$$

Now Try Exercise 23.

OBJECTIVE 4 **Find the mode.** Another statistical measure is the **mode,** which is the number that occurs *most often* in a list of numbers. For example, the test scores for ten students are shown below.

74, 81, 39, **74**, 82, 80, 100, 92, **74**, 85

The mode is 74. Three students earned a score of 74, so 74 appears more times on the list than any other score. It is *not* necessary to place the numbers in numerical order when looking for the mode, although that may help you find it more easily.

A list can have two modes; such a list is sometimes called *bimodal*. If no number occurs more frequently than any other number in a list, the list has *no mode*.

EXAMPLE 7 Finding the Mode

Find the mode for each list of numbers.

(a) 51, 32, 49, 51, 49, 90, 49, 60, 17, 60
The number 49 occurs three times, which is more often than any other number. Therefore, **49** is the mode.

(b) 482, 485, 483, 485, 487, 487, 489, 486
Because both **485** and **487** occur twice, each is a mode. This list is *bimodal*.

(c) $10,708; $11,519; $10,972; $12,546; $13,905; $12,182
No price occurs more than once. This list has *no mode*.

Measures of Central Tendency

The **mean** is the sum of all the values divided by the number of values. It is the mathematical *average*.

The **median** is the *middle number* in a group of values that are listed from smallest to largest. It divides a group of numbers in half.

The **mode** is the value that occurs *most often* in a group of values.

Now Try Exercise 31.

OBJECTIVE 5 **Evaluate the variability of a set of data by finding the range of values.**   If two students each have a mean (average) score of 60 on their math tests, you might think they have done work of equal quality. However, a closer look at the test scores below gives a different impression.

	Test 1	Test 2	Test 3	Test 4	Test 5	Mean	Median
Student Y	55	60	60	60	65	60	60
Student Z	20	40	60	80	100	60	60

The means and medians are all 60. But student Y's scores are all clustered around 60. Student Z's scores, however, have greater *variability* because they are spread over a wider range, from 20 to 100. The **variability** of a set of data is the spread of the data around the mean. A quick way to evaluate the variability is to look at the *range of values*. To find the range, subtract the lowest value from the highest value.

EXAMPLE 8 **Finding the Range**

(a) Find the range of test scores for student Y, shown above.
   Subtract student Y's lowest test score of 55 from the highest test score of 65.

$$65 - 55 = 10 \longleftarrow \text{Range of scores for student Y}$$

(b) Find the range of test scores for student Z, shown above.

Highest score

$$100 - 20 = 80 \longleftarrow \text{Range of scores for student Z}$$

Lowest score

Now Try Exercise 43(a).

Generally, the greater the *range* in a set of values, the greater the *variability*. When analyzing data, look at the mean and/or median and also look at the variability. For example, if you were the teacher of students Y and Z in Example 8 above, you would analyze their test data quite differently. Student Y has consistently gotten fairly low scores and may need some extra help. Student Z's scores, on the other hand, have improved dramatically, and he or she will probably do well on future tests.

EXAMPLE 9 **Evaluating the Variability of Data**

(a) Which set of data shows greater variability in the points scored in a basketball game? Show your work.

	Game 1	Game 2	Game 3	Game 4	Game 5	Mean
Player A	18	9	27	3	25	16.4
Player B	18	16	17	17	15	16.6

Range for player A = 27 − 3 = 24

Range for player B = 18 − 15 = 3

Player A's data has a greater range and greater variability.

Continued on Next Page

(b) If you are a coach, which player would you like on your team, player A or player B? Explain why.

Answers may vary. Some things to consider are: Player A had some very good games, and the means for the two players are nearly the same. But player B is more consistent and perhaps more reliable. You would want to know more details about why player A scored so few points in games 2 and 4.

Now Try Exercises 43(b) and 43(c).

5.7 Exercises

Find the mean for each list of numbers. Round answers to the nearest tenth when necessary. See Example 1.

1. Ages of infants at the child care center (in months): 4, 9, 6, 4, 7, 10, 9

2. Monthly electric bills: $53, $77, $38, $29, $49, $48

3. Final exam scores: 92, 51, 59, 86, 68, 73, 49, 80

4. Quiz scores: 18, 25, 21, 8, 16, 13, 23, 19

5. Annual salaries: $31,900; $32,850; $34,930; $39,712; $38,340, $60,000

6. Numbers of people attending baseball games: 27,500; 18,250; 17,357; 14,298; 33,110

Solve each application problem. See Examples 2 and 3.

7. The Athletic Shoe Store sold shoes at the following prices: $75.52, $36.15, $58.24, $21.86, $47.68, $106.57, $82.72, $52.14, $28.60, $72.92. Find the mean shoe sales amount.

8. In one evening, a waitress collected the following checks from her dinner customers: $30.10, $42.80, $91.60, $51.20, $88.30, $21.90, $43.70, $51.20. Find the mean dinner check amount.

9. The table below shows the face value (policy amount) of life insurance policies and the number of policies sold for each amount by the New World Life Company during one week. Find the weighted mean amount for the policies sold.

Policy Amount	Number of Policies Sold
$ 10,000	6
$ 20,000	24
$ 25,000	12
$ 30,000	8
$ 50,000	5
$100,000	3
$250,000	2

10. Detroit Metro-Sales Company prepared the table below showing the gasoline mileage obtained by each of the cars in their automobile fleet. Find the weighted mean to determine the miles per gallon for the fleet of cars.

Miles per Gallon	Number of Autos
15	5
20	6
24	10
30	14
32	5
35	6
40	4

Find the weighted mean. Round answers to the nearest tenth when necessary. See Example 3.

11.

Quiz Score	Frequency
3	4
5	2
6	5
8	5
9	2

12.

Credits per Student	Frequency
9	3
12	5
13	2
15	6
18	1

13.

Hours Worked	Frequency
12	4
13	2
15	5
19	3
22	1
23	5

14.

Students per Class	Frequency
25	1
26	2
29	5
30	4
32	3
33	5

Find the GPA (grade point average) for students earning the following grades. Assume A = 4, B = 3, C = 2, D = 1, and F = 0. Round answers to the nearest hundredth. See Example 4.

15.

Course	Credits	Grade
Biology	4	B
Biology lab	2	A
Mathematics	5	C
Health	1	F
Psychology	3	B

16.

Course	Credits	Grade
Chemistry	3	A
English	3	B
Mathematics	4	B
Theater	2	C
Astronomy	3	C

17. Look again at the grades in Exercise 15. Find the student's GPA in each of these situations.

  (a) The student earned a B instead of an F in the 1-credit class.

  (b) The student earned a B instead of a C in the 5-credit class.

  (c) Both (a) and (b) happened.

18. List the credits for the courses you're taking at this time. List the lowest grades you think you will earn in each class and find your GPA. Then list the highest grades you think you will earn and find your GPA.

Find the median for each list of numbers. See Examples 5 and 6.

19. Number of e-mail messages received:
9, 12, 14, 15, 23, 24, 28

20. Patients seen each day at a clinic:
99, 108, 109, 123, 126, 129, 146, 168, 170

21. Students enrolled in algebra each semester:
328, 549, 420, 592, 715, 483

22. Number of cars in the parking lot each day:
520, 523, 513, 1283, 338, 509, 290, 420

23. Pounds of shrimp sold each day:
51, 48, 96, 40, 47, 40, 95, 56, 34, 49

24. Number of gallons of paint sold per week:
1072, 1068, 1093, 1042, 1056, 205, 1009, 1081

The table lists the cruising speed and distance flown without refueling for several types of larger airplanes used to carry passengers. Use the table to answer Exercises 25–28.

Type of Airplane	Cruising Speed (miles per hour)	Distance without Refueling (miles)
747-400	565	7650
747-200	558	6450
DC-9	505	1100
DC-10	550	5225
727	530	1550
757	530	2875

Source: Northwest Airlines *World Traveler.*

25. What is the average distance flown without refueling, to the nearest mile?

26. Find the average cruising speed, to the nearest mile per hour.

27. (a) Find the median distance flown.

    (b) Explain how you found the median.

28. (a) Find the median cruising speed.

    (b) Is the median similar to the average speed from Exercise 26? Explain why or why not.

Find the mode or modes for each list of numbers. See Example 7.

29. Number of samples taken each hour:
3, 8, 5, 1, 7, 6, 8, 4, 5, 8

30. Monthly water bills:
$21, $32, $46, $32, $49, $32, $49, $25, $32

31. Ages of senior residents (in years):
74, 68, 68, 68, 75, 75, 74, 74, 70, 77

32. Patients admitted to the hospital each week:
30, 19, 25, 78, 36, 20, 45, 85, 38

33. The number of boxes of candy sold by each child:
5, 9, 17, 3, 2, 8, 19, 1, 4, 20, 10, 6

34. The weights of soccer players (in pounds):
158, 161, 165, 162, 165, 157, 163, 162

The table lists monthly normal temperatures from November through April for some of the coldest and warmest U.S. cities. Use the table to answer Exercises 35–40. Round answers to the nearest whole degree. For help in determining variability, see Examples 8 and 9.

Normal Monthly Temperatures (in Degrees Fahrenheit)						
City	Nov.	Dec.	Jan.	Feb.	Mar.	Apr.
Barrow, Alaska	-2	-11	-13	-18	-15	-2
Fairbanks, Alaska	3	-7	-10	-4	11	31
Honolulu, Hawaii	77	74	73	73	74	76
Miami, Florida	74	69	67	69	72	75

Source: National Climate Data Center.

35. (a) Find Barrow's mean temperature and Fairbanks' mean temperature for the six-month period.

    (b) How much warmer is Fairbanks' mean than Barrow's mean?

36. (a) Find Barrow's median temperature and Fairbanks' median temperature for the six-month period.

    (b) How much cooler is Barrow's median than Fairbanks' median?

37. Which set of temperatures has greater variability, Barrow or Fairbanks? Show how to calculate the range for each city.

38. (a) Find the mean and median for Honolulu's temperatures during the six-month period.

    (b) Why are the mean and median so similar?

39. (a) Find the mean and median for Miami's temperatures during the six-month period.

    (b) What do you notice about the mean and median? Explain how this is possible.

40. Which set of temperatures has less variability, Honolulu or Miami? Show how to calculate the range for each city.

RELATING CONCEPTS (EXERCISES 41–44) For Individual or Group Work

*Use your knowledge of statistics as you **work Exercises 41–44 in order.***

41. (a) Find the mean, median, and range for each student's test data. Round to the nearest tenth when necessary.
    Student P: 92, 80, 61, 49, 82, 53
    Student Q: 70, 76, 77, 60, 67, 72

    (b) Which student's data has greater variability?

    (c) If you were the teacher, what advice would you give to each student?

42. (a) Find the mean, median, and range for each golfer's scores. Round to the nearest tenth when necessary.
    Golfer G: 84, 87, 83, 89, 88
    Golfer H: 94, 88, 76, 89, 85

    (b) Which golfer had scores with less variability?

    (c) If you were the coach, what advice would you give to each golfer?

43. (a) Calculate the range for the ages of tenants in each building. Show your work.
    Building A: 63, 70, 66, 62, 74, 65, 74
    Building B: 23, 20, 26, 18, 21, 23, 19
    Building C: 45, 32, 69, 25, 50, 37, 41

    (b) Which building has the least variability in the tenants' ages?

    (c) Based on your work above, which building would you like to live in? Explain why.

44. (a) Find the set of data on monthly salaries of computer technicians that has the greatest variability. Show your work.
    Company P: $4900, $5500, $5000, $7200, $6100
    Company Q: $5800, $5900, $5500, $6000, $5900
    Company R: $6200, $6400, $5800, $5200, $5600

    (b) Based on your work in part (a), which company would you like to work for? Explain why.

5.8 Geometry Applications: Pythagorean Theorem and Square Roots

In **Section 3.2,** you used this formula for area of a square, $A = s^2$. The blue square below has an area of 25 cm² because (5 cm)(5 cm) = 25 cm².

5 cm

5 cm

Area = 25 cm²
Area = (5 cm)(5 cm)

side = ? cm

Area = 49 cm²
Area = (? cm)(? cm)

The red square above has an area of 49 cm². To find the length of a side, ask yourself, "What number can be multiplied by itself to give 49?" Because (7)(7) = 49, the length of each side is 7 cm. Also, because (7)(7) = 49 we say that 7 is the *square root* of 49, or $\sqrt{49} = 7$.

Square Root

The positive **square root** of a positive number is one of two identical positive factors of that number.

For example, $\sqrt{36} = 6$ because (6)(6) = 36.

> **NOTE**
> There is another square root for 36. We know that
>
> $$(6)(6) = 36 \quad \text{and} \quad (-6)(-6) = 36$$
>
> so the *positive* square root of 36 is 6 and the *negative* square root of 36 is −6. In this section we will work only with positive square roots. You will learn more about negative square roots in other math courses.

A number that has a whole number as its square root is called a *perfect square*. For example, 9 is a perfect square because $\sqrt{9} = 3$, and 3 is a whole number. The first few perfect squares are listed below.

The First Twelve Perfect Squares

$\sqrt{1} = 1$	$\sqrt{16} = 4$	$\sqrt{49} = 7$	$\sqrt{100} = 10$
$\sqrt{4} = 2$	$\sqrt{25} = 5$	$\sqrt{64} = 8$	$\sqrt{121} = 11$
$\sqrt{9} = 3$	$\sqrt{36} = 6$	$\sqrt{81} = 9$	$\sqrt{144} = 12$

OBJECTIVE 1 Find square roots using the square root key on a calculator. If a number is *not* a perfect square, then you can find its *approximate* square root by using a calculator with a square root key.

⊞ **Calculator Tip** To find a square root on a *scientific* calculator, use the $\sqrt{\ }$ or the $\sqrt{x}$ key. (On some models you may have to press the ②ⁿᵈ key to access the square root function.) You do *not* need to use the ⊜ key. Try these. (Your calculator may show only eight digits in the answers.)

To find $\sqrt{16}$ press: 16 $\sqrt{x}$ Answer is 4

To find $\sqrt{7}$ press: 7 $\sqrt{x}$ Answer is 2.645751311

For $\sqrt{7}$, your calculator shows 2.645751311, which is an *approximate* answer. We will round to the nearest thousandth, so $\sqrt{7} \approx 2.646$. To check, multiply 2.646 times 2.646. Do you get 7 as the result? No, you get 7.001316, which is very close to 7. The difference is due to rounding.

EXAMPLE 1 **Finding the Square Root of Numbers**

⊞ Use a calculator to find each square root. Round to the nearest thousandth.

(a) $\sqrt{35}$ Calculator shows 5.916079783; round to 5.916.

(b) $\sqrt{124}$ Calculator shows 11.13552873; round to 11.136.

◀◀◀ **Now Try Exercises 9 and 13.**

OBJECTIVE 2 **Find the unknown length in a right triangle.** One place you will use square roots is when working with the *Pythagorean Theorem*. This theorem applies only to *right* triangles (triangles with a 90° angle). The longest side of a right triangle is called the **hypotenuse**. It is opposite the right angle. The other two sides are called *legs*. The legs form the right angle. Here are some right triangles.

Examples of right triangles

Pythagorean Theorem

$$(\text{hypotenuse})^2 = (\text{leg})^2 + (\text{leg})^2$$

In other words, square the length of each side. After you have squared all the sides, the sum of the squares of the two legs will equal the square of the hypotenuse. An example is shown below.

$$
\begin{aligned}
(\text{hypotenuse})^2 &= (\text{leg})^2 + (\text{leg})^2 \\
5^2 &= 4^2 + 3^2 \\
25 &= 16 + 9 \\
25 &= 25
\end{aligned}
$$

The theorem is named after Pythagoras, a Greek mathematician who lived about 2500 years ago. He and his followers may have used floor tiles to prove the theorem, as shown on the next page.

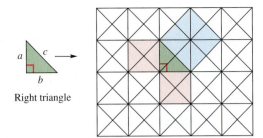

Right triangle

The green right triangle in the center of the floor tiles has sides a, b, and c. The pink square drawn on side a contains four triangular tiles. The pink square on side b contains four tiles. The blue square on side c contains eight tiles. The number of tiles in the square on side c equals the sum of the number of tiles in the squares on sides a and b, that is, 8 tiles = 4 tiles + 4 tiles. As a result, you often see the Pythagorean Theorem written as $c^2 = a^2 + b^2$.

   If you know the lengths of any two sides in a right triangle, you can use the Pythagorean Theorem to find the length of the third side.

> ### Formulas Based on the Pythagorean Theorem
>
> To find the hypotenuse: **hypotenuse** $= \sqrt{(\textbf{leg})^2 + (\textbf{leg})^2}$
>
> To find a leg: **leg** $= \sqrt{(\textbf{hypotenuse})^2 - (\textbf{leg})^2}$

EXAMPLE 2   **Finding the Unknown Length in Right Triangles**

Find the unknown length in each right triangle. Round your answers to the nearest tenth when necessary.

(a)

3 ft
4 ft

The unknown length is the side opposite the right angle, which is the hypotenuse. Use the formula for finding the hypotenuse.

$$\text{hypotenuse} = \sqrt{(\text{leg})^2 + (\text{leg})^2} \qquad \text{Find the hypotenuse.}$$
$$\text{hypotenuse} = \sqrt{(3)^2 + (4)^2} \qquad \text{Legs are 3 and 4}$$
$$= \sqrt{9 + 16} \qquad (3)(3) \text{ is } 9 \quad \text{and} \quad (4)(4) \text{ is } 16$$
$$= \sqrt{25}$$
$$= 5$$

The hypotenuse is 5 ft long.

(b)

15 cm
7 cm

You *do* know the length of the hypotenuse (15 cm), so it is the length of one of the legs that is unknown. Use the formula for finding a leg.

$$\text{leg} = \sqrt{(\text{hypotenuse})^2 - (\text{leg})^2} \qquad \text{Find a leg.}$$
$$\text{leg} = \sqrt{(15)^2 - (7)^2} \qquad \text{Hypotenuse is 15; one leg is 7}$$
$$= \sqrt{225 - 49} \qquad (15)(15) \text{ is } 225 \quad \text{and} \quad (7)(7) \text{ is } 49$$
$$= \sqrt{176} \qquad \text{Use a calculator to find } \sqrt{176}$$
$$\approx 13.3 \qquad \text{Round } 13.26649916 \text{ to } 13.3$$

The length of the leg is approximately 13.3 cm.

Now Try Exercises 21 and 29.

OBJECTIVE 3 Solve application problems involving right triangles.
The next example shows an application of the Pythagorean Theorem.

EXAMPLE 3 Using the Pythagorean Theorem

A television antenna is on the roof of a house, as shown. Find the length of the support wire. Round your answer to the nearest tenth of a meter if necessary.

A right triangle is formed. The total length of the leg on the left is 3 m + 2 m = 5 m.

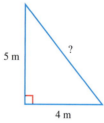

Notice that the support wire is opposite the right angle, so it is the *hypotenuse* of the right triangle.

$$\text{hypotenuse} = \sqrt{(\text{leg})^2 + (\text{leg})^2} \qquad \text{Find the hypotenuse.}$$

$$\text{hypotenuse} = \sqrt{(5)^2 + (4)^2} \qquad \text{Legs are 5 and 4}$$

$$= \sqrt{25 + 16} \qquad 5^2 \text{ is 25 and } 4^2 \text{ is 16}$$

$$= \sqrt{41} \qquad \text{Use a calculator to find } \sqrt{41}$$

$$\approx 6.4 \qquad \text{Round 6.403124237 to 6.4}$$

The length of the support wire is approximately 6.4 m.

Now Try Exercise 37.

5.8 Exercises

Find each square root. Starting with Exercise 5, find the square root using a calculator. Round your answers to the nearest thousandth when necessary. See Example 1.

1. $\sqrt{16}$ **2.** $\sqrt{4}$ **3.** $\sqrt{64}$ **4.** $\sqrt{81}$

5. $\sqrt{11}$ **6.** $\sqrt{23}$ **7.** $\sqrt{5}$ **8.** $\sqrt{2}$

9. $\sqrt{73}$ **10.** $\sqrt{80}$ **11.** $\sqrt{101}$ **12.** $\sqrt{125}$

13. $\sqrt{361}$ **14.** $\sqrt{729}$ **15.** $\sqrt{1000}$ **16.** $\sqrt{2000}$

17. You know that $\sqrt{25} = 5$ and $\sqrt{36} = 6$. Using just that information (no calculator), describe how you could *estimate* $\sqrt{30}$. How would you estimate $\sqrt{26}$ or $\sqrt{35}$? Now check your estimates using a calculator.

18. Explain the relationship between *squaring* a number and finding the *square root* of a number. Include two examples to illustrate your explanation.

Find the unknown length in each right triangle. Use a calculator to find square roots. Round your answers to the nearest tenth when necessary. See Example 2.

19.
15 ft 90° 36 ft

20.
9 cm 12 cm

21.
8 in. 90° 15 in.

22.
30 in. 72 in.

23.
16 mm 20 mm

24.
5 m 13 m

25.

3 in.

8 in.

26.

5 cm

11 cm

27.

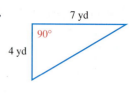

7 yd

90°

4 yd

28.

7 km

10 km

29.

22 cm

17 cm

30.

16 cm

9 cm

90°

31.

1.3 m

90°

2.5 m

32.

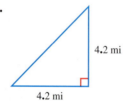

4.2 mi

4.2 mi

33.

11.5 cm

8.2 cm

34.

9.1 mm

10.8 mm

35.

13.2 km   90°

21.6 km

36.

26.5 ft

37.4 ft

Solve each application problem. Round your answers to the nearest tenth when necessary.
See Example 3.

37. Find the length of this loading ramp.

4 ft

?

7 ft

38. Find the unknown length in this roof plan.

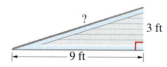

?

3 ft

9 ft

39. How high is the airplane above the ground?

40. Find the height of this farm silo.

41. How long is the diagonal brace on this rectangular gate?

42. Find the height of this rectangular television screen. (*Source:* Sears.)

43. To reach his ladylove, a knight placed a 12 ft ladder against the castle wall. If the base of the ladder is 3 ft from the building, how high on the castle will the top of the ladder reach? Draw a sketch of the castle and ladder and solve the problem.

44. William drove his car 15 miles north, then made a right turn and drove 7 miles east. How far is he, in a straight line, from his starting point? Draw a sketch to illustrate the problem and then solve it.

45. Explain the *two* errors made by a student in solving this problem. Also find the correct answer. Round to the nearest tenth.

$$? = \sqrt{(13)^2 + (20)^2}$$

$$= \sqrt{169 + 400}$$

$$= \sqrt{569} \approx 23.9 \text{ m}^2$$

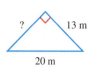

13 m

20 m

46. Explain the *two* errors made by a student in solving this problem. Also find the correct answer. Round to the nearest tenth.

$$? = \sqrt{(9)^2 + (7)^2}$$

$$= \sqrt{18 + 14}$$

$$= \sqrt{32} \approx 5.657 \text{ in.}$$

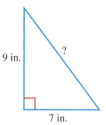

9 in.

?

7 in.

RELATING CONCEPTS (EXERCISES 47–50)   For Individual or Group Work

Use your knowledge of the Pythagorean Theorem to **work Exercises 47–50 in order.** *Round answers to the nearest tenth.*

47. A major league baseball diamond is a square shape measuring 90 ft on each side. If the catcher throws a ball from home plate to second base, how far is he throwing the ball? (*Source:* American League of Professional Baseball Clubs.)

Second base

90 ft

Third base   ?   First base

90 ft

Home plate

48. A softball diamond is only 60 ft on each side. (*Source:* Amateur Softball Association.)

(a) Draw a sketch of the softball diamond and label the bases and the lengths of the sides.

(b) How far is it to throw a ball from home plate to second base?

49. Look back at your answer to Exercise 47. Explain how you can tell the distance from third base to first base without doing any further calculations.

50. (a) Look back at your answer to Exercise 48. Suppose you measured the distance from home plate to second base on a softball diamond and found it was 80 ft. What would this tell you about the length of each side of the diamond? (Assume the diamond is still a square.)

(b) Bonus question: For the diamond in part (a), find the length of each side, to the nearest tenth.

5.9 Problem Solving: Equations Containing Decimals

OBJECTIVE **1** **Solve equations containing decimals using the addition property of equality.** In **Section 2.3** you used the addition property of equality to solve an equation like $c + 5 = 30$. The addition property says that you can add the *same* number to *both* sides of an equation and still keep it balanced. You can also use this property when an equation contains decimal numbers.

OBJECTIVES

1 Solve equations containing decimals using the addition property of equality.

2 Solve equations containing decimals using the division property of equality.

3 Solve equations containing decimals using both properties of equality.

4 Solve application problems involving equations with decimals.

EXAMPLE 1 **Using the Addition Property of Equality**

Solve each equation and check each solution.

(a) $w + 2.9 = -0.6$

The first step is to get the variable term (w) by itself on the left side of the equal sign. Use the addition property to "get rid of" the 2.9 on the left side by adding its opposite, -2.9, to both sides.

$$
\begin{array}{rcl}
w + 2.9 & = & -0.6 \\
\underline{-2.9 \quad\; -2.9} & & \text{Add } -2.9 \text{ to both sides.} \\
w + \underbrace{0}_{} & = & -3.5 \\
\underbrace{w} & = & -3.5
\end{array}
$$

The solution is -3.5. To check the solution, go back to the *original* equation.

$$
\begin{array}{rcll}
Check & w + 2.9 & = & -0.6 \qquad \text{Original equation} \\
& \underbrace{-3.5 + 2.9} & = & -0.6 \qquad \text{Replace } w \text{ with } -3.5 \\
& -0.6 & = & -0.6 \qquad \text{Balances}
\end{array}
$$

When w is replaced with -3.5, the equation balances, so -3.5 is the correct solution (***not*** -0.6).

(b) $7 = -4.3 + x$

To get x by itself on the right side of the equal sign, add 4.3 to both sides.

$$
\begin{array}{rcl}
7 & = & -4.3 + x \\
\underline{+4.3 \qquad +4.3} & & \text{Add 4.3 to both sides.} \\
11.3 & = & \underbrace{0 \; + x} \\
11.3 & = & x
\end{array}
$$

The solution is 11.3. To check the solution, go back to the *original* equation.

$$
\begin{array}{rcll}
Check & 7 = -4.3 + & x & \text{Original equation} \\
& 7 = \underbrace{-4.3 + 11.3} & & \text{Replace } x \text{ with } 11.3 \\
& 7 = \qquad 7 & & \text{Balances}
\end{array}
$$

When x is replaced with 11.3, the equation balances, so 11.3 is the correct solution (***not*** 7).

Now Try Exercises 1 and 3. ▶▶▶

OBJECTIVE 2 Solve equations containing decimals using the division property of equality. You can also use the division property of equality (from **Section 2.4**) when an equation contains decimals.

> **EXAMPLE 2** Using the Division Property of Equality
>
> Solve each equation and check each solution.
>
> **(a)** $5x = 12.4$
>
> On the left side of the equation, the variable is multiplied by 5. To undo the multiplication, divide both sides by 5.
>
> 5x means 5 · x.　　　$5x = 12.4$
>
> $$\frac{5 \cdot x}{5} = \frac{12.4}{5} \qquad \text{Divide both sides by 5}$$
>
> On the left side, divide out the common factor of 5.
>
> $$\frac{\overset{1}{\cancel{5}} \cdot x}{\underset{1}{\cancel{5}}} = 2.48 \qquad \text{On the right side, } 12.4 \div 5 \text{ is } 2.48$$
>
> $$x = 2.48$$
>
> The solution is 2.48. To check the solution, go back to the original equation.
>
> *Check*　　　$5x = 12.4$　　　Original equation
>
> 　　　　　　　$5(2.48) = 12.4$　　　Replace x with 2.48
>
> 　　　　　　　$12.4 = 12.4$　　　Balances
>
> When x is replaced with 2.48, the equation balances, so 2.48 is the correct solution (***not*** 12.4).
>
> **(b)** $-9.3 = 1.5t$
>
> Signs are different, so quotient is negative.
>
> $$\frac{-9.3}{1.5} = \frac{\overset{1}{\cancel{1.5}}t}{\underset{1}{\cancel{1.5}}} \qquad \begin{array}{l}\text{Divide both sides by the}\\ \text{coefficient of the}\\ \text{variable term, 1.5}\end{array}$$
>
> $$-6.2 = t$$
>
> The solution is -6.2. To check the solution, go back to the original equation.
>
> *Check*　　　$-9.3 = 1.5t$　　　Original equation
>
> 　　　　　　　$-9.3 = 1.5(-6.2)$　　Replace t with -6.2
>
> 　　　　　　　$-9.3 = -9.3$　　　Balances
>
> When t is replaced with -6.2, the equation balances, so -6.2 is the correct solution (***not*** -9.3).
>
> ◀◀◀ **Now Try Exercise 11.**

OBJECTIVE 3 Solve equations containing decimals using both properties of equality. Sometimes you need to use both the addition and division properties to solve an equation, as shown in Example 3.

EXAMPLE 3 **Solving Equations with Several Steps**

(a) $2.5b + 0.35 = -2.65$

The first step is to get the variable term, $2.5b$, by itself on the left side of the equal sign.

$$
\begin{array}{rl}
2.5b + 0.35 &= -2.65 \\
\underline{-0.35 \quad -0.35} & \qquad \text{Add } -0.35 \text{ to both sides.} \\
2.5b + 0 &= -3.00 \\
2.5b &= -3
\end{array}
$$

The next step is to divide both sides by the coefficient of the variable term. In $2.5b$, the coefficient is 2.5.

$$
\frac{\cancel{2.5}b}{\cancel{2.5}} = \frac{-3}{2.5} \qquad
\begin{array}{l}
\text{On the right side, signs do} \\
\textit{not } \text{match, so the quotient} \\
\text{is } \textit{negative.}
\end{array}
$$

$$
b = -1.2
$$

The solution is -1.2. To check the solution, go back to the original equation.

Check
$$
\begin{array}{rll}
2.5b + 0.35 &= -2.65 & \text{Original equation} \\
2.5\,(-1.2) + 0.35 &= -2.65 & \text{Replace } b \text{ with } -1.2 \\
-3 + 0.35 &= -2.65 & \\
-2.65 &= -2.65 & \begin{array}{l}\text{Balances, so } -1.2 \text{ is the} \\ \text{correct solution.}\end{array}
\end{array}
$$

(b) $5x - 0.98 = 2x + 0.4$

There is a variable term on both sides of the equation. You can choose to keep the variable term on the left side, or to keep the variable term on the right side. Either way will work. Just pick the left side or the right side.

Suppose that you decide to keep the variable term, $5x$, on the left side. Use the addition property to "get rid of" $2x$ on the right side by adding its opposite, $-2x$, to both sides.

$$
\begin{array}{rl}
5x - 0.98 = & 2x + 0.4 \\
\underline{-2x -2x} & \qquad \text{Add } -2x \text{ to both sides.} \\
3x - 0.98 = & 0 + 0.4
\end{array}
$$

Change subtraction to adding the opposite.

$$
\begin{array}{rl}
3x + (-0.98) = & 0.4 \\
\underline{+0.98 +0.98} & \qquad \text{Add } 0.98 \text{ to both sides.} \\
3x + 0 = & 1.38
\end{array}
$$

$$
\frac{\cancel{3}x}{\cancel{3}} = \frac{1.38}{3} \qquad \text{Divide both sides by 3}
$$

$$
x = 0.46
$$

The solution is 0.46. To check the solution, go back to the original equation.

Check
$$
\begin{array}{rll}
5x - 0.98 = 2x + 0.4 & \qquad & \text{Original equation} \\
5(0.46) - 0.98 = 2(0.46) + 0.4 & \qquad & \text{Replace } x \text{ with } 0.46 \\
2.3 - 0.98 = 0.92 + 0.4 & & \\
1.32 = 1.32 & & \begin{array}{l}\text{Balances, so } 0.46 \text{ is the} \\ \text{correct solution.}\end{array}
\end{array}
$$

Now Try Exercises 13 and 19. ▶▶▶

OBJECTIVE **4** **Solve application problems involving equations with decimals.** Use the six problem-solving steps from **Section 3.3.**

EXAMPLE 4 Solving an Application Problem

Many larger airplanes have phones that can be used to call people on the ground. In January 2004, the cost of using the air-to-ground cell phone was $3.28 per minute plus a $2.99 connection charge. (*Source:* AT&T.) Hernando was billed $19.39 for one call. How many minutes did the call last?

Step 1   **Read** the problem. It is about the cost of a telephone call.

> **Unknown:** number of minutes the call lasted
> **Known:** Costs are $3.28 per minute plus $2.99; total cost was $19.39.

Step 2   **Assign a variable:** There is only one unknown, so let m be the number of minutes.

Step 3   **Write an equation.**

Cost per minute		Number of minutes		Connection charge		Total cost
3.28	•	m	+	2.99	=	19.39

Step 4   **Solve** the equation.

$$3.28m + 2.99 = 19.39$$

$$\underline{ -2.99 \quad -2.99} \qquad \text{Add } -2.99 \text{ to both sides.}$$

$$3.28m + 0 = 16.40$$

$$\frac{\cancel{3.28}m}{\cancel{3.28}} = \frac{16.40}{3.28} \qquad \text{Divide both sides by 3.28}$$

$$m = 5$$

Step 5   **State the answer.** The call lasted 5 minutes.

Step 6   **Check** the solution by putting it back into the original problem.

> $3.28 per minute times 5 minutes = $16.40
> $16.40 plus $2.99 connection charge = $19.39
> Hernando was billed $19.39. ← Matches ⤴

Because 5 minutes "works" when put back into the original problem, it is the correct solution.

◀◀◀ Now Try Exercise 25.

5.9 Exercises

Solve each equation and check each solution. See Examples 1 and 2.

1. $h + 0.63 = 5.1$ *Check*

2. $-0.2 = k - 0.7$ *Check*

3. $-20.6 + n = -22$ *Check*

4. $g - 5 = 6.03$ *Check*

5. $0 = b - 0.008$ *Check*

6. $0.18 + m = -4.5$ *Check*

7. $2.03 = 7a$ *Check*

8. $-6.2c = 0$ *Check*

9. $0.8p = -96$ *Check*

10. $-10.16 = -4r$ *Check*

11. $-3.3t = -2.31$ *Check*

12. $8.3w = -49.8$ *Check*

Solve each equation. Show your work. See Example 3.

13. $7.5x + 0.15 = -6$

14. $0.8 = 0.2y + 3.4$

15. $-7.38 = 2.05z - 7.38$

16. $6.2h - 0.4 = 2.7$

17. $3c + 10 = 6c + 8.65$

18. $2.1b + 5 = 1.6b + 10$

19. $0.8w - 0.4 = -6 + w$

20. $7r + 9.64 = -2.32 + 5r$ **21.** $-10.9 + 0.5p = 0.9p + 5.3$ **22.** $0.7x - 4.38 = x - 2.16$

Solve each application problem using the six problem-solving steps. See Example 4.

23. Most adult medication doses are for a person weighing 150 pounds. For a 45-pound child, the adult dose should be multiplied by 0.3. If the child's dose of a decongestant is 9 milligrams, what is the adult dose?

24. For a 30-pound child, an adult dose of medication should be multiplied by 0.2. If the child's dose of a cough suppressant is 3 milliliters, find the adult dose.

25. A storm blew down many trees. Several neighbors rented a chain saw for $275.80 and helped each other cut up and stack the wood. The rental company charges $65.95 per day plus a $12 sharpening fee. How many days was the saw rented? (*Source:* Central Rental.)

26. A 20-inch chain saw can be rented for $29.95 for the first two hours, and $9 for each additional hour. Steve's rental charge was $56.95. How many hours did he rent the saw? (*Source:* Central Rental.)

RELATING CONCEPTS (EXERCISES 27–30) For Individual or Group Work

When doing aerobic exercises, it is important to increase your heart rate (the number of beats per minute) so that you get the maximum benefit from the exercise. But you don't want your heart rate to be so fast that it is dangerous. Here is an expression for finding a safe maximum heart rate for a healthy person with no heart disease: $0.7(220 - a)$ *where a is the person's age.* **Work Exercises 27–30 in order:** *Write an equation and solve it to find each person's age. Assume all the people are healthy.*

27. How old is a person who has a maximum safe heart rate of 140 beats per minute? *Hint:* Use the distributive property to simplify $0.7(220 - a)$.

28. How old is a person who has a maximum safe heart rate of 126 beats per minute?

29. If a person's maximum safe heart rate is 134 (rounded to the nearest whole number), how old is the person, to the nearest whole year?

30. If a person's maximum safe heart rate is 117 (rounded to the nearest whole number), how old is the person, to the nearest whole year?

5.10 Geometry Applications: Circles, Cylinders, and Surface Area

OBJECTIVE **1** **Find the radius and diameter of a circle.** Suppose you start with one dot on a piece of paper. Then you draw many dots that are each 2 cm away from the first dot. If you draw enough dots (points) you'll end up with a circle, as shown below on the left. Each point on the circle is exactly 2 cm away from the *center* of the circle. The 2 cm distance is called the *radius, r*, of the circle. The distance across the circle (passing through the center) is called the *diameter, d*, of the circle. In this circle, the diameter is 4 cm.

OBJECTIVES

1 Find the radius and diameter of a circle.

2 Find the circumference of a circle.

3 Find the area of a circle.

4 Find the volume of a cylinder.

5 Find the surface area of a rectangular solid.

6 Find the surface area of a cylinder.

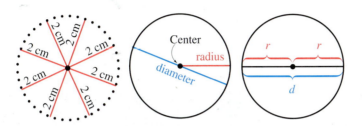

Circle, Radius, and Diameter

A **circle** is a two-dimensional (flat) figure with all points the same distance from a fixed center point.

The **radius** (*r*) is the distance from the center of the circle to any point on the circle.

The **diameter** (*d*) is the distance across the circle passing through the center.

Using the circle above on the right as a model, you can see some relationships between the radius and diameter.

Finding the Diameter and Radius of a Circle

$$\text{diameter} = 2 \cdot \text{radius}$$
$$d = 2r$$

$$\text{and} \qquad r = \frac{d}{2}$$

EXAMPLE 1 **Finding the Diameter and Radius of Circles**

Find the unknown length of the diameter or radius in each circle.

(a)

Because the radius is 9 cm, the diameter is twice as long.

$$d = 2 \cdot \quad r$$
$$d = 2 \cdot 9 \text{ cm}$$
$$d = 18 \text{ cm}$$

Continued on Next Page

(b)

$r = ?$

17 m

The radius is half the diameter.

$$r = \frac{d}{2} \quad \text{so} \quad r = \frac{17 \text{ m}}{2}$$

$$r = 8.5 \text{ m} \quad \text{or} \quad 8\frac{1}{2} \text{ m}$$

◀◀◀ **Now Try Exercises 1 and 3.**

OBJECTIVE 2 Find the circumference of a circle. The perimeter of a circle is called its **circumference.** Circumference is the distance around the edge of a circle.

The diameter of the can in the drawing is about 10.6 cm, and the circumference of the can is about 33.3 cm. Dividing the circumference of the circle by the diameter gives an interesting result.

$$\frac{\text{Circumference}}{\text{diameter}} = \frac{33.3}{10.6} \approx 3.14 \qquad \textcolor{blue}{\text{Rounded to the nearest hundredth}}$$

Dividing the circumference of *any* circle by its diameter *always* gives an answer close to 3.14. This means that going around the edge of any circle is a little more than 3 times as far as going straight across the circle.

This ratio of circumference to diameter is called π (the Greek letter **pi,** pronounced PIE). There is no decimal that is exactly equal to π, but here is the *approximate* value.

$$\pi \approx 3.14159265359$$

Rounding the Value of *Pi* (π)

We usually round π to 3.14. Therefore, calculations involving π will give approximate answers and should be written using the $\approx$ symbol.

Use the following formulas to find the *circumference* of a circle.

Finding the Circumference (Distance Around a Circle)

$$\text{Circumference} = \pi \cdot \text{diameter}$$

$$C = \pi d$$

or, because $d = 2r$ then $C = \pi \cdot 2r$ usually written $C = 2\pi r$

Remember to use linear units such as ft, yd, m, and cm when measuring circumference (**not** square units).

EXAMPLE 2 **Finding the Circumference of Circles**

Find the circumference of each circle. Use 3.14 as the approximate value for π. Round answers to the nearest tenth.

(a)

The *diameter* is 38 m, so use the formula with *d* in it.

$$C = \pi \cdot d$$
$$C \approx 3.14 \cdot 38 \text{ m}$$
$$C \approx 119.3 \text{ m} \quad \leftarrow \text{Rounded}$$

38 m

(b)

In this example, the length of the *radius* is labeled, so it is easier to use the formula with *r* in it.

$$C = 2 \cdot \pi \cdot r$$
$$C \approx 2 \cdot 3.14 \cdot 11.5 \text{ cm}$$
$$C \approx 72.2 \text{ cm} \quad \leftarrow \text{Rounded}$$

11.5 cm

▦ **Calculator Tip** Most *scientific* calculators have a ⓟ key. Try pressing it. With a 10-digit display, you'll see the value of π to the nearest billionth.

$$\boxed{3.141592654}$$

But this is still an approximate value, although it is more precise than rounding π to 3.14. Try finding the circumference in Example 2(a) above using the ⓟ key.

ⓟ ⓧ 38 ⓔ **119.3805208**   Rounds to 119.4

When you used 3.14 as the approximate value of π, the result rounded to 119.3, so the answers are slightly different. In this book we will use 3.14. Our measurements of radius and diameter are given as whole numbers or with tenths, so it is acceptable to round π to hundredths. Also, some students may be using a calculator without a ⓟ key.

Now Try Exercises 5a and 7a. ▶▶▶

OBJECTIVE 3 **Find the area of a circle.**   To find the formula for the area of a circle, start by cutting two circles into many pie-shaped pieces.

Circumference (distance around) is $2 \cdot \pi \cdot r$

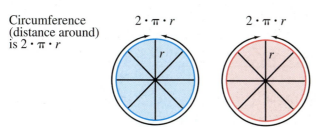

$2 \cdot \pi \cdot r$ $2 \cdot \pi \cdot r$

Unfold the circles, much as you might "unfold" a peeled orange, and put them together as shown here.

$2 \cdot \pi \cdot r$

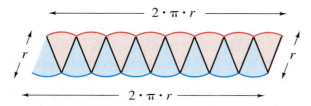

$2 \cdot \pi \cdot r$

The figure is approximately a rectangle with width r (the radius of the original circle) and length $2 \cdot \pi \cdot r$ (the circumference of the original circle). The area of the "rectangle" is length times width.

$$\text{Area of "rectangle"} = \overset{l}{\overbrace{}} \cdot w$$
$$\text{Area of "rectangle"} = 2 \cdot \pi \cdot r \cdot r$$
$$\text{Area of "rectangle"} = 2 \cdot \pi \cdot r^2 \quad \leftarrow \text{Recall that } r \cdot r \text{ is } r^2$$

Because the "rectangle" was formed from *two* circles, the area of *one* circle is half as much.

$$\frac{1}{\underset{1}{\cancel{2}}} \cdot \overset{1}{\cancel{2}} \cdot \pi \cdot r^2 = 1 \cdot \pi \cdot r^2 \quad \text{or simply} \quad \pi r^2$$

> ### Finding the Area of a Circle
>
> $$\text{Area of a circle} = \pi \cdot \text{radius} \cdot \text{radius}$$
> $$A = \pi r^2$$
>
> Remember to use *square units* when measuring area.

EXAMPLE 3 **Finding the Area of Circles**

Find the area of each circle. Use 3.14 for π. Round your answers to the nearest tenth.

(a) A circle with a radius of 8.2 cm

Use the formula $A = \pi r^2$, which means $A = \pi \cdot r \cdot r$.

$$A = \pi \cdot r \cdot r$$
$$A \approx 3.14 \cdot 8.2 \text{ cm} \cdot 8.2 \text{ cm}$$
$$A \approx 211.1 \text{ cm}^2 \quad \leftarrow \text{Square units for area}$$

(b)

To use the area formula $A = \pi r^2$, you need to know the radius (r). In this circle, the *diameter* is 10 ft. First find the radius.

$$r = \frac{d}{2}$$
$$r = \frac{10 \text{ ft}}{2} = 5 \text{ ft}$$

Now find the area.

$$A \approx 3.14 \cdot 5 \text{ ft} \cdot 5 \text{ ft}$$
$$A \approx 78.5 \text{ ft}^2 \quad \leftarrow \text{Square units for area}$$

> **CAUTION**
> When finding *circumference,* you can start with either the radius or the diameter. When finding *area,* you must use the *radius.* If you are given the diameter, divide it by 2 to find the radius. Then find the area.

◀◀◀ **Now Try Exercises 5(b) and 7(b).**

▦ **Calculator Tip**   You can find the area of the circle in Example 3(a) on the previous page using your calculator. The first method works on all types of calculators.

$$3.14 \; \text{⊗} \; 8.2 \; \text{⊗} \; 8.2 \; \text{=} \; \textbf{211.1336}$$

You round the answer to 211.**1** (nearest tenth).

On a *scientific* or *graphing* calculator you can also use the ⓧ² key, which automatically squares the number you enter (that is, multiplies the number times itself).

$$3.14 \; \text{⊗} \; 8.2 \; \text{⊗} \; \text{=} \; \textbf{211.1336}$$

In the next example we will find the area of a *semicircle,* which is half the area of a circle.

EXAMPLE 4   **Finding the Area of a Semicircle**

Find the area of the semicircle below. Use 3.14 for π. Round your answer to the nearest tenth.

First, find the area of a whole circle with a radius of 12 ft.

$$A = \pi \cdot r \cdot r$$
$$A \approx 3.14 \cdot 12 \text{ ft} \cdot 12 \text{ ft}$$
$$A \approx 452.16 \text{ ft}^2 \quad \text{Do not round yet.}$$

Divide the area of the whole circle by 2 to find the area of the semicircle.

$$\frac{452.16 \text{ ft}^2}{2} = 226.08 \text{ ft}^2$$

The *last* step is rounding 226.08 to the nearest tenth.

$$\text{Area of semicircle} \approx 226.1 \text{ ft}^2 \leftarrow \text{Rounded}$$

Now Try Exercise 15. ▶▶▶

EXAMPLE 5   **Applying the Concept of Circumference**

A circular rug is 8 feet in diameter. The cost of fringe for the edge is $2.25 per foot. What will it cost to add fringe to the rug? Use 3.14 for π.

$$\text{Circumference} = \pi \cdot d$$
$$C \approx 3.14 \cdot 8 \text{ ft}$$
$$C \approx 25.12 \text{ ft}$$

$$\text{cost} = \text{cost per foot} \cdot \text{Circumference}$$
$$\text{cost} = \frac{\$2.25}{1 \text{ ft}} \cdot \frac{25.12 \text{ ft}}{1}$$
$$\text{cost} = \$56.52$$

The cost of adding fringe to the rug is $56.52.

Now Try Exercise 23. ▶▶▶

EXAMPLE 6 Applying the Concept of Area

Find the cost of covering the underside of the rug in Example 5 with a non-slip rubber backing. The rubber backing costs $1.50 per square foot. Use 3.14 for π.

First find the radius.

$$r = \frac{d}{2} = \frac{8 \text{ ft}}{2} = 4 \text{ ft}$$

Then find the area.

$$A = \pi \cdot r^2$$
$$A \approx 3.14 \cdot 4 \text{ ft} \cdot 4 \text{ ft}$$
$$A \approx 50.24 \text{ ft}^2$$

$$\text{cost} = \frac{\$1.50}{1 \text{ ft}^2} \cdot \frac{50.24 \text{ ft}^2}{1} = \$75.36$$

Now Try Exercise 21.

OBJECTIVE 4 Find the volume of a cylinder. Several *cylinders* are shown below.

The height must be perpendicular to the circular top and bottom of the cylinder.

These are called *right circular cylinders* because the top and bottom are circles, and the side makes a right angle with the top and bottom. Examples of cylinders are a soup can, a home water heater, and a piece of pipe.

Use the following formula to find the *volume* of a *cylinder*. Notice that the first part of the formula, $\pi \cdot r \cdot r$, is the *area* of the circular base.

Finding the Volume of a Cylinder

$$\text{Volume of a cylinder} = \pi \cdot r \cdot r \cdot h$$
$$V = \pi r^2 h$$

Remember to use **cubic units** when measuring volume.

EXAMPLE 7 Finding the Volume of Cylinders

Find the volume of each cylinder. Use 3.14 as the approximate value of π. Round your answers to the nearest tenth when necessary.

(a)

The diameter is 20 m, so the radius is 20 m ÷ 2 = 10 m. The height is 9 m. Use the formula.

$$V = \pi \cdot r \cdot r \cdot h$$
$$V \approx 3.14 \cdot 10 \text{ m} \cdot 10 \text{ m} \cdot 9 \text{ m}$$
$$V \approx 2826 \text{ m}^3 \quad \leftarrow \text{Cubic units for volume}$$

(b)

$$V \approx 3.14 \cdot 6.2 \text{ cm} \cdot 6.2 \text{ cm} \cdot 38.4 \text{ cm}$$
$$V \approx 4634.94144 \quad \text{Now round to tenths.}$$
$$V \approx 4634.9 \text{ cm}^3 \quad \leftarrow \text{Cubic units for volume}$$

Now Try Exercise 55.

OBJECTIVE 5 Find the surface area of a rectangular solid. You have just learned how to find the *volume* of a cylinder. In **Section 4.8** you found the *volume* of a rectangular solid. For example, the volume of the cereal box shown below is $V = lwh = (7 \text{ in.})(2 \text{ in.})(10 \text{ in.}) = 140 \text{ in.}^3$ But if your company makes cereal boxes, you also need to know how much cardboard is needed for each box. You need to find the *surface area* of the box.

The unfolded box is made up of six rectangles:
front, back, top, bottom, left side, right side.

Surface area is the area on the surface of a three-dimensional object (a solid). For a rectangular solid like the cereal box, the surface area is the sum of the areas of the six rectangular sides. Notice that the top and bottom have the same area, the front and back have the same area, and the left side and right side have the same area.

$$SA = l \cdot w + l \cdot w + l \cdot h + l \cdot h + w \cdot h + w \cdot h$$

$$SA = \underbrace{2lw}_{} + \underbrace{2lh}_{} + \underbrace{2wh}_{}$$

> **Finding the Surface Area of a Rectangular Solid**
>
> $$\text{Surface Area} = (2 \cdot l \cdot w) + (2 \cdot l \cdot h) + (2 \cdot w \cdot h)$$
> $$SA = \quad 2lw \quad + \quad 2lh \quad + \quad 2wh$$
>
> Remember that area is measured in square units, so use **square units** when measuring *surface* area.

EXAMPLE 8 **Finding the Volume and Surface Area of a Rectangular Solid**

Find the volume and surface area of this shipping carton.

First find the volume.

$V = lwh$

$V = 5 \text{ ft} \cdot 3 \text{ ft} \cdot 4 \text{ ft}$

$V = 60 \text{ ft}^3$ ← Cubic units for volume

Next find the surface area.

$SA = \quad 2lw \quad + \quad 2lh \quad + \quad 2wh$

$SA = (2 \cdot 5 \text{ ft} \cdot 3 \text{ ft}) + (2 \cdot 5 \text{ ft} \cdot 4 \text{ ft}) + (2 \cdot 3 \text{ ft} \cdot 4 \text{ ft})$

$SA = \quad 30 \text{ ft}^2 \quad + \quad 40 \text{ ft}^2 \quad + \quad 24 \text{ ft}^2$

$SA = 94 \text{ ft}^2$ ← Square units for area

Now Try Exercise 47. ▶▶▶

OBJECTIVE **6** **Find the surface area of a cylinder.** You can use the same idea of "unfolding" a shape to find the surface area of a cylinder, such as the soup can shown below. Finding the surface area will tell you how much aluminum you need to make the can.

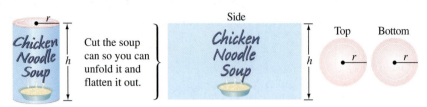

The unfolded soup can is made up of a rectangular side, a circular top, and a circular bottom.

Remember that the formula for the area of a circle is πr^2.

$$SA = \underbrace{2\pi r \cdot h}_{} + \underbrace{\pi r^2 + \pi r^2}_{}$$

$$SA = 2\pi rh + 2\pi r^2$$

Finding the Surface Area of a Right Circular Cylinder

$$\text{Surface Area} = (2 \cdot \pi \cdot r \cdot h) + (2 \cdot \pi \cdot r \cdot r)$$
$$SA = 2\pi rh + 2\pi r^2$$

Remember that area is measured in square units, so use **square units** when measuring *surface* area.

EXAMPLE 9 **Finding the Volume and Surface Area of a Right Circular Cylinder**

Find the volume and surface area of this water tank. Use 3.14 as the approximate value for π. Round your answers to the nearest tenth when necessary.

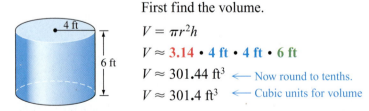

First find the volume.

$V = \pi r^2 h$

$V \approx \textbf{3.14} \cdot \textbf{4 ft} \cdot \textbf{4 ft} \cdot \textbf{6 ft}$

$V \approx 301.44 \text{ ft}^3$ ← Now round to tenths.

$V \approx 301.4 \text{ ft}^3$ ← Cubic units for volume

Now find the surface area.

$SA = 2\pi rh + 2\pi r^2$

$SA \approx (2 \cdot \textbf{3.14} \cdot \textbf{4 ft} \cdot \textbf{6 ft}) + (2 \cdot \textbf{3.14} \cdot \textbf{4 ft} \cdot \textbf{4 ft})$

$SA \approx 150.72 \text{ ft}^2 + 100.48 \text{ ft}^2$

$SA \approx 251.2 \text{ ft}^2$ ← Square units for area

Now Try Exercise 45.

5.10 Exercises

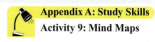

Appendix A: Study Skills
Activity 9: Mind Maps

Find the unknown length in each circle. See Example 1.

1.

2.

3.

4.

Find the circumference and area of each circle. Use 3.14 as the approximate value for π. Round your answers to the nearest tenth. See Examples 2 and 3.

5.

6.

7.

8.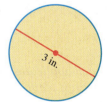

Find the circumference and area of circles having the following diameters. Use 3.14 for π. Round your answers to the nearest tenth. See Examples 2 and 3.

9. $d = 15$ cm

10. $d = 39$ ft

11. $d = 7\frac{1}{2}$ ft

12. $d = 4\frac{1}{2}$ yd

13. $d = 8.65$ km

14. $d = 19.5$ mm

Find each shaded area. Note that Exercises 17 and 18 contain semicircles. Use 3.14 as the approximate value of π. Round your answers to the nearest tenth when necessary. See Example 4.

15.

16.

17.

10 cm

20 cm

18.

20 in.

18 in.

19. How would you explain π to a friend who is not in your math class? Write an explanation. Then make up a test question that requires the use of π, and show how to solve it.

20. Explain how circumference and perimeter are alike. How are they different? Make up two problems, one involving perimeter, the other circumference. Then show how to solve your problems.

Solve each application problem. Use **3.14** *as the approximate value of* π. *Round answers to the nearest tenth. See Examples 5 and 6.*

21. An irrigation system moves around a center point to water a circular area for crops. If the irrigation system is 50 yd long, how large is the watered area?

50 yd

22. If you swing a ball held at the end of a string 20 cm long, how far will the ball travel on each turn?

20 cm

23. A Michelin Cross Terrain SUV tire has an overall diameter of 29.10 inches. How far will a point on the tire tread move in one complete turn? (*Source:* Michelin.)

Bonus question: How many revolutions does the tire make per mile? Round to the nearest whole number.

24. In September 2005, hurricane Katrina slammed into coastal Mississippi and flooded New Orleans. The diameter of the circular storm was 210 miles. In July 2005, hurricane Dennis, with a diameter of only 80 miles, hit the Florida Panhandle. What was the area of each storm, to the nearest hundred square miles? (*Source:* Associated Press.)

For Exercises 25–30, first draw a circle and label the radius or diameter. Then solve the problem. For all Exercises 25–38, use 3.14 for π and round answers to the nearest tenth.

25. A radio station can be heard 150 miles in all directions during evening hours. How many square miles are in the station's broadcast area?

26. An earthquake was felt by people 900 km away in all directions from the epicenter (the source of the earthquake). How much area was affected by the quake?

27. The diameter of Diana Hestwood's wristwatch is 1 in. and the radius of the clock face on her kitchen wall is 3 in. Find the circumference and the area of each clock face.

28. The diameter of the largest known ball of twine is 12 ft 9 in. The sign posted near the ball says it has a circumference of 40 ft. Is the sign correct? *Hint:* First change 9 in. to feet and add it to 12 ft. (*Source: Guinness Book of World Records.*)

29. Blaine Fenstad wants to buy a pair of two-way radios. Some models have a range of 2 miles under ideal conditions. More expensive models have a range of 5 miles. What is the difference in the area covered by the 2-mile and 5-mile models? (*Source: Best Buy.*)

30. The National Audubon Society holds an end-of-year bird count. Volunteers count all the birds they see in a circular area during a 24-hour period. Each circle has a diameter of 15 miles. About 1700 circular areas are counted across the United States each December. What is the total area covered by the count?

31. A forester measures the circumference of a living tree at chest height, then calculates the diameter.

(a) If the circumference of one tree is 144 cm, what is the diameter?

(b) Explain how you solved part (a).

32. In Atlanta, Interstate 285 circles the city and is known as the "perimeter." If the circumference of the circle made by the highway is 62.8 miles, find:

(a) the diameter of the circle

(b) the area inside the circle.

(*Source: Greater Atlanta Newcomer's Guide.*)

33. The Mormons traveled west to Utah by covered wagon in 1847. They tied a rag to a wagon wheel to keep track of the distance they traveled. The radius of the wheel was 2.33 ft. How far did the rag travel each time the wheel made a complete revolution?
(*Source: Trail of Hope.*)

34. First work Exercise 33. Then find how many wheel revolutions equaled one mile. There are 5280 ft in one mile.

35. Find the cost of sod, at $1.76 per square foot, for this playing field that has a semicircle on each end.

29 ft

8 ft

16 ft

29 ft

36. Find the area of this skating rink.

60 ft

25 ft

25 ft

25 ft

37. Find the shaded area.

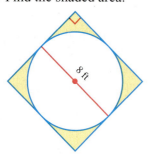

8 ft

38. Find the shaded area.

9 cm

12 cm

RELATING CONCEPTS (EXERCISES 39–44)   For Individual or Group Work

 *Use the table below to **work Exercises 39–44 in order.***

Find the best buy for each type of pizza. The best buy is the lowest cost per square inch of pizza. All the pizzas are circular in shape, and the measurement given on the menu board is the diameter of the pizza in inches. Use 3.14 as the approximate value of π. Round the area to the nearest tenth. Round cost per square inch to the nearest thousandth.

Pizza Menu	Small $7\frac{1}{2}"$	Medium 13"	Large 16"
Cheese only	$2.80	$ 6.50	$ 9.30
"The Works"	$3.70	$ 8.95	$14.30
Deep-dish combo	$4.35	$10.95	$15.65

39. Find the area of a small pizza.

40. Find the area of a medium pizza.

41. Find the area of a large pizza.

42. What is the cost per square inch for each size of cheese pizza? Which size is the best buy?

43. What is the cost per square inch for each size of "The Works" pizza? Which size is the best buy?

44. You have a coupon for 95¢ off any small pizza. What is the cost per square inch for each size of deep-dish combo pizza? Which size is the best buy?

Find the volume and surface area of each cylinder or rectangular solid. Use 3.14 as the approximate value of π. Round your answers to the nearest tenth when necessary. See Examples 7–9.

45.

5 ft

6 ft

46.

12 in.

21 in.

47.

10 m

16.5 m

9.8 m

48.

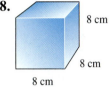

8 cm

8 cm

8 cm

49.

18 in.   3 in.

50.

2 ft

8 ft

51.

37 mm

15 mm

10 mm

52.

0.7 m

3 m

3.5 m

53. Explain the *two* errors made by a student in finding the volume of a cylinder with a diameter of 7 cm and a height of 5 cm. Find the correct answer.

$$V \approx 3.14 \cdot 7 \cdot 7 \cdot 5$$

$$V \approx 769.3 \text{ cm}^2$$

54. Look again at Exercise 48. The figure is a *cube*.

(a) What is special about the measurements of a cube?

(b) Find a shortcut you can use to calculate the surface area of a cube.

Solve each application problem. Use 3.14 as the approximate value of π. Round your answers to the nearest tenth when necessary.

55. A city sewer pipe has a diameter of 5 ft and a length of 200 ft. Find the volume of the pipe.

56. A cylindrical woven basket made by a Northwest Coast tribe is 8 cm high and has a diameter of 11 cm. What is the volume of the basket?

57. A box for graham crackers measures 5.5 in. by 2.8 in. by 8 in. high. Find the amount of cardboard needed to make the box.

58. A soda can is 12.5 cm tall and has a diameter of 6.7 cm. How much aluminum is needed to make the can? (Assume the can has a flat top and bottom.)

Chapter 5 Group Activity: LAWN FERTILIZER AND WEED KILLER

Objective Divide with decimal quotients; round whole numbers and decimals; decide how to round quotients based on the situation.

Gotta Be Green

A lot's being said about personal responsibility these days, and the idea seems to be ending up on the front lawn—literally! Each spring, homeowners across the country gear up to green up their lawns, and the increased use of fertilizer has a lot of environmentalists concerned about the potential effects of chemical runoff into nearby rivers and streams.

Each year, according to a study conducted by the University of Minnesota's Department of Agriculture, each household in the Minneapolis/St. Paul metro area uses an average of 36 pounds of lawn fertilizer. That adds up to 25,529,295 pounds, or 12,765 tons. Add to that another 193,000 pounds of weed killer and you're looking at the total picture for keeping it green in the Twin Cities.

Source: Minneapolis Star Tribune.

1. According to the article,
   (a) How many pounds of lawn fertilizer are used each year in the *entire metro area?*

   (b) Do a division on your calculator to find the number of *households* in the metro area.

   (c) Would it make sense to round your answer to part (b)? If so, how would you round it?

   (d) How many pounds of *weed killer* are used each year in the entire metro area*?*

   (e) Do a division on your calculator to find the number of pounds of *weed killer* used by each household in the metro area. Round your answer to the nearest hundredth.

2. There are 2000 pounds in one ton.
   (a) Find the number of tons equivalent to 25,529,295 pounds of fertilizer.

   (b) Does your answer match the figure given in the article? If not, what did the author of the article do to get 12,765 tons?

   (c) Is the author's figure accurate? Why or why not?

   (d) Find the number of tons equivalent to 193,000 pounds of weed killer.

   (e) How can you write the division problem 193,000 ÷ 2,000 in a simpler way? Explain what you did.

3. According to the article,

(a) "each household in the Minneapolis/St. Paul metro area uses an average of 36 pounds of lawn fertilizer" each year. What mathematical operation do you do to find an average?

(b) When the calculations were done to find the average, the answer was probably not *exactly* 36 pounds. List six different values that are *less than* 36 that would round to 36. List two values with one decimal place, two values with two decimal places, and two values with three decimal places.

(c) List six different values that are *greater than* 36 that would round to 36. List two values each with one, two, and three decimal places.

(d) What is the *smallest* number that can be rounded to 36? What is the *largest* number?

5.1   **decimals**   Decimals, like fractions, are used to show parts of a whole.

         decimal point   A decimal point is the dot that is used to separate the whole number part from the fractional part of a decimal number.

         place value   Place value is the value assigned to each place to the right or left of the decimal point. Whole numbers, such as ones and tens, are to the *left* of the decimal point. Fractional parts, such as tenths and hundredths, are to the *right* of the decimal point.

5.2   **round**   To round is to "cut off" a number after a certain place, such as to round to the nearest hundredth. The rounded number is less accurate than the original number. You can use the symbol " $\approx$ " to mean "approximately equal to."

         decimal places   Decimal places are the number of digits to the *right* of the decimal point; for example, 6.37 has two decimal places, 4.706 has three decimal places.

5.5   **repeating decimal**   A repeating decimal (like the 6 in 0.166 . . .) is a decimal number with one or more digits that repeat forever; it never ends. Use three dots to indicate that it is a repeating decimal. Or write the number with a bar above the repeating digits, as in $0.1\overline{6}$. (Use the dots or the bar, but not both.)

5.7   **mean**   The mean is the sum of all the values divided by the number of values. It is often called the *average*.

         weighted mean   The weighted mean is a mean calculated so that each value is multiplied by its frequency.

         median   The median is the middle number in a group of values that are listed from smallest to largest. It divides a group of values in half. If there is an even number of values, the median is the mean (average) of the two middle values.

         mode   The mode is the value that occurs most often in a group of values.

         variability   The variability of a set of data is the spread of the data around the mean. A quick way to evaluate variability is to find the range of values.

5.8   **square root**   A positive square root of a positive number is one of two equal positive factors of the number.

         hypotenuse   The hypotenuse is the side of a right triangle opposite the 90° angle; it is the longest side.

         Example: See the red side in the triangle at the right.

5.10   **circle**   A circle is a two-dimensional (flat) figure with all points the same distance from a fixed center point.

         Example: See figure at the right.

         radius   Radius is the distance from the center of a circle to any point on the circle.

         Example: See the red radius in the circle at the right.

         diameter   Diameter is the distance across a circle, passing through the center.

         Example: See the blue diameter in the circle at the right.

         circumference   Circumference is the distance around a circle.

         π (pi) π is the ratio of the circumference to the diameter of any circle. It is approximately equal to 3.14.

         surface area   Surface area is the area on the surface of a three-dimensional object (a solid). Surface area is measured in square units.

NEW SYMBOLS

$3.8\overline{6}$ Bar above repeating digit(s) in a decimal number

$3.866 . . .$ Three dots indicate a repeating decimal

$\approx$ is approximately equal to

$\sqrt{}$ square root

π Greek letter pi (pronounced PIE); ratio of the circumference of a circle to its diameter

NEW FORMULAS

$$\text{mean} = \frac{\text{sum of all values}}{\text{number of values}}$$

$$\text{hypotenuse} = \sqrt{(\text{leg})^2 + (\text{leg})^2}$$

$$\text{leg} = \sqrt{(\text{hypotenuse})^2 - (\text{leg})^2}$$

diameter of a circle: $d = 2r$

radius of a circle: $r = \dfrac{d}{2}$

Circumference of a circle: $C = \pi d$
            or $C = 2\pi r$

Area of a circle: $A = \pi r^2$

Volume of a cylinder: $V = \pi r^2 h$

Surface Area of a rectangular solid:
$$SA = 2lw + 2lh + 2wh$$

Surface Area of a cylinder: $SA = 2\pi rh + 2\pi r^2$

TEST YOUR WORD POWER

See how well you have learned the vocabulary in this chapter. Answers follow the Quick Review.

1. **Decimal numbers** are like fractions in that they both
   A. must be written in lowest terms
   B. need common denominators
   C. have decimal points
   D. represent parts of a whole.

2. **Decimal places** refer to
   A. the digits from 0 to 9
   B. digits to the left of the decimal point
   C. digits to the right of the decimal point
   D. the number of zeros in a decimal number.

3. The **hypotenuse** is
   A. the long base in a rectangle
   B. the height in a parallelogram
   C. the longest side in a right triangle
   D. the distance across a circle, passing through the center.

4. The **decimal point**
   A. separates the whole number part from the fractional part
   B. is always moved when finding a quotient
   C. separates tenths from hundredths
   D. is at the far left side of a whole number.

5. The number $0.\overline{3}$ is an example of
   A. an estimate
   B. a repeating decimal
   C. a rounded number
   D. a truncated number.

6. π is the ratio of
   A. the diameter to the radius of a circle
   B. the circumference to the diameter of a circle
   C. the circumference to the radius of a circle
   D. the diameter to the circumference of a circle.

7. The **median** for a set of values is
   A. the mathematical average
   B. the value that occurs most often
   C. the sum of each value times its frequency
   D. the middle value when the values are listed from smallest to largest.

8. The **circumference** of a circle is
   A. the perimeter of the circle
   B. found using the expression πr^2
   C. the distance across the circle
   D. measured in square units.

QUICK REVIEW

Concepts	Examples

5.1 *Reading and Writing Decimals*

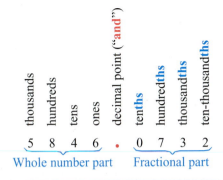

Whole number part   Fractional part

Write each decimal in words.

15.38

8 is in hundredths place.

fifteen **and** thirty-eight **hundredths**

0.0103

3 is in ten-thousandths place.

one hundred three **ten-thousandths**

5.1 *Writing Decimals as Fractions*
The digits to the right of the decimal point are the numerator. The place value of the rightmost digit determines the denominator.

Always write the fraction in lowest terms.

Write 0.45 as a fraction in lowest terms.

The numerator is 45. The rightmost digit, 5, is in the hundredths place, so the denominator is 100. Then write the fraction in lowest terms.

$$\frac{45}{100} = \frac{45 \div 5}{100 \div 5} = \frac{9}{20} \leftarrow \text{Lowest terms}$$

5.2 *Rounding Decimals*
Find the place to which you are rounding. Draw a cut-off line to the right of that place; the rest of the digits will be dropped. Look *only* at the first digit being cut. If it is *4 or less,* the part you are keeping stays the same. If it is *5 or more,* the part you are keeping rounds up. Do not move the decimal point when rounding. Use the sign "≈" to mean "approximately equal to."

Round 0.17952 to the nearest thousandth.

First digit cut is *5 or more* so round up.

$$0.179 \,|\, 52$$

$$
\begin{array}{r}
0.179 \quad \leftarrow \text{Keep this part.}\\
+\ 0.001 \quad \leftarrow \text{To round up, add 1 thousandth.}\\
\hline
0.180
\end{array}
$$

0.17952 rounds to 0.180. Write $0.17952 \approx 0.180$.

5.3 *Adding Positive and Negative Decimals*
Estimate the answer by using front end rounding: round each number to the highest possible place.

To find the exact answer, line up the decimal points. If needed, write in zeros as placeholders. Add or subtract the absolute values as if they were whole numbers. Line up the decimal point in the answer.

If the numbers have the same sign, use the common sign as the sign of the sum. If the numbers have different signs, the sign of the sum is the sign of the number with the larger absolute value.

Add $5.68 + 785.3 + 12 + 2.007$.

Estimate: *Exact:*

$$
\begin{array}{r}
6\\
800\\
10\\
+\ 2\\
\hline
818
\end{array}
\qquad
\begin{array}{r}
5.680\\
785.300\\
12.000\\
+\ 2.007\\
\hline
804.987
\end{array}
$$

Use zeros as placeholders so all numbers have three decimal places.

Line up decimal points.

All addends are positive, so the sign of the sum is positive. The estimate and exact answer are both in hundreds, so the decimal point is probably in the correct place.

Concepts	*Examples*

5.3 *Subtracting Positive and Negative Decimals*

Rewrite subtracting as adding the opposite of the second number. Then follow the rules for adding positive and negative decimals.

Subtract. $4.2 -$ **12.91**

$$4.2 + (-12.91)$$

$|4.2|$ is 4.2 and $|-12.91|$ is 12.91.
Subtract $12.91 - 4.2$ to get 8.71.
Because -12.91 has the larger absolute value and is negative, the answer will be negative.

$$4.2 + (-12.91) = -8.71$$

5.4 *Multiplying Positive and Negative Decimals*

Step 1   Multiply as you would for whole numbers.

Step 2   Count the total number of decimal places in both factors.

Step 3   Write the decimal point in the answer so it has the same number of decimal places as the total from Step 2. You may need to write extra zeros on the left side of the product in order to get enough decimal places in the answer.

Step 4   If two factors have the *same sign,* the product is *positive.* If two factors have *different signs,* the product is *negative.*

Multiply $(0.169)(-0.21)$

$$
\begin{array}{r}
0.169 \quad \leftarrow \text{3 decimal places} \\
\times \quad 0.21 \quad \leftarrow \text{2 decimal places} \\
\hline
169 \qquad \text{5 total decimal places} \\
338 \qquad\qquad \\
\hline
.03549 \quad \leftarrow \text{5 decimal places in product}
\end{array}
$$

Write in a 0 so you can count over 5 decimal places.
   The factors have different signs, so the product is negative.

$$(0.169)(-0.21) = -0.03549$$

5.5 *Dividing by a Decimal*

Step 1   Change the divisor to a whole number by moving the decimal point to the right.

Step 2   Move the decimal point in the dividend the same number of places to the right.

Step 3   Write the decimal point in the quotient directly above the decimal point in the dividend. Then divide as with whole numbers.

Step 4   If the numbers have the *same sign,* the quotient is *positive.* If they have *different signs,* the quotient is *negative.*

Divide -52.8 by -0.75.
First consider $52.8 \div 0.75$.

$$
\begin{array}{r}
70.4 \\
0.75\overline{)52.800} \\
525 \\
\hline
300 \\
300 \\
\hline
0
\end{array}
$$

Move decimal point two places to the right in divisor and dividend.
Write zeros in the dividend so you can move the decimal point and continue dividing until the remainder is 0.

The quotient is positive because both the divisor and dividend were negative (same signs means positive quotient).

5.6 *Writing Fractions as Decimals*

Divide the numerator by the denominator. If necessary, round to the place indicated.

Write $\frac{1}{8}$ as a decimal.

$\frac{1}{8}$ means $1 \div 8$. Write it as $8\overline{)1}$.
The decimal point is on the right side of 1.

$$
\begin{array}{r}
0.125 \\
8\overline{)1.000} \\
8 \\
\hline
20 \\
16 \\
\hline
40 \\
40 \\
\hline
0
\end{array}
$$

$\leftarrow$ Write the decimal point and three zeros so you can continue dividing.

Therefore, $\frac{1}{8}$ is equivalent to 0.125.

Concepts	Examples

Concepts

5.6 *Comparing the Size of Fractions and Decimals*

Step 1   Write any fractions as decimals.
Step 2   Write zeros so that all the numbers being compared have the same number of decimal places.
Step 3   Use < to mean "is less than," > to mean "is greater than," or list the numbers from smallest to largest.

Examples

Arrange in order from smallest to largest. $0.505 \quad \frac{1}{2} \quad 0.55$

$0.505 = 505$ thousandths ← 505 is in the middle

$\frac{1}{2} = 0.5 = 0.500 = 500$ thousandths ← 500 is smallest.

$0.55 = 0.550 = 550$ thousandths ← 550 is largest.

(smallest) $\frac{1}{2} \quad 0.505 \quad 0.55$ (largest)

5.7 *Finding the Mean (Average) of a Set of Values*

Step 1   Add all values to obtain a total.
Step 2   Divide the total by the number of values.

Here are Heather Hall's test scores in her math course.

$$93 \quad 76 \quad 83 \quad 93$$
$$78 \quad 82 \quad 87 \quad 85$$

Find Heather's mean score to the nearest tenth.

$$\text{mean} = \frac{93 + 76 + 83 + 93 + 78 + 82 + 87 + 85}{8}$$

$$= \frac{677}{8} \approx 84.6 \leftarrow \text{Mean test score}$$

5.7 *Finding the Median of a Set of Values*

Step 1   Arrange the data from smallest to largest.
Step 2   If there is an odd number of values, select the middle value. If there is an even number of values, find the average of the two middle values.

Find the median for Heather Hall's scores from the previous example.

List the scores from smallest to largest.

$$76 \quad 78 \quad 82 \quad \mathbf{83 \quad 85} \quad 87 \quad 93 \quad 93$$

Middle values

The middle two values are 83 and 85. Find the average of these two values.

$$\frac{83 + 85}{2} = 84 \quad \leftarrow \text{Median test score}$$

5.7 *Finding the Mode of a Set of Values*

Find the value that appears most often in the list of values. This is the mode.

If no value appears more than once, there is no mode. If two different values appear the same number of times, the list is bimodal.

Find the mode for Heather's scores in the previous example.

The most frequently occurring score is 93 (it occurs twice). Therefore, the mode is 93.

5.7 *Evaluating the Variability of Data*

The variability of a set of data is the spread of the data around the mean. A quick way to evaluate the variability of data is to look at the range of values. Subtract the lowest value from the highest value. Generally, the greater the range in a set of values, the greater the variability.

Which set of test scores has greater variability?

Student M: 78, 85, 82, 84, 90, 86

Student N: 87, 89, 65, 95, 68, 88

Range for student M = 90 − 78 = 12

Range for student N = 95 − 65 = 30

Student N's scores have greater range and greater variability.

Concepts	Examples

5.8 Finding the Square Root of a Number

Use the square root key on a calculator, $\sqrt{}$ or $\sqrt{x}$.
Round to the nearest thousandth when
necessary.

$$\sqrt{64} = 8$$

$$\sqrt{43} \approx 6.557 \quad \leftarrow \text{6.557438524 is rounded}$$
$$\text{to the nearest thousandth.}$$

5.8 Finding the Unknown Length in a Right Triangle

To find the hypotenuse, use:

$$\text{hypotenuse} = \sqrt{(\text{leg})^2 + (\text{leg})^2}$$

The hypotenuse is the side opposite the right angle.
It is the longest side in a right triangle.

Find the unknown length in this right triangle. Round to the
nearest tenth.

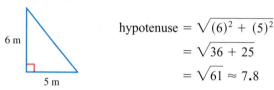

$$\text{hypotenuse} = \sqrt{(6)^2 + (5)^2}$$
$$= \sqrt{36 + 25}$$
$$= \sqrt{61} \approx 7.8$$

The hypotenuse is about 7.8 m long.

To find a leg, use:

$$\text{leg} = \sqrt{(\text{hypotenuse})^2 - (\text{leg})^2}$$

The legs are the sides that form the right angle.

Find the unknown length in this right triangle. Round to the
nearest tenth.

$$\text{leg} = \sqrt{(25)^2 - (16)^2}$$
$$= \sqrt{625 - 256}$$
$$= \sqrt{369} \approx 19.2$$

The leg is about 19.2 cm long.

5.9 Solving Equations Containing Decimals

Solve the equation and check the solution.

$$
\begin{array}{ll}
4.5x + 0.7 = -5.15 & \\
\underline{ -0.7 \quad -0.7} & \text{Add } -0.7 \text{ to both sides.} \\
4.5x + 0 \;\;= -5.85 &
\end{array}
$$

Step 1   Use the addition property of equality to get the
variable term by itself on one side of the equal
sign.

$$
\frac{4.5x}{4.5} = \frac{-5.85}{4.5} \qquad \text{Divide both sides by 4.5}
$$
$$x = -1.3$$

Step 2   Divide both sides by the coefficient of the variable
term to find the solution.

The solution is -1.3. To check the solution, go back to the
original equation.

Step 3   Check the solution by going back to the *original*
equation. Replace the variable with the solution.
If the equation balances, the solution is correct.

Check $4.5x + 0.7 = -5.15$ Original equation

$4.5(-1.3) + 0.7 = -5.15$ Replace x with -1.3

$-5.85 + 0.7 = -5.15$

$-5.15 = -5.15$ Balances

When x is replaced with -1.3, the equation balances, so
-1.3 is the correct solution (***not*** -5.15).

Concepts	**Examples**
5.10 *Circles* Use this formula to find the *diameter* of a circle when you are given the radius. diameter = 2 • radius or $d = 2r$	Find the diameter of a circle if the radius is 3 ft. $d = 2 • r$ $d = 2 • \textbf{3 ft} = 6$ ft
Use this formula to find the *radius* of a circle when you are given the diameter. radius = $\dfrac{\text{diameter}}{2}$ or $r = \dfrac{d}{2}$	Find the radius of a circle if the diameter is 5 cm. $r = \dfrac{d}{2}$ $r = \dfrac{\textbf{5 cm}}{2} = 2.5$ cm
Use these formulas to find the *circumference* of a circle. When you know the *radius,* use this formula. $C = 2 • \pi •$ radius or $C = 2\pi r$ When you know the *diameter*, use this formula. $C = \pi •$ diameter or $C = \pi d$ Use 3.14 as the approximate value for π.	Find the circumference of a circle with a radius of 7 yd. Round your answer to the nearest tenth. Circumference = $2 • \pi • r$ $C \approx 2 • \textbf{3.14} • \textbf{7 yd}$ $C \approx \textbf{44.0} \text{ yd} \leftarrow$ Rounded
Use this formula to find the *area* of a circle. $A = \pi • r • r$ or $A = \pi r^2$ Use 3.14 as the approximate value of π. Area is measured in **square units**.	Find the area of the circle. Round your answer to the nearest tenth. Area = $\pi • r • r$ $A \approx \textbf{3.14} • \textbf{3 cm} • \textbf{3 cm}$ $A \approx 28.3 \text{ cm}^2 \leftarrow \begin{cases}\text{Rounded;}\\\text{square units for area}\end{cases}$
5.10 *Volume of a Cylinder* Use this formula to find the volume of a cylinder. Volume = $\pi • r • r • h$ or $V = \pi r^2 h$ where r is the radius of the circular base and h is the height of the cylinder. Volume is measured in **cubic units**.	Find the volume of this cylinder. First, find the radius. $r = \dfrac{8 \text{ m}}{2} = 4$ m $V = \pi • r • r • h$ $V \approx \textbf{3.14} • \textbf{4 m} • \textbf{4 m} • \textbf{10 m}$ $V \approx 502.4 \text{ m}^3 \leftarrow$ Cubic units for volume

Concepts	*Examples*

5.10 *Surface Area of a Rectangular Solid*

Use this formula to find the surface area of a rectangular solid.

Surface Area = $(2 \cdot l \cdot w) + (2 \cdot l \cdot h) + (2 \cdot w \cdot h)$

or $SA = 2lw + 2lh + 2wh$

where *l* is the length, *w* is the width, and *h* is the height of the solid.

Surface area is measured in **square units**.

Find the surface area of this packing crate.

5 m

3 m

6 m

$SA = 2lw + 2lh + 2wh$

$SA = (2 \cdot 5\ m \cdot 3\ m) + (2 \cdot 5\ m \cdot 6\ m) + (2 \cdot 3\ m \cdot 6\ m)$

$SA = 30\ m^2 + 60\ m^2 + 36\ m^2$

$SA = 126\ m^2$ ← Square units for surface area

5.10 *Surface Area of a Cylinder*

Use this formula to find the surface area of a cylinder.

Surface Area = $(2 \cdot \pi \cdot r \cdot h) + (2 \cdot \pi \cdot r \cdot r)$

or $SA = 2\pi rh + 2\pi r^2$

where *r* is the radius of the circular base and *h* is the height of the cylinder. Use 3.14 as the approximate value of π.

Surface area is measured in **square units**.

Find the surface area of a hot water tank with a height of 4.5 ft and a diameter of 1.8 ft. Round your answer to the nearest tenth.

First, find the radius. $r = \dfrac{1.8\ ft}{2} = 0.9\ ft$

$SA = 2\pi rh + 2\pi r^2$

$SA \approx (2 \cdot 3.14 \cdot 0.9\ ft \cdot 4.5\ ft) + (2 \cdot 3.14 \cdot 0.9\ ft \cdot 0.9\ ft)$

$SA \approx 25.434\ ft^2 + 5.0868\ ft^2$

$SA \approx 30.5208\ ft^2$ Now round to tenths.

$SA \approx 30.5\ ft^2$ ← Square units for area

ANSWERS TO TEST YOUR WORD POWER

1. D; *Example:* For 0.7, the whole is cut into ten parts, and you are interested in 7 of the parts.
2. C; *Examples:* The number 6.87 has two decimal places; 0.309 has three decimal places.
3. C; *Example:* In triangle *ABC*, side *AC* (the red side) is the hypotenuse.

C

A   B

4. A; *Example:* In 5.42, the decimal point separates the whole number part, 5 ones, from the decimal part, 42 hundredths.
5. B; *Example:* The bar above the 3 in $0.\overline{3}$ indicates that the 3 repeats forever.

6. B; *Example:* The ratio of a circumference of 12.57 cm to a diameter of 4 cm is $\dfrac{12.57}{4} \approx 3.14$ (rounded).

7. D; *Example:* For this set of prices—\$4, \$3, \$7, \$2, \$5, \$6, \$7—the median is \$5.
8. A; *Example:* If the radius of a circle is 4 ft, then $C \approx 2 \cdot 3.14 \cdot 4 \approx 25.1$ ft. So the distance around the circle is about 25.1 ft.

Chapter **5**
REVIEW EXERCISES

[5.1] *Name the digit that has the given place value.*

1. 243.059

tenths 0

hundredths 5

2. 0.6817

ones 0

tenths 6

3. $5824.39

hundreds 8

hundredths 9

4. 896.503

tenths 5

tens 9

5. 20.73861

tenths 7

ten-thousandths 8

Write each decimal as a fraction or mixed number in lowest terms.

6. 0.5 5/10

7. 0.75

8. 4.05

9. 0.875

10. 0.027

11. 27.8

Write each decimal in words.

12. 0.8 *zero point eight*

13. 400.29

14. 12.007

15. 0.0306

Write each decimal in numbers.

16. Eight and three tenths

8.3

17. Two hundred five thousandths

200.005

18. Seventy and sixty-six ten-thousandths

70.0066

19. Thirty hundredths

30

[5.2] *Round to the place indicated.*

20. 275.635 to the nearest tenth

276

21. 72.789 to the nearest hundredth

73

22. 0.1604 to the nearest thousandth

0.160

23. 0.0905 to the nearest thousandth

10

24. 0.98 to the nearest tenth

10

Round each money amount to the nearest cent.

25. $15.8333

26. $0.698

27. $17,625.7906

Round each income or expense item to the nearest dollar.

28. The income from the pancake breakfast was $350.48.

29. Each member paid $129.50 in dues.

30. The refreshments cost $99.61.

31. The bank charges were $29.37.

[5.3] *Find each sum or difference.*

32. $0.4 - 6.07$

-5.67

33. $-20 + 19.97$

-0.03

34. $-1.35 + 7.229$

-8.579

35. $0.005 + (3 - 9.44)$

-6.435

First, use front end rounding to estimate each answer. Then find the exact answer.

36. Americans' favorite recreational activity is exercise walking. About 81.3 million people go walking at least once during the year. Swimming is second with 56.3 million people, and camping is third with 49.9 million people. How many more people go walking than camping? (*Source:* National Sporting Goods Association.)

Estimate:

Exact:

37. Today, Jasmin had $306 in her checking account. She wrote a check to the day care center for $215.53 and a check for $44.67 at the grocery store. What is the new balance in her account?

Estimate:

Exact:

38. Joey spent $1.59 for toothpaste, $5.33 for a gift, and $18.94 for a toaster. He gave the clerk three $10 bills. How much change did he get?

Estimate:

Exact:

39. Roseanne is training for a wheelchair race. She raced 2.3 kilometers on Monday, 4 kilometers on Wednesday, and 5.25 kilometers on Friday. How far did she race altogether?

Estimate:

Exact:

[5.4] *First, use front end rounding to estimate each answer. Then find the exact answer.*

40. *Estimate:*   *Exact:*

$\begin{array}{r} 40 \\ \times\ 3 \\ \hline \end{array}$ $\begin{array}{r} 6.138 \\ \times\ 3.7 \\ \hline \end{array}$

41. *Estimate:*   *Exact:*

$\begin{array}{r} 40 \\ \times\ 3 \\ \hline \end{array}$ $\begin{array}{r} 42.9 \\ \times\ 3.3 \\ \hline \end{array}$

Multiply.

42. $(-5.6)(-0.002)$

43. $(0.071)(-0.005)$

[5.5] *Decide if each answer is reasonable by rounding the numbers and estimating the answer. If the exact answer is not reasonable, find and correct the error.*

44. $706.2 \div 12 = 58.85$

Estimate:

45. $26.6 \div 2.8 = 0.95$

Estimate:

Divide. Round quotients to the nearest thousandth when necessary.

46. $3\overline{)43.4}$

47. $\dfrac{-72}{-0.06}$

48. $-0.00048 \div 0.0012$

[5.4–5.5] *Solve each application problem.*

49. Adrienne worked 46.5 hours this week. Her hourly wage is $14.24 for the first 40 hours and 1.5 times that rate over 40 hours. Find her total earnings to the nearest dollar.

50. A book of 12 tickets costs $35.89 at the State Fair midway. What is the cost per ticket, to the nearest cent?

51. Stock in Math Tronic sells for $3.75 per share. Kenneth is thinking of investing $500. How many whole shares could he buy?

52. Ground beef is on sale at $0.99 per pound. How much will Ms. Lee pay for 3.5 pounds of hamburger, to the nearest cent?

Simplify.

53. $3.5^2 + 8.7(-1.95)$

54. $11 - 3.06 \div (3.95 - 0.35)$

[5.6] *Write each fraction as a decimal. Round to the nearest thousandth when necessary.*

55. $3\dfrac{4}{5}$

56. $\dfrac{16}{25}$

57. $1\dfrac{7}{8}$

58. $\dfrac{1}{9}$

Arrange each group of numbers in order from smallest to largest.

59. $3.68, 3.806, 3.6008$

60. $0.215, 0.22, 0.209, 0.2102$

61. $0.17, \dfrac{3}{20}, \dfrac{1}{8}, 0.159$

[5.7] *Find the mean and the median for each set of data.*

62. Digital cameras sold each day:
18, 12, 15, 24, 9, 42, 54, 87, 21, 3

63. Number of insurance claims processed:
54, 28, 35, 43, 17, 37, 68, 75, 39

64. Find the weighted mean.

Dollar Value	Frequency
$42	3
$47	7
$53	2
$55	3
$59	5

65. (a) Find the mode or modes for each set of data.
Hiking boots at store J priced at $107, $69, $139, $107, $160, $84, $160

Hiking boots at store K priced at $119, $136, $99, $119, $139, $119, $95

(b) Which store's prices have greater variability? Show your work.

[5.8] ▦ *Find the unknown length in each right triangle. Use a calculator to find square roots. Round your answers to the nearest tenth when necessary.*

66.

8 in.   90°   15 in.

67.

24 cm   25 cm

68.

15 cm   90°   11 cm

69.

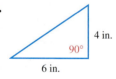

4 in.   90°   6 in.

70.

2.2 m   1.3 m

71.

12 km   8.5 km

[5.9] *Solve each equation. Show your work.*

72. $-0.1 = b - 0.35$

73. $-3.8x = 0$

74. $6.8 + 0.4n = 1.6$

75. $-0.375 + 1.75a = 2a$

76. $0.3y - 5.4 = 2.7 + 0.8y$

[5.10] *Find the unknown radius or diameter.*

77. The radius of a circular irrigation field is 68.9 m. What is the diameter of the field?

78. The diameter of a juice can is 3 in. What is the radius of the can?

Find the circumference and area of each circle. Use **3.14** *as the approximate value for* π. *Round your answers to the nearest tenth.*

79.

1 cm

80.

17.4 m

81.

12 in.

Find the volume and surface area of each solid. Use **3.14** *as the approximate value for* π. *Round your answers to the nearest tenth when necessary.*

82.

7 cm

5 cm

83.

24 m   4 m

84. A rectangular cooler that measures 3.5 ft by 1.5 ft and is 1.5 ft high

---

MIXED REVIEW EXERCISES

Simplify.

85. $89.19 + 0.075 + 310.6 + 5$

86. $72.8(-3.5)$

87. $1648.3 \div 0.46$ Round to thousandths.

88. $30 - 0.9102$

89. $(4.38)(0.007)$

90. $0.005\overline{)0.047}$

91. $72.105 + 8.2 - 95.37$

92. $\dfrac{-81.36}{9}$

93. $(0.6 - 1.22) + 4.8(-3.15)$

94. $0.455(18)$

95. $(-1.6)(-0.58)$

96. $0.218\overline{)7.63}$

97. $-21.059 - 20.8$

98. $18.3 - 3^2 \div 0.5$

Use the information in the ad to solve Exercises 99–103. Round money answers to the nearest cent. (Disregard any sales tax.)

99. How much would one pair of men's socks cost?

100. How much more would one pair of men's socks cost than one pair of children's socks?

101. How much would Fernando pay for a dozen pairs of men's socks?

102. How much would Akiko pay for five pairs of teen jeans and four pairs of women's jeans?

103. What is the difference between the cheapest sale price for athletic shoes and the highest regular price?

Solve each equation.

104. $4.62 = -6.6y$

105. $1.05x - 2.5 = 0.8x + 5$

Solve each application problem. Round answers to the nearest tenth when necessary.

106. A circular table has a diameter of 5 ft. What length of a rubber strip is needed to go around the edge of the table? What is the area of the tabletop?

107. Jerry missed one math test, so his test scores are 82, 0, 78, 93, 85. Find his mean and median scores.

108. LaRae drove 16 miles south, then made a 90° right turn and drove 12 miles west. How far is she, in a straight line, from her starting point?

109. A juice can that is 7 in. tall has a diameter of 4 in. What is the volume of the can?

Chapter 5

TEST

Write each decimal as a fraction or mixed number in lowest terms.

1. 18.4   **2.** 0.075

Write each decimal in words.

3. 60.007   **4.** 0.0208

First, use front end rounding to estimate each answer. Then find the exact answer.

5. $7.6 + 82.0128 + 39.59$ **6.** $-5.79(1.2)$

7. $-79.1 - 3.602$ **8.** $-20.04 \div (-4.8)$

Find the exact answer.

9. $670 - 0.996$ **10.** $0.15\overline{)72}$ **11.** $(-0.006)(-0.007)$

12. Pat bought 3.4 meters of fabric. She paid $15.47. What was the cost per meter?

13. Davida ran a race in 3.059 minutes. Angela ran the race in 3.5 minutes. Who won? By how much?

14. Mr. Yamamoto bought 1.85 pounds of cheese at $2.89 per pound. How much did he pay for the cheese, to the nearest cent?

Solve each equation. Show your work.

15. $-5.9 = y + 0.25$ **16.** $-4.2x = 1.47$

17. $3a - 22.7 = 10$ **18.** $-0.8n + 1.88 = 2n - 6.1$

1. _____

2. _____

3. _____

4. _____

5. Estimate: _____
   Exact: _____

6. Estimate: _____
   Exact: _____

7. Estimate: _____
   Exact: _____

8. Estimate: _____
   Exact: _____

9. _____

10. _____

11. _____

12. _____

13. _____

14. _____

15. _____

16. _____

17. _____

18. _____

19. _____

19. Arrange in order from smallest to largest: $0.44, 0.451, \dfrac{9}{20}, 0.4506$

20. _____

20. Simplify: $6.3^2 - 5.9 + 3.4(-0.5)$

21. _____

21. Find the mean number of books loaned: 52, 61, 68, 69, 73, 75, 79, 84, 91, 98.

22. _____

22. Find the mode for hot tub temperatures (Fahrenheit) of 96°, 104°, 103°, 104°, 103°, 104°, 91°, 74°, 103°.

23. _____

23. Find the weighted mean.

Cost	Frequency
$ 6	7
$10	3
$11	4
$14	2
$19	3
$24	1

24. (a) _____

   (b) _____

   (c) _____

24. (a) Find the median cost of a math textbook: $54.50, $48, $39.75, $89, $56.25, $49.30, $46.90, $51.80.

  (b) Find the median cost of a biology textbook: $75, $50.45, $49.80, $30.95, $63, $97.55, $47.30.

  (c) Which set of costs has less variability? Show your work.

Find the unknown lengths. Round your answers to the nearest tenth when necessary.

25. _____

25.

?   6 cm   7 cm

26.

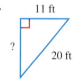

11 ft   ?   20 ft

26. _____

Use 3.14 as the approximate value for π and round your answers to the nearest tenth when necessary.

27. _____

27. Find the radius.

25 in.

28. Find the circumference.

0.9 km

29. Find the area.

16.2 cm

28. _____

29. _____

30. Find the volume.

18 ft   5 ft

30. _____

31. Find the surface area of the solid in Problem 30.

31. _____

Cumulative Review Exercises

CHAPTERS 1–5

1. Write these numbers in words.

   (a) 45.0203

   (b) 30,000,650,008

2. Write these numbers using digits.

   (a) One hundred sixty million, five hundred

   (b) Seventy-five thousandths

Round each number as indicated.

3. 46,908 to the nearest hundred

4. 6.197 to the nearest hundredth

5. 0.66148 to the nearest thousandth

6. 9951 to the nearest hundred

Simplify.

7. $-5 - 8$

8. $-0.003(0.02)$

9. $\dfrac{-7}{0}$

10. $8 + 4(2 - 5)^2$

11. $-\dfrac{3}{8}(-48)$

12. $|4| - |-10|$

13. $0.721 + 55.9$

14. $3\dfrac{1}{3} - 1\dfrac{5}{6}$

15. $\left(\dfrac{3}{b^2}\right)\left(\dfrac{b}{8}\right)$

16. $12 - 0.853$

17. $\dfrac{3}{10} - \dfrac{3}{4}$

18. $\dfrac{\frac{5}{16}}{-10}$

19. $-3.75 \div (-2.9)$ Round answer to nearest tenth.

20. $\dfrac{x}{2} + \dfrac{3}{5}$

21. $5 \div \left(-\dfrac{5}{8}\right)$

22. $2^5 - 4^3$

23. $\dfrac{-36 \div (-2)}{-6 - 4(0 - 6)}$

24. $(0.8)^2 - 3.2 + 4(-0.8)$

25. $\dfrac{3}{4} \div \dfrac{3}{10}\left(\dfrac{1}{4} + \dfrac{2}{3}\right)$

The ages (in years) of the students in a math class are: 19, 23, 24, 19, 20, 29, 26, 35, 20, 22, 26, 23, 25, 26, 20, 30. *Use this data for Exercises 26–28.*

26. Find the mean age, to the nearest tenth.

27. Find the median age.

28. Find the mode.

Solve each equation. Show your work.

29. $3h - 4h = 16 - 12$

30. $-2x = x - 15$

31. $20 = 6r - 45.4$

32. $-3(y + 4) = 7y + 8$

33. $-0.8 + 1.4n = 2 + 0.7n$

Find the unknown length, perimeter, area, or volume. Use 3.14 as the approximate value of π and round answers to the nearest tenth.

34. Find the perimeter and the area.

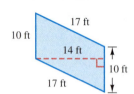

35. Find the circumference and the area.

36. Find the length of the third side and the area.

37. Find the volume and surface area.

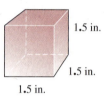

First, use front end rounding to estimate each answer. Then find the exact answer.

38. Lameck had two $20 bills. He spent $17.96 on gasoline and $0.87 for a candy bar at the convenience store. How much money does he have left?

Estimate:

Exact:

39. Toshihiro bought $2\frac{1}{3}$ yards of cotton fabric and $3\frac{7}{8}$ yards of wool fabric. How many yards did he buy in all?

Estimate:

Exact:

40. Paulette bought 2.7 pounds of grapes for $6.18. What was the cost per pound, to the nearest cent?

Estimate:

Exact:

41. Carter Community College received a $78,000 grant from a local hospital to help students pay tuition for nursing classes. How much money could be given to each of 107 students? Round to the nearest dollar.

Estimate:

Exact:

Solve each application problem using the six problem-solving steps.

42. A $30,000 scholarship is being divided between two students so that one student receives three times as much as the other. How much will each student receive?

43. The perimeter of a rectangular game board is 48 in. The length is 4 in. more than the width. Find the length and width of the game board.

Ratio, Proportion, and Line/Angle/ Triangle Relationships

6

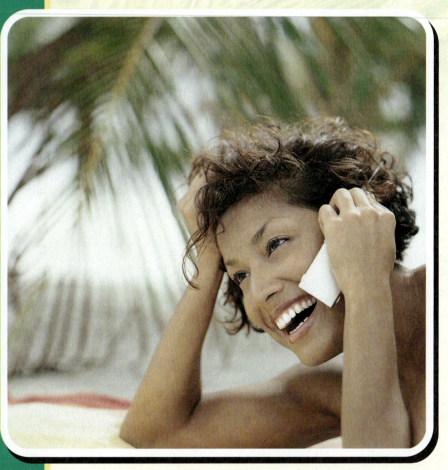

Seven out of every ten American households have cellular phones. Worldwide, about 1 out of 5 people use a cell phone! (*Source:* Consumer Electronics Association.)

You can use unit rates to find the best deal on cell phone service plans. But most plans do not cover long-distance calls to certain U.S. locations or to other countries. Then you can use unit rates to find the cheapest long-distance calling card. (See **Section 6.2,** Exercises 37–38 for calling cards and Exercises 47–50 for cell phone plans; and **Cumulative Review,** Exercises 30–33 for international calling cards.)

6.1 Ratios

OBJECTIVES

1. Write ratios as fractions.

2. Solve ratio problems involving decimals or mixed numbers.

3. Solve ratio problems after converting units.

A **ratio** compares two quantities. You can compare two numbers, such as 8 and 4, or two measurements that have the *same* type of units, such as 3 days and 12 days. (*Rates* compare measurements with different types of units and are covered in the next section.)

Ratios can help you see important relationships. For example, if the ratio of your monthly expenses to your monthly income is 10 to 9, then you are spending $10 for every $9 you earn and going deeper into debt.

OBJECTIVE 1 Write ratios as fractions.   A ratio can be written in three ways.

> **Writing a Ratio**
>
> The ratio of $7 **to** $3 can be written as follows.
>
> $$7 \text{ to } 3 \quad \text{or} \quad 7{:}3 \quad \text{or} \quad \frac{7}{3} \leftarrow \text{Fraction bar indicates "to."}$$
>
> "**:**" indicates "**to.**"

Writing a ratio as a fraction is the most common method, and the one we will use here. All three ways are read, "the ratio of 7 **to** 3." The word **to** separates the quantities being compared.

> **Writing a Ratio as a Fraction**
>
> Order is important when you're writing a ratio. The quantity mentioned **first** is the **numerator**. The quantity mentioned **second** is the **denominator**. For example:
>
> The ratio of **5** to **12**   is written $\dfrac{5}{12}$

EXAMPLE 1 Writing Ratios

Ancestors of the Pueblo Indians built multistory apartment towns in New Mexico about 1100 years ago. A room might measure 14 feet long, 11 feet wide, and 15 feet high.

Continued on Next Page

Write each ratio as a fraction, using the room measurements.

(a) Ratio of length to width

$$\text{The ratio of \textbf{length} to \textbf{width} is } \frac{14 \text{ feet}}{11 \text{ feet}} = \frac{14}{11}$$

Numerator
(mentioned first)

Denominator
(mentioned second)

You can divide out common *units* just as you divided out common *factors* when writing fractions in lowest terms. (See **Section 4.2**.) However, do *not* rewrite the fraction as a mixed number. Keep it as the ratio of 14 to 11.

(b) Ratio of width to height

$$\text{The ratio of width \textbf{to} height is } \frac{11 \text{ feet}}{15 \text{ feet}} = \frac{11}{15}$$

> **CAUTION**
> Remember, the *order* of the numbers is important in a ratio. Look for the words "ratio of *a* to *b*." Write the ratio as $\frac{a}{b}$, *not* $\frac{b}{a}$. The quantity mentioned first is the numerator.

Now Try Exercise 1. ▶▶▶

Any ratio can be written as a fraction. Therefore, you can write a ratio in *lowest terms,* just as you do with any fraction.

EXAMPLE 2 **Writing Ratios in Lowest Terms**

Write each ratio as a fraction in lowest terms.

(a) 60 days to 20 days
The ratio is $\frac{60}{20}$. Write this ratio in lowest terms by dividing the numerator and the denominator by 20.

$$\frac{60}{20} = \frac{60 \div 20}{20 \div 20} = \frac{3}{1} \quad \begin{cases} \text{Ratio in} \\ \text{lowest terms} \end{cases}$$

So, the ratio of 60 days to 20 days is 3 to 1, or, written as a fraction, $\frac{3}{1}$.

> **CAUTION**
> In the fractions chapter (**Chapter 4**), you would have rewritten $\frac{3}{1}$ as 3. But a *ratio* compares *two* quantities, so you need to keep both parts of the ratio and write it as $\frac{3}{1}$.

(b) 50 ounces of medicine to 120 ounces of medicine
The ratio is $\frac{50}{120}$. Divide the numerator and the denominator by 10.

$$\frac{50}{120} = \frac{50 \div 10}{120 \div 10} = \frac{5}{12} \quad \begin{cases} \text{Ratio in} \\ \text{lowest terms} \end{cases}$$

So, the ratio of 50 ounces to 120 ounces is $\frac{5}{12}$.

Continued on Next Page

(c) 15 people in a large van to 6 people in a small van

$$\text{The ratio is } \frac{15}{6} = \frac{15 \div 3}{6 \div 3} = \frac{5}{2} \quad \left\{ \begin{array}{l} \text{Ratio in} \\ \text{lowest terms} \end{array} \right.$$

> **NOTE**
> Although $\frac{5}{2} = 2\frac{1}{2}$, ratios are *not* written as mixed numbers. Nevertheless, in Example 2(c) above, the ratio $\frac{5}{2}$ does mean the large van holds $2\frac{1}{2}$ times as many people as the small van.

Now Try Exercise 7.

OBJECTIVE 2 Solve ratio problems involving decimals or mixed numbers. Sometimes a ratio compares two decimal numbers or two fractions. It is easier to understand if we rewrite the ratio as a ratio of two whole numbers.

EXAMPLE 3 Using Decimal Numbers in a Ratio

The price of a Sunday newspaper increased from $1.50 to $1.75. Find the ratio of the <u>increase in price</u> **to** <u>the original price</u>.

The words <u>increase in price</u> are mentioned first, so the increase will be the numerator. How much did the price go up? Use subtraction.

$$\text{new price} - \text{original price} = \text{increase}$$
$$\$1.75 \quad - \quad \$1.50 \quad = \quad \$0.25$$

The words <u>original price</u> are mentioned second, so the original price of $1.50 is the denominator.

The ratio of <u>increase in price</u> **to** <u>original price</u> is shown below.

$$\frac{0.25}{1.50} \quad \begin{array}{l} \leftarrow \text{increase} \\ \leftarrow \text{original price} \end{array}$$

Now rewrite the ratio as a ratio of whole numbers. Recall that if you multiply both the numerator and denominator of a fraction by the same number, you get an equivalent fraction. The decimals in this example are hundredths, so multiply by 100 to get whole numbers. (If the decimals are tenths, multiply by 10. If thousandths, multiply by 1000.) Then write the ratio in lowest terms.

$$\frac{0.25}{1.50} = \frac{(0.25)(\mathbf{100})}{(1.50)(\mathbf{100})} = \frac{25}{150} = \frac{25 \div \mathbf{25}}{150 \div \mathbf{25}} = \frac{1}{6} \quad \left\{ \begin{array}{l} \text{Ratio in} \\ \text{lowest terms} \end{array} \right.$$

Ratio as two whole numbers

Now Try Exercise 11.

EXAMPLE 4 Using Mixed Numbers in Ratios

Write each ratio as a comparison of whole numbers in lowest terms.

(a) 2 days to $2\frac{1}{4}$ days

Write the ratio as follows. Divide out the common units.

$$\frac{2 \text{ days}}{2\frac{1}{4} \text{ days}} = \frac{2}{2\frac{1}{4}}$$

Continued on Next Page

Next, write 2 as $\frac{2}{1}$ and $2\frac{1}{4}$ as the improper fraction $\frac{9}{4}$.

$$\frac{2}{2\frac{1}{4}} = \frac{\frac{2}{1}}{\frac{9}{4}}$$

Now rewrite the problem in horizontal format, using the "÷" symbol for division. Finally, multiply by the reciprocal of the divisor, as you did in **Chapter 4.**

$$\frac{\frac{2}{1}}{\frac{9}{4}} = \frac{2}{1} \div \frac{9}{4} = \frac{2}{1} \cdot \frac{4}{9} = \frac{8}{9}$$

Reciprocals

The ratio, in lowest terms, is $\frac{8}{9}$.

(b) $3\frac{1}{4}$ to $1\frac{1}{2}$

Write the ratio as $\dfrac{3\frac{1}{4}}{1\frac{1}{2}}$. Then write $3\frac{1}{4}$ and $1\frac{1}{2}$ as improper fractions.

$$3\frac{1}{4} = \frac{13}{4} \quad \text{and} \quad 1\frac{1}{2} = \frac{3}{2}$$

The ratio is shown below.

$$\frac{3\frac{1}{4}}{1\frac{1}{2}} = \frac{\frac{13}{4}}{\frac{3}{2}}$$

Rewrite as a division problem in horizontal format, using the "÷" symbol. Then multiply by the reciprocal of the divisor.

$$\frac{13}{4} \div \frac{3}{2} = \frac{13}{4} \cdot \frac{2}{3} = \frac{13 \cdot \overset{1}{\cancel{2}}}{\underset{1}{\cancel{2}} \cdot 2 \cdot 3} = \frac{13}{6} \quad \leftarrow \begin{cases} \text{Ratio in} \\ \text{lowest terms} \end{cases}$$

Reciprocals

> **NOTE**
> We can also work Examples 4(a) and 4(b) above by using decimals.
> **(a)** $2\frac{1}{4}$ is equivalent to 2.25, so we have the ratio shown below.
>
> $$\frac{2}{2\frac{1}{4}} = \frac{2}{2.25} = \frac{(2)(100)}{(2.25)(100)} = \frac{200}{225} = \frac{200 \div 25}{225 \div 25} = \frac{8}{9} \quad \leftarrow \text{Same result}$$
>
> **(b)** $3\frac{1}{4}$ is equivalent to 3.25 and $1\frac{1}{2}$ is equivalent to 1.5.
>
> $$\frac{3\frac{1}{4}}{1\frac{1}{2}} = \frac{3.25}{1.5} = \frac{(3.25)(100)}{(1.5)(100)} = \frac{325}{150} = \frac{325 \div 25}{150 \div 25} = \frac{13}{6} \quad \leftarrow \text{Same result}$$
>
> This method would *not* work for fractions that are repeating decimals, such as $\frac{1}{3}$ or $\frac{5}{6}$.

Now Try Exercise 13. ▶▶▶

O B J E C T I V E 3 Solve ratio problems after converting units. When a ratio compares measurements, both measurements must be in the *same* units. For example, *feet* must be compared to *feet, hours* to *hours, pints* to *pints,* and *inches* to *inches.*

EXAMPLE 5 Ratio Applications Using Measurement

(a) Write the ratio of the length of the shorter board on the left to the length of the longer board on the right. Compare in inches.

First, express 2 feet in inches. Because 1 foot has 12 inches, 2 feet is

$$2 \cdot \textbf{12 inches} = 24 \text{ inches.}$$

The length of the board on the left is 24 inches, so the ratio of the lengths is shown below. The common units divide out.

$$\frac{2 \text{ ft}}{30 \text{ in.}} = \frac{24 \text{ inches}}{30 \text{ inches}} = \frac{24}{30}$$

Write the ratio in lowest terms.

$$\frac{24}{30} = \frac{24 \div \textbf{6}}{30 \div \textbf{6}} = \frac{4}{5} \left\{ \begin{array}{l} \text{Ratio in} \\ \text{lowest terms} \end{array} \right.$$

The shorter board on the left is $\frac{4}{5}$ the length of the longer board on the right.

NOTE

Notice in the example above that we wrote the ratio using the smaller unit (inches are smaller than feet). Using the smaller unit will help you avoid working with fractions. If we wrote the ratio using feet, then:

$$30 \text{ inches} = 2\frac{1}{2} \text{ feet.}$$

The ratio in feet is shown below.

$$\frac{2 \text{ feet}}{2\frac{1}{2} \text{ feet}} = \frac{2}{1} \div \frac{5}{2} = \frac{2}{1} \cdot \frac{2}{5} = \frac{4}{5} \quad \leftarrow \text{Same result}$$

The ratio is the same, but it takes more steps to get the answer. Using the smaller unit is usually easier.

(b) Write the ratio of 28 days to 3 weeks.

Since it is usually easier to write the ratio using the smaller measurement, compare in *days* because days are shorter than weeks.

First express 3 weeks in days. Because 1 week has 7 days, 3 weeks is

$$3 \cdot \textbf{7 days} = 21 \text{ days.}$$

So the ratio in days is shown below.

$$\frac{28 \text{ days}}{3 \text{ weeks}} = \frac{28 \text{ days}}{21 \text{ days}} = \frac{28}{21} = \frac{28 \div 7}{21 \div 7} = \frac{4}{3} \longleftarrow \left\{ \begin{array}{l} \text{Ratio in} \\ \text{lowest terms} \end{array} \right.$$

The following table will help you set up ratios that compare measurements. You will work with these measurements again in **Chapter 8.**

Measurement Comparisons

Length	Capacity (Volume)
12 inches = 1 foot	2 cups = 1 pint
3 feet = 1 yard	2 pints = 1 quart
5280 feet = 1 mile	4 quarts = 1 gallon

Weight	Time
16 ounces = 1 pound	60 seconds = 1 minute
2000 pounds = 1 ton	60 minutes = 1 hour
	24 hours = 1 day
	7 days = 1 week

Now Try Exercise 19. ▶▶▶

6.1 Exercises

Write each ratio as a fraction in lowest terms. See Examples 1 and 2.

1. 8 days to 9 days

2. $11 to $15

3. $100 to $50

4. 35¢ to 7¢

5. 30 minutes to 90 minutes

6. 9 pounds to 36 pounds

7. 80 miles to 50 miles

8. 300 people to 450 people

9. 6 hours to 16 hours

10. 45 books to 35 books

Write each ratio as a ratio of whole numbers in lowest terms. See Examples 3 and 4.

11. $4.50 to $3.50

12. $0.08 to $0.06

13. 15 to $2\frac{1}{2}$

14. 5 to $1\frac{1}{4}$

15. $1\frac{1}{4}$ to $1\frac{1}{2}$

16. $2\frac{1}{3}$ to $2\frac{2}{3}$

Write each ratio as a fraction in lowest terms. For help, use the table of measurement relationships on page 427. See Example 5.

17. 4 feet to 30 inches

18. 8 feet to 4 yards

19. 5 minutes to 1 hour

20. 8 quarts to 5 pints

21. 15 hours to 2 days

22. 3 pounds to 6 ounces

23. 5 gallons to 5 quarts

24. 3 cups to 3 pints

The table shows the number of greeting cards that Americans buy for various occasions. Use the information to answer Exercises 25–30. Write each ratio as a fraction in lowest terms.

Holiday/Event	Cards Sold
Valentine's Day	900 million
Mother's Day	150 million
Father's Day	95 million
Graduation	60 million
Thanksgiving	30 million
Halloween	25 million

Source: Hallmark Cards.

25. Find the ratio of Thanksgiving cards to graduation cards.

26. Find the ratio of Halloween cards to Mother's Day cards.

27. Find the ratio of Valentine's Day cards to Halloween cards.

28. Find the ratio of Mother's Day cards to Father's Day cards.

29. Explain how you might use the information in the table if you owned a shop selling gifts and greeting cards.

30. Why is the ratio of Valentine's Day cards to graduation cards $\frac{15}{1}$? Give two possible reasons.

The bar graph shows worldwide sales of the most popular songs of all time. Use the graph to complete Exercises 31–34. Write each ratio as a fraction in lowest terms.

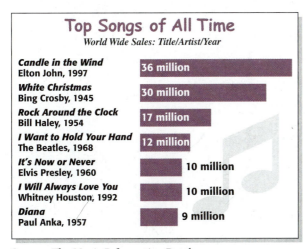

Top Songs of All Time
World Wide Sales: Title/Artist/Year

Candle in the Wind
Elton John, 1997 — 36 million

White Christmas
Bing Crosby, 1945 — 30 million

Rock Around the Clock
Bill Haley, 1954 — 17 million

I Want to Hold Your Hand
The Beatles, 1968 — 12 million

It's Now or Never
Elvis Presley, 1960 — 10 million

I Will Always Love You
Whitney Houston, 1992 — 10 million

Diana
Paul Anka, 1957 — 9 million

Source: The Music Information Database.

31. Write two ratios that compare the top-selling song to the second best seller and the top-selling song to the third best seller.

32. Write two ratios that compare sales of Elvis Presley's song with sales of the songs just ahead and just behind it in the graph.

33. Sales of which two songs give a ratio of $\frac{3}{1}$? There may be more than one correct answer.

34. Sales of which two songs give a ratio of $\frac{5}{6}$? There may be more than one correct answer.

Use the circle graph of one person's monthly budget to complete Exercises 35–38.
Write each ratio as a fraction in lowest terms.

35. Find the ratio of taxes to transportation.

36. Find the ratio of rent to food.

37. Find the ratio of

   (a) rent to total budget.

   (b) rent and utilities to taxes and miscellaneous.

38. Find the ratio of

   (a) utilities to total budget.

   (b) food and taxes to transportation.

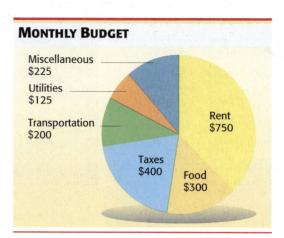

MONTHLY BUDGET

Miscellaneous $225
Utilities $125
Transportation $200
Rent $750
Taxes $400
Food $300

For each figure, find the ratio of the length of the longest side to the length of the
shortest side. Write each ratio as a fraction in lowest terms. See Examples 2–4.

39.

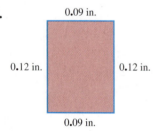

7 ft   6 ft

5 ft

40.

25 kilometers
5 kilometers
24.5 kilometers

41.

1.8 meters
0.3 meter   0.3 meter
1.8 meters

42.

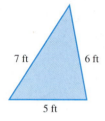

0.09 in.
0.12 in.   0.12 in.
0.09 in.

43.

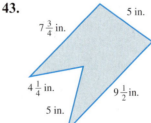

5 in.
$7\frac{3}{4}$ in.
$4\frac{1}{4}$ in. $9\frac{1}{2}$ in.
5 in.

44.

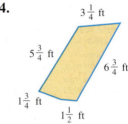

$3\frac{1}{4}$ ft
$5\frac{3}{4}$ ft $6\frac{3}{4}$ ft
$1\frac{3}{4}$ ft $1\frac{1}{2}$ ft

Write each ratio as a fraction in lowest terms.

45. The price of automobile engine oil recently went from $10.80 to $13.50 per case of 12 quarts. Find the ratio of the increase in price to the original price.

46. The price of an antibiotic decreased from $8.80 to $5.60 for 10 tablets. Find the ratio of the decrease in price to the original price.

47. The first time a movie was made in Minnesota, the cast and crew spent $59\frac{1}{2}$ days filming winter scenes. The next year, another movie was filmed in $8\frac{3}{4}$ weeks. Find the ratio of the first movie's filming time to the second movie's time. Compare in weeks.

48. The percheron, a large draft horse, measures about $5\frac{3}{4}$ feet at the shoulder. The prehistoric ancestor of the horse measured only $15\frac{3}{4}$ inches at the shoulder. Find the ratio of the percheron's height to its prehistoric ancestor's height. Compare in inches. (*Source: Eyewitness Books: Horse.*)

$5\frac{3}{4}$ ft $15\frac{3}{4}$ in.

RELATING CONCEPTS (EXERCISES 49–52) For Individual or Group Work

Use your knowledge of ratios to **work Exercises 49–52 in order.**

49. In this painting, what is the ratio of the length of the longest side to the length of the shortest side? What other measurements could the painting have and still maintain the same ratio?

$2\frac{7}{12}$ ft

$2\frac{7}{12}$ ft $2\frac{7}{12}$ ft

$2\frac{7}{12}$ ft

50. The ratio of my son's age to my daughter's age is 4 to 5. One possibility is that my son is 4 years old and my daughter is 5 years old. Find six other possibilities that fit the 4 to 5 ratio.

51. Amelia said that the ratio of her age to her mother's age is 5 to 3. Is this possible? Explain your answer.

52. Would you prefer that the ratio of your income to your friend's income be 1 to 3 or 3 to 1? Explain your answer.

6.2 Rates

OBJECTIVES

1 Write rates as fractions.
2 Find unit rates.
3 Find the best buy based on cost per unit.

A *ratio* compares two measurements with the same type of units, such as 9 feet to 12 feet (both length measurements). But many of the comparisons we make use measurements with different types of units, such as shown here.

160 dollars for 8 hours (money to time)

450 miles on 15 gallons (distance to capacity)

This type of comparison is called a **rate.**

OBJECTIVE 1 Write rates as fractions. Suppose that yesterday you hiked 18 miles **in** 4 hours. The *rate* at which you hiked can be written as a fraction in lowest terms.

$$\frac{18 \text{ miles}}{4 \text{ hours}} = \frac{18 \text{ miles} \div 2}{4 \text{ hours} \div 2} = \frac{9 \text{ miles}}{2 \text{ hours}} \quad \leftarrow \text{ Rate in lowest terms}$$

In a rate, you often find one of these words separating the quantities you are comparing.

in   for   on   per   from

CAUTION
When writing a rate, always include the units, such as miles, hours, dollars, and so on. Because the units in a rate are different, the units do *not* divide out.

EXAMPLE 1 Writing Rates in Lowest Terms

Write each rate as a fraction in lowest terms.

(a) 5 gallons of chemical **for** $60

$$\frac{5 \text{ gallons} \div 5}{60 \text{ dollars} \div 5} = \frac{1 \text{ gallon}}{12 \text{ dollars}} \quad \gtrdot \text{ Write the units: gallons and dollars.}$$

(b) $1500 wages **in** 10 weeks

$$\frac{1500 \text{ dollars} \div 10}{10 \text{ weeks} \div 10} = \frac{150 \text{ dollars}}{1 \text{ week}}$$

(c) 2225 miles **on** 75 gallons of gas

$$\frac{2225 \text{ miles} \div 25}{75 \text{ gallons} \div 25} = \frac{89 \text{ miles}}{3 \text{ gallons}}$$

◄◄◄ Now Try Exercise 9.

OBJECTIVE 2 Find unit rates. When the *denominator* of a rate is 1, it is called a **unit rate.** We use unit rates frequently. For example, you earn $12.75 for *1 hour* of work. This unit rate is written:

$12.75 **per** hour   or $12.75 / hour.

You drive 28 miles on *1 gallon* of gas. This unit rate is written

28 miles **per** gallon   or   28 miles / gallon.

Use **per** or a / mark when writing unit rates.

EXAMPLE 2 **Finding Unit Rates**

Find each unit rate.

(a) 337.5 miles on 13.5 gallons of gas
Write the rate as a fraction.

$$\frac{337.5 \text{ miles}}{13.5 \text{ gallons}} \longleftarrow \text{The fraction bar indicates division.}$$

Divide 337.5 by 13.5 to find the unit rate.

$$13.5\overline{)337.5} \quad \overset{2\,5.}{}$$

$$\frac{337.5 \text{ miles} \div \textbf{13.5}}{13.5 \text{ gallons} \div \textbf{13.5}} = \frac{25 \text{ miles}}{1 \text{ gallon}}$$

The unit rate is 25 miles **per** gallon, or 25 miles/gallon.

(b) 549 miles in 18 hours

$$\frac{549 \text{ miles}}{18 \text{ hours}} \quad \text{Divide.} \quad 18\overline{)549.0} \quad \overset{30.5}{}$$

The unit rate is 30.5 miles per hour, or 30.5 miles/hour.

(c) $810 in 6 days

$$\frac{810 \text{ dollars}}{6 \text{ days}} \quad \text{Divide.} \quad 6\overline{)810} \quad \overset{135}{}$$

The unit rate is $135 per day, or $135/day.

Now Try Exercise 17. ▶▶▶

OBJECTIVE 3 **Find the best buy based on cost per unit.** When shopping for groceries, household supplies, and health and beauty items, you will find many different brands and package sizes. You can save money by finding the lowest *cost per unit*.

Cost per Unit

Cost per unit is a rate that tells how much you pay for *one* item or *one* unit. Examples are $2.85 per gallon, $47 per shirt, and $4.98 per pound.

EXAMPLE 3 **Determining the Best Buy**

The local store charges the following prices for pancake syrup. Find the best buy.

Continued on Next Page

The best buy is the container with the *lowest* cost per unit. All the containers are measured in *ounces* (oz), so you first need to find the *cost per ounce* for each one. Divide the price of the container by the number of ounces in it. Round to the nearest thousandth if necessary.

Let the *order* of the *words* help you set up the rate.

Size	Cost per Unit (Rounded)
12 ounces	$\dfrac{\$1.28}{12\ \text{ounces}} \approx \$0.107$ per ounce (highest)
24 ounces	$\dfrac{\$1.81}{24\ \text{ounces}} \approx \$0.075$ per ounce (lowest)
36 ounces	$\dfrac{\$2.73}{36\ \text{ounces}} \approx \$0.076$ per ounce

The lowest cost per ounce is $0.075, so the 24-ounce container is the best buy.

> **NOTE**
> Earlier we rounded money amounts to the nearest hundredth (nearest cent). But when comparing unit costs, rounding to the nearest thousandth will help you see the difference between very similar unit costs. Notice that the 24-ounce and 36-ounce syrup containers above would both have rounded to $0.08 per ounce if we had rounded to hundredths.

Now Try Exercise 27.

Calculator Tip   When using a calculator to find unit prices, remember that division is *not* commutative. In Example 3 above you wanted to find cost per ounce. Let the *order* of the *words* help you enter the numbers in the correct order.

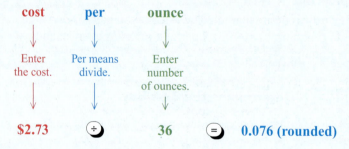

If you reversed the order and entered 36 ÷ 2.73 =, the result is the number of *ounces* per *dollar*. How could you use that information to find the best buy? (*Answer:* The best buy would be the greatest number of ounces per dollar.)

Finding the best buy is sometimes a complicated process. Things that affect the cost per unit can include "cents off" coupons and differences in how much use you'll get out of each unit.

EXAMPLE 4   **Solving Best Buy Applications**

Solve each application problem.

(a) There are many brands of liquid laundry detergent. If you feel they all do a good job of cleaning your clothes, you can base your purchase on cost per unit. But some brands are "concentrated" so you can use less detergent for each load of clothes. Which of the choices shown below is the best buy?

To find Sudzy's unit cost, divide $3.99 by 64 ounces, not 50 ounces. You're getting as many clothes washed as if you bought 64 ounces. Similarly, to find White-O's unit cost, divide $9.89 by 256 ounces (twice 128 ounces, or 2 • 128 ounces = 256 ounces).

$$\text{Sudzy} \quad \frac{\$3.99}{64 \text{ ounces}} \approx \$0.062 \quad \text{per ounce}$$

$$\text{White-O} \quad \frac{\$9.89}{256 \text{ ounces}} \approx \$0.039 \quad \text{per ounce}$$

White-O has the lower cost per ounce and is the best buy. (However, if you try it and it doesn't get out all the stains, Sudzy may be worth the extra cost.)

(b) "Cents-off" coupons also affect the best buy. Suppose you are looking at these choices for "extra-strength" pain reliever. Both brands have the same amount of pain reliever in each tablet.

Brand X is $2.29 for 50 tablets.

Brand Y is $10.75 for 200 tablets.

You have a 40¢ coupon for Brand X and a 75¢ coupon for Brand Y. Which choice is the best buy?

To find the best buy, first subtract the coupon amounts, then divide to find the lowest cost per ounce.

$$\text{Brand X costs} \quad \$2.29 - \$0.40 = \$1.89$$

$$\frac{\$1.89}{50 \text{ tablets}} \approx \$0.038 \text{ per tablet}$$

$$\text{Brand Y costs} \quad \$10.75 - \$0.75 = \$10.00$$

$$\frac{\$10.00}{200 \text{ tablets}} = \$0.05 \text{ per tablet}$$

Brand X has the lower cost per tablet and is the best buy.

Now Try Exercise 45. ▶▶▶

6.2 Exercises

FOR
EXTRA
HELP

 Addison-Wesley
Math Tutor Center

 MathXL

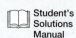 Digital Video Tutor CD 4
Videotape 6

Student's
Solutions
Manual

 MyMathLab

 MathXL
Tutorials on CD

Write each rate as a fraction in lowest terms. See Example 1.

1. 10 cups for 6 people

2. $12 for 30 pens

3. 15 feet in 35 seconds

4. 100 miles in 30 hours

5. 14 people for 28 dresses

6. 12 wagons for 48 horses

7. 25 letters in 5 minutes

8. 68 pills for 17 people

9. $63 for 6 visits

10. 25 doctors for 310 patients

11. 72 miles on 4 gallons

12. 132 miles on 8 gallons

Find each unit rate. See Example 2.

13. $60 in 5 hours

14. $2500 in 20 days

15. 50 eggs from 10 chickens

16. 36 children from 12 families

17. 7.5 pounds for 6 people

18. 44 bushels from 8 trees

19. $413.20 for 4 days

20. $74.25 for 9 hours

Earl kept the following record of the gas he bought for his car. For each entry, find the number of miles he traveled and the unit rate. Round your answers to the nearest tenth.

	Date	Odometer at Start	Odometer at End	Miles Traveled	Gallons Purchased	Miles per Gallon
21.	2/4	27,432.3	27,758.2		15.5	
22.	2/9	27,758.2	28,058.1		13.4	
23.	2/16	28,058.1	28,396.7		16.2	
24.	2/20	28,396.7	28,704.5		13.3	

Source: Author's car records.

Find the best buy (based on the cost per unit) for each item. See Example 3. (Source: Cub Foods.)

25. Black pepper

26. Shampoo

27. Cereal
13 ounces for $2.80
15 ounces for $3.15
18 ounces for $3.98

28. Soup (same size cans)
2 cans for $1.90
3 cans for $2.89
5 cans for $4.58

29. Chunky peanut butter
12 ounces for $1.29
18 ounces for $1.79
28 ounces for $3.39
40 ounces for $4.39

30. Baked beans
8 ounces for $0.59
16 ounces for $0.77
21 ounces for $0.99
28 ounces for $1.36

31. Suppose you are choosing between two brands of chicken noodle soup. Brand A is $0.88 per can and Brand B is $0.98 per can. But Brand B has more chunks of chicken in it. Which soup is the best buy? Explain your choice.

32. A small bag of potatoes costs $0.19 per pound. A large bag costs $0.15 per pound. But there are only two people in your family, so half the large bag would probably rot before you used it up. Which ba is the best buy? Explain.

Solve each application problem. See Examples 2–4.

33. Makesha lost 10.5 pounds in six weeks. What was her rate of loss in pounds per week?

34. Enrique's taco recipe uses thr feed 10 people. Give the ra

35. Russ works 7 hours to earn $85.82. What is his pay rate per hour?

36. Find the cost of 1 ga $55.62.

The table lists information about three long-distance calling cards. The connection fee is charged each time you make a call, no matter how long the call lasts. Use the information in the table to answer Exercises 37–40. Round answers to the nearest thousandth when necessary.

LONG-DISTANCE CALLING CARDS (U.S.)

Card Name	Cost per Minute	Connection Fee
Radiant Penny	$0.01	$0.39
IDT Special	$0.022	$0.14
Access America	$0.047	$0.00

Source: www.1callcard.com

37. (a) Find the *actual* total cost, including the connection charge, for a five-minute call using each card.

(b) Find the cost per minute for this call using each card and select the best buy.

38. (a) Find the *actual* total cost, including the connection charge, for a 30-minute call using each card.

(b) Find the cost per minute for this call using each card and select the best buy.

39. Find the *actual* total cost per minute for a 15-minute call and a 20-minute call using each card. Then select the best buy for each call.

40. All the cards round calls up to the next full minute.

(a) Suppose you call the wrong number. How much would you pay for this 40-*second* call on each card?

(b) How much would you save on this call by using Access America instead of Radiant Penny?

41. In the 2000 Olympics, Michael Johnson ran the 400-meter event in a record time of approximately 44 seconds (actually 43.84 seconds). Give his rate in seconds per meter and in meters per second. Use 44 seconds as the time. (*Source:* www.Olympics.com)

42. Sofia can clean and adjust five hearing aids in four hours. Give her rate in hearing aids per hour and in hours per hearing aid.

44. Which is the best buy, assuming these laundry detergents both clean equally well?

45. Three brands of cornflakes are available. Brand G is priced at $2.39 for 10 ounces. Brand K is $3.99 for 20.3 ounces and Brand P is $3.39 for 16.5 ounces. You have a coupon for 50¢ off Brand P and a coupon for 60¢ off Brand G. Which cereal is the best buy based on cost per unit?

46. Two brands of facial tissue are available. Brand K has a special price of three boxes of 175 tissues each for $5. Brand S is priced at $1.29 per box of 125 tissues. You have a coupon for 20¢ off one box of Brand S and a coupon for 45¢ off one box of Brand K. How can you get the best buy on one box of tissue?

RELATING CONCEPTS (EXERCISES 47–50)   For Individual or Group Work

On the first page of this chapter, we said that unit rates can help you get the best deal on cell phone service. Use the information in the table to **work Exercises 47–50 in order.**

Company	Anytime Minutes	One-Time Activation Fee	Monthly Charge	Termination Fee
Verizon	400	$35	$59.99	$175
T-Mobile	600	$35	$39.99	$200
Nextel	500	$35	$45.99	$200
Sprint	500	$36	$55	$150

Source: Advertisements appearing in *Minneapolis Star Tribune*.

Notes:

1. All companies require that you sign a contract for one year of service and charge a one-time termination fee if you quit early.

2. Unused minutes cannot be carried over to the next month.

47. How much will each company's activation fee cost you on a monthly basis during the one-year contract?

48. All the plans allow unlimited calls on nights and weekends, so you would be using the "anytime minutes" on weekdays. Figure out the average number of weekdays per month. Then, for each plan, how many minutes could you use per weekday? Round to the nearest whole minute.

49. Find the actual average cost per "anytime minute" during the one-year contract for each company, including the activation fee. Assume you use all the minutes and no more. Decide how to round your answers so you can find the best buy.

50. Suppose that after two months you can̲ service because you found that you o 100 "anytime minutes" per month conditions, find the actual cost r minute" for each company, to

6.3 Proportions

OBJECTIVE 1 Write proportions. A **proportion** states that two ratios (or rates) are equivalent. For example,

$$\frac{\$20}{4 \text{ hours}} = \frac{\$40}{8 \text{ hours}}$$

is a proportion that says the rate $\dfrac{\$20}{4 \text{ hours}}$ is equivalent to the rate $\dfrac{\$40}{8 \text{ hours}}$. As the amount of money doubles, the number of hours also doubles. This proportion is read:

20 dollars **is to** 4 hours **as** 40 dollars **is to** 8 hours.

EXAMPLE 1 Writing Proportions

Write each proportion.

(a) 6 feet is to 11 feet **as** 18 feet is to 33 feet.

$$\frac{6 \text{ feet}}{11 \text{ feet}} = \frac{18 \text{ feet}}{33 \text{ feet}} \quad \text{so} \quad \frac{6}{11} = \frac{18}{33} \qquad \text{The common units (feet) divide out and are not written.}$$

(b) $9 is to 6 liters **as** $3 is to 2 liters.

$$\frac{\$9}{6 \text{ liters}} = \frac{\$3}{2 \text{ liters}} \qquad \text{Units must be written.}$$

Now Try Exercise 1.

OBJECTIVE 2 Determine whether proportions are true or false. There are two ways to see whether a proportion is true. One way is to *write both of the ratios in lowest terms.*

EXAMPLE 2 Writing Both Ratios in Lowest Terms

Determine whether each proportion is true or false by writing both ratios in lowest terms.

(a) $\dfrac{5}{9} = \dfrac{18}{27}$

Write each ratio in lowest terms.

$$\frac{5}{9} \quad \leftarrow \text{ Already in lowest terms} \qquad\qquad \frac{18 \div 9}{27 \div 9} = \frac{2}{3} \quad \leftarrow \text{ Lowest terms}$$

is *not* equivalent to $\frac{2}{3}$, the proportion is *false*. The ratios are *not* ...al.

$$\frac{28}{21}$$

...te each ratio in lowest terms.

$$\frac{16 \div 4}{12 \div 4} = \frac{4}{3} \quad \text{and} \quad \frac{28 \div 7}{21 \div 7} = \frac{4}{3}$$

...ratios are equivalent to $\frac{4}{3}$, so the proportion is *true*. The ratios are ...ortional.

Now Try Exercise 13.

A second way to see whether a proportion is true is to find *cross products*.

Using Cross Products to Determine Whether a Proportion Is True

To see whether a proportion is true, first multiply along one diagonal, then multiply along the other diagonal, as shown here.

$$5 \cdot 4 = \mathbf{20}$$

$$\frac{2}{5} = \frac{4}{10}$$

Cross products are equal.

$$2 \cdot 10 = \mathbf{20}$$

In this case the **cross products** are both 20. When cross products are *equal,* the proportion is *true.* If the cross products are *unequal,* the proportion is *false.*

NOTE

Why does the cross products test work? It is based on rewriting both fractions with a common denominator of $5 \cdot 10$ or 50. (We do not search for the *lowest* common denominator. We simply use the product of the two given denominators.)

$$\frac{2 \cdot \mathbf{10}}{5 \cdot \mathbf{10}} = \frac{20}{50} \quad \text{and} \quad \frac{4 \cdot \mathbf{5}}{10 \cdot \mathbf{5}} = \frac{20}{50}$$

We see that $\frac{2}{5}$ and $\frac{4}{10}$ are equivalent because both can be rewritten as $\frac{20}{50}$. The cross products test takes a shortcut by comparing only the two numerators ($20 = 20$).

EXAMPLE 3 **Using Cross Products**

Use cross products to see whether each proportion is true or false.

(a) $\frac{3}{5} = \frac{12}{20}$ Multiply along one diagonal and then multiply along the other diagonal.

$$5 \cdot 12 = \mathbf{60}$$

$$\frac{3}{5} = \frac{12}{20}$$

Equal

$$3 \cdot 20 = \mathbf{60}$$

The cross products are *equal,* so the proportion is *true.*

CAUTION

Use cross products *only* when working with *proportions*. Do ***not*** use cross products when multiplying fractions, adding fractions, or writing fractions in lowest terms.

Continued on Next Page

(b) $\dfrac{2\frac{1}{3}}{3\frac{1}{3}} = \dfrac{9}{16}$ Find the cross products.

Changed to improper fractions

$$3\frac{1}{3} \cdot 9 = \dfrac{10}{\cancel{3}} \cdot \dfrac{\cancel{9}^{3}}{1} = \dfrac{30}{1} = \mathbf{30} \leftarrow$$

$$\dfrac{2\frac{1}{3}}{3\frac{1}{3}} = \dfrac{9}{16} \qquad \text{Unequal}$$

$$2\frac{1}{3} \cdot 16 = \dfrac{7}{3} \cdot \dfrac{16}{1} = \dfrac{112}{3} = \mathbf{37\frac{1}{3}} \leftarrow$$

The cross products are *unequal,* so the proportion is *false.*

> **NOTE**
> The numbers in a proportion do *not* have to be whole numbers. They may be fractions, mixed numbers, decimal numbers, and so on.

Now Try Exercises 17 and 21.

OBJECTIVE 3 Find the unknown number in a proportion. Four numbers are used in a proportion. If any three of these numbers are known, the fourth can be found. For example, find the unknown number that will make this proportion true.

$$\dfrac{3}{5} = \dfrac{x}{40}$$

The variable x represents the unknown number. Start by finding the cross products.

$$\dfrac{3}{5} = \dfrac{x}{40} \qquad \begin{matrix} 5 \cdot x \\ 3 \cdot 40 \end{matrix} \text{ Cross products}$$

To make the proportion true, the cross products must be equal. This gives us the following equation.

$$5 \cdot x = 3 \cdot 40$$
$$5x = 120$$

Recall from **Section 2.4** that we can solve an equation of this type by dividing both sides by the coefficient of the variable term. In this case, the coefficient of $5x$ is 5.

$$\dfrac{5x}{5} = \dfrac{120}{5} \quad \leftarrow \text{Divide both sides by 5}$$

On the left side, divide out the common factor of 5; slashes indicate the divisions.

$$\dfrac{\cancel{5}^{1} \cdot x}{\cancel{5}_{1}} = 24 \qquad \text{On the right side, divide 120 by 5 to get 24}$$

$$x = 24$$

The unknown number in the proportion is 24. The complete proportion is shown below.

$$\frac{3}{5} = \frac{24}{40} \leftarrow x \text{ is } 24$$

Check by finding the cross products. If they are equal, you solved the problem correctly. If they are unequal, rework the problem.

$$5 \cdot 24 = \mathbf{120} \leftarrow$$

$$\frac{3}{5} = \frac{24}{40} \qquad \text{Equal; proportion is true.}$$

$$3 \cdot 40 = \mathbf{120} \leftarrow$$

The cross products are equal, so the solution, $x = 24$, is correct.

CAUTION
The solution is 24, which is the unknown number in the proportion. 120 is *not* the solution; it is the cross product you get when *checking* the solution.

Solve a proportion for an unknown number by using the following steps.

Solving a Proportion to Find an Unknown Number

Step 1 Find the cross products.

Step 2 Show that the cross products are equivalent.

Step 3 Divide both sides of the equation by the coefficient of the variable term.

Step 4 Check by writing the solution in the *original* proportion and finding the cross products.

EXAMPLE 4 **Solving Proportions**

Find the unknown number in each proportion. Round answers to the nearest hundredth when necessary.

(a) $\dfrac{16}{x} = \dfrac{32}{20}$

Recall that ratios can be rewritten in lowest terms. If desired, you can do that *before* finding the cross products. In this example, write $\frac{32}{20}$ in lowest terms as $\frac{8}{5}$, which gives the proportion $\dfrac{16}{x} = \dfrac{8}{5}$.

$x \cdot 8 \leftarrow$

Step 1 $\qquad \dfrac{16}{x} = \dfrac{8}{5} \qquad$ Find the cross products.

$16 \cdot 5 \leftarrow$

Step 2 $\quad x \cdot 8 = 16 \cdot 5 \quad$ Show that the cross products are equivalent.

$\qquad x \cdot 8 = \quad 80$

Continued on Next Page

Step 3 $\dfrac{x \cdot \cancel{8}}{\cancel{8}} = \dfrac{80}{8}$ ← Divide both sides by 8

$x = 10$ ← Find x. (No rounding is necessary.)

Step 4 Write the solution in the *original* proportion and check by finding the cross products.

x is 10 → $\dfrac{16}{10} = \dfrac{32}{20}$

$10 \cdot 32 = \mathbf{320}$

$16 \cdot 20 = \mathbf{320}$

Equal; proportion is true.

The cross products are equal, so 10 is the correct solution (***not*** 320).

> **NOTE**
> It is not necessary to write the ratios in lowest terms before solving. However, if you do, you will have smaller numbers to work with.

(b) $\dfrac{7}{12} = \dfrac{15}{x}$

Step 1 $\dfrac{7}{12} = \dfrac{15}{x}$

$12 \cdot 15 = 180$

$7 \cdot x$

Find the cross products.

Step 2 $7 \cdot x = 180$ ← Show that cross products are equivalent.

Step 3 $\dfrac{\cancel{7} \cdot x}{\cancel{7}} = \dfrac{180}{7}$ ← Divide both sides by 7

$x \approx 25.71$ ← Rounded to nearest hundredth

When the division does not come out even, check for directions on how to round your answer. Divide out one more place, then round.

$$7)\overline{180.000} \quad \dfrac{25.714}{} \leftarrow \text{Divide out to thousandths so you can round to hundredths.}$$

Step 4 Write the solution in the original proportion and check by finding the cross products.

$\dfrac{7}{12} \approx \dfrac{15}{25.71}$

$(12)(15) = \mathbf{180}$

$(7)(25.71) = \mathbf{179.97}$

Very close, but not equal

The cross products are slightly different because you rounded the value of x. However, they are close enough to see that the problem was done correctly and 25.71 is the approximate solution.

◀◀◀ Now Try Exercises 33 and 41.

The next example shows how to solve for the unknown number in a proportion with fractions or decimals.

EXAMPLE 5 Solving Proportions with Mixed Numbers and Decimals

Find the unknown number in each proportion.

(a) $\dfrac{2\frac{1}{5}}{6} = \dfrac{x}{10}$

Step 1 $\quad \dfrac{2\frac{1}{5}}{6} = \dfrac{x}{10}$

$6 \cdot x \leftarrow$
$2\frac{1}{5} \cdot 10 \leftarrow$ Find the cross products.

Find $2\frac{1}{5} \cdot 10$.

$$2\frac{1}{5} \cdot 10 = \frac{11}{5} \cdot \frac{10}{1} = \frac{11 \cdot 2 \cdot \overset{1}{\cancel{5}}}{\cancel{5} \cdot 1} = \frac{22}{1} = 22$$

Changed to improper fraction

Step 2 $\quad 6 \cdot x = 22 \leftarrow$ Show that cross products are equivalent.

Step 3 $\quad \dfrac{\overset{1}{\cancel{6}} \cdot x}{\underset{1}{\cancel{6}}} = \dfrac{22}{6} \leftarrow$ Divide both sides by 6.

Write the solution as a mixed number in lowest terms.

$$x = \frac{22 \div 2}{6 \div 2} = \frac{11}{3} = 3\frac{2}{3}$$

Step 4 Write the solution in the original proportion and check by finding the cross products.

$$6 \cdot 3\frac{2}{3} = \frac{2 \cdot \overset{1}{\cancel{3}}}{1} \cdot \frac{11}{\underset{1}{\cancel{3}}} = \frac{22}{1} = \textcolor{red}{22} \leftarrow$$

$$\frac{2\frac{1}{5}}{6} = \frac{3\frac{2}{3}}{10} \qquad \text{Equal}$$

$$2\frac{1}{5} \cdot 10 = \frac{11}{\underset{1}{\cancel{5}}} \cdot \frac{2 \cdot \overset{1}{\cancel{5}}}{1} = \frac{22}{1} = \textcolor{red}{22} \leftarrow$$

The cross products are equal, so $3\frac{2}{3}$ is the correct solution (*not* 22).

Continued on Next Page

NOTE

You can use decimal numbers and your calculator to solve Example 5(a) on the previous page. $2\frac{1}{5}$ is equivalent to 2.2, so the cross products are

$$6 \cdot x = (2.2)(10)$$

$$\frac{\overset{1}{\cancel{6}} \cdot x}{\underset{1}{\cancel{6}}} = \frac{22}{6}$$

When you divide 22 by 6 on your calculator, it shows 3.666666667. Write the answer using a bar to show the repeating digit: $3.\overline{6}$. Or round the answer to 3.67 (nearest hundredth).

(b) $\dfrac{1.5}{0.6} = \dfrac{2}{x}$

$$(1.5)(x) = (0.6)(2) \quad \longleftarrow \text{ Show that cross products are equivalent.}$$

$$(1.5)(x) = 1.2$$

$$\frac{(\overset{1}{\cancel{1.5}})(x)}{\underset{1}{\cancel{1.5}}} = \frac{1.2}{1.5} \quad \longleftarrow \text{ Divide both sides by 1.5.}$$

$$x = \frac{1.2}{1.5} \qquad\qquad \text{Complete the division.}$$

$$x = 0.8 \qquad\qquad \overset{\textstyle .8}{1.5\overline{)1.20}}$$

So the unknown number is 0.8. Write the solution in the original proportion and check by finding the cross products.

$$(0.6)(2) = \textbf{1.2}$$

$$\frac{\textbf{1.5}}{\textbf{0.6}} = \frac{\textbf{2}}{\textbf{0.8}} \qquad \Big] \text{ Equal}$$

$$(1.5)(0.8) = \textbf{1.2}$$

The cross products are equal, so 0.8 is the correct solution (*not* 1.2).

Now Try Exercises 43 and 47.

6.3 **Exercises**

Write each proportion. See Example 1.

1. $9 is to 12 cans as $18 is to 24 cans.

2. 28 people is to 7 cars as 16 people is to 4 cars.

3. 200 adults is to 450 children as 4 adults is to 9 children.

4. 150 trees is to 1 acre as 1500 trees is to 10 acres.

5. 120 feet is to 150 feet as 8 feet is to 10 feet.

6. $6 is to $9 as $10 is to $15.

7. 2.2 hours is to 3.3 hours as 3.2 hours is to 4.8 hours.

8. 4 meters is to 4.75 meters as 6 meters is to 7.125 meters.

Determine whether each proportion is true *or* false *by writing the ratios in lowest terms. Show the simplified ratios and then write* true *or* false*. See Example 2.*

9. $\dfrac{6}{10} = \dfrac{3}{5}$

10. $\dfrac{1}{4} = \dfrac{9}{36}$

11. $\dfrac{5}{8} = \dfrac{25}{40}$

12. $\dfrac{2}{3} = \dfrac{20}{27}$

13. $\dfrac{150}{200} = \dfrac{200}{300}$

14. $\dfrac{100}{120} = \dfrac{75}{100}$

Use cross products to determine whether each proportion is true *or* false*. Show the cross products and circle* True *or* False*. See Example 3.*

15. $\dfrac{2}{9} = \dfrac{6}{27}$

    True   False

16. $\dfrac{20}{25} = \dfrac{4}{5}$

    True   False

17. $\dfrac{20}{28} = \dfrac{12}{16}$

    True   False

18. $\dfrac{16}{40} = \dfrac{22}{55}$

True   False

19. $\dfrac{110}{18} = \dfrac{160}{27}$

True   False

20. $\dfrac{600}{420} = \dfrac{20}{14}$

True   False

21. $\dfrac{3.5}{4} = \dfrac{7}{8}$

True   False

22. $\dfrac{36}{23} = \dfrac{9}{5.75}$

True   False

23. $\dfrac{18}{15} = \dfrac{2\frac{5}{6}}{2\frac{1}{2}}$

True   False

24. $\dfrac{1\frac{3}{10}}{3\frac{1}{5}} = \dfrac{4}{9}$

True   False

25. $\dfrac{6}{3\frac{2}{3}} = \dfrac{18}{11}$

True   False

26. $\dfrac{16}{13} = \dfrac{2}{1\frac{5}{8}}$

True   False

27. Suppose Ichiro Suzuki of the Seattle Mariners had 16 hits in 50 times at bat, and Sammy Sosa of the Chicago Cubs was at bat 400 times and got 128 hits. Paul is trying to convince Jamie that the two men hit equally well. Show how you could use a proportion and cross products to see whether Paul is correct.

28. Jay worked 3.5 hours and packed 91 cartons. Craig packed 126 cartons in 5.25 hours. To see if the men worked equally fast, Barry set up this proportion:

$$\frac{3.5}{91} = \frac{126}{5.25}$$

Explain what is wrong with Barry's proportion and write a correct one. Is the correct proportion true or false?

Solve each proportion to find the unknown number. Round your answers to hundredths when necessary. Check your answers by finding cross products. See Examples 4 and 5.

29. $\dfrac{1}{3} = \dfrac{x}{12}$

30. $\dfrac{x}{6} = \dfrac{15}{18}$

31. $\dfrac{15}{10} = \dfrac{3}{x}$

32. $\dfrac{5}{x} = \dfrac{20}{8}$

33. $\dfrac{x}{11} = \dfrac{32}{4}$

34. $\dfrac{12}{9} = \dfrac{8}{x}$

35. $\dfrac{42}{x} = \dfrac{18}{39}$

36. $\dfrac{49}{x} = \dfrac{14}{18}$

37. $\dfrac{x}{25} = \dfrac{4}{20}$

38. $\dfrac{6}{x} = \dfrac{4}{8}$

39. $\dfrac{8}{x} = \dfrac{24}{30}$

40. $\dfrac{32}{5} = \dfrac{x}{10}$

41. $\dfrac{99}{55} = \dfrac{44}{x}$

42. $\dfrac{x}{12} = \dfrac{101}{147}$

43. $\dfrac{0.7}{9.8} = \dfrac{3.6}{x}$

44. $\dfrac{x}{3.6} = \dfrac{4.5}{6}$

45. $\dfrac{250}{24.8} = \dfrac{x}{1.75}$

46. $\dfrac{4.75}{17} = \dfrac{43}{x}$

Find the unknown number in each proportion. Write your answers as whole or mixed numbers when possible. See Example 5.

47. $\dfrac{15}{1\frac{2}{3}} = \dfrac{9}{x}$

48. $\dfrac{x}{\frac{3}{10}} = \dfrac{2\frac{2}{9}}{1}$

49. $\dfrac{2\frac{1}{3}}{1\frac{1}{2}} = \dfrac{x}{2\frac{1}{4}}$

50. $\dfrac{1\frac{5}{6}}{x} = \dfrac{\frac{3}{14}}{\frac{6}{7}}$

Solve each proportion two different ways. First change all the numbers to decimal form and solve. Then change all the numbers to fraction form and solve; write your answers in lowest terms.

51. $\dfrac{\frac{1}{2}}{x} = \dfrac{2}{0.8}$

52. $\dfrac{\frac{3}{20}}{0.1} = \dfrac{0.03}{x}$

53. $\dfrac{x}{\frac{3}{50}} = \dfrac{0.15}{1\frac{4}{5}}$

54. $\dfrac{8\frac{4}{5}}{1\frac{1}{10}} = \dfrac{x}{0.4}$

RELATING CONCEPTS (EXERCISES 55–56) For Individual or Group Work

*Work Exercises 55–56 in order. First prove that the proportions are **not** true. Then create four true proportions for each exercise by changing only one number at a time.*

55. $\dfrac{10}{4} = \dfrac{5}{3}$

56. $\dfrac{6}{8} = \dfrac{24}{30}$

Summary Exercises on Ratios, Rates, and Proportions

Use the circle graph of one college's enrollment to complete Exercises 1–4. Write each ratio as a fraction in lowest terms.

1. Write the ratio of freshmen to juniors.

2. What is the ratio of freshmen to the total college enrollment?

3. Find the ratio of seniors and sophomores to juniors.

4. Write the ratio of freshmen and sophomores to juniors and seniors.

COLLEGE ENROLLMENT

Seniors 1850

Freshmen 2400

Juniors 2000

Sophomores 2150

The bar graph shows the number of Americans who play various instruments. Use the graph to complete Exercises 5 and 6. Write each ratio as a fraction in lowest terms.

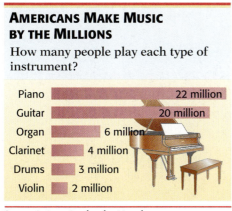

AMERICANS MAKE MUSIC BY THE MILLIONS

How many people play each type of instrument?

Piano	22 million
Guitar	20 million
Organ	6 million
Clarinet	4 million
Drums	3 million
Violin	2 million

Source: America by the Numbers.

5. Write six ratios that compare the least popular instrument to each of the other instruments.

6. Which two instruments give each of these ratios: **(a)** $\frac{5}{1}$; **(b)** $\frac{2}{1}$? There may be more than one correct answer.

The table lists data on the top three individual scoring NBA basketball games of all time. Use the data to answer Exercises 7 and 8. Round answers to the nearest tenth.

7. What was Wilt Chamberlain's scoring rate in points per minute and in minutes per point?

Player	Date	Points	Min
Wilt Chamberlain	3/2/62	100	48
David Thompson	4/9/78	73	43
David Robinson	4/24/94	71	44

Source: www.NBA.com

8. Find David Robinson's scoring rate in points per minute and minutes per point.

Wilt Chamberlain (#13)

9. Lucinda's paycheck showed gross pay of $652.80 for 40 hours of work and $195.84 for 8 hours of overtime work. Find her regular hourly pay rate and her overtime rate.

10. Satellite TV is being offered to new subscribers at $25 per month for 50 channels, $39 per month for 100 channels, or $48 per month for 150 channels. What is the monthly cost per channel under each plan? (*Source:* Dish1Up Satellites.)

11. Find the best buy on gourmet coffee beans.

11 ounces for $6.79

12 ounces for $7.24

16 ounces for $10.99

(*Source:* Cub Foods.)

12. Which brand of cat food is the best buy? You have a coupon for $2 off on Brand P, and another for $1 off on Brand N.

Brand N is $3.75 for 3.5 pounds

Brand P is $5.99 for 7 pounds

Brand R is $10.79 for 18 pounds

Use either the method of writing in lowest terms or the method of finding cross products to decide whether each proportion is true *or* false. *Show your work and then write* true *or* false.

13. $\dfrac{28}{21} = \dfrac{44}{33}$

14. $\dfrac{2.3}{8.05} = \dfrac{0.25}{0.9}$

15. $\dfrac{2\frac{5}{8}}{3\frac{1}{4}} = \dfrac{21}{26}$

Solve each proportion to find the unknown number. Round your answers to hundredths when necessary.

16. $\dfrac{7}{x} = \dfrac{25}{100}$

17. $\dfrac{15}{8} = \dfrac{6}{x}$

18. $\dfrac{x}{84} = \dfrac{78}{36}$

19. $\dfrac{10}{11} = \dfrac{x}{4}$

20. $\dfrac{x}{17} = \dfrac{3}{55}$

21. $\dfrac{2.6}{x} = \dfrac{13}{7.8}$

22. $\dfrac{0.14}{1.8} = \dfrac{x}{0.63}$

23. $\dfrac{\frac{1}{3}}{8} = \dfrac{x}{24}$

24. $\dfrac{6\frac{2}{3}}{4\frac{1}{6}} = \dfrac{\frac{6}{5}}{x}$

6.4 Problem Solving with Proportions

OBJECTIVE 1 Use proportions to solve application problems.
Proportions can be used to solve a wide variety of problems. Watch for problems in which you are given a ratio or rate and then asked to find part of a corresponding ratio or rate. Remember that a ratio or rate compares two quantities and often includes one of these indicator words.

in   for   on   per   from   to

Use the six problem-solving steps from **Section 3.3.** When setting up the proportion, use a variable to represent the unknown number. We have used the letter x, but you may use any letter you like.

EXAMPLE 1 Solving a Proportion Application

Mike's car can travel 163 **miles on** 6.4 **gallons** of gas. How far can it travel on a full tank of 14 **gallons** of gas? Round to the nearest mile.

Step 1   **Read** the problem. It is about how far a car can travel on a certain amount of gas.

> Unknown: miles traveled on 14 gallons of gas
> Known: 163 miles traveled on 6.4 gallons of gas

Step 2   **Assign a variable.** There is only one unknown, so let x be the number of miles traveled on 14 gallons.

Step 3   **Write an equation.** The equation is in the form of a proportion. Decide what is being compared. This problem compares **miles** to **gallons**. Write the two rates described in the problem. Be sure that *both* rates compare miles to gallons in the same order. In other words, miles is in both numerators and gallons is in both denominators.

Matching units

This rate compares **miles** to **gallons**. $\dfrac{163\ \textbf{miles}}{6.4\ \textbf{gallons}} = \dfrac{x\ \textbf{miles}}{14\ \textbf{gallons}}$ This rate compares **miles** to **gallons**.

Matching units

Step 4   **Solve** the equation. Ignore the units while finding the cross products.

$$\frac{163\ \text{miles}}{6.4\ \text{gallons}} = \frac{x\ \text{miles}}{14\ \text{gallons}}$$

$(6.4)(x) = (163)(14)$ Show that cross products are equivalent.

$(6.4)(x) = 2282$

$$\frac{\overset{1}{\cancel{(6.4)}}(x)}{\underset{1}{\cancel{6.4}}} = \frac{2282}{6.4}$$ Divide both sides by 6.4

$x = 356.5625$ Round to 357

Step 5   **State the answer.** Mike's car can travel 357 miles, rounded to the nearest mile, on a full tank of gas.

Continued on Next Page

Check the solution by putting it back into the original problem. The car traveled 163 miles on 6.4 gallons of gas; 14 gallons is a little more than *twice as much* gas, so the car should travel a little more than *twice as far*.

$$(2)(163 \text{ miles}) = 326 \text{ miles} \quad \leftarrow \text{Estimate}$$

The solution, 357 miles, is a little more than the estimate of 326 miles, so it is reasonable.

CAUTION

When setting up the proportion, do *not* mix up the units in the rates.

$$\left.\begin{array}{c}\text{compares } \textbf{miles} \\ \text{to } \textbf{gallons}\end{array}\right\} \quad \frac{163 \textbf{ miles}}{6.4 \textbf{ gallons}} = \frac{14 \textbf{ gallons}}{x \textbf{ miles}} \quad \left\{\begin{array}{c}\text{compares } \textbf{gallons} \\ \text{to } \textbf{miles}\end{array}\right.$$

These rates do *not* compare things in the same order and *cannot* be set up as a proportion.

Now Try Exercise 9.

EXAMPLE 2 **Solving a Proportion Application**

A newspaper report says that 7 out of 10 people surveyed watch the news on TV. At that rate, how many of the 3200 people in town would you expect to watch the news?

Step 1 **Read** the problem. It is about people watching the news on TV.

Unknown: how many people in town are expected to watch the news on TV
Known: 7 out of 10 people surveyed watched the news on TV.

Step 2 **Assign a variable.** There is only one unknown, so let x be the number of people in town who watch the news on TV.

Step 3 **Write an equation.** Set up the two rates as a proportion. You are comparing people who watch the news to people surveyed. Write the two rates described in the example. Be sure that both rates make the same comparison. "People who watch the news" is mentioned first, so it should be in the numerator of *both* rates.

$$\begin{array}{c}\text{People who watch the news} \rightarrow \\ \text{Total group} \rightarrow \\ \text{(people surveyed)}\end{array} \quad \frac{7}{10} = \frac{x}{3200} \quad \begin{array}{c}\leftarrow \text{People who watch the news} \\ \leftarrow \text{Total group} \\ \text{(people in town)}\end{array}$$

Step 4 **Solve** the equation.

$$\frac{7}{10} = \frac{x}{3200}$$

$$(10)(x) = (7)(3200) \qquad \text{Show that cross products are equivalent.}$$

$$(10)(x) = 22{,}400$$

$$\frac{(\cancel{10})(x)}{\cancel{10}} = \frac{22{,}400}{10} \qquad \text{Divide both sides by 10}$$

$$x = 2240$$

Continued on Next Page

Step 5   **State the answer.** You would expect 2240 people in town to
watch the news on TV.

Step 6   **Check** the solution by putting it back into the original problem.
Notice that 7 out of 10 people is more than half the people,
but less than all the people. Half of the 3200 people in town is
$3200 \div 2 = 1600$, so between 1600 and 3200 people would be
expected to watch the news on TV. The solution, 2240 people, is
between 1600 and 3200, so it is reasonable.

CAUTION
Always check that your answer is reasonable. If it isn't, look at the
way your proportion is set up. Be sure you have matching units in the
numerators and matching units in the denominators.

For example, suppose you set up the last proportion *incorrectly,* as
shown below.

$$\frac{7}{10} = \frac{3200}{x} \quad \leftarrow \textbf{Incorrect setup}$$

$$(7)(x) = (10)(3200)$$

$$\frac{\overset{1}{\cancel{(7)}}(x)}{\underset{1}{\cancel{7}}} = \frac{32,000}{7}$$

$$x \approx 4571 \text{ people} \quad \leftarrow \textbf{Unreasonable answer}$$

This answer is *unreasonable* because there are only 3200 people in the
town; it is *not* possible for 4571 people to watch the news.

Now Try Exercise 21. ▶▶▶

6.4 Exercises

FOR EXTRA HELP

 Addison-Wesley Math Tutor Center

 MathXL

 Digital Video Tutor CD 4 Videotape 6

 Student's Solutions Manual

MyMathLab MyMathLab

 MathXL Tutorials on CD

Set up and solve a proportion for each application problem. See Example 1.

1. Caroline can sketch four cartoon strips in five hours. How long will it take her to sketch 18 strips?

2. The Cosmic Toads recorded eight songs on their first CD in 26 hours. How long will it take them to record 14 songs for their second CD?

3. Sixty newspapers cost $27. Find the cost of 16 newspapers.

4. Twenty-two guitar lessons cost $396. Find the cost of 12 lessons.

5. If three pounds of fescue grass seed cover about 350 square feet of ground, how many pounds are needed for 4900 square feet?

6. Anna earns $1242.08 in 14 days. How much does she earn in 260 days?

7. Tom makes $455.75 in five days. How much does he make in three days?

8. If 5 ounces of a medicine must be mixed with 8 ounces of water, how many ounces of medicine would be mixed with 20 ounces of water?

9. The bag of rice noodles shown below makes 7 servings. At that rate, how many ounces of noodles do you need for 12 servings, to the nearest ounce?

10. This can of sweet potatoes is enough for four servings. How many ounces are needed for nine servings, to the nearest ounce?

11. Three quarts of a latex enamel paint will cover about 270 square feet of wall surface. How many quarts will you need to cover 350 square feet of wall surface in your kitchen and 100 square feet of wall surface in your bathroom?

12. One gallon of clear gloss wood finish covers about 550 square feet of surface. If you need to apply three coats of finish to 400 square feet of surface, how many gallons do you need, to the nearest tenth?

Use the floor plan shown to answer Exercises 13–16. On the plan, one inch represents four feet.

13. What is the actual length and width of the kitchen?

14. What is the actual length and width of the family room?

15. What is the actual length and width of the dining area?

16. What is the actual length and width of the entire floor plan?

The table below lists recommended amounts of food to order for 25 party guests. Use the table to answer Exercises 17 and 18. (Source: Cub Foods.)

FOOD FOR 25 GUESTS

Item	Amount
Fried chicken	40 pieces
Lasagna	14 pounds
Deli meats	4.5 pounds
Sliced cheese	$2\frac{1}{3}$ pounds
Bakery buns	3 dozen
Potato salad	6 pounds

17. How much of each food item should Nathan and Amanda order for a graduation party with 60 guests?

18. Taisha is having 20 neighbors over for a Fourth of July picnic. How much food should she buy?

In Exercises 19–24, set up a proportion to solve each problem. Check to see whether your answer is reasonable. Then flip one side of your proportion to see what answer you get with an incorrect setup. Explain why the second answer is unreasonable. See Example 2.

19. About 7 out of 10 people entering our community college need to take a refresher math course. If we have 2950 entering students, how many will probably need refresher math? (*Source:* Minneapolis Community and Technical College.)

20. In a survey, only 3 out of 100 people like their eggs poached. At that rate, how many of the 60 customers who ordered eggs at Soon-Won's restaurant this morning asked to have them poached? Round to the nearest whole person.

21. About 1 out of 8 people choose vanilla as their favorite ice cream flavor. If 238 people attend an ice cream social, how many would you expect to choose vanilla? Round to the nearest whole person.

22. In a test of 200 sewing machines, only one had a defect. At that rate, how many of the 5600 machines shipped from the factory have defects?

23. About 98 out of 100 U.S. households have at least one TV set. There were 109,300,000 U.S. households in 2003. How many households had one or more TVs? (*Source:* Nielsen Media Research.)

24. In a survey, 3 out of 100 dog owners washed their pets by having the dogs go into the shower with them. If the survey is accurate, how many of the 31,200,000 dog owners in the United States use this method? (*Source:* Teledyne Water Pik; American Veterinary Medical Association.)

Set up and solve a proportion for each problem.

25. The stock market report says that five stocks went up for every six stocks that went down. If 750 stocks went down yesterday, how many went up?

26. The human body contains 90 pounds of water for every 100 pounds of body weight. How many pounds of water are in a child who weighs 80 pounds?

27. The ratio of the length of an airplane wing to its width is 8 to 1. If the length of a wing is 32.5 meters, how wide must it be? Round to the nearest hundredth.

28. The Rosebud School District wants a student-to-teacher ratio of 19 to 1. How many teachers are needed for 1850 students? Round to the nearest whole number.

29. The number of calories you burn is proportional to your weight. A 150-pound person burns 222 calories during 30 minutes of tennis. How many calories would a 210-pound person burn, to the nearest whole number? (*Source: Wellness Encyclopedia.*)

30. (Complete Exercise 29 first.) A 150-pound person burns 189 calories during 45 minutes of grocery shopping. How many calories would a 115-pound person burn, to the nearest whole number? (*Source: Wellness Encyclopedia.*)

31. At 3 P.M., Coretta's shadow is 1.05 meters long. Her height is 1.68 meters. At the same time, a tree's shadow is 6.58 meters long. How tall is the tree? Round to the nearest hundredth.

32. (Complete Exercise 31 first.) Later in the day, Coretta's shadow was 2.95 meters long. How long a shadow did the tree have at that time? Round to the nearest hundredth.

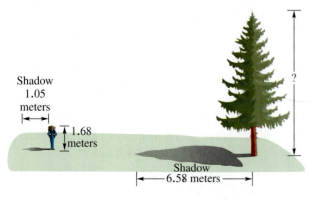

Shadow 1.05 meters

1.68 meters

Shadow 6.58 meters

Shadow 2.95 meters

Shadow ?

33. Can you set up a proportion to solve this problem? Explain why or why not. Jim is 25 years old and weighs 180 pounds. How much will he weigh when he is 50 years old?

34. Write your own application problem that can be solved by setting up a proportion. Also show the proportion and the steps needed to solve your problem.

35. A survey of college students shows that 4 out of 5 drink coffee. Of the students who drink coffee, 1 out of 8 adds cream to it. How many of the 50,700 students at Ohio State University would be expected to use cream in their coffee?

36. About 9 out of 10 adults think it is a good idea to exercise regularly. But of the ones who think it is a good idea, only 1 in 6 actually exercises at least three times a week. At this rate, how many of the 300 employees in our company exercise regularly?

37. The nutrition information on a bran cereal box says that a $\frac{1}{3}$ cup serving provides 80 calories and 8 grams of dietary fiber. At that rate, how many calories and grams of fiber are in a $\frac{1}{2}$ cup serving? (*Source:* Kraft Foods, Inc.)

38. A $\frac{2}{3}$ cup serving of penne pasta has 210 calories and 2 grams of dietary fiber. How many calories and grams of fiber would be in a 1 cup serving? (*Source:* Borden Foods.)

RELATING CONCEPTS (EXERCISES 39–42) For Individual or Group Work

A box of instant mashed potatoes has the list of ingredients shown in the table. Use this information to **work Exercises 39–42 in order.**

Ingredient	For 12 Servings
Water	$3\frac{1}{2}$ cups
Margarine	6 tablespoons
Milk	$1\frac{1}{2}$ cups
Potato flakes	4 cups

Source: General Mills.

39. Find the amount of each ingredient needed for six servings. Show *two* different methods for finding the amounts. One method should use proportions.

40. Find the amount of each ingredient needed for 18 servings. Show *two* different methods for finding the amounts, one using proportions and one using your answers from Exercise 39.

41. Find the amount of each ingredient needed for three servings, using your answers from either Exercise 39 or Exercise 40.

42. Find the amount of each ingredient needed for nine servings, using your answers from either Exercise 40 or Exercise 41.

6.5 Geometry: Lines and Angles

Geometry starts with the idea of a point. A **point** can be described as a location in space. It has no length or width. A point is represented by a dot and is named by writing a capital letter next to the dot.

Point P

OBJECTIVE 1 Identify and name lines, line segments, and rays. A **line** is a straight row of points that goes on forever in both directions. A line is drawn by using arrowheads to show that it never ends. The line is named by using the letters of any two points on the line.

Line AB, written $\overleftrightarrow{AB}$

A piece of a line that has two endpoints is called a **line segment.** A line segment is named for its endpoints. The segment with endpoints P and Q is shown below. It can be named $\overline{PQ}$ or $\overline{QP}$.

Line segment PQ, written $\overline{PQ}$

A **ray** is a part of a line that has only one endpoint and goes on forever in one direction. A ray is named by using the endpoint and some other point on the ray. The endpoint is always mentioned first.

Ray RS, written $\overrightarrow{RS}$

EXAMPLE 1 Identifying and Naming Lines, Rays, and Line Segments

Identify each figure below as a line, line segment, or ray and name it using the appropriate symbol.

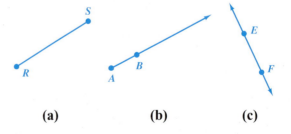

(a)   (b)   (c)

Figure **(a)** has two endpoints, so it is a *line segment* named $\overline{RS}$ or $\overline{SR}$.

Figure **(b)** starts at point A and goes on forever in one direction, so it is a *ray* named $\overrightarrow{AB}$.

Figure **(c)** goes on forever in both directions, so it is a *line* named $\overleftrightarrow{EF}$ or $\overleftrightarrow{FE}$.

Now Try Exercise 5.

OBJECTIVE **2** **Identify parallel and intersecting lines.**   A *plane* is an infinitely large, flat surface. A floor or a wall is part of a plane. Lines that are in the *same plane,* but that never intersect (never cross), are called **parallel lines,** while lines that cross are called **intersecting lines.** (Think of an intersection, where two streets cross each other.)

EXAMPLE 2   **Identifying Parallel and Intersecting Lines**

Label each pair of lines as appearing to be parallel or as intersecting.

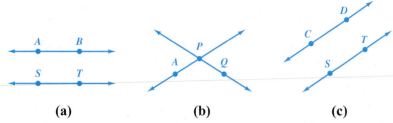

(a)   (b)   (c)

The lines in Figures **(a)** and **(c)** do not intersect; they appear to be *parallel lines.*

The lines in Figure **(b)** cross at *P,* so they are *intersecting lines.*

> **CAUTION**
> Appearances may be deceiving! Do not assume that lines are parallel unless it is stated that they are parallel.

Now Try Exercise 9. ▶▶▶

OBJECTIVE **3** **Identify and name angles.**   An **angle** is made up of two rays that start at a common endpoint. This common endpoint is called the *vertex.*

$\overrightarrow{PQ}$ and $\overrightarrow{PR}$ are called the *sides* of the angle. The angle can be named in four different ways, as shown below.

$$\angle 1 \qquad \angle P \qquad \angle QPR \qquad \angle RPQ$$

Vertex alone   Vertex in the middle

Naming an Angle
To name an angle, write the vertex alone or write the vertex in the middle of two other points, one from each side. If two or more angles have the *same vertex,* as in Example 3 on the next page, do *not* use the vertex alone to name an angle.

EXAMPLE 3   **Identifying and Naming an Angle**

Name the highlighted angle.

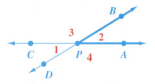

The angle can be named ∠*BPA*, ∠*APB*, or ∠2. It cannot be named ∠*P*, using the vertex alone, because four different angles have *P* as their vertex.

Now Try Exercise 13.

OBJECTIVE 4   Classify angles as right, acute, straight, or obtuse. Angles can be measured in **degrees.** The symbol for degrees is a small, raised circle °. Think of the minute hand on a clock as a ray of an angle. Suppose it is at 12:00. During one hour of time, the minute hand moves around in a complete circle. It moves 360 *degrees*, or 360°. In half an hour, at 12:30, the minute hand has moved halfway around the circle, or 180°. An angle of 180° is called a **straight angle.** When two rays go in opposite directions and form a straight line, then the rays form a straight angle.

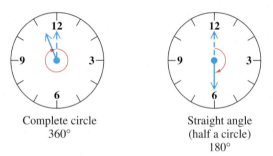

Complete circle
360°

Straight angle
(half a circle)
180°

In a quarter of an hour, at 12:15, the minute hand has moved $\frac{1}{4}$ of the way around the circle, or 90°. An angle of 90° is called a **right angle.** The rays of a right angle form one corner of a square. So, to show that an angle is a **right angle**, we draw a **small square** at the vertex.

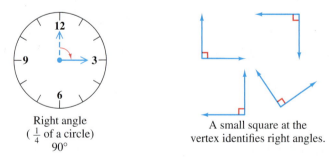

Right angle
($\frac{1}{4}$ of a circle)
90°

A small square at the
vertex identifies right angles.

An angle that measures 1° is shown below. You can see that an angle of 1° is very small.

1° angle

Some other terms used to describe angles are shown below.

Acute angles measure less than 90°.

Examples of acute angles

Obtuse angles measure more than 90° but less than 180°.

Examples of obtuse angles

Section 9.2 shows you how to use a tool called a *protractor* to measure the number of degrees in an angle.

Classifying Angles

Acute angles measure less than 90°.

Right angles measure *exactly* 90°.

Obtuse angles measure more than 90° but less than 180°.

Straight angles measure *exactly* 180°.

NOTE
Angles can also be measured in radians, which you will learn about in a later math course.

EXAMPLE 4 **Classifying an Angle**

Label each angle as acute, right, obtuse, or straight.

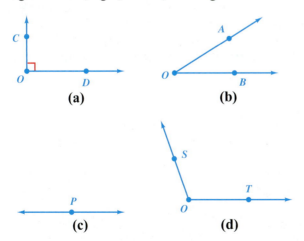

(a)   (b)   (c)   (d)

Figure **(a)** shows a *right angle* (exactly 90° and identified by a small square at the vertex).

Figure **(b)** shows an *acute angle* (less than 90°).

Figure **(c)** shows a *straight angle* (exactly 180°).

Figure **(d)** shows an *obtuse angle* (more than 90° but less than 180°).

Now Try Exercises 21 and 23.

OBJECTIVE **5** **Identify perpendicular lines.** Two lines are called **perpendicular lines** if they intersect to form a right angle.

$\overleftrightarrow{CB}$ and $\overleftrightarrow{ST}$ are **perpendicular** lines because they intersect at right angles, indicated by the small red square in the figure. Perpendicular lines can be written in the following way: $\overleftrightarrow{CB} \perp \overleftrightarrow{ST}$.

EXAMPLE 5 **Identifying Perpendicular Lines**

Which pairs of lines are perpendicular?

(a) (b) (c)

The lines in Figures **(b)** and **(c)** are *perpendicular* to each other, because they intersect at right angles.

The lines in Figure **(a)** are *intersecting lines,* but they are *not* perpendicular because they do not form a right angle.

> **Now Try Exercises 7 and 11.**

OBJECTIVE **6** **Identify complementary angles and supplementary angles and find the measure of a complement or supplement of a given angle.** Two angles are called **complementary angles** if the sum of their measures is 90°. If two angles are complementary, each angle is the *complement* of the other.

EXAMPLE 6 **Identifying Complementary Angles**

Identify each pair of complementary angles.

$\angle MPN$ (40°) and $\angle NPC$ (50°) are complementary angles because

$$40° + 50° = \mathbf{90°}$$

$\angle CAB$ (30°) and $\angle FHG$ (60°) are complementary angles because

$$30° + 60° = \mathbf{90°}$$

> **Now Try Exercise 25.**

EXAMPLE 7 Finding the Complement of Angles

Find the complement of each angle.

(a) $30°$
Find the complement of $30°$ by subtracting. $90° - 30° = \mathbf{60°}$ ← Complement

(b) $75°$
Find the complement of $75°$ by subtracting. $90° - 75° = \mathbf{15°}$ ← Complement

Now Try Exercise 29. ▶▶▶

Two angles are called **supplementary angles** if the sum of their measures is $180°$. If two angles are supplementary, each angle is the *supplement* of the other.

EXAMPLE 8 Identifying Supplementary Angles

Identify each pair of supplementary angles.

$\angle BOA$ and $\angle BOC$, because $65° + 115° = \mathbf{180°}$
$\angle BOA$ and $\angle ERF$, because $65° + 115° = \mathbf{180°}$
$\angle BOC$ and $\angle MPN$, because $115° + \ \ 65° = \mathbf{180°}$
$\angle MPN$ and $\angle ERF$, because $65° + 115° = \mathbf{180°}$

Now Try Exercise 27. ▶▶▶

EXAMPLE 9 Finding the Supplement of Angles

Find the supplement of each angle.

(a) $70°$
Find the supplement of $70°$ by subtracting. $180° - 70° = \mathbf{110°}$ ← Supplement

(b) $140°$
Find the supplement of $140°$ by subtracting. $180° - 140° = \mathbf{40°}$ ← Supplement

Now Try Exercise 33. ▶▶▶

OBJECTIVE 7 Identify congruent angles and vertical angles and use this knowledge to find the measures of angles.   Two angles are called **congruent angles** if they measure the same number of degrees. If two angles are congruent, this is written as $\angle A \cong \angle B$ and read as, "angle A **is congruent to** angle B." Here is an example.

$\angle A \cong \angle B$

Example of congruent angles

The figure below has two pairs of angles that are congruent.

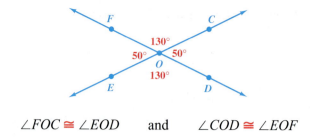

$$\angle FOC \cong \angle EOD \quad \text{and} \quad \angle COD \cong \angle EOF$$

Angles that share a common side and a common vertex are called *adjacent* angles, such as $\angle FOC$ and $\angle COD$ in the figure above. Angles that do *not* share a common side are called *nonadjacent* angles. Two nonadjacent angles formed by two intersecting lines are called **vertical angles.**

> **EXAMPLE 10** Identifying Vertical Angles
>
> Identify the vertical angles in this figure.
>
>
>
> | |
> |---|
>
> $\angle AOF$ and $\angle COE$ are vertical angles because they do *not* share a common side and they are formed by two intersecting lines ($\overleftrightarrow{CF}$ and $\overleftrightarrow{EA}$).
>
> $\angle COA$ and $\angle EOF$ are also vertical angles.
>
>

Look back at the figure above Example 10. Notice that the two *congruent* angles that measure 130° are also *vertical* angles. Also, the two congruent angles that measure 50° are vertical angles. This illustrates the following property.

> **Vertical Angles Are Congruent**
>
> If two angles are *vertical* angles, they are *congruent;* that is, they measure the same number of degrees.

Now Try Exercise 37.

EXAMPLE 11 **Finding the Measures of Vertical Angles**

In the figure below, find the measure of each unlabeled angle.

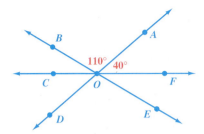

(a) $\angle COD$

$\angle COD$ and $\angle AOF$ are vertical angles, so they are congruent. This means they measure the same number of degrees.

The measure of $\angle AOF$ is 40°   so   the measure of $\angle COD$ is **40°** also.

(b) $\angle DOE$

$\angle DOE$ and $\angle BOA$ are vertical angles, so they are congruent.

The measure of $\angle BOA$ is 110°   so   the measure of $\angle DOE$ is **110°** also.

(c) $\angle COB$

Look at $\angle COB$, $\angle BOA$, and $\angle AOF$. Notice that $\overrightarrow{OC}$ and $\overrightarrow{OF}$ go in opposite directions. Therefore, $\angle COF$ is a straight angle and measures 180°. To find the measure of $\angle COB$, subtract the sum of the other two angles from 180°.

$$180° - (110° + 40°) = 180° - (150°) = 30°$$

The measure of $\angle COB$ is **30°**.

(d) $\angle EOF$

$\angle EOF$ and $\angle COB$ are vertical angles, so they are congruent. We know from part (c) above that the measure of $\angle COB$ is 30° so the measure of $\angle EOF$ is **30°** also.

Now Try Exercise 39.

OBJECTIVE 8 Identify corresponding angles and alternate interior angles and use this knowledge to find the measures of angles.   We can also find congruent angles (angles with the same measure) when two *parallel lines* are crossed by a third line, called a *transversal*. When a transversal crosses two *parallel* lines, eight angles are formed, as shown below. There are special names for certain pairs of angles.

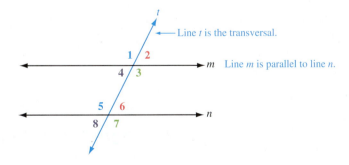

$\angle 1$ and $\angle 5$ are called **corresponding angles.** Notice that they are both on the same side of the transversal (line t) and in the same relative position. *Corresponding angles are congruent*, so $\angle 1$ and $\angle 5$ measure the same number of degrees. There are four pairs of corresponding angles.

∠1 and ∠5 are corresponding angles, so ∠1 ≅ ∠5.

∠2 and ∠6 are corresponding angles, so ∠2 ≅ ∠6.

∠3 and ∠7 are corresponding angles, so ∠3 ≅ ∠7.

∠4 and ∠8 are corresponding angles, so ∠4 ≅ ∠8.

When a transversal crosses two parallel lines, angles 3, 4, 5, and 6 are called *interior angles*. You can see that they are "inside" the *parallel* lines.

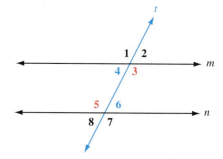

∠3 and ∠5 are alternate interior angles.

∠4 and ∠6 are alternate interior angles.

When two lines are *parallel*, then **alternate interior angles** *are congruent* (they have the same measure). Notice that alternate interior angles are on opposite sides of the transversal.

$$∠3 ≅ ∠5 \qquad \text{and} \qquad ∠4 ≅ ∠6$$

Angles Formed by Parallel Lines and a Transversal

When two parallel lines are crossed by a transversal:
1. Corresponding angles are congruent, and
2. Alternate interior angles are congruent.

EXAMPLE 12 Identifying Corresponding Angles and Alternate Interior Angles

In each figure, line *m* is parallel to line *n*. Identify all pairs of corresponding angles and all pairs of alternate interior angles.

(a)

(b)

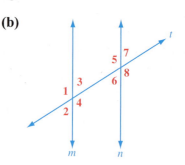

There are four pairs of corresponding angles:

∠5 and ∠3   ∠6 and ∠4

∠1 and ∠7   ∠2 and ∠8

Alternate interior angles:

∠1 and ∠4   ∠2 and ∠3

Corresponding angles:

∠1 and ∠5   ∠3 and ∠7

∠2 and ∠6   ∠4 and ∠8

Alternate interior angles:

∠3 and ∠6   ∠4 and ∠5

Now Try Exercise 47.

Recall that two angles are supplementary angles if the sum of their measures is 180°. Also remember that the two rays that form a 180° angle form a straight line. Now you can combine your knowledge about supplementary angles with the information on parallel lines.

EXAMPLE 13 Working with Parallel Lines

In the figure at the right, line *m* is parallel to line *n* and the measure of ∠4 is 70°. Find the measures of the other angles.

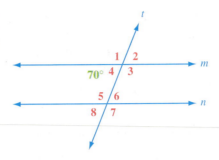

As you find the measure of each angle, write it on the figure.

∠4 ≅ ∠8 (corresponding angles), so the measure of ∠8 is also 70°.
∠4 ≅ ∠6 (alternate interior angles), so the measure of ∠6 is also 70°.
∠6 ≅ ∠2 (corresponding angles), so the measure of ∠2 is also 70°.

Notice that the exterior sides of ∠4 and ∠3 form a straight line, that is, a straight angle of 180°. Therefore, ∠4 and ∠3 are supplementary angles and the sum of their measures is 180°. If ∠4 is 70° then ∠3 must be 110° because 180° − 70° = 110°. So the measure of ∠3 is 110°.

∠3 ≅ ∠7 (corresponding angles), so the measure of ∠7 is also 110°.
∠3 ≅ ∠5 (alternate interior angles), so the measure of ∠5 is also 110°.
∠5 ≅ ∠1 (corresponding angles), so the measure of ∠1 is also 110°.

With the measures of all the angles labeled, you can double-check that each pair of angles that forms a straight angle also adds up to 180°.

Now Try Exercise 49. ▶▶▶

6.5 Exercises

 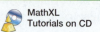
Identify each figure as a line, line segment, *or* ray *and name it using the appropriate symbol. See Example 1.*

1.

2.

3.

4.

5.

6.

Label each pair of lines as appearing to be parallel, *as* perpendicular, *or as* intersecting. *See Examples 2 and 5.*

7.

8.

9.

10.

11.

12.

Name each highlighted angle by using the three-letter form of identification. See Example 3.

13.

14.

15.

16.

17.

18.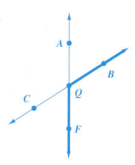

Label each angle as acute, right, obtuse, *or* straight. *For right angles and straight angles, indicate the number of degrees in the angle. See Example 4.*

19.

20.

21.

22.

23.

24.

Identify each pair of complementary angles. See Example 6.

25.

26.

Identify each pair of supplementary angles. See Example 8.

27.

28.

Find the complement of each angle. See Example 7.

29. 40°   **30.** 35°   **31.** 86°   **32.** 59°

Find the supplement of each angle. See Example 9.

33. 130°   **34.** 75°   **35.** 90°   **36.** 5°

In Exercises 37 and 38, identify the angles that are congruent. See Example 10.

37.

38.

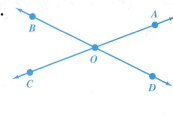

In Exercises 39 and 40, find the measure of each of the angles. See Example 11.

39. In the figure below, ∠*AOH* measures 37° and ∠*COE* measures 63°. Find the measure of each of the other angles.

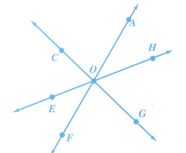

40. In the figure below, ∠*POU* measures 105° and ∠*UOT* measures 40°. Find the measure of each of the other angles.

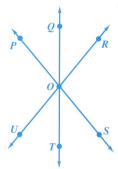

RELATING CONCEPTS (EXERCISES 41–46) For Individual or Group Work

*Use the figure to **work Exercises 41–46 in order.** Decide whether each statement is* true *or* false. *If it is true, explain why. If it is false, rewrite it to make a true statement.*

41. ∠*UST* is 90°.

42. $\overleftrightarrow{SQ}$ and $\overleftrightarrow{PQ}$ are perpendicular.

43. The measure of ∠*USQ* is less than the measure of ∠*PQR*.

44. $\overleftrightarrow{ST}$ and $\overleftrightarrow{PR}$ are intersecting.

45. $\overleftrightarrow{QU}$ and $\overleftrightarrow{TS}$ are parallel.

46. ∠*UST* and ∠*UQR* measure the same number of degrees.

In each figure, line m is parallel to line n. Identify all pairs of corresponding angles and all pairs of alternate interior angles. See Example 12.

47.

48.

In each figure, line m is parallel to line n. Find the measure of each angle. See Example 13.

49. ∠8 measures 130°.

50. ∠2 measures 80°.

51. ∠6 measures 47°.

52. ∠2 measures 108°.

53. ∠6 measures 114°.

54. ∠3 measures 59°.

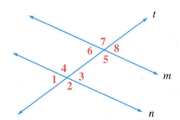

In each figure, $\overrightarrow{BA}$ is parallel to $\overrightarrow{CD}$. Find the measure of each numbered angle.

55.

56.

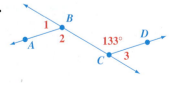

6.6 Geometry Applications: Congruent and Similar Triangles

Two useful concepts in geometry are *congruence* and *similarity*. If two figures are *identical*, both in *shape* and in *size*, we say the figures are **congruent.** In other words, the figures are perfect duplicates of each other, like getting two prints from the same negative on a roll of film. If two figures have the *same shape* but are *different sizes,* we say the figures are **similar,** like getting a print from a roll of film and then an enlargement of the same print. We'll explore the ideas of congruence and similarity using triangles.

OBJECTIVE 1 Identify corresponding parts of congruent triangles. The two triangles shown below are *congruent* because they are the *same shape* and the *same size.*

One way to name this triangle is △ABC.

One way to name this triangle is △DEF.

Suppose you picked up △ABC and slid it over on top of △DEF. You would see that the two triangles are a perfect match. ∠1 would be on top of ∠4, so they are called *corresponding angles.* Similarly, ∠2 and ∠5 are corresponding angles, and ∠3 and ∠6 are corresponding angles. You would see that corresponding angles have the same measure, as indicated below.

$$m\angle 1 = m\angle 4 \qquad m\angle 2 = m\angle 5 \qquad m\angle 3 = m\angle 6$$

The abbreviation for measure is *m*, so $m\angle 1$ is read, "the measure of angle 1."

When you put △ABC on top of △DEF, you would also see that side AB is on top of side DE. We say that $\overline{AB}$ and $\overline{DE}$ are *corresponding sides.* Similarly, $\overline{BC}$ and $\overline{EF}$ are corresponding sides, and $\overline{AC}$ and $\overline{DF}$ are corresponding sides. You would see that corresponding sides have the same length.

$$AB = DE \qquad BC = EF \qquad AC = DF$$

Because corresponding angles have the same measure, and corresponding sides have the same length, we know that △ABC **is congruent to** △DEF. We can write this as △ABC ≅ △DEF

> **Congruent Triangles**
>
> If two triangles are congruent, then
> 1. Corresponding angles have the same measure, and
> 2. Corresponding sides have the same length.

EXAMPLE 1 Identifying Corresponding Parts in Congruent Triangles

Each pair of triangles is congruent. List the corresponding angles and corresponding sides.

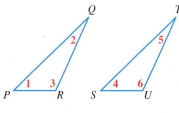

(a) If you picked up △PQR and slid it over on top of △STU, the two triangles would match.
The corresponding parts are:

∠1 and ∠4 $\overline{PQ}$ and $\overline{ST}$

∠2 and ∠5 $\overline{PR}$ and $\overline{SU}$

∠3 and ∠6 $\overline{QR}$ and $\overline{TU}$

Continued on Next Page

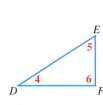

(b) If you picked up $\triangle ABC$ and slid it over on top of $\triangle DEF$, it wouldn't match. But if you *rotate* $\triangle ABC$ before sliding it on top of $\triangle DEF$, it *will* match.

The corresponding parts are:

$\angle 1$ and $\angle 6$	$\overline{BC}$ and $\overline{DE}$
$\angle 2$ and $\angle 4$	$\overline{BA}$ and $\overline{DF}$
$\angle 3$ and $\angle 5$	$\overline{CA}$ and $\overline{EF}$

Now Try Exercise 1. ▶▶▶

OBJECTIVE 2 Prove that triangles are congruent using SAS, SSS, and ASA.   One way to prove that two triangles are congruent would be to measure all the angles and all the sides. If the measures of the corresponding angles and sides are equal, then the triangles are congruent. But here are three quicker methods to prove that two triangles are congruent.

Proving That Two Triangles Are Congruent

1. Angle–Side–Angle (ASA) Method
If two angles and the side between them on one triangle measure the same as the corresponding parts on another triangle, the triangles are congruent.

If $m\angle 1 = m\angle 3$ and $m\angle 2 = m\angle 4$ and $a = x$ then the two triangles are congruent.

2. Side–Side–Side (SSS) Method
If three sides of one triangle measure the same as the corresponding sides of another triangle, the triangles are congruent.

If $a = x$ and $b = y$ and $c = z$ then the two triangles are congruent.

3. Side–Angle–Side (SAS) Method
If two sides and the angle between them on one triangle measure the same as the corresponding parts on another triangle, the triangles are congruent.

If $a = x$ and $b = y$ and $m\angle 1 = m\angle 2$ then the two triangles are congruent.

EXAMPLE 2 Proving That Two Triangles Are Congruent

Explain which method can be used to prove that each pair of triangles is congruent. Choose from ASA, SSS, and SAS.

(a)

(b)

(c)

(a) On both triangles, two corresponding sides and the angle between them measure the same, so the Side–Angle–Side (SAS) method can be used to prove that the triangles are congruent.

(b) On both triangles, two corresponding angles and the side between them measure the same, so the Angle–Side–Angle (ASA) method can be used to prove that the triangles are congruent.

(c) Each pair of corresponding sides has the same length, so the Side–Side–Side (SSS) method can be used to prove that the triangles are congruent.

> **Now Try Exercise 11.**

OBJECTIVE 3 Identify corresponding parts of similar triangles. Now that you've worked with *congruent* triangles, let's look at *similar* triangles. Remember that congruent triangles match exactly, both in shape and in size. Similar triangles, on the other hand, have the same shape but are *different sizes*. Three pairs of similar triangles are shown here.

Each pair of triangles has the same shape because the corresponding angles have the same measure. But the corresponding sides are *not* the same length, so the triangles are of *different sizes*.

Two similar triangles are shown to the right. Notice that corresponding angles have the same measure, but corresponding sides have different lengths.

$\overline{CB}$ corresponds to $\overline{RQ}$. Similarly, $\overline{CA}$ corresponds to $\overline{RP}$, and $\overline{BA}$ corresponds to $\overline{QP}$. Notice that each side in the larger triangle is *twice* the length of the corresponding side in the smaller triangle.

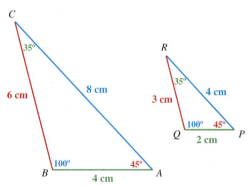

OBJECTIVE **4** **Find the unknown lengths of sides in similar triangles.**
Similar triangles are useful because of the following definition.

> **Similar Triangles**
>
> If two triangles are similar, then
> 1. Corresponding angles have the same measure, and
> 2. The ratios of the lengths of corresponding sides are equal.

EXAMPLE 3 **Finding the Unknown Lengths of Sides in Similar Triangles**

Find the length of $\overline{DF}$ in the smaller triangle. Assume the triangles are similar.

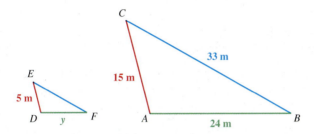

The length you want to find in the smaller triangle is $\overline{DF}$, and it corresponds to $\overline{AB}$ in the larger triangle. Then, notice that $\overline{ED}$ in the smaller triangle corresponds to $\overline{CA}$ in the larger triangle, and you know both of their lengths. Since the *ratios* of the lengths of corresponding sides are equal, you can set up a proportion. (Recall that a proportion states that two ratios are equal.)

$$\text{Corresponding sides} \begin{cases} DF \to \dfrac{y}{24} \leftarrow AB \end{cases} = \dfrac{5}{15} \begin{cases} \leftarrow ED \\ \leftarrow CA \end{cases} \text{Corresponding sides}$$

$$\frac{y}{24} = \frac{1}{3} \qquad \text{Write } \tfrac{5}{15} \text{ in lowest terms as } \tfrac{1}{3}$$

Find the cross products.

$$\frac{y}{24} = \frac{1}{3} \qquad \begin{array}{l} 24 \cdot 1 = 24 \\ \\ y \cdot 3 \end{array}$$

$$y \cdot 3 = 24 \qquad \text{Show that cross products are equivalent.}$$

$$\frac{y \cdot \overset{1}{\cancel{3}}}{\underset{1}{\cancel{3}}} = \frac{24}{3} \qquad \text{Divide both sides by 3.}$$

$$y = 8$$

$\overline{DF}$ has a length of 8 m.

Now Try Exercise 21. ▶▶▶

EXAMPLE 4 **Finding an Unknown Length and the Perimeter**

Find the perimeter of the smaller triangle. Assume the triangles are similar.

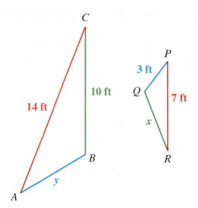

First find x, the length of $\overline{RQ}$ in the smaller triangle, then add the lengths of all three sides to find the perimeter.

The smaller triangle is turned "upside down" compared to the larger triangle, so be careful when identifying corresponding sides. $\overline{PR}$ is the longest side in the smaller triangle, and $\overline{AC}$ is the longest side in the larger triangle. So $\overline{PR}$ and $\overline{AC}$ are corresponding sides and you know both of their lengths. $\overline{QR}$, the length you want to find in the smaller triangle, corresponds to $\overline{BC}$ in the larger triangle. The ratios of the lengths of corresponding sides are equal, so you can set up a proportion.

$$\begin{array}{c} QR \rightarrow \\ BC \rightarrow \end{array} \dfrac{x}{10} = \dfrac{7}{14} \begin{array}{c} \leftarrow PR \\ \leftarrow AC \end{array}$$

$$\dfrac{x}{10} = \dfrac{1}{2} \quad \text{Write } \tfrac{7}{14} \text{ in lowest terms as } \tfrac{1}{2}$$

Find the cross products.

$$\dfrac{x}{10} \bowtie \dfrac{1}{2} \qquad \begin{array}{l} 10 \cdot 1 = 10 \\ \\ x \cdot 2 \end{array}$$

$$x \cdot 2 = 10 \qquad \text{Show that cross products are equivalent.}$$

$$\dfrac{x \cdot \overset{1}{\cancel{2}}}{\cancel{2}_{1}} = \dfrac{10}{2} \qquad \text{Divide both sides by 2.}$$

$$x = 5$$

$\overline{RQ}$ has a length of 5 ft.

Now add the lengths of all three sides to find the perimeter of the smaller triangle.

$$\text{Perimeter} = 5 \text{ ft} + 3 \text{ ft} + 7 \text{ ft} = 15 \text{ ft}$$

Now Try Exercise 23.

OBJECTIVE **5** **Solve application problems involving similar triangles.**
The next example shows an application of similar triangles.

EXAMPLE 5 **Using Similar Triangles in an Application**

A flagpole casts a shadow 99 m long at the same time that a pole 10 m tall casts a shadow 18 m long. Find the height of the flagpole.

The triangles shown are similar, so write a proportion to find h.

Height in larger triangle → $\dfrac{h}{10} = \dfrac{99}{18}$ ← Shadow in larger triangle
Height in smaller triangle → ← Shadow in smaller triangle

Find the cross products and show that they are equivalent.

$$h \cdot 18 = 10 \cdot 99$$
$$h \cdot 18 = 990$$

$$\frac{h \cdot \cancel{18}}{\cancel{18}} = \frac{990}{18} \qquad \text{Divide both sides by 18.}$$

$$h = 55$$

The flagpole is 55 m high.

> **NOTE**
> There are several other correct ways to set up the proportion in Example 5 above. One way is to simply flip the ratios on *both* sides of the equal sign.
>
> $$\frac{10}{h} = \frac{18}{99}$$
>
> But there is another option, shown below.
>
> Height in larger triangle → $\dfrac{h}{99} = \dfrac{10}{18}$ ← Height in smaller triangle
> Shadow in larger triangle → ← Shadow in smaller triangle
>
> Notice that both ratios compare *height* to *shadow* in the same order. The ratio on the left describes the larger triangle, and the ratio on the right describes the smaller triangle.

Now Try Exercise 27. ▷▷▷

6.6 Exercises

Each pair of triangles is congruent. List the corresponding angles and the corresponding sides. See Example 1.

1.

2.

3.

4.

5.

6.
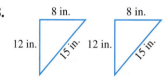

Determine which of these methods can be used to prove that each pair of triangles is congruent: Angle–Side–Angle (ASA), Side–Side–Side (SSS), or Side–Angle–Side (SAS). See Example 2.

7.

8.

9.

10.

11.

12.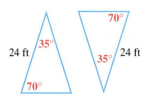

Work Exercises 13–16 in order. *Given the information in each exercise, explain how you can prove that the indicated triangles are congruent. Note: A midpoint divides a segment into two congruent parts.*

13.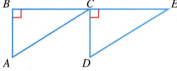

C is the midpoint of $\overline{BE}$ and $CD = BA$.
Prove that $\triangle ABC \cong \triangle DCE$.

14.

P is the midpoint of both $\overline{WY}$ and $\overline{XZ}$; $WZ = XY$.
Prove that $\triangle WPZ \cong \triangle YPX$.

15.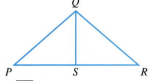

$\overline{QS} \perp \overline{PR}$ and S is the midpoint of $\overline{PR}$. Prove that
$\triangle PQS \cong \triangle RQS$.

16.

M is the midpoint of both $\overline{LO}$ and $\overline{PN}$.
Prove that $\triangle PLM \cong \triangle NOM$.

Find the unknown lengths in each pair of similar triangles. See Example 3.

17.

18.

19.

20.

21.

22.

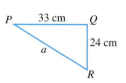

In Exercises 23 and 24, find the perimeter of each triangle. Assume the triangles are similar. See Example 4.

23.

24.

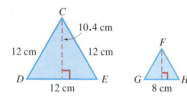

25. Triangles *CDE* and *FGH* are similar. Find the perimeter and area of triangle *FGH*. *Note:* The heights of similar triangles have the same ratio as corresponding sides. Round to the nearest tenth when necessary.

26. Triangles *JKL* and *MNO* are similar. Find the perimeter and area of triangle *MNO*.

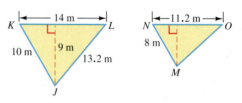

Solve each application problem. See Example 5.

27. The height of the house shown here can be found by comparing its shadow to the shadow cast by a 3-foot stick. Find the height of the house by writing a proportion and solving it.

28. A fire lookout tower provides an excellent view of the surrounding countryside. The height of the tower can be found by lining up the top of the tower with the top of a 2-meter stick. Use similar triangles to find the height of the tower.

29. Look up the word *similar* in a dictionary. What is the nonmathematical definition of this word? Describe two examples of similar objects at home, school, or work.

30. Look up the word *congruent* in a dictionary. What is the nonmathematical definition of this word? Describe two examples of congruent objects at home, school, or work.

Find the unknown length in Exercises 31–34. Round your answers to the nearest tenth. Note: When a line is drawn parallel to one side of a triangle, the smaller triangle that is formed will be similar to the original triangle. In Exercises 31–32, the red segments are parallel.

31.

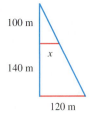

100 m

140 m

x

120 m

Hint: Redraw the two triangles and label the sides.

100 m

x

100 m + 140 m = 240 m

120 m

32.

c

50 in.

5 in.   45 in.

Hint: Redraw the two triangles and label the sides.

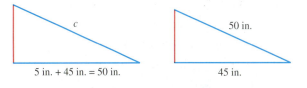

c

5 in. + 45 in. = 50 in.

50 in.

45 in.

33. Use similar triangles and a proportion to find the length of the lake shown here. (*Hint:* The side 100 m long in the smaller triangle corresponds to the side of 100 m + 120 m = 220 m in the larger triangle.)

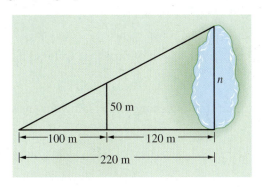

n

50 m

100 m   120 m

220 m

34. To find the height of the tree, find *y* and then add $5\frac{1}{2}$ ft for the distance from the ground to the eye level of the person.

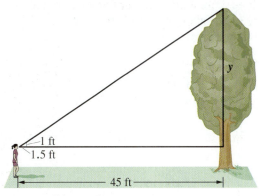

y

1 ft

1.5 ft

45 ft

Distance from person to tree.

Chapter 6   Group Activity: FEEDING HUMMINGBIRDS

📠 **Objective**   **Apply ratios and solve proportions**

A recipe can be used to make as much of a mixture as you might need as long as the ingredients are kept proportional. Use the recipe for a homemade mixture of sugar water for hummingbird feeders to answer these problems.

1. What is the ratio of sugar to water in the recipe?

   What is the ratio of water to sugar in the recipe?

2. Complete each table.

Sugar	Water
1 cup	4 cups
	5 cups
	6 cups
	7 cups
2 cups	8 cups

Sugar	Water
1 cup	4 cups
	3 cups
	2 cups
	1 cup

3. How much water would you need
   (a) if you wanted to use 3 cups of sugar?

   (b) if you wanted to use 4 cups of sugar?

   (c) if you wanted to use $\frac{1}{3}$ cup of sugar?

4. One cup of sugar weighs about 200 grams and one cup of water weighs about 235 grams. If you mix 1 cup of sugar and 4 cups of water, what is the approximate weight of the resulting mixture?

Feeding Hummingbirds

After getting a hummingbird feeder, the next step is to fill it! You have two choices at this point: you can either buy one of the commercial mixtures or you can make your own solution:

> **Recipe for Homemade Mixture:**
> **1 part sugar (not honey)**
> **4 parts water**
> **Boil for 1 to 2 minutes. Cool.**
> **Store extra in refrigerator.**

The concentration of the sugar is important. A 1 to 4 ratio of sugar to water is recommended because it approximates the ratio of sugar to water found in the nectar of many hummingbird flowers. A recent study of native California wildflowers visited by hummingbirds showed that their nectar had an average sugar concentration of 21 percent. This is sweet enough to attract the hummers without being too sweet. If you increase the concentration of sugar, it may be harder for the birds to digest; if you decrease the concentration, they may lose interest.

Boiling the solution helps retard fermentation. Sugar-and-water solutions are subject to rapid spoiling, especially in hot weather.

Source: The Hummingbird Book.

---

5. The article says that the nectar from wildflowers visited by hummingbirds has an average sugar concentration of 21 percent. That represents a ratio of 21 to 100. If you wanted to mix a sugar solution in the same proportion as the wildflower nectar, how much water should you use for 1 cup of sugar? What proportion did you set up to solve this problem?

6. As you change the amounts of water and sugar, should you change the length of time that you boil the mixture? Explain your answer.

7. Will the length of time it takes to get the water hot enough to start boiling change? Explain your answer.

Chapter 6

SUMMARY

KEY TERMS

6.1	ratio	A ratio compares two quantities that have the same type of units. For example, the ratio of 6 apples to 11 apples is written in fraction form as $\frac{6}{11}$. The common units (apples) divide out.
6.2	rate	A rate compares two measurements with different types of units. Examples are 96 dollars for 8 hours, or 450 miles on 18 gallons.
	unit rate	A unit rate has 1 in the denominator.
	cost per unit	Cost per unit is a rate that tells how much you pay for one item or one unit. The lowest cost per unit is the best buy.
6.3	proportion	A proportion states that two ratios or rates are equivalent.
	cross products	Multiply along one diagonal and then along the other diagonal to find the cross products of a proportion. If the cross products are equal, the proportion is true.
6.5	point	A point is a location in space. *Example:* Point *P* at the right.
	line	A line is a straight row of points that goes on forever in both directions. *Example:* Line *AB*, written $\overleftrightarrow{AB}$, at the right.
	line segment	A line segment is a piece of a line with two endpoints. *Example:* Line segment *PQ*, written $\overline{PQ}$, at the right.
	ray	A ray is a part of a line that has one endpoint and extends forever in one direction. *Example:* Ray *RS*, written $\overrightarrow{RS}$, at the right.
	parallel lines	Parallel lines are two lines in the same plane that never intersect (never cross). *Example:* $\overleftrightarrow{AB}$ is parallel to $\overleftrightarrow{ST}$ at the right.
	intersecting lines	Intersecting lines cross. *Example:* $\overleftrightarrow{RQ}$ intersects $\overleftrightarrow{AB}$ at point *P* at the right.
	angle	An angle is made up of two rays that have a common endpoint called the vertex. *Example:* Angle 1 at the right.
	degrees	Degrees are used to measure angles; a complete circle is 360 degrees, written 360°.
	straight angle	A straight angle is an angle that measures exactly 180°; its sides form a straight line. *Example:* Angle *G* at the right.
	right angle	A right angle is an angle that measures exactly 90°. *Example:* Angle *AOB* at the right.
	acute angle	An acute angle is an angle that measures less than 90°. *Example:* Angle *E* at the right.
	obtuse angle	An obtuse angle is an angle that measures more than 90° but less than 180°. *Example:* Angle *F* at the right.
	perpendicular lines	Perpendicular lines are two lines that intersect to form a right angle. *Example:* $\overleftrightarrow{PQ}$ is perpendicular to $\overleftrightarrow{RS}$ at the right.

(continued)

KEY TERMS

complementary angles	Complementary angles are two angles whose measures add up to 90°.
supplementary angles	Supplementary angles are two angles whose measures add up to 180°.
congruent angles	Congruent angles are angles that measure the same number of degrees.
vertical angles	Vertical angles are two nonadjacent congruent angles formed by intersecting lines. *Example:* $\angle COA$ and $\angle EOF$ are vertical angles at the right.
corresponding angles	Corresponding angles are formed when two parallel lines are crossed by a transversal; corresponding angles are congruent and are on the same side of the transversal and in the same relative position. *Example:* In the figure at the right, line m is parallel to line n. The pairs of corresponding angles are $\angle 1$ and $\angle 5$, $\angle 2$ and $\angle 6$, $\angle 3$ and $\angle 7$, $\angle 4$ and $\angle 8$.
alternate interior angles	When two parallel lines are crossed by a transversal, there are two pairs of alternate interior angles and each pair is congruent. They are on opposite sides of the transversal. *Example:* In the figure at the right, line m is parallel to line n. The pairs of alternate interior angles are $\angle 3$ and $\angle 5$, $\angle 4$ and $\angle 6$.

6.6 congruent figures	Congruent figures are identical both in shape and in size.
similar figures	Similar figures have the same shape but are different sizes.
congruent triangles	Congruent triangles are triangles with the same shape and the same size; corresponding angles measure the same number of degrees and corresponding sides have the same length.
similar triangles	Similar triangles are triangles with the same shape but not necessarily the same size; corresponding angles measure the same number of degrees and the *ratios* of the lengths of corresponding sides are equal.

NEW SYMBOLS

$\overleftrightarrow{AB}$	line AB		$\perp$	is perpendicular to		$\cong$	is congruent to
$\overline{EF}$	line segment EF		$\angle MRN$	angle MRN		$m\angle 2$	the measure of angle 2
$\overrightarrow{RS}$	ray RS		$1°$	one degree		$\triangle ABC$	triangle ABC

TEST YOUR WORD POWER

See how well you have learned the vocabulary in this chapter. Answers follow the Quick Review.

1. A **ratio**
   - **A.** can be written only as a fraction
   - **B.** compares two quantities that have the same type of units
   - **C.** compares two quantities that have different types of units
   - **D.** is the reciprocal of a rate.

2. A **rate**
   - **A.** can be written only as a decimal
   - **B.** compares two quantities that have the same type of units
   - **C.** compares two quantities that have different types of units
   - **D.** is the reciprocal of a ratio.

3. Cost per unit is
   - **A.** the best buy
   - **B.** a ratio written in lowest terms
   - **C.** found by comparing cross products
   - **D.** the price of one item or one unit.

4. A **proportion**
   - **A.** shows that two ratios or rates are equivalent
   - **B.** contains only whole numbers or decimals
   - **C.** always has one unknown number
   - **D.** states that two improper fractions are equivalent.

5. An **obtuse angle**
   - **A.** is formed by perpendicular lines
   - **B.** is congruent to a right angle
   - **C.** measures more than 90° but less than 180°
   - **D.** measures less than 90°.

6. Two angles that are **complementary**
   - **A.** have measures that add up to 180°
   - **B.** are always congruent
   - **C.** form a straight angle
   - **D.** have measures that add up to 90°.

(continued)

7. Perpendicular lines
 A. intersect to form a right angle
 B. intersect to form an acute angle
 C. never intersect
 D. have a common endpoint called the vertex.

8. In a pair of similar triangles
 A. corresponding sides have the same length
 B. all the angles have the same measure
 C. the perimeters are equal
 D. the ratios of the lengths of corresponding sides are equal.

QUICK REVIEW

Concepts	**Examples**

6.1 *Writing a Ratio*

A ratio compares two quantities that have the same type of units. A ratio is usually written as a fraction with the number that is mentioned first in the numerator. The common units divide out and are not written in the answer. Check that the fraction is in lowest terms.

Write this ratio as a fraction in lowest terms.

60 ounces of medicine **to** 160 ounces of medicine

$$\frac{60 \text{ ounces}}{160 \text{ ounces}} = \frac{60 \div 20}{160 \div 20} = \frac{3}{8} \quad \left\{ \begin{array}{l} \text{Ratio in} \\ \text{lowest terms} \end{array} \right.$$

Divide out common units.

6.1 *Using Mixed Numbers in a Ratio*

If a ratio has mixed numbers, change the mixed numbers to improper fractions. Rewrite the problem in horizontal form, using the "÷" symbol for division. Finally, multiply by the reciprocal of the divisor.

Write as a ratio of whole numbers in lowest terms. $\quad 2\frac{1}{2} \quad \text{to} \quad 3\frac{3}{4}$

$$\frac{2\frac{1}{2}}{3\frac{3}{4}} \quad \text{Ratio with mixed numbers}$$

$$= \frac{\frac{5}{2}}{\frac{15}{4}} \quad \text{Ratio with improper fractions}$$

$$= \frac{5}{2} \div \frac{15}{4} = \frac{5}{2} \cdot \frac{4}{15} = \frac{\overset{1}{\cancel{5}} \cdot \overset{1}{\cancel{2}} \cdot 2}{\cancel{2} \cdot 3 \cdot \cancel{5}} = \frac{2}{3} \quad \left\{ \begin{array}{l} \text{Ratio in} \\ \text{lowest} \\ \text{terms} \end{array} \right.$$

Reciprocals

6.1 *Using Measurements in Ratios*

When a ratio compares measurements, both measurements must be in the *same* units. It is usually easier to compare the measurements using the smaller unit, for example, inches instead of feet.

Write 8 inches to 6 feet as a ratio in lowest terms.

Compare using the smaller unit, inches. Because 1 foot has 12 inches, 6 feet is

$$6 \cdot \textbf{12 inches} = 72 \text{ inches}.$$

The ratio is shown below.

$$\frac{8 \text{ inches}}{6 \text{ feet}} = \frac{8 \text{ inches}}{72 \text{ inches}} = \frac{8 \div 8}{72 \div 8} = \frac{1}{9} \quad \left\{ \begin{array}{l} \text{Ratio in} \\ \text{lowest terms} \end{array} \right.$$

6.2 *Writing Rates*

A rate compares two measurements with different types of units. The units do *not* divide out, so you must write them as part of the rate.

Write the rate as a fraction in lowest terms.

475 miles in 10 hours

$$\frac{475 \text{ miles} \div 5}{10 \text{ hours} \div 5} = \frac{95 \text{ miles}}{2 \text{ hours}} \quad \left. \begin{array}{l} \text{Must write units:} \\ \text{miles and hours} \end{array} \right.$$

Concepts	Examples

6.2 *Finding a Unit Rate*

A unit rate has 1 in the denominator. To find the unit rate, divide the numerator by the denominator. Write unit rates using the word **per** or a / mark.

Write as a unit rate: $1278 in 9 days.

$$\frac{\$1278}{9 \text{ days}} \leftarrow \text{Fraction bar indicates division.}$$

$$\begin{array}{c} 142 \\ 9\overline{)1278} \end{array} \quad \text{so} \quad \frac{\$1278 \div 9}{9 \text{ days} \div 9} = \frac{\$142}{1 \text{ day}}$$

Write the answer as $142 **per** day or $142/day.

6.2 *Finding the Best Buy*

The best buy is the item with the lowest cost per unit. Divide the price by the number of units. Round to thousandths when necessary. Then compare to find the lowest cost per unit.

Find the best buy on cheese.

2 pounds for $2.25

3 pounds for $3.40

Find cost per unit (cost per pound).

$$\frac{\$2.25}{2} = \$1.125 \text{ per pound}$$

$$\frac{\$3.40}{3} \approx \$1.133 \text{ per pound}$$

The lower cost per pound is $1.125, so 2 pounds of cheese is the best buy.

6.3 *Writing Proportions*

A proportion states that two ratios or rates are equivalent. The proportion "5 is to 6 as 25 is to 30" is written as shown below.

$$\frac{5}{6} = \frac{25}{30}$$

To see whether a proportion is true or false, multiply along one diagonal, then multiply along the other diagonal. If the two cross products are equal, the proportion is true. If the two cross products are unequal, the proportion is false.

Write as a proportion: 8 is to 40 as 32 is to 160

$$\frac{8}{40} = \frac{32}{160}$$

Is this proportion true or false?

$$\frac{6}{8\frac{1}{2}} = \frac{24}{34}$$

Find the cross products.

$$8\frac{1}{2} \cdot 24 = \frac{17}{2} \cdot \frac{\overset{1}{2} \cdot 12}{1} = \mathbf{204}$$

$$\frac{6}{8\frac{1}{2}} = \frac{24}{34}$$

$$6 \cdot 34 = \mathbf{204} \quad \text{Equal}$$

The cross products are equal, so the proportion is true.

Concepts	**Examples**
6.3 *Solving Proportions*	Find the unknown number.
Solve for an unknown number in a proportion by using these steps.	$$\frac{12}{x} = \frac{6}{8}$$ Write $\frac{6}{8}$ in lowest terms. $$\frac{12}{x} = \frac{3}{4}$$
Step 1 Find the cross products. (If desired, you can rewrite the ratios in lowest terms before finding the cross products.)	*Step 1* $$\frac{12}{x} = \frac{3}{4}$$ $x \cdot 3$ — Find the cross products. $12 \cdot 4$
Step 2 Show that the cross products are equivalent.	*Step 2* $x \cdot 3 = \underbrace{12 \cdot 4}$ Show that cross products are equivalent. $x \cdot 3 = 48$
Step 3 Divide both sides of the equation by the coefficient of the variable term.	*Step 3* $$\frac{x \cdot \overset{1}{\cancel{3}}}{\underset{1}{\cancel{3}}} = \frac{48}{3}$$ Divide both sides by 3 $x = 16$
Step 4 Check by writing the solution in the *original* proportion and finding the cross products.	*Step 4* x is 16 → $$\frac{12}{16} = \frac{6}{8}$$ $16 \cdot 6 = \mathbf{96}$ $12 \cdot 8 = \mathbf{96}$ Equal The cross products are equal, so 16 is the correct solution.
6.4 *Applications of Proportions*	If 3 pounds of grass seed cover 450 square feet of lawn, how much seed is needed for 1500 square feet of lawn?
Use the six problem-solving steps from **Section 3.3.**	
Step 1 **Read** the problem.	*Step 1* The problem is about the amount of grass seed needed for a lawn. Unknown: pounds of seed needed for 1500 square feet of lawn Known: 3 pounds cover 450 square feet of lawn
Step 2 **Assign a variable.**	*Step 2* There is only one unknown, so let x be the pounds of seed needed for 1500 square feet of lawn.
Step 3 **Write an equation.**	*Step 3* The equation is in the form of a proportion. Be sure that both rates in the proportion compare pounds to square feet in the same order. 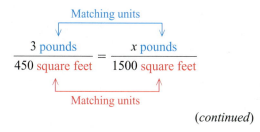
	(continued)

Concepts	Examples

6.4 *Applications of Proportions* *(continued)*

Step 4 **Solve** the equation.

Step 4 Ignore the units while finding the cross products and solving for x.

$$450 \cdot x = 3 \cdot 1500 \quad \text{Show that cross products}$$
$$450 \cdot x = 4500 \quad \text{are equivalent.}$$

$$\frac{\overset{1}{\cancel{450}} \cdot x}{\underset{1}{\cancel{450}}} = \frac{4500}{450} \quad \text{Divide both sides by 450}$$

$$x = 10$$

Step 5 **State the answer.**

Step 5 10 pounds of grass seed are needed for 1500 square feet of lawn.

Step 6 **Check** the solution by putting it back into the original problem.

Step 6 450 square feet of lawn needs 3 pounds of seed; 1500 square feet is about *three times as much* lawn, so about *three times as much* seed is needed.

$$(3)(3 \text{ pounds}) = 9 \text{ pounds} \leftarrow \text{Estimate}$$

The solution, 10 pounds, is close to the estimate of 9 pounds, so it is reasonable.

6.5 *Lines*

A *line* is straight row of points that goes on forever in both directions. If part of a line has one endpoint, it is a *ray*. If it has two endpoints, it is a *line segment*.

Identify each figure as a line, line segment, or ray, and name it using the appropriate symbol.

Figure **(a)** shows a ray named $\overrightarrow{OS}$. Figure **(b)** shows a line named $\overleftrightarrow{PQ}$ or $\overleftrightarrow{QP}$. Figure **(c)** shows a line segment named $\overline{ST}$ or $\overline{TS}$.

If two lines intersect at right angles, they are *perpendicular*. If two lines in the same plane never intersect, they are *parallel*.

Label each pair of lines as appearing to be parallel or as perpendicular.

Figure **(a)** shows perpendicular lines (they intersect at 90°), and figure **(b)** shows lines that appear to be parallel (they never intersect).

Concepts

6.5 Angles

If the sum of the measures of two angles is 90°, they are *complementary*.

If the sum of the measures of two angles is 180°, they are *supplementary*.

If two angles measure the same number of degrees, the angles are *congruent*. The symbol for congruent is ≅.

Two nonadjacent angles formed by two intersecting lines are called *vertical angles*. Vertical angles are congruent.

6.5 Parallel Lines

When two parallel lines are crossed by a transversal, corresponding angles are congruent, and alternate interior angles are congruent. Use this information to find the measures of the other angles.

Read $m\angle 1$ as "the measure of angle 1."

6.6 Proving That Two Triangles Are Congruent

Congruent triangles are identical both in shape and in size. This means that corresponding angles have the same measure and corresponding sides have the same length.

Here are three ways to prove that two triangles are congruent.

1. Angle–Side–Angle (ASA) method: If two angles and the side between them on one triangle measure the same as the corresponding parts on another triangle, the triangles are congruent.

2. Side–Side–Side (SSS) method: If three sides of one triangle measure the same as the corresponding sides of another triangle, the triangles are congruent.

3. Side–Angle–Side (SAS) method: If two sides and the angle between them on one triangle measure the same as the corresponding parts on another triangle, the triangles are congruent.

Examples

Find the complement and supplement of a 35° angle.

$$90° - 35° = 55° \text{ (the complement)}$$
$$180° - 35° = 145° \text{ (the supplement)}$$

Identify the vertical angles in this figure. Which angles are congruent?

∠1 and ∠3 are vertical angles.

∠2 and ∠4 are vertical angles.

Vertical angles are congruent, so ∠1 ≅ ∠3 and ∠2 ≅ ∠4.

Line m is parallel to line n and the measure of ∠4 is 125°. Find the measures of the other angles.

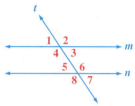

∠4 ≅ ∠8 (corresponding angles), so $m\angle 8 = 125°$.

∠4 ≅ ∠6 (alternate interior angles), so $m\angle 6 = 125°$.

∠6 ≅ ∠2 (corresponding angles), so $m\angle 2 = 125°$.

∠4 and ∠3 are supplements, so $m\angle 3 = 180° - 125° = 55°$.

∠3 ≅ ∠7 (corresponding angles), so $m\angle 7 = 55°$.

∠3 ≅ ∠5 (alternate interior angles), so $m\angle 5 = 55°$.

∠5 ≅ ∠1 (corresponding angles), so $m\angle 1 = 55°$.

Determine which method can be used to prove that each pair of triangles is congruent.

(a)   **(b)**

(c)

(a) On both triangles, two corresponding angles and the side between them measure the same, so use ASA.

(b) Each pair of corresponding sides has the same length, so use SSS.

(c) On both triangles, two corresponding sides and the angle between them measure the same, so use SAS.

Concepts	Examples

6.6 Finding the Unknown Lengths in Similar Triangles

Use the fact that in similar triangles, the ratios of the lengths of corresponding sides are equal. Write a proportion. Then find the cross products and show that they are equivalent. Finish solving for the unknown length.

Find the unknown lengths in this pair of similar triangles.

$$\frac{x}{8} = \frac{5}{10} \qquad\qquad \frac{y}{12} = \frac{5}{10}$$

$$x \bullet 10 = 8 \bullet 5 \qquad\qquad y \bullet 10 = 12 \bullet 5$$

$$\frac{x \bullet \overset{1}{\cancel{10}}}{\underset{1}{\cancel{10}}} = \frac{40}{10} \qquad\qquad \frac{y \bullet \overset{1}{\cancel{10}}}{\underset{1}{\cancel{10}}} = \frac{60}{10}$$

$$x = 4 \text{ m} \qquad\qquad y = 6 \text{ m}$$

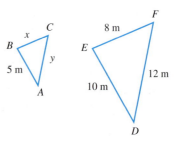

ANSWERS TO TEST YOUR WORD POWER

1. B; *Example:* The ratio of 3 miles to 4 miles is $\frac{3}{4}$; the common units (miles) divide out.

2. C; *Example:* $4.50 for 3 pounds is a rate comparing dollars to pounds.

3. D; *Example:* $3.95 per gallon tells the price of one gallon (one unit).

4. A; *Example:* The proportion $\frac{5}{6} = \frac{25}{30}$ says that $\frac{5}{6}$ is equivalent to $\frac{25}{30}$.

5. C; *Examples:* Angles that measure 91°, 120°, and 175° are all obtuse angles.

6. D; *Example:* If $\angle 1$ measures 35° and $\angle 2$ measures 55°, the angles are complementary because 35° + 55° = 90°.

7. A; *Example:* $\overleftrightarrow{EF}$ is perpendicular to $\overleftrightarrow{GH}$.

8. D; *Example:* Triangle *ABC* is similar to triangle *DEF*, so the ratios of corresponding sides are equal.

$$\frac{AB}{DE} = \frac{3 \text{ m}}{6 \text{ m}} = \frac{1}{2} \qquad \frac{BC}{EF} = \frac{2 \text{ m}}{4 \text{ m}} = \frac{1}{2} \qquad \frac{AC}{DF} = \frac{3.5 \text{ m}}{7 \text{ m}} = \frac{1}{2}$$

Chapter 6
REVIEW EXERCISES

[6.1] *Write each ratio as a fraction in lowest terms. Change to the same units when necessary, using the table of measurement comparisons in* **Section 6.1.** *Use the information in the graph to answer Exercises 1–3.*

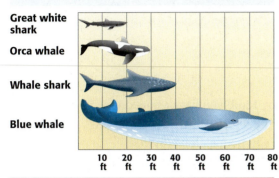

AVERAGE LENGTH OF SHARKS AND WHALES

Great white shark

Orca whale

Whale shark

Blue whale

10 ft 20 ft 30 ft 40 ft 50 ft 60 ft 70 ft 80 ft

Source: Grolier Multimedia Encyclopedia.

1. Orca whale's length to whale shark's length

2. Blue whale's length to great white shark's length

3. Which two animals' lengths give a ratio of $\frac{1}{2}$? There are several answers.

4. $2.50 to $1.25

5. $0.30 to $0.45

6. $1\frac{2}{3}$ cups to $\frac{2}{3}$ cup

7. $2\frac{3}{4}$ miles to $16\frac{1}{2}$ miles

8. 5 hours to 100 minutes

9. 9 in. to 2 ft

10. 1 ton to 1500 pounds

11. 8 hours to 3 days

12. Jake sold $350 worth of his kachina figures. Ramona sold $500 worth of her pottery. What is the ratio of Ramona's sales to Jake's sales?

13. Ms. Wei's new car gets 35 miles per gallon. Her old car got 25 miles per gallon. Find the ratio of the new car's mileage to the old car's mileage.

14. This fall, 6000 students are taking math courses and 7200 students are taking English courses. Find the ratio of math students to English students.

[6.2] *Write each rate as a fraction in lowest terms.*

15. $88 for 8 dozen

16. 96 children in 40 families

17. In his keyboarding class, Patrick can type four pages in 20 minutes. Give his rate in pages per minute and minutes per page.

18. Elena made $24 in three hours. Give her earnings in dollars per hour and hours per dollar.

Find the best buy.

19. Minced onion

      13 ounces for $2.29

       8 ounces for $1.45

       3 ounces for $0.95

20. Dog food; you have a coupon for $1 off on 25 pounds or more.

      50 pounds for $19.95

      25 pounds for $10.40

       8 pounds for $3.40

[6.3] *Use either the method of writing in lowest terms or of finding cross products to decide whether each proportion is* true *or* false. *Show your work and then write* true *or* false.

21. $\dfrac{6}{10} = \dfrac{9}{15}$

22. $\dfrac{6}{48} = \dfrac{9}{36}$

23. $\dfrac{47}{10} = \dfrac{98}{20}$

24. $\dfrac{1.5}{2.4} = \dfrac{2}{3.2}$

25. $\dfrac{3\frac{1}{2}}{2\frac{1}{3}} = \dfrac{6}{4}$

Find the unknown number in each proportion. Round answers to the nearest hundredth when necessary.

26. $\dfrac{4}{42} = \dfrac{150}{x}$

27. $\dfrac{16}{x} = \dfrac{12}{15}$

28. $\dfrac{100}{14} = \dfrac{x}{56}$

29. $\dfrac{5}{8} = \dfrac{x}{20}$

30. $\dfrac{x}{24} = \dfrac{11}{18}$

31. $\dfrac{7}{x} = \dfrac{18}{21}$

32. $\dfrac{x}{3.6} = \dfrac{9.8}{0.7}$

33. $\dfrac{13.5}{1.7} = \dfrac{4.5}{x}$

34. $\dfrac{0.82}{1.89} = \dfrac{x}{5.7}$

[6.4] *Set up and solve a proportion for each application problem.*

35. The ratio of cats to dogs at the animal shelter is 3 to 5. If there are 45 dogs, how many cats are there?

36. Danielle had 8 hits in 28 times at bat during last week's games. If she continues to hit at the same rate, how many hits will she get in 161 times at bat?

37. If 3.5 pounds of ground beef cost $9.77, what will 5.6 pounds cost? Round to the nearest cent.

38. About 4 out of 10 students are expected to vote in campus elections. There are 8247 students. How many are expected to vote? Round to the nearest whole number.

39. The scale on Brian's model railroad is 1 inch to 16 feet. One of the scale model boxcars is 4.25 inches long. What is the length of a real boxcar in feet?

40. Marvette makes necklaces to sell at a local gift shop. She made 2 dozen necklaces in $16\frac{1}{2}$ hours. How long will it take her to make 40 necklaces?

41. A 180-pound person burns 284 calories playing basketball for 25 minutes. At this rate, how many calories would the person burn in 45 minutes, to the nearest whole number? (*Source: Wellness Encyclopedia.*)

42. In the hospital pharmacy, Michiko sees that a medicine is to be given at the rate of 3.5 milligrams for every 50 pounds of body weight. How much medicine should be given to a patient who weighs 210 pounds?

[6.5] *Identify each figure as a* line, line segment, *or* ray *and name it using the appropriate symbol.*

43.

44.

45.

Label each pair of lines as appearing to be parallel, *as* perpendicular, *or as* intersecting.

46.

47.

48.

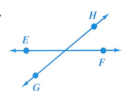

Label each angle as an acute, right, obtuse, *or* straight angle. *For right and straight angles, indicate the number of degrees in the angle.*

49.

50.

51.

52.

Find the complement or supplement of each angle.

53. Find each complement.

   (a) 80°

   (b) 45°

   (c) 7°

54. Find each supplement.

   (a) 155°

   (b) 90°

   (c) 33°

55. In the figure below, $\angle 2$ measures 60°. Find the measure of each of the other angles.

56. Line m is parallel to line n and $\angle 8$ measures 160°. Find the measures of the other angles.

[6.6] *Determine which method can be used to prove that each pair of triangles is congruent.*

57.

58.

59.

Find the unknown lengths in each pair of similar triangles. Then find the perimeter of the larger triangle in each pair.

60.

61.

62.

MIXED REVIEW EXERCISES

Find the unknown number in each proportion. Round to hundredths when necessary.

63. $\dfrac{x}{45} = \dfrac{70}{30}$

64. $\dfrac{x}{52} = \dfrac{0}{20}$

65. $\dfrac{64}{10} = \dfrac{x}{20}$

66. $\dfrac{15}{x} = \dfrac{65}{100}$

67. $\dfrac{7.8}{3.9} = \dfrac{13}{x}$

68. $\dfrac{34.1}{x} = \dfrac{0.77}{2.65}$

Write each ratio as a fraction in lowest terms. Change to the same units when necessary.

69. 4 dollars to 10 quarters

70. $4\dfrac{1}{8}$ inches to 10 inches

71. 10 yards to 8 feet

72. $3.60 to $0.90

73. 12 eggs to 15 eggs

74. 37 meters to 7 meters

75. 3 pints to 4 quarts

76. 15 minutes to 3 hours

77. $4\dfrac{1}{2}$ miles to $1\dfrac{3}{10}$ miles

Set up and solve a proportion for each application problem.

78. Nearly 7 out of 8 fans buy something to drink at rock concerts. How many of the 28,500 fans at today's concert would be expected to buy a beverage? Round to the nearest hundred fans.

79. Emily spent $150 on car repairs and $400 on car insurance. What is the ratio of the amount spent on insurance to the amount spent on repairs?

80. Antonio is choosing among three packages of plastic wrap. Is the best buy 25 feet for $0.78; 75 feet for $1.99; or 100 feet for $2.59? He has a coupon for 50¢ off that is good for either of the larger two packages.

81. On this scale drawing of a backyard patio, 0.5 inch represents 6 feet. If the patio measures 1.75 inches long and 1.25 inches wide on the drawing, what will be the actual length and width of the patio when it is built?

0.5 in. = 6 ft

82. A vitamin supplement for cats is to be given at the rate of 1000 milligrams for a 5-pound cat. (*Source:* St. Jon Pet Care Products.)

  (a) How much should be given to a 7-pound cat?

  (b) How much should be given to an 8-ounce kitten?

83. Charles made 251 points during 169 minutes of playing time last year. If he plays 14 minutes in tonight's game, how many points would you expect him to make? Round to the nearest whole number.

84. An antibiotic is to be given at the rate of $1\dfrac{1}{2}$ teaspoons for every 24 pounds of body weight. How much should be given to an infant who weighs 8 pounds?

Label each figure. Choose from these labels: line segment, ray, parallel lines, perpendicular lines, intersecting lines, acute angle, right angle, straight angle, obtuse angle. *Indicate the number of degrees in the right angle and the straight angle.*

85.

86.

87.

88.

89.

90.

91.

92.

93.

94. Explain what is happening in each sentence.
   (a) The road was so slippery that my car did a 360.

   (b) After the election, the governor's view on new taxes took a 180° turn.

95. (a) Can two obtuse angles be supplementary? Explain why or why not.

   (b) Can two acute angles be complementary? Explain why or why not.

96. In the figure below, ∠2 measures 45° and ∠7 measures 55°. Find the measure of each of the other angles.

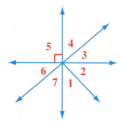

97. In the figure below, line *m* is parallel to line *n* and ∠5 measures 75°. Find the measures of the other angles.

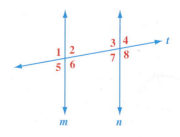

Chapter **6**
TEST

Write each rate or ratio as a fraction in lowest terms. Change to the same units when necessary.

1. $15 for 75 minutes

2. 3 hours to 40 minutes

3. The little theater at our college has 320 seats. The auditorium has 1200 seats. Find the ratio of auditorium seats to theater seats.

4. Find the best buy on spaghetti sauce. You have a coupon for 75¢ off Brand X and a coupon for 25¢ off Brand Y.
   28 ounces of Brand X for $3.89
   18 ounces of Brand Y for $1.89
   13 ounces of Brand Z for $1.29

Find the unknown number in each proportion. Round answers to the nearest hundredth when necessary.

5. $\dfrac{5}{9} = \dfrac{x}{45}$

6. $\dfrac{3}{1} = \dfrac{8}{x}$

7. $\dfrac{x}{20} = \dfrac{6.5}{0.4}$

8. $\dfrac{2\frac{1}{3}}{x} = \dfrac{\frac{8}{9}}{4}$

Set up and solve a proportion for each application problem.

9. Pedro typed 240 words in five minutes in his keyboarding class. At that rate, how many words could he type in 12 minutes?

10. About 2 out of every 15 people are left-handed. How many of the 650 students in our school would you expect to be left-handed? Round to the nearest whole number.

11. A medication is given at the rate of 8.2 grams for every 50 pounds of body weight. How much should be given to a 145-pound person? Round to the nearest tenth.

12. On a scale model, 1 inch represents 8 feet. If a building in the model is 7.5 inches tall, what is the actual height of the building in feet?

1. _____

2. _____

3. _____

4. _____

5. _____

6. _____

7. _____

8. _____

9. _____

10. _____

11. _____

12. _____

13. _____

14. _____

15. _____

16. _____

17. _____

18. _____

19. _____

20. _____

21. _____

22. _____

23. _____

24. _____

25. _____

Choose the figure that matches each label. For right and straight angles, indicate the number of degrees in the angle.

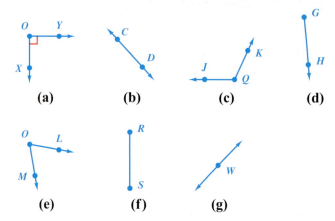

(a)   **(b)**   **(c)**   **(d)**

(e)   **(f)**   **(g)**

13. Acute angle is figure _____.

14. Right angle is figure _____ and its measure is _____.

15. Ray is figure _____.

16. Straight angle is figure _____ and its measure is _____.

17. Write a definition of parallel lines and a definition of perpendicular lines. Make a sketch to illustrate each definition.

18. Find the complement of an 81° angle.

19. Find the supplement of a 20° angle.

20. In the figure below, ∠4 measures 50° and ∠6 measures 95°. Find the measures of the other angles.

21. In the figure below, line *m* is parallel to line *n* and ∠3 measures 65°. Find the measures of the other angles.

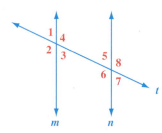

Determine which method can be used to prove that each pair of triangles is congruent.

22.

23.

24. Find the unknown lengths in these similar triangles.

25. Find the perimeter of each of these similar triangles.

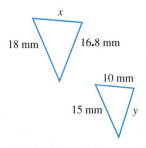

Cumulative Review Exercises

CHAPTERS 1–6

1. Write these numbers in words.

   (a) 77,001,000,805

   (b) 0.02

2. Write these numbers using digits.

   (a) Three and forty thousandths

   (b) Five hundred million, thirty-seven thousand

Simplify.

3. $\dfrac{0}{-16}$

4. $|0| + |-6 - 8|$

5. $4\dfrac{3}{4} - 1\dfrac{5}{6}$

6. $\dfrac{h}{5} - \dfrac{3}{10}$

7. $100 - 0.0095$

8. $\dfrac{-4 + 7}{9 - 3^2}$

9. $-6 + 3(0 - 4)$

10. $\dfrac{-8}{\frac{4}{7}}$

11. $\dfrac{5n}{6m^3} \div \dfrac{10}{3m^2}$

12. $(0.06)(-0.007)$

13. $\dfrac{x}{14y} \cdot \dfrac{7}{xy}$

14. $\dfrac{9}{n} + \dfrac{2}{3}$

15. $-40 + 8(-5) + 2^4$

16. $5.8 - (-0.6)^2 \div 0.9$

17. $\left(-\dfrac{1}{2}\right)^3 \left(\dfrac{2}{3}\right)^2$

Solve each equation. Show your work.

18. $7y + 5 = -3 - y$

19. $-2 + \dfrac{3}{5}x = 7$

20. $\dfrac{4}{x} = \dfrac{14}{35}$

*Solve each application problem using the six problem-solving steps from **Chapter 3**.*

21. Tuyen heard on the radio that the temperature had risen 15 degrees by noon, then dropped 23 degrees due to a storm, then risen 5 degrees and was now at 71 degrees. What was the starting temperature?

22. The width of a rectangular swimming pool is 14 ft less than the length. The perimeter of the pool is 100 ft. Find the length and the width.

Find the unknown length, perimeter, area, or volume. When necessary, use 3.14 as the approximate value of π and round answers to the nearest tenth.

23. Find the perimeter and the area.

2.8 km
0.7 km   0.7 km
2.8 km

 24. Find the diameter, circumference, and area.

8.5 m

 25. Find the volume and surface area.

3 ft
12 ft

Solve each application problem. Show your work.

26. The honor society has a goal of collecting 1500 pounds of food to fill Thanksgiving baskets. So far they've collected $\frac{5}{6}$ of their goal. How many more pounds do they need?

27. The directions on a can of plant food call for $\frac{1}{2}$ teaspoon in two quarts of water. How much plant food is needed for five quarts?

28. The distance around Dunning Pond is $1\frac{1}{10}$ miles. Norma ran around the pond four times in the morning and $2\frac{1}{2}$ times in the afternoon. How far did she run in all?

 29. Rodney used 49.8 gallons of gas for his SUV while driving 896.5 miles on a vacation. How many miles per gallon did he get, rounded to the nearest tenth?

Use the information in the table on international long-distance calling card rates to answer Exercises 30–33. Round answers to the nearest whole minute when necessary.

$10 International Phone Card *No Connnection Fee!*	
Place a call to	**Cost per minute**
Mexico City	$0.10
Philippines	$0.333
Japan	$0.14
Hong Kong	$0.111
Canada	$0.08
Ghana	$0.172
South Korea	$0.104

Source: Walgreens; IDT.

30. List the per-minute rates from least to greatest.

31. If you buy a $10 calling card, how long a call could you make to Japan?

 32. How many $10 cards would you have to buy to make two hours' worth of calls to the Philippines?

33. What is the ratio of minutes you can call Canada for $10 to minutes you can call Mexico City for $10?

Percent

Everyone likes to buy things that are "on sale." It helps if you can estimate the sale price *before* deciding to buy. Most states add on a sales tax, so that is part of your final cost as well. Find out how to calculate discounts and sales tax in **Section 7.5**, Examples 1 and 4. Then get some practice in Exercises 1–10 and 17–24.

In **Section 7.5** you'll also learn how to use percents to estimate restaurant tips and calculate the interest on loans.

7.1 The Basics of Percent

OBJECTIVE 1 Learn the meaning of percent. You have probably seen percents frequently in daily life. The symbol for percent is **%**. For example, during one day you may leave a 15% tip for the waitress at dinner, pay 7% sales tax on a CD player, and buy shoes at 25% off the regular price. The next day your score on a math test may be 89% correct.

The Meaning of Percent

A **percent** is a ratio with a denominator of 100. So percent means "per 100" or "how many out of 100." The symbol for percent is **%**. Read 15% as "fifteen percent."

Here are some everyday situations using percents.

- If you left a $15 tip when the restaurant bill was $100, then you left $15 per $100 or $\frac{15}{100}$ or 15%.

- If you pay $7 in tax on a $100 CD player, then the tax rate is $7 per $100 or $\frac{7}{100}$ or 7%.

- If you earn 89 points on a 100-point math test, then your score is 89 out of 100 or $\frac{89}{100}$ or 89%.

OBJECTIVE 2 Write percents as decimals. In order to work with percents, you will need to write them as decimals or as fractions. We'll start by writing percents as equivalent decimal numbers. Twenty-five percent, or 25%, means 25 parts out of 100 parts, or $\frac{25}{100}$. Remember that the fraction bar indicates division. So we can write $\frac{25}{100}$ as $25 \div 100$. When you do the division, $25 \div 100$ is 0.25.

Indicates division $\longrightarrow$ $\dfrac{25}{100}$ can be written as $25 \div 100 = 0.25$
                                                         $\uparrow$
                                              Indicates division

Another way to remember how to write a percent as a decimal is to use the meaning of the word *percent*. The first part of the word, *per*, is an indicator word for *division*. The last part of the word, *cent*, comes from the Latin word for *hundred*. (Recall that there are 100 *cent*s in a dollar, and 100 years in a *cent*ury.)

25%   is   25 **percent**

25 $\div$ **100** = 0.25

Writing a Percent as an Equivalent Decimal

To write a percent as a decimal, drop the % symbol and divide by 100.

EXAMPLE 1   **Writing Percents as Decimals**

Write each percent as a decimal.

(a) 47% $47\% = 47 \div 100 = 0.47$ Decimal form

(b) 3% $3\% = 3 \div 100 = 0.03$ Decimal form

(c) 28.2% $28.2\% = 28.2 \div 100 = 0.282$ Decimal form

(d) 100% $100\% = 100 \div 100 = 1.00$ Decimal form

(e) 135% $135\% = 135 \div 100 = 1.35$ Decimal form

> **CAUTION**
> In Example 1(d) above, notice that 100% is 1.00, or 1, which is a whole number. Whenever you have a percent that is *100% or greater,* the equivalent decimal number will be *1 or greater*. Notice in Example 1(e) above that 135% is 1.35 (greater than 1).

Now Try Exercises 5 and 7. ▶▶▶

In the exercise set for **Section 5.5,** you discovered a shortcut for *dividing* by 100: Move the decimal point *two* places to the *left*. You can use this shortcut when writing percents as decimals.

EXAMPLE 2   **Changing Percents to Decimals by Moving the Decimal Point**

Write each percent as a decimal by dropping the percent symbol and moving the decimal point two places to the left.

(a) 17%

          ⌐—— Decimal point starts at far right side.
$17\% = 17.\%$

          .17 ← Percent symbol is dropped.
          ↑
          └—— Decimal point is moved two places to the left. This is a quick way to divide by 100

$17\% = 0.17$

(b) 160%

$160\% = 160.\% = 1.60 \text{ or } 1.6$ Decimal point starts at far right side. 1.60 is equivalent to 1.6

(c) 4.9%

          04.9%   0 is attached so the decimal point can be moved two places to the left.

$4.9\% = 0.049$

(d) 0.6%

          00.6% = 0.006   0 is attached so the decimal point can be moved two places to the left.

Continued on Next Page

NOTE

In Example 2(d), notice that 0.6% is less than 1%. Because 1% is equivalent to 0.01 or $\frac{1}{100}$, any fraction of a percent smaller than 1% is less than 0.01. The decimal equivalent of 0.6% is 0.006, which is less than 0.01.

Now Try Exercise 13.

OBJECTIVE **3** **Write decimals as percents.**   You can write a decimal as a percent. For example, the decimal 0.25 is the same as the fraction $\frac{25}{100}$. This fraction means 25 out of 100 parts, or 25%. Notice that multiplying 0.25 by 100 gives the same result.

$$(0.25)(100) = 25\%$$

This result makes sense because we are doing the opposite of what we did to change a percent to a decimal.

To change a percent to a decimal, we *drop* the % symbol and *divide* by 100. So, to *reverse* the process and change a decimal to a percent, we *multiply* by 100 and *attach* a % symbol.

Writing a Decimal as a Percent

To write a decimal as a percent, multiply by 100 and attach a % symbol.

NOTE

A quick way to *multiply* a number by 100 is to move the decimal point *two* places to the *right*. Notice that this is the opposite of the shortcut for *dividing* by 100.

Drop % symbol; *divide* by 100.
(Move decimal point two places *left*.)

Decimal   **Percent**

Multiply by 100; attach % symbol.
(Move decimal point two places *right*.)

EXAMPLE 3   **Changing Decimals to Percents by Moving the Decimal Point**

Write each decimal as a percent.

(a) 0.21

0.21

Decimal point is moved two places to the right.

0.21 = 21%   ⟵ Percent symbol is attached after decimal point is moved.

Decimal point is *not* written with whole number percents.

(b) 0.529 = 52.9%   ⟵ Percent symbol is attached after decimal point is moved.

(c) 1.92 = 192%   ⟵ Percent symbol is attached after decimal point is moved.

Continued on Next Page

(d) 2.5

$$2.5\underset{\frown}{0} \qquad \text{0 is attached so the decimal point can be moved two places to the right.}$$

$$2.5 = 250\%$$

(e) 3

$$3. = 3.\underset{\frown}{00} \quad \text{so} \quad 3 = 300\%$$

> **CAUTION**
> In Examples 3(c), 3(d), and 3(e) above, notice that 1.92, 2.5, and 3 are greater than 1. Because the number 1 is equivalent to 100%, all *numbers greater than 1* will be equivalent to *percents greater than 100%.*

Now Try Exercises 17 and 29. ▶▶▶

OBJECTIVE 4 Write percents as fractions. Percents can also be written as fractions. Recall that a percent is a ratio with a denominator of 100. For example, 89% is $\frac{89}{100}$. Because the fraction bar indicates division, we are dividing the percent by 100, just as we did when writing a percent as a decimal.

> **Writing a Percent as a Fraction**
> To write a percent as a fraction, drop the % symbol and write the number over 100. Then write the fraction in lowest terms.

EXAMPLE 4 **Writing Percents as Fractions**

Write each percent as a fraction or mixed number in lowest terms.

(a) 25%   Drop the % symbol and write 25 over 100.

$$25\% = \frac{25}{100} \quad \leftarrow \text{25 per 100}$$

$$= \frac{25 \div 25}{100 \div 25} = \frac{1}{4} \quad \leftarrow \text{Lowest terms}$$

As a check, write 25% as a decimal.

$$25\% = 25 \div 100 = 0.25 \quad \leftarrow \text{Percent sign dropped}$$

Recall that 0.25 means 25 hundredths.

$$0.25 = \frac{25}{100} = \frac{25 \div 25}{100 \div 25} = \frac{1}{4} \quad \leftarrow \text{Same result as above}$$

(b) 76%   Drop the % symbol and write 76 over 100.

The percent becomes the numerator.

$$76\% = \frac{76}{100} \quad \leftarrow \text{The } \textit{denominator} \text{ is always 100 because percent means parts per 100}$$

Write $\frac{76}{100}$ in lowest terms. $\dfrac{76 \div 4}{100 \div 4} = \dfrac{19}{25} \quad \leftarrow \text{Lowest terms}$

Continued on Next Page

(c) 150%

$$150\% = \frac{150}{100} = \frac{150 \div 50}{100 \div 50} = \frac{3}{2} = 1\frac{1}{2} \quad \leftarrow \text{Mixed number}$$

(d) $100\% = \dfrac{100}{100} = 1 \leftarrow$ Whole number

> **NOTE**
> Remember that percent means *per 100.*

◀◀◀ Now Try Exercises 37 and 45.

Example 5 below shows how to write decimal percents and fraction percents as fractions.

EXAMPLE 5 **Writing Decimal Percents or Fraction Percents as Fractions**

Write each percent as a fraction in lowest terms.

(a) 15.5%

Drop the % symbol and write 15.5 over 100.

$$15.5\% = \frac{15.5}{100}$$

To get a whole number in the numerator, multiply the numerator and denominator by 10. (Multiplying by $\frac{10}{10}$ is the same as multiplying by 1.)

$$\frac{15.5}{100} = \frac{(15.5)(10)}{(100)(10)} = \frac{155}{1000}$$

Write the fraction in lowest terms.

$$\frac{155 \div 5}{1000 \div 5} = \frac{31}{200} \quad \leftarrow \text{Lowest terms}$$

(b) $33\frac{1}{3}\%$

Drop the % symbol and write $33\frac{1}{3}$ over 100.

$$33\frac{1}{3}\% = \frac{33\frac{1}{3}}{100}$$

When there is a mixed number in the numerator, write it as an improper fraction. So $33\frac{1}{3}$ is $\frac{100}{3}$.

$$\frac{33\frac{1}{3}}{100} = \frac{\frac{100}{3}}{100}$$

Continued on Next Page

Now you have a complex fraction (see **Section 4.6**). Rewrite the complex fraction using the ÷ symbol for division. Then follow the steps for dividing fractions.

$$\frac{\frac{100}{3}}{100} = \frac{100}{3} \div 100 = \frac{100}{3} \div \frac{100}{1} = \frac{100}{3} \cdot \frac{1}{100} = \frac{\overset{1}{\cancel{100}} \cdot 1}{3 \cdot \underset{1}{\cancel{100}}} = \frac{1}{3}$$

Reciprocals

> **NOTE**
> In Example 5(a), we could have changed 15.5% to $15\frac{1}{2}$% and then written it as the improper fraction $\frac{31}{2}$ over 100. But it is usually easier to work with decimal percents as they are.

Now Try Exercises 39 and 43. ▶▶▶

OBJECTIVE 5 Write fractions as percents. Recall that to write a percent as a fraction, you *drop* the percent symbol and *divide* by 100. So, to reverse the process and change a fraction to a percent, you *multiply* by 100 and *attach* a percent symbol.

> **Writing a Fraction as a Percent**
> To write a fraction as a percent, multiply by 100 and attach a % symbol. This is the same as multiplying by 100%.

> **NOTE**
> Look back at Example 4(d) to see that 100% = 1. Recall that multiplying a number by 1 does *not* change the value of the number. So multiplying by 100% does not change the value of a number; it just gives us an *equivalent percent*.

EXAMPLE 6 Writing Fractions as Percents

Write each fraction as a percent. Round to the nearest tenth of a percent, if necessary.

(a) $\frac{2}{5}$ Multiply $\frac{2}{5}$ by 100%.

$$\frac{2}{5} = \left(\frac{2}{5}\right)(\mathbf{100\%}) = \left(\frac{2}{5}\right)\left(\frac{100}{1}\%\right) = \left(\frac{2}{5}\right)\left(\frac{5 \cdot 20}{1}\%\right) = \frac{2 \cdot \overset{1}{\cancel{5}} \cdot 20}{\underset{1}{\cancel{5}} \cdot 1}\%$$

$$= \frac{40}{1}\% = 40\%$$

To check the result, write 40% as $\frac{40}{100}$ and simplify to lowest terms.

$$40\% = \frac{40}{100} = \frac{40 \div \mathbf{20}}{100 \div \mathbf{20}} = \frac{2}{5} \quad \leftarrow \text{Original fraction}$$

Continued on Next Page

(b) $\frac{5}{8}$ Multiply $\frac{5}{8}$ by 100%.

$$\frac{5}{8} = \left(\frac{5}{8}\right)(100\%) = \left(\frac{5}{8}\right)\left(\frac{100}{1}\%\right) = \left(\frac{5}{2 \cdot 4}\right)\left(\frac{4 \cdot 25}{1}\%\right) = \frac{5 \cdot \overset{1}{4} \cdot 25}{2 \cdot 4 \cdot 1}\%$$

$$= \frac{125}{2}\% = 62\frac{1}{2}\%$$

You can also do the last step of simplifying $\frac{125}{2}$ on your calculator. Enter $\frac{125}{2}$ as 125 ÷ 2 = . The result is 62.5, so $\frac{5}{8} = 62\frac{1}{2}\%$ or $\frac{5}{8} = 62.5\%$.

(c) $\frac{1}{6}$ Multiply $\frac{1}{6}$ by 100%.

$$\frac{1}{6} = \left(\frac{1}{6}\right)(100\%) = \left(\frac{1}{6}\right)\left(\frac{100}{1}\%\right) = \left(\frac{1}{2 \cdot 3}\right)\left(\frac{2 \cdot 50}{1}\%\right) = \frac{1 \cdot \overset{1}{2} \cdot 50}{2 \cdot 3 \cdot 1}\%$$

$$= \frac{50}{3}\% = 16\frac{2}{3}\%$$

To simplify $\frac{50}{3}$ on your calculator, enter 50 ÷ 3 = . The result is 16.66666666, with the 6 continuing to repeat. The directions say to round to the nearest *tenth* of a percent.

Next digit is *5 or more.*

1 6.6|6 6 6 6 6 6%   rounds to   16.**7%**

Tenths place

So $\frac{1}{6} = 16\frac{2}{3}\%$ or $\frac{1}{6} \approx 16.7\%$.

Calculator Tip   In Example 6(a), you can use your calculator to write $\frac{2}{5}$ as a percent.

Step 1   Enter $\frac{2}{5}$ as 2 ÷ 5 = . Your calculator shows   **0.4**

Decimal equivalent of $\frac{2}{5}$

Step 2   Change the decimal number 0.4 to a percent by moving the decimal point two places to the right (multiplying by 100).

0.40 = 40**%** ← Attach % symbol.

Try this technique on Examples 6(b) and 6(c).

For $\frac{5}{8}$, enter 5 ÷ 8 = . Your calculator shows   **0.625**

Move the decimal point in 0.625 two places to the right.

0.625 = 62.5**%** ← Attach % symbol.

Continued on Next Page

For $\frac{1}{6}$, enter 1 ⊝ 6 ⊜. Your calculator shows | **0.166666666** |

Some calculators show 7 in the last place. ⟶

Move the decimal point two places to the right. Then round to the nearest tenth.

$$0.166666666 = 16.66666666\% \approx 16.7\% \; \leftarrow \text{Attach \% symbol.}$$

With a fraction such as $\frac{1}{6}$, your calculator gives only an *approximate* answer. You can't get the exact answer of $16\frac{2}{3}\%$ by using your calculator.

Now Try Exercises 57 and 61. ▶▶▶

OBJECTIVE 6 Use 100% and 50%. When working with percents, it is helpful to have several reference points. 100% and 50% are two helpful reference points.

100% means 100 parts out of 100 parts. That's *all* of the parts. If you pay 100% of a $245 dentist bill, you pay $245 (*all* of it).

EXAMPLE 7 Finding 100% of a Number

Fill in the blanks.

(a) 100% of $34 is _____.

100% is *all* of the money.

So 100% of $34 is $34_____.

(b) 100% of 4 miles is _____.

100% is *all* of the miles.

So 100% of 4 miles is 4 miles____.

Now Try Exercise 103(a). ▶▶▶

50% means 50 parts out of 100 parts, which is *half* of the parts because $\frac{50}{100} = \frac{1}{2}$. So 50% of $12 is $6 (*half* of the money).

EXAMPLE 8 Finding 50% of a Number

Fill in the blanks.

(a) 50% of $20 is _____.

50% is *half* of the money.

So 50% of $20 is $10_____.

(b) 50% of 280 miles is _____.

50% is *half* of the miles.

So 50% of 280 miles is 140 miles.

Now Try Exercise 103(b). ▶▶▶

7.1   Exercises

FOR EXTRA HELP

 Addison-Wesley Math Tutor Center

 MathXL

Digital Video Tutor CD 4 Videotape 7

Student's Solutions Manual

 MyMathLab MymathLab

MathXL Tutorials on CD

Write each percent as a decimal. See Examples 1 and 2.

1. 25%   **2.** 35%   **3.** 30%   **4.** 20%

5. 6%   **6.** 3%   **7.** 140%   **8.** 250%

9. 7.8%   **10.** 6.7%   **11.** 100%   **12.** 600%

13. 0.5%   **14.** 0.2%   **15.** 0.35%   **16.** 0.076%

Write each decimal as a percent. See Example 3.

17. 0.5   **18.** 0.6   **19.** 0.62   **20.** 0.18

21. 0.03   **22.** 0.09   **23.** 0.125   **24.** 0.875

25. 0.629   **26.** 0.494   **27.** 2   **28.** 5

29. 2.6   **30.** 1.8   **31.** 0.0312   **32.** 0.0625

Write each percent as a fraction or mixed number in lowest terms. See Examples 4 and 5.

33. 20%   **34.** 40%   **35.** 50%   **36.** 75%

37. 55%   **38.** 35%   **39.** 37.5%   **40.** 87.5%

41. 6.25%   **42.** 43.75%   **43.** $16\frac{2}{3}$%   **44.** $83\frac{1}{3}$%

45. 130%   **46.** 175%   **47.** 250%   **48.** 325%

Write each fraction as a percent. If you're using a calculator, first work each one by hand. Then use your calculator and round to the nearest tenth of a percent, if necessary. See Example 6.

49. $\dfrac{1}{4}$ **50.** $\dfrac{1}{5}$ **51.** $\dfrac{3}{10}$ **52.** $\dfrac{9}{10}$

53. $\dfrac{3}{5}$ **54.** $\dfrac{3}{4}$ **55.** $\dfrac{37}{100}$ **56.** $\dfrac{63}{100}$

57. $\dfrac{3}{8}$ **58.** $\dfrac{1}{8}$ **59.** $\dfrac{1}{20}$ **60.** $\dfrac{1}{50}$

61. $\dfrac{5}{9}$ **62.** $\dfrac{7}{9}$ **63.** $\dfrac{1}{7}$ **64.** $\dfrac{5}{7}$

In each statement, write percents as decimals and decimals as percents. See Examples 1–3.

65. In 1900, only 8% of U.S. homes had a telephone. (*Source: Harper's Index.*)

66. In 1900, only 14% of homes in the United States had a bathtub. (*Source: Harper's Index.*)

67. Tornadoes can occur on any day of the year, but 42% of them appear in May and June. (*Source: National Severe Storms Laboratory.*)

68. Only 2.1% of tornadoes occur in December, making it the least likely month for twisters. (*Source: National Severe Storms Laboratory.*)

69. The property tax rate in Alpine County is 0.035.

70. A church building fund has 0.49 of the money needed.

71. The number of people taking CPR training this session is 2 times that of the last session.

72. Attendance at this year's company picnic is 3 times last year's attendance.

Write a fraction and a percent for the shaded part of each figure. Then write a fraction and a percent for the unshaded part of each figure.

73.

74.

75.

76.

77.

78.

Complete this table. Write fractions and mixed numbers in lowest terms.

	Fraction	**Decimal**	**Percent**
79.	$\dfrac{1}{100}$	_____	_____
80.	$\dfrac{1}{10}$	_____	_____
81.	_____	0.2	_____
82.	_____	0.25	_____
83.	_____	_____	30%
84.	_____	_____	40%
85.	$\dfrac{1}{2}$	_____	_____
86.	$\dfrac{3}{4}$	_____	_____
87.	_____	_____	90%
88.	_____	_____	100%
89.	_____	1.5	_____
90.	_____	2.25	_____

This bar graph shows recent trends in employee benefits. It includes the most frequently offered "work perks" and the percent of the responding companies offering them. The survey was conducted on-line and included 4800 companies ranging in size from 2 to 5000 employees. Use this graph to answer Exercises 91–94. If the answer is a percent, also write it as a fraction in lowest terms. (Source: Work Perks Survey, Ceridian Employer Services.)

91. (a) What was the fourth-most-common perk?

(b) What portion of the companies offer that perk?

92. (a) What was the top work perk?

(b) What portion of the companies offer that perk?

WORK PERKS

- **82%** Casual dress
- **60.5%** Flexible hours
- **49%** Personal development training
- **40%** Employee entertainment/ Company product discounts
- **36%** Free food/Beverages
- **27%** Telecommuting
- **15%** Fitness centers
- **1%** On-site child care

Percent of Companies

Type of Perk

93. (a) Which perk was offered by *about* $\frac{1}{2}$ of the companies?

(b) Which perk was offered by *about* $\frac{1}{4}$ of the companies?

94. (a) If free food/beverages had been offered by $33\frac{1}{3}$% instead of 36% of the companies, what fraction would that be?

(b) What percent of the companies would have to offer flexible hours for the fraction to be $\frac{2}{3}$?

The diagram shows the different types of teeth in an adult's mouth. Use the diagram to answer Exercises 95–100. Write each answer in Exercises 95–98 as a fraction in lowest terms, as a decimal, and as a percent.

95. What portion of an adult's teeth are incisors, designed to bite and cut?

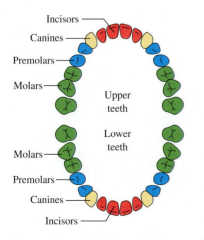

Incisors
Canines
Premolars
Molars

Upper teeth

Lower teeth

Molars
Premolars
Canines
Incisors

Source: Time-Life *Human Body.*

96. The pointy canine teeth tear and rip food. They are what portion of an adult's teeth?

97. The molars, which grind up food, are what portion of an adult's teeth?

98. What portion of an adult's teeth are premolars?

99. Some people have four fewer molars than shown in the tooth diagram. For these people, the canines are what portion of their teeth? Write your answer as a fraction in lowest terms and as a percent.

100. For the adults who have four fewer molars than shown, the incisors are what portion of their teeth? Write your answer as a fraction in lowest terms and as a percent.

101. Explain and correct the errors made by students when they used their calculators on these problems.

(a) Write $\frac{7}{20}$ as a percent.

Student entered 7 $\div$ 20 $=$, and the result was 0.35. So $\frac{7}{20} = 0.35\%$

(b) Write $\frac{16}{25}$ as a percent.

Student entered 25 $\div$ 16 $=$, and the result was 1.5625. So $\frac{16}{25} = 156.25\%$

102. Explain and correct the errors made by students who moved decimal points to solve these problems.

(a) Write 3.2 as a percent.

$03.2 = 0.032$ so $3.2 = 0.032\%$

(b) Write 60% as a decimal.

$00.60 = 0.0060$ so $60\% = 0.0060$

Fill in the blanks in Exercises 103–116. Remember that 100% *is* all *of something and* 50% *is* half *of it. See Examples 7 and 8.*

103. (a) 100% of $78 is _____.

(b) 50% of $78 is _____.

104. (a) 100% of 5 hours is _____.

(b) 50% of 5 hours is _____.

105. (a) 100% of 15 inches is _____.

(b) 50% of 15 inches is _____.

106. (a) 100% of $6000 is _____.

(b) 50% of $6000 is _____.

107. There are 20 children in the preschool class. 100% of the children are served breakfast and lunch. How many children are served both meals?

_____

108. The Speedy Delivery company owns 345 vans. 100% of the vans are painted white with blue lettering. How many company vans are painted white with blue lettering?

_____

109. (a) 100% of 2.8 miles is _____.

(b) 50% of 2.8 miles is _____.

110. (a) 100% of $2.50 is _____.

(b) 50% of $2.50 is _____.

111. (a) John owes $285 for tuition. Financial aid will pay 50% of the cost. Financial aid will pay

_____.

(b) What percent of the tuition will John have to pay?

_____

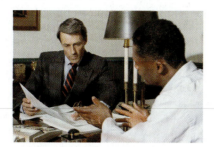

112. (a) The Animal Humane Society took in 20,000 animals last year. About 50% of them were dogs. The number of dogs taken in was about how many?

_____

(b) What percent of the animals were _not_ dogs?

_____

113. (a) About 50% of the 8200 students at our college work more than 20 hours per week. How many students is this?

_____

(b) What percent of the students work 20 hours or less per week?

_____

114. (a) Shalayna's cell phone plan includes 1600 minutes of "off-peak" calling time per month. Last month she used 50% of her "off-peak" minutes. How many minutes did she use?

_____

(b) What percent of her "off-peak" minutes were not used?

_____

115. (a) Dylan's test score was 100%. How many of the 35 problems did he work correctly?

_____

(b) How many problems did Dylan miss?

_____

116. (a) Latrell made 100% of his 12 free throws during a practice game today. How many free throws did he make?

_____

(b) How many free throws did Latrell miss?

_____

117. Describe a shortcut way to find 50% of a number. Include two examples to illustrate the shortcut.

118. Describe a shortcut way to find 100% of a number. Include two examples to illustrate the shortcut.

7.2 The Percent Proportion

OBJECTIVES

1. Identify the percent, whole, and part.
2. Solve percent problems using the percent proportion.

We will show you two ways to solve percent problems. One is the proportion method, which we discuss in this section. The other is the percent equation method, which we explain in **Section 7.3.**

OBJECTIVE 1 Identify the percent, whole, and part. You have learned that a statement of two equivalent ratios is called a proportion (see **Section 6.3**). For example, the fraction $\frac{3}{5}$ is the same as the ratio 3 to 5, and 60% is the ratio 60 to 100. As the figure below shows, these two ratios are equivalent and make a proportion.

The **percent proportion** can be used to solve percent problems.

The Percent Proportion

Percent is to *100* **as** *part* is to *whole*.

$$\frac{\text{percent}}{100} = \frac{\text{part}}{\text{whole}}$$

Always 100 ⟶ because percent means "per 100"

In some textbooks the percent proportion is written using the words *amount* and *base*.

$$\frac{\text{percent}}{100} = \frac{\text{amount}}{\text{base}}$$

Here is the proportion for the figure at the top of the page.

60% means 60 parts out of 100 parts. $\quad \dfrac{60}{100} = \dfrac{3}{5}$ ← Shaded (3 parts) ← Whole (5 parts)

If we write $\frac{60}{100}$ in lowest terms, it is equal to $\frac{3}{5}$, so the proportion is true.

$$\frac{60}{100} = \frac{60 \div 20}{100 \div 20} = \frac{3}{5} \quad \begin{cases} \text{Matches ratio on} \\ \text{right side of proportion} \end{cases}$$

In order to use the percent proportion to solve problems, you must be able to pick out the *percent,* the *whole,* and the *part.* Look for the percent first, because it is the easiest to identify.

Identifying the Percent

The **percent** is a ratio of a part to a whole, with 100 as the denominator. In a problem, the percent appears with the word *percent* or with the symbol % after it.

EXAMPLE 1 **Identifying the Percent in Percent Problems**

Identify the percent in each problem.

(a) 32% of the 900 women were retired. How many were retired?

Percent

The percent is 32. The number 32 appears with the symbol %.

(b) $150 is 25 percent of what number?

Percent

The percent is 25 because 25 appears with the word *percent*.

(c) If 7 students failed, what percent of the 350 students failed?

Percent (unknown)

The word *percent* has no number with it, so the percent is the unknown part of the problem.

 Now Try Exercises 7(a) and 9(a).

The second thing to look for in a percent problem is the *whole* (sometimes called the *base*).

> ### Identifying the Whole
> The **whole** is the entire quantity. In a problem, the *whole* often appears after the word **of**.

EXAMPLE 2 **Identifying the Whole in Percent Problems**

These problems are the same as those in Example 1 above. Now identify the *whole*.

(a) 32% **of** the 900 women were retired. How many were retired?

Percent   Whole

The whole is 900 women. The number 900 appears after the word *of*.

(b) $150 is 25 percent **of** what number?

Percent   Whole (unknown; follows *of*)

The whole is unknown.

(c) If 7 students failed, what percent **of** the 350 students failed?

Percent (unknown)   Whole (follows *of*)

The whole is 350 students. The number 350 appears after the word *of*.

 Now Try Exercises 7(b) and 9(b).

The third and final thing to identify in a percent problem is the *part* (sometimes called the *amount*).

> ### Identifying the Part
> The **part** is the number being compared to the whole.

> **NOTE**
> If you have trouble identifying the *part,* find the *whole* and the *percent* first. The remaining number is the *part.*

EXAMPLE 3 Identifying the Part in Percent Problems

These problems are the same as those in Examples 1 and 2. Identify the *part.*

(a) 32% **of** the 900 women were retired. How many were retired?

Percent   Whole   Part (unknown)

The part of the women who were retired is unknown. In other words, some part of 900 women were retired.

(b) $150 is 25 percent **of** what number?

Part   Percent   Whole (unknown)

150 is the remaining number, so 150 is the part.

(c) If 7 students failed, what percent **of** 350 students failed?

Part   Percent (unknown)   Whole

The part of the students who failed is 7 students.

Now Try Exercises 7(c) and 9(c).

OBJECTIVE 2 Solve percent problems using the percent proportion.

EXAMPLE 4 Using the Percent Proportion to Find the Part

Use the percent proportion to answer this question.

15% of $165 is how much money?

Recall that the percent proportion is $\dfrac{\text{percent}}{100} = \dfrac{\text{part}}{\text{whole}}$. First identify the percent by looking for the % symbol or the word *percent.* Then look for the *whole* (usually follows the word *of*). Finally, identify the *part.*

15% **of** $165 is how much money?

Percent   Whole   Part (unknown)
            (follows *of*)

Set up the percent proportion. Here we use *n* as the variable representing the unknown part. You may use any letter you like.

$$\text{Percent} \rightarrow \ \frac{15}{100} = \frac{n}{165} \ \begin{array}{l} \leftarrow \text{Part (unknown)} \\ \leftarrow \text{Whole} \end{array}$$
$$\text{Always } 100 \rightarrow$$

Recall from **Section 6.3** that the first step in solving a proportion is to find the cross products.

Step 1 $\dfrac{15}{100} = \dfrac{n}{165}$ $\begin{array}{l} 100 \cdot n \\ 15 \cdot 165 \end{array}$ ⎤ Find the cross products.

Continued on Next Page

Step 2 $100 \cdot n = \underline{15 \cdot 165}$ Show that the cross products are equivalent.

$100 \cdot n = \quad 2475$

Step 3 $\dfrac{\overset{1}{\cancel{100}} \cdot n}{\cancel{100}_{\,1}} = \dfrac{2475}{100}$ Divide both sides by 100, the coefficient of the variable term. On the left side, divide out the common factor of 100.

$n = 24.75$ On the right side, $2475 \div 100$ is 24.75

The part is **$24.75**, so 15% of $165 is $24.75.

> **CAUTION**
> When you use the percent proportion, do ***not*** move the decimal point in the percent or in the answer.

Now Try Exercise 3. ▶▶▶

EXAMPLE 5 **Using the Percent Proportion to Find the Percent**

Use the percent proportion to answer this question.

8 pounds is what percent **of** 160 pounds?

↓	↓	↓
Part	Percent (unknown)	Whole (follows *of*)

The percent proportion is $\dfrac{\text{percent}}{100} = \dfrac{\text{part}}{\text{whole}}$. Set up the proportion using p as the variable representing the unknown percent. Then find the cross products.

Percent (unknown) → $\dfrac{p}{100} = \dfrac{8}{160}$ ← Part
Always 100 → $\qquad\qquad$ ← Whole

$$\dfrac{p}{100} \bowtie \dfrac{8}{160} \qquad \begin{array}{l} 100 \cdot 8 = 800 \\ p \cdot 160 \end{array} \Big\} \text{Cross products}$$

$p \cdot 160 = 800$ Show that the cross products are equivalent.

$\dfrac{p \cdot \overset{1}{\cancel{160}}}{\cancel{160}_{\,1}} = \dfrac{800}{160}$ Divide both sides by 160

$p = 5$

The percent is **5%**. So 8 pounds is 5% of 160 pounds.

> **CAUTION**
> When you're finding an unknown percent, as in Example 5 above, be careful to label your answer with the % symbol. Do ***not*** add a decimal point or move the decimal point in your answer.

Now Try Exercise 5. ▶▶▶

EXAMPLE 6 **Using the Percent Proportion to Find the Whole**

Use the percent proportion to answer this question.

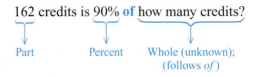

162 credits is 90% **of** how many credits?

Part   Percent   Whole (unknown);
                       (follows *of*)

Percent → $\dfrac{90}{100} = \dfrac{162}{n}$ ← Part
Always 100 →   ← Whole (unknown)

$$\dfrac{90}{100} = \dfrac{162}{n} \qquad 100 \cdot 162 = 16{,}200$$
$$90 \cdot n \qquad\qquad\qquad \text{Cross products}$$

$90 \cdot n = 16{,}200$ Show that the cross products are equivalent.

$$\dfrac{\overset{1}{\cancel{90}} \cdot n}{\underset{1}{\cancel{90}}} = \dfrac{16{,}200}{90}$$ Divide both sides by 90

$$n = 180$$

The whole is **180 credits**. So 162 credits is 90% of 180 credits.

Now Try Exercise 11.

So far in all the examples, the part has been *less* than the whole. This is because all the percents have been less than 100%. Recall that 100% of something is *all* of it. When the percent is *less* than 100%, you have *less* than all of it.

Now let's look at percents *greater* than 100%. For example,

100% of $20 is all of the money, or $20.

150% of $20 is *more* than $20.

100%	+	**50%**	=	**150%**
of the money is		of the money is		of the money is
$20	+	**$10**	=	**$30**

So 150% of $20 is $30.

When the percent is *greater* than 100%, the part is *larger* than the whole, as you'll see in Example 7.

EXAMPLE 7 **Working with Percents Greater Than 100%**

Use the percent proportion to answer each question.

(a) How many students is 210% **of** 40 students?

Part (unknown)   Percent   Whole (follows *of*)

Percent → $\dfrac{210}{100} = \dfrac{n}{40}$ ← Part (unknown)
Always 100 →   ← Whole

$\dfrac{210}{100} = \dfrac{n}{40}$

$100 \cdot n$ ←
$210 \cdot 40 = 8400$ ←   Cross products

$100 \cdot n = 8400$ Show that the cross products
                         are equivalent.

$\dfrac{\overset{1}{\cancel{100}} \cdot n}{\underset{1}{\cancel{100}}} = \dfrac{8400}{100}$ Divide both sides by 100

$n = 84$

The part is **84 students**, which is *more* than the whole of 40 students. This result makes sense because the percent is 210%. If it was exactly 200%, we would have *2 times the whole,* and 2 times 40 students is 80 students. So 210% should be even a little more than 80 students. Our answer of 84 students is reasonable.

(b) $68 is what percent **of** $50?

Part   Percent   Whole
         (unknown)   (follows *of*)

Percent (unknown) → $\dfrac{p}{100} = \dfrac{68}{50}$ ← Part
Always 100 →   ← Whole

$\dfrac{p}{100} = \dfrac{68}{50}$

$100 \cdot 68 = 6800$ ←
$p \cdot 50$ ←   Cross products

$p \cdot 50 = 6800$ Show that the cross products
                        are equivalent.

$\dfrac{p \cdot \overset{1}{\cancel{50}}}{\underset{1}{\cancel{50}}} = \dfrac{6800}{50}$ Divide both sides by 50

$p = 136$

The percent is **136%**. This result makes sense because $68 is *more* than $50, so $68 has to be *more than 100%* of $50.

Now Try Exercise 13. ▶▶▶

7.2 Exercises

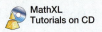
*In Exercises 1–12, **(a)** identify the percent; **(b)** identify the whole; **(c)** identify the part;*
(d) *write a percent proportion and solve it to answer the question. See Examples 1–6.*

1. What is 10% of 3000 runners?

2. What is 35% of 2340 volunteers?

3. 4% of 120 feet is how many feet?

4. 9% of $150 is how much money?

5. 16 pepperoni pizzas is what percent of 32 pizzas?

6. 35 hours is what percent of 140 hours?

7. What percent of 200 calories is 16 calories?

8. What percent of 350 parking spaces is 7 handicapped parking spaces?

9. 495 successful students is 90% of what number of students?

10. 84 e-mails is 28% of what number of e-mails?

11. $12\frac{1}{2}$% of what amount is $3.50?

12. $5\frac{1}{2}$% of what amount is $17.60?

Write a percent proportion and solve it to answer Exercises 13–24. If necessary, round money answers to the nearest cent and percent answers to the nearest tenth of a percent. See Examples 4–7.

13. 250% of 7 hours is how long?

14. What is 130% of 60 trees?

15. What percent of $172 is $32?

16. $14 is what percent of $398?

17. 748 books is 110% of what number of books?

18. 145% of what number of inches is 11.6 inches?

19. What is 14.7% of $274?

20. 8.3% of $43 is how much?

21. 105 employees is what percent of 54 employees?

22. What percent of 46 animals is 100 animals?

23. $0.33 is 4% of what amount?

24. 6% of what amount is $0.03?

25. A student turned in the following answers on a test. You can see that *two* of the answers are incorrect *without working the problems*. Find the incorrect answers and explain how you identified them (without actually solving the problems).

$$50\% \text{ of } \$84 \text{ is } \underline{\quad \$42 \quad}$$
$$150\% \text{ of } \$30 \text{ is } \underline{\quad \$20 \quad}$$
$$25\% \text{ of } \$16 \text{ is } \underline{\quad \$32 \quad}$$
$$100\% \text{ of } \$217 \text{ is } \underline{\quad \$217 \quad}$$

26. Name the three parts in a percent problem. For each of these three parts, write a sentence telling how you identify it.

27. Explain and correct the *two* errors that a student made when solving this problem:
$14 is what percent of $8?

$$\frac{p}{100} = \frac{8}{14} \qquad p \cdot 14 = 100 \cdot 8$$

$$\frac{p \cdot \overset{1}{\cancel{14}}}{\underset{1}{\cancel{14}}} = \frac{800}{14}$$

$$p \approx 57.1$$

The answer is 57.1 (rounded).

28. Explain and correct the *two* errors that a student made when solving this problem: 9 children is 30% of what number of children?

$$\frac{30}{100} = \frac{n}{9} \qquad 100 \cdot n = 30 \cdot 9$$

$$\frac{\overset{1}{\cancel{100}} \cdot n}{\underset{1}{\cancel{100}}} = \frac{270}{100}$$

$$n = 2.7$$

The answer is 2.7%.

7.3 The Percent Equation

OBJECTIVE 1 Estimate answers to percent problems involving 25%. Before showing you the percent equation, we need to do some more estimation. As you have learned when working with integers, fractions, and decimals, it is always a good idea to estimate the answer. Doing so helps you catch mistakes. Also, when you're out shopping or eating in a restaurant, you will be able to estimate the sales tax, discount, or tip.

In **Section 7.1** we used shortcuts for 100% of a number (all of the number) and 50% of a number (divide the number by 2). Now let's look at a quick way to work with 25%.

25% means 25 parts out of 100 parts, or $\frac{25}{100}$, which is the same as $\frac{1}{4}$

25% of $40 would be $\frac{1}{4}$ of $40, or $10

A quick way to find $\frac{1}{4}$ of a number is to *divide it by 4*. Recall that the denominator, 4, tells you that the whole is divided into 4 equal parts.

EXAMPLE 1 Estimating 25% of a Number

Estimate the answer to each question.

(a) What is 25% of $817?
Use front end rounding to round $817 to $800. Then divide $800 by 4. The estimate is **$200**.

(b) Find 25% of 19.7 miles.
Use front end rounding to round 19.7 miles to 20 miles. Then divide 20 miles by 4. The estimate is **5 miles**.

(c) 25% of 49 days is how long?
You could round 49 days to 50 days, using front end rounding. Then divide 50 by 4 to get an estimate of **12.5 days**.
However, the division step is simpler if you notice that 48 is a multiple of 4. You can round 49 days to 48 days and divide by 4 to get an estimate of **12 days**. Either way gives you a fairly good idea of the correct answer.

Now Try Exercise 3.

OBJECTIVE 2 Find 10% and 1% of a number by moving the decimal point. There are also helpful shortcuts for finding 10% or 1% of a number.

Ten percent, or 10%, means 10 parts out of 100 parts or $\frac{10}{100}$, which is the same as $\frac{1}{10}$. A quick way to find $\frac{1}{10}$ of a number is to *divide it by 10*. The denominator, 10, tells you that the whole is divided into 10 equal parts. The shortcut for dividing by 10 is to move the decimal point *one* place to the *left*.

EXAMPLE 2 Finding 10% of a Number by Moving the Decimal Point

Find the *exact* answer to each question by moving the decimal point.

(a) What is 10% of $817?
To find 10% of $817, divide $817 by 10. Do the division by moving the decimal point *one* place to the *left*. The decimal point starts at the far right side of $817.

10% of $817. = $81.70 ← Exact answer

↑
Write this 0 because it's money.

So 10% of $817 is **$81.70**.

Continued on Next Page

(b) Find 10% of 19.7 miles.

To find 10% of 19.7 miles, divide 19.7 by 10. Move the decimal point *one* place to the *left*.

$$10\% \text{ of } 19.7 \text{ miles} = 1.97 \text{ miles} \longleftarrow \text{Exact answer}$$

So 10% of 19.7 miles is **1.97 miles**.

> **Now Try Exercise 5.** ▶▶▶

One percent, or 1%, is 1 part out of 100 parts or $\frac{1}{100}$. This time the denominator of 100 tells you that the whole is divided into 100 parts. Recall that a quick way to divide by 100 is to move the decimal point *two* places to the *left*.

EXAMPLE 3 **Finding 1% of a Number by Moving the Decimal Point**

Find the *exact* answer to each question by moving the decimal point.

(a) What is 1% of $817?

To find 1% of $817, divide $817 by 100. Do the division by moving the decimal point *two* places to the *left*.

$$1\% \text{ of } \$817. = \$8.17 \longleftarrow \text{Exact answer}$$

So 1% of $817 is **$8.17**.

(b) Find 1% of 19.7 miles.

To find 1% of 19.7 miles, divide 19.7 by 100. Move the decimal point *two* places to the *left*.

$$1\% \text{ of } 19.7 \text{ miles} = 0.197 \text{ mile} \longleftarrow \text{Exact answer}$$

So 1% of 19.7 miles is **0.197 mile**.

> **Now Try Exercise 9.** ▶▶▶

Here is a summary of the shortcuts you can use with percents.

Percent Shortcuts

200% of a number is 2 times the number; **300% of a number** is 3 times the number; and so on.

100% of a number is the entire number.

To find **50% of a number,** divide the number by 2.

To find **25% of a number,** divide the number by 4.

To find **10% of a number,** divide the number by 10. To do the division, move the decimal point in the number *one* place to the *left*.

To find **1% of a number,** divide the number by 100. To do the division, move the decimal point in the number *two* places to the *left*.

OBJECTIVE **3** **Solve basic percent problems using the percent equation.** In **Section 7.2** you used a proportion to solve percent problems. Now you will learn how to solve these problems using the percent equation.

> **Percent Equation**
>
> $$\text{percent } \textbf{of} \text{ whole} = \text{part}$$
>
> The word **of** indicates multiplication, so the **percent equation** becomes
>
> $$\text{percent} \cdot \text{whole} = \text{part}$$
>
> Be sure to write the percent as a decimal or fraction before using the equation.

The percent equation is just a rearrangement of the percent proportion. Recall that in the proportion you wrote the percent over 100. Because there is no 100 in the equation, you have to change the percent to a decimal or fraction by dividing by 100 *before* using the equation.

> **NOTE**
> Once you have set up a percent equation, we encourage you to use your calculator to do the multiplying or dividing needed to solve the equation. For this reason, we will always write the percent as a decimal. If you're doing the problems by hand, changing the percent to a fraction may be easier at times. Either method will work.

Examples 4, 5, and 6 below are the same percent questions that were in the examples in **Section 7.2**. There we used a proportion to answer each question. Now we will use an equation to answer them. You can then compare the equation method with the proportion method.

EXAMPLE 4 **Using the Percent Equation to Find the Part**

Write and solve a percent equation to answer each question.

(a) 15% of $165 is how much money?

Translate the sentence into an equation. Recall that *of* indicates multiplication and *is* translates to the equal sign. The percent must be written in decimal form. Use any letter you like to represent the unknown quantity. (We will use n for an unknown number and p for an unknown percent.)

$$15.\% \text{ of } \$165 \text{ is how much money?}$$

Write 15% as the decimal 0.15

$$0.15 \quad \cdot \quad 165 \quad = \quad n$$

To solve the equation, simplify the left side, multiplying 0.15 by 165.

$$(0.15)(165) = n$$
$$24.75 \quad = n$$

So 15% of $165 is **$24.75**, which matches the answer obtained using a proportion (see Example 4 in **Section 7.2**).

Continued on Next Page

Check Use estimation to check that the solution is reasonable. First find 10% of $165 by moving the decimal point.

10% of 165. is $16.50   and

5% of $165 would be half as much, that is, half of $16.50 or about $8.

So the *estimate* for 15% of $165 is $16.50 + $8 = $24.50. The exact answer of $24.75 is very close to this estimate, so it is reasonable.

(b) How many students is 210% of 40 students?
   Translate the sentence into an equation. Write the percent in decimal form.

$$\text{How many students is } 210.\% \text{ of } 40 \text{ students?}$$

Write 210% as the decimal 2.10

$$n = 2.10 \cdot 40$$

This time the two sides of the percent equation are reversed, so

$$\text{part} = \text{percent} \cdot \text{whole.}$$

Recall that the variable may be on either side of the equal sign. To solve the equation, simplify the right side, multiplying 2.10 by 40.

$$n = (2.10)(40)$$
$$n = 84$$

So **84 students** is 210% of 40 students. This matches the answer obtained by using a proportion (see Example 7(a) in **Section 7.2**).

Check Use estimation to check that the solution is reasonable. 210% is close to 200%.

200% of 40 students is 2 times 40 students = 80 students ← Estimate

The exact answer of 84 students is close to the estimate, so it is reasonable.

Now Try Exercise 27. ▷▷▷

EXAMPLE 5 **Using the Percent Equation to Find the Percent**

Write and solve a percent equation to answer each question.

(a) 8 pounds is what percent of 160 pounds?
   Translate the sentence into an equation. This time the percent is unknown. Do ***not*** move the decimal point in the other numbers.

$$\text{8 pounds is what percent of 160 pounds?}$$

$$8 = p \cdot 160$$

To solve the equation, divide both sides by 160.

On the left side, divide 8 by 160 $\dfrac{8}{160} = \dfrac{p \cdot \overset{1}{\cancel{160}}}{\underset{1}{\cancel{160}}}$ On the right side, divide out the common factor of 160

Solution in *decimal* form $0.05 = p$

Now ***multiply the solution by 100*** to change it from a *decimal* to a *percent*.

$$0.05 = 5\%$$

So 8 pounds is **5%** of 160 pounds. This matches the answer obtained by using a proportion (see Example 5 in **Section 7.2**).

Continued on Next Page

Check The solution makes sense because 10% of 160 pounds would be 16 pounds.

$$\text{10\% of 160. pounds is 16 pounds}\quad\text{so}$$

5% of 160 pounds is half as much, that is, half of 16 pounds, or 8 pounds.

8 pounds matches the number given in the original problem, so 5% is the correct solution.

(b) What percent of $50 is $68?
    Translate the sentence into an equation and solve it.

$$\underbrace{\text{What percent}}\;\text{of }\$50\text{ is }\$68?$$

$$p \quad\cdot\quad 50 = 68$$

$$\frac{p \cdot 50}{50} = \frac{68}{50}\qquad \text{Divide both sides by 50}$$

$$p = 1.36 \;\leftarrow \text{Solution in } decimal \text{ form}$$

Now ***multiply the solution by 100*** to change it from a *decimal* to a *percent*.

$$1.36 = 136\%$$

So **136%** of $50 is $68. This matches the answer obtained by using a proportion (see Example 7(b) in **Section 7.2**).

Check The solution makes sense because 100% of $50 would be $50 (all of it), and 200% of $50 would be 2 times $50, or $100. So $68 has to be between 100% and 200%.

136% is between   100% of $50 = $50 $68 is between
100% and 200%.   200% of $50 = $100 $50 and $100

The solution of 136% fits the conditions.

> **CAUTION**
> When you use an equation to solve for an unknown percent, *the solution will be in decimal form.* Remember to ***multiply the solution by 100*** to change it from decimal form to a percent. The shortcut is to move the decimal point in the solution *two* places to the *right* and attach the % symbol.

Now Try Exercise 31.

EXAMPLE 6 **Using the Percent Equation to Find the Whole**

Write and solve a percent equation to answer each question.

(a) 162 credits is 90% of how many credits?
Translate the sentence into an equation. Write the percent in decimal form.

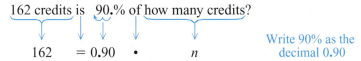

162 credits is 90.% of how many credits?

$$162 \quad = 0.90 \quad \cdot \quad n \qquad \text{Write 90\% as the decimal 0.90}$$

Recall that 0.90 is equivalent to 0.9, so use 0.9 in the equation.

$$\frac{162}{0.9} = \frac{(0.9)(n)}{0.9} \qquad \text{Divide both sides by 0.9}$$

$$180 = n$$

So 162 credits is 90% of **180 credits**.

Check The solution makes sense because 90% of 180 credits should be 10% less than 100% of the credits, and 10% of 180. credits is 18 credits.

100% **−** **10%** **=** **90%**
of 180 credits **of 180. credits** **of 180 credits**

180 credits **−** **18 credits** **=** **162 credits** ← Matches the number given in the original problem

(b) 250% of what amount is $75?
Translate the sentence into an equation. Write the percent in decimal form.

250.% of what amount is $75?

Write 250% as the decimal 2.5

$$2.5 \quad \cdot \quad n \quad = \quad 75$$

$$\frac{(2.5)(n)}{2.5} = \frac{75}{2.5} \qquad \text{Divide both sides by 2.5}$$

$$n = 30$$

So 250% of **$30** is $75.

Check The solution makes sense because 200% of $30 is 2 times $30 = $60, and 50% of $30 is $30 ÷ 2 = $15.

200% **+** **50%** **=** **250%**
of $30 **of $30** **of $30**

(2)($30) **$30 ÷ 2**

$60 **+** **$15** **=** **$75** ← Matches the number given in the original problem

Now Try Exercise 41. ▸▸▸

7.3 Exercises

FOR EXTRA HELP

 Addison-Wesley Math Tutor Center

 MathXL

 Digital Video Tutor CD 4 Videotape 7

Student's Solutions Manual

 MyMathLab

MathXL Tutorials on CD

*Use your estimation skills and the percent shortcuts to select the most reasonable answers. Circle your choices. Do **not** write an equation or proportion and solve it. See Examples 1–3.*

1. Find 50% of 3000 patients.

150 patients   1500 patients   300 patients

2. What is 50% of 192 pages?

48 pages   384 pages   96 pages

3. 25% of $60 is how much?

$15 $6 $30

4. Find 25% of $2840.

$28.40 $710 $284

5. What is 10% of 45 pounds?

0.45 pound   22.5 pounds   4.5 pounds

6. 10% of 7 feet is how many feet?

0.7 foot   3.5 feet   14 feet

7. Find 200% of $3.50.

$0.35 $1.75 $7.00

8. What is 300% of $12?

$4 $36 $1.20

9. 1% of 5200 students is how many students?

520 students   52 students   2600 students

10. Find 1% of 460 miles.

0.46 mile   46 miles   4.6 miles

11. Find 10% of 8700 cell phones.

8700 phones   4350 phones   870 phones

12. 25% of 128 CDs is how many CDs?

64 CDs   32 CDs   1280 CDs

13. What is 25% of 19 hours?

4.75 hours   1.9 hours   2.5 hours

14. What is 1% of $37?

$370 $3.70 $0.37

15. (a) Describe a shortcut for finding 10% of a number and explain *why* your shortcut works.

16. (a) Describe a shortcut for finding 1% of a number and explain *why* it works.

(b) Once you know 10% of a certain number, explain how you could use that information to find 20% and 30% of the same number.

(b) Once you know 1% of a certain number, explain how you could use that information to find 2% and 3% of the same number.

Write and solve an equation to answer each question in Exercises 17–48. See Examples 4–6.

17. 35% of 660 programs is how many programs?

18. 55% of 740 canisters is how many canisters?

19. 70 truckloads is what percent of 140 truckloads?

20. 30 crew members is what percent of 75 crew members?

21. 476 circuits is 70% of what number of circuits?

22. 621 tons is 45% of what number of tons?

23. $12\frac{1}{2}$% of what number of people is 135 people?

24. $6\frac{1}{2}$% of what number of bottles is 130 bottles?

25. What is 65% of 1300 species?

26. What is 75% of 360 dosages?

27. 4% of $520 is how much?

28. 7% of $480 is how much?

29. 38 styles is what percent of 50 styles?

30. 75 offices is what percent of 125 offices?

31. What percent of $264 is $330?

32. What percent of $480 is $696?

33. 141 employees is 3% of what number of employees?

34. 16 books is 8% of what number of books?

35. 32% of 260 quarts is how many quarts?

36. 44% of 430 liters is how many liters?

37. $1.48 is what percent of $74?

38. $0.51 is what percent of $8.50?

39. How many tablets is 140% of 500 tablets?

40. How many patients is 175% of 540 patients?

41. 40% of what number of salads is 130 salads?

42. 75% of what number of wrenches is 675 wrenches?

43. What percent of 160 liters is 2.4 liters?

44. What percent of 600 miles is 7.5 miles?

45. 225% of what number of gallons is 11.25 gallons?

46. 180% of what number of ounces is 6.3 ounces?

47. What is 12.4% of 8300 meters?

48. What is 13.2% of 9400 acres?

49. Explain and correct the error in each of these solutions.

(a) 3 hours is what percent of 15 hours?

$$3 = p \cdot 15$$

$$\frac{3}{15} = \frac{p \cdot \cancel{15}^{1}}{\cancel{15}_{1}}$$

$$0.2 = p$$

The answer is 0.2%.

(b) $50 is what percent of $20?

$$50 \cdot p = 20$$

$$\frac{\cancel{50}^{1} \cdot p}{\cancel{50}_{1}} = \frac{20}{50}$$

$$p = 0.40 = 40\%$$

The answer is 40%.

50. Explain and correct the error in each of these solutions.

(a) 12 inches is 5% of what number of inches?

$$(12)(0.05) = n$$

$$0.6 = n$$

The answer is 0.6 inch.

(b) What is 4% of 30 pounds?

$$n = (4)(30)$$

$$n = 120$$

The answer is 120 pounds.

RELATING CONCEPTS (EXERCISES 51–52) For Individual or Group Work

Use your knowledge of fractions to **work Exercises 51 and 52 in order.**

51. Suppose that you have this problem:
$33\frac{1}{3}\%$ of $162 is how much?

(a) First, change $33\frac{1}{3}\%$ to a fraction. (See **Section 7.1** for help.) Then, write an equation and solve it.

(b) Now solve the problem by changing $33\frac{1}{3}\%$ to a decimal. (*Hint:* Look at part (a) to see what fraction is equivalent to $33\frac{1}{3}\%$. Change the fraction to a decimal, using your calculator. Keep *all* the decimal places shown in the calculator's display window. Now write the equation and solve it, using your calculator.)

(c) Compare your answers from part (a) and part (b). How different are they?

52. Now suppose that you have this problem:
22 cans is $66\frac{2}{3}\%$ of what number of cans?

(a) First, change $66\frac{2}{3}\%$ to a fraction. (See **Section 7.1** for help.) Then, write an equation and solve it. (See **Section 4.7** for help.)

(b) Now solve the problem by changing $66\frac{2}{3}\%$ to a decimal. Use your calculator and keep all the decimal places shown. Now write the equation and solve it.

(c) Compare your answers from part (a) and part (b). How different are they?

Summary Exercises on Percent

1. Complete this table. Write fractions in lowest terms and as whole or mixed numbers when possible.

	Fraction	Decimal	Percent
(a)	$\dfrac{3}{100}$		
(b)			30%
(c)		0.375	
(d)			160%
(e)	$\dfrac{1}{16}$		
(f)			5%
(g)		2.0	
(h)	$\dfrac{4}{5}$		
(i)		0.072	

2. Use percent shortcuts to answer these questions.

   (a) 10% of 35 ft is _____.

   (b) 100% of 19 miles is _____.

   (c) 50% of 210 cows is _____.

   (d) 1% of $8 is _____.

   (e) 25% of 2000 women is _____.

   (f) 300% of $15 is _____.

   (g) 10% of $875 is _____.

   (h) 25% of 48 pounds is _____.

   (i) 1% of 9500 students is _____.

Use the percent proportion or percent equation to answer each question. If necessary, round money amounts to the nearest cent and percent answers to the nearest tenth of a percent.

3. 9 Web sites is what percent of 72 Web sites?

4. 30 DVDs is 40% of what number of DVDs?

5. 6% of $8.79 is how much?

6. 945 students is what percent of 540 students?

7. $3\frac{1}{2}$% of 168 pounds is how much, to the nearest tenth of a pound?

8. 1.25% of what number of hours is 7.5 hours?

9. What percent of 80,000 deer is 40,000 deer?

10. 465 camp sites is 93% of what number of camp sites?

11. What number of golf balls is 280% of 35 golf balls?

12. What percent of $66 is $1.80?

13. 9% of what number of apartments is 207 apartments?

14. What weight is 84% of 0.75 ounce?

15. $1160 is what percent of $800?

16. Find 3.75% of 6500 voters, to the nearest whole number.

17. What is 300% of 0.007 inch?

18. 24 minutes is what percent of 6 minutes?

19. What percent of 60 yards is 4.8 yards?

20. $0.17 is 25% of what amount?

The circle graph shows the average costs for various wedding expenses. Use the graph to answer Exercises 21–24. Round money answers to the nearest cent, if necessary.

WHAT IT COSTS TO RING THE WEDDING BELLS
The average wedding has 200 guests and costs $20,500. Here is how couples are spending the money.

Invitations, clergy, limousine 6%
Rings for bride and groom 20%
Reception 50%
Flowers $4\frac{1}{2}$%
Photography/Videography 14.6%
Bride's dress 5%
Percents are rounded.

Source: Bride's magazine.

21. Which item is least expensive, and how much is spent on it?

22. What amount is paid for invitations, clergy, and limousine?

23. Find the cost of photography and videography.

24. How much, on average, is spent for each guest at the reception?

7.4 Problem Solving with Percent

OBJECTIVE 1 Solve percent application problems. Solving percent problems involves identifying three items: the *percent,* the *whole,* and the *part.* Then you can write a percent equation or percent proportion and solve it to answer the question in the problem. Use the six problem-solving steps from **Section 3.3.**

EXAMPLE 1 Finding the Part

A new low-income housing project charges 30% of a family's income as rent. The Smiths' family income is $1260 per month. How much will the Smiths pay for rent?

Step 1 **Read the problem.** It is about a family paying part of its income for rent.

> **Unknown:** amount of rent
> **Known:** 30% of income paid for rent; $1260 monthly income

Step 2 **Assign a variable.** There is only one unknown, so let n be the amount paid for rent.

Step 3 **Write an equation.** Use the percent equation.

$$\text{percent} \cdot \text{whole} = \text{part}$$

Recall that the *whole* often follows the word *of.*

The percent is given in the problem: 30%. The key word **of** appears *right after* 30%, which means that you can use the phrase "30% **of** a family's income" to help you write one side of the equation. Write 30% as the decimal 0.30.

30.% **of** a family's income is the rent.

$$0.30 \quad \cdot \quad \$1260 \quad = \quad n$$

Step 4 **Solve the equation.** Simplify the left side, multiplying 0.30 by 1260.

$$\underbrace{(0.30)(1260)}_{378} = n$$
$$378 = n$$

Step 5 **State the answer.** The Smith family will pay **$378** for rent.

Step 6 **Check** the solution. The solution of $378 makes sense because 10% of $1260. is $126, so 30% would be 3 times $126, or $378.

> **NOTE**
> You could also use the percent proportion in *Step 3* above.
>
> $$\text{Percent} \rightarrow \frac{30}{100} = \frac{n}{1260} \begin{array}{l} \leftarrow \text{Part (unknown)} \\ \leftarrow \text{Whole} \end{array}$$
> Always 100 $\rightarrow$
>
> The answer will be the same, $378.
>
> Throughout the rest of this chapter, we will use the percent equation. You may, if you wish, use the percent proportion instead. The final answers will be the same.

Now Try Exercise 1.

EXAMPLE 2 Finding the Part

When Britta received her first $180 paycheck as a math tutor, $12\frac{1}{2}\%$ was withheld for federal income tax. How much was withheld?

Step 1   **Read** the problem. It is about part of Britta's pay being withheld for taxes.

> **Unknown:** amount withheld for taxes
>
> **Known:** $12\frac{1}{2}\%$ of earnings withheld; $180 in pay

Step 2   **Assign a variable.** There is only one unknown, so let n be the amount withheld for taxes.

Step 3   **Write an equation.** Use the percent equation. The *percent* is given: $12\frac{1}{2}\%$. Write $12\frac{1}{2}\%$ as 12.5% and then move the decimal point two places to the left. So 12.5% becomes the decimal 0.125.

   The key word *of* doesn't appear after $12\frac{1}{2}\%$. Instead, think about whether you know the *whole* or the *part*. You know Britta's *whole* paycheck is $180, but you do *not* know what *part* of it was withheld.

$$\text{percent} \cdot \text{whole} = \text{part}$$
$$12.5\% \cdot \$180 = n$$

Step 4   **Solve.** $(0.125)\ (180) = n$
$$22.5 = n$$

Step 5   **State the answer.** **$22.50** was withheld from Britta's paycheck.

Step 6   **Check the solution.** The solution of $22.50 makes sense because 10% of $180. is $18, so a little more than $18 should be withheld.

> **Now Try Exercise 5.** ▶▶▶

EXAMPLE 3 Finding the Percent

On a 15-point quiz, Zenitia earned 13 points. What percent correct is this, to the nearest whole percent?

Step 1   **Read** the problem. It is about points earned on a quiz.

> **Unknown:** percent correct
>
> **Known:** earned 13 out of 15 points

Step 2   **Assign a variable.** Let p be the unknown percent.

Step 3   **Write an equation.** Use the percent equation. There is no number with a % symbol in the problem. The question, "What percent is this?" tells you that the *percent* is unknown. The *whole* is all the points on the quiz (15 points), and 13 points is the *part* of the quiz that Zenitia did correctly.

$$\text{percent} \cdot \text{whole} = \text{part}$$
$$p \cdot 15 \text{ points} = 13 \text{ points}$$

Continued on Next Page

Step 4   **Solve** the equation.

$$\frac{p \cdot \cancel{15}^{\,1}}{\cancel{15}_{\,1}} = \frac{13}{15} \quad \text{Divide both sides by 15}$$

$$p = 0.8\overline{6} \quad \leftarrow \text{The solution is a repeating decimal.}$$

Multiply the solution by 100 to change it from a *decimal* to a *percent*.

$$0.866666667 \approx 86.6666667\% \approx 87\% \quad \leftarrow \text{Rounded}$$

Step 5   **State the answer.** Zenitia had **87%** correct, rounded to the nearest whole percent.

Step 6   **Check the solution.** The solution of 87% makes sense because she earned most of the possible points, so the percent should be fairly close to 100%.

◀◀◀ Now Try Exercise 7.

EXAMPLE 4   **Finding the Percent**

The rainfall in the Red River Valley was 33 inches this year. The average rainfall is 30 inches. This year's rainfall is what percent of the average rainfall?

Step 1   **Read the problem.** It is about comparing this year's rainfall to the average rainfall.

> **Unknown:** This year's rain is what percent of the average?
>
> **Known:** 33 inches this year; 30 inches is average

Step 2   **Assign a variable.** Let *p* be the unknown percent.

Step 3   **Write an equation.** The percent is unknown. The key word **of** appears *right after* the word *percent,* so you can use that sentence to help you write the equation.

This year's rainfall is what percent **of** the average rainfall?

$$33 \text{ inches} \quad = \quad p \quad \cdot \quad 30 \text{ inches}$$

Step 4   **Solve the equation.**

$$33 = p \cdot 30$$

$$\frac{33}{30} = \frac{p \cdot \cancel{30}^{\,1}}{\cancel{30}_{\,1}} \quad \text{Divide both sides by 30}$$

Solution in *decimal* form → $1.1 = p$

Multiply the solution by 100 to change it from a *decimal* to a *percent*.

$$1.10 = 110\%$$

Step 5   **State the answer.** This year's rainfall is **110%** of the average rainfall.

Step 6   **Check the solution.** The solution of 110% makes sense because 33 inches is *more* than 30 inches (more than 100% of the average rainfall), so 33 inches must be *more* than 100% of 30 inches.

◀◀◀ Now Try Exercise 11.

EXAMPLE 5 Finding the Whole

A newspaper article stated that 648 pints of blood were donated at the blood bank last month, which was only 72% of the number of pints needed. How many pints of blood were needed?

Step 1   **Read the problem.** It is about blood donations.

>   **Unknown:** number of pints needed
>
>   **Known:** 648 pints were donated;
>   648 pints is 72% of the number needed.

Step 2   **Assign a variable.** Let n be the number of pints needed.

Step 3   **Write an equation.** The percent is given in the problem: 72%. The key word **of** appears *right after* 72%, so you can use the phrase "72% **of** the number of pints needed" to help you write one side of the equation.

$$72.\% \text{ of the number of pints needed is } 648 \text{ pints.}$$

$$0.72 \quad \cdot \qquad n \qquad = 648$$

Step 4   **Solve.**
$$\frac{(0.\overset{1}{\cancel{72}})(n)}{\underset{1}{\cancel{0.72}}} = \frac{648}{0.72} \qquad \text{Divide both sides by } 0.72$$

$$n = 900$$

Step 5   **State the answer. 900 pints** of blood were needed.

Step 6   **Check the solution.** The solution of 900 pints makes sense because 10% of 900 pints is 90 pints, so 70% would be 7 times 90 pints, or 630 pints, which is close to the number given in the problem (648 pints).

> **Now Try Exercise 13.** ▶▶▶

OBJECTIVE 2 Solve problems involving percent of increase or decrease.   We are often interested in looking at increases or decreases in prices, earnings, population, and many other numbers. This type of problem involves finding the percent of change. Use the following steps to find the **percent of increase.**

Finding the Percent of Increase

Step 1   Use subtraction to find the *amount* of increase.

Step 2   Use a form of the percent equation to find the *percent* of increase.

$$\text{percent } \textbf{of} \quad \text{whole} \quad = \quad \text{part}$$

$$\text{percent } \textbf{of} \text{ original value} = \text{amount of increase}$$

EXAMPLE 6 Finding the Percent of Increase

Brad's hourly wage as assistant manager of a fast-food restaurant was raised from \$9.40 to \$9.87. What was the percent of increase?

Step 1 **Read the problem.** It is about an increase in wages.

> **Unknown:** percent of increase
>
> **Known:** Original hourly wage was \$9.40; new hourly wage is \$9.87.

Step 2 **Assign a variable.** Let p be the percent of increase.

Step 3 **Write an equation.** First subtract \$9.87 − \$9.40 to find how much Brad's wage went up. That is the *amount* of increase. Then write an equation to find the unknown *percent* of increase. Be sure to use his *original* wage (\$9.40) in the equation because we are looking for the change from his *original* wage. Do **not** use the new wage of \$9.87 in the equation.

$$\$9.87 - 9.40 = \$0.47 \quad \leftarrow \textit{Amount of increase}$$

$$\underbrace{\text{percent } \mathbf{of} \text{ original wage}} = \underbrace{\text{amount of increase}}$$

$$p \quad \cdot \quad \$9.40 \quad = \quad \$0.47$$

Step 4 **Solve.**

$$\frac{(p)(9.40)}{9.40} = \frac{0.47}{9.40} \qquad \text{Divide both sides by 9.40}$$

$$p = 0.05 \quad \leftarrow \text{Solution in } \textit{decimal} \text{ form}$$

Multiply the solution by 100 to change it from a *decimal* to a *percent*.

$$0.05 = 5\%$$

Step 5 **State the answer.** Brad's hourly wage increased **5%**.

Step 6 **Check the solution.** 10% of \$9.40 would be a \$0.94 raise. A raise of \$0.47 is half as much, and half of 10% is 5%, so the solution checks.

> **CAUTION**
> When writing the percent equation in *Step 3* above, be sure to multiply the percent (p) by the **original** wage. Do **not** use the *new* wage in the percent equation.

Now Try Example 25.

Use a similar procedure to find the **percent of decrease.**

Finding the Percent of Decrease

Step 1 Use subtraction to find the *amount* of decrease.

Step 2 Use a form of the percent equation to find the *percent* of decrease.

$$\text{percent } \mathbf{of} \quad \text{whole} \quad = \quad \text{part}$$

$$\text{percent } \mathbf{of} \text{ original value} = \text{amount of decrease}$$

EXAMPLE 7   **Finding the Percent of Decrease**

Rozenia trained for six months to run in a marathon. Her weight dropped from 137 pounds to 122 pounds. Find the percent of decrease. Round to the nearest whole percent.

Step 1   **Read the problem.** It is about a decrease in weight.

> **Unknown:** percent of decrease
> **Known:** Original weight was 137 pounds;
> new weight is 122 pounds.

Step 2   **Assign a variable.** Let p be the percent of decrease.

Step 3   **Write an equation.** First subtract 137 pounds − 122 pounds to find how much Rozenia's weight went down. That is the *amount* of decrease. Then write an equation to find the unknown *percent* of decrease. Be sure to use her *original* weight (137 pounds) in the equation because we are looking for the change from her *original* weight. Do **not** use the new weight of 122 pounds in the equation.

137 pounds − 122 pounds = 15 pounds ← *Amount* of decrease

percent **of** original weight = amount of decrease

$$p \quad \cdot \quad 137 \text{ pounds} \quad = \quad 15 \text{ pounds}$$

Step 4   **Solve.**
$$\frac{p \cdot \overset{1}{\cancel{137}}}{\underset{1}{\cancel{137}}} = \frac{15}{137} \qquad \text{Divide both sides by 137}$$

$$p = 0.109489051 \quad \leftarrow \text{Solution in } decimal \text{ form}$$

Multiply the solution by 100 to change it from a *decimal* to a *percent*.

$$0.109489051 = 10.9489051\% \approx 11\%$$

> Rounded to nearest whole percent

Step 5   **State the answer.** Rozenia's weight decreased approximately **11%**.

Step 6   **Check the solution.** A 10% decrease would be 13.7. = 13.7 pounds, so an 11% decrease is a reasonable solution.

CAUTION

When writing the percent equation in *Step 3* above, be sure to multiply the percent (p) by the ***original*** weight. Do ***not*** use the *new* weight in the percent equation.

Now Try Example 27. ▶▶▶

7.4 Exercises

Use the six problem-solving steps in Exercises 1–22. Round percent answers to the nearest tenth of a percent. See Examples 1–5.

1. Robert Garrett, who works part-time, earns $210 per week and has 18% of this amount withheld for taxes, Social Security, and Medicare. Find the amount withheld.

2. Most shampoos contain 75% to 90% water. If a 16-ounce bottle of shampoo contains 78% water, find the number of ounces of water in the bottle, to the nearest tenth. (*Source: Consumer Reports.*)

3. An ATM machine charges $2 for any size cash withdrawal. The $2 fee is what percent of a

   (a) $20 withdrawal

   (b) $40 withdrawal

   (c) $100 withdrawal

   (d) $200 withdrawal.

4. A mail-order company charges $8 for shipping and handling on any order of $50 or less. The $8 shipping charge is what percent of a

   (a) $10 order

   (b) $16 order

   (c) $25 order

   (d) $50 order.

The figure below shows, on average, what portion of the human body is made up of water, protein, fat, and so on. Use the figure to answer Exercises 5 and 6. Round answers to the nearest tenth.

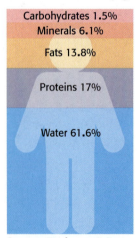

Carbohydrates 1.5%
Minerals 6.1%
Fats 13.8%
Proteins 17%
Water 61.6%

Source: Beakman & Jax, Universal Press Syndicate.

5. If an adult weighs 165 pounds, how much of that weight is

   (a) water

   (b) minerals.

6. If a teenager weighs 92 pounds, how much of that weight is

   (a) fat

   (b) carbohydrates.

7. The guided-missile destroyer USS *Sullivans* has a 335-person crew of which 44 are female. What percent of the crew is female? What percent of the crew is male? (*Source:* U.S. Navy.)

8. In a test by *Consumer Reports,* 6 of the 123 cans of tuna analyzed contained more than the 30-microgram intake limit of mercury. What percent of the cans contained an excessive level of mercury? What percent of the cans contained less than or equal to 30 micrograms of mercury?

9. The U.S. Bureau of the Census reported that Americans who are 65 years of age or older made up 12.7% of the total population in the 2000 census. It said that there were 35.7 million Americans in this group. Find the total U.S. population. (Round to the nearest million.)

10. Julie Ward has 8.5% of her earnings deposited into the credit union. If this amounts to $263.50 per month, find her monthly and annual earnings.

11. The campus honor society hoped to raise $50,000 in donations from businesses for scholarships. It actually raised $69,000. This amount was what percent of the goal?

12. Doug had budgeted $220 for textbooks but ended up spending $316.80. The amount he spent was what percent of his budget?

13. Alfonso earned a score of 95% on his test. He did 38 problems correctly. How many problems were on the test?

14. In a telephone survey, 467 U.S. women said they had done a home improvement project in the past two years. This was 45% of the women in the survey. How many women were surveyed, to the nearest whole number? (*Source:* Opinion Research Corp.)

15. During the 2004–2005 NBA season, Kevin Garnett made 683 field goals, which was 50.22% of the shots he tried. How many shots did he try, to the nearest whole number?

16. Chauncey Billups, who plays basketball for the NBA Detroit Pistons, attempted 382 free throws during the 2004–2005 season. He made 89.8% of his shots. How many free throws did Billups make, to the nearest whole number?

17. An ad for steel-belted radial tires promises 15% better mileage. If Sheera's SUV has gotten 20.6 miles per gallon in the past, what mileage can she expect after the new tires are installed? (Round to the nearest tenth of a mile.)

18. Spam and Spam Lite have yearly sales of $92 million, which is about 62% of all canned lunch meat sales. Find the total yearly sales of canned lunch meat, to the nearest million. (*Source:* Hormel Foods.)

The graph (pictograph) shows the percent of chicken noodle soup sold during the cold and flu season. Use this information to answer Exercises 19–22.

SOUP'S ON

350 million cans of chicken noodle soup are sold each year. More than half are bought during cold-and-flu season, with January being the number one month. The percent sold during each flu-season month is shown.

15%

10% 9% 8% 11% 7%

October November December January February March

Source: *USA Today.*

19. Which of the flu season months had the lowest sales of chicken noodle soup? How many cans were bought that month?

20. What percent of the chicken noodle soup sales take place during the flu months of October through March? What percent of sales take place in the *non-flu* season months?

21. Find the number of cans of soup sold in the highest sales month and in the second-highest sales month.

22. How many more cans of soup were sold in October than in November? How many more were sold in November than December?

Use the six problem-solving steps to find the percent increase or decrease. Round your answers to the nearest tenth of a percent. See Examples 6 and 7.

23. Henry Ford started the Ford Motor Company. The Model T first appeared in 1908 and cost $825. In 1913, Ford started using an assembly line and the price of the Model T dropped to $290. What was the percent of decrease? (*Source:* Kenneth C. Davis.)

24. In 1914, Henry Ford became a hero to his workers because he doubled the daily minimum wage. He also cut the workday from nine hours to eight hours. Find the percent of decrease in the number of hours in the workday. (*Source:* Kenneth C. Davis.)

25. Students at Lane College were charged $1449 as tuition this semester. If the tuition was $1328 last semester, find the percent of increase.

26. Americans are eating more fish. This year the average American will eat 15.5 pounds compared to only 12.5 pounds per year a decade ago. Find the percent of increase. (*Source: Consumer Reports.*)

27. Jordan's part-time work schedule has been reduced to 18 hours per week. He had been working 30 hours per week. What is the percent decrease?

28. Janis works as a hair stylist. During January, she cut her price on haircuts from $28 to $25.50 to try to get more customers. By what percent did she decrease the price?

29. In 1967 there were 78 animal species listed as threatened with extinction in the United States. In 2004 there were 519 animals on the list. What was the percent increase, to the nearest whole percent? (*Source:* U.S. Fish and Wildlife Service.)

30. The world population was estimated at 6,377,642,000 people in 2004. It is projected to reach 7,824,487,000 people by 2025. By what percent will the world's population increase in those 21 years? (*Source:* United Nations.)

31. You can have an *increase* of 150% in the price of something. Could there be a 150% *decrease* in its price? Explain why or why not.

32. Show how to use a shortcut to find 25% of $80. Then explain how to use the result to find 75% of $80 and 125% of $80 *without* solving a proportion or equation.

RELATING CONCEPTS (EXERCISES 33–36) For Individual or Group Work

*As you **work Exercises 33–36 in order**, explain why each solution does **not** make sense. Then find and correct the error.*

33. The recommended maximum daily amount of dietary fat is 65 grams. George ate 78 grams of fat today. He ate what percent of the recommended amount?

$$p \cdot 78 = 65$$

$$\frac{p \cdot \cancel{78}^{1}}{\cancel{78}_{1}} = \frac{65}{78}$$

$$p = 0.833 = 83.3\%$$

34. The Goblers soccer team won 18 of its 25 games this season. What percent did the team win?

$$p \cdot 25 = 18$$

$$\frac{p \cdot \cancel{25}^{1}}{\cancel{25}_{1}} = \frac{18}{25}$$

$$p = 0.72\%$$

35. The human brain is $2\frac{1}{2}\%$ of total body weight. How much would the brain of a 150-pound person weigh?

$$2\frac{1}{2}\% \text{ of } 150 = n$$

$$(2.5)(150) = n$$

$$375 = n$$

36. Yesterday, because of an ice storm, 80% of the students were absent. How many of the 800 students made it to class?

$$80\% \text{ of } 800 = n$$

$$(0.80)(800) = n$$

$$640 = n$$

The answer is 640 students.

7.5 Consumer Applications: Sales Tax, Tips, Discounts, and Simple Interest

OBJECTIVES

1. Find sales tax and total cost.
2. Estimate and calculate restaurant tips.
3. Find the discount and sale price.
4. Calculate simple interest and the total amount due on a loan.

Four of the more common uses of percent in daily life are sales taxes, tips, discounts, and simple interest on loans.

OBJECTIVE 1 **Find sales tax and total cost.** Most states collect **sales taxes** on the purchases you make in stores. Your county or city may also add on a small amount of sales tax. For example, your state may charge $6\frac{1}{2}\%$ on purchases and your city may add on another $\frac{1}{2}\%$ for a total of 7%. The exact percent varies from place to place but is usually from 4% to 8%. The stores collect the tax and send it to the city or state government where it is used to pay for things like road repair, public schools, parks, police and fire protection, and so on.

You can use a form of the percent equation to calculate sales tax. The **tax rate** is the *percent*. The cost of the item(s) you are buying is the *whole*. The amount of tax you pay is the *part*.

Finding Sales Tax and Total Cost

Use a form of the percent equation to find sales tax.

$$\text{percent} \quad \bullet \quad \text{whole} \quad = \quad \text{part}$$

$$\text{tax rate} \quad \bullet \quad \text{cost of item} \quad = \quad \text{amount you pay in sales tax}$$

Then add to find how much you will pay in all.

$$\text{cost of item} + \text{sales tax} = \text{total cost paid by you}$$

EXAMPLE 1 Finding Sales Tax and Total Cost

Suppose that you buy a DVD player for $289 from A-1 Electronics. The sales tax rate in your state is $6\frac{1}{2}\%$. How much is the tax? What is the total cost of the DVD player?

Step 1 **Read the problem.** It asks for the sales tax on a DVD player and the total cost.

Step 2 **Assign a variable.** Let n be the amount of tax.

Step 3 **Write an equation.** Use the sales tax equation. Write $6\frac{1}{2}\%$ as 6.5% and then move the decimal point two places to the left.

$$\text{tax rate} \bullet \text{cost of item} = \text{sales tax}$$

$$06.5\% \bullet \quad \$289 \quad = \quad n$$

Step 4 **Solve.**
$$(0.065)(289) = n$$
$$18.785 = n$$

The store will round the tax to the nearest cent, so $18.785 rounds to **$18.79**.

Now add the sales tax to the cost of the DVD player to find your total cost.

$$\text{cost of item} + \text{sales tax} = \text{total cost}$$

$$\$289 \quad + \quad \$18.79 \quad = \quad \textbf{\$307.79}$$

Continued on Next Page

Step 5   **State the answer.** The tax is **$18.79** and the total cost of the DVD
player, including tax, is **$307.79**.

Step 6   **Check:** Use estimation to check that the amount of sales tax is
reasonable. Round $289 to $300. Then 1% of $300. is $3. Round
$6\frac{1}{2}$% to 7%. Then 7% would be 7 times $3 or $21 for sales tax.
Our solution of $18.79 is close to $21, so it is reasonable.

Now Try Exercise 7. ▶▶▶

EXAMPLE 2   **Finding the Sales Tax Rate**

Ms. Ortiz bought a $21,950 pickup truck. She paid an additional $1646.25
in sales tax. What was the sales tax rate?

Step 1   **Read** the problem. It asks for the sales tax rate on a truck purchase.

Step 2   **Assign a variable.** Let p be the tax rate (the percent).

Step 3   **Write an equation.** Use the sales tax equation.

tax rate • cost of item = sales tax

$$p \quad • \quad \$21{,}950 \quad = \$1646.25$$

Step 4   **Solve.**
$$\frac{(p)\,\overset{1}{(21{,}950)}}{\underset{1}{21{,}950}} = \frac{1646.25}{21{,}950} \qquad \text{Divide both sides by 21,950}$$

$$p = 0.075 \quad \longleftarrow \text{Solution in } \textit{decimal} \text{ form}$$

Multiply the solution by 100 to change it from a decimal to a
percent: $0.075 = 7.5\%$.

Step 5   **State the answer.** The sales tax rate is **7.5%** (or $7\frac{1}{2}\%$).

Step 6   **Check.** Use estimation to check that the solution is reasonable. If
the tax rate was 1%, then 1% of $21950. = $219.50, or about $200.

Round 7.5% to 8%. Then 8% would be 8 times $200, or $1600.

The tax amount given in the original problem, $1646.25, is close
to the estimate, so our solution of 7.5% is reasonable.

Now Try Exercise 9. ▶▶▶

OBJECTIVE 2   **Estimate and calculate restaurant tips.**   Waiters and
waitresses rely on tips as a major part of their income. The general rule
of thumb is to leave 15% of your bill for food and beverages as a tip for
the server. If you receive exceptional service or are eating in an upscale
restaurant, consider leaving a 20% tip.

EXAMPLE 3   **Estimating 15% and 20% Tips**

First estimate each tip. Then calculate the exact amount.

(a) Kirby took his wife to dinner at a nice restaurant to celebrate her
promotion at work. The bill came to $77.85. How much should he
leave for a 20% tip?

Estimate: Round $77.85 to $80. Then 10% of $80. is $8.

20% would be 2 times $8, or **$16**. ← Estimate

Continued on Next Page

Exact: Use the percent equation. Write 20% as a decimal. The bill for food and beverages is the *whole* and the tip is the *part*.

$$\text{percent} \cdot \text{whole} = \text{part}$$

$$20.\% \cdot \$77.85 = n$$

$$(0.20)(77.85) = n$$

$$15.57 = n$$

20% of $77.85 is **$15.57**, which is close to the estimate of $16.

A tip is usually rounded off to a convenient amount, such as the nearest quarter or nearest dollar, so Kirby left $16.

(b) Linda, Peggy, and Mary ordered similarly priced lunches and agreed to split the bill plus a 15% tip. How much should each woman pay if the bill is $21.63?

Estimate: Round $21.63 to $20. Then 10% of $20, is $2.

5% of $20 would be half as much, that is, half of $2, or $1.

So an estimate of the 15% tip is $2 + $1 = **$3**. ← Estimates ⌐

An estimate of the amount each woman should pay is ($20 + $3) ÷ 3 ≈ **$8**.

Exact: Use the percent equation to calculate the 15% tip. Add the tip to the bill. Then divide the total by 3 to find the amount each woman should pay.

$$\text{percent} \cdot \text{whole} = \text{part}$$

$$15.\% \cdot \$21.63 = n$$

$$(0.15)(21.63) = n$$

$$3.2445 = n$$

Round $3.2445 to **$3.24** (nearest cent), which is close to the estimate of $3 for the tip.

Add: $21.63 + $3.24 = $24.87 ← Total cost of lunch and tip

Divide: $24.87 ÷ 3 = **$8.29** ← Amount paid by each woman

Now Try Exercise 11.

OBJECTIVE 3 Find the discount and sale price.   Most people prefer buying things when they are on sale. A store will reduce prices, or **discount,** to attract additional customers. You can use a form of the percent equation to calculate the discount. The *rate* of discount is the *percent*. The original price is the *whole*. The amount that will be discounted (subtracted from the original price) is the *part*.

Finding the Discount and Sale Price

Use a form of the percent equation to find the discount.

$$\text{percent} \quad \cdot \quad \text{whole} \quad = \quad \text{part}$$

rate of discount · original price = amount of discount

Then subtract to find the sale price.

$$\text{original price} - \text{amount of discount} = \text{sale price}$$

EXAMPLE 4 **Finding the Discount and Sale Price**

The Oak Mill Furniture Store has an oak entertainment center with an original price of $840 on sale at 15% off. Find the sale price of the entertainment center.

Step 1 **Read** the problem. It asks for the sale price on an entertainment center.

Step 2 **Assign a variable.** Let n be the amount of discount.

Step 3 **Write an equation.** Use a form of the percent equation to find the discount. Write 15% as the decimal 0.15.

rate of discount • original price = amount of discount

Step 4 **Solve.**

$$(0.15)(840) = n$$
$$126 = n$$

The amount of discount is **$126**. Find the sale price of the entertainment center by subtracting the amount of the discount ($126) from the original price.

original price − amount of discount = sale price

$$\$840 \quad - \quad \$126 \quad = \quad \$714$$

Step 5 **State the answer.** During the sale, you can buy the entertainment center for **$714**.

Step 6 **Check.** Round $840 to $800. Then 10% of $800 is $80 and 5% is half as much, or $40. So 15% of $800 is $80 + $40 = $120. An estimate of the sale price is $800 − $120 = $680, so the exact answer of $714 is reasonable.

Now Try Exercise 41. ▶▶▶

Calculator Tip In Example 4 above, you can use a *scientific* calculator to find the amount of discount and subtract the discount from the original price.

840 ⊖ .15 ⊗ 840 ⊜ 714

Original price — Amount of discount — Sale price

Your *scientific* calculator observes the order of operations, so it will automatically do the multiplication before the subtraction. (Recall that simple, four-function calculators *may not* follow the order of operations; they might give an incorrect result.)

OBJECTIVE 4 **Calculate simple interest and the total amount due on a loan.** **Interest** is a fee paid, or a charge made, for lending or borrowing money. The amount of money borrowed is called the **principal.** The charge for interest is usually given as a percent, called the **interest rate.** The interest rate is assumed to be *per year* (for *one* year) unless stated otherwise.

In most cases, interest is calculated on the original principal and is called **simple interest.** Use the following **interest formula** to find simple interest.

Formula for Simple Interest

$$\text{Interest} = \text{principal} \cdot \text{rate} \cdot \text{time}$$

The formula is usually written using variables.

$$I = p \cdot r \cdot t \qquad \text{or} \qquad I = prt$$

When you repay a loan, the interest is added to the original principal to find the total amount due.

Finding the Total Amount Due

$$\text{amount due} = \text{principal} + \text{interest}$$

NOTE
Simple interest calculations are used for most short-term business, automobile, and consumer loans.

EXAMPLE 5   **Finding Simple Interest for 1 Year**

Find the simple interest and total amount due on a $2000 loan at 6% for 1 year.

    The amount borrowed, or principal (p), is $2000. The interest rate (r) is 6%, which is 0.06 as a decimal, and the time of the loan (t) is 1 year. Use the interest formula.

$$I = \quad p \; \cdot \; r \; \cdot t$$

$$I = (2000)(0.06)(1)$$

$$I = 120$$

The interest is **$120**. Now add the principal and the interest to find the total amount due.

$$\text{\textcolor{red}{amount due}} = \text{principal} + \text{\textcolor{blue}{interest}}$$
$$= \quad \$2000 \; + \; \textcolor{blue}{\$120}$$
$$= \$2120$$

The total amount due is **$2120**.

> **Now Try Exercise 25.**

EXAMPLE 6   **Finding Simple Interest for More Than 1 Year**

Find the simple interest and total amount due on a $4200 loan at $8\frac{1}{2}$% for $3\frac{1}{2}$ years.

    The principal (p) is $4200. The rate ($r$) is $8\frac{1}{2}$%, which is the same as 8.5%. Move the decimal point two places to the left to change 8.5% to a decimal.

$$8\tfrac{1}{2}\% = 8.5\% = 08.5 = 0.085$$

The time (t) is $3\frac{1}{2}$ or 3.5 years. Use the formula.

$$I = \quad p \; \cdot \; r \; \cdot \; t$$

$$I = (4200)(0.085)(3.5)$$

$$I = 1249.50$$

Continued on Next Page

The interest is $1249.50. Now add the principal and the interest to find the total amount due.

$$\textbf{amount due} = \text{principal} + \textbf{interest}$$
$$= \ \ \$4200 \ \ + \$1249.50$$
$$= \$5449.50$$

The total amount due is **$5449.50**.

> **CAUTION**
> Be careful when changing a mixed number percent, like $8\frac{1}{2}\%$, to a decimal. Writing $8\frac{1}{2}\%$ as 8.5% is only the first step. There is a decimal point in 8.5% but there is still a **%** sign. You must divide by 100 before dropping the % sign. ***Remember to move the decimal point two places to the left,*** as shown in Example 6 above. So $8\frac{1}{2}\% = 8.5\% = 0.085$.

Now Try Exercise 29. ▶▶▶

Interest rates are given *per year*. For loan periods of less than one year, be careful to express the time as a fraction of a year.

If the time is given in months, use a denominator of 12, because there are 12 months in a year. A loan of 9 months would be for $\frac{9}{12}$ of a year, a loan of 7 months would be for $\frac{7}{12}$ of a year, and so on.

EXAMPLE 7 **Finding Simple Interest for Less Than 1 Year**

Find the simple interest and total amount due on $840 at $9\frac{3}{4}\%$ for 7 months.
   The principal is $840. The rate is $9\frac{3}{4}\%$ or 0.0975.

$$9\frac{3}{4}\% = 9.75\ \% = 09.75 = 0.0975$$

The time is $\frac{7}{12}$ of a year. Use the formula ***I = prt.***

$$I = \underbrace{(840)(0.0975)}\left(\frac{7}{12}\right) \qquad \text{7 months is } \tfrac{7}{12} \text{ of a year.}$$

$$= \ \ (81.9) \ \ \ \left(\frac{7}{12}\right)$$

$$= \left(\frac{81.9}{1}\right)\left(\frac{7}{12}\right) \qquad \text{Multiply numerators.}$$
$$\qquad\qquad\qquad\qquad\qquad \text{Multiply denominators.}$$

$$= \frac{573.3}{12} = 47.775 \qquad \text{Divide 573.3 by 12}$$

The interest is **$47.78**, rounded to the nearest cent.
The total amount due is $840 + **$47.78** = **$887.78**.

🖩 **Calculator Tip** The calculator solution for finding the interest in Example 7 above uses chain calculations.

$$840 \ \otimes \ .0975 \ \otimes \ 7 \ \oplus \ 12 \ \ominus \ 47.775 \ \leftarrow \text{Round to \$47.78}$$

Now Try Exercise 31. ▶▶▶

7.5 Exercises

Find the amount of the sales tax or the tax rate and the total cost. Round money answers to the nearest cent. See Examples 1 and 2.

Cost of Item	Tax Rate	Amount of Tax	Total Cost
1. $100	6%	_____	_____
2. $200	4%	_____	_____
3. $68	_____	$2.04	_____
4. $185	_____	$9.25	_____
5. $365.98	8%	_____	_____
6. $28.49	7%	_____	_____
7. $2.10	$5\frac{1}{2}\%$	_____	_____
8. $7.00	$7\frac{1}{2}\%$	_____	_____
9. $12,600	_____	$567	_____
10. $21,800	_____	$1417	_____

For each restaurant bill, estimate a 15% tip and a 20% tip. Then find the exact amounts for a 15% tip and a 20% tip. Round exact amounts to the nearest cent, if necessary. See Example 3.

Bill	Estimate of 15% Tip	Exact 15% Tip	Estimate of 20% Tip	Exact 20% Tip
11. $32.17				
12. $21.94				
13. $78.33				
14. $67.85				
15. $9.55				
16. $52.61				

Find the amount or rate of discount and the sale price. Round money answers to the nearest cent if necessary. See Example 4.

Original Price	Rate of Discount	Amount of Discount	Sale Price
17. $100	15%		
18. $200	20%		
19. $180		$54	
20. $38		$9.50	
21. $17.50	25%		
22. $76	60%		
23. $37.88	10%		
24. $59.99	40%		

Find the simple interest and total amount due on each loan. See Examples 5–7.

	Principal	Rate	Time	Interest	Total Amount Due
25.	$300	14%	1 year	_____	_____
26.	$600	11%	6 months	_____	_____
27.	$740	6%	9 months	_____	_____
28.	$1180	9%	2 years	_____	_____
29.	$1500	$9\frac{1}{2}$%	$1\frac{1}{2}$ years	_____	_____
30.	$3000	$6\frac{1}{2}$%	$2\frac{1}{2}$ years	_____	_____
31.	$17,800	$7\frac{3}{4}$%	8 months	_____	_____
32.	$20,500	$8\frac{1}{4}$%	5 months	_____	_____

Solve each application problem. Round money answers to the nearest cent, if necessary.

33. Diamonds at Discounts sells diamond engagement rings at 40% off the regular price. Find the sale price of a $\frac{1}{2}$-carat diamond ring normally priced at $1950.

34. An Apple iPod mini MP3 player originally priced at $219 is marked down 9%. Find the price of the iPod after the markdown.

35. Evelina Jones lends $7500 to her son Rick, the owner of Rick's Limousine Service. He will repay the loan at the end of 9 months at $8\frac{1}{2}\%$ simple interest. What is the total amount that Rick will owe his mother?

36. The owners of Delta Trucking purchased four diesel-powered tractors for cross-country hauling at a cost of $87,500 per tractor. If they borrowed the purchase price for $1\frac{1}{2}$ years at 11% simple interest, find the total amount due.

37. A Nokia wireless phone with instant messaging and a built-in camera is priced at $99.99. The sales tax rate is $6\frac{1}{2}\%$. Find the total cost of the phone. (*Source:* CompUSA)

38. A weekday "golf/breakfast special" includes breakfast, 18 holes of golf, and use of a cart for $37.95 plus tax per person. If the sales tax rate is $7\frac{1}{2}\%$, find the total cost per person. How much will three friends pay to play golf?

39. An Anderson wood frame French door is priced at $1980 with a sales tax of $99. Find the sales tax rate.

40. Textbooks for two classes cost $135 plus sales tax of $8.10. Find the sales tax rate.

41. A "super 45% off sale" begins today. What is the sale price of a ski parka normally priced at $135?

42. A discontinued Whirlpool model side-by-side refrigerator with in-door icemaker originally sold for $1197. What is the sale price with a 35% discount?

43. Ricia and Seitu split a $43.70 dinner bill plus 15% tip. How much did each person pay?

44. Marvette took her brother out to dinner for his birthday. The bill for food was $58.36 and for wine was $15.44. How much was her 20% tip, rounded to the nearest dollar?

45. A 36" flat-tube TV normally priced at $590 is on sale for 18% off. Find the discount and the sale price.

46. This week SUVs are offered at 15% off manufacturers' suggested price. Find the discount and the sale price of an SUV originally priced at $23,500.

47. Ms. Henderson owes $1900 in taxes. She is charged a penalty of $12\frac{1}{4}\%$ annual interest and pays the taxes and penalty after 6 months. Find the total amount she must pay.

48. Norell Di Loreto, owner of Sunset Realtors, borrows $27,000 to update her office computer system. If the loan is for 24 months at $7\frac{3}{4}\%$, find the total amount due on the loan.

49. Vincente and Samuel ordered a large deep-dish pizza for $17.98. How much did they give the delivery person to pay for the pizza and a 15% tip, rounded to the nearest dollar?

50. Cher, Maya, and Adara shared a $25.50 bill for a buffet lunch. Because the server only brought their beverages, they left a 10% tip instead of the usual 15%. How much did each person pay?

Use the information in the store ad to answer Exercises 51–54. Round sale prices and sales tax to the nearest cent if necessary.

STORE CLOSE-OUT!

All clothing is now 45% off!

All jewelry is now 30% off!

All electronics are now 65% off!

6% sales tax added to jewelry and electronics purchases.

CASH ONLY! ALL SALES ARE FINAL!

51. Danika bought a computer modem originally priced at $129 and a $60 pair of earrings. What was her bill for the two items?

52. Find David's total bill for a $189 jacket and a $75 graphing calculator.

53. Sergei purchased a television originally priced at $287.95, two pairs of $48 jeans, and a $95 ring. Find his total bill.

54. Richard picked out three pairs of $15 running shorts, two $28 shirts, and a 4.1 megapixel digital camera originally priced at $99.99. How much did he pay in all?

RELATING CONCEPTS (EXERCISES 55–56) For Individual or Group Work

Use your knowledge of percent to work Exercises 55 and 56 in order.

55. (a) College students are offered a 6% discount on a dictionary that sells for $18.50. If the sales tax is 6%, find the cost of the dictionary, including the sales tax, to the nearest cent.

(b) In part (a) the rate of discount and the sales tax rate are the same percent. Explain why the answer did *not* end up back at $18.50.

56. (a) A combination printer, copier, scanner, and FAX machine priced at $398 is marked down 7% to promote the new model. If the sales tax is also 7%, find the cost of the machine, including sales tax, to the nearest cent.

(b) What rate of sales tax would have made the answer in part (a) end up back at $398? Round your answer to the nearest hundredth of a percent.

Chapter **7** | Group Activity: MAKE YOUR INVESTMENTS GROW—COMPOUND INTEREST

Objective Calculate compound interest on investments using a compound interest table

Simple interest is paid only on the original principal. But savings accounts and most investments earn *compound interest*. In that case, interest is paid on the principal *and* the interest earned. Calculating compound interest can be quite tedious. For this reason compound interest tables have been developed.

Suppose you deposit $1 in a savings account today that earns 4% compounded annually and you allow the deposit to remain for three years. The diagram below shows the compound amount in your account at the end of each of the 3 years.

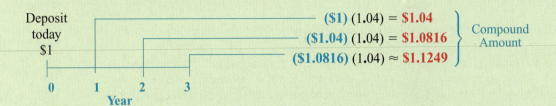

Deposit today $1

($1) (1.04) = **$1.04**
($1.04) (1.04) = **$1.0816** } Compound Amount
($1.0816) (1.04) ≈ **$1.1249**

0   1   2   3
Year

Using the compound amounts for $1, a table can be formed. Look at the table below and find the column headed 4%. The first three numbers (for years 1, 2, and 3) are the same as those we have calculated for $1 at 4% for 3 years.

COMPOUND INTEREST TABLE

Years	3.00%	3.50%	4.00%	4.50%	5.00%	5.50%	6.00%	8.00%
1	1.0300	1.0350	1.0400	1.0450	1.0500	1.0550	1.0600	1.0800
2	1.0609	1.0712	1.0816	1.0920	1.1025	1.1130	1.1236	1.1664
3	1.0927	1.1087	**1.1249**	1.1412	1.1576	1.1742	1.1910	1.2597
4	1.1255	1.1475	1.1699	1.1925	1.2155	1.2388	1.2625	1.3605
5	1.1593	1.1877	1.2167	1.2462	1.2763	1.3070	1.3382	1.4693
6	1.1941	1.2293	1.2653	1.3023	1.3401	1.3788	1.4185	1.5869
7	1.2299	1.2723	1.3159	1.3609	1.4071	1.4547	1.5036	1.7138
8	1.2668	1.3168	1.3686	1.4221	1.4775	1.5347	1.5938	1.8509
9	1.3048	1.3629	1.4233	1.4861	1.5513	1.6191	1.6895	1.9990
10	1.3439	1.4106	1.4802	1.5530	1.6289	1.7081	1.7908	2.1589

To find the compound amount for $1000 deposited at 4% interest for 3 years, multiply the number from the table (1.1249) times the principal ($1000).

($1000) (**1.1249**) = $1124.90 ← Amount in account after 3 years

1. (a) Use the table to find the compound amount for $1000 deposited at 6% for 3 years, and the compound amount at 8% for 3 years.

(b) How much more do you earn in 3 years on $1000 if the interest rate is 6% instead of 4%? How much more if the interest rate is 8% instead of 4%?

2. (a) Use the table to find the compound amount on $15,000 invested for 10 years at 4%, at 6%, and at 8%.

(b) How much more do you earn at 6% than 4%? How much more at 8% than 4%?

3. Compound interest makes a tremendous difference when you invest money for retirement. Suppose a person plans to retire at age 65. Use the part of the compound interest table shown at the right. Find the compound amount on $5000 invested at 8% when the person is 45 years old, and left in the account until age 65. Then find the compound amount if the investment had been made at age 35, if the investment had been made at age 25, and if the investment had been made at age 15.

Years	8.00%
20	4.6610
30	10.0627
40	21.7245
50	46.9016

Chapter 7

SUMMARY

KEY TERMS

7.1 **percent** Percent means "per one hundred." A percent is a ratio with a denominator of 100.

7.2 **percent proportion** The proportion used to solve percent problems is $\dfrac{percent}{100} = \dfrac{part}{whole}$.

whole The *whole* in a percent problem is the entire quantity or the total. It is sometimes called the *base*.

part The *part* in a percent problem is the number being compared to the *whole*.

7.3 **percent equation** The percent equation is: percent • whole = part. The equation can be used instead of the percent proportion to solve percent problems.

7.4 **percent of increase or decrease** Percent of increase or decrease is the amount of change (increase or decrease) expressed as a percent of the original value.

7.5 **sales tax** Sales tax is a percent of the total sales charged as a tax.

tax rate The tax rate is the percent used when calculating the amount of tax.

discount Discount is often expressed as a percent of the original price; it is then deducted from the original price, resulting in the sale price.

interest Interest is a fee paid or a charge made for lending or borrowing money.

principal Principal is the amount of money on which interest is charged.

interest rate Often referred to as *rate*, it is the charge for interest and is given as a percent.

simple interest When interest is calculated on the original principal, it is called simple interest.

interest formula The interest formula is used to calculate simple interest. It is:

Interest = principal • rate • time or $I = prt$.

NEW FORMULAS

Finding simple interest: $I = prt$

NEW SYMBOLS

% percent symbol (meaning "per 100")

TEST YOUR WORD POWER

See how well you have learned the vocabulary in this chapter. Answers follow the Quick Review.

1. **Percent** means
   A. per one thousand
   B. part divided by whole
   C. per one hundred
   D. part times whole.

2. The **whole** in a percent problem is
   A. the entire quantity or total
   B. a ratio with a denominator of 100
   C. the amount of change
   D. always 100.

3. When calculating sales tax, the **rate** is
   A. the part
   B. the whole
   C. the total cost
   D. the percent.

4. The **interest formula** is
   A. tax rate • cost of item = sales tax
   B. $I = prt$
   C. percent • whole = part
   D. $\dfrac{percent}{100} = \dfrac{part}{whole}$.

5. When calculating interest on a loan, the **principal** is the
   A. amount of money borrowed
   b. total amount due
   C. rate charged for borrowing money
   D. amount of simple interest.

6. The **percent of increase or decrease** compares the amount of change to
   A. 100
   B. the percent

C. the original value before the change
D. the new value after the change.

7. The **percent equation** is
   A. tax rate • cost of item = sales tax
   B. $I = prt$
   C. percent • whole = part
   D. $\dfrac{percent}{100} = \dfrac{part}{whole}$.

8. A **discount** is
   A. added to the original price
   B. divided by 100
   C. multiplied by the percent
   D. subtracted from the original price.

QUICK REVIEW

Concepts	Examples

7.1 Basics of Percent

Writing a Percent as a Decimal
To write a percent as a decimal, move the decimal point *two* places to the *left* and drop the % sign.

$$50\% = 50.\% = 0.50 \text{ or } 0.5$$
$$3\% = 03.\% = 0.03$$

Writing a Decimal as a Percent
To write a decimal as a percent, move the decimal point *two* places to the *right* and attach a % sign.

$$0.75 = 0.75 = 75\%$$
$$3.6 = 3.60 = 360\%$$

Writing a Percent as a Fraction
To write a percent as a fraction, drop the % symbol and write the number over 100. Then write the fraction in lowest terms.

$$35\% = \frac{35}{100} = \frac{35 \div 5}{100 \div 5} = \frac{7}{20} \leftarrow \text{Lowest terms}$$
$$125\% = \frac{125}{100} = \frac{125 \div 25}{100 \div 25} = \frac{5}{4} = 1\frac{1}{4}$$

Writing a Fraction as a Percent
To write a fraction as a percent, multiply by 100 and attach a % symbol. This is the same as multiplying by 100%.

$$\frac{3}{5} = \frac{3}{5} \cdot \frac{100}{1}\% = \frac{3 \cdot \overset{1}{5} \cdot 20}{\underset{1}{5} \cdot 1}\% = \frac{60}{1}\% = 60\%$$

$$\frac{7}{8} = \frac{7}{8} \cdot \frac{100}{1}\% = \frac{7 \cdot \overset{1}{4} \cdot 25}{2 \cdot \underset{1}{4} \cdot 1}\% = \frac{175}{2}\% = 87\frac{1}{2}\% \text{ or } 87.5\%$$

7.2 Using the Percent Proportion

The percent proportion is shown below.

$$\underset{\text{Always } 100 \longrightarrow}{\frac{\text{percent}}{100}} = \frac{\text{part}}{\text{whole}}$$

Identify the percent first. It appears with the word *percent* or the % symbol.

The *whole* is the entire quantity or total. It often appears after the word **of**.

The *part* is the number being compared to the whole.

30 children is what percent **of** 75 children?

$$\underset{\text{Always } 100 \longrightarrow}{\overset{\text{Percent (unknown)} \longrightarrow}{\frac{p}{100}}} = \frac{30}{75} \begin{matrix} \leftarrow \text{Part} \\ \leftarrow \text{Whole (follows } of \text{)} \end{matrix}$$

To solve the proportion, find the cross products.

$$\frac{p}{100} \bowtie \frac{30}{75} \quad \begin{matrix} 100 \cdot 30 = 3000 \\ p \cdot 75 \end{matrix} \quad \text{Cross products}$$

$$p \cdot 75 = 3000 \quad \text{Show that the cross products are equivalent.}$$

$$\frac{p \cdot \overset{1}{75}}{\underset{1}{75}} = \frac{3000}{75} \quad \text{Divide both sides by 75}$$

$$p = 40$$

30 children is **40%** of 75 children.

7.3 Percent Shortcuts

200% of a number is 2 times the number.

100% of a number is the entire number.

To find 50% of a number, divide the number by 2.

To find 25% of a number, divide the number by 4.

To find 10% of a number, move the decimal point *one* place to the *left*.

To find 1% of a number, move the decimal point *two* places to the *left*.

200% of $35 is 2 times $35 = $70.

100% of 600 women is *all* the women (600 women).

50% of $8000 is $8000 ÷ 2 = $4000.

25% of 40 pens is 40 ÷ 4 = 10 pens.

10% of $92.40 is $9.24.

1% of 62. miles is 0.62 mile.

Concepts	Examples

7.3–7.4 *Using the Percent Equation*
Use the six problem-solving steps from **Chapter 3.**

Step 1 **Read** the problem.

Todd's regular pay is $540 per week but $43.20 is taken out of each paycheck for medical insurance. What percent is that?

Step 1 The problem asks for the percent of Todd's pay that is taken out for insurance.
  Unknown: percent taken out
  Known: Whole paycheck is $540; $43.20 taken out.

Step 2 **Assign a variable.**

Step 2 Let p be the unknown percent.

Step 3 **Write an equation.**

Step 3 Use the percent equation. The *whole* is Todd's entire pay of $540 and the *part* is $43.20 (the part taken out of his paycheck).

$$\text{percent} \cdot \text{whole} = \text{part}$$
$$p \quad \cdot \quad 540 \quad = 43.20$$

Step 4 **Solve the equation.**

Step 4
$$\frac{p \cdot \overset{1}{\cancel{540}}}{\underset{1}{\cancel{540}}} = \frac{43.20}{540} \quad \text{Divide both sides by 540}$$
$$p = 0.08 \quad \leftarrow \text{Decimal form}$$

Multiply 0.08 by 100 so that $0.08 = 8\%$.

Step 5 **State the answer.**

Step 5 8% of Todd's pay is taken out for medical insurance.

Step 6 **Check** the solution.

Step 6 Use estimation. If 10% of Todd's pay were withheld, then 10% of $540. = $54, which is a little more than the $43.20 actually withheld. So 8% is a reasonable solution.

7.4 *Finding Percent of Increase or Decrease*
Use subtraction to find the *amount* of increase or decrease. When writing the equation, be careful to use the *original* value as the whole.

Enrollment rose from 3820 students to 5157 students. Find the percent of increase.

$$5157 \text{ students} - 3820 \text{ students} = 1337 \text{ students}$$
↑
Amount of increase

$$\text{percent } \textbf{of} \text{ original value} = \text{amount of increase}$$
$$p \quad \cdot \quad 3820 \quad = \quad 1337$$
$$p \cdot 3820 = 1337$$
$$\frac{p \cdot \overset{1}{\cancel{3820}}}{\underset{1}{\cancel{3820}}} = \frac{1337}{3820} \quad \text{Divide both sides by 3820}$$
$$p = 0.35 \quad \leftarrow \text{Decimal form}$$
$$0.35 = \textbf{35\%} \quad \leftarrow \text{Percent increase}$$

The enrollment increased **35%**.

Concepts	*Examples*
7.5 *Consumer Applications* **Finding Sales Tax** To find sales tax, use this equation. tax rate • cost of item = sales tax Write the percent as a decimal.	Find the sales tax on an $89 pair of binoculars if the sales tax rate is 6%. tax rate • cost of item = sales tax 06.% • $89 = n (0.06) (89) = n 5.34 = n The sales tax is **$5.34**.

Estimating and Calculating Restaurant Tips To estimate a 15% tip, first find 10% of the food bill by moving the decimal point *one* place to the *left*. Then add half of that amount for the other 5%.
 	
To estimate a 20% tip, first find 10% by moving the decimal point *one* place to the *left*. Then double the amount.
 	
To find the exact tip, write the percent as a decimal. The bill for food and beverages is the *whole* and the tip is the *part*.
	Estimate a 15% tip on a restaurant bill of $38.72. Then find the exact tip.
	
	Estimate: Round $38.72 to $40.
	
	  10% of $40. is $4.
	  Half of $4 is $2.
	
	So an estimate of the 15% tip is $4 + $2 = $6.
	
	Exact: percent • whole = part
	
	15.% • 38.72 = n
	
	(0.15) (38.72) = n
	
	5.808 = n
	
	The exact tip is **$5.81** (rounded to nearest cent).

Finding a Discount To find a discount, use this formula. rate of discount • original price = amount of discount Then subtract to find the sale price. original price − amount of discount = sale price	All calculators are on sale at 20% off. Find the sale price of a calculator originally marked $35. 20% • $35 = n (0.20) (35) = n 7 = n The amount of discount is $7. Then $35 − $7 = **$28** ← Sale price
Finding Simple Interest and Total Amount Due To find the simple interest on a loan, use the formula $I = prt$. Interest = principal • rate • time Time (t) is in years. When the time is given in months, use a fraction with 12 in the denominator because there are 12 months in a year. Write the rate (the percent) as a decimal. To find the total amount due on the loan, add the interest to the original principal.	$2800 is borrowed at 8% for 5 months. Find the amount of interest and the total amount due. $$I = p \cdot r \cdot t$$ $$= (2800)(0.08)\left(\frac{5}{12}\right)$$ $$= (224)\left(\frac{5}{12}\right) = \frac{(224)(5)}{12} \approx \textbf{\$93.33}$$ Total amount due = $2800 + **$93.33** = **$2893.33**

ANSWERS TO TEST YOUR WORD POWER

1. C; *Example:* 7% means 7 per 100, or, 7 out of 100.

2. A; *Example:* In the problem "15 computers is what percent of 75 computers," the whole is 75 computers, which is the total group of computers.

3. D; *Example:* Houston has a sales tax rate of 8.25%; Minneapolis has a sales tax rate of 7%.

4. B; *Example:* The interest formula is Interest = principal • rate • time. If you borrow $4000 for 2 years at a rate of 9%, then $I = (4000)(0.09)(2) = \$720$.

5. A; *Example:* If you borrow $4000 for 2 years at a rate of 9%, the principal is $4000.

6. C; *Example:* If your rent increased from $800 to $850, the *percent of increase* compares the amount of change ($50) to the *original* rent ($800).

7. C; *Example:* To answer the question, "What percent of 40 Web pages is 12 Web pages," let p be the unknown percent and write the equation as: $p • 40 = 12$.

8. D; *Example:* If sunglasses are on sale at 10% off, then a pair of sunglasses regularly priced at $25 will have a discount of $2.50 subtracted from the price; you will pay $22.50. To find the amount of discount, multiply $(0.10)(\$25)$ to get $2.50.

Chapter 7
REVIEW EXERCISES

[7.1] *Write each percent as a decimal and each decimal as a percent.*

1. 25%

2. 180%

3. 12.5%

4. 7%

5. 2.65

6. 0.02

7. 0.3

8. 0.002

Write each percent as a fraction or mixed number in lowest terms. Write each fraction as a percent.

9. 12%

10. 37.5%

11. 250%

12. 5%

13. $\dfrac{3}{4}$

14. $\dfrac{5}{8}$

15. $3\dfrac{1}{4}$

16. $\dfrac{3}{50}$

Complete this table.

Fraction	Decimal	Percent
$\dfrac{1}{8}$	**17.** _____	**18.** _____
19. _____	0.15	**20.** _____
21. _____	**22.** _____	180%

Use percent shortcuts to fill in the blanks.

23. 100% of $46 is _____.

24. 50% of $46 is _____.

25. 100% of 9 hours is _____.

26. 50% of 9 hours is _____ .

[7.2] *Use a percent proportion to answer each question. If necessary, round percent answers to the nearest tenth of a percent.*

27. 338.8 meters is 140% of what number of meters?

28. 2.5% of what number of cases is 425 cases?

29. What is 6% of 450 cellular phones?

30. 60% of 1450 reference books is how many books?

31. What percent of 380 pairs is 36 pairs?

32. 1440 cans is what percent of 640 cans?

[7.3] *Use the percent equation to answer each question. Round money answers to the nearest cent, if necessary.*

33. 11% of $23.60 is how much?

34. What is 125% of 64 days?

35. 1.28 ounces is what percent of 32 ounces?

36. $46 is 8% of what number of dollars?

37. 8 people is 40% of what number of people?

38. What percent of 174 ft is 304.5 ft?

[7.4] *Use the six problem-solving steps to answer each question. If necessary, round percent answers to the nearest tenth of a percent.*

39. (a) A medical clinic found that 16.8% of the patients were late for their appointments in January. The number of patients who were late was 504. Find the total number of patients in January.

(b) In February, only 345 patients were late. Find the percent of decrease in late patients from January to February.

40. (a) Coreen budgeted $280 for food on her vacation. She actually spent 130% of that amount. How much did she spend on food?

(b) Coreen's vacation budget included $50 for gifts and souvenirs, but she spent $112. What percent of the budgeted amount did she spend?

41. (a) In the first part of a tree-planting project, 640 of the 800 trees planted were still living one year later. What percent of the trees planted were still living?

(b) In the second part of the project, 850 trees were planted. What was the percent of increase in trees planted?

42. Scientists tell us that worldwide there are 9600 species of birds and that 1000 of these species are in danger of extinction. What percent of the bird species are in danger of extinction, to the nearest tenth of a percent?

The ivory-billed woodpecker, thought to be extinct, was recently found again in Arkansas.

[7.5] *Find the amount of sales tax or the tax rate and the total cost. Round to the nearest cent, if necessary.*

	Amount of Sale	Tax Rate	Amount of Tax	Total Cost
43.	$2.79	4%	_____	_____
44.	$780	_____	$58.50	_____

For each restaurant bill, estimate a 15% tip and a 20% tip. Then find the exact amount for a 15% tip and a 20% tip. Round to the nearest cent, if necessary.

	Bill	Estimated 15%	Exact 15%	Estimated 20%	Exact 20%
45.	$42.73	_____	_____	_____	_____
46.	$8.05	_____	_____	_____	_____

Find the amount or rate of discount and the sale price.

	Original Price	Rate of Discount	Amount of Discount	Sale Price
47.	$37.50	10%	_____	_____
48.	$252	_____	$63	_____

Find the simple interest and total amount due on each loan.

	Principal	Rate	Time	Interest	Total Amount Due
49.	$350	$6\frac{1}{2}\%$	3 years	_____	_____
50.	$1530	16%	9 months	_____	_____

MIXED REVIEW EXERCISES

The bar graph shows the types of electronic/computer games that adults like to play. Use the information in the graph to answer Exercises 51–54.

CARE FOR A GAME OF CARDS?
One in three adults say they play electronic or computer games. Types of games they play:

Cards 68%

Sports/driving 46%

Action/combat 43%

Board 39%

Sci-fi/simulation 37%

Note: People may play more than one type of game, so the percents add up to more than 100%.

Source: Cable & Telecommunications Association for Marketing.

51. Write the portion of adults who like to play electronic/computer games as a fraction and as a percent.

52. What type of game is most popular? Write the portion of players who picked this type as a percent, a decimal, and a fraction in lowest terms.

53. If 830 adults were surveyed, how many of them play electronic/computer games, to the nearest whole number?

54. Using your answer from Exercise 53, how many of the playing adults chose the most popular and least popular types of game?

The table shows the number of animals received during the first nine months of the year by the local Animal Humane Society and the number placed in new homes. Use the table to answer Exercises 55–60. Round percent answers to the nearest tenth of a percent.

Animals Received	Placed in New Homes
5371 dogs	2599 dogs
6447 cats	2346 cats
2223 other*	406 other*

*"Other" includes rabbits, birds, gerbils, mice, etc.

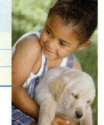

55. What percent of the dogs were placed in new homes?

56. The number of dogs received so far this year is 75% of the total number expected. How many dogs are expected, to the nearest whole number?

57. What percent of the cats were placed in new homes?

58. During the first nine months of last year, 2300 cats were received. What is the percent increase in cats received from last year to this year?

59. What percent of all the animals received were placed in new homes?

60. The Society's goal was to place 40% of all animals received. So far they have missed their goal by how many animals? Round to the nearest whole number.

Chapter 7

TEST

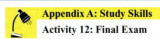

Appendix A: Study Skills
Activity 12: Final Exam

Write each percent as a decimal and each decimal as a percent.

1. 75%   **2.** 0.6   **3.** 1.8

4. 0.075   **5.** 300%   **6.** 2%

Write each percent as a fraction or mixed number in lowest terms.

7. 62.5%   **8.** 240%

Write each fraction or mixed number as a percent.

9. $\dfrac{1}{20}$ **10.** $\dfrac{7}{8}$ **11.** $1\dfrac{3}{4}$

Write and solve a proportion or an equation to answer each question. Show your work.

12. 16 laptops is 5% of what number of laptops?

13. $192 is what percent of $48?

14. Erica Green has saved 75% of the amount needed for a down payment on a condominium. If she has saved $14,625, find the total down payment needed.

15. The price of a used car is $7950 plus sales tax of $6\frac{1}{2}$%. Find the total cost of the car including sales tax.

16. Enrollment in mathematics courses increased from 1440 students last semester to 1925 students this semester. Find the percent of increase, to the nearest whole percent.

1. _____

2. _____

3. _____

4. _____

5. _____

6. _____

7. _____

8. _____

9. _____

10. _____

11. _____

12. _____

13. _____

14. _____

15. _____

16. _____

17. _____

17. Explain a shortcut for finding 50% of a number and a shortcut for finding 25% of a number. Show an example of how to use each shortcut.

18. _____

18. Explain how you would *estimate* a 15% tip on a restaurant bill of $31.94. Then explain how you would *estimate* a 20% tip on the same bill.

19. _____

19. Find the exact 15% tip, to the nearest cent, for the restaurant bill in Problem 18. If you and two friends are sharing the bill and the exact tip, how much will each person pay?

Find the amount of discount and the sale price of each item in Problems 20 and 21. Round answers to the nearest cent, if necessary. Show your work.

20. _____

20. Jeremy plans to use his 8% employee discount to buy a $48 clock radio so he can get to work on time.

21. _____

21. A Sony progressive scan multiformat DVD recorder and VCR combination regularly priced at $229.95 is on sale at 18% off.

22. _____

22. Jamal found a $1089 computer on sale at 30% off because it was a "discontinued" model. The store will let him pay for it over 6 months with no interest charge. How much will each monthly payment need to be to cover the discounted price plus 7% sales tax?

23. _____

23. What is the simple interest and total amount due on a four-year loan of $5000 at $8\frac{1}{4}\%$?

24. _____

24. Kendra borrowed $860 to pay medical expenses. The loan is for 6 months at 12% simple interest. Find the simple interest and the total amount due on the loan.

Cumulative Review Exercises

CHAPTERS 1–7

1. Write these numbers in words.

(a) 90.105

(b) 125,000,670

2. Write these numbers in digits.

(a) Thirty billion, five million

(b) Seventy-eight ten-thousandths

3. Round each number as indicated.

(a) 49,617 to the nearest thousand

(b) 0.7039 to the nearest hundredth

(c) 8945 to the nearest hundred

4. Name the property illustrated by each example.

(a) $(-7)(0.8) = (0.8)(-7)$

(b) $\left(\dfrac{2}{3} + \dfrac{3}{4}\right) + \dfrac{1}{2} = \dfrac{2}{3} + \left(\dfrac{3}{4} + \dfrac{1}{2}\right)$

5. Write $>$ or $<$ between each pair of numbers to make true statements.

-18 _____ -8 0 _____ -5

6. Arrange in order from smallest to largest.

0.705 0.755 $\dfrac{3}{4}$ 0.7005

7. Find the mean and median for this set of data on rent prices: $710, $780, $650, $785, $1125, $695, $740, $685.

8. Find the best buy on cereal.

17 ounces of Brand A for $2.89

21 ounces of Brand B for $3.59

15 ounces of Brand C for $2.79

Simplify.

9. $50 - 1.099$

10. $(-3)^2 + 2^3$

11. $\dfrac{3b}{10a} \cdot \dfrac{15ab}{4}$

12. $3\dfrac{3}{10} - 2\dfrac{4}{5}$

13. $0 + 2(-6 + 1)$

14. $\dfrac{4}{5} + \dfrac{x}{4}$

15. $-20 - 20$

16. $(-0.5)(0.002)$

17. $\dfrac{-\dfrac{10}{11}}{-\dfrac{5}{6}}$

18. $\dfrac{3}{8}$ of 328

19. $\dfrac{-16 + 2^4}{-3 - 2}$

20. $\dfrac{7}{8} - \dfrac{2}{m}$

21. $\dfrac{4.8}{-0.16}$ **22.** $\dfrac{8}{9} \div 2n$ **23.** $1\dfrac{5}{6} + 1\dfrac{2}{3}$ **24.** $5 - 1\dfrac{7}{9}$

25. $|10 - 30| + (-4)^3$ **26.** $0.6 \div 12(3.6 - 4)$ **27.** $\dfrac{3}{10} - \left(\dfrac{1}{4} - \dfrac{3}{4}\right)^2 + \left(\dfrac{1}{2}\right)^2$

Evaluate each expression when w is -4, x is -2, and y is 3.

28. $-6w - 5$ **29.** $5x + 3y$ **30.** x^3y **31.** $-5w^2x$

Simplify each expression.

32. $-2x^2 + 5x - 7x^2$ **33.** $ab - ab$ **34.** $-10(4w^3)$ **35.** $3(h - 4) + 2$

Solve each equation. Show your work.

36. $2n - 3n = 0 - 5$ **37.** $12 - h = -3h$ **38.** $5a - 0.6 = 10.4$

39. $-\dfrac{7}{8} = \dfrac{3}{16}y$ **40.** $3 + \dfrac{1}{10}b = 5$ **41.** $\dfrac{0.2}{3.25} = \dfrac{10}{x}$

42. $32 - 3h = 5h + 8$ **43.** $3 + 2(x + 4) = -4x + 7 + 2x$

Translate each sentence into an equation and solve it.

44. If 5 is subtracted from four times a number, the result is −17. What is the number?

45. The sum of a number and 31 is three times the number plus 1. Find the number.

Solve each application problem using the six problem-solving steps from **Chapter 3.**

46. Lawrence brought home three packages of disposable diapers. He used 17 diapers the first day and 19 the next day. There were 12 diapers left. How many diapers were originally in each package?

47. Susanna and Neoka are splitting $1620 for painting a house. Neoka worked twice as many hours, so she should earn twice as much. How much should each woman receive?

Find the unknown length, perimeter, circumference, area, or volume. When necessary, use **3.14** *as the approximate value of π and round answers to the nearest tenth.*

48. Find the perimeter and the area.

49. Find the circumference and the area.

50. Find the unknown length in these similar triangles.

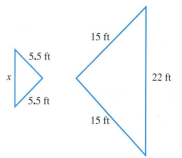

51. Find the volume and surface area of this rectangular solid.

52. Find the volume.

53. Find the unknown length and the area of the figure.

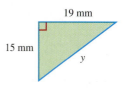

Solve each application problem.

54. A survey of the 5600 students on our campus found that $\frac{3}{8}$ of the students work 20 hours or more per week. How many students is that?

55. The Americans With Disabilities Act provides the single parking space design below. Find the perimeter and area of this parking space, including the accessible aisle.

56. The Jackson family is making three kinds of holiday cookies that require brown sugar. The recipes call for $2\frac{1}{4}$ cups, $1\frac{1}{2}$ cups, and $\frac{3}{4}$ cup, respectively. They bought two packages of brown sugar, each holding $2\frac{1}{3}$ cups. The amount bought is how much more or less than the amount needed?

57. Leather jackets are on sale at 30% off the regular price. Tracy likes a jacket with a regular price of $189. Find the amount of discount and the sale price she will pay.

58. On the Illinois map, one centimeter represents 12 kilometers. The center of Springfield is 7.8 cm from the center of Bloomington on the map. What is the actual distance in kilometers?

59. On a 35-problem math test, Juana solved 31 problems correctly. What percent of the problems were correct? Round to the nearest tenth of a percent.

60. Daniel uses an on-line company to make prints from his digital camera. He ordered 18 prints at $0.12 per print plus $1.97 for shipping. What is the cost per print, including shipping, to the nearest cent? (*Source:* snapfish.com)

The bar graph below shows the average number of vacation days (including holidays) received each year by workers in various countries. Use the information in the graph to answer Exercises 61–64. Round to the nearest tenth when necessary.

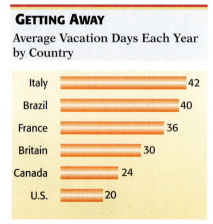

GETTING AWAY

Average Vacation Days Each Year by Country

Country	Days
Italy	42
Brazil	40
France	36
Britain	30
Canada	24
U.S.	20

Source: World Tourism Organization.

61. Assuming that a workweek is 5 days long, which country has an average of:

  (a) six weeks of vacation;

  (b) eight weeks of vacation.

62. Find the ratio of vacation days when comparing:

  (a) Brazil to the United States;

  (b) Canada to France.

63. (a) U.S. vacation days are what percent of Italy's vacation days?

  (b) France's vacation days are what percent of U.S. vacation days?

64. By what percent would U.S. vacation days need to increase to match the vacation days in:

  (a) Canada;   **(b)** Britain;   **(c)** Brazil.

Measurement

8

A greatly enlarged photo of a computer microchip.

We are constantly measuring things, from very large to very small.

For example, on August 27, 2003, the planet Mars was closer to Earth than at any time in the last 60,000 years. Mars was "only" 34,646,437 miles away! (See **Section 8.1,** Exercises 63–66.)

On the other hand, computer microchips often measure a tiny 5 mm long by 1 mm wide. (Learn about millimeters in **Section 8.2** and then try Exercise 39.)

8.1 Problem Solving with English Measurement

OBJECTIVES

1. Learn the basic measurement units in the English system.

2. Convert among measurement units using multiplication or division.

3. Convert among measurement units using unit fractions.

4. Solve application problems using English measurement.

We measure things all the time: the distance traveled on vacation, the floor area we want to cover with carpet, the amount of milk in a recipe, the weight of the bananas we buy at the store, the number of hours we work, and many more.

In the United States we still use the **English system** of measurement for many everyday activities. Examples of English units are inches, feet, quarts, ounces, and pounds. However, science, medicine, sports, and manufacturing use the **metric system** (meters, liters, and grams). And, because the rest of the world uses *only* the metric system, U.S. businesses have been changing to the metric system in order to compete internationally.

OBJECTIVE 1 Learn the basic measurement units in the English system. Until the switch to the metric system is complete, we still need to know how to use the English system of measurement. The table below lists the relationships you should memorize. The time relationships are used in both the English and metric systems.

English Measurement Relationships

Length	Weight
12 inches (in.) = 1 foot (ft)	16 ounces (oz) = 1 pound (lb)
3 feet (ft) = 1 yard (yd)	2000 pounds (lb) = 1 ton (T)
5280 feet (ft) = 1 mile (mi)	

Capacity	Time
8 fluid ounces = 1 cup (c) (fl oz)	60 seconds (sec) = 1 minute (min)
2 cups (c) = 1 pint (pt)	60 minutes (min) = 1 hour (hr)
2 pints (pt) = 1 quart (qt)	24 hours (hr) = 1 day
4 quarts (qt) = 1 gallon (gal)	7 days = 1 week (wk)

As you can see, there is no simple way to convert among these various measures. The units evolved over hundreds of years and were based on a variety of "standards." For example, one yard was the distance from the tip of a king's nose to his thumb when his arm was outstretched. An inch was three dried barleycorns laid end to end.

EXAMPLE 1 Knowing English Measurement Units

Memorize the English measurement conversions shown above. Then answer these questions.

(a) _____ day = 24 hr   Answer: 1 day

(b) 1 yd = _____ ft   Answer: 3 ft

Now Try Exercise 7.

OBJECTIVE 2 Convert among measurement units using multiplication or division. You often need to convert from one unit of measure to another. Two methods of converting measurements are shown here. Study each way and use the method you prefer. The first method involves deciding whether to multiply or divide.

Converting among Measurement Units

1. *Multiply* when converting from a larger unit to a smaller unit.
2. *Divide* when converting from a smaller unit to a larger unit.

EXAMPLE 2 Converting from One Unit of Measure to Another

Convert each measurement.

(a) 7 ft to inches

You are converting from a *larger* unit to a *smaller* unit (a *foot* is longer than an *inch*), so multiply.

Because *1 ft = **12** in.,* multiply by 12.

$$7 \text{ ft} = 7 \cdot \mathbf{12} = 84 \text{ in.}$$

(b) $3\frac{1}{2}$ lb to ounces

You are converting from a *larger* unit to a *smaller* unit (a *pound* is heavier than an *ounce*), so multiply.

Because *1 lb = **16** oz,* multiply by 16.

$$3\frac{1}{2} \text{ lb} = 3\frac{1}{2} \cdot \mathbf{16} = \frac{7}{\underset{1}{\cancel{2}}} \cdot \frac{\overset{8}{\cancel{16}}}{1} = \frac{56}{1} = 56 \text{ oz}$$

(c) 20 qt to gallons

You are converting from a *smaller* unit to a *larger* unit (a *quart* is smaller than a *gallon*), so divide.

Because *4 qt = 1 gal,* divide by 4.

$$20 \text{ qt} = \frac{20}{4} = 5 \text{ gal}$$

Divide by 4 ↑

(d) 45 min to hours

You are converting from a *smaller* unit to a *larger* unit (a *minute* is less than an *hour*), so divide.

Because ***60** min = 1 hr,* divide by 60 and write the fraction in lowest terms.

$$45 \text{ min} = \frac{45}{\mathbf{60}} = \frac{45 \div \mathbf{15}}{60 \div \mathbf{15}} = \frac{3}{4} \text{ hr} \quad \leftarrow \text{ Lowest terms}$$

Divide by 60 ↑

Now Try Exercise 11. ▶▶▶

OBJECTIVE 3 Convert among measurement units using unit fractions. If you have trouble deciding whether to multiply or divide when converting measurements, use *unit fractions* to solve the problem. You'll also use this method in science classes. A **unit fraction** is equivalent to 1. Here is an example.

$$\frac{12 \text{ in.}}{12 \text{ in.}} = \frac{\overset{1}{\cancel{12 \text{ in.}}}}{\underset{1}{\cancel{12 \text{ in.}}}} = 1$$

Use the table of measurement relationships on the previous page to find that 12 in. is the same as 1 ft. So in the numerator you can substitute 1 ft for 12 in., or in the denominator you can substitute 1 ft for 12 in. This makes two useful unit fractions.

$$\frac{\mathbf{1 \text{ ft}}}{12 \text{ in.}} = 1 \quad \text{or} \quad \frac{12 \text{ in.}}{\mathbf{1 \text{ ft}}} = 1$$

To convert from one measurement unit to another, just multiply by the appropriate unit fraction. Remember, a unit fraction is equivalent to 1. Multiplying something by 1 does *not* change its value.

Use these guidelines to choose the correct unit fraction.

> **Choosing a Unit Fraction**
>
> The *numerator* should use the measurement unit you want in the *answer.*
> The *denominator* should use the measurement unit you want to *change.*

EXAMPLE 3 **Using Unit Fractions with Length Measurements**

(a) Convert 60 in. to feet.

Use a unit fraction with feet (the unit for your answer) in the numerator, and inches (the unit being changed) in the denominator. Because *1 ft = 12 in.,* the necessary unit fraction is

$$\frac{1 \text{ ft}}{12 \text{ in.}} \quad \leftarrow \text{Unit for your answer is feet.}$$
$$\leftarrow \text{Unit being changed is inches.}$$

Next, multiply 60 in. times this unit fraction. Write 60 in. as the fraction $\dfrac{60 \text{ in.}}{1}$ and divide out common units and factors wherever possible.

$$60 \text{ in.} \cdot \frac{1 \text{ ft}}{12 \text{ in.}} = \frac{\overset{5}{\cancel{60} \text{ in.}}}{1} \cdot \frac{1 \text{ ft}}{\underset{1}{\cancel{12} \text{ in.}}} = \frac{5 \cdot 1 \text{ ft}}{1} = 5 \text{ ft}$$

These units should match.

Divide out inches.

Divide 60 and 12 by 12

(b) Convert 9 ft to inches.

Select the correct unit fraction to change 9 ft to inches.

$$\frac{12 \text{ in.}}{1 \text{ ft}} \quad \leftarrow \text{Unit for your answer is inches.}$$
$$\leftarrow \text{Unit being changed is feet.}$$

Multiply 9 ft times the unit fraction.

$$9 \text{ ft} \cdot \frac{12 \text{ in.}}{1 \text{ ft}} = \frac{9 \cancel{\text{ ft}}}{1} \cdot \frac{12 \text{ in.}}{1 \cancel{\text{ ft}}} = \frac{9 \cdot 12 \text{ in.}}{1} = 108 \text{ in.}$$

These units should match.

Divide out feet.

> **CAUTION**
> If no units will divide out, you made a mistake in choosing the unit fraction.

Now Try Exercise 17.

EXAMPLE 4 **Using Unit Fractions with Capacity and Weight Measurements**

(a) Convert 9 pt to quarts.

First select the correct unit fraction.

$$\frac{1 \text{ qt}}{2 \text{ pt}} \quad \leftarrow \text{Unit for your answer is quarts.}$$
$$\leftarrow \text{Unit being changed is pints.}$$

Continued on Next Page

Now multiply.

Write as mixed number.

$$9 \text{ pt} \cdot \frac{1 \text{ qt}}{2 \text{ pt}} = \frac{9 \text{ pt}}{1} \cdot \frac{1 \text{ qt}}{2 \text{ pt}} = \frac{9}{2} \text{ qt} = 4\frac{1}{2} \text{ qt}$$

These units should match.   Divide out pints.

(b) Convert $7\frac{1}{2}$ gal to quarts.

Write as an improper fraction.

$$\frac{7\frac{1}{2} \text{ gal}}{1} \cdot \frac{4 \text{ qt}}{1 \text{ gal}} = \frac{15}{2} \cdot \frac{4}{1} \text{ qt}$$

Divide out gallons.

$$= \frac{15}{\underset{1}{\cancel{2}}} \cdot \frac{\overset{2}{\cancel{4}}}{1} \text{ qt}$$

$$= 30 \text{ qt}$$

(c) Convert 36 oz to pounds.

$$\frac{\overset{9}{\cancel{36}} \text{ oz}}{1} \cdot \frac{1 \text{ lb}}{\underset{4}{\cancel{16}} \text{ oz}} = \frac{9}{4} \text{ lb} = 2\frac{1}{4} \text{ lb}$$

NOTE

In Example 4(c) above, you get $\frac{9}{4}$ lb. Recall that $\frac{9}{4}$ means $9 \div 4$. If you do $9 \div 4$ on your calculator, you get **2.25** lb. English measurements usually use fractions or mixed numbers, like $2\frac{1}{4}$ lb. However, 2.25 lb is also correct and is the way grocery stores often show weights of produce, meat, and cheese.

Now Try Exercises 27 and 35. ▶▶▶

EXAMPLE 5 **Using Several Unit Fractions**

Sometimes you may need to use two or three unit fractions to complete a conversion.

(a) Convert 63 in. to yards.

Use the unit fraction $\frac{1 \text{ ft}}{12 \text{ in.}}$ to change inches to feet and the unit fraction $\frac{1 \text{ yd}}{3 \text{ ft}}$ to change feet to yards. Notice how all the units divide out except yards, which is the unit you want in the answer.

$$\frac{63 \text{ in.}}{1} \cdot \frac{1 \text{ ft}}{12 \text{ in.}} \cdot \frac{1 \text{ yd}}{3 \text{ ft}} = \frac{63}{36} \text{ yd} = \frac{63 \div 9}{36 \div 9} \text{ yd} = \frac{7}{4} \text{ yd} = 1\frac{3}{4} \text{ yd}$$

Continued on Next Page

You can also divide out common factors in the numbers.

$$\frac{\overset{7}{\cancel{\overset{21}{\cancel{63}}}}}{1} \cdot \frac{1}{\underset{4}{\cancel{12}}} \cdot \frac{1}{\underset{1}{\cancel{3}}} = \frac{7}{4} = 1\frac{3}{4} \text{ yd}$$

Instead of changing $\frac{7}{4}$ to $1\frac{3}{4}$, you can enter $7 \div 4$ on your calculator to get 1.75 yd. Both answers are correct because 1.75 is equivalent to $1\frac{3}{4}$.

(b) Convert 2 days to seconds.

Use three unit fractions. The first one changes days to hours, the next one changes hours to minutes, and the last one changes minutes to seconds. All the units divide out except seconds, which is what you want in your answer.

$$\frac{2 \text{ \cancel{days}}}{1} \cdot \frac{24 \text{ \cancel{hr}}}{1 \text{ \cancel{day}}} \cdot \frac{60 \text{ \cancel{min}}}{1 \text{ \cancel{hr}}} \cdot \frac{60 \text{ seconds}}{1 \text{ \cancel{min}}} = 172{,}800 \text{ seconds}$$

Divide out **days**.

Divide out **hr**.

Divide out **min**.

◄◄◄ Now Try Exercise 45.

OBJECTIVE 4 Solve application problems using English measurement. To solve measurement application problems, we will use the steps summarized here.

Step 1 **Read** the problem.

Step 2 **Work out a plan.**

Step 3 **Estimate** a reasonable answer.

Step 4 **Solve** the problem.

Step 5 **State the answer.**

Step 6 **Check** your work.

Because measurement applications often involve conversions, writing an equation may not be the most helpful way to solve the problem. Therefore, Steps 2 and 3 are different from the ones you learned in **Chapter 3;** the other steps are the same.

EXAMPLE 6 Solving English Measurement Applications

(a) A 36 oz can of coffee is on sale at Jerry's Foods for $7.89. What is the cost per pound, to the nearest cent? (*Source:* Jerry's Foods.)

Step 1 **Read** the problem. The problem asks for the cost per *pound* of coffee.

Step 2 **Work out a plan.** The weight of the coffee is given in *ounces* but the answer must be cost per *pound*. Convert ounces to pounds. The word *per* indicates division. You need to divide the cost by the number of pounds.

Step 3 **Estimate** a reasonable answer. To estimate, round $7.89 to $8. Then, there are 16 oz in a pound, so 36 oz is a little more than 2 pounds. So, $8 \div 2 = $4 per pound as our estimate.

Continued on Next Page

Step 4   **Solve** the problem. Use a unit fraction to convert 36 oz to pounds.

$$\frac{\overset{9}{\cancel{36\ \cancel{oz}}}}{1} \cdot \frac{1\ \text{lb}}{\underset{4}{\cancel{16\ \cancel{oz}}}} = \frac{9}{4}\ \text{lb} = 2.25\ \text{lb}$$

Then divide to find the *cost* per *pound.*

$$\begin{matrix}\text{cost} \longrightarrow \\ \text{per} \longrightarrow \\ \text{pound} \longrightarrow\end{matrix}\ \frac{\$7.89}{2.25\ \text{lb}} = 3.50\overline{6} \approx 3.51 \qquad \text{Rounded}$$

Step 5   **State the answer.** The coffee costs $3.51 per pound (to the nearest cent).

Step 6   **Check** your work. The exact answer of $3.51 is close to our estimate of $4.

(b) Bilal's favorite cake recipe uses $1\frac{2}{3}$ cups of milk. If he makes six cakes for a bake sale at his son's school, how many quarts of milk will he need?

Step 1   **Read** the problem. The problem asks for the number of *quarts* of milk needed for six cakes.

Step 2   **Work out a plan.** Multiply to find the number of *cups* of milk for six cakes. Then convert *cups* to *quarts* (the unit required in the answer).

Step 3   **Estimate** a reasonable answer. To estimate, round $1\frac{2}{3}$ cups to 2 cups. Then, 2 cups times 6 = 12 cups. There are 4 cups in a quart, so 12 cups ÷ 4 = 3 quarts as our estimate.

Step 4   **Solve** the problem. First multiply. Then use unit fractions to convert.

$$1\frac{2}{3} \cdot 6 = \frac{5}{\underset{1}{\cancel{3}}} \cdot \frac{\overset{2}{\cancel{6}}}{1} = \frac{10}{1} = 10\ \text{cups} \quad \left\{\begin{matrix}\text{Milk needed} \\ \text{for six cakes}\end{matrix}\right.$$

$$\frac{\overset{5}{\cancel{10\ \cancel{\text{cups}}}}}{1} \cdot \frac{1\ \cancel{\text{pt}}}{\underset{1}{\cancel{2\ \cancel{\text{cups}}}}} \cdot \frac{1\ \text{qt}}{2\ \cancel{\text{pt}}} = \frac{5}{2}\ \text{qt} = 2\frac{1}{2}\ \text{qt}$$

Step 5   **State the answer.** Bilal needs $2\frac{1}{2}$ qt (or 2.5 qt) of milk.

Step 6   **Check** your work. The exact answer of $2\frac{1}{2}$ qt is close to our estimate of 3 qt.

NOTE

In *Step 2* above, we *first multiplied* $1\frac{2}{3}$ cups times 6 to find the number of cups needed, then *converted* 10 cups to $2\frac{1}{2}$ quarts. It would also work to *first convert* $1\frac{2}{3}$ cups to $\frac{5}{12}$ qt, then *multiply* $\frac{5}{12}$ qt times 6 to get $2\frac{1}{2}$ qt.

Now Try Exercise 55. ▶▶▶

8.1 Exercises

Fill in the blanks with the measurement relationships you have memorized. See Example 1.

1. (a) 1 yd = _____ ft

   (b) _____ in. = 1 ft

2. (a) 1 ft = _____ in.

   (b) _____ ft = 1 mi

3. (a) _____ fl oz = 1 c

   (b) 1 qt = _____ pt

4. (a) _____ qt = 1 gal

   (b) 1 pt = _____ c

5. (a) 1 mi = _____ ft

   (b) _____ ft = 1 yd

6. (a) 1 wk = _____ days

   (b) _____ sec = 1 min

7. (a) _____ lb = 1 T

   (b) 1 lb = _____ oz

8. (a) _____ oz = 1 lb

   (b) 1 T = _____ lb

9. (a) 1 min = _____ sec

   (b) _____ min = 1 hr

10. (a) 1 day = _____ hr

   (b) _____ sec = 1 min

Convert each measurement by multiplying or dividing. See Example 2.

11. (a) 120 sec = _____ min

   (b) 4 hr = _____ min

12. (a) 180 min = _____ hr

   (b) 5 min = _____ sec

13. (a) 2 qt = _____ gal

   (b) $6\frac{1}{2}$ ft = _____ in.

14. (a) $4\frac{1}{2}$ gal = _____ qt

   (b) 12 oz = _____ lb

15. An adult African elephant could weigh 7 to 8 tons. How many pounds could it weigh? (*Source: The Top 10 of Everything.*)

16. A reticulated python is the world's longest snake. It grows to a length of 18 to 33 feet. How many yards long can the snake be?

Convert each measurement in Exercises 17–38 using unit fractions. See Examples 3 and 4.

17. 9 yd = _____ ft

18. 20,000 lb = _____ T

19. 7 lb = _____ oz

20. 96 oz = _____ lb

21. 5 qt = _____ pt

22. 26 pt = _____ qt

23. 90 min = _____ hr

24. 45 sec = _____ min

25. 3 in. = _____ ft

26. 30 in. = _____ ft

27. 24 oz = _____ lb

28. 36 oz = _____ lb

29. 5 c = _____ pt

30. 15 qt = _____ gal

Use the information in the bar graph below to answer Exercises 31–32.

Thickness of Lake Ice Needed for Safe Walking/Driving

15 in. Pickup truck
12 in. Car
5 in. Snowmobile or ATV
4 in. Person walking

Source: Wisconsin DNR.

31. If the ice on a lake is $\frac{1}{2}$ ft thick, what will it safely support?

32. How many feet of ice are needed to safely drive a pickup truck on a lake?

33. $2\frac{1}{2}$ T = _____ lb

34. $4\frac{1}{2}$ pt = _____ c

35. $4\frac{1}{4}$ gal = _____ qt

36. $2\frac{1}{4}$ hr = _____ min

37. After 15 years a saguaro cactus is still only one-third to two-thirds of a foot tall, depending upon rainfall. How tall could the cactus be in inches? (*Source: Ecology of the Saguaro III.*)

38. Yao Ming, an NBA basketball player originally from China, is $7\frac{1}{2}$ ft tall. What is his height in inches? (*Source: www.NBA.com*)

Use two or three unit fractions to make each conversion. See Example 5.

39. 6 yd = _____ in.

40. 2 T = _____ oz

41. 112 c = _____ qt

42. 336 hr = _____ wk

43. 6 days = _____ sec

44. 5 gal = _____ c

45. $1\frac{1}{2}$ T = _____ oz

46. $3\frac{1}{3}$ yd = _____ in.

47. The statement $8 = 2$ is *not* true. But with appropriate measurement units, it *is* true.

$$8 \text{ } quarts = 2 \text{ } gallons$$

Attach measurement units to these numbers to make the statements true.

(a) 1 _____ = 16 _____

(b) 10 _____ = 20 _____

(c) 120 _____ = 2 _____

(d) 2 _____ = 24 _____

(e) 6000 _____ = 3 _____

(f) 35 _____ = 5 _____

48. Explain in your own words why you can add 2 feet + 12 inches to get 3 feet, but you cannot add 2 feet + 12 pounds.

Convert each measurement. See Example 5.

49. $2\frac{3}{4}$ mi = _____ in.

50. $5\frac{3}{4}$ tons = _____ oz

51. $6\frac{1}{4}$ gal = _____ fl oz

52. $3\frac{1}{2}$ days = _____ sec

53. 24,000 oz = _____ T

54. 57,024 in. = _____ mi

Solve each application problem in Exercises 55–62. Show your work. See Example 6.

55. Geralyn bought 20 oz of strawberries for $2.29. What was the price per pound for the strawberries, to the nearest cent?

56. Zach paid $0.79 for a 1.6 oz candy bar. What was the cost per pound?

57. Dan orders supplies for the science labs. Each of the 24 stations in the chemistry lab needs 2 ft of rubber tubing. If rubber tubing sells for $8.75 per yard, how much will it cost to equip all the stations?

58. In 2001, Marquette, Michigan, had 219 in. of snowfall, while Detroit, Michigan, had 18 in. What was the difference in snowfall between the two cities, in feet? Round to the nearest tenth. (*Source: World Almanac.*)

59. Tropical cockroaches are the fastest land insects. They can run about 5 feet per second. At this rate, how long would it take the cockroach to travel one mile? (*Source: Guinness Book of World Records.*) Give your answer

(a) in seconds

(b) In minutes, to the nearest tenth.

60. A snail moves at an average speed of 2 feet every 3 minutes. At that rate, how long would it take the snail to travel one mile? (*Source: Beakman and Jax.*) Give your answer

(a) in hours

(b) in days.

61. At the day care center, each of the 15 toddlers drinks about $\frac{2}{3}$ cup of milk with lunch. The center is open 5 days a week.

(a) How many quarts of milk will the center need for one week of lunches?

(b) If the center buys milk in gallon containers, how many containers should be ordered for one week?

62. Bob's Candies in Albany, Georgia, makes 135,000 pounds of candy canes each day. (*Source:* Bob's Candies, Inc.)

(a) How many tons of candy canes are produced during a 5-day workweek?

(b) The plant operates 24 hours per day. How many tons of candy canes are produced each hour, to the nearest tenth?

RELATING CONCEPTS (EXERCISES 63–66) For Individual or Group Work

*On the first page of this chapter, we said that the planet Mars was only 34,646,437 miles from Earth in August 2003. Use this information and learn more about Mars as you **work Exercises 63–66 in order.*** (*Sources:* www.NASA.com *and* Associated Press.)

63. In order to better understand how far away Mars was in August 2003, we need to compare the distance to something more familiar.

(a) The distance around Earth is about 24,907 miles. How many times would you travel around Earth before going 34,646,437 miles? Round your answer to the nearest whole number.

(b) The driving distance from Los Angeles to New York City is 2786 miles. How many times would you have to make that drive to match the distance from Earth to Mars? Round to the nearest whole number.

64. The divisions you did in Exercise 63 were long and messy, so a calculator was very helpful. Now let's see how we can do it *without* a calculator.

(a) Round the distance from Mars to Earth to the nearest million. Round the distance around Earth to the nearest thousand. Now do the division *without* a calculator, using the shortcut for division by multiples of 10 in **Section 1.7.**

(b) Round the driving distance from Los Angeles to New York City to the nearest hundred. Now do the division by hand using the shortcut.

(c) Compare your answers above to those from Exercise 63. Are they just as useful as the answers in Exercise 63?

65. The tallest volcano on any planet in our solar system is on Mars. It is called Olympic Mons and is 16.7 miles high.

(a) How many feet tall is Olympic Mons?

(b) Mount Everest, the tallest place on Earth, is 29,035 ft above sea level. How many miles tall is it, to the nearest tenth?

(c) About how many *times* taller is Olympic Mons than Mount Everest?

66. The pull of gravity on Mars is less than it is on Earth. Because of this, a person who weighs 100 pounds on Earth would weigh only 36 pounds on Mars.

(a) Set up a proportion to find out how much a 180-pound person would weigh on Mars, to the nearest pound.

(b) Set up a proportion to figure out how much a 9-pound baby would weigh on Mars, to the nearest pound.

(c) How much would *you* weigh on Mars?

8.2 The Metric System—Length

OBJECTIVES

1. Learn the basic metric units of length.
2. Use unit fractions to convert among units.
3. Move the decimal point to convert among units.

Around 1790, a group of French scientists developed the metric system of measurement. It is an organized system based on multiples of 10, like our number system and our money. After you are familiar with metric units, you will see that they are easier to use than the hodgepodge of English measurement relationships you used in **Section 8.1.**

> **NOTE**
> The metric system information in this text is consistent with usage guidelines from the National Institute of Standards and Technology, www.nist.gov/metric.

OBJECTIVE 1 Learn the basic metric units of length. The basic unit of length in the metric system is the **meter** (also spelled *metre*). Use the symbol **m** for meter; do not put a period after it. If you put five of the pages from this textbook side by side, they would measure about 1 meter. Or, look at a yardstick—a meter is just a little longer. A yard is 36 inches long; a meter is about 39 inches long.

book page	book page	book page	book page	book page

|← 1 meter (about 39 in.) →|
|← 1 yard (36 in.) →|

In the metric system you use meters for things like buying fabric for sewing projects, measuring the length of your living room, talking about heights of buildings, or describing track and field athletic events.

Buy 2 m
of fabric
(about 2 yd)

6 m
(about 20 ft)

15 m
(about
49 ft)

To make longer or shorter length units in the metric system, **prefixes** are written in front of the word *meter*. For example, the prefix *kilo* means **1000**, so a *kilo*meter is **1000** meters. The table below shows how to use the prefixes for length measurements. It is helpful to memorize the prefixes because they are also used with weight and capacity measurements. The blue boxes are the units you will use most often in daily life.

Prefix	kilo-meter	hecto-meter	deka-meter	meter	deci-meter	centi-meter	milli-meter
Meaning	1000 meters	100 meters	10 meters	1 meter	$\frac{1}{10}$ of a meter	$\frac{1}{100}$ of a meter	$\frac{1}{1000}$ of a meter
Symbol	km	hm	dam	m	dm	cm	mm

Length units that are used most often

Here are some comparisons to help you get acquainted with the commonly used length units: km, m, cm, mm.

*Kilo*meters are used instead of miles. A kilometer is 1000 meters. It is about 0.6 mile (a little more than half a mile) or about 5 to 6 city blocks. If you participate in a 10 km run, you'll run about 6 miles.

A meter is divided into 100 smaller pieces called *centi*meters. Each centimeter is $\frac{1}{100}$ of a meter. Centimeters are used instead of inches. A centimeter is a little shorter than $\frac{1}{2}$ inch. The cover of this textbook is about 22 cm wide. A nickel is about 2 cm across. Measure the width and length of your little finger on this centimeter ruler. The width of your little finger is probably about 1 cm, or a little more.

A meter is divided into 1000 smaller pieces called *milli*meters. Each millimeter is $\frac{1}{1000}$ of a meter. It takes 10 mm to equal 1 cm, so it is a very small length. The thickness of a dime is about 1 mm. Measure the width of your pen or pencil and the width of your little finger on this millimeter ruler.

EXAMPLE 1 Using Metric Length Units

Write the most reasonable metric unit in each blank. Choose from km, m, cm, and mm.

(a) The distance from home to work is 20 _____ .

20 **km** because kilometers are used instead of miles.

20 km is about 12 miles.

(b) My wedding ring is 4 _____ wide.

4 **mm** because the width of a ring is very small.

(c) The newborn baby is 50 _____ long.

50 **cm**, which is half of a meter; a meter is about 39 inches so half a meter is around 20 inches.

Now Try Exercises 17 and 19.

OBJECTIVE **2** **Use unit fractions to convert among units.** You can convert among metric length units using unit fractions. Keep these relationships in mind when setting up the unit fractions.

Metric Length Relationships

1 km = 1000 m so the unit fractions are:	**1 m = 1000 mm** so the unit fractions are:
$\dfrac{1 \text{ km}}{1000 \text{ m}}$ or $\dfrac{1000 \text{ m}}{1 \text{ km}}$	$\dfrac{1 \text{ m}}{1000 \text{ mm}}$ or $\dfrac{1000 \text{ mm}}{1 \text{ m}}$
1 m = 100 cm so the unit fractions are:	**1 cm = 10 mm** so the unit fractions are:
$\dfrac{1 \text{ m}}{100 \text{ cm}}$ or $\dfrac{100 \text{ cm}}{1 \text{ m}}$	$\dfrac{1 \text{ cm}}{10 \text{ mm}}$ or $\dfrac{10 \text{ mm}}{1 \text{ cm}}$

EXAMPLE 2 **Using Unit Fractions to Convert Length Measurements**

Convert each measurement using unit fractions.

(a) 5 km to m

Put the unit for the answer (meters) in the numerator of the unit fraction; put the unit you want to change (km) in the denominator.

Unit fraction equivalent to 1 $\left\{\dfrac{1000 \text{ m}}{1 \text{ km}}\right.$ ← Unit for answer
← Unit being changed

Multiply. Divide out common units where possible.

$$5 \text{ km} \cdot \frac{1000 \text{ m}}{1 \text{ km}} = \frac{5 \text{ km}}{1} \cdot \frac{1000 \text{ m}}{1 \text{ km}} = \frac{5 \cdot 1000 \text{ m}}{1} = 5000 \text{ m}$$

These units should match.

Thus, 5 km = 5000 m.

The answer makes sense because a kilometer is much longer than a meter, so 5 km will contain many meters.

(b) 18.6 cm to m

Multiply by a unit fraction that allows you to divide out centimeters.

Unit fraction

$$\frac{18.6 \text{ cm}}{1} \cdot \frac{1 \text{ m}}{100 \text{ cm}} = \frac{18.6}{100} \text{ m} = 0.186 \text{ m}$$

Thus, 18.6 cm = 0.186 m.

There are 100 cm in a meter, so 18.6 cm will be a small part of a meter. The answer makes sense.

Now Try Exercise 25. (Use a unit fraction.)

OBJECTIVE 3 **Move the decimal point to convert among units.**   By
now you have probably noticed that conversions among metric units are
made by multiplying or dividing by 10, by 100, or by 1000. A quick way to
multiply by 10 is to move the decimal point one place to the *right*. Move it
two places to the right to multiply by 100, three places to multiply by 1000.
Dividing is done by moving the decimal point to the *left* in the same manner.

An alternate conversion method to unit fractions is moving the decimal
point using this **metric conversion line.**

Here are the steps for using the conversion line.

> **Using the Metric Conversion Line**
>
> *Step 1*   Find the unit you are given on the metric conversion line.
>
> *Step 2*   Count the number of places to get from the unit you are given
> to the unit you want in the answer.
>
> *Step 3*   Move the decimal point the **same number of places** and in the
> **same direction** as you did on the conversion line.

EXAMPLE 3 **Using the Metric Conversion Line**

Use the metric conversion line to make the following conversions.

(a) 5.702 km to m

Find **km** on the metric conversion line. To get to **m** , you move *three
places* to the *right*. So move the decimal point in 5.702 *three places* to the
right.

Thus, 5.702 km = 5702 m.

(b) 69.5 cm to m

Find **cm** on the conversion line. To get to **m** , move *two places* to the *left*.

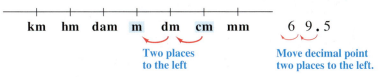

Thus, 69.5 cm = 0.695 m.

───── **Continued on Next Page**

(c) 8.1 cm to mm

From **cm** to **mm** is *one place* to the *right*.

One place to the right

Move decimal point one place to the right.

Thus, 8.1 cm = 81 mm.

Now Try Exercise 31.

EXAMPLE 4 **Practicing Length Conversions**

Convert using the metric conversion line.

(a) 1.28 m to mm

Moving from **m** to **mm** is going *three places to the right*. In order to move the decimal point in 1.28 three places to the right, you must write a 0 as a placeholder.

1.28**0**   Zero is written in as a placeholder.

Move decimal point three places to the right.

Thus, 1.28 m = 1280 mm.

(b) 60 cm to m

From **cm** to **m** is two places to the left. The decimal point in 60 starts at the *far right side* because 60 is a whole number. Then move it two places to the left.

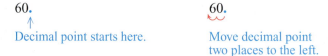

60.   60.

Decimal point starts here.   Move decimal point two places to the left.

Thus, 60 cm = 0.60 m, which is equivalent to 0.6 m.

(c) 8 m to km

From **m** to **km** is three places to the left. The decimal point in 8 starts at the far right side. In order to move it three places to the left, you must write two zeros as placeholders.

Two zeros are written in as placeholders.

8.   008.

Decimal point starts here.   Move decimal point three places to the left.

Thus, 8 m = 0.008 km.

Now Try Exercise 29.

8.2 Exercises

FOR EXTRA HELP   Tutor Center Addison-Wesley Math Tutor Center   *Math XL* MathXL   Digital Video Tutor CD 5 Videotape 8   Student's Solutions Manual   *MyMathLab* MyMathLab   MathXL Tutorials on CD

Use your knowledge of the meaning of metric prefixes to fill in the blanks.

1. *kilo* means _____ so

1 km = _____ m

2. *deka* means _____ so

1 dam = _____ m

3. *milli* means _____ so

1 mm = _____ m

4. *deci* means _____ so

1 dm = _____ m

5. *centi* means _____ so

1 cm = _____ m

6. *hecto* means _____ so

1 hm = _____ m

Use this ruler to measure the width of your hand and thumb for Exercises 7–10.

7. The width of your hand in centimeters

8. The width of your hand in millimeters

9. The width of your thumb in millimeters

10. The width of your thumb in centimeters

Write the most reasonable metric length unit in each blank. Choose from km, m, cm, and mm. See Example 1.

11. The child was 91 _____ tall.

12. The cardboard was 3 _____ thick.

13. Ming-Na swam in the 200 _____ backstroke race.

14. The bookcase is 75 _____ wide.

15. Adriana drove 400 _____ on her vacation.

16. The door is 2 _____ high.

17. An aspirin tablet is 10 _____ across.

18. Lamard jogs 4 _____ every morning.

19. A paper clip is about 3 _____ long.

20. My pen is 145 _____ long.

21. Dave's truck is 5 _____ long.

22. Wheelchairs need doorways that are at least 80 _____ wide.

23. Describe at least three examples of metric length units that you have come across in your daily life.

24. Explain one reason the metric system would be easier for a child to learn than the English system.

Convert each measurement. Use unit fractions or the metric conversion line. See Examples 2–4.

25. 7 m to cm

26. 18 m to cm

27. 40 mm to m

28. 6 mm to m

29. 9.4 km to m

30. 0.7 km to m

31. 509 cm to m

32. 30 cm to m

33. 400 mm to cm

34. 25 mm to cm

35. 0.91 m to mm

36. 4 m to mm

37. Is 82 cm greater than or less than 1 m? What is the difference in the lengths?

38. Is 1022 m greater than or less than 1 km? What is the difference in the lengths?

39. On the first page of this chapter we said that computer micro-chips may be only 5 mm long and 1 mm wide. Using the ruler from Exercises 7–10, draw a rectangle that measures 5 mm by 1 mm. Then convert each measurement to centimeters.

A greatly enlarged photo of a computer microchip.

40. Some cameras use film that is 35 mm wide. Film for movie theaters may be 70 mm wide. Using the ruler from Exercises 7–10, draw a line that is 35 mm long and a line 70 mm long. Then convert each measurement to centimeters.

41. The Roe River near Great Falls, Montana, is the shortest river in the world, with a north fork that is just under 18 m long. How many kilometers long is the north fork of the river? (*Source: Guinness Book of Amazing Nature.*)

42. There are 60,000 km of blood vessels in the human body. How many meters of blood vessels are in the body? (*Source: Big Book of Knowledge.*)

43. The median height for U.S. females who are 20 to 29 years old is about 1.64 m. Convert this height to centimeters and to millimeters. (*Source: U.S. National Center for Health Statistics.*)

44. The median height for 20- to 29-year-old males in the United States is about 177 cm. Convert this height to meters and to millimeters. (*Source: U.S. National Center for Health Statistics.*)

45. Use two unit fractions to convert 5.6 mm to km.

46. Use two unit fractions to convert 16.5 km to mm.

8.3 The Metric System—Capacity and Weight (Mass)

We use capacity units to measure liquids, such as the amount of milk in a recipe, the gasoline in our car tank, and the water in an aquarium. (The English capacity units used in the United States are cups, pints, quarts, and gallons.) The basic metric unit for capacity is the **liter** (also spelled *litre*). The capital letter **L** is the symbol for liter, to avoid confusion with the numeral 1.

OBJECTIVES

1. Learn the basic metric units of capacity.
2. Convert among metric capacity units.
3. Learn the basic metric units of weight (mass).
4. Convert among metric weight (mass) units.
5. Distinguish among basic metric units of length, capacity, and weight (mass).

OBJECTIVE 1 Learn the basic metric units of capacity.   The liter is related to metric length in this way: A box that measures 10 cm on every side holds exactly one liter. (The volume of the box is 10 cm • 10 cm • 10 cm = 1000 cubic centimeters. Volume is discussed in **Section 4.8.**) A liter is just a little more than 1 quart.

10 cm 10 cm 10 cm
Holds exactly 1 liter (L)

About 1 liter of milk

About 1 liter of oil for your car

A liter is a little more than one quart (just $\frac{1}{4}$ cup more).

In the metric system you use liters for things like buying milk and soda at the store, filling a pail with water, and describing the size of your home aquarium.

Buy a 2 L bottle of soda.

Use a 12 L pail to wash floors.

Watch the fish in your 40 L aquarium.

To make larger or smaller capacity units, we use the same **prefixes** as we did with length units. For example, *kilo* means **1000** so a *kilo*meter is **1000** meters. In the same way, a *kilo*liter is **1000** liters.

Prefix	kilo- liter	hecto- liter	deka- liter	liter	deci- liter	centi- liter	milli- liter
Meaning	1000 liters	100 liters	10 liters	1 liter	$\frac{1}{10}$ of a liter	$\frac{1}{100}$ of a liter	$\frac{1}{1000}$ of a liter
Symbol	kL	hL	daL	L	dL	cL	mL

Capacity units used most often

The capacity units you will use most often in daily life are liters (L) and *milli*liters (mL). A tiny box that measures 1 cm on every side holds exactly one milliliter. (In medicine, this small amount is also called 1 cubic centimeter, or 1 cc for short.) It takes 1000 mL to make 1 L. Here are some useful comparisons.

1 cm 1 cm 1 cm Holds exactly 1 milliliter (mL)	Teaspoon holds 5 mL	One cup holds about 250 mL

EXAMPLE 1 Using Metric Capacity Units

Write the most reasonable metric unit in each blank. Choose from L and mL.

(a) The bottle of shampoo held 500 _____ .
500 **mL** because 500 L would be about 500 quarts, which is too much.

(b) I bought a 2 _____ carton of orange juice.
2 **L** because 2 mL would be less than a teaspoon.

◀◀◀ **Now Try Exercises 1 and 3.**

OBJECTIVE 2 Convert among metric capacity units. Just as with length units, you can convert between milliliters and liters using unit fractions.

Metric Capacity Relationships

1 L = 1000 mL, so the unit fractions are:

$$\frac{1 \text{ L}}{1000 \text{ mL}} \quad \text{or} \quad \frac{1000 \text{ mL}}{1 \text{ L}}$$

Or you can use a metric conversion line to decide how to move the decimal point.

1000	100	10	1	$\frac{1}{10}$	$\frac{1}{100}$	$\frac{1}{1000}$
kL	**hL**	**daL**	**L**	**dL**	**cL**	**mL**

EXAMPLE 2 Converting among Metric Capacity Units

Convert using the metric conversion line or unit fractions.

(a) 2.5 L to mL

Using the metric conversion line:
From **L** to **mL** is *three places* to the *right*.

2.5̲0̲0̲   Write two zeros as placeholders.

Thus, 2.5 L = 2500 mL.

Using unit fractions:

Multiply by a unit fraction that allows you to divide out liters.

$$\frac{2.5 \text{ L}}{1} \cdot \frac{1000 \text{ mL}}{1 \text{ L}} = 2500 \text{ mL}$$

Thus, 2.5 L = 2500 mL.

Continued on Next Page

(b) 80 mL to L

Using the metric conversion line:	Using unit fractions:

Using the metric conversion line:

From **mL** to **L** is *three places* to the *left*.

80.   080.
↑

Decimal point   Move decimal point
starts here.   three places to the left.

Thus, 80 mL = 0.080 L or 0.08 L.

Using unit fractions:

Multiply by a unit fraction that allows you to divide out mL.

$$\frac{80 \text{ mL}}{1} \cdot \frac{1 \text{ L}}{1000 \text{ mL}}$$

$$= \frac{80}{1000} \text{ L} = 0.08 \text{ L}$$

Thus, 80 mL = 0.08 L.

Now Try Exercises 37 and 49. ▶▶▶

OBJECTIVE 3 Learn the basic metric units of weight (mass). The **gram** is the basic metric unit for *mass*. Although we often call it "weight," there is a difference. Weight is a measure of the pull of gravity; the farther you are from the center of the earth, the less you weigh. In outer space you become weightless, but your mass, the amount of matter in your body, stays the same regardless of where you are. In science courses, it will be important to distinguish between the weight of an object and its mass. But for everyday purposes, we will use the word *weight*.

The gram is related to metric length in this way: The weight of the water in a box measuring 1 cm on every side is 1 gram. This is a very tiny amount of water (1 mL) and a very small weight. One gram is also the weight of a dollar bill or a single raisin. A nickel weighs 5 grams. A regular-sized hamburger and bun weighs from 175 to 200 grams.

The 1 mL of water in this tiny box weighs 1 gram.

A nickel weighs 5 grams.

A dollar bill weighs 1 gram.

A hamburger weighs 175 to 200 grams.

To make larger or smaller weight units, we use the same **prefixes** as we did with length and capacity units. For example, *kilo* means **1000** so a *kilo*meter is **1000** meters, a *kilo*liter is **1000** liters, and a *kilo*gram is **1000** grams.

Prefix	kilo-gram	hecto-gram	deka-gram	gram	deci-gram	centi-gram	milli-gram
Meaning	1000 grams	100 grams	10 grams	1 gram	$\frac{1}{10}$ of a gram	$\frac{1}{100}$ of a gram	$\frac{1}{1000}$ of a gram
Symbol	kg	hg	dag	g	dg	cg	mg

Weight (mass) units that are used most often

The units you will use most often in daily life are kilograms (kg), grams (g), and milligrams (mg). *Kilo*grams are used instead of pounds. A kilogram is 1000 grams. It is about 2.2 pounds. Two packages of butter plus one stick of butter weigh about 1 kg. An average newborn baby weighs 3 to 4 kg; a college football player might weigh 100 to 130 kg.

1 kilogram is about 2.2 pounds.   100 to 130 kg   3 to 4 kg

Extremely small weights are measured in *milli*grams. It takes 1000 mg to make 1 g. Recall that a dollar bill weighs about 1 g. Imagine cutting it into 1000 pieces; the weight of one tiny piece would be 1 mg. Dosages of medicine and vitamins are given in milligrams. You will also use milligrams in science classes.

Cut a dollar bill into 1000 pieces.
One tiny piece weighs 1 milligram.

EXAMPLE 3 Using Metric Weight Units

Write the most reasonable metric unit in each blank. Choose from kg, g, and mg.

(a) Ramon's suitcase weighed 20 _____.
20 **kg** because kilograms are used instead of pounds.
20 kg is about 44 pounds.

(b) LeTia took a 350 _____ aspirin tablet.
350 **mg** because 350 g would be more than the weight of a hamburger, which is too much.

(c) Jenny mailed a letter that weighed 30 _____.
30 **g** because 30 kg would be much too heavy and 30 mg is less than the weight of a dollar bill.

Now Try Exercises 11 and 15.

OBJECTIVE 4 Convert among metric weight (mass) units.   As with length and capacity, you can convert among metric weight units by using unit fractions. The unit fractions you need are shown below.

Metric Weight (Mass) Relationships

1 kg = 1000 g so the unit fractions are:

$$\frac{1 \text{ kg}}{1000 \text{ g}} \quad \text{or} \quad \frac{1000 \text{ g}}{1 \text{ kg}}$$

1 g = 1000 mg so the unit fractions are:

$$\frac{1 \text{ g}}{1000 \text{ mg}} \quad \text{or} \quad \frac{1000 \text{ mg}}{1 \text{ g}}$$

Or you can use a metric conversion line to decide how to move the decimal point.

1000	100	10	1	$\frac{1}{10}$	$\frac{1}{100}$	$\frac{1}{1000}$
kg	**hg**	**dag**	**g**	**dg**	**cg**	**mg**

EXAMPLE 4 Converting among Metric Weight Units

Convert using the metric conversion line or unit fractions.

(a) 7 mg to g

Using the metric conversion line:
From **mg** to **g** is *three places* to the *left*.

7.   007.
↑   Move decimal
Decimal point   point three places
starts here.   to the left.

Thus, 7 mg = 0.007 g.

Using unit fractions:

Multiply by a unit fraction that allows you to divide out mg.

$$\frac{7 \text{ mg}}{1} \cdot \frac{1 \text{ g}}{1000 \text{ mg}} = \frac{7}{1000} \text{ g}$$
$$= 0.007 \text{ g}$$

Thus, 7 mg = 0.007 g.

(b) 13.72 kg to g

Using the metric conversion line:
From **kg** to **g** is *three places* to the *right*.

13.720   Decimal point moves
          three places to the
          right.

Thus, 13.72 kg = 13,720 g.
                      ↑
                 A comma
            (not a decimal point)

Using unit fractions:

Multiply by a unit fraction that allows you to divide out kg.

$$\frac{13.72 \text{ kg}}{1} \cdot \frac{1000 \text{ g}}{1 \text{ kg}} = 13,720 \text{ g}$$
                                          ↑
                                     A comma
                              (not a decimal point)

Thus, 13.72 kg = 13,720 g.

Now Try Exercises 43 and 51.

OBJECTIVE 5 Distinguish among basic metric units of length, capacity, and weight (mass). As you encounter things to be measured at home, on the job, or in your classes at school, be careful to use the correct type of measurement unit.

Use *length units* (kilometers, meters, centimeters, millimeters) to measure:

how long	how high	how far away
how wide	how tall	how far around (perimeter)
how deep	distance	

Use *capacity units* (liters, milliliters) to measure liquids (things that can be poured) such as:

water	shampoo	gasoline
milk	perfume	oil
soft drinks	cough syrup	paint

Also use liters and milliliters to describe how much liquid something can hold, such as an eyedropper, measuring cup, pail, or bathtub.

Use *weight units* (kilograms, grams, milligrams) to measure:

the weight of something   how heavy something is

In **Chapters 3 and 4** you used square units (such as cm^2 and m^2) to measure area, and cubic units (such as cm^3 and m^3) to measure volume.

EXAMPLE 5 Using a Variety of Metric Units

First decide which type of units are needed: length, capacity, or weight. Then write the most appropriate metric unit in the blank. Choose from km, m, cm, mm, L, mL, kg, g, and mg.

(a) The letter needs another stamp because it weighs 40 _____.

Use _____ units.

The letter weighs 40 **g** because 40 mg is less than the weight of a dollar bill and 40 kg would be about 88 pounds.

Use **weight** units because of the word "weighs."

(b) The swimming pool is 3 _____ deep at the deep end.

Use _____ units.

The pool is 3 **m** deep because 3 cm is only about an inch and 3 km is more than a mile.

Use **length** units because of the word "deep."

(c) This is a 340 _____ can of juice.

Use _____ units.

It is a 340 **ml** can because 340 liters would be more than 340 quarts.

Use **capacity** units because juice is a liquid.

Now Try Exercises 57 and 61.

8.3 Exercises

Write the most reasonable metric unit in each blank. Choose from L, mL, kg, g, and mg.
See Examples 1 and 3.

1. The glass held 250 _____ of water.

2. Hiromi used 12 _____ of water to wash the kitchen floor.

3. Dolores can make 10 _____ of soup in that pot.

4. Jay gave 2 _____ of vitamin drops to the baby.

5. Our yellow Labrador dog grew up to weigh 40 _____.

6. A small safety pin weighs 750 _____.

7. Lori caught a small sunfish weighing 150 _____.

8. One dime weighs 2 _____.

9. Andre donated 500 _____ of blood today.

10. Barbara bought the 2 _____ bottle of cola.

11. The patient received a 250 _____ tablet of medication each hour.

12. The 8 people on the elevator weighed a total of 500 _____.

13. The gas can for the lawn mower holds 4 _____.

14. Kevin poured 10 _____ of vanilla into the mixing bowl.

15. Pam's backpack weighs 5 _____ when it is full of books.

16. One grain of salt weighs 2 _____.

Today, medical measurements are usually given in the metric system. Since we convert among metric units of measure by moving the decimal point, it is possible that mistakes can be made. Examine the following dosages and indicate whether they are reasonable or unreasonable. If a dose is unreasonable, indicate whether it is too much *or* too little.

17. Drink 4.1 L of Kaopectate after each meal.

18. Drop 1 mL of solution into the eye twice a day.

19. Soak your feet in 5 kg of Epsom salts per liter of water.

20. Inject 0.5 L of insulin each morning.

21. Take 15 mL of cough syrup every four hours.

22. Take 200 mg of vitamin C each day.

23. Take 350 mg of aspirin three times a day.

24. Buy a tube of ointment weighing 0.002 g.

25. Describe at least two examples of metric capacity units and two examples of metric weight units that you have come across in your daily life.

26. Explain in your own words how the meter, liter, and gram are related.

27. Describe how you decide which unit fraction to use when converting 6.5 kg to grams.

28. Write an explanation of each step you would use to convert 20 mg to grams using the metric conversion line.

Convert each measurement. Use unit fractions or the metric conversion line. See Examples 2 and 4.

29. 15 L to mL

30. 6 L to mL

31. 3000 mL to L

32. 18,000 mL to L

33. 925 mL to L

34. 200 mL to L

35. 8 mL to L

36. 25 mL to L

37. 4.15 L to mL

38. 11.7 L to mL

39. 8000 g to kg

40. 25,000 g to kg

41. 5.2 kg to g

42. 12.42 kg to g

43. 0.85 g to mg

44. 0.2 g to mg

45. 30,000 mg to g

46. 7500 mg to g

47. 598 mg to g

48. 900 mg to g

49. 60 mL to L

50. 6.007 kg to g

51. 3 g to kg

52. 12 mg to g

53. 0.99 L to mL

54. 13,700 mL to L

Write the most appropriate metric unit in each blank. Choose from km, m, cm, mm, L, mL, kg, g, and mg. See Example 5.

55. The masking tape is 19 _____ wide.

56. The roll has 55 _____ of tape on it.

57. Buy a 60 _____ jar of acrylic paint for art class.

58. One onion weighs 200 _____.

59. My waist measurement is 65 _____.

60. Add 2 _____ of windshield washer fluid to your car.

61. A single postage stamp weighs 90 _____.

62. The hallway is 10 _____ long.

Solve each application problem. Show your work. (*Source for Exercises 63–68*: *Top 10 of Everything.*)

63. Human skin has about 3 million sweat glands, which release an average of 300 mL of sweat per day. How many liters of sweat are released each day?

64. In hot climates, the sweat glands in a person's skin may release up to 3.5 L of sweat in one day. How many milliliters is that?

65. The average weight of a human brain is 1.34 kg. How many grams is that?

66. A healthy human heart pumps about 70 mL of blood per beat. How many liters of blood does it pump per beat?

67. On average, we breathe in and out roughly 900 mL of air every 10 seconds. How many liters of air is that?

68. In the Victorian era, people believed that heavier brains meant greater intelligence. They were impressed that Otto von Bismarck's brain weighed 1907 g, which is how many kilograms?

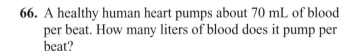

69. A small adult cat weighs from 3000 g to 4000 g. How many kilograms it that? (*Source: Lyndale Animal Hospital.*)

70. If the letter you are mailing weighs 29 g, you must put additional postage on it. How many kilograms does the letter weigh? (*Source: U.S. Postal Service.*)

71. Is 1005 mg greater than or less than 1 g? What is the difference in the weights?

72. Is 990 mL greater than or less than 1 L? What is the difference in the amounts?

73. One nickel weighs 5 g. How many nickels are in 1 kg of nickels?

74. The ratio of the total length of all the fish to the amount of water in an aquarium can be 3 cm of fish for every 4 L of water. What is the total length of all fish you can put in a 40 L aquarium? (*Source: Tropical Aquarium Fish.*)

RELATING CONCEPTS (EXERCISES 75–78) For Individual or Group Work

Recall that the prefix **kilo** *means 1000, so a* **kilo***meter is 1000 meters. You'll learn about other prefixes for numbers greater than 1000 as you* **work Exercises 75–78 in order.**

75. (a) The prefix *mega* (abbreviated with a capital M) means one million. So a *mega*meter (Mm) is how many meters?

1 Mm = _____ m

(b) Figure out a unit fraction that you can use to convert megameters to meters. Then use it to convert 3.5 Mm to meters.

76. (a) The prefix *giga* (abbreviated with a capital G) means one billion. So a *giga*meter (Gm) is how many meters?

1 Gm = _____ m

(b) Figure out a unit fraction you can use to convert meters to gigameters. Then use it to convert 2500 m to gigameters.

77. (a) The prefix *tera* (abbreviated with a capital T) means one trillion. So a *tera*meter (Tm) is how many meters?

1 Tm = _____ m

(b) Think carefully before you fill in the blanks.

1 Tm = _____ Gm

1 Tm = _____ Mm

78. A computer's memory is measured in *bytes*. A byte can represent a single letter, a digit, or a punctuation mark. The memory for a desktop computer may be measured in megabytes (abbreviated MB) or gigabytes (abbreviated GB). Using the meanings of *mega* and *giga,* it would seem that

1 MB = _____ bytes and

1 GB = _____ bytes.

However, because computers use a base 2 or binary system, 1 MB is actually 2^{20} and 1 GB is 2^{30}. Use your calculator to find the actual values.

2^{20} = _____ 2^{30} = _____

8.4 Problem Solving with Metric Measurement

OBJECTIVE 1 Solve application problems involving metric measurements. One advantage of the metric system is the ease of comparing measurements in application situations. Just be sure that you are comparing similar units: mg to mg, km to km, and so on.

We will use the same problem-solving steps you used for English measurement applications in **Section 8.1.**

OBJECTIVE

1 Solve application problems involving metric measurements.

EXAMPLE 1 Solving a Metric Application

Cheddar cheese is on sale at $8.99 per kilogram. Jake bought 350 g of the cheese. How much did he pay, to the nearest cent?

Step 1 **Read** the problem. The problem asks for the cost of 350 g of cheese.

Step 2 **Work out a plan.** The price is $8.99 per *kilogram*, but the amount Jake bought is given in *grams*. Convert grams to kilograms (the unit in the price). Then multiply the weight times the cost per kilogram.

Step 3 **Estimate** a reasonable answer. Round the cost of 1 kg from $8.99 to $9. There are 1000 g in a kilogram, so 350 g is about $\frac{1}{3}$ of a kilogram. Jake is buying about $\frac{1}{3}$ of a kilogram, so $\frac{1}{3}$ of $9 = $3 as our estimate.

Step 4 **Solve** the problem. Use a unit fraction to convert 350 g to kilograms.

$$\frac{350 \text{ g}}{1} \cdot \frac{1 \text{ kg}}{1000 \text{ g}} = \frac{350}{1000} \text{ kg} = 0.35 \text{ kg}$$

Now multiply 0.35 kg times the cost per kilogram.

$$\frac{\$8.99}{1 \text{ kg}} \cdot \frac{0.35 \text{ kg}}{1} = \$3.1465 \approx \$3.15 \quad \text{Rounded}$$

Step 5 **State the answer.** Jake paid $3.15, rounded to the nearest cent.

Step 6 **Check** your work. The exact answer of $3.15 is close to our estimate of $3.

Now Try Exercise 1. ▶▶▶

EXAMPLE 2 Solving a Metric Application

Olivia has 2.6 m of lace. How many centimeters of lace can she use to trim each of six hair ornaments? Round to the nearest tenth of a centimeter.

Step 1 **Read** the problem. The problem asks for the number of centimeters of lace for each of six hair ornaments.

Step 2 **Work out a plan.** The given amount of lace is in *meters,* but the answer must be in *centimeters.* Convert meters to centimeters, then divide by 6 (the number of hair ornaments).

Continued on Next Page

Step 3 **Estimate** a reasonable answer. To estimate, round 2.6 m of lace to 3 m. Then, 3 m = 300 cm, and 300 cm ÷ 6 = 50 cm as our estimate.

Step 4 **Solve** the problem. On the metric conversion line, moving from **m** to **cm** is two places to the right, so move the decimal point in 2.6 m two places to the right. Then divide by 6.

$$2.60 \text{ m} = 260 \text{ cm} \qquad \frac{260 \text{ cm}}{6 \text{ ornaments}} \approx 43.3 \text{ cm per ornament}$$

Step 5 **State the answer.** Olivia can use about 43.3 cm of lace on each ornament.

Step 6 **Check** your work. The exact answer of 43.3 cm is close to our estimate of 50 cm.

 Now Try Exercise 5.

> **NOTE**
> In Example 1, we used a unit fraction to convert the measurement. In Example 2, we moved the decimal point. Use whichever method you prefer. Also, there is more than one way to solve an application problem. Another way to solve Example 2 is to divide 2.6 m by 6 to get 0.4333 m of lace for each ornament. Then convert 0.4333 m to 43.3 cm (rounded to the nearest tenth).

EXAMPLE 3 Solving a Metric Application

Rubin measured a board and found that the length was 3 m plus an additional 5 cm. He cut off a piece measuring 1 m 40 cm for a shelf. Find the length in meters of the remaining piece of board.

Step 1 **Read** the problem. Part of a board is cut off. The problem asks what length of board, in meters, is left over.

Step 2 **Work out a plan.** The lengths involve two units, m and cm. Rewrite both lengths in meters (the unit called for in the answer), and then subtract.

Step 3 **Estimate** a reasonable answer. To estimate, 3 m 5 cm can be rounded to 3 m, because 5 cm is less than half of a meter (less than 50 cm). Round 1 m 40 cm down to 1 m. Then, 3 m − 1 m = 2 m as our estimate.

Step 4 **Solve** the problem. Rewrite the lengths in meters. Then subtract.

$$
\begin{array}{ll}
3\text{m} \Rightarrow \quad 3.00 \text{ m} & 1\text{ m} \Rightarrow \quad 1.0 \text{ m} \\
\text{plus 5 cm} \Rightarrow \underline{+\ 0.05 \text{ m}} & \text{plus 40 cm} \Rightarrow \underline{+\ 0.4 \text{ m}} \\
\qquad\qquad 3.05 \text{ m} & \qquad\qquad 1.4 \text{ m}
\end{array}
$$

Subtract to find leftover length.
$$
\begin{array}{ll}
3.05 \text{ m} & \leftarrow \text{ Board} \\
\underline{-\ 1.40 \text{ m}} & \leftarrow \text{ Shelf} \\
1.65 \text{ m} & \leftarrow \text{ Leftover piece}
\end{array}
$$

Step 5 **State the answer.** The length of the remaining piece of board is 1.65 m.

Step 6 **Check** your work. The exact answer of 1.65 m is close to our estimate of 2 m.

 Now Try Exercise 9.

8.4 Exercises

Solve each application problem. Show your work. Round money answers to the nearest cent. See Examples 1–3.

1. Bulk rice at the food co-op is $0.98 per kilogram. Pam scooped some rice into a bag and put it on the scale. How much will she pay for 850 g of rice?

2. Lanh is buying a piece of plastic tubing measuring 315 cm for the science lab. The price is $4.75 per meter. How much will Lanh pay?

3. A miniature Yorkshire terrier, one of the smallest dogs, may weigh only 500 g. But a St. Bernard, the heaviest dog, could easily weigh 90 kg. What is the difference in the weights of the two dogs, in kilograms? (*Source: Big Book of Knowledge.*)

4. The world's longest insect is the giant stick insect of Indonesia, measuring 33 cm. The fairy fly, the smallest insect, is just 0.2 mm long. How much longer is the giant stick insect, in millimeters? (*Source: Big Book of Knowledge.*)

5. An adult human body contains about 5 L of blood. If each beat of the heart pumps 70 mL of blood, how many times must the heart beat to pass all the blood through the heart? Round to the nearest whole number of beats. (*Source: Harper's Index.*)

6. A floor tile measures 30 cm by 30 cm and weighs 185 g. How many kilograms would a stack of 24 tiles weigh? How much would five stacks of tile weigh? (*Source: The Tile Shop.*)

7. Each piece of lead for a mechanical pencil has a thickness of 0.5 mm and is 60 mm long. Find the total length in centimeters of the lead in a package with 30 pieces. If the price of the package is $3.29, find the cost per centimeter for the lead. (*Source: Pentel.*)

60 mm

8. The apartment building caretaker puts 750 mL of chlorine into the swimming pool every day. How many liters should he order to have a one-month (30-day) supply on hand? If chlorine is sold in containers that hold 4 L, how many containers should be ordered for one month? How much chlorine will be left at the end of the month?

9. Rosa is building a bookcase. She has one board that is 2 m 8 cm long and another that is 2 m 95 cm long. How long are the two boards together, in meters?

10. Janet has 10 m 30 cm of fabric. She wants to make curtains for three windows that are all the same size. How much fabric is available for each window, to the nearest tenth of a meter?

11. In a chemistry lab, each of the 45 students needs 85 mL of acid. How many one-liter bottles of acid need to be ordered? How much acid will be left over?

Use the bar graph below to answer Exercises 13 and 14.

Caffeine Meter
Average milligrams of caffeine per 8 oz cup or equivalent

Double espresso 160 mg

Drip coffee 90 mg

Cola 45 mg

25 mg Chocolate bar

5 mg Decaffeinated coffee

Source: Celestial Seasonings.

15. During August 2003, Mars moved closer to Earth at a rate of about 10,000 meters per second. How much closer, in kilometers, did Mars get to Earth

(a) in one second,

(b) in one minute,

(c) in one hour.

(*Source:* www.NASA.com)

12. James needs two 1.3 m pieces and two 85 cm pieces of wood molding to frame a picture. The price is $5.89 per meter plus 7% sales tax. How much will James pay?

13. If Agnete usually drinks three 8 oz cups of drip coffee each day, how many grams of caffeine will she consume in one week?

14. Lorenzo's doctor has suggested that he cut down on caffeine. So Lorenzo switched from drinking four 8 oz cups of cola every day to drinking two 8 oz cups of decaffeinated coffee. How many fewer grams of caffeine is he consuming each week?

16. Some of the newest football stadiums have FieldTurf instead of grass. Use the drawing below to find the total thickness in centimeters of the top two layers of FieldTurf.

64 mm fiber grass

4.5 cm rubber and sand filler

asphalt base

Source: Sports Facilities Commission.

RELATING CONCEPTS (EXERCISES 17–21) For Individual or Group Work

It is difficult to weigh very light objects, such as a single sheet of paper or a single staple (unless you have a very expensive scientific scale). One way around this problem is to weigh a large number of the items and then divide to find the weight of one item. Of course, before dividing, you must subtract the weight of the box or wrapper that the items are packaged in to find the net weight. To complete the table, **work Exercises 17–20 in order.** *Then answer Exercise 21.*

	Item	Total Weight	Weight of Packaging	Net Weight	Weight of One Item in Grams	Weight of One Item in Milligrams
17.	Box of 50 envelopes	255 g	40 g	_____	_____	_____
18.	Box of 1000 staples	350 g	20 g	_____	_____	_____
19.	Ream of paper (500 sheets)	_____	50 g	_____	_____	3000 mg
20.	Box of 100 small paper clips	_____	5 g	_____	_____	500 mg

21. Bonus question: One million seeds from an aspen tree weigh about 0.45 kg. How much does one seed weigh in milligrams? (*Source:* Scenic State Park.)

8.5 Metric–English Conversions and Temperature

OBJECTIVES

1 Use unit fractions to convert between metric and English units.

2 Learn common temperatures on the Celsius scale.

3 Use formulas to convert between Celsius and Fahrenheit temperatures

OBJECTIVE 1 Use unit fractions to convert between metric and English units. Until the United States has switched completely from the English system to the metric system, it will be necessary to make conversions from one system to the other. *Approximate* conversions can be made with the help of the table below, in which the values have been rounded to the nearest hundredth or thousandth. (The only value that is exact, not rounded, is 1 inch = 2.54 cm.)

Metric to English	English to Metric
1 kilometer $\approx$ 0.62 mile	1 mile $\approx$ 1.61 kilometers
1 meter $\approx$ 1.09 yards	1 yard $\approx$ 0.91 meter
1 meter $\approx$ 3.28 feet	1 foot $\approx$ 0.30 meter
1 centimeter $\approx$ 0.39 inch	1 inch = 2.54 centimeters
1 liter $\approx$ 0.26 gallon	1 gallon $\approx$ 3.78 liters
1 liter $\approx$ 1.06 quarts	1 quart $\approx$ 0.95 liter
1 kilogram $\approx$ 2.20 pounds	1 pound $\approx$ 0.45 kilogram
1 gram $\approx$ 0.035 ounce	1 ounce $\approx$ 28.35 grams

EXAMPLE 1 Converting between Metric and English Length Units

Convert 10 m to yards using unit fractions. Round your answer to the nearest tenth, if necessary.

We're changing from a *metric* unit to an *English* unit. In the "Metric to English" side of the table above, you see that 1 meter $\approx$ 1.09 yards. Two unit fractions can be written using that information.

$$\frac{1 \text{ m}}{1.09 \text{ yd}} \qquad \text{or} \qquad \frac{1.09 \text{ yd}}{1 \text{ m}}$$

Multiply by the unit fraction that allows you to divide out meters (that is, meters is in the denominator).

$$10 \text{ m} \cdot \frac{1.09 \text{ yd}}{1 \text{ m}} = \frac{10 \text{ m}}{1} \cdot \frac{1.09 \text{ yd}}{1 \text{ m}} = \frac{(10)(1.09 \text{ yd})}{1} = 10.9 \text{ yd}$$

These units should match.

Thus, 10 m $\approx$ 10.9 yd.

NOTE
In Example 1 above, you could also use the numbers from the "English to Metric" side of the table that involve meters and yards:
1 yard $\approx$ 0.91 meter.

$$\frac{10 \text{ m}}{1} \cdot \frac{1 \text{ yd}}{0.91 \text{ m}} = \frac{10}{0.91} \text{ yd} \approx 10.99 \text{ yd}$$

The answer is slightly different because the values in the table are rounded. Also, you have to divide instead of multiply, which is usually more difficult to do without a calculator. We will use the first method in this chapter.

Now Try Exercise 1. ▶▶▶

EXAMPLE 2 **Converting between Metric and English Weight and Capacity Units**

Convert using unit fractions. Round your answers to the nearest tenth.

(a) 3.5 kg to pounds

Look in the "Metric to English" side of the table on the previous page to see that 1 kilogram ≈ 2.20 pounds. Use this information to write a unit fraction that allows you to divide out kilograms.

$$\frac{3.5 \ \cancel{kg}}{1} \cdot \frac{2.20 \ lb}{1 \ \cancel{kg}} = \frac{(3.5)(2.20 \ lb)}{1} = 7.7 \ lb$$

Thus, 3.5 kg ≈ 7.7 lb.

(b) 18 gal to liters

Look in the "English to Metric" side of the table to see that 1 gallon ≈ 3.78 liters. Write a unit fraction that allows you to divide out gallons.

$$\frac{18 \ \cancel{gal}}{1} \cdot \frac{3.78 \ L}{1 \ \cancel{gal}} = \frac{(18)(3.78 \ L)}{1} = 68.04 \ L$$

68.04 rounded to the nearest tenth is 68.0.
Thus, 18 gal ≈ 68.0 L.

(c) 300 g to ounces

In the "Metric to English" side of the table, 1 gram ≈ 0.035 ounce.

$$\frac{300 \ \cancel{g}}{1} \cdot \frac{0.035 \ oz}{1 \ \cancel{g}} = \frac{(300)(0.035 \ oz)}{1} = 10.5 \ oz$$

Thus, 300 g ≈ 10.5 oz.

CAUTION
Because the metric and English systems were developed independently, almost all comparisons are approximate. Your answers should be written with the "≈" symbol to show they are approximate.

Now Try Exercise 7.

OBJECTIVE 2 Learn common temperatures on the Celsius scale.
In the metric system, temperature is measured on the **Celsius** scale. On the Celsius scale, water freezes at 0 °C and boils at 100 °C. The small raised circle stands for "degrees" and the capital **C** is for Celsius. Read the temperatures like this:

Water freezes at 0 degrees Celsius (0 °C).

Water boils at 100 degrees Celsius (100 °C).

The English temperature system, used only in the United States, is measured on the **Fahrenheit** scale. On this scale:

Water freezes at 32 degrees Fahrenheit (32 °F).

Water boils at 212 degrees Fahrenheit (212 °F).

The thermometer below shows some typical temperatures in both Celsius and Fahrenheit. For example, comfortable room temperature is about 20 °C or 68 °F, and normal body temperature is about 37 °C or 98.6 °F.

	Celsius (Metric system)	Fahrenheit (English system)
Water boils	100 °C	212 °F
Hot coffee	60 °C	140 °F
Hot bath water	50 °C	122 °F
Normal body temperature	37 °C	98.6 °F
Summer day	30 °C	86 °F
Room temperature	20 °C	68 °F
Spring day	10 °C	50 °F
Water freezes	0 °C	32 °F
Very cold winter day	−18 °C	0 °F

> **NOTE**
> The freezing and boiling temperatures are exact. The other temperatures are approximate. Even normal body temperature varies slightly from person to person.

EXAMPLE 3 Using Celsius Temperatures

Circle the Celsius temperature that is most reasonable for each situation.

(a) Warm summer day   29 °C   64 °C   90 °C

29 °C is reasonable. 64 °C and 90 °C are too hot; they're both above the temperature of hot bath water (above 122 °F).

(b) Inside a freezer   −10 °C   3 °C   25 °C

−10 °C is the reasonable temperature because it is the only one below the freezing point of water (0 °C). Your frozen foods would start thawing at 3 °C or 25 °C.

Now Try Exercise 27. ▶▶▶

OBJECTIVE 3 **Use formulas to convert between Celsius and Fahrenheit temperatures.** You can use these formulas to convert between Celsius and Fahrenheit temperatures.

Celsius–Fahrenheit Conversion Formulas

Converting from Fahrenheit (F) to Celsius (C)

$$C = \frac{5(F - 32)}{9}$$

Converting from Celsius (C) to Fahrenheit (F)

$$F = \frac{9C}{5} + 32$$

As you use these formulas, be sure to follow the order of operations from **Section 1.8.**

Order of Operations

Step 1   Work inside *parentheses* or *other grouping symbols.*

Step 2   Simplify any expressions with *exponents.*

Step 3   Do the remaining *multiplications and divisions* as they occur from left to right.

Step 4   Do the remaining *additions and subtractions* as they occur from left to right.

EXAMPLE 4 **Converting Fahrenheit to Celsius**

Convert 10 °F to Celsius. Round your answer to the nearest degree.

Use the formula and the order of operations.

$$C = \frac{5(\mathbf{F} - 32)}{9}$$ Replace F with 10

$$= \frac{5(\mathbf{10} - 32)}{9}$$ Work inside parentheses first. $10 - 32$ becomes $10 + (-32)$.

$$= \frac{5(-22)}{9}$$ Multiply in the numerator; positive times negative gives a negative product.

$$= \frac{-110}{9}$$ Divide; negative divided by positive gives a negative quotient.

$$= -12.\overline{2}$$ Round to -12 (nearest degree).

Thus, 10 °C ≈ −12 °F.

Now Try Exercise 31.

EXAMPLE 5 **Converting Celsius to Fahrenheit**

Convert 15 °C to Fahrenheit.

Use the formula and follow the order of operations.

$$F = \frac{9\mathbf{C}}{5} + 32$$ Replace C with 15

$$= \frac{9 \cdot \mathbf{15}}{5} + 32$$

$$= \frac{9 \cdot 3 \cdot \overset{1}{\cancel{5}}}{\underset{1}{\cancel{5}}} + 32$$ Divide out the common factor. Multiply in the numerator.

$$= 27 + 32$$ Add.

$$= 59$$

Thus, 15 °C = 59 °F.

Now Try Exercise 37.

8.5 Exercises

FOR EXTRA HELP

 Addison-Wesley Math Tutor Center

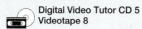 MathXL

Digital Video Tutor CD 5 Videotape 8

Student's Solutions Manual

 MyMathLab

 MathXL Tutorials on CD

Use the table on the first page of this section and unit fractions to make approximate conversions from metric to English or English to metric. Round your answers to the nearest tenth. See Examples 1 and 2.

1. 20 m to yards

2. 8 km to miles

3. 80 m to feet

4. 85 cm to inches

5. 16 ft to meters

6. 3.2 yd to meters

7. 150 g to ounces

8. 2.5 oz to grams

9. 248 lb to kilograms

10. 7.68 kg to pounds

11. 28.6 L to quarts

12. 15.75 L to gallons

13. For the 2000 Olympics, the 3M Company used 5 g of pure gold to coat Michael Johnson's track shoes. (*Source:* 3M Company.)

   (a) How many ounces of gold were used, to the nearest tenth?

   (b) Was this enough extra weight to slow him down?

14. The label on a Van Ness auto feeder for cats and dogs says it holds 1.4 kg of dry food. How many pounds of food does it hold, to the nearest tenth?

Source: Van Ness Plastics.

15. The heavy-duty wash cycle in a dishwasher uses 8.4 gal of water. How many liters does it use, to the nearest tenth? (*Source:* Frigidaire.)

16. The rinse-and-hold cycle in a dishwasher uses only 4.5 L of water. How many gallons does it use, to the nearest tenth? (*Source:* Frigidaire.)

17. The smallest pet fish are dwarf gobies, which are half an inch long. How many centimeters long is a dwarf gobie, to the nearest tenth? (*Hint:* Write half an inch in decimal form.)

18. The fastest nerve signals in the human body travel 120 m per second. How many feet per second do the signals travel? (*Source: Big Book of Knowledge.*)

19. On Northwest Airlines flights, a piece of carry-on luggage cannot exceed 40 lb in weight or measure more than 22 in. by 14 in. by 9 in. What measurements would be given to Northwest passengers from countries outside the United States? Round to the nearest tenth of a kilogram and nearest centimeter, if necessary. (*Source:* www.nwa.com)

20. Northwest Airlines has a limit of 50 lb for each checked piece of luggage. The maximum size is 62 in., which is length + width + height. Convert these measurements to metric for Northwest flights from countries outside the United States. Round to the nearest tenth of a kilogram and nearest centimeter, if necessary. (*Source:* www.nwa.com)

21. Part of a tea bag packet imported from Scotland is shown below. Are the two weights on the packet equivalent? Explain your answer.

Source: Heath and Heather.

22. Part of the wrapper from a ball of twine is shown below. The manufacturer has designed the label so the twine can be sold worldwide. Are the two lengths listed on the wrapper equivalent? Explain your answer.

Source: Lehigh Group.

The BabyBjörn is a popular baby carrier imported from Sweden. Use the information from the instruction sheet that comes with the carrier to answer Exercises 23–24. (*Source:* BabyBjörn, Sweden.)

Infant facing adult position

Minimum baby size:
53 cm and 3.5 kg

Maximum baby size: 15 kg

23. Can the carrier be safely used for a newborn infant who weighs 8 lb and is 19.5 in. long? Explain your answer.

24. Can the carrier be safely used for a baby who weighs 30 lb?

Circle the most reasonable Celsius temperature for each situation. See Example 3.

25. A snowy day

    12 °C   28 °C   −8 °C

26. Brewing coffee

    80 °C   180 °C   15 °C

27. A high fever

    21 °C   40 °C   103 °C

28. Swimming pool water

    90 °C   78 °C   25 °C

29. Oven temperature

    150 °C   50 °C   30 °C

30. Light jacket weather

    0 °C   10 °C   −10 °C

Use the conversion formulas from this section and the order of operations to convert Fahrenheit temperatures to Celsius and Celsius temperatures to Fahrenheit. Round your answers to the nearest degree if necessary. See Examples 4 and 5.

31. 60 °F

32. 80 °F

33. −4 °F

34. 15 °F

35. 8 °C

36. 18 °C

37. −5 °C

38. 0 °C

Solve each application problem. Round your answers to the nearest degree, if necessary.

39. The highest temperature ever recorded on Earth was 136 °F at E1 Aziza, Libya, in 1922. The lowest was −129 °F in Antarctica. Convert these temperatures to Celsius. (*Source: The World Almanac.*)

40. Hummingbirds have a normal body temperature of 107 °F. But on cold nights they go into a state of torpor where their body temperature drops to 39 °F. What are these temperatures in Celsius? (*Source: Wildbird.*)

41. The directions for a self-stick clothes hook with adhesive on the back are as follows: "Apply to surfaces above 50 °F. Adhesive could soften and lose adhesion above 105 °F." What are these temperatures in the metric system? (*Source:* 3M Company.)

42. A box of imported Belgian chocolates carries a warning to keep the box dry and at <18 °C. Translate this warning into the English system. (*Source:* Chocolaterie Guylian N.V.)

43. Would a drop in temperature of 20 Celsius degrees be more or less than a drop of 20 Fahrenheit degrees? Explain your answer.

44. Describe one advantage of switching from the Fahrenheit temperature scale to the Celsius scale. Describe one disadvantage.

45. **(a)** Here is the tag on a pair of hiking boots. In what kind of weather would you be most comfortable wearing these boots?

Comfort
range
24 °C to 4 °C

Source: Sorel.

(b) For what Fahrenheit temperatures are the boots designed?

(c) What range of metric temperatures do you have in January where you live? Would you be comfortable in these boots?

46. Sleeping bags made by Eddie Bauer are sold around the world. Each type of sleeping bag is designed for outdoor camping in certain temperatures.

Junior bag	5 °C or warmer
Removable liner bag	0 °C to 15 °C
Conversion bag	−7 °C to 0 °C

Source: Eddie Bauer.

(a) At what Fahrenheit temperatures should you use the Junior bag?

(b) What Fahrenheit temperatures is the removable liner bag designed for?

(c) The conversion bag is designed for what Fahrenheit temperatures?

RELATING CONCEPTS (EXERCISES 47–54) For Individual or Group Work

The article below appeared in American newspapers. However, both Newfoundland (part of Canada) and Ireland use the metric system. Their newspapers would have reported all the measurements in metric units. Complete the conversions to metric, rounding answers to the nearest tenth.

Q: A recent news brief reported on some men who flew a model airplane from Newfoundland to Ireland. Can you provide some details of the flight?

A: The model plane is 6 feet long and weighs 11 pounds. Made of balsa wood and mylar, it crossed the Atlantic—the flight path took it 1,888.3 miles—in 38 hours, 23 minutes. It soared at a cruising altitude of 1,000 feet. The plane used a souped-up piston engine and carried less than a gallon of fuel, as mandated by rules of the Federation Aeronautique Internationale, the governing body of model airplane building. When it landed in County Galway, Ireland, it had less than 2 fluid ounces of fuel left. The plane was built by Maynard Hill of Silver Spring, Maryland.

(*Sources: New York Times* and *Cox News Service.*)

47. Length of model plane

48. Weight of plane

49. Length of flight path

50. Time of flight

51. Cruising altitude

52. Fuel at the start in milliliters

53. Fuel left after landing, in milliliters (*Hint:* First convert 2 fl oz to quarts.)

54. What *percent* of the fuel was left at the end of the flight?

Objective   Convert among English and metric length units; use decimals and percents.

Growing Sunflowers

The front and back of a seed packet for sunflowers are shown below. Look at the top of the packet first.

Sunflower, Mammoth Grey Stripe

Tall Plants, Huge Flowers

Net Wt. 3.5 g $1.19

Sunflower, Mammoth Grey Stripe

The stalk of this sunflower will grow to 12'/4 m. Flowers will range from 6"/15 cm to 15"/38 cm in diameter. Sunflowers can thrive in poor soil with little moisture.

Type	Height	Planting Depth	Seed Spacing	Thinning Height	Spacing After Thinning	Days to Germination
Annual	8-10' 2.4-3 m	1/2" 13 mm	6" 15 cm	3" 8 cm	2' 61 cm	10-20

Select a sunny or lightly shaded location and plant outdoors, where plants are to remain, after all danger of frost is past. For tallest plants, sow in good soil with moderate moisture.

Stock #1185

7 18964 98119 7

Source: Olds Seed Solutions.

1. There were 42 seeds in the packet. If 40 of the seeds sprouted, what was the cost per sprout, to the nearest cent?

2. If vegetable and flower seeds were on sale at 30% off, what was the cost per sprout, to the nearest cent?

3. What percent of the seeds sprouted, to the nearest whole percent?

4. How many seeds would weigh 1 gram?

5. The table at the bottom of the packet uses the symbol (') for feet, and the symbol (") for inches.

    (a) How tall will the plants grow, in feet?

    (b) How tall will they grow, in inches?

    (c) How tall will they grow, in yards?

6. How many inches tall should the plants be when you thin them (remove less vigorous plants to give others room to grow)? How tall is that in feet?

7. What is the range in the diameter of the flowers, in inches, and in feet? Diameter is the distance across the circular flower.

8. Convert the height, planting depth, and seed spacing from English units to metric. How do your answers compare to the metric measurements on the packet?

Hair and Nail Growth

Q How fast do hair and nails grow? Do they grow faster in the summer?

A Fingernails grow, on average, about one-tenth of a millimeter per day, although there is considerable variation among individuals. Fingernails grow faster than toenails, and nails on the longest fingers appear to grow the fastest.

Fingernails, as well as hair and skin, grow faster in the summer, presumably under the influence of sunlight, which expands blood vessels, bringing more oxygen and nutrients to the area and allowing for faster growth.

The rate the scalp hair grows is 0.3 to 0.4 millimeter per day, or about 6 inches a year.

Source: Minneapolis Star Tribune.

9. How much do nails grow in one week? one month? one year?

10. How much does scalp hair grow in one week? one month? one year? (Use metric units.)

11. Is the statement about hair growing 6 inches a year accurate? Explain your answer.

Chapter 8
SUMMARY

KEY TERMS

8.1 **English system** The English system of measurement (U.S. system of units) is used for many daily activities only in the United States. Commonly used units in this system include quarts, pounds, feet, miles, and degrees Fahrenheit.

metric system The metric system of measurement is an international system used in manufacturing, science, medicine, sports, and other fields. Commonly used units in this system include meters, liters, grams, and degrees Celsius.

unit fraction A unit fraction involves measurement units and is equivalent to 1. Unit fractions are used to convert among different measurements.

8.2 **meter** The meter is the basic unit of length in the metric system. The symbol **m** is used for meter. One meter is a little longer than a yard.

prefixes Attaching a prefix to meter, liter, or gram produces larger or smaller units. For example, the prefix *kilo* means 1000, so a *kilo*meter is 1000 meters.

metric conversion line The metric conversion line is a line showing the various metric measurement prefixes and their size relationship to each other.

8.3 **liter** The liter is the basic unit of capacity in the metric system. The symbol **L** is used for liter. One liter is a little more than one quart.

gram The gram is the basic unit of weight (mass) in the metric system. The symbol **g** is used for gram. One gram is the weight of 1 milliliter of water or one dollar bill.

8.5 **Celsius** The Celsius scale is used to measure temperature in the metric system. Water freezes at 0 °C and boils at 100 °C.

Fahrenheit The Fahrenheit scale is used to measure temperature in the English system. Water freezes at 32 °F and boils at 212 °F.

NEW SYMBOLS

Frequently used length units
km kilometer
m meter
cm centimeter
mm millimeter

Frequently used capacity units
L liter
mL milliliter

Frequently used weight (mass) units
kg kilogram
g gram
mg milligram

°C degrees Celsius (metric temperature unit)

°F degrees Fahrenheit (English temperature unit)

NEW FORMULAS

Converting from Celsius to Fahrenheit: $F = \dfrac{9C}{5} + 32$

Converting from Fahrenheit to Celsius: $C = \dfrac{5(F - 32)}{9}$

See how well you have learned the vocabulary in this chapter. Answers follow the Quick Review.

1. The **metric system**
  A. uses meters, liters, and degrees Fahrenheit
  B. is based on multiples of 10
  C. is used only in the United States
  D. has evolved over centuries.

2. The **English system** of measurement
  A. is used throughout the world
  B. is based on multiples of 12
  C. uses feet, inches, quarts, and pounds
  D. was developed by a group of scientists in 1790.

3. A **unit fraction**
  A. has the unit you want to change in the numerator
  B. has a denominator of 1
  C. must be written in lowest terms
  D. is equivalent to 1.

4. A **gram** is
  A. the weight of 1 mL of water
  B. abbreviated gm
  C. equivalent to 1000 kg
  D. approximately equal to 2.2 pounds.

5. A **meter** is
  A. equivalent to 1000 cm
  B. approximately equal to $\frac{1}{2}$ inch
  C. abbreviated m with no period after it
  D. the basic unit of capacity in the metric system.

6. The **Celsius** temperature scale
  A. shows water freezing at 32°
  B. is used in the English system of measurement
  C. shows water boiling at 100°
  D. cannot be converted to the Fahrenheit temperature scale.

QUICK REVIEW

Concepts	*Examples*
8.1 *The English System of Measurement* Memorize the basic measurement relationships. Then, to convert units, multiply when changing from a larger unit to a smaller unit; divide when changing from a smaller unit to a larger unit.	Convert each measurement. **(a)** 5 ft to inches $$5 \text{ ft} = 5 \cdot \mathbf{12} = 60 \text{ in.}$$ **(b)** 3 lb to ounces $$3 \text{ lb} = 3 \cdot \mathbf{16} = 48 \text{ oz}$$ **(c)** 15 qt to gallons $$15 \text{ qt} = \frac{15}{4} = 3\frac{3}{4} \text{ gal}$$
8.1 *Using Unit Fractions* Another, more useful, conversion method is multiplying by a unit fraction. The unit you want in the answer should be in the numerator. The unit you want to change should be in the denominator.	Convert 32 oz to pounds. $32 \text{ oz} \cdot \dfrac{1 \text{ lb}}{16 \text{ oz}}$ } Unit fraction These units should match. $= \dfrac{\overset{2}{\cancel{32} \text{ oz}}}{1} \cdot \dfrac{1 \text{ lb}}{\underset{1}{\cancel{16} \text{ oz}}}$ Divide out **ounces**. Divide out **common factors**. $= 2 \text{ lb}$

Concepts	Examples

8.1 *Solving English Measurement Application Problems*
To solve measurement application problems, use these problem-solving steps.

Use the six steps to solve this problem.

Mr. Green has 10 yd of rope. He is cutting it into eight pieces so his sailing class can practice knot tying. How many feet of rope will each of his eight students get?

Step 1   **Read** the problem.

Step 1   The problem asks how many feet of rope can be given to each of eight students.

Step 2   **Work out a plan.**

Step 2   Convert 10 yd to feet (the unit required in the answer). Then divide by eight students.

Step 3   **Estimate** a reasonable answer.

Step 3   There are 3 ft in one yard, so there are 30 ft in 10 yd. Then 30 ft ÷ 8 ≈ 4 ft as our estimate.

Step 4   **Solve** the problem.

Step 4   Use a unit fraction to convert 10 yd to feet, then divide.

$$\frac{10 \text{ yd}}{1} \cdot \frac{3 \text{ ft}}{1 \text{ yd}} = 30 \text{ ft}$$

$$\frac{30 \text{ ft}}{8 \text{ students}} = 3\frac{3}{4} \text{ ft or } 3.75 \text{ ft per student}$$

Step 5   **State the answer.**

Step 5   Each student gets $3\frac{3}{4}$ ft or 3.75 ft of rope.

Step 6   **Check** your work.

Step 6   The exact answer of 3.75 ft is close to our estimate of 4 ft.

8.2 *Basic Metric Length Units*
Use approximate comparisons to judge which length units are appropriate:

1 mm is the thickness of a dime.

1 cm is about $\frac{1}{2}$ inch.

1 m is a little more than 1 yard.

1 km is about 0.6 mile.

Write the most reasonable metric unit in each blank. Choose from km, m, cm, and mm.

The room is 6 __**m**__ long.

A paper clip is 30 __**mm**__ long.

He drove 20 __**km**__ to work.

8.2 and 8.3 *Converting within the Metric System*
Using Unit Fractions
One conversion method is to multiply by a unit fraction. Use a fraction with the unit you want in the answer in the numerator and the unit you want to change in the denominator.

Convert.

(a) 9 g to kg

$$\frac{9 \text{ g}}{1} \cdot \frac{1 \text{ kg}}{1000 \text{ g}} = \frac{9}{1000} \text{ kg} = 0.009 \text{ kg}$$

Thus, 9g = 0.009 kg.

(b) 3.6 m to cm

$$\frac{3.6 \text{ m}}{1} \cdot \frac{100 \text{ cm}}{1 \text{ m}} = 360 \text{ cm}$$

Thus, 3.6 m = 360 cm.

(continued)

Concepts	**Examples**

8.2 *and* **8.3** *Converting within the Metric System* (*continued*)

Using the Metric Conversion Line
Another conversion method is to find the unit you are given on the metric conversion line. Count the number of places to get from the unit you are given to the unit you want. Move the decimal point the same number of places and in the same direction.

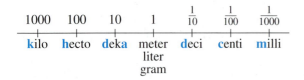

$$1000 \quad 100 \quad 10 \quad 1 \quad \tfrac{1}{10} \quad \tfrac{1}{100} \quad \tfrac{1}{1000}$$

kilo   hecto   deka   meter   deci   centi   milli
                              liter
                              gram

Convert.

(a) 68.2 kg to g

From **kg** to **g** is three places to the right.

6 8.200   Decimal point is moved three places to the right.

Thus, 68.2 kg = 68,200 g.

(b) 300 mL to L

From **mL** to **L** is three places to the left.

300.   Decimal point is moved three places to the left.

Thus, 300 mL = 0.3 L.

(c) 825 cm to m

From **cm** to **m** is two places to the left.

825.   Decimal point is moved two places to the left.

Thus, 825 cm = 8.25 m.

8.3 *Basic Metric Capacity Units*
Use approximate comparisons to judge which capacity units are appropriate:

> 1 L is a little more than 1 quart.
>
> 1 mL is the amount of water in a cube 1 cm on each side.
>
> 5 mL is about one teaspoon.
>
> 250 mL is about one cup.

Write the most reasonable metric unit in each blank. Choose from L and mL.

The pail holds 12 __L__ .

The milk carton from the vending machine holds 250 __mL__ .

8.3 *Basic Metric Weight (Mass) Units*
Use approximate comparisons to judge which weight units are appropriate:

> 1 kg is about 2.2 pounds.
>
> 1 g is the weight of 1 mL of water or one dollar bill.
>
> 1 mg is $\dfrac{1}{1000}$ of a gram; very tiny!

Write the most reasonable metric unit in each blank. Choose from kg, g, and mg.

The wrestler weighed 95 __kg__ .

She took a 500 __mg__ aspirin tablet.

One banana weighs 150 __g__ .

Concepts	Examples

8.4 *Solving Metric Application Problems*

Convert units so you are comparing kg to kg, cm to cm, and so on. When a measurement involves two units, such as 6 m 20 cm, write it in terms of the unit called for in the answer (6.2 m or 620 cm).

Use these problem-solving steps.

Step 1 **Read** the problem.

Step 2 **Work out a plan.**

Step 3 **Estimate** a reasonable answer.

Step 4 **Solve** the problem.

Step 5 **State the answer.**

Step 6 **Check** your work.

Use the six steps to solve this problem.

George cut 1 m 35 cm off of a 3 m board. How long was the leftover piece, in meters?

Step 1 The problem asks for the length of the leftover piece in meters.

Step 2 Convert the cut-off measurement to meters (the unit required in the answer), and then subtract to find the "leftover."

Step 3 To estimate, round 1 m 35 cm to 1 m, because 35 cm is less than half a meter. Then, 3 m − 1 m = 2 m as our estimate.

Step 4 Convert the cut-off measurement to meters, then subtract.

$$
\begin{array}{rl}
1\text{ m} \rightarrow & 1.00\text{ m} \\
\text{plus }35\text{ cm} \rightarrow & +\ 0.35\text{ m} \\
\hline
& 1.35\text{ m}
\end{array}
\qquad
\begin{array}{rl}
& 3.00\text{ m} \leftarrow \text{Board} \\
& -\ 1.35\text{ m} \leftarrow \text{Cut off} \\
\hline
& 1.65\text{ m} \leftarrow \text{Left}
\end{array}
$$

Step 5 The leftover piece is 1.65 m long.

Step 6 The exact answer of 1.65 m is close to our estimate of 2 m.

8.5 *Converting between Metric and English Units*

Write a unit fraction using the values in the table of conversion factors on the first page of **Section 8.5.** Because the values in the table are rounded, your answers will be approximate.

Convert. Round answers to the nearest tenth.

(a) 23 m to yards

From the table, 1 meter ≈ 1.09 yards.

$$
\frac{23 \ \cancel{m}}{1} \cdot \frac{1.09 \text{ yd}}{1 \ \cancel{m}} = 25.07 \text{ yd}
$$

25.07 rounds to 25.1, so 23 m ≈ 25.1 yd.

(b) 4 oz to grams

From the table, 1 ounce ≈ 28.35 grams.

$$
\frac{4 \ \cancel{oz}}{1} \cdot \frac{28.35 \text{ g}}{1 \ \cancel{oz}} = 113.4 \text{ g}
$$

Thus, 4 oz ≈ 113.4 g.

Concepts	Examples

8.5 *Common Celsius Temperatures*

Use approximate and exact comparisons to judge which temperatures are appropriate.

Exact comparisons:

0 °C is the freezing point of water (32 °F).

100 °C is the boiling point of water (212 °F).

Approximate comparisons:

10 °C for a spring day (50 °F)

20 °C for room temperature (68 °F)

30 °C for summer day (86 °F)

37 °C for normal body temperature (98.6 °F)

Circle the Celsius temperature that is most reasonable.

(a) Hot summer day:

(35 °C)   90 °C   110 °C

(b) The first snowy day in winter:

−20 °C   (0 °C)   15 °C

8.5 *Converting between Fahrenheit and Celsius Temperatures (continued)*

Use this formula to convert from Fahrenheit (F) to Celsius (C).

$$C = \frac{5(F - 32)}{9}$$

Convert 100 °F to Celsius. Round your answer to the nearest degree if necessary.

$$C = \frac{5(\mathbf{100} - 32)}{9} \quad \text{Replace F with 100}$$

$$= \frac{5(68)}{9}$$

$$= \frac{340}{9}$$

$$= 37.\overline{7} \quad \text{Round to 38}$$

Thus, 100 °F ≈ 38 °C.

Use this formula to convert from Celsius (C) to Fahrenheit (F).

$$F = \frac{9C}{5} + 32$$

Convert −8 °C to Fahrenheit. Round your answer to the nearest degree, if necessary.

$$F = \frac{9(\mathbf{-8})}{5} + 32 \quad \text{Replace C with } -8$$

$$= \frac{-72}{5} + 32$$

$$= -14.4 + 32$$

$$= 17.6 \quad \text{Round to 18}$$

Thus, −8 °C ≈ 18 °F.

ANSWERS TO TEST YOUR WORD POWER

1. B; *Examples:* 10 meters = 1 dekameter; 100 meters = 1 hectometer; 1000 meters = 1 kilometer.

2. C; *Examples:* Feet and inches are used to measure length, quarts to measure capacity, pounds to measure weight.

3. D; *Example:* Because 12 in. = 1 ft, the unit fraction $\frac{12 \text{ in.}}{1 \text{ ft}}$ is equivalent to $\frac{12 \text{ in.}}{12 \text{ in.}} = 1$.

4. A; *Example:* A small box measuring 1 cm on every edge holds exactly 1 mL of water, and the water weighs 1 g.

5. C; *Example:* A measurement of 16 meters is written 16 m (without a period).

6. C; *Example:* In the metric system, water freezes at 0 °C and boils at 100 °C. The English system uses the Fahrenheit temperature scale where water freezes at 32 °F and boils at 212 °F.

Chapter **8**
REVIEW EXERCISES

[8.1] *Fill in the blanks with the measurement relationships you have memorized.*

1. 1 lb = _____ oz

2. _____ ft = 1 yd

3. 1 T = _____ lb

4. _____ qt = 1 gal

5. 1 hr = _____ min

6. 1 c = _____ fl oz

7. _____ sec = 1 min

8. _____ ft = 1 mi

9. _____ in. = 1 ft

Convert using unit fractions.

10. 4 ft = _____ in.

11. 6000 lb = _____ T

12. 64 oz = _____ lb

13. 18 hr = _____ day

14. 150 min = _____ hr

15. $1\frac{3}{4}$ lb = _____ oz

16. $6\frac{1}{2}$ ft = _____ in.

17. 7 gal = _____ c

18. 4 days = _____ sec

19. The average depth of the world's oceans is 12,460 ft. (*Source: Handy Ocean Answer Book.*)

  (a) What is the average depth in yards?

  (b) What is the average depth in miles, to the nearest tenth?

20. During the first year of a program to recycle office paper, a company recycled 123,260 pounds of paper. The company received $40 per ton for the paper. How much money did the company make? (*Source: I. C. System.*)

[8.2] *Write the most reasonable metric length unit in each blank. Choose from km, m, cm, and mm.*

21. My thumb is 20 _____ wide.

22. Her waist measurement is 66 _____.

23. The two towns are 40 _____ apart.

24. A basketball court is 30 _____ long.

25. The height of the picnic bench is 45 _____.

Height

26. The eraser on the end of my pencil is 5 _____ long.

Convert using unit fractions or the metric conversion line.

27. 5 m to cm

28. 8.5 km to m

29. 85 mm to cm

30. 370 cm to m   **31.** 70 m to km   **32.** 0.93 m to mm

[8.3] *Write the most reasonable metric unit in each blank. Choose from L, mL, kg, g, and mg.*

33. The eyedropper holds 1 _____.

34. I can heat 3 _____ of water in this saucepan.

35. Loretta's hammer weighed 650 _____.

36. Yongshu's suitcase weighed 20 _____ when it was packed.

37. My fish tank holds 80 _____ of water.

38. I'll buy the 500 _____ bottle of mouthwash.

39. Mara took a 200 _____ antibiotic pill.

40. This piece of chicken weighs 100 _____.

Convert using unit fractions or the metric conversion line.

41. 5000 mL to L   **42.** 8 L to mL   **43.** 4.58 g to mg

44. 0.7 kg to g   **45.** 6 mg to g   **46.** 35 mL to L

[8.4] *Solve each application problem. Show your work.*

47. Each serving of punch at the wedding reception will be 180 mL. How many liters of punch are needed for 175 servings?

48. Jason is serving a 10 kg turkey to 28 people. How many grams of meat is he allowing for each person? Round to the nearest whole gram.

49. Yerald weighed 92 kg. Then he lost 4 kg 750 g. What is his weight now, in kilograms?

50. Young-Mi bought 750 g of onions. The price was $1.49 per kilogram. How much did she pay, to the nearest cent?

[8.5] *Use the table on the first page of **Section 8.5** and unit fractions to make approximate conversions. Round your answers to the nearest tenth, if necessary.*

51. 6 m to yards   **52.** 30 cm to inches

53. 108 km to miles

54. 800 mi to kilometers

55. 23 qt to liters

56. 41.5 L to quarts

*Write the appropriate **metric** temperature in each blank.*

57. Water freezes at _____.

58. Water boils at _____.

59. Normal body temperature is about _____.

60. Comfortable room temperature is about _____.

*Use the conversion formulas in **Section 8.5** to convert each temperature to Fahrenheit or to Celsius. Round to the nearest degree, if necessary.*

61. 77 °F

62. 5 °F

63. −2 °C

64. Water coming into a dishwasher should be at least 49 °C to clean the dishes properly. What Fahrenheit temperature is that? (*Source:* Frigidaire.)

MIXED REVIEW EXERCISES

Write the most reasonable metric unit in each blank. Choose from km, m, cm, mm, L, mL, kg, g, and mg.

65. I added 1 _____ of oil to my car.

66. The box of books weighed 15 _____.

67. Larry's shoe is 30 _____ long.

68. Jan used 15 _____ of shampoo on her hair.

69. My fingernail is 10 _____ wide.

70. I walked 2 _____ to school.

71. The tiny bird weighed 15 _____.

72. The new library building is 18 _____ wide.

73. The cookie recipe uses 250 _____ of milk.

74. Renee's pet mouse weighs 30 _____.

75. One postage stamp weighs 90 _____.

76. I bought 30 _____ of gas for my car.

Convert the following using unit fractions, the metric conversion line, or the temperature conversion formulas.

77. 10.5 cm to millimeters

78. 45 min to hours

79. 90 in. to feet

80. 1.3 m to centimeters

81. 25 °C to Fahrenheit

82. $3\frac{1}{2}$ gal to quarts

83. 700 mg to grams

84. 0.81 L to milliliters

85. 5 lb to ounces

86. 60 kg to grams

87. 1.8 L to milliliters

88. 30 °F to Celsius

89. 0.36 m to centimeters

90. 55 mL to liters

Solve each application problem. Show your work.

91. Peggy had a board measuring 2 m 4 cm. She cut off 78 cm. How long is the board now, in meters?

2 m 4 cm

78 cm

92. During the 12-day Minnesota State Fair, one of the biggest in the United States, Sweet Martha's booth sells an average of 3000 pounds of cookies per day. How many tons of cookies are sold in all? (*Source: Minneapolis Star Tribune.*)

93. Olivia is sending a recipe to her mother in Mexico. Among other things, the recipe calls for 4 oz of rice and a baking temperature of 350 °F. Convert these measurements to metric, rounding to the nearest gram and nearest degree.

94. While on vacation in Canada, Jalo became ill and went to a health clinic. They said he weighed 80.9 kg and was 1.83 m tall. Find his weight in pounds and height in feet. Round to the nearest tenth.

The largest two-axle trucks in the world are mining trucks used to haul huge loads of rock and iron ore. Some information about these $1.5 million trucks is given below. Fill in the blank spaces in the table. Then use the information to answer Exercises 101 and 102. Round answers to the nearest tenth.

WORLD'S LARGEST TWO-AXLE TRUCK

| | Mining Truck | Measurements | |
		Metric Units	English Units
95.	Length of truck	_____ m	44 ft
96.	Height of truck	6.7 m	_____ ft
97.	Height of tire	_____ m	4 yd
98.	Width of tire tread	102 cm	_____ in.
99.	Weight of load truck can carry	220,000 kg	_____ lb
100.	Fuel needed to travel 1 mile	_____ to _____ L	5 to 6 gal

Source: Hull Rust Mahoning Mine.

101. Using English measurements:

(a) The truck can carry a load weighing how many tons?

(b) How many inches high is each tire?

102. Using metric measurements:

(a) What is the width of the tire tread in meters?

(b) How tall is the truck in centimeters?

Chapter **8**
TEST

Convert each measurement.

1. 9 gal = _____ qt

2. 45 ft = _____ yd

3. 135 min = _____ hr

4. 9 in. = _____ ft

5. $3\frac{1}{2}$ lb = _____ oz

6. 5 days = _____ min

Write the most reasonable metric unit in each blank. Choose from km, m, cm, mm, L, mL, kg, g, and mg.

7. My husband weighs 75 _____.

8. I hiked 5 _____ this morning.

9. She bought 125 _____ of cough syrup.

10. This apple weighs 180 _____.

11. This page is about 21 _____ wide.

12. My watch band is 10 _____ wide.

13. I bought 10 _____ of soda for the picnic.

14. The bracelet is 16 _____ long.

Convert the following measurements. Show your work.

15. 250 cm to meters

16. 4.6 km to meters

17. 5 mm to centimeters

18. 325 mg to grams

19. 16 L to milliliters

20. 0.4 kg to grams

21. 10.55 m to centimeters

22. 95 mL to liters

1. _____
2. _____
3. _____
4. _____
5. _____
6. _____
7. _____
8. _____
9. _____
10. _____
11. _____
12. _____
13. _____
14. _____
15. _____
16. _____
17. _____
18. _____
19. _____
20. _____
21. _____
22. _____

23. _____

23. The rainiest place in the world is Mount Waialeale in Hawaii, which receives 460 inches of rain each year. What is the average rainfall per month, in feet, to the nearest tenth? (*Source:* National Geographic Society.)

24. (a) _____

(b) _____

24. A 6-inch Subway Veggie Delite sandwich has 590 mg of sodium. A 6-inch Super Subway Melt sandwich has 2.9 g of sodium. (*Source:* Subway.)

   (a) How much more sodium is in the Super Melt sandwich, in milligrams, than in the Veggie Delite?

   (b) The recommended amount of sodium is less than 2400 mg daily. How much more or less sodium does the Super Melt have than the recommended daily amount?

Pick the metric temperature that is most appropriate in each situation.

25. _____

25. The water is almost boiling.

   210 °C   155 °C   95 °C

26. The tomato plants may freeze tonight.

   30 °C   20 °C   0 °C

26. _____

*Use the table from **Section 8.5** and unit fractions to convert each measurement. Round your answers to the nearest tenth, if necessary.*

27. _____

27. 6 ft to meters

28. 125 lb to kilograms

28. _____

29. _____

29. 50 L to gallons

30. 8.1 km to miles

30. _____

Use the conversion formulas to convert each temperature. Round your answers to the nearest degree, if necessary.

31. _____

31. 74 °F to Celsius

32. −12 °C to Fahrenheit

32. _____

Solve this application problem. Show your work.

33. _____

33. Denise is making five matching pillows. She needs 1 m 20 cm of braid to trim each pillow. If the braid costs $3.98 per yard, how much will she spend to trim the pillows, to the nearest cent? (First find the number of meters of braid Denise needs.)

34. _____

34. Describe two benefits the United States would achieve by switching entirely to the metric system.

Cumulative Review Exercises

CHAPTERS 1–8

1. Write these numbers in words.

   (a) 603,005,040,000

   (b) 9.040

2. Write these numbers in digits.

   (a) Eighty and eight hundredths

   (b) Two hundred million, sixty-five thousand, four

3. Round each number as indicated.

   (a) 0.9802 to the nearest tenth

   (b) 495 to the nearest ten

   (c) 306,472,000 to the nearest million

4. Find the mean and median, to the nearest tenth, for this set of data on hours worked: 22, 18, 40, 18, 20, 21, 45, 25.

Simplify.

5. $(-2)^3 - 3^2$

6. $2\dfrac{2}{5} - \dfrac{3}{4}$

7. $\dfrac{-4}{0.16}$

8. $\dfrac{7}{6x^2} \cdot \dfrac{9x}{14y}$

9. $(-0.003)(-0.05)$

10. $0.083 - 42$

11. $\dfrac{3}{c} - \dfrac{5}{6}$

12. $-15 - 15$

13. $10 - 3(6 - 7)$

14. $\dfrac{-3(-4)}{27 - 3^3}$

15. $\dfrac{3}{5}$ of (-400)

16. $3\dfrac{7}{12} - 4$

17. $\dfrac{9y^2}{8x} \div \dfrac{y}{6x^2}$

18. $\dfrac{2}{3} + \dfrac{n}{m}$

19. $1\dfrac{1}{6} + 1\dfrac{2}{3}$

20. $\dfrac{\frac{14}{15}}{-6}$

21. $\dfrac{12 \div (2 - 5) + 12(-1)}{2^3 - (-4)^2}$

22. $(-0.8)^2 \div (0.8 - 1)$

23. $\left(-\dfrac{1}{3}\right)^2 - \dfrac{1}{4}\left(\dfrac{4}{9}\right)$

Evaluate each expression when $k = -6$, $m = -1$, and $n = 2$.

24. $-3k + 4$ **25.** $4k - 5n$ **26.** $3m^3n$

Simplify each expression.

27. $3p - 3p^2 - 4p$ **28.** $-5(x + 2) - 4$ **29.** $7(-2r^3)$

Solve each equation. Show your work.

30. $2b + 2 = -5 + 5$ **31.** $12.92 - a = 4.87$ **32.** $7(t + 6) = 42$

33. $-5n = n - 12$ **34.** $2 = \dfrac{1}{4}w - 3$ **35.** $\dfrac{1.5}{45} = \dfrac{x}{12}$

36. $4y - 3 = 7y + 12$ **37.** $3(k - 6) - 4 = -2(k + 1)$

Translate each sentence into an equation and solve it.

38. If eight times a number is subtracted from eleven times the number, the result is -9. Find the number.

39. When twice a number is decreased by 8, the result is the number increased by 7. Find the number.

*Solve each application problem using the six problem-solving steps from **Chapter 3**.*

40. The perimeter of a rectangle is 124 cm. The width is 25 cm. Find the length.

41. A 90 ft pipe is cut into two pieces so that one piece is 6 ft shorter than the other. Find the length of each piece.

Find the unknown length, perimeter, circumference, area, or volume. When necessary, use 3.14 as the approximate value of π and round answers to the nearest tenth.

42. Find the perimeter and area.

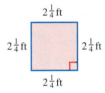

43. Find the circumference and the area.

44. Find the perimeter and the area.

45. Find the unknown length and the area of the figure.

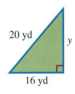

46. Find the volume and the surface area of this rectangular solid.

47. Find the unknown length in these similar triangles.

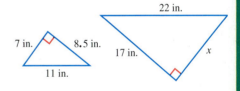

Convert the following using unit fractions, the metric conversion line, or the temperature conversion formulas.

48. $4\frac{1}{2}$ ft to inches

49. 72 hours to days

50. 3.7 kg to grams

51. 60 cm to meters

52. 7 mL to liters

53. $-20\,°C$ to Fahrenheit

Solve each application problem. Show your work.

54. Mei Ling must earn 60 credits to receive an associate of arts degree. She has 35 credits. What percent of the necessary credits does she have? Round to the nearest whole percent.

55. Which bag of chips is the best buy: Brand T is $15\frac{1}{2}$ ounces for \$2.99, Brand F is 14 ounces for \$2.49, and Brand H is 18 ounces for \$3.89. You have a coupon for 40¢ off Brand H and another for 30¢ off Brand T.

56. On the Illinois map, one centimeter represents 12 km. The center of Springfield is 7.8 cm from the center of Bloomington on the map.

   (a) What is the actual distance in kilometers?

   (b) What is the actual distance in miles, to the nearest tenth?

57. Bags of slivered almonds weigh 115 g each. They are packed in a carton that weighs 450 g. How many kilograms would a carton containing 48 bags weigh?

58. Steven bought $4\frac{1}{2}$ yards of canvas material to repair the tents used by the scout troop. He used $1\frac{2}{3}$ yards on one tent and $1\frac{3}{4}$ yards on another. How much material is left?

59. Mark bought 650 grams of maple sugar candy on his vacation in Montreal. The candy is priced at \$14.98 per kilogram. How much did Mark pay, to the nearest cent?

60. Calbert works at a Wal-Mart store and used his employee discount to buy a \$189.94 DVD player at 10% off. Find the amount of his discount, to the nearest cent, and the sale price. (*Source:* Wal-Mart.)

61. Akuba is knitting a scarf. Six rows of knitting result in 5 cm of scarf. At that rate, how many rows will she have to knit to make a 100 cm scarf?

62. A loan of \$3500 will be paid back with $7\frac{1}{2}$% simple interest at the end of six months. Find the total amount due.

63. A shelter for homeless people received a \$400 donation to buy blankets. If the blankets cost \$17 each, how many blankets can be purchased? How much money will be left over?

64. A spray-on bandage applies a transparent film over a wound that protects it from bacteria. The cost is \$6 for about 40 applications. What is the cost per application? (*Source:* Curad.)

65. The average hospital stay is now about 5 days, compared to about 8 days in 1970. What is the percent decrease in the length of hospital stays? (*Source:* National Center for Health Statistics.)

Graphs

9

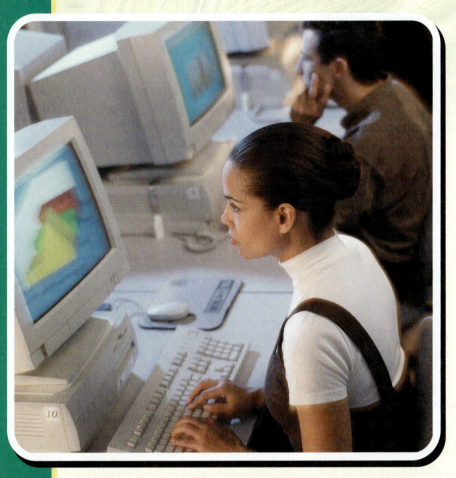

The Internet can be a great place to do research. Whether you're writing term papers and speeches for college courses, or planning on a career as a writer or reporter, you'll need to find information at various Web sites, interpret the data, and communicate it to other people. The data will often be presented in the form of a table (see **Section 9.1,** Examples 1 and 2) or a graph (see **Sections 9.2 and 9.3**). Interpreting the data accurately is critical to the success of your report, article, or speech. (See **Section 9.3,** Exercises 37–44.)

9.1 Problem Solving with Tables and Pictographs

OBJECTIVES

1. Read and interpret data presented in a table.

2. Read and interpret data from a pictograph.

Throughout this book you have used numbers, expressions, formulas, and equations to communicate rules or information. In this chapter, you'll see how *tables, pictographs, circle graphs, bar graphs,* and *line graphs* are also used to communicate information.

OBJECTIVE 1 Read and interpret data presented in a table. A **table** presents data organized into rows and columns. The advantage of a table is that you can find very specific, exact values. The disadvantages are that you may have to spend some time searching through the table to find what you want, and it may not be easy to see trends or patterns.

The table below shows information about the performance of eight U.S. airlines during the year 2003 (January–December).

PERFORMANCE DATA FOR SELECTED U.S. AIRLINES
JANUARY–DECEMBER 2003

Airline	On-Time Performance	Luggage Handling*
Airtran	78%	2.8
American	82%	4.5
Continental	82%	3.1
Delta	82%	3.8
Northwest	83%	3.4
Southwest	88%	3.6
United	83%	3.9
US Airways	80%	3.6

*Luggage problems per 1000 passengers.
Source: Department of Transportation Air Travel Consumer Report.

For example, by reading from left to right along the row marked Delta, you first see that 82% of Delta's flights were on time during 2003. The next number is 3.8 and the heading at the top of that column is *Luggage Handling**. The little star (asterisk) tells you to look below the table for more information. Next to the asterisk below the table it says "Luggage problems per 1000 passengers." So, almost 4 passengers out of every 1000 passengers on Delta flights had some sort of problem with their luggage.

EXAMPLE 1 Reading and Interpreting Data from a Table

Use the table above on airline performance to answer these questions.

(a) What percent of US Airways' flights were on time?

Look across the row labeled US Airways to see that 80% of its flights were on time.

(b) Which airline had the worst luggage handling record?

Look down the column headed Luggage Handling. To find the *worst* record, look for the *highest* number of luggage problems. The highest number is 4.5. Then look to the left to find the airline, which is American.

(c) What was the average percent of on-time flights for the three airlines with the best performance, to the nearest whole percent?

Look down the column headed On-Time Performance to find the three highest numbers: 88%, 83%, and 83%. To find the average, add the values and divide by 3.

Continued on Next Page

$$\frac{88 + 83 + 83}{3} = \frac{254}{3} \approx 85 \text{ (rounded to nearest whole percent)}$$

The average on-time performance for the three best airlines was about 85%.

Now Try Exercise 7.

EXAMPLE 2   **Interpreting Data from a Table**

The table below shows the maximum cab fares in five different cities in February 2004. The "flag drop" charge is made when the driver starts the meter. "Wait time" is the charge for having to wait in the middle of a ride.

MAXIMUM TAXICAB FARES ALLOWED IN SELECTED CITIES IN FEBRUARY 2004

City	Flag Drop	Price per Mile	Wait Time (per Hour)
Chicago	$1.90	$1.60	$20
Denver	$1.60	$1.80	$18
Miami	$1.70	$2.20	$21
New York	$2.50	$2	$12
Portland	$2.50	$1.50	$20

Source: Web sites for the various cities or their airports.

Use the table to answer these questions.

(a) What is the maximum fare for a 9-mile ride in New York that includes having the cab wait 15 minutes while you pick up a package at a store?

The price per mile in New York is $2, so the cost for 9 miles is $9(\$2) = \$18$. Then, add the flag drop charge of $2.50. Finally, figure out the cost of the wait time. One way to do that is to set up a proportion. Recall that 1 hour is 60 minutes, so each side of the proportion compares the cost to the number of minutes of wait time.

$$\text{Cost} \rightarrow \frac{\$12}{60 \text{ min}} = \frac{\$x}{15 \text{ min}} \leftarrow \text{Cost} \atop \leftarrow \text{Wait time}$$

$$60 \cdot x = 12 \cdot 15 \qquad \text{Find the cross products.}$$

$$\frac{\overset{1}{\cancel{60}} x}{\underset{1}{\cancel{60}}} = \frac{180}{60} \qquad \text{Divide both sides by 60}$$

$$x = \$3 \longleftarrow \text{Charge for waiting 15 minutes}$$

Total fare = $18 + $2.50 + $3 = $23.50

(b) It is customary to give the cab driver a tip. Find the total cost of the cab ride in part (a) above if the passenger added a 15% tip, rounded to the nearest quarter (nearest $0.25).

Use the percent equation to find the exact tip.

$$\text{percent} \cdot \text{whole} = \text{part} \qquad \text{Percent equation}$$

$$(15.\%)(\$23.50) = n \qquad \text{Write 15\% as a decimal.}$$

$$(0.15)(\$23.50) = n$$

$$\$3.525 = n \qquad \text{Round to \$3.50 (nearest \$0.25).}$$

The total cost of the cab ride is $23.50 + $3.50 tip = $27.00.

Now Try Exercise 17.

OBJECTIVE **2** **Read and interpret data from a pictograph.** Tables show numbers in rows and columns. Graphs, on the other hand, are a *visual* way to communicate data; that is, they show a *picture* of the information rather than a list of numbers. In this section you'll work with one type of graph, *pictographs*, and in the next two sections you'll learn about circle graphs, bar graphs, and line graphs.

The advantage of a graph is that you can easily make comparisons or see trends just by looking. The disadvantage is that the graph may not give you the specific, more exact numbers you need in some situations.

EXAMPLE 3 Reading and Interpreting a Pictograph

A **pictograph** uses symbols or pictures to represent various amounts. The pictograph below shows the population of five U.S. metropolitan areas (cities with their surrounding suburbs) in 2000. The *key* at the bottom of the graph tells you that each symbol of a person represents 2 million people. A fractional part of a symbol represents a fractional part of 2 million people.

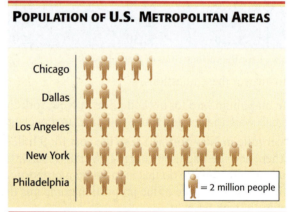

POPULATION OF U.S. METROPOLITAN AREAS

Source: U.S. Bureau of the Census, U.S. Department of Commerce.

Use the pictograph to answer these questions.

(a) What is the approximate population of Chicago?

The population of Chicago is represented by 4 whole symbols (4 • 2 million = 8 million) plus half of a symbol ($\frac{1}{2}$ of 2 million is 1 million) for a total of 9 million.

Recall that 2 million can be written as 2,000,000, so another way to get the answer is to multiply **4.5** (the number of symbols) times 2,000,000 for a total of 9,000,000. So the approximate population of Chicago is 9 million people (or 9,000,000).

(b) How much greater is the population of Los Angeles than Philadelphia?

Los Angeles shows 8 symbols and Philadelphia shows 3 symbols. So Los Angeles has 5 more symbols than Philadelphia, and 5 • 2 million = 10 million. Thus, Los Angeles has about 10 million (or 10,000,000) more people than Philadelphia.

Continued on Next Page

> **NOTE**
> One disadvantage of a pictograph is that it is difficult to draw and interpret fractional parts of a symbol. The data on metropolitan area populations were rounded a great deal in order to use only whole symbols and half symbols.
>
> For example, the actual metropolitan area population of Los Angeles is 16,402,079 people. We rounded the population to the *nearest million*. The actual population of Los Angeles is closer to 16 million than to 17 million, so there are 8 whole symbols in the pictograph. That means our pictograph is off by about 400,000 people. Therefore, if you need fairly accurate numbers for a particular situation, it's better to use the information from a table than a pictograph.

Now Try Exercise 25.

9.1 Exercises

FOR EXTRA HELP

 Addison-Wesley Math Tutor Center

 MathXL

 Digital Video Tutor CD 5 Videotape 9

Student's Solutions Manual

 MyMathLab

MathXL Tutorials on CD

This table lists the basketball players in the NBA with the highest scoring average at the end of the 2003–2004 season. (Players must have a minimum of 10,000 points or 400 games.) Use the table to answer Exercises 1–10. See Examples 1 and 2.

ALL-TIME NBA STATISTICAL LEADERS—SCORING AVERAGE (AT THE END OF THE 2003–2004 SEASON)

Player	Games	Points	Average Points Per Game
Michael Jordan	1072	32,292	30.1
Wilt Chamberlain	1045	31,419	30.1
Elgin Baylor	846	23,419	27.4
Shaquille O'Neal*	809	21,914	27.1
Jerry West	932	25,192	
Allen Iverson*	535	14,436	
Bob Pettit	792	20,880	

*Player active in 2003–2004 season.
Source: National Basketball Association.

1. **(a)** How many points did Wilt Chamberlain score during his NBA career?

   (b) Which player(s) scored more points than Chamberlain?

2. **(a)** How many games did Elgin Baylor play in during his career in the NBA?

   (b) Which player is closest to Baylor in number of games?

3. **(a)** Which player has been in the greatest number of games?

   (b) Which player has been in the fewest number of games?

4. Which players have scored more than 25,000 points? List them in order, starting with the player with the greatest number of points.

5. What is the difference in points scored between the player with the greatest number of points and the player with the least number of points?

6. How many fewer games has Shaquille O'Neal played in than Michael Jordan?

7. Complete the table by finding the average number of points scored per game by Jerry West, by Allen Iverson, and by Bob Pettit. Look at the other averages in the table to decide how to round your answers.

8. Find the overall scoring average (points per game) for all seven players listed in the table. Use the numbers in the *Games* column and the *Points* column to calculate your answer.

9. According to the table, what is different about O'Neal and Iverson from the rest of the players?

10. Michael Jordan and Wilt Chamberlain are tied in first place for average points per game. Explain what happens if you round their average points per game to the nearest hundredth instead of the nearest tenth.

This table shows the number of calories burned during 30 minutes of various types of exercise. The table also shows how the number of calories burned varies according to the weight of the person doing the exercise. Use the table to answer Exercises 11–20. See Examples 1 and 2.

CALORIES BURNED DURING 30 MINUTES OF EXERCISE BY PEOPLE OF DIFFERENT WEIGHTS

Activity	Calories Burned in 30 Minutes		
	110 Pounds	140 Pounds	170 Pounds
Moderate jogging	322	410	495
Moderate walking	110	140	170
Moderate bicycling	140	180	220
Aerobic dance	200	255	310
Racquetball	210	268	325
Tennis	160	205	250

Source: Fairview Health Services.

11. A person weighing 140 pounds is looking at the table.

   (a) How many calories will be burned during 30 minutes of aerobic dance?

   (b) Which activity burns the most calories?

12. A person weighing 170 pounds is looking at the table.

   (a) How many calories are burned during 30 minutes of tennis?

   (b) Which activity burns the fewest calories?

13. (a) Which activities can a 110-pound person do to burn at least 200 calories in 30 minutes?

   (b) Which activities can a 170-pound person do to burn at least 200 calories in 30 minutes?

14. (a) Which activities would burn fewer than 200 calories in 30 minutes for a 140-pound person?

   (b) Which activities would burn fewer than 300 calories in 30 minutes for a 170-pound person?

15. How many total calories will a 140-pound person burn during 15 minutes of bicycling and 60 minutes of walking?

16. How many total calories will a 110-pound person burn during 90 minutes of tennis and 15 minutes of aerobic dance?

Set up and solve proportions to answer Exercises 17–20.

17. How many calories would you expect a 125-pound person to burn

   (a) during 30 minutes of moderate jogging?

   (b) during 30 minutes of racquetball? Round to the nearest whole number.

18. How many calories would you expect a 185-pound person to burn

   (a) during 30 minutes of walking?

   (b) during 30 minutes of bicycling? Round to the nearest ten.

19. How many more calories would a 158-pound person burn during 15 minutes of aerobic dance than during 20 minutes of walking? Round to the nearest whole number.

20. How many fewer calories would a 196-pound person burn during 25 minutes of walking than during 20 minutes of tennis? Round to the nearest whole number.

This pictograph shows the approximate number of passenger arrivals and departures at selected U.S. airports in 2003. Use the pictograph to answer Exercises 21–28. See Example 3.

PASSENGER ARRIVALS AND DEPARTURES AT SELECTED U.S. AIRPORTS

Atlanta
Los Angeles
Chicago
Houston
Seattle

🧳 = 10 million passenger arrivals and departures

Source: Airports Council International—North America.

21. Approximately how many passenger arrivals and departures took place at the

(a) Chicago airport?

(b) Houston airport?

22. Approximately how many passenger arrivals and departures took place at the

(a) Atlanta airport?

(b) Seattle airport?

23. What is the approximate total number of arrivals and departures at the three busiest airports?

24. What is the difference in the number of arrivals and departures at the busiest airport and the least-busy airport?

25. How many fewer arrivals and departures did Los Angeles' airport have compared to Chicago's airport?

26. Find the approximate total number of arrivals and departures for the three least-busy airports.

27. What is the approximate total number of arrivals and departures for all five airports?

28. Find the average number of arrivals and departures for the five airports.

RELATING CONCEPTS (EXERCISES 29–34) For Individual or Group Work

Look back at the first table in this section, Performance Data for Selected U.S. Airlines.
*Use the table as you **work Exercises 29–34 in order.***

29. Suppose you are planning a business trip where you will fly to a new city each day on a tight schedule. You'll travel light, carrying one small bag on the plane rather than checking it. If you can choose any one of the airlines in the table, which one would you pick? Explain why.

30. Suppose you are planning the business trip described in Exercise 29 and the only airline that goes to the cities you want is US Airways. What could you do to minimize the problems caused by the possibility of late flights?

31. Now you are planning a two-week vacation trip to a beachfront resort. You'll be checking several bags and your expensive golf clubs. If you can choose any one of the airlines in the table, which one would you pick? Explain why.

32. Suppose you are planning the vacation trip described in Exercise 31 and American is the only airline that goes to the city you want. What could you do to minimize possible luggage handling problems?

33. Think of three possible reasons an airline might have a lower percentage of on-time flights during a particular year than they usually do. What, if anything, could the airline do to resolve each of the problems you listed?

34. Describe three other factors you might consider when selecting an airline, other than on-time performance and luggage handling problems.

9.2 Reading and Constructing Circle Graphs

A *circle graph* is another way to show a *picture* of a set of data. This picture can often be understood faster and more easily than a formula or a list of numbers.

OBJECTIVE 1 **Read a circle graph.** A **circle graph** is used to show how a total amount is divided into parts. The circle graph below shows you how 24 hours in the life of a college student are divided among different activities.

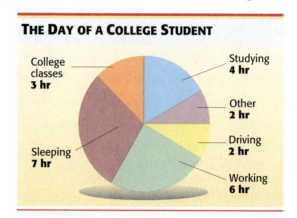

THE DAY OF A COLLEGE STUDENT

College classes 3 hr

Studying 4 hr

Other 2 hr

Driving 2 hr

Working 6 hr

Sleeping 7 hr

> Now Try Exercise 1.

OBJECTIVE 2 **Use a circle graph.** The circle graph above uses pie-shaped pieces called *sectors* to show the amount of time spent on each activity (the total must be 24 hours). The circle graph can therefore be used to compare the time spent on one activity to the total number of hours in the day.

EXAMPLE 1 Using a Circle Graph

Find the ratio of time spent in college classes to the total number of hours in a day. Write the ratio as a fraction in lowest terms. (See **Section 6.1.**)

The circle graph shows that 3 of the 24 hours in a day are spent in class. The ratio of class time to the hours in a day is shown below.

$$\frac{3 \text{ hours (college classes)}}{24 \text{ hours (whole day)}} = \frac{3 \text{ hours}}{24 \text{ hours}} = \frac{\overset{1}{\cancel{3}}}{\underset{1}{\cancel{3} \cdot 8}} = \frac{1}{8} \quad \left\{ \text{Ratio in lowest terms} \right.$$

> Now Try Exercise 5.

The circle graph above can also be used to find the ratio of the time spent on one activity to the time spent on any other activity. See the next example.

EXAMPLE 2 Finding a Ratio from a Circle Graph

Find the ratio of working time to class time.

The circle graph shows 6 hours spent working and 3 hours spent in class. The ratio of working time to class time is shown below.

$$\frac{6 \text{ hours (working)}}{3 \text{ hours (class)}} = \frac{6 \text{ hours}}{3 \text{ hours}} = \frac{\overset{1}{\cancel{3}} \cdot 2}{\underset{1}{\cancel{3}}} = \frac{2}{1} \quad \leftarrow \text{Ratio in lowest terms}$$

> Now Try Exercise 3(b)

A circle graph often shows data as percents. For example, total U.S. sales of frozen pizza are $2 billion each year. The circle graph below shows how sales are divided among various companies that make frozen pizza. The entire circle represents $2 billion in sales. Each sector represents the sales of one company as a percent of the total sales. The total in a circle graph must be 100%, although it may be slightly more or less due to rounding the percent for each sector.

HOT SALES OF FROZEN PIZZA

Americans eat $2 billion worth of frozen pizzas each year. The percent of total sales for each company is rounded to the nearest whole percent.

Van De Kamps **4%**

All other **10%**

Tony's Pizza Service **30%**

Nestle **5%**

Private Label **5%**

Pillsbury Corp. **9%**

Kraft **37%**

Source: Information Resources, Inc.

EXAMPLE 3 **Calculating an Amount Using a Circle Graph**

Use the circle graph on frozen pizza sales to find the amount of sales for Nestle.

Recall the percent equation.

$$\text{percent} \cdot \text{whole} = \text{part}$$

The percent for Nestle is 5%. Rewrite 5% as the decimal 0.05. The *whole* is the total sales of $2 billion (the entire circle).

$$\begin{array}{ccc} \text{percent} \cdot & \text{whole} & = \text{part} \\ \downarrow & \downarrow & \downarrow \\ 05.\% \cdot & \$2 \text{ billion} & = n \qquad \text{Write 5\% as a decimal.} \\ \downarrow & \downarrow & \\ (0.05)(\$2,000,000,000) = & n \qquad \text{Write \$2 billion as} \\ & & \$2,000,000,000 \\ \$100,000,000 & = n \end{array}$$

The sales for Nestle are $100,000,000.

Now Try Exercise 21. ▶▶▶

OBJECTIVE 3 **Use a protractor to draw a circle graph.** The coordinator of the Fair Oaks Youth Soccer League organizes teams in five age groups. She counts the number of registered players in each age group as shown in the table on the next page. Then she calculates what percent of the total each group represents. For example, there are 59 players in the "Under 8" group, out of 298 total players.

$$\text{percent} \cdot \text{whole} = \text{part}$$
$$p \cdot 298 = 59$$
$$\frac{p \cdot 298}{298} = \frac{59}{298}$$
$$p \approx 0.197 \approx 20\%$$

FAIR OAKS YOUTH SOCCER LEAGUE

Age Group	Number of Players	Percent of Total (rounded to nearest whole percent)
Under 8 years	59	20% ← 59 players ≈ 20% of 298
Ages 8–9	46	15%
Ages 10–11	75	25%
Ages 12–13	74	25%
Ages 14–15	44	15%
Total	**298**	**100%**

You can show these percents in a circle graph. Recall that a circle has 360 degrees (written 360°). The 360° represents the entire league, or 100% of the soccer players.

EXAMPLE 4 **Drawing a Circle Graph**

Using the data on *age groups,* find the number of degrees in the sector that would represent the "Under 8" group, and begin constructing a circle graph.

A complete circle has 360°. Because the "Under 8" group makes up 20% of the total number of players, the number of degrees needed for the "Under 8" sector of the circle graph is 20% of 360°.

$$20.\% \text{ of } 360° = n$$
$$(0.20)(360°) = n$$
$$72° = n$$

Use the circle drawn below and a tool called a **protractor** to make a circle graph. First, using a ruler or straightedge, draw a line from the center of the circle to the left edge. Place the hole in the protractor over the center of the circle, making sure that the 0 mark and the black line on the protractor are right over the line you drew. Find 72° and make a mark as shown in the illustration. Then remove the protractor and use the straightedge to draw a line from the 72° mark to the center of the circle. This sector is 72° and represents the "Under 8" group.

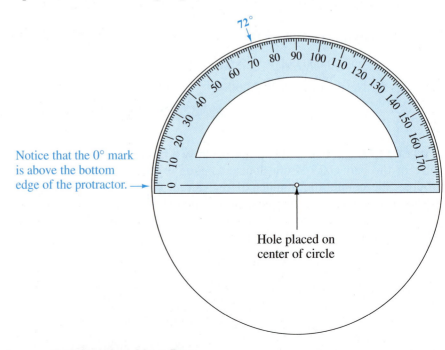

Notice that the 0° mark is above the bottom edge of the protractor. ⟶

Hole placed on center of circle

Continued on Next Page

To draw the "Ages 8–9" sector, begin by finding the number of degrees in the sector, which is 15% of the total circle, or 15% of 360°.

$$15.\% \text{ of } 360° = n$$
$$(0.15)(360°) = n$$
$$54° \qquad = n$$

Again, place the hole of the protractor at the center of the circle, but this time align 0 with the previous 72° mark. Make a mark at 54° and draw a line from the mark to the center of the circle. This sector is 54° and represents the "Ages 8–9" group.

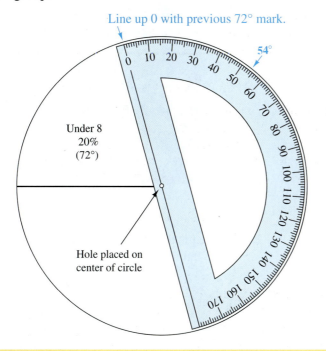

CAUTION
You must be certain that the hole in the protractor is placed on the exact center of the circle each time you measure the size of a sector.

Now Try Exercise 29(a)–(g). ▶▶▶

9.2 Exercises

This circle graph shows the cost of adding a family room to an existing home. Use this circle graph to answer Exercises 1–6. Write ratios as fractions in lowest terms. See Examples 1 and 2.

COST OF A FAMILY ROOM ADDITION

Carpentry $12,100
Materials $9800
Electrical $2000
Window coverings $900
Floor $2400
Windows $3000
Painting $1800

1. (a) Find the total cost of adding the family room.

   (b) What is the largest single expense?

2. (a) What is the second-largest expense in adding the family room?

   (b) What is the smallest expense?

3. (a) Find the ratio of the cost of materials to the total remodeling cost.

   (b) Find the ratio of the cost of windows to the cost of electrical.

4. (a) Find the ratio of the cost of painting to the total remodeling cost.

   (b) Find the ratio of the cost of windows to the cost of window coverings.

5. Find the ratio of the cost of carpentry, windows, and window coverings to the total remodeling cost.

6. Find the ratio of the cost of windows and electrical to the cost of the floor and painting.

This circle graph, adapted from USA Today, *shows the number of people in a survey who gave various reasons for eating dinner at restaurants. Use this circle graph to answer Exercises 7–14. See Examples 1 and 2.*

ON THE TOWN

When asked in a survey why they ate dinner in restaurants, a group of people gave these reasons.

Wanted food they couldn't cook at home **1740**

Atmosphere **1200**

Enjoy eating out **1140**

Quicker **720**

Less work/ no cleanup **1020**

180 Don't know

Source: Market Facts for Tyson Foods.

7. (a) Which reason was given by the least number of people?

(b) Which reason was given by the second-fewest number of people?

8. (a) Which reason was given by the highest number of people?

(b) Which reason was given by the second-highest number of people?

Find each ratio in Exercises 9–14. Write the ratios as fractions in lowest terms.

9. Those who said dining out is "Quicker" to total people in the survey

10. Those who said "Enjoy eating out" to the total people in the survey

11. Those who said "Less work/no cleanup" to those who said "Atmosphere"

12. Those who said "Don't know" to those who said "Quicker"

13. Those who said "Wanted food they couldn't cook at home" to those who said "Don't know"

14. Those who said "Atmosphere" to those who said "Enjoy eating out"

This circle graph shows the costs necessary to comply with the Americans with Disabilities Act (ADA) at the Dos Pueblos College. The cost of each item is expressed as a percent of the total cost of $1,740,000. Use the graph to find the dollar amount spent for each item in Exercises 15–20. See Example 3.

ADA COMPLIANCE COSTS

Ramps and handrails 25%
Restrooms 30%
Doors and thresholds 10%
Other 10%
Parking 10%
Walkways and curbs 15%

15. Restrooms

16. Ramps and handrails

17. Doors and thresholds

18. Parking

19. Walkways and curbs

20. Other

This circle graph shows the results of a survey on favorite hot dog toppings in the United States. Each topping is expressed as a percent of the 3200 people in the survey. Use the graph to find the number of people favoring each of the toppings in Exercises 21–26.

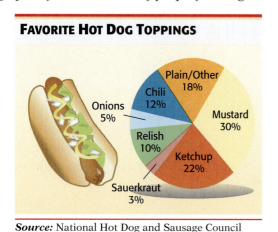

FAVORITE HOT DOG TOPPINGS

Plain/Other 18%
Chili 12%
Onions 5%
Mustard 30%
Relish 10%
Ketchup 22%
Sauerkraut 3%

Source: National Hot Dog and Sausage Council

21. Onions

22. Sauerkraut

23. The most popular topping

24. The second-most-popular topping

25. How many more people chose chili than chose relish?

26. How many fewer people chose onions than chose mustard?

27. Describe the procedure for determining how large each sector must be to represent each of the items in a circle graph.

28. A protractor is the tool used to draw a circle graph. Give a brief explanation of what the protractor does and how you would use it to measure and draw each sector in the circle graph.

29. During one semester Kara Diano spent $5460 for school expenses as shown in this table. Find all numbers missing from the table.

Item	Dollar Amount	Percent of Total	Degrees of a Circle
(a) Rent	$1365	25%	_____
(b) Food	$1092	_____	72°
(c) Clothing	$ 546	_____	_____
(d) Books and supplies	$ 546	_____	_____
(e) Tuition and fees	$ 819	_____	_____
(f) Savings	$ 273	_____	_____
(g) Entertainment	$ 819	_____	_____

(h) Draw a circle graph using the budget information from the table. Label each sector with the budget item and the percent of total for that item. See Example 4.

30. The Pathfinder Research Group asked 4488 Americans how they fall asleep. The results are shown in the figure on the right.

Use this information to complete the table. Round to the nearest whole percent and to the nearest degree.

SET TO SLEEP

Number of Americans surveyed who fall asleep on their:

Source: Pathfinder Research Group.

	Sleeping Position	Number of Americans	Percent of Total	Number of Degrees
(a)	Side	_____	_____	_____
(b)	Back	_____	_____	_____
(c)	Stomach	_____	_____	_____
(d)	Varies	_____	_____	_____
(e)	Not sure	_____	_____	_____

(f) Add up the percents. Is the total 100%? Explain why or why not.

(g) Add up the degrees. Is the total 360°? Explain why or why not.

(h) Draw a circle graph using the information from the table above. Label each sector with the sleeping position and the percent of total for that position.

31. White Water Rafting Company divides its annual sales into five categories as shown in the table.

Category	Annual Sales	Percent of Total
Adventure classes	$12,500	_____
Grocery/provision sales	$40,000	_____
Equipment rentals	$60,000	_____
Rafting tours	$50,000	_____
Equipment sales	$37,500	_____

(a) Find the total sales for the year.

(b) Find the percent of the total for each category and write it in the table.

(c) Find the number of degrees in a circle graph for each item.

(d) Make a circle graph showing the sales information. Label each sector with the category and the percent of total for that category.

32. A book publisher had 25% of total sales in mysteries, 10% in biographies, 15% in cookbooks, 15% in romance novels, 20% in science, and the rest in travel books.

(a) Find the number of degrees in a circle graph for each type of book.

(b) Draw a circle graph, using the information given. Label each sector with the type of book and the percent of total for that type.

9.3 Bar Graphs and Line Graphs

OBJECTIVES

Read and understand
1 a bar graph;
2 a double-bar graph;
3 a line graph;
4 a comparison line graph.

OBJECTIVE 1 **Read and understand a bar graph.** A **bar graph** is useful for showing comparisons. For example, the bar graph below compares the number of college graduates who continued taking advanced courses in their major field during each of five years.

GRADUATES TAKING ADVANCED COURSES

EXAMPLE 1 **Using a Bar Graph**

How many college graduates took advanced classes in their major field in 2003?

The bar for 2003 rises to 25. Notice the label along the left side of the graph that says "Number of Graduates (in thousands)." The phrase *in thousands* means you have to multiply 25 by 1000 to get 25,000. So, 25,000 (not 25) graduates took advanced classes in their major field in 2003.

Now Try Exercise 1.

OBJECTIVE 2 **Read and understand a double-bar graph.**
A **double-bar graph** can be used to compare two sets of data. The graph below shows the number of DSL (digital subscriber line) installations each quarter for two different years.

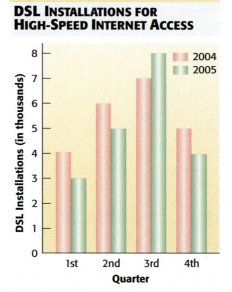

DSL INSTALLATIONS FOR HIGH-SPEED INTERNET ACCESS

Note:
1st quarter = Jan–Mar
2nd quarter = Apr–June
3rd quarter = July–Sept
4th quarter = Oct–Dec

EXAMPLE 2 **Reading a Double-Bar Graph**

Use the double-bar graph on DSL installations to find the following.

(a) The number of DSL installations in the second quarter of 2004.

There are two bars for the second quarter. The color code in the upper right-hand corner of the graph tells you that the **red bars** represent 2004. So the **red bar** on the *left* is for the 2nd quarter of 2004. It rises to 6. Multiply 6 by 1000 because the label on the left side of the graph says *in thousands*. So there were 6000 DSL installations for the second quarter in 2004.

(b) The number of DSL installations in the second quarter of 2005.

The **green bar** for the second quarter rises to 5 and 5 times 1000 is 5000. So, in the second quarter of 2005, there were 5000 DSL installations.

> **CAUTION**
> Use a ruler or straightedge to line up the top of the bar with the number on the left side of the graph.

Now Try Exercise 7. ▶▶▶

OBJECTIVE 3 Read and understand a line graph. A **line graph** is often useful for showing a trend. The line graph below shows the number of trout stocked along the Feather River during five months. Each dot indicates the number of trout stocked during the month directly below that dot.

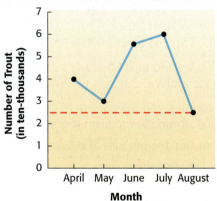

TROUT STOCKED IN FEATHER RIVER

EXAMPLE 3 **Understanding a Line Graph**

Use the line graph on trout to answer each question.

(a) In which month were the least number of trout stocked?

The lowest point on the graph is the dot directly over August, so the least number of trout were stocked in August.

(b) How many trout were stocked in August?

Use a ruler or straightedge to line up the August dot with the numbers along the left edge of the graph. (See the red dashed line on the graph.)

The August dot is halfway between the 2 and 3. Notice that the label on the left side says *in ten thousands*. So August is halfway between $(2 \cdot 10{,}000)$ and $(3 \cdot 10{,}000)$. It is halfway between 20,000 and 30,000. That means 25,000 trout were stocked in August.

Now Try Exercise 21. ▶▶▶

OBJECTIVE **4** **Read and understand a comparison line graph.** Two sets of data can also be compared by drawing two line graphs together as a **comparison line graph.** For example, the line graph below compares the number of minivans sold and the number of Sport Utility Vehicles (SUVs) sold during each of five years.

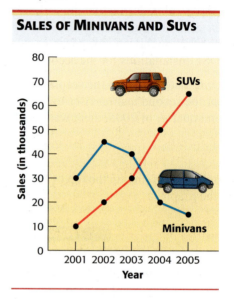

SALES OF MINIVANS AND SUVS

EXAMPLE 4 **Interpreting a Comparison Line Graph**

Use the comparison line graph above to find the following.

(a) The number of minivans sold in 2002

Find the dot on the **blue line** above 2002. Use a ruler or straightedge to line up the dot with the numbers along the left edge. The dot is halfway between 40 and 50, which is 45. Then, 45 times 1000 is 45,000 minivans sold in 2002.

(b) The number of SUVs sold in 2005

The **red line** on the graph shows that 65,000 SUVs were sold in 2005.

(c) The amount of decrease and the percent of decrease in minivan sales from 2002 to 2003. Round to the nearest whole percent.

Use subtraction to find the *amount* of decrease.

$$\begin{array}{ccccc} \textbf{minivan} & & \textbf{minivan} & & \textbf{amount} \\ \textbf{sales in 2002} & - & \textbf{sales in 2003} & = & \textbf{of decrease} \\ 45{,}000 & - & 40{,}000 & = & 5000 \end{array}$$

Now use the percent equation to find the *percent* of decrease.

$$\text{percent } \textbf{of} \text{ original sales} = \text{amount of decrease}$$
$$p \quad \bullet \quad 45{,}000 \quad = \quad 5000$$

$$\frac{p \bullet 45{,}000}{45{,}000} = \frac{5000}{45{,}000} \qquad \text{\color{blue}{Divide both sides by 45,000}}$$

$$p = 0.11\overline{1} \quad \leftarrow \text{\color{blue}{Solution in \textit{decimal} form}}$$

Multiply by 100 to change the decimal to a percent, then round.

$$p = 0.11\overline{1} = 11.\overline{1}\% \approx 11\% \text{ to nearest whole percent}$$

The amount of decrease is 5000 minivans. The percent of decrease is 11% (rounded).

Now Try Exercise 25.

This bar graph shows the top seven reasons people say they shop on-line. Use the graph to answer Exercises 1–6. See Example 1.

CONVENIENCE SIMPLIFIED

Top Reasons People Shop On-Line

I can shop during off hours.	74%
It saves time.	68%
It saves trips to the mall or the store.	67%
I can shop at stores I cannot visit where I live.	66%
I find better prices on-line.	57%
I can shop without salesperson pressure.	54%
I can compare products more easily.	50%

Source: EMARKETER.

1. What is the top reason people shop on-line? What percent gave this reason?

2. What is the second most popular reason for shopping on-line? What percent gave this reason?

3. What percent of the people say they find better prices on-line? If 600 people were surveyed, how many gave this answer?

4. What percent say it saves trips to the mall or store? If 600 people were surveyed, how many gave this answer?

5. Which reason(s) were given by $\frac{1}{2}$ of the people? Which reason(s) were given by nearly $\frac{3}{4}$ of the people?

6. Which reason(s) were given by about $\frac{2}{3}$ of the people?

This double-bar graph shows the number of workers who were unemployed in a city during the first six months of 2004 and 2005. Use this graph to answer Exercises 7–12. Round percents to the nearest whole number. See Example 2.

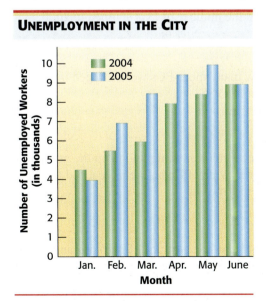

UNEMPLOYMENT IN THE CITY

■ 2004
■ 2005

Number of Unemployed Workers (in thousands)

Month: Jan. Feb. Mar. Apr. May June

7. In the first half of 2005, which month had the greatest number of unemployed workers? What was the total number unemployed in that month?

8. How many workers were unemployed in January of 2004?

9. How many more workers were unemployed in February of 2005 than in February of 2004?

10. How many fewer workers were unemployed in March of 2004 than in March of 2005?

11. Find the amount of increase and the percent of increase in the number of unemployed workers from February 2004 to April 2004.

12. Find the amount of increase and the percent of increase in the number of unemployed workers from January 2005 to June 2005.

This double-bar graph shows sales of super unleaded and supreme unleaded gasoline at a service station for each of five years. Use this graph to answer Exercises 13–18. See Example 2.

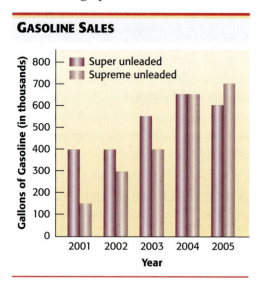

GASOLINE SALES

13. How many gallons of supreme unleaded gasoline were sold in 2001?

14. How many gallons of super unleaded gasoline were sold in 2004?

15. In which year did the greatest difference in sales between super unleaded and supreme unleaded gasoline occur? Find the difference.

16. In which year did the sales of supreme unleaded gasoline surpass the sales of super unleaded gasoline?

17. Find the amount of increase and percent of increase in supreme unleaded gasoline sales from 2001 to 2005. Round to the nearest whole percent.

18. Find the amount of increase and percent of increase in super unleaded gasoline sales from 2001 to 2005.

This line graph shows how sales of personal computers (PCs) have increased since they first became widely available in 1985. Use the line graph to answer Exercises 19–24. Round percents to the nearest whole percent. See Example 3.

WORLDWIDE SHIPMENTS OF PERSONAL COMPUTERS (PCs)

Source: Gartner Dataquest

19. How many PCs were shipped in 1990?

20. Find the number of PCs shipped in 2000.

21. How many more PCs were shipped in 2005 than in 1985?

22. Place a ruler or straightedge on the graph to find the year in which PC shipments reached **(a)** 40 million; **(b)** 120 million.

23. Find the amount of increase in shipments from 1995 to 2000. What was the percent of increase, to the nearest whole percent?

24. What was the amount of increase and the percent of increase in PC shipments from 1985 to 1990?

This comparison line graph shows the number of compact discs (CDs) sold by two different chain stores during each of five years. Use this graph to find the annual number of CDs sold in each year listed in Exercises 25–28. See Example 4.

COMPACT DISC SALES

25. (a) Chain Store A in 2001

 (b) Chain Store B in 2001

26. (a) Chain Store A in 2002

 (b) Chain Store B in 2002

27. (a) Chain Store A in 2004

 (b) Chain Store A in 2005

28. (a) Chain Store B in 2003

 (b) Chain Store B in 2004

29. Describe the pattern(s) or trend(s) you see in the graph.

30. Store B used to have lower sales than Store A. What might have happened to cause this change? Give four possible explanations.

This comparison line graph shows the sales and profits of Tacos-To-Go for each of four years. Use the graph to answer Exercises 31–36.

ANNUAL TACOS-TO-GO SALES

31. Find the year of lowest sales and the amount of sales.

32. Find the year of greatest profit and the amount of profit.

33. For each of the four years, find what percent the profit is of sales. Round to the nearest whole percent.

34. Find the total sales for all four years and the total profit for all four years. Total profit is what percent of total sales? Round to the nearest whole percent.

35. Give two possible explanations for the decrease in sales from 2002 to 2003 and two possible explanations for the increase in sales from 2003 to 2005.

36. *Based on the graph,* what conclusion can you make about the relationship between sales and profits for Tacos-To-Go?

RELATING CONCEPTS (EXERCISES 37–44) For Individual or Group Work

Use the line graph on Worldwide Shipments of PCs on page 658 and the double-line graph on Compact Disc Sales on page 659 as you **work Exercises 37–44 in order.** *Round percent answers to the nearest whole percent.*

37. What overall trend do you see in the graph on worldwide shipments of Personal Computers (PCs)?

38. Give at least two possible explanations for the increase in PC shipments.

39. Find the percent of increase in PC shipments from

(a) 1985 to 1990

(b) 1990 to 1995

(c) 1995 to 2000

(d) 2000 to 2005.

40. Look at your answers in Exercise 39. What trend do you see?

41. What future conditions could result in a decrease in PC shipments? Give at least two possibilities.

42. Look at the graph of Compact Disc Sales. For each store, compare the sales of compact discs in 2005 to sales in 2001. Find the percent of increase or the percent of decrease for:

(a) Chain Store A

(b) Chain Store B.

43. *Based on the graph,* what amount of CD sales would you predict for

(a) Store A in 2006?

(b) Store B in 2006?

Explain how you arrived at your predictions.

44. (a) *Based on the graph,* which store would you like to own? Explain why.

(b) Name at least three other things you would want to know before deciding which store to buy.

9.4 The Rectangular Coordinate System

OBJECTIVES

1 Plot a point, given the coordinates, and find the coordinates, given a point.

2 Identify the four quadrants and determine which points lie within each one.

OBJECTIVE 1 Plot a point, given the coordinates, and find the coordinates, given a point. A bar graph or line graph shows the relationship between two things. The line graph below is from Example 3 in **Section 9.3.** It shows the relationship between the month of the year and the number of trout stocked in the Feather River.

TROUT STOCKED IN FEATHER RIVER

Each black dot on the graph represents a particular month paired with a particular number of trout. This is an example of **paired data.** We write each pair inside parentheses, with a comma separating the two items. To be consistent, we will always list the item on the *horizontal axis* first. In this case, the months are shown on the **horizontal axis** (the line that goes "left and right"), and the number of trout is shown along the **vertical axis** (the line that goes "up and down").

Paired Data from Line Graph on Trout Stocked in Feather River

(Apr, 40,000) (May, 30,000) (June, 55,000) (July, 60,000) (Aug, 25,000)

Each data pair gives you the location of a particular spot on the graph, and that spot is marked with a dot. This idea of paired data can be used to locate particular places on any flat surface.

Think of a small town laid out in a grid of square blocks, as shown below. To tell a taxi driver where to go, you could say, "the corner of 4th Avenue and 2nd Street" or just "4th and 2nd." As an *ordered pair,* it would be (4, 2). Of course, both you and the taxi driver need to know that the avenue is mentioned first (the number on the horizontal axis) and that the street is mentioned second (the number on the vertical axis). If the driver goes to (2, 4) instead, you'll be at the wrong corner.

Corner of 2nd Avenue and 4th Street; (2, 4) as an ordered pair

Corner of 4th Avenue and 2nd Street; (4, 2) as an ordered pair

Move along horizontal axis first.

Here are two examples of points plotted on the grid at the right.

(a) (3, 5)

Start at 0. Move *to the right* along the horizontal axis until you reach 3. Then move *up* 5 units so that you are aligned with 5 on the vertical axis. Make a dot. This is the plot, or graph, of the point (3, 5).

(b) (5, 3)

Start at 0. Move *to the right* along the horizontal axis until you reach 5. Then move *up* 3 units so that you are aligned with 3 on the vertical axis. Make a dot. This is the plot, or graph, of the point (5, 3).

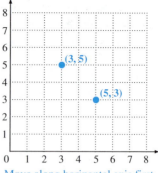

Move along horizontal axis first.

CAUTION
Notice that the points (3, 5) and (5, 3) are ***not*** the same. The "address" of a point is called an **ordered pair** because the *order* within the pair is important. Always move along the horizontal axis first.

You have been using both positive and negative numbers throughout this book. We can extend our grid system to include negative numbers, as shown below. The horizontal axis is now a number line with 0 at the center, positive numbers extending to the right and negative numbers to the left. This horizontal number line is called the ***x*-axis.**

The vertical axis is also a number line, with positive numbers extending upward from 0 and negative numbers extending downward from 0. The vertical axis is called the ***y*-axis.** Together, the *x*-axis and the *y*-axis form a rectangular **coordinate system.** The center point (0, 0) is where the *x*-axis crosses the *y*-axis; it is called the **origin.**

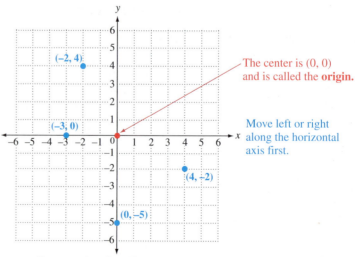

The center is (0, 0) and is called the **origin.**

Move left or right along the horizontal axis first.

Rectangular Coordinate System

EXAMPLE 1 **Plotting Points on a Rectangular Coordinate System**

Plot each point on the rectangular coordinate system shown at the bottom of the previous page.

(a) $(4, -2)$

Start at 0. Then move left or right along the horizontal x-axis first.

Because 4 is *positive,* move *to the right* until you reach 4. Now, because the 2 is *negative,* move *down* 2 units so that you are aligned with -2 on the y-axis. Make a dot and label it $(4, -2)$.

(b) $(-2, 4)$

Starting at 0, move left or right along the horizontal x-axis first. In this case, move *to the left* until you reach -2. Then move *up* 4 units. Make a dot and label it $(-2, 4)$. Notice that $(-2, 4)$ is ***not*** the same as $(4, -2)$.

(c) $(0, -5)$

Move left or right along the horizontal x-axis first. However, because the first number is 0, stay right at the center of the coordinate system. Then move *down* 5 units. Make a dot and label it $(0, -5)$.

(d) $(-3, 0)$

Starting at 0, move *to the left* along the horizontal x-axis to -3. Then, because the second number is 0, stay on -3. Do *not* move up or down. Make a dot and label it $(-3, 0)$.

> **NOTE**
> When the *first* number in an ordered pair is 0, the point is on the y-axis, as in Example 2(c) above. When the *second* number in an ordered pair is 0, the point is on the x-axis, as in Example 2(d) above.

Now Try Exercise 1.

We can use a coordinate system and an ordered pair to show the location of any point. The numbers in the ordered pair are called the **coordinates** of the point.

EXAMPLE 2 **Finding the Coordinates of Points**

Find the coordinates of points $A, B, C,$ and D.

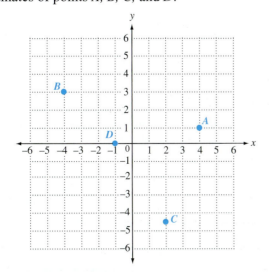

To reach point A from the origin, move 4 units *to the right;* then move *up* 1 unit. The coordinates are $(4, 1)$.

Continued on Next Page

To reach point *B* from the origin, move 4 units *to the left;* then move *up* 3 units. The coordinates are (−4, 3).

To reach point *C* from the origin, move 2 units *to the right;* then move *down* approximately $4\frac{1}{2}$ units. The approximate coordinates are $(2, -4\frac{1}{2})$.

To reach point *D* from the origin, move 1 unit *to the left;* then do *not* move either up or down. The coordinates are (−1, 0).

> **NOTE**
> If a point is between the lines on the coordinate system, you can use fractions to give the approximate coordinates. For example, the approximate coordinates of point *C* above are $(2, -4\frac{1}{2})$.

Now Try Exercise 5.

OBJECTIVE 2 Identify the four quadrants and determine which points lie within each one. The *x*-axis and *y*-axis divide the coordinate system into four regions, called **quadrants.** These quadrants are numbered with Roman numerals, as shown below. Points on the axes themselves are not in any quadrant.

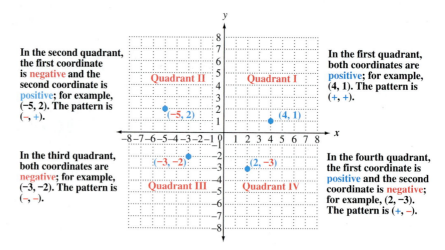

In the second quadrant, the first coordinate is negative and the second coordinate is positive; for example, (−5, 2). The pattern is (−, +).

In the first quadrant, both coordinates are positive; for example, (4, 1). The pattern is (+, +).

In the third quadrant, both coordinates are negative; for example, (−3, −2). The pattern is (−, −).

In the fourth quadrant, the first coordinate is positive and the second coordinate is negative; for example, (2, −3). The pattern is (+, −).

EXAMPLE 3 Working with Quadrants

(a) All points in the third quadrant are similar in what way? Give two examples of points in the third quadrant.

For all points in quadrant III, both coordinates are negative. The pattern is (−, −). There are many possible examples, such as (−2, −5) and (−4, −4). Just be sure that both numbers are negative.

(b) In which quadrant is each point located: (3, 5); (1, −6); (−4, 0)?

For (3, 5) the pattern is (+, +), so the point is in **quadrant I.**

For (1, −6) the pattern is (+, −), so the point is in **quadrant IV.**

The point corresponding to (−4, 0) is on the *x*-axis, so it isn't in any quadrant.

Now Try Exercise 7.

9.4 Exercises

FOR EXTRA HELP

 Addison-Wesley Math Tutor Center

 MathXL

 Digital Video Tutor CD 5 Videotape 9

Student's Solutions Manual

 MyMathLab

MathXL Tutorials on CD

Plot each point on the rectangular coordinate system. Label each point with its coordinates. See Example 1.

1. $(3, 7)$ $(-2, 2)$ $(-3, -7)$ $(2, -2)$ $(0, 6)$
$(6, 0)$ $(0, -4)$ $(-4, 0)$

2. $(5, 2)$ $(-3, -3)$ $(4, -1)$ $(-4, 1)$ $(-1, 0)$
$(0, 3)$ $(2, 0)$ $(0, -5)$

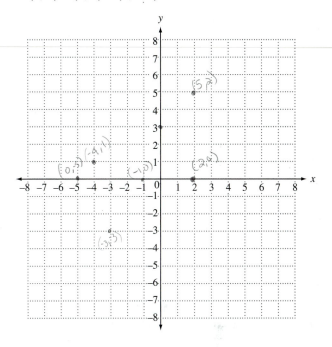

3. $(-5, 3)$ $(4, 4)$ $(-2\frac{1}{2}, 0)$ $(3, -5)$ $(0, 0)$
$(2, \frac{1}{2})$ $(-7, -5)$ $(-1, -6)$

4. $(1, 7)$ $(0, 3\frac{1}{2})$ $(-5, -1)$ $(6, -2)$ $(-2, 6)$
$(0, 0)$ $(-3, 3)$ $(-\frac{1}{2}, -2)$

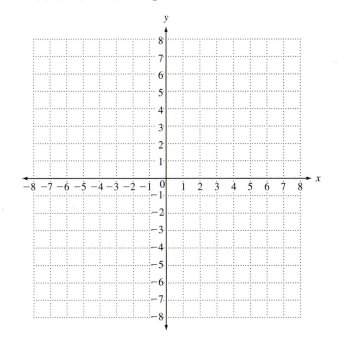

Give the coordinates of each point. See Example 2.

5.

6.

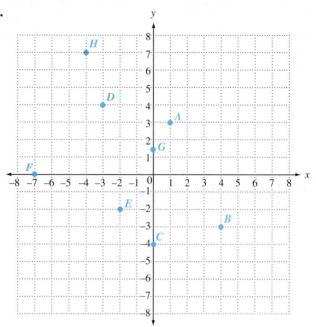

Identify the quadrant in which each point is located. See Example 3.

7. In which quadrant is each point located?
$(-3, -7)$ $(0, 4)$ $(10, -16)$ $(-9, 5)$

8. In which quadrant is each point located?
$(1, 12)$ $(20, -8)$ $(-5, 0)$ $(-14, 14)$

Complete each ordered pair with a number that will make the point fall in the specified quadrant.

9. (a) Quadrant II $(-4, \underline{\hspace{0.5cm}})$

(b) Quadrant IV $(7, \underline{\hspace{0.5cm}})$

(c) No quadrant $(\underline{\hspace{0.5cm}}, -2)$

(d) Quadrant III $(\underline{\hspace{0.5cm}}, -1\frac{1}{2})$

(e) Quadrant I $(3\frac{1}{4}, \underline{\hspace{0.5cm}})$

10. (a) Quadrant III $(-5, \underline{\hspace{0.5cm}})$

(b) Quadrant I $(\underline{\hspace{0.5cm}}, 3)$

(c) Quadrant IV $(\underline{\hspace{0.5cm}}, -\frac{1}{2})$

(d) No quadrant $(6, \underline{\hspace{0.5cm}})$

(e) Quadrant II $(\underline{\hspace{0.5cm}}, 1\frac{3}{4})$

11. Explain how to graph the ordered pair (a, b), where a and b are positive or negative integers.

12. Explain how to graph the ordered pair (a, b) where a is 0 and b is an integer. Explain how to graph (a, b) where a is an integer and b is 0.

9.5   Introduction to Graphing Linear Equations

In **Chapters 2–7** you solved equations that had only one variable, such as $2n - 3 = 7$ or $\frac{1}{3}x = 10$. Each of these equations had exactly one solution; n is 5 in the first equation, and x is 30 in the second equation. In other words, there was only *one* number that could replace the variable and make the equation balance. As you take more algebra courses, you will work with equations that have two variables and many different numbers that will make the equation balance. This section will get you started.

OBJECTIVE 1 Graph linear equations in two variables. Suppose that you have 6 hours of study time available during a weekend. You plan to study math and psychology. For example, you could spend 4 hours on math and then 2 hours on psychology, for a total of 6 hours. Or you could spend $1\frac{1}{2}$ hours on math and then $4\frac{1}{2}$ hours on psychology, for a total of 6 hours. Here is a list of *some* of the possible combinations.

Hours on Math	+	Hours on Psychology	=	Total Hours Studying
0	+	6	=	6
1	+	5	=	6
$1\frac{1}{2}$	+	$4\frac{1}{2}$	=	6
3	+	3	=	6
4	+	2	=	6
$5\frac{1}{2}$	+	$\frac{1}{2}$	=	6
6	+	0	=	6

We can write an equation to represent this situation.

$$\underset{m}{\text{hours studying math}} + \underset{p}{\text{hours studying psychology}} = \underset{6}{\text{total of 6 hours}}$$

This equation, $m + p = 6$, has *two* variables. The hours spent on math (m) can vary, and the hours spent on psychology (p) can vary.

As you can see, there is more than one solution for this equation. We can list possible solutions as *ordered pairs*. The first number in the pair is the value of m (math), and the second number in the pair is the corresponding value of p (psychology).

$$(m, p) \quad (m, p) \quad (m, p) \quad (m, p) \quad (m, p) \quad (m, p) \quad (m, p)$$

$$(0, 6) \quad (1, 5) \quad \left(1\frac{1}{2}, 4\frac{1}{2}\right) \quad (3, 3) \quad (4, 2) \quad \left(5\frac{1}{2}, \frac{1}{2}\right) \quad (6, 0)$$

Another way to show the solutions is to plot the ordered pairs, as you learned to do in **Section 9.4.** This method will give us a "picture" of the solutions that we listed.

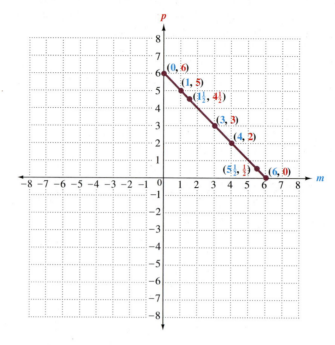

$m + p = 6$

The variables m and p represent hours, and hours can be 0 or positive numbers, but *not* negative numbers.

Notice that all the solutions (all the ordered pairs) lie on a straight line. When you draw a line connecting the ordered pairs, you have graphed the solutions. **Every point on the line is a solution.** You can use the line to find additional solutions besides the ones that we listed. For example, the point (5, 1) is on the line. This point tells you that another solution is 5 hours on math and 1 hour on psychology. The fact that the line is a *straight* line tells you that $m + p = 6$ is a *linear equation.* (The word *line* is part of the word *line*ar.) Later on in algebra you will work with equations whose solutions form a curve rather than a straight line when you graph them.

To draw the line for $m + p = 6$, we really needed only two solutions (two ordered pairs). But it's a good idea to use a third ordered pair as a check. If the three ordered pairs are *not* in a straight line, there is an error in your work.

Graphing a Linear Equation

To **graph a linear equation,** find at least three ordered pairs that satisfy the equation. Then plot the ordered pairs on a coordinate system and connect them with a straight line. *Every* point on the line is a solution of the equation.

EXAMPLE 1 **Graphing a Linear Equation**

Graph $x + y = 3$ by finding three solutions and plotting the ordered pairs. Then use the graph to find a fourth solution of the equation.

There are many possible solutions. Start by picking three different values for x. You can choose any numbers you like, but 0 and small numbers usually are easy to use. Then find the value of y that will make the sum equal to 3. Set up a table to organize the information.

Continued on Next Page

x	y	Check that $x + y = 3$	Ordered Pair (x, y)
0	3	$0 + 3 = 3$	$(0, 3)$
1	2	$1 + 2 = 3$	$(1, 2)$
2	1	$2 + 1 = 3$	$(2, 1)$

Plot the ordered pairs and draw a line through the points, extending it in both directions as shown below.

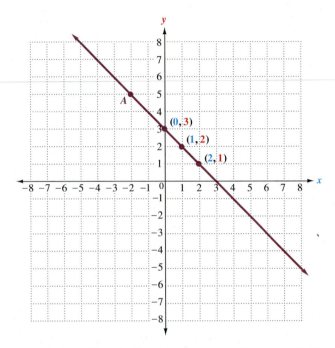

Now you can use the graph to find more solutions of $x + y = 3$. *Every* point on the line is a solution. Suppose that you pick **point A**. The coordinates are $(-2, 5)$.

To check that $(-2, 5)$ is a solution, substitute **-2** for **x** and **5** for **y** in the original equation.

$$x + y = 3 \qquad \text{Original equation}$$
$$-2 + 5 = 3$$
$$3 \quad = 3 \qquad \text{Balances}$$

The equation balances, so $(-2, 5)$ is another solution of $x + y = 3$.

NOTE
The line in Example 1 above was extended in both directions because *every* point on the line is a solution of $x + y = 3$. However, when we graphed the line for the hours spent studying, $m + p = 6$, we did *not* extend the line. That is because the variables m and p represented hours, and hours can only be 0 or positive numbers; all the solutions had to be in the first quadrant.

Now Try Exercise 1.

EXAMPLE 2 Graphing a Linear Equation

Graph $y = -2x$ by finding three solutions and plotting the ordered pairs. Then use the graph to find a fourth solution of the equation.

You can choose any three values for x, but small numbers such as 0, 1, and 2 are easy to use. Then $y = -2x$ tells you that y is -2 times the value of x.

$$y = -2x$$
$$y \text{ is } -2 \text{ times } x$$

First set up a table.

x	$y = -2 \cdot x$	Ordered Pair (x, y)
0	$-2 \cdot 0$ is 0	(0, 0)
1	$-2 \cdot 1$ is -2	(1, -2)
2	$-2 \cdot 2$ is -4	(2, -4)

Plot the ordered pairs and draw a line through the points. Be sure to draw arrows on both ends of the line to show that it continues in both directions.

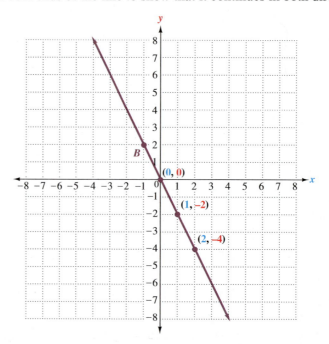

Now use the graph to find more solutions. *Every* point on the line is a solution.

Suppose that you pick **point B**. The coordinates are $(-1, 2)$. To check that $(-1, 2)$ is a solution, substitute **-1** for **x** and **2** for **y** in the original equation.

$$y = -2x \qquad \text{Original equation}$$
$$2 = -2(-1)$$
$$2 = 2 \qquad \text{Balances}$$

The equation balances, so $(-1, 2)$ is another solution of $y = -2x$.

Now Try Exercise 11.

EXAMPLE 3 **Graphing a Linear Equation**

Graph $y = \dfrac{1}{2}x$ by finding three solutions and plotting the ordered pairs.
Then use the graph to find a fourth solution of the equation.

Complete the table. The coefficient of x is $\frac{1}{2}$, so choose even numbers like 2, 4, and 6 as values for x because they are easy to divide in half. The equation $y = \frac{1}{2}x$ tells you that y is $\frac{1}{2}$ *times* the value of x.

x	$y = \frac{1}{2} \cdot x$	Ordered Pair (x, y)
2	$\frac{1}{2} \cdot 2$ is **1**	$(2, 1)$
4	$\frac{1}{2} \cdot 4$ is **2**	$(4, 2)$
6	$\frac{1}{2} \cdot 6$ is **3**	$(6, 3)$

Plot the ordered pairs and draw a line through the points.

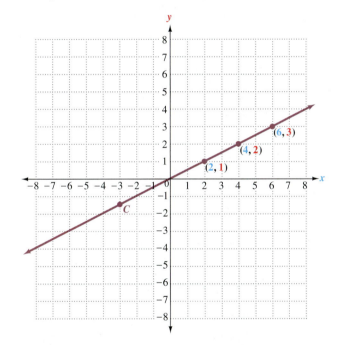

Now use the graph to find more solutions. *Every* point on the line is a solution.

Suppose that you pick **point C**. The coordinates are $(-3, -1\frac{1}{2})$. Check that $(-3, -1\frac{1}{2})$ is a solution by substituting -3 for x and $-1\frac{1}{2}$ for y.

$$y = \frac{1}{2}x \qquad \text{Original equation}$$

$$-1\frac{1}{2} = \frac{1}{2}(-3)$$

$$-1\frac{1}{2} = -\frac{3}{2} \qquad \text{Balances; } -1\frac{1}{2} \text{ is equivalent to } -\frac{3}{2}.$$

The equation balances, so $(-3, -1\frac{1}{2})$ is another solution of $y = \frac{1}{2}x$.

Now Try Exercise 15. ▶▶▶

EXAMPLE 4 **Graphing a Linear Equation**

Graph the equation $y = x + 4$ by finding three solutions and plotting the ordered pairs. Then use the graph to find two more solutions of the equation.

First set up a table. The equation $y = x + 4$ tells you that y must be 4 more than the value of x.

x	$y = x + 4$	Ordered Pair (x, y)
0	$0 + 4$ is 4	$(0, 4)$
1	$1 + 4$ is 5	$(1, 5)$
2	$2 + 4$ is 6	$(2, 6)$

Plot the ordered pairs and draw a line through the points.

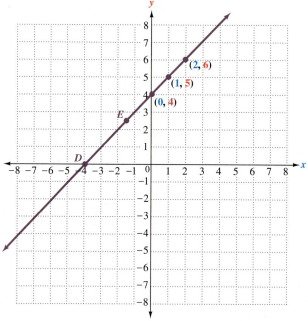

Now use the graph to find two more solutions. *Every* point on the line is a solution. Suppose that you pick **point D** at $(-4, 0)$ and **point E** at $\left(-1\frac{1}{2}, 2\frac{1}{2}\right)$. Check that both ordered pairs are solutions.

Check $(-4, 0)$

$$y = x + 4$$
$$0 = \underbrace{-4 + 4}$$
$$0 = \quad 0 \quad \text{Balances}$$

Check $\left(-1\frac{1}{2}, 2\frac{1}{2}\right)$

$$y = x + 4$$
$$2\frac{1}{2} = -1\frac{1}{2} + 4$$
$$\frac{5}{2} = \underbrace{-\frac{3}{2} + \frac{8}{2}}$$
$$\frac{5}{2} = \quad \frac{5}{2} \quad \text{Balances}$$

Both equations balance, so $(-4, 0)$ and $\left(-1\frac{1}{2}, 2\frac{1}{2}\right)$ are also solutions of $y = x + 4$.

Now Try Exercise 7.

OBJECTIVE **2** **Identify the slope of a line as positive or negative.**
Let's look again at some of the lines that we graphed for various equations.
All are straight lines, but some are almost flat and some tilt steeply upward
or downward.

Graph from Example 1: $x + y = 3$

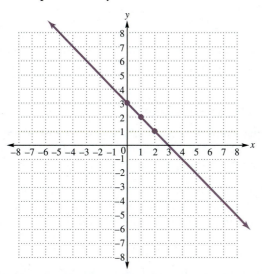

As you move from *left to right,* the line slopes downward, as if you were walking down a hill. When a line tilts downward, we say that it has a *negative slope*.
   Now look at the table of solutions we used to draw the line.

The value of	x	y	The value of
x is *increasing*	**0**	**3**	y is *decreasing*
from 0 to 1 to 2	**1**	**2**	from 3 to 2 to 1
	2	**1**	

As the value of *x increases,* the value of *y* does the *opposite*—it *decreases.*
Whenever one variable increases while the other variable decreases, the line
will have a negative slope.

Graph from Example 3: $y = \dfrac{1}{2}x$

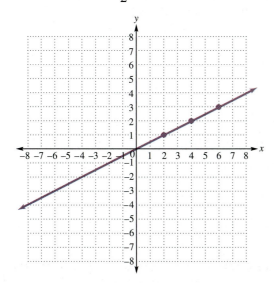

As you move from *left to right,* this line slopes upward, as if you were walking up a hill. When a line tilts upward, we say that it has a *positive slope.*

Now look at the table of solutions we used to draw the line.

The value of
x is *increasing*
from 2 to 4 to 6

x	y
2	1
4	2
6	3

The value of
y is *increasing*
from 1 to 2 to 3

As the value of *x* increases, the value of *y* does the *same* thing—it also *increases*. Whenever both variables do the same thing (both increase or both decrease), the line will have a positive slope.

Positive and Negative Slopes

As you move from left to right, a line with a *positive* slope tilts *upward* or rises. As the value of one variable increases, the value of the other variable also increases (does the same).

As you move from left to right, a line with a *negative* slope tilts *downward* or falls. As the value of one variable increases, the value of the other variable decreases (does the opposite).

EXAMPLE 5   **Identifying Positive or Negative Slope in a Line**

Look back at the graph of $y = -2x$ in Example 2. Then complete these sentences.

The graph of $y = -2x$ has a _____ slope.

As the value of *x* increases, the value of *y* _____.

The graph of $y = -2x$ has a <u>negative</u> slope (because it tilts downward).

As the value of *x* increases, the value of *y* <u>decreases</u> (does the opposite).

◀◀◀ **Now Try Exercises 13(a) and 13(c).**

9.5 Exercises

FOR EXTRA HELP

 Addison-Wesley Math Tutor Center

 MathXL

 Digital Video Tutor CD 5 Videotape 9

Student's Solutions Manual

MyMathLab

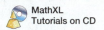 MathXL Tutorials on CD

Graph each equation by completing the table to find three solutions and plotting the ordered pairs. Then use the graph to find two *other solutions. See Example 1.*

1. $x + y = 4$

x	y	Ordered Pair (x, y)
0		
1		
2		

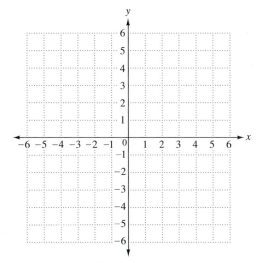

Two other solutions are (__, __) and (__, __).

2. $x + y = -4$

x	y	Ordered Pair (x, y)
0		
1		
2		

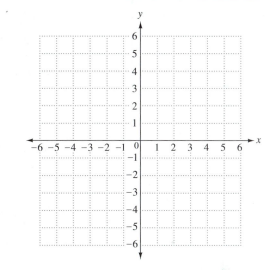

Two other solutions are (__, __) and (__, __).

3. $x + y = -1$

x	y	Ordered Pair (x, y)
0		
1		
2		

Two other solutions are (__, __) and (__, __).

4. $x + y = 1$

x	y	Ordered Pair (x, y)
0		
1		
2		

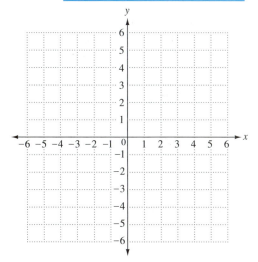

Two other solutions are (__, __) and (__, __).

5. The line in Exercise 1 crosses the y-axis at what point? _____. The line in Exercise 3 crosses the y-axis at what point? _____. Based on these examples, where would the graph of $x + y = -6$ cross the y-axis? _____. Where would the graph of $x + y = 99$ cross the y-axis? _____

6. Look at where the line crosses the x-axis and where it crosses the y-axis in Exercises 2 and 4. What pattern do you see?

Graph each equation. Make your own table using the listed values of x. See Examples 2 and 4.

7. $y = x - 2$

Use 1, 2, and 3 as the values of x.

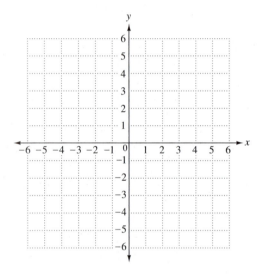

8. $y = x + 1$

Use 1, 2, and 3 as the values of x.

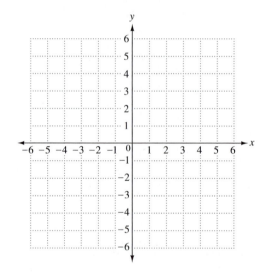

9. $y = x + 2$

Use 0, -1, and -2 as the values of x.

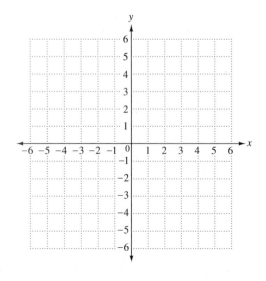

10. $y = x - 1$

Use 0, -1, and -2 as the values of x.

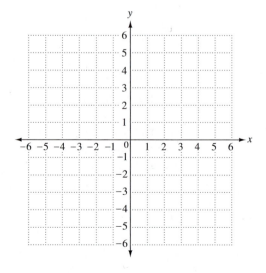

11. $y = -3x$

Use 0, 1, and 2 as the values of x.

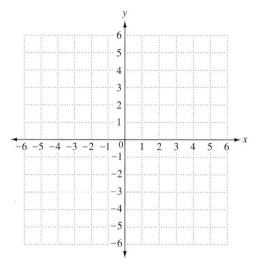

12. $y = -2x$

Use 0, 1, and 2 as the values of x.

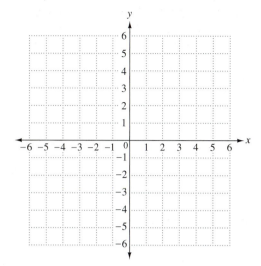

13. Look back at the graphs in Exercises 7, 9, and 11. Then complete these sentences. See Example 5.

   (a) The graph of $y = x - 2$ in Exercise 7 has a
_____ slope. As the value of x increases, the value of y _____.

   (b) The graph of $y = x + 2$ in Exercise 9 has a
_____ slope. As the value of x decreases, the value of y _____.

   (c) The graph of $y = -3x$ in Exercise 11 has a
_____ slope. As the value of x increases, the value of y _____.

14. Look back at the graphs in Exercises 8, 10, and 12. Then complete these sentences. See Example 5.

   (a) The graph of $y = x + 1$ in Exercise 8 has a
_____ slope. As the value of x increases, the value of y _____.

   (b) The graph of $y = x - 1$ in Exercise 10 has a
_____ slope. As the value of x decreases, the value of y _____.

   (c) The graph of $y = -2x$ in Exercise 12 has a
_____ slope. As the value of x increases, the value of y _____.

Graph each equation. Make your own table using the listed values of x. See Examples 1–4.

15. $y = \dfrac{1}{3}x$

Use 0, 3, and 6 as the values of x.

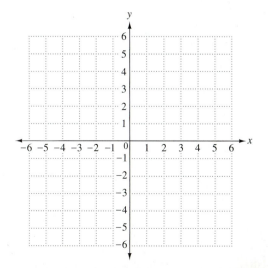

16. $y = \dfrac{1}{2}x$

Use 0, 2, and 4 as the values of x.

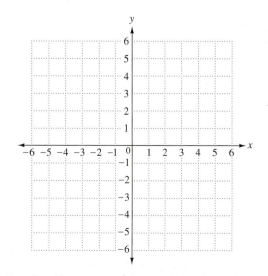

17. $y = x$

Use -1, -2, and -3 as the values of x.

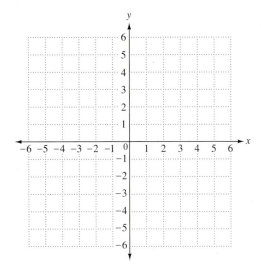

18. $x + y = 0$

Use 1, 2, and 3 as the values of x.

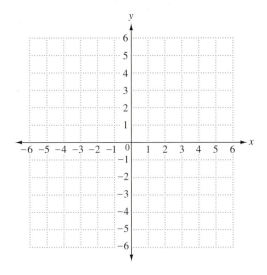

19. $y = -2x + 3$

Use 0, 1, and 2 as the values of x.

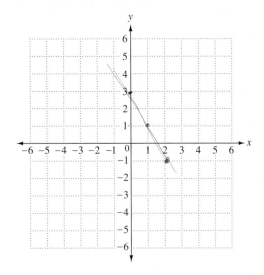

20. $y = 3x - 4$

Use 0, 1, and 2 as the values of x.

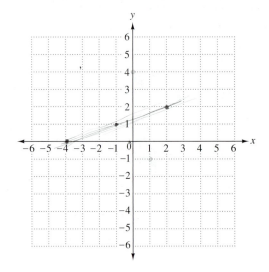

Graph each equation. Choose three values for x. Make a table showing your x values and the corresponding y values. After graphing the equation, state whether the line has a positive or negative slope. See Examples 1–5.

21. $x + y = -3$

22. $x + y = 2$

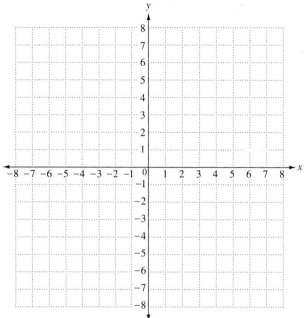

23. $y = \dfrac{1}{4}x$ (*Hint:* Try using 0 and multiples of 4 as the values of x.)

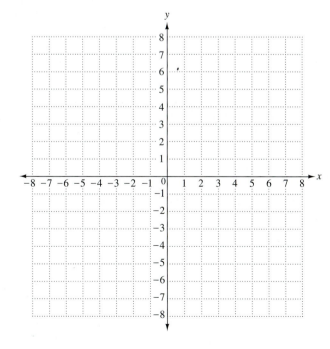

24. $y = -\dfrac{1}{3}x$ (*Hint:* Try using 0 and multiples of 3 as the values of x.)

25. $y = x - 5$

26. $y = x + 4$

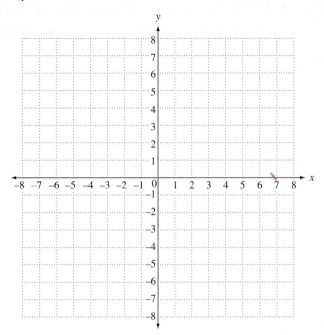

27. $y = -3x + 1$

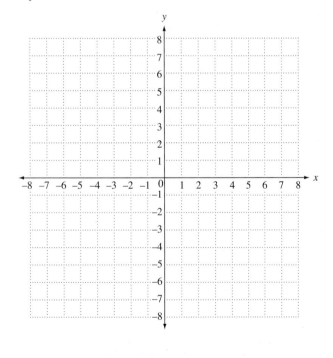

28. $y = 2x - 2$

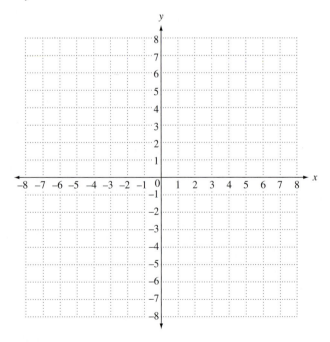

Chapter 9 **Group Activity: SURFING THE NET**

Objective   **Interpret a bar graph; use percents; collect data and construct a bar graph.**

1. Look at the "Source" information at the bottom of the graph. How were the numbers in the graph obtained?

CAUGHT IN THE NET

Web users have cut back on the following activities to get more online time:

Activity	Percent
Television	52%
Video/computer games	36%
Sleeping	18%
Reading	12%
Seeing friends	12%
Work/school	11%
Other	10%
Exercising	3%

Note: Respondents could choose more than one activity.

2. A researcher seeks information about members of a *population*. The individuals who are polled must be representative of the *population*. Describe the population that was targeted by this survey.

Source: NUKE InterNETWORK poll of 500 regular users.

3. How many people in the poll said they cut back on television viewing to find time to use the Internet?

4. Find the number of people in the poll who cut back on each of the other activities listed in the graph.

5. Add up all the responses to the poll from Problems 3 and 4. Why is the total more than the 500 people that were in the poll?

6. Suppose you took a similar poll of 100 students at your school. Would you expect the results to be similar to those shown in the graph? Why or why not?

7. Suppose you first asked students if they regularly used the Internet, and then took a similar poll of 100 of those students. Would you expect the results to be similar to those shown in the graph? Why or why not?

8. Conduct a survey of your class members. First find out if they regularly use the Internet. Ask those who regularly use the Internet which of the activities they cut back on to have more surfing time. Each person polled can select more than one activity.

(a) How many students are in your class poll? _____

(b) Complete the table below using the responses from those who regularly use the Internet.

Activity	Number Who Cut Back on the Activity	Percent Who Cut Back on the Activity
Television		
Video/computer games		
Sleeping		
Reading		
Seeing friends		
Work/school		
Other		
Exercising		

9. Make a bar graph showing your survey data.

10. How is your data similar to the graph shown on the previous page? How is it different?

		KEY TERMS
9.1	**table**	A table presents data organized into rows and columns.
	pictograph	A pictograph uses symbols or pictures to represent various amounts.
9.2	**circle graph**	A circle graph shows how a total amount is divided into parts or sectors. It is based on percents of 360°.
	protractor	A protractor is a tool (usually in the shape of a half-circle) used to measure the number of degrees in an angle or part of a circle.
9.3	**bar graph**	A bar graph uses bars of various heights to show quantity or frequency.
	double-bar graph	A double-bar graph compares two sets of data by showing two sets of bars.
	line graph	A line graph uses dots connected by line segments to show trends.
	comparison line graph	A comparison line graph shows how two or more sets of data relate to each other by showing a line graph for each set of data.
9.4	**paired data**	When each number in a set of data is matched with another number by some rule of association, we call it paired data.
	horizontal axis	The horizontal axis is the number line in a coordinate system that goes "left and right."
	vertical axis	The vertical axis is the number line in a coordinate system that goes "up and down."
	ordered pair	An ordered pair is the "address" of a point in a coordinate system. The *order* of the numbers is important. The first number tells how far to move left or right from 0 along the horizontal axis. The second number tells how far to move up or down from 0 along the vertical axis.
	***x*-axis**	The horizontal axis is called the *x*-axis.
	***y*-axis**	The vertical axis is called the *y*-axis.
	coordinate system	Together, the *x*-axis and the *y*-axis form a rectangular coordinate system. *Example:* See figure at the right.
	origin	The center point of a rectangular coordinate system is (0, 0) and is called the origin. *Example:* See red dot marking (0, 0) in figure at the right.
	coordinates	Coordinates are the numbers in the ordered pair that specify the location of a point on a rectangular coordinate system. *Example:* See the point $(-4, 2)$ at the right.
	quadrants	The *x*-axis and the *y*-axis divide the coordinate system into four regions called quadrants; they are designated with Roman numerals. *Example:* The point $(-4, 2)$ at the right is in quadrant II.
9.5	**graph a linear equation**	All the solutions of a linear equation (all the ordered pairs that satisfy the equation) lie along a straight line. When you draw the line, you have graphed the equation.

		NEW SYMBOLS
(x, y)	ordered pair	

TEST YOUR WORD POWER

See how well you have learned the vocabulary in this chapter. Answers follow the Quick Review.

1. A **circle graph**
   - A. uses symbols to represent various amounts
   - B. is useful for showing trends
   - C. shows how a total amount is divided into parts
   - D. compares two sets of data.

2. A **line graph**
   - A. uses symbols to represent various amounts
   - B. is useful for showing trends
   - C. shows how a total amount is divided into parts
   - D. compares two sets of data.

3. A **protractor** is used to
   - A. measure degrees in an angle
   - B. draw circles of various sizes
   - C. measure lengths
   - D. calculate circumference.

4. Two sets of data can be compared using a
   - A. circle graph
   - B. pictograph
   - C. line graph
   - D. double-bar graph.

5. The **x-axis** in a rectangular coordinate system is the
   - A. number line that goes "up and down"
   - B. number line that goes "left and right"
   - C. center point of the grid
   - D. vertical axis.

6. **Quadrants** are the
   - A. numbers in an ordered pair
   - B. solutions of a linear equation
   - C. four regions in a coordinate system
   - D. paired data shown on a graph.

7. **Coordinates** are
   - A. points in a straight line on a coordinate system
   - B. designated with Roman numerals
   - C. the solutions of a linear equation
   - D. numbers used to locate a point in a coordinate system.

8. A **rectangular coordinate system**
   - A. is formed by the x-axis and y-axis
   - B. is divided into eight quadrants
   - C. is the solution of a linear equation
   - D. has only positive numbers.

QUICK REVIEW

Concepts

Examples

9.1 *Reading a Table*

The data in a table is organized into rows and columns. As you read from left to right along each row, check the heading at the top of each column.

Use the table below to answer the question.

PER CAPITA CONSUMPTION OF SELECTED BEVERAGES IN GALLONS

	1985	1990	1995	2000
Milk	26.7	25.7	24.3	22.6
Coffee	27.4	26.9	20.5	26.3
Bottled water	4.5	8.0	11.6	17.7
Soft drinks	35.7	46.2	47.5	49.3

Source: U.S. Dept. of Agriculture Economic Research Service.

In which years was the consumption of milk greater than the consumption of coffee?

Read across the rows labeled "milk" and "coffee" from left to right, comparing the numbers in each column. In 1995 the figure for milk (24.3 gallons) is greater than for coffee (20.5 gallons).

Concepts	Examples

9.1 *Reading a Pictograph*

A pictograph uses symbols or pictures to represent various amounts. The *key* tells you how much each symbol represents. A fractional part of a symbol represents a fractional part of the symbol's value.

Use the pictograph to answer these questions.

SALES OF FOOD AND DRINK AT U.S. RESTAURANTS

1985
1990
1995
2000

= $50 billion

Source: National Restaurant Association.

(a) How much was spent at U.S. restaurants in 1985?

Sales for 1985 are represented by 3 whole symbols (3 • $50 billion = $150 billion) plus half of a symbol ($\frac{1}{2}$ of $50 billion = $25 billion) for a total of $175 billion.

(b) How much did restaurant expenditures increase from 1985 to 2000?

The year 2000 shows four more symbols than 1985, so the increase is 4 • $50 billion = $200 billion.

9.2 *Constructing a Circle Graph*

Step 1   Determine the percent of the total for each item.

Step 2   Find the number of degrees out of 360° that each percent represents.

Step 3   Use a protractor to measure the number of degrees for each item in the circle.

Step 4   Label each sector in the circle with the item name and percent of total for that item.

Construct a circle graph for these expenses from a business trip.

Item	Amount
Transportation	$200
Lodging	$300
Food	$250
Entertainment	$150
Other	$100
Total	**$1000**

9.2 *Constructing a Circle Graph (continued)*

Item	Amount	Percent of Total		Sector Size
Transportation	$200	$\dfrac{\$200}{\$1000} = \dfrac{1}{5} = \mathbf{20\%}$	so 20% of 360° $= (0.20)(360)$	$= 72°$
Lodging	$300	$\dfrac{\$300}{\$1000} = \dfrac{3}{10} = \mathbf{30\%}$	so 30% of 360° $= (0.30)(360)$	$= 108°$
Food	$250	$\dfrac{\$250}{\$1000} = \dfrac{1}{4} = \mathbf{25\%}$	so 25% of 360° $= (0.25)(360)$	$= 90°$
Entertainment	$150	$\dfrac{\$150}{\$1000} = \dfrac{3}{20} = \mathbf{15\%}$	so 15% of 360° $= (0.15)(360)$	$= 54°$
Other	$100	$\dfrac{\$100}{\$1000} = \dfrac{1}{10} = \mathbf{10\%}$	so 10% of 360° $= (0.10)(360)$	$= 36°$

BUSINESS TRIP EXPENSES

Lodging **30%** (108°)
Transportation **20%** (72°)
Food **25%** (90°)
Other **10%** (36°)
Entertainment **15%** (54°)

9.3 *Reading a Bar Graph*

The height of the bar is used to show the quantity or frequency (number) in a specific category. Use a ruler or straightedge to line up the top of each bar with the numbers on the left side of the graph.

Use the bar graph below to determine the number of students who earned each letter grade.

STUDENT LETTER GRADES

Number of Students

Grade

The number of students earning each letter grade were
A: 3 students; B: 7 students; C: 4 students; D: 2 students.

Concepts	Examples

9.3 *Reading a Line Graph*

A dot is used to show the number or quantity in a specific class. The dots are connected with line segments. This kind of graph is used to show a trend.

The line graph below shows the annual sales for the Fabric Supply Center for each of four years.

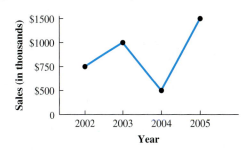

Which year had the lowest sales? What was the amount of sales that year?

2004; sales were $500 • 1000 = $500,000

9.4 *Plotting Points*

Start at the center of the coordinate system (the origin). The first number in an ordered pair tells you how far to move *left* or *right* along the horizontal axis; *positive* numbers are to the *right,* and *negative* numbers are to the *left.* The second number in an ordered pair tells you how far to move *up* or *down; positive* numbers are *up,* and *negative* numbers are *down.*

To plot $(3, -2)$, start at 0, move to the *right* 3 units, and then move *down* 2 units.

To plot $(-2, 3)$, start at 0, move to the *left* 2 units, and then move *up* 3 units.

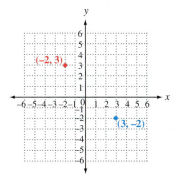

9.4 *Identifying Quadrants*

The *x*-axis and *y*-axis divide the coordinate system into four regions called quadrants. The quadrants are designated with Roman numerals.

Points in the first quadrant fit the pattern $(+, +)$.

Points in the second quadrant fit the pattern $(-, +)$.

Points in the third quadrant fit the pattern $(-, -)$.

Points in the fourth quadrant fit the pattern $(+, -)$.

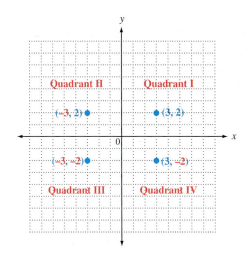

Concepts

9.5 *Graphing Linear Equations*

Choose any three values for x. Then find the corresponding values of y. Plot the three ordered pairs on a coordinate system. Draw a line through the points, extending it in both directions. If the three points do *not* lie on a straight line, there is an error in your work. *Every point on the line* is a solution of the given equation.

Examples

Graph $y = -2x$.

x	$y = -2 \cdot x$	Ordered Pair (x, y)
1	$-2 \cdot 1$ is -2	$(1, -2)$
2	$-2 \cdot 2$ is -4	$(2, -4)$
3	$-2 \cdot 3$ is -6	$(3, -6)$

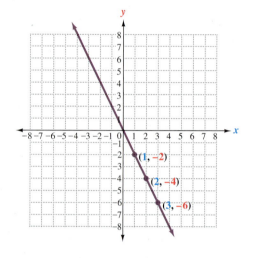

---

1. C; *Example:* A circle graph can show how a 24-hour day is divided among various activities.

2. B; *Example:* A line graph can show changes in the amount of sales over several months or several years.

3. A; *Example:* A protractor is a tool, usually in the shape of a half-circle, used to measure or draw angles of a certain number of degrees. A tool called a compass is used to draw circles of various sizes.

4. D; *Example:* Two sets of bars in different colors can compare monthly unemployment figures for two different years.

5. B; *Example:* The x-axis is the horizontal number line with 0 at the center, negative numbers extending to the left, and positive numbers extending to the right.

6. C; *Example:* The x-axis and the y-axis divide the coordinate system into four regions; each region is designated by a Roman numeral.

7. D; *Example:* To locate the point $(2, -3)$, start at the origin and move 2 units to the right along the x-axis, then move down 3 units.

8. A; *Example:* Together, a horizontal number line (x-axis) and vertical number line (y-axis) form a rectangular coordinate system.

Chapter **9**
R E V I E W   E X E R C I S E S

[9.1] *Use the table at the right to answer Exercises 1–4.*

1. Which sport had the
   (a) fewest men's teams?

   (b) second-greatest number of women's teams?

2. Which sport had
   (a) about 10,000 female athletes?

   (b) about 7000 male athletes?

PARTICIPATION IN SELECTED NCAA SPORTS

Sport	Males			Females		
	Teams	Athletes	Average Squad	Teams	Athletes	Average Squad
Basketball	950	15,141	15.9	966	13,392	13.9
Cross country	792	10,271	13.0	838	10,141	12.1
Golf	678	7197	10.6	282	2323	8.2
Gymnastics	28	413	14.8	91	1311	14.4
Volleyball	74	1052		923	12,284	

Source: The National Collegiate Athletic Association (NCAA).

3. (a) How many more men participated in basketball than cross country?

   (b) How many fewer women participated in gymnastics than basketball?

4. Find the average squad size for men's and women's volleyball teams and complete the table. Round to the nearest tenth.

Use the pictograph to answer Exercises 5–8.

5. What is the average yearly snowfall in
   (a) Juneau?
   (b) Washington, D.C.?

6. What is the average yearly snowfall in
   (a) Minneapolis?
   (b) Cleveland?

7. Find the difference in the average yearly snowfall between
   (a) Buffalo and Cleveland.
   (b) Memphis and Minneapolis.

8. Find the difference in the average yearly snowfall between the city with the greatest amount and the city with the least amount.

AVERAGE YEARLY SNOWFALL IN SELECTED U.S. CITIES

Buffalo, New York

Cleveland, Ohio

Juneau, Alaska

Memphis, Tennessee

Minneapolis, Minnesota

Washington, D.C.

= 10 inches

Source: National Climatic Data Center.

[9.2] *Use this circle graph for Exercises 9–13.*

9. **(a)** What was the largest single expense of the vacation? How much was that item?

   (b) What was the second-most-expensive item? How much was that item?

   (c) What was the total cost of the vacation?

COST OF A FAMILY VACATION

Use the family vacation circle graph to find each ratio. Write the ratio as a fraction in lowest terms.

10. Cost of lodging to the cost of food

11. Cost of gasoline to the total cost of the vacation

12. Cost of sightseeing to the total cost of the vacation

13. Cost of gasoline to the cost of the Other category

[9.3] *This bar graph shows the top six home improvement projects that people do themselves. Use the graph to answer Exercises 14–19.*

CAN YOU FIX IT?

Projects homeowners take on themselves according to a recent survey:

Painting, wallpapering — 63%
Landscaping, gardening — 54%
Interior decorating — 51%
Carpentry — 43%
Window treatments — 34%
Construction work — 33%

Note: Multiple responses allowed.
Source: American Express Home Improvement Telephone Survey.

14. What is the most popular project? What percent of the homeowners in the survey gave that answer?

15. Which project was selected the least? What percent of the homeowners in the survey gave that answer?

16. What percent of the homeowners in the survey selected carpentry projects? There were 341 people surveyed. How many of them selected carpentry, to the nearest whole number?

17. Of the 341 homeowners surveyed, how many selected landscaping or gardening projects? Round to the nearest whole number.

18. Name the project selected by about **(a)** $\frac{1}{2}$ of the homeowners **(b)** $\frac{1}{3}$ of the homeowners.

19. Give two reasons painting and wallpapering might be selected so much more than construction work.

This double-bar graph shows the number of acre-feet of water in the Lake Natoma reservoir for each of the first six months of 2004 and 2005. Use this graph to answer Exercises 20–25.

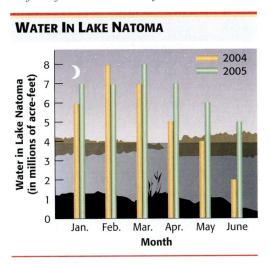

WATER IN LAKE NATOMA

20. From January through June 2005, which month had the greatest amount of water in the lake? How much was there?

21. During which month in the first half of 2004 was the least amount of water in the lake? How much was there?

22. How many acre-feet of water were in the lake in June of 2005?

23. How many acre-feet of water were in the lake in January of 2004?

24. Find the amount of decrease and the percent of decrease in the acre-feet of water in the lake from March 2005 to June 2005.

25. Find the amount of decrease and the percent of decrease in the acre-feet of water in the lake from April 2004 to June 2004.

This comparison line graph shows the annual floor-covering sales of two different home improvement centers during each of five years. Use this graph to find the amount of sales in each year listed in Exercises 26–29 and to answer Exercises 30 and 31.

ANNUAL FLOOR-COVERING SALES

26. Center A in 2002

27. Center A in 2004

28. Center B in 2003

29. Center B in 2005

30. What trend do you see in Center A's sales from 2002 to 2005? Why might this have happened?

31. What trend do you see in Center B's sales starting in 2003? Why might this have happened?

[9.2] *The Broadway Hair Salon spent a total of $22,400 to open a new shop. The breakdown of expenditures for various items is shown. Find all the missing numbers in Exercises 32–36.*

Item	Dollar Amount	Percent of Total	Degrees of Circle
32. Plumbing and electrical changes	$2240	10%	_____
33. Work stations	$7840	_____	_____
34. Small appliances	$4480	_____	_____
35. Interior decoration	$5600	_____	_____
36. Supplies	_____	_____	36°

37. Draw a circle graph using the information in Exercises 32–36. Label each sector with the item and the percent of total for that item.

[9.4] *In Exercise 38, plot each point on the rectangular coordinate system and label it with its coordinates. In Exercise 39, give the coordinates of each point.*

38. $(1, 7)$ $(-1\frac{1}{2}, 0)$ $(-4, -2)$ $(0, 3)$ $(2, -\frac{1}{2})$ **39.**
  $(-7, 1)$ $(5, -5)$ $(0, -4)$

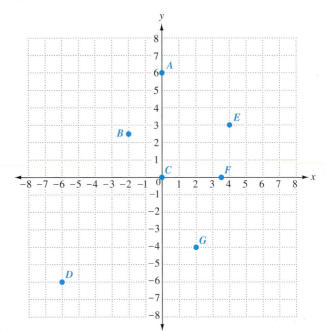

[9.5] *Graph each equation. Complete the table using the listed values of x. State whether each line has a positive or negative slope. Use the graph to find two other solutions of the equation.*

40. $x + y = -2$

Use 0, 1, and 2 as the values of x.

x	y	(x, y)

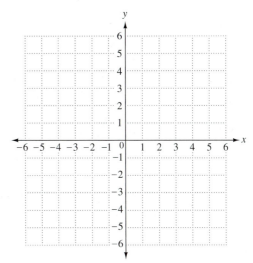

The graph of $x + y = -2$ has a _____ slope.

Two other solutions of $x + y = -2$ are

(____, ____) and (____, ____).

41. $y = x + 3$

Use 0, 1, and 2 as the values of x.

x	y	(x, y)

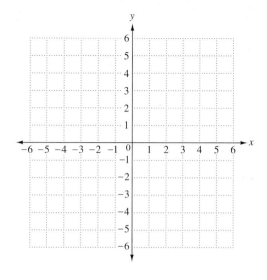

The graph of $y = x + 3$ has a _____ slope.

Two other solutions for $y = x + 3$ are

(_____, _____) and (_____, _____).

42. $y = -4x$

Use -1, 0, and 1 as the values of x.

x	y	(x, y)

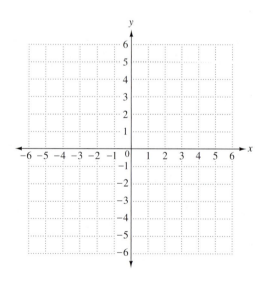

The graph of $y = -4x$ has a _____ slope.

Two other solutions of $y = -4x$ are

(_____, _____) and (_____, _____).

Chapter **9**

T E S T

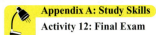

Appendix A: Study Skills
Activity 12: Final Exam

Use this table to answer Problems 1–3.

AMOUNT OF CALCIUM IN SELECTED FOODS

Food	Measure	Calories	Calcium (mg)
Swiss cheese	1 oz	95	219
Cream cheese	1 oz	100	23
Yogurt, fruit flavor	8 oz	230	345
Skim milk	1 cup	85	302
Sardines	3 oz	175	371

Source: U.S. Department of Agriculture.

1. Which food has **(a)** the greatest amount of calcium? **(b)** the least amount of calcium?

   1. (a) _____

      (b) _____

2. A packet of Swiss cheese contains 1.75 oz. How many calories are in that amount of Swiss cheese? Round to the nearest whole number.

   2. _____

3. Find the amount of calcium in a 6 oz container of fruit-flavored yogurt. Round to the nearest whole number.

   3. _____

Use this pictograph to answer Problems 4–6.

NUMBER OF U.S. ENDANGERED WILDLIFE SPECIES

Mammals

Birds

Reptiles

Fish

Insects

= 10 Species

Source: U.S. Fish and Wildlife Service.

4. **(a)** Which group has the greatest number of endangered species?
   (b) How many endangered species are in that group?

   4. (a) _____

      (b) _____

5. How many more species of fish are endangered than reptiles?

   5. _____

6. What is the total number of endangered species shown in the pictograph?

   6. _____

This circle graph shows the advertising budget for Lakeland Amusement Park. Find the dollar amount budgeted for each category listed in Exercises 7–10. The total advertising budget is $2,800,000.

7. _____

7. Which form of advertising has the largest budget? What amount is this?

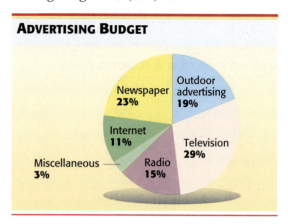

ADVERTISING BUDGET

Newspaper 23%
Outdoor advertising 19%
Internet 11%
Television 29%
Miscellaneous 3%
Radio 15%

8. _____

8. Which form of advertising has the smallest budget? How much is this?

9. _____

9. How much is budgeted for Internet advertising?

10. _____

10. How much is budgeted for newspaper ads?

This graph shows one student's income and expenses for four years.

11. _____

11. In what year did the student's expenses exceed income? By how much?

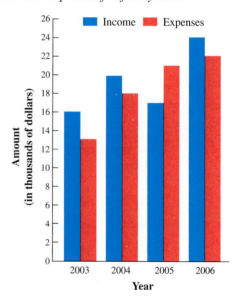

12. _____

12. Find the amount of increase and the percent of increase in the student's expenses from 2003 to 2004. Round to the nearest whole percent.

13. _____

13. In what year did the student's income decline? Give two possible explanations for the decline.

This graph shows enrollment at two community colleges.

14. _____

14. What was College A's enrollment in 2003? What was College B's enrollment in 2004?

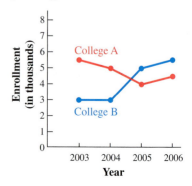

15. _____

15. Which college had higher enrollment in 2006? How much higher?

16. _____

16. Give two possible explanations for the fact that College B's enrollment passed College A's.

During a one-year period, Oak Mill Furniture Sales had a total of $480,000 in expenses. Find all the numbers missing from the table.

Item	Dollar Amount	Percent of Total	Degrees of a Circle	
17. Salaries	$168,000	_____	_____	17. _____
18. Delivery expense	$24,000	_____	_____	18. _____
19. Advertising	$96,000	_____	_____	19. _____
20. Rent	$144,000	_____	_____	20. _____
21. Other	_____	_____	36°	21. _____

22. Draw a circle graph using a protractor and the information in Problems 17–21. Label each sector with the item and the percent of total for that item.

22.

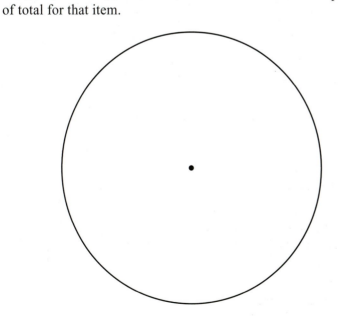

Plot each point on the coordinate system below. Label each point with its coordinates.

23. $(-5, 3)$

24. $(1, -4)$

25. $(0, 6)$

26. $(2, 0)$

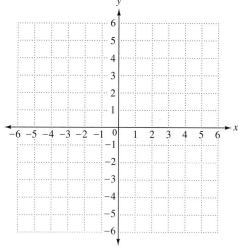

Give the coordinates of each lettered point shown below, and state which quadrant the point is in.

27. _____

28. _____

29. _____

30. _____

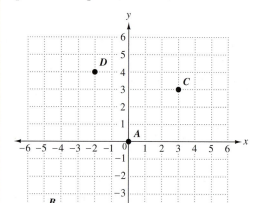

27. Point A

28. Point B

29. Point C

30. Point D

31. Graph $y = x - 4$ on the coordinate system below. Make your own table using 0, 1, and 2 as the values of x.

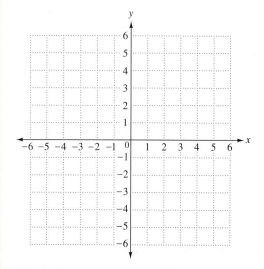

32. (a) _____

    (b) _____

32. (a) Use the graph in Problem 31 to find *two* other solutions of $y = x - 4$.

    (b) State whether the graph of $y = x - 4$ has a positive or negative slope.

Cumulative Review Exercises

CHAPTERS 1-9

1. Write these numbers in words.

   (a) 0.0602

   (b) 300,000,560

2. Write these numbers in digits.

   (a) Seventy billion, five million, forty-three

   (b) Eighteen and nine hundredths

3. Round each number as indicated.

   (a) 3.049 to the nearest tenth

   (b) 0.7982 to the nearest hundredth

   (c) 68,592,000 to the nearest million

4. Find the mean, median, and mode for this set of data on student ages in years: 23, 29, 18, 23, 36, 62, 23, 19, 27, 30.

Simplify.

5. $\dfrac{4}{5} + 2\dfrac{1}{3}$

6. $\dfrac{0.8}{-3.2}$

7. $5^2 + (-4)^3$

8. $(0.002)(-0.05)$

9. $\dfrac{4a}{9} \cdot \dfrac{6b}{2a^3}$

10. $1\dfrac{1}{4} - 3\dfrac{5}{6}$

11. $-13 + 2.993$

12. $\dfrac{2}{7} - \dfrac{8}{x}$

13. $-3 - 33$

14. $\dfrac{10m}{3n^2} \div \dfrac{2m^2}{5n}$

15. $\dfrac{-3}{-\dfrac{9}{10}}$

16. $\dfrac{3(-7)}{2^4 - 16}$

17. $10 - 2\dfrac{5}{8}$

18. $\dfrac{3}{w} + \dfrac{x}{6}$

19. $8 + 4(2 - 5)$

20. $\dfrac{7}{8}$ of 960

21. $\dfrac{(-4)^2 + 8(0 - 2)}{8 \div 2(-3 + 5) - 10}$

22. $0.5 - 0.25(3.2)^2$

23. $6\left(-\dfrac{1}{2}\right)^3 + \dfrac{2}{3}\left(\dfrac{3}{5}\right)$

Evaluate each expression when $a = 2$, $b = 3$, and $c = -5$.

24. $20 + 4c$

25. $7b - 4c$

26. $-2ac^2$

Simplify each expression.

27. $-4x + x^2 - x$

28. $3(y - 4) + 2y$

29. $-5(8h^2)$

Solve each equation. Show your work.

30. $2x - 3 = -20 + 3$

31. $-12 = 3(y + 2)$

32. $6x = 14 - x$

33. $-8 = \dfrac{2}{3}m + 2$

34. $\dfrac{2}{13.5} = \dfrac{2.4}{n}$

35. $-20 = -w$

36. $3.4x - 6 = 8 + 1.4x$

37. $2(h - 1) = -3(h + 12) - 11$

Translate each sentence into an equation and solve it.

38. If five times a number is subtracted from 12, the result is the number. Find the number.

39. When -8 is added to twice a number, the result is -28. What is the number?

*Solve each application problem using the six problem-solving steps from **Chapter 3**.*

40. An $1800 scholarship is to be split between two people so that one person gets $500 more than the other. How much will each person receive?

41. The length of a rectangular movie theater is three times the width. The perimeter is 280 ft. Find the width and the length.

*Find the unknown length, perimeter, circumference, area, or volume. Use **3.14** as the approximate value of π, and round answers to the nearest tenth.*

42. Find the perimeter and area.

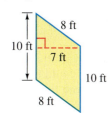

43. Find the perimeter and area.

44. Find the unknown length, the perimeter, and the area.

45. A circular hot tub has a diameter of 6 ft. Find its circumference and the area of the tub's floor.

46. A cylindrical oil tank is 12 m tall and has a diameter of 20 m. Find its volume.

47. Find the volume of a rectangular storage shed that is 4 yd long, 3 yd wide, and $2\frac{1}{2}$ yd high.

Convert each measurement using unit fractions, the metric conversion line, or the temperature conversion formulas.

48. $2\frac{1}{4}$ hours to minutes

49. 54 in. to feet

50. 1.85 L to milliliters

51. 35 mm to centimeters

52. 10 g to kilograms

53. 25 °F to Celsius; round to the nearest whole degree.

Solve each application problem.

54. A survey found that 19 out of 25 adults are nonsmokers. If Mathtronic has 732 employees, how many would be expected to be nonsmokers? Round to the nearest whole number.

55. Esther bought a cellular phone at a 15% discount. The regular price was $129. She also paid $6\frac{1}{2}$% sales tax. What was her total cost for the phone, to the nearest cent?

56. Abiola earned 167 points out of 180 on the prealgebra final exam. What percent of the points did she earn, to the nearest tenth?

57. Century College received a $30,000 technology grant.

   (a) How many $957 computers were they able to buy with the grant money? How much was left over?

   (b) How many $19 calculators could they buy with the money left over from the computer purchase? How much was left then?

58. Wayne bought $2\frac{1}{4}$ pounds of lunch meat. He used $\frac{1}{6}$ pound of meat on each of five sandwiches. How much meat is left?

59. Plot the following points and label each one with its coordinates. Which of the points are in the second quadrant? Which are in the third quadrant?

$(3, -5)$ $(0, -4)$ $(-2, 3\frac{1}{2})$ $(1, 0)$ $(4, 2)$

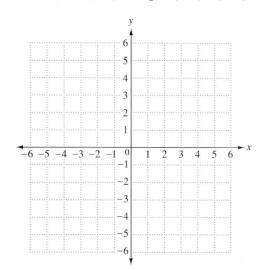

60. Graph $y = x + 6$. Use 0, -1, and -2 as the values of x. After graphing the equation, state whether the line has a positive or negative slope, and find two other solutions of $y = x + 6$.

Exponents and Polynomials

10

The expression $100t - 13t^2$ gives the distance in feet that a car going approximately 68 mph will skid in t seconds. This expression in t is an example of a *polynomial*. Accident investigators use polynomials like this to determine the length of a skid or the elapsed time during a skid. (See **Section 10.4**, Exercises 65–68, where we use this polynomial to approximate skidding distance.)

703

10.1 The Product Rule and Power Rules for Exponents

OBJECTIVE 1 Review the use of exponents. In **Section 1.8** we used exponents to write repeated products. Recall that in the expression 5^2, the number 5 is called the *base* and 2 is called the *exponent* or *power*. The expression 5^2 is written in *exponential form*. Usually we do not write an exponent of 1, but sometimes it is convenient to do so. In general, for any quantity x, we can write x as x^1.

EXAMPLE 1 Review of Using Exponents

Write $3 \cdot 3 \cdot 3 \cdot 3 \cdot 3$ in exponential form, and find the value of the exponential expression.

Since 3 occurs as a factor five times, the base is 3 and the exponent is 5. The exponential expression is 3^5, read "3 to the fifth power" or simply "3 to the fifth." To find the value of 3^5, actually do the multiplication.

$$3^5 = 3 \cdot 3 \cdot 3 \cdot 3 \cdot 3 = 243$$

Now Try Exercise 9.

EXAMPLE 2 Evaluating Exponential Expressions

Evaluate each exponential expression. Name the base and the exponent.

	Base	**Exponent**
(a) $5^4 = 5 \cdot 5 \cdot 5 \cdot 5 = 625$	5	4
(b) $-5^4 = -(5 \cdot 5 \cdot 5 \cdot 5) = -625$	5	4
(c) $(-5)^4 = (-5)(-5)(-5)(-5) = 625$	-5	4

CAUTION
It is important to understand the difference between parts (b) and (c) of Example 2 above. In -5^4 the lack of parentheses shows that the exponent 4 applies only to the base **5**. In $(-5)^4$ the parentheses show that the exponent 4 applies to the base **-5**. In summary, $-a^n$ and $(-a)^n$ *mean different things.* The exponent applies only to what is *immediately* to the left of it.

Expression	**Base**	**Exponent**	**Example**
$-a^n$	a	n	$-3^2 = -(3 \cdot 3) = -9$
$(-a)^n$	$-a$	n	$(-3)^2 = (-3)(-3) = 9$

Now Try Exercise 13.

OBJECTIVE 2 Use the product rule for exponents. To develop the product rule, we use the definition of an exponential expression.

$$2^4 \cdot 2^3 = \overbrace{(2 \cdot 2 \cdot 2 \cdot 2)}^{\text{4 factors}} \overbrace{(2 \cdot 2 \cdot 2)}^{\text{3 factors}}$$

$$= \underbrace{2 \cdot 2 \cdot 2 \cdot 2 \cdot 2 \cdot 2 \cdot 2}_{\text{4 factors + 3 factors = 7 factors}}$$

$$= 2^7$$

Here is another example.

$$6^2 \cdot 6^3 = (6 \cdot 6)(6 \cdot 6 \cdot 6)$$
$$= 6 \cdot 6 \cdot 6 \cdot 6 \cdot 6$$
$$= 6^5 \leftarrow \text{2 factors + 3 factors = 5 factors}$$

Generalizing from these examples, $2^4 \cdot 2^3 = 2^{4+3} = 2^7$ and $6^2 \cdot 6^3 = 6^{2+3} = 6^5$. In each case, adding the exponents gives the exponent of the product, suggesting the **product rule for exponents.**

Product Rule for Exponents

If m and n are positive integers, then $\quad a^m \cdot a^n = a^{m+n}$
(Keep the same base and add the exponents.)

Example: $\quad 6^2 \cdot 6^5 = 6^{2+5} = 6^7$

CAUTION
Avoid the common error of multiplying the *bases* when using the product rule.

$$6^2 \cdot 6^5 \neq 36^7 \leftarrow \text{Error} \qquad 6^2 \cdot 6^5 = 6^7 \leftarrow \text{Correct}$$

Keep the *same base* and add the exponents.

EXAMPLE 3 **Using the Product Rule**

Use the product rule for exponents to find each product, if possible.

(a) $6^3 \cdot 6^5 = 6^{3+5} = 6^8 \quad$ by the product rule.

(b) $(-4)^7(-4)^2 = (-4)^{7+2} = (-4)^9 \quad$ by the product rule.

(c) $x^2 \cdot x = x^2 \cdot x^1 = x^{2+1} = x^3 \quad$ Recall that x can be written as x^1.

(d) $m^4 \cdot m^3 \cdot m^4 = m^{4+3+4} = m^{11} \quad$ by the product rule.

(e) $2^3 \cdot 3^2$

The product rule does *not* apply to the product $2^3 \cdot 3^2$ because the *bases are different.*

$$2^3 \cdot 3^2 = (2 \cdot 2 \cdot 2)(3 \cdot 3) = (8)(9) = 72$$

CAUTION
The *bases* must be the *same* before we can apply the product rule for exponents.

(f) $2^3 + 2^4$

The product rule does *not* apply to $2^3 + 2^4$ because it is a *sum,* not a *product.*

$$2^3 + 2^4 = (2 \cdot 2 \cdot 2) + (2 \cdot 2 \cdot 2 \cdot 2) = (8) + (16) = 24$$

Now Try Exercises 29 and 31.

EXAMPLE 4 Using the Product Rule

Multiply $2x^3$ and $4x^3$.

Recall that $2x^3$ means $2 \cdot x^3$ and $4x^3$ means $4 \cdot x^3$. Now use the associative and commutative properties and the product rule.

$$2x^3 \cdot 4x^3 = (2 \cdot 4) \cdot (x^3 \cdot x^3) = 8x^{3+3} = 8x^6$$

CAUTION

Be sure you understand the difference between *adding* and *multiplying* exponential expressions. Here is a comparison.

Adding expressions $2x^3 + 4x^3 = (2 + 4)x^3 = 6x^3$

Multiplying expressions $(2x^3)(4x^3) = (2 \cdot 4)x^{3+3} = 8x^6$

◀◀◀ Now Try Exercise 35.

OBJECTIVE 3 **Use the rule $(a^m)^n = a^{mn}$.** We can simplify an expression such as $(8^3)^2$ with the product rule for exponents, as follows.

$$(8^3)^2 = (8^3)(8^3) = 8^{3+3} = 8^6$$

Notice that the *product* of the original exponents, $3 \cdot 2$, gives 6, the exponent in 8^6. Here is another example.

$$(5^2)^4 = 5^2 \cdot 5^2 \cdot 5^2 \cdot 5^2 \quad \text{Definition of exponent}$$
$$= 5^{2+2+2+2} \quad \text{By the product rule}$$
$$= 5^8$$

Notice that the *product* of the original exponents gives the exponent of the result: $2 \cdot 4 = 8$. These examples suggest **power rule (a) for exponents.**

Power Rule (a) for Exponents

If m and n are positive integers, then $(a^m)^n = a^{mn}$
(Raise a power to a power by multiplying exponents.)

Example: $(3^2)^4 = 3^{2 \cdot 4} = 3^8$

EXAMPLE 5 Using Power Rule (a)

Use power rule (a) for exponents to simplify each expression. Write answers in exponential form.

(a) $(2^5)^3 = 2^{5 \cdot 3} = 2^{15}$

(b) $(5^7)^2 = 5^{7 \cdot 2} = 5^{14}$

(c) $(x^5)^5 = x^{5 \cdot 5} = x^{25}$

(d) $(n^3)^2 = n^{3 \cdot 2} = n^6$

◀◀◀ Now Try Exercise 43.

OBJECTIVE **4** **Use the rule** $(ab)^m = a^m b^m$. We can use the definition of an exponential expression and the commutative and associative properties to develop two more rules for exponents. Here is an example.

$$(4x)^3 = (4x)(4x)(4x) \qquad \text{Definition of exponent}$$
$$= 4 \cdot 4 \cdot 4 \cdot x \cdot x \cdot x \qquad \text{Commutative and associative properties}$$
$$= 4^3 x^3 \qquad \text{Definition of exponent}$$

This example suggests **power rule (b) for exponents.**

Power Rule (b) for Exponents

If m is a positive integer, then $(ab)^m = a^m b^m$
(Raise a product to a power by raising each factor to the power.)

Example: $(2p)^5 = 2^5 p^5$

EXAMPLE 6 **Using Power Rule (b)**

Use power rule (b) to simplify each expression.

(a) $(3xy)^2 = 3^2 x^2 y^2$ Power rule (b)

$\qquad\qquad = 9x^2 y^2$ $3^2 = 3 \cdot 3 = 9$

(b) $9(pq)^2 = 9(p^2 q^2)$ Power rule (b); notice that the 9 is *outside* the parentheses, so the exponent applies *only* to pq,

$\qquad\qquad = 9p^2 q^2$ which is *inside* the parentheses.

(c) $5(2m^2 p^3)^4 = 5[2^4 (m^2)^4 (p^3)^4]$ Power rule (b); notice that the 5 is *outside* the parentheses so the exponent does *not* apply to the 5.

$\qquad\qquad = 5(2^4 m^8 p^{12})$ Power rule (a)

$\qquad\qquad = 5 \cdot 2^4 m^8 p^{12}$ $5 \cdot 2^4 = 5 \cdot 16 = 80$

$\qquad\qquad = 80 m^8 p^{12}$

Now Try Exercises 47 and 49.

CAUTION
Power rule (b) *does not* apply to a *sum.*

$$(x + 4)^2 \neq x^2 + 4^2 \quad \leftarrow \text{Error}$$

You will learn how to work with $(x + 4)^2$ in **Section 10.5.**

OBJECTIVE 5 Use the rule $\left(\dfrac{a}{b}\right)^m = \dfrac{a^m}{b^m}$. Since the quotient $\dfrac{a}{b}$ can be written as $a \cdot \dfrac{1}{b}$, we can use power rule (b), together with some of the properties of real numbers, to get **power rule (c) for exponents.**

> **Power Rule (c) for Exponents**
>
> If m is a positive integer, then $\left(\dfrac{a}{b}\right)^m = \dfrac{a^m}{b^m}$ as long as $b \neq 0$.
>
> (Raise a quotient to a power by raising both the numerator and the denominator to the power. The denominator cannot be 0.)
>
> *Example:* $\left(\dfrac{5}{3}\right)^2 = \dfrac{5^2}{3^2}$

EXAMPLE 7 Using Power Rule (c)

Simplify each expression.

(a) $\left(\dfrac{2}{3}\right)^5 = \dfrac{2^5}{3^5} = \dfrac{32}{243}$

(b) $\left(\dfrac{m}{n}\right)^4 = \dfrac{m^4}{n^4}$ when $n \neq 0$ Recall that division by 0 is undefined, so the denominator cannot be equal to 0.

> **Now Try Exercise 53.**

Here is a list of the rules for exponents discussed in this section. These rules are basic to the study of algebra and should be *memorized.*

Rules for Exponents

If m and n are positive integers, then:

		Examples
Product rule	$a^m \cdot a^n = a^{m+n}$	$6^2 \cdot 6^5 = 6^{2+5} = 6^7$
Power rule (a)	$(a^m)^n = a^{mn}$	$(3^2)^4 = 3^{2 \cdot 4} = 3^8$
Power rule (b)	$(ab)^m = a^m b^m$	$(2p)^5 = 2^5 p^5$
Power rule (c)	$\left(\dfrac{a}{b}\right)^m = \dfrac{a^m}{b^m}$ when $b \neq 0$	$\left(\dfrac{5}{3}\right)^2 = \dfrac{5^2}{3^2}$

10.1 Exercises

1. What is the understood exponent for x in the expression xy^2?

2. How are the expressions 3^2, 5^3, and 7^4 read?

Decide whether each statement is true *or* false. *If a statement is* false, *correct it.*

3. $3^3 = 9$

4. $2^4 + 2^3 = 2^7$

5. $(a^2)^3 = a^5$

6. $\left(\dfrac{1}{4}\right)^2 = \dfrac{1}{4^2}$

Write each expression using exponents. See Example 1.

7. $(-2)(-2)(-2)(-2)(-2)$

8. $w \cdot w \cdot w \cdot w \cdot w \cdot w$

9. $\left(\dfrac{1}{2}\right)\left(\dfrac{1}{2}\right)\left(\dfrac{1}{2}\right)\left(\dfrac{1}{2}\right)\left(\dfrac{1}{2}\right)\left(\dfrac{1}{2}\right)$

10. $\left(-\dfrac{1}{4}\right)\left(-\dfrac{1}{4}\right)\left(-\dfrac{1}{4}\right)\left(-\dfrac{1}{4}\right)\left(-\dfrac{1}{4}\right)$

11. $(-8p)(-8p)$

12. $(-7x)(-7x)(-7x)(-7x)$

13. Explain how the expressions $(-3)^4$ and -3^4 are different.

14. Explain how the expressions $(5x)^3$ and $5x^3$ are different.

Identify the base and the exponent for each exponential expression. In Exercises 15–18, also evaluate the expression. See Example 2.

15. 3^5

16. 2^7

17. $(-3)^5$

18. $(-2)^7$

19. $(-6x)^4$

20. $(-8x)^4$

21. $-6x^4$

22. $-8x^4$

23. Explain why the product rule does *not* apply to the expression $5^2 + 5^3$. Then evaluate the expression by finding the individual powers and adding the results.

24. Explain why the product rule does *not* apply to the expression $3^2 \cdot 4^3$. Then evaluate the expression by finding the individual powers and multiplying the results.

Use the product rule to simplify each expression. Write each answer in exponential form.
See Examples 3 and 4.

25. $5^2 \cdot 5^6$

26. $3^6 \cdot 3^7$

27. $4^2 \cdot 4^7 \cdot 4^3$

28. $5^3 \cdot 5^8 \cdot 5^2$

29. $(-7)^3(-7)^6$

30. $(-9)^8(-9)^5$

31. $t^3 \cdot t \cdot t^{13}$

32. $n \cdot n^6 \cdot n^9$

33. $(-8r^4)(7r^3)$

34. $(10a^7)(-4a^3)$

35. $(-6p^5)(-7p^5)$

36. $(-5w^8)(-9w^8)$

For each group of terms, first add the given terms. Then start over and multiply them.

37. $5x^4, 9x^4$

38. $8t^5, 3t^5$

39. $-7a^2, 2a^2, 10a^2$

40. $6x^3, 9x^3, -2x^3$

Use the power rules for exponents to simplify each expression. Write each answer in exponential form. See Examples 5–7.

41. $(4^3)^2$

42. $(8^3)^6$

43. $(t^4)^5$

44. $(y^6)^5$

45. $(7r)^3$

46. $(11x)^4$

47. $(5xy)^5$

48. $(9pq)^6$

49. $8(qr)^3$

50. $4(vw)^5$

51. $\left(\dfrac{1}{2}\right)^3$

52. $\left(\dfrac{1}{3}\right)^5$

53. $\left(\dfrac{a}{b}\right)^3$ when $b \neq 0$

54. $\left(\dfrac{r}{t}\right)^4$ when $t \neq 0$

55. $\left(\dfrac{9}{5}\right)^8$

56. $\left(\dfrac{12}{7}\right)^3$

57. $(-2x^2y)^3$

58. $(-5m^4p^2)^3$

59. $(3a^3b^2)^2$

60. $(4x^3y^5)^4$

Find the area of each figure. For help, refer to the formulas found on the inside back cover of this textbook.

61.

$3x^2$
$4x^3$

62.

m^2
$3m^4$

10.2 Integer Exponents and the Quotient Rule

OBJECTIVES

1. Use 0 as an exponent.
2. Use negative numbers as exponents.
3. Use the quotient rule for exponents.
4. Use the product rule with negative exponents.

In the last section we studied the product rule for exponents. In all our earlier work, exponents were positive integers. Now we want to develop meaning for exponents that are not positive integers.

Look at this list of exponential expressions.

Exponent is decreasing by 1 each time.
$$2^4 = 16$$
$$2^3 = 8$$ 16 ÷ 2 is 8
$$2^2 = 4$$ 8 ÷ 2 is 4

Do you see the pattern in the values? Each time we reduce the exponent by 1, the value is divided by 2 (the base). Using this pattern, we can continue the list to smaller and smaller integer exponents.

$$2^1 = 2$$ 2 ÷ 2 is 1
$$2^0 = 1$$ 1 ÷ 2 is $\frac{1}{2}$
$$2^{-1} = \frac{1}{2}$$ $\frac{1}{2}$ ÷ 2 is $\frac{1}{4}$
$$2^{-2} = \frac{1}{4}$$ $\frac{1}{4}$ ÷ 2 is $\frac{1}{8}$
$$2^{-3} = \frac{1}{8}$$

From the list above, it appears that we should define 2^0 as 1 and define negative exponents as reciprocals.

OBJECTIVE 1 Use 0 as an exponent.   We want the definitions of 0 and negative exponents to satisfy the rules for exponents from **Section 10.1.** For example, if $6^0 = 1$, we can check the product rule as shown below.

$$6^0 \cdot 6^2 = 1 \cdot 6^2 = 6^2 \quad \text{and} \quad 6^0 \cdot 6^2 = 6^{0+2} = 6^2$$

The results match, so the product rule is satisfied. If you check the power rules, you will see that they are also valid for a base raised to an exponent of 0. Thus, we define an exponent of 0 as follows.

Zero Exponent

If a is any *nonzero* real number, then, $a^0 = 1$

Example: $17^0 = 1$

EXAMPLE 1 Using Zero Exponents

Evaluate each exponential expression.

(a) $60^0 = 1$

(b) $(-60)^0 = 1$

(c) $-60^0 = -(1) = -1$

(d) $y^0 = 1$, if $y \neq 0$

(e) $6y^0 = 6(1) = 6$, if $y \neq 0$

(f) $(6y)^0 = 1$, if $y \neq 0$

CAUTION
Notice the difference between parts (b) and (c) of Example 1 above. In Example 1(b) the base is -60 and the exponent is 0. But in Example 1(c), the base is 60.

Now Try Exercises 9 and 11.

OBJECTIVE 2 Use negative numbers as exponents. In the list at the start of this section, we saw that $2^{-2} = \frac{1}{4}$, $2^{-3} = \frac{1}{8}$ and $2^{-4} = \frac{1}{16}$. If we write the denominators as powers of 2, we see an interesting pattern in the exponents.

$$2^{-2} = \frac{1}{4} = \frac{1}{2^2} \qquad 2^{-3} = \frac{1}{8} = \frac{1}{2^3} \qquad 2^{-4} = \frac{1}{16} = \frac{1}{2^4}$$

So it seems that 2^{-n} should equal $\frac{1}{2^n}$. Is the product rule valid in such cases? For example, if we use the product rule to multiply 6^{-2} by 6^2, we get the following.

$$6^{-2} \cdot 6^2 = 6^{-2+2} = 6^0 = 1$$

The expression 6^{-2} behaves as if it were the reciprocal of 6^2, because their product is 1. The reciprocal of 6^2 may be written $\frac{1}{6^2}$, leading us to define 6^{-2} as $\frac{1}{6^2}$. This example illustrates the definition of negative exponents.

Negative Exponents

If a is any *nonzero* real number and n is any integer, then $\quad a^{-n} = \dfrac{1}{a^n}$

Example: $\quad 3^{-2} = \dfrac{1}{3^2}$

EXAMPLE 2 Using Negative Exponents

Simplify by writing each expression with positive exponents. Then evaluate the expression, if possible.

(a) $4^{-2} = \dfrac{1}{4^2} = \dfrac{1}{16}$

(b) $5^{-3} = \dfrac{1}{5^3} = \dfrac{1}{125}$

(c) $4^{-1} - 2^{-1} = \dfrac{1}{4} - \dfrac{1}{2}$ Apply the exponents first.

$$= \dfrac{1}{4} - \dfrac{2}{4} \qquad \text{Then subtract.}$$

$$= -\dfrac{1}{4}$$

(d) $p^{-2} = \dfrac{1}{p^2}$ when $p \neq 0$

CAUTION

A negative exponent does ***not*** indicate a negative number; negative exponents lead to reciprocals.

Expression	**Example**
a^{-n}	$3^{-2} = \dfrac{1}{3^2} = \dfrac{1}{9}$ ← **Not** a negative number

---

◀◀◀ Now Try Exercises 25 and 27.

OBJECTIVE 3 Use the quotient rule for exponents. What about the quotient of two exponential expressions with the same base? We know the following.

$$\frac{6^5}{6^3} = \frac{\overset{1}{\cancel{6}} \cdot \overset{1}{\cancel{6}} \cdot \overset{1}{\cancel{6}} \cdot 6 \cdot 6}{\underset{1}{\cancel{6}} \cdot \underset{1}{\cancel{6}} \cdot \underset{1}{\cancel{6}}} = \frac{6^2}{1} = 6^2$$

Notice that the difference between the exponents, $5 - 3 = 2$, results in the exponent in the quotient. Here is another example.

$$\frac{6^2}{6^4} = \frac{\overset{1}{\cancel{6}} \cdot \overset{1}{\cancel{6}}}{\underset{1}{\cancel{6}} \cdot \underset{1}{\cancel{6}} \cdot 6 \cdot 6} = \frac{1}{6^2} = 6^{-2}$$

In this case, the difference between the exponents is $2 - 4 = -2$. These examples suggest the quotient rule for exponents.

Quotient Rule for Exponents

If a is any nonzero real number, and m and n are integers, then

$$\frac{a^m}{a^n} = a^{m-n}.$$

(Keep the base and subtract the exponents.)

Example: $\dfrac{5^8}{5^4} = 5^{8-4} = 5^4$

CAUTION

A common ***error*** is to write $\dfrac{5^8}{5^4} = 1^{8-4} = 1^4$. When using the quotient rule, the quotient should have the ***same base.*** The base here is 5.

$$\frac{5^8}{5^4} = 5^{8-4} = 5^4$$

If you are not sure, use the definition of an exponent to write out the factors.

$$\frac{5^8}{5^4} = \frac{\overset{1}{\cancel{5}} \cdot \overset{1}{\cancel{5}} \cdot \overset{1}{\cancel{5}} \cdot \overset{1}{\cancel{5}} \cdot 5 \cdot 5 \cdot 5 \cdot 5}{\underset{1}{\cancel{5}} \cdot \underset{1}{\cancel{5}} \cdot \underset{1}{\cancel{5}} \cdot \underset{1}{\cancel{5}}} = \frac{5^4}{1} = 5^4$$

⌐ **EXAMPLE 3 Using the Quotient Rule for Exponents**

Simplify, using the quotient rule for exponents. Write answers with positive exponents.

(a) $\dfrac{5^8}{5^6} = 5^{8-6} = 5^2$

(c) $\dfrac{5^{-3}}{5^{-7}} = 5^{-3-(-7)} = 5^4$

(b) $\dfrac{4^2}{4^9} = 4^{2-9} = 4^{-7} = \dfrac{1}{4^7}$

(d) $\dfrac{x^5}{x^{-3}} = x^{5-(-3)} = x^8$ when $x \neq 0$

> **Now Try Exercises 33, 37 and 45.** ▶▶▶

OBJECTIVE 4 Use the product rule with negative exponents. As shown in the next example, we may use the product rule to simplify expressions with negative exponents.

EXAMPLE 4 **Using the Product Rule with Negative Exponents**

Simplify each expression. Assume all variables represent nonzero real numbers. Write answers with positive exponents.

(a) $10^5(10^{-4}) = 10^{5+(-4)}$ Using the product rule, add the exponents.

$$= 10^1 \quad \text{or} \quad 10$$

(b) $(10^{-5})(10^{-8}) = 10^{-5+(-8)}$ Product rule

$$= 10^{-13} = \frac{1}{10^{13}} \quad \text{Definition of negative exponent}$$

(c) $y^{-2} \cdot y^5 \cdot y^{-8} = \underbrace{y^{-2+5}} \cdot y^{-8}$ Product rule

$$= y^3 \cdot y^{-8}$$

$$= y^{3+(-8)} \quad \text{Product rule}$$

$$= y^{-5}$$

$$= \frac{1}{y^5} \quad \text{Definition of negative exponent}$$

Now Try Exercises 63, 65, and 73.

The definitions and rules for exponents are summarized below.

Definitions and Rules for Exponents

If m and n are any integers,   **Examples**

Product rule	$a^m \cdot a^n = a^{m+n}$	$7^4 \cdot 7^5 = 7^{4+5} = 7^9$
Zero exponent	$a^0 = 1$ when $a \neq 0$	$(-3)^0 = 1$
Negative exponent	$a^{-n} = \dfrac{1}{a^n}$ when $a \neq 0$	$5^{-3} = \dfrac{1}{5^3}$
Quotient rule	$\dfrac{a^m}{a^n} = a^{m-n}$ when $a \neq 0$	$\dfrac{2^2}{2^5} = 2^{2-5} = 2^{-3} = \dfrac{1}{2^3}$
Power rule (a)	$(a^m)^n = a^{mn}$	$(4^2)^3 = 4^{2\cdot3} = 4^6$
Power rule (b)	$(ab)^m = a^m b^m$	$(3k)^4 = 3^4 k^4$
Power rule (c)	$\left(\dfrac{a}{b}\right)^m = \dfrac{a^m}{b^m}$ when $b \neq 0$	$\left(\dfrac{2}{3}\right)^2 = \dfrac{2^2}{3^2}$

CAUTION

Read directions carefully. In the examples above, we **simplified** each expression and wrote the answer using positive exponents.

If you are told to **evaluate** an expression, go one step further and do all possible operations.

Example: $\dfrac{2^2}{2^5}$ $\xrightarrow[\text{by using the quotient rule.}]{\textbf{Simplify}}$ $2^{2-5} = 2^{-3} = \dfrac{1}{2^3}$ $\xrightarrow[\text{by finding } 2^3 \text{ in the denominator.}]{\textbf{Evaluate}}$ $\dfrac{1}{8}$

10.2 Exercises

FOR EXTRA HELP

 Addison-Wesley Math Tutor Center

 MathXL

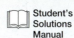 Digital Video Tutor CD 6 Videotape 10

Student's Solutions Manual

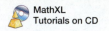 MyMathLab

MathXL Tutorials on CD

Decide whether each expression is positive, negative, or 0.

1. $(-2)^{-3}$ **2.** $(-3)^{-2}$ **3.** -2^4 **4.** -3^6

5. $(-2)^6$ **6.** $(-2)^3$ **7.** $1 - 5^0$ **8.** $1 - 7^0$

Each expression is equal to either 0, 1, or -1. Decide which is correct. See Example 1.

9. $(-4)^0$ **10.** $(-10)^0$ **11.** -9^0 **12.** -5^0

13. $(-2)^0 - 2^0$ **14.** $(-8)^0 - 8^0$ **15.** $\dfrac{0^{10}}{10^0}$ **16.** $\dfrac{0^5}{5^0}$

Evaluate each expression. See Examples 1 and 2.

17. $7^0 + 9^0$ **18.** $8^0 + 6^0$ **19.** b^0 when $b \neq 0$ **20.** x^0 when $x \neq 0$

21. $15x^0$ when $x \neq 0$ **22.** $12g^0$ when $g \neq 0$ **23.** 4^{-3} **24.** 5^{-4}

25. $5^{-1} + 3^{-1}$ **26.** $6^{-1} + 2^{-1}$ **27.** x^{-4} **28.** b^{-3}

RELATING CONCEPTS (EXERCISES 29–32)   For Individual or Group Work

In Objective 1, we used the product rule to develop a definition for an exponent of 0. We can also use the quotient rule. To see this, **work Exercises 29–32 in order.**

29. Consider the expression $\frac{25}{25}$. What is its simplest form?

30. Rewrite $\frac{25}{25}$ using the fact that $25 = 5^2$.

31. Apply the quotient rule for exponents to your answer for Exercise 30. Give the answer as a power of 5.

32. Because your answers for Exercises 29 and 31 both represent $\frac{25}{25}$, they must be equal. Write this equality. What definition does it support?

Use the quotient rule to simplify each expression. Write each expression with positive exponents. Assume that all variables represent nonzero real numbers. See Example 3.

33. $\dfrac{6^7}{6^2}$

34. $\dfrac{8^5}{8^3}$

35. $\dfrac{10^4}{10}$

36. $\dfrac{7^8}{7}$

37. $\dfrac{y^2}{y^6}$

38. $\dfrac{d^9}{d^5}$

39. $\dfrac{c^6}{c^5}$

40. $\dfrac{x^4}{x^3}$

41. $\dfrac{5^3}{5^7}$

42. $\dfrac{9^2}{9^8}$

43. $\dfrac{m^7}{m^8}$

44. $\dfrac{n^2}{n^3}$

45. $\dfrac{3^{-4}}{3^{-8}}$

46. $\dfrac{5^{-3}}{5^{-10}}$

47. $\dfrac{a^{-2}}{a^{-5}}$

48. $\dfrac{b^{-4}}{b^{-6}}$

49. $\dfrac{2^{-10}}{2^{-2}}$

50. $\dfrac{4^{-7}}{4^{-3}}$

51. $\dfrac{r^{-12}}{r^{-8}}$

52. $\dfrac{v^{-10}}{v^{-5}}$

53. $\dfrac{10^6}{10^{-4}}$

54. $\dfrac{10^4}{10^{-2}}$

55. $\dfrac{10^{-2}}{10^3}$

56. $\dfrac{10^{-1}}{10^7}$

57. $\dfrac{10^3}{10^{-8}}$

58. $\dfrac{10^5}{10^{-7}}$

59. $\dfrac{10^{-4}}{10^4}$

60. $\dfrac{10^{-6}}{10^6}$

Simplify each expression. Write answers with only positive exponents. Assume that all variables represent nonzero real numbers. See Example 4.

61. $10^6(10^{-2})$

62. $10^{-7}(10^4)$

63. $a^{-4}(a^3)$

64. $w^9(w^{-8})$

65. $(2^{-4})(2^{-4})$

66. $(2^{-1})(2^{-5})$

67. $(x^{-10})(x^{-1})$

68. $(h^{-5})(h^{-5})$

69. $10^8 \cdot 10^{-2} \cdot 10^{-4}$

70. $10^{-2} \cdot 10^7 \cdot 10^{-2}$

71. $(y^{-3})(y^5)(y^{-4})$

72. $(n^{-8})(n^{-2})(n^5)$

73. $m^{-6} \cdot m^3 \cdot m^{12}$

74. $b^{-3} \cdot b^3 \cdot b^8$

75. $(10^{-1})(10^{-2})(10^{-3})$

76. $(10^{-4})(10^{-5})(10^{-6})$

Summary Exercises on Exponents

Evaluate each expression. Assume all variables represent nonzero real numbers.

1. $(-3)^2$

2. $5^3 + 5^2$

3. 15^0

4. $3^{-2} - 5^{-1}$

5. p^0

6. 4^{-3}

7. -3^2

8. $2^3 \cdot 3^2$

9. -2^3

10. $-(8)^0$

11. $6^{-1} + 2^{-2}$

12. $9g^0$

13. 2^{-5}

14. $(-12)^0$

15. $(-2)^3$

16. $(-7n)^0$

Simplify. Write all answers in exponential form. Assume all variables represent nonzero real numbers.

17. $(6^4)(6^4)$

18. x^{-7}

19. $10^2(10^{-10})$

20. $(2^3)^3$

21. $(y^{-12})(y^6)(y^6)$

22. $6(2m^2n)^5$

23. $(-5)^2(-5)^8$

24. $\dfrac{w^{-4}}{w^{-7}}$

25. $(m^6)^4$

26. $\dfrac{n^{-5}}{n^{-3}}$

27. $(10^{-2})(10^{-2})(10^{-2})$

28. $d^5 \cdot d \cdot d^3$

29. $4x^4 \cdot 4x^4$

30 $\dfrac{c^4}{c^5}$

31. $(10^4)^2$

32. $\dfrac{4^{-1}}{4^{-6}}$

33. $\left(\dfrac{x}{y}\right)^6$

34. $3^3 \cdot 3^3 \cdot 3^5$

35. $\dfrac{7^{12}}{7^{11}}$

36. $h^{-6}(h^7)$

37. $-8a^3(2a)$

38. $\left(\dfrac{1}{2}\right)^4$

39. $(5xy)^3$

40. $(m^{13})(m^{-2})(m^{-6})$

41. $\dfrac{10^{-3}}{10^4}$

42. $b^9(b^{-10})$

43. $(b^3)^5$

44. $\dfrac{5^3}{5^{10}}$

45. $10^{-8}(10^5)$

46. $(-7y^4)(-3y^4)$

47. $\dfrac{10^5}{10^{-8}}$

48. $(k^{-4})(k^{-4})$

49. $10(ab)^4$

50. $x^{-5} \cdot x \cdot x^{-5}$

10.3 An Application of Exponents: Scientific Notation

OBJECTIVE 1 Express numbers in scientific notation. An important example of the use of exponents comes from science. The numbers occurring in science are often extremely large (such as the distance from Earth to the sun, 93,000,000 miles) or extremely small (the wavelength of yellow-green light, approximately 0.0000006 meter). Because of the difficulty of working with many zeros, scientists often express such numbers with exponents. Each number is written as $a \times 10^n$, where n is an integer and a is greater than or equal to 1 and less than 10 (that is, $1 \le |a| < 10$). This form is called **scientific notation.** There is always one nonzero digit before the decimal point. Here are some examples.

OBJECTIVES

1 Express numbers in scientific notation.
2 Convert numbers in scientific notation to numbers without exponents.
3 Use scientific notation in calculations.
4 Solve application problems using scientific notation.

35 is written 3.5×10^1

because $35 = 3.5 \times 10 = 3.5 \times 10^1$

56,200 is written 5.62×10^4

because $56,200 = 5.62 \times 10,000 = 5.62 \times 10^4$

0.09 is written 9×10^{-2}

because $0.09 = \dfrac{9}{100} = 9 \times \dfrac{1}{100} = 9 \times \dfrac{1}{10^2} = 9 \times 10^{-2}$

To write a number in scientific notation, follow the steps listed below.

Writing a Number in Scientific Notation

Step 1 Move the decimal point to the right of the first nonzero digit.

Step 2 To figure out the exponent on 10, count the number of places you moved the decimal point. The number of places is the absolute value of the exponent on 10.

Step 3 To decide on the sign of the exponent, look at the *original* number. If the *original* number was "large" (10 or more), the exponent on 10 is positive. If the *original* number was "small" (between 0 and 1), the exponent is negative.

EXAMPLE 1 Writing Numbers in Scientific Notation

Write each number in scientific notation.

(a) 93,000,000
Move the decimal point to the right of the first nonzero digit. Count the number of places the decimal point was moved.

9.3 000 000   Move the decimal point 7 places.

We will write the number as 9.3×10^n. The *original* number (93,000,000) was "large" so the exponent n will be **positive 7**.

$$93,000,000 = 9.3 \times 10^7$$

(b) $302,100 = 3.021 \times 10^5$ Move the decimal point 5 places. *Original* number was "large" so the exponent is *positive* 5.

Continued on Next Page

(c) 0.00462

Move the decimal point to the right of the first nonzero digit and count the number of places it was moved.

$$0\ 004.62 \quad \text{Move the decimal point 3 places.}$$

The *original* number (0.00462) was "small" so the exponent will be **negative 3**.

$$0.00462 = 4.62 \times 10^{-3}$$

(d) $0.0000762 = 7.62 \times 10^{-5}$ Move the decimal point 5 places; *original* number was "small" so the exponent is **negative 5**.

>>> **Now Try Exercises 15 and 19.**

OBJECTIVE 2 Convert numbers in scientific notation to numbers without exponents. To convert a number written in scientific notation to a number without exponents, remember that multiplying a number by a positive power of 10 will make the number larger; multiplying by a negative power of 10 will make the number smaller.

EXAMPLE 2 Writing Numbers without Exponents

Write each number without exponents.

(a) 6.2×10^3

Since the exponent is positive, make 6.2 larger by moving the decimal point 3 places to the *right,* inserting zeros as needed.

$$6.2 \times \mathbf{10^3} = 6.200 = 6200$$

(b) $4.283 \times 10^5 = 4.28300 = 428{,}300$ Move 5 places to the *right.*

(c) $9.73 \times 10^{-2} = 09.73 = 0.0973$ Move 2 places to the *left;* multiplying by a negative power of 10 makes the number *smaller.*

As shown in Example 2 above, the exponent tells the number of places and the direction that the decimal point is moved.

>>> **Now Try Exercises 25 and 29.**

OBJECTIVE 3 Use scientific notation in calculations. The next example uses scientific notation with products and quotients.

EXAMPLE 3 Multiplying and Dividing with Scientific Notation

Simplify. Write the answers both in scientific notation and as numbers without exponents.

(a) $(6 \times 10^3)(5 \times 10^{-4})$

$$= (6 \times 5)(10^3 \times 10^{-4}) \quad \text{Commutative and associative properties}$$

$$= 30 \times 10^{-1} \quad \text{Product rule for exponents}$$

$$= 30. = 3.0 \quad \text{or} \quad 3 \quad \text{Write without exponents.}$$

(b) $\dfrac{6 \times 10^{-5}}{2 \times 10^3} = \dfrac{6}{2} \times \dfrac{10^{-5}}{10^3} = 3 \times 10^{-8} = 0.00000003$

>>> **Now Try Exercises 37 and 39.**

OBJECTIVE 4 Solve application problems using scientific notation.
Sometimes application problems may involve one or more numbers written in scientific notation and other numbers written without exponents. By writing all the numbers in scientific notation, we can use the product and quotient rules for exponents to help solve the problem.

EXAMPLE 4 Solving an Application Problem

There are approximately 3.03×10^{23} molecules in one gram of hydrogen gas. How many molecules are in 2000 g of hydrogen gas? Write your answer both in scientific notation and as a number without exponents.

First write 2000 in scientific notation.

$$2000 = 2.000 \qquad \text{Move 3 places to the left; } \textit{original} \text{ number}$$
$$\text{was "large" so the exponent is } \textit{positive.}$$
$$= 2.0 \times 10^3$$

Now multiply the number of molecules in one gram times the number of grams.

$$(3.03 \times 10^{23})(2.0 \times 10^3)$$
$$= (3.03 \times 2.0)(10^{23} \times 10^3) \qquad \text{Use the product rule for exponents.}$$
$$= 6.06 \times 10^{26}$$

Written in scientific notation, there are 6.06×10^{26} molecules in 2000 g of hydrogen gas. Written without an exponent, the number is 606,000,000,000,000,000,000,000,000 molecules.

Now Try Exercise 47.

Calculator Tip Most scientific calculators have a key that allows you to enter numbers in scientific notation. Usually this key is labeled (EE) or (EXP). Then you could solve Example 4, above, by pressing:

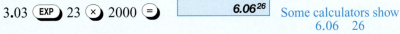

3.03 (EXP) 23 (×) 2000 (=) 6.06^{26} Some calculators show
6.06   26

The exponent in the answer is 26. **You have to write the × 10 part of the answer** to get 6.06×10^{26}.

If you need to enter a number in scientific notation that has a negative exponent, use the *change of sign* key. For example, to enter 8.45×10^{-6}, press:

8.45 (EXP) 6 (+/−) 8.45^{-06}

EXAMPLE 5 Solving an Application Problem

At a printing shop, a stack of 5×10^3 sheets of paper measures 0.38 meter high. What is the thickness of one sheet of paper? Write your answer both in scientific notation and as a number without exponents.

First write 0.38 in scientific notation.

$$0.38 = 0\,3.8 \qquad \text{Move 1 place to the right; } \textit{original} \text{ number}$$
$$\text{was "small" so exponent is } \text{negative.}$$
$$= 3.8 \times 10^{-1}$$

Continued on Next Page

Now divide the total height of the stack by the number of sheets in the stack.

$$\frac{3.8 \times 10^{-1}}{5 \times 10^{3}} = \frac{3.8}{5} \times \frac{10^{-1}}{10^{3}} = 0.76 \times 10^{-4}$$ Use the quotient rule for exponents.

The answer of 0.76×10^{-4} is *not* in proper scientific notation yet, because 0.76 is not between 1 and 10. The decimal point in 0.76 must be moved *1 place* to get 7.6. Then, because the original number (0.76) was "small," add *negative* 1 to the exponent.

$$\overset{(-4)+(-1)}{0.76 \times 10^{-4}} = 7.6 \times 10^{-5}$$

Move decimal point 1 place.

The thickness of one sheet of paper is 7.6×10^{-5} meter. Written without an exponent, the thickness is 0.000076 meter.

Now Try Exercise 45.

10.3 Exercises

Write the numbers (other than dates) in these statements in scientific notation. See Example 1.

1. The mass of Pluto, the smallest planet, is 0.0021 times that of Earth; the mass of Jupiter, the largest planet, is 317.83 times that of Earth. (*Source: World Almanac.*)

2. NASA budgeted $13,750,000,000 in each of the years 2003 and 2004 for the international space station. (*Source:* NASA.)

3. There are 831,200,000 pounds of pumpkins grown every year in the United States. Illinois leads the nation with 319,000,000 pounds.

4. The human body contains trace amounts of several minerals. For example, a 150-pound body has about 0.00023 pound of copper in it.

State whether or not the given number is written in scientific notation. If it is not, write it as such.

5. 4.56×10^3 **6.** 7.34×10^5 **7.** 5,600,000   **8.** 34,000

9. 0.004   **10.** 0.0007   **11.** 0.8×10^2 **12.** 0.9×10^3

13. Explain in your own words how to write a very large number in scientific notation, and how to write a very small number in scientific notation.

14. Explain how to multiply a number by a positive power of ten. Then explain how to multiply a number by a negative power of ten.

Write each number in scientific notation. See Example 1.

15. 5,876,000,000   **16.** 9,994,000,000   **17.** 82,350   **18.** 78,330

19. 0.000007   **20.** 0.0000004   **21.** 0.00203   **22.** 0.0000578

Write each number without exponents. See Example 2.

23. 7.5×10^5 **24.** 8.8×10^6 **25.** 5.677×10^{12} **26.** 8.766×10^9

27. 6.21×10^0

28. 8.56×10^0

29. 7.8×10^{-4}

30. 8.9×10^{-5}

31. 5.134×10^{-9}

32. 7.123×10^{-10}

Simplify. Write the answers both in scientific notation and as numbers without exponents. See Example 3.

33. $(2 \times 10^8)(3 \times 10^3)$

34. $(4 \times 10^7)(3 \times 10^3)$

35. $(5 \times 10^4)(3 \times 10^2)$

36. $(8 \times 10^5)(2 \times 10^3)$

37. $(3.15 \times 10^{-4})(2.04 \times 10^8)$

38. $(4.92 \times 10^{-3})(2.25 \times 10^7)$

39. $\dfrac{9 \times 10^{-5}}{3 \times 10^{-1}}$

40. $\dfrac{12 \times 10^{-4}}{4 \times 10^{-3}}$

41. $\dfrac{8 \times 10^3}{2 \times 10^2}$

42. $\dfrac{5 \times 10^4}{1 \times 10^3}$

43. $\dfrac{(2.6 \times 10^{-3})(7.0 \times 10^{-1})}{(2 \times 10^2)(3.5 \times 10^{-3})}$

44. $\dfrac{(9.5 \times 10^{-1})(2.4 \times 10^4)}{(5 \times 10^3)(1.2 \times 10^{-2})}$

Solve each problem. Write the answers both in scientific notation and without exponents. See Examples 4 and 5.

45. The population of China in 2003 was about 1.304×10^9 people. The country's land area is 3,700,000 square miles. What is the average number of people per square mile in China? (*Source: United Nations World Population Prospects.*)

46. The mass of Earth is approximately 5.98×10^{24} kg, and the mass of the moon is approximately 7.36×10^{22} kg. Earth's mass is how many times greater than the moon's mass? (*Source:* Time Life *The Universe.*)

47. The average diameter of a hydrogen atom is 1×10^{-10} meter. (*Source: FactFinder.*)

   (a) If one million hydrogen atoms are lined up in a row, how long would the row be?

   (b) Find the length of a row of one billion hydrogen atoms.

48. The speed of light is about 1.86×10^5 miles per second. (*Source:* Time Life *The Universe.*)

   (a) How far does light travel in one minute?

   (b) How far does light travel in one hour?

49. Americans mailed about 2×10^{10} cards, letters, and packages during December 2002. The U.S. population is about 2.88×10^8 people. On average, how many items did each person mail during that month? (*Source:* United States Postal Service and Bureau of the Census.)

50. American golfers buy more than 5×10^8 golf balls each year. On average, how many golf balls are sold each month? (*Source:* National Sporting Goods Association.)

10.4 Adding and Subtracting Polynomials

Recall from **Section 2.2** that in an expression such as

$$4x^3 + 6x^2 + 5x + 8,$$

the quantities that are added, $4x^3$, $6x^2$, $5x$, and 8, are called *terms*. In the term $4x^3$, the number 4 is called the *numerical coefficient,* or simply the *coefficient,* of x^3. In the same way, 6 is the coefficient of x^2 in the term $6x^2$, and 5 is the coefficient of x in the term $5x$. The 8 is the constant term.

OBJECTIVE 1 Review combining like terms. In **Section 2.2,** we saw that *like terms* are terms with exactly the same variables, and with the same exponents on the variables. Only the coefficients may differ. Like terms are combined, or *added,* by adding their coefficients using the distributive property.

> **OBJECTIVES**
>
> **1** Review combining like terms.
>
> **2** Use the vocabulary for polynomials.
>
> **3** Evaluate polynomials.
>
> **4** Add polynomials.
>
> **5** Subtract polynomials.

EXAMPLE 1 Review of Adding Like Terms

Simplify each expression by adding like terms.

(a) $-4x^3 + 6x^3 = (-4 + 6)\,x^3$ Distributive property
$$= 2x^3$$

(b) $9x^6 - 14x^6 + x^6 = (9 - 14 + 1)\,x^6 = -4x^6$

(c) $12m^2 + 5m + 4m^2 = (12 + 4)\,m^2 + 5m$
$$= 16m^2 + 5m$$

(d) $3x^2y + 4x^2y - x^2y = (3 + 4 - 1)\,x^2y = 6x^2y$

> **CAUTION**
> In Example 1(c) above, we *cannot* add $16m^2$ and $5m$. These two terms are *unlike* because the exponents on the variables are different. *Unlike terms* have different variables or different exponents on the same variables.

Now Try Exercises 15 and 19. ▸▸▸

OBJECTIVE 2 Use the vocabulary for polynomials. A **polynomial in x** is a term, or the sum of several terms, of the form ax^n, where a is any real number and n is any whole number. Here is an example.

$$16x^8 - 7x^6 + 5x^4 - 3x^2 + 4 \longleftarrow \begin{cases} \text{Example of a} \\ \text{polynomial in } x \end{cases}$$

This polynomial is written in **descending powers,** because the exponents on x decrease from left to right. On the other hand,

$$2x^3 - x^2 + \frac{4}{x} \longleftarrow \begin{cases} Not \text{ a polynomial because there} \\ \text{is a variable in the denominator} \end{cases}$$

is *not* a polynomial, since a variable appears in a denominator. Of course, we could define a *polynomial* using any variable, not just x, as in Example 1(c) above. In fact, polynomials may have terms with more than one variable, as in Example 1(d) above.

The **degree of a term** is the *sum* of the exponents on the variables. For example, $3x^4$ has degree 4, while $6x^{17}$ has degree 17. The term $5x$ has degree 1, and $2x^2y$ has degree $2 + 1 = 3$ (because y has an understood exponent of 1).

The **degree of a polynomial** is the highest degree of any nonzero term in the polynomial. For example, $3x^4 - 5x^2 + 6$ is a polynomial of degree 4, the polynomial $5x + 7$ is of degree 1, and $x^2y + xy - 5xy^2$ is of degree 3.

Three types of polynomials are very common and are given special names. A polynomial with only one term is called a **monomial** because *mon(o)-* means "one," as in *mono*rail. Some examples are shown below.

$$9m \qquad -6y^5 \qquad a^2 \qquad 6 \qquad \text{Monomials}$$

A polynomial with exactly two terms is called a **binomial** because *bi-* means "two," as in *bi*cycle. Examples are shown below.

$$-9x^4 + 9x^3 \qquad 8m^2 + 6m \qquad 3m^5 - 9m^2 \qquad \text{Binomials}$$

A polynomial with exactly three terms is called a **trinomial** because *tri-* means "three," as in *tri*angle. Examples are shown below.

$$9m^3 - 4m^2 + 6 \qquad -3m^5 - 9m^2 + 2 \qquad \text{Trinomials}$$

EXAMPLE 2 Classifying Polynomials

Simplify each polynomial, if possible, by combining like terms. Then give the degree and tell whether the simplified polynomial is a monomial, a binomial, a trinomial, or none of these.

(a) $2x^3 + 5$

The polynomial cannot be simplified. The degree is 3. The polynomial is a binomial because it has exactly two terms.

(b) $4x - 5x + 2x$

Add like terms to simplify: $4x - 5x + 2x = x$. The degree is 1 (since $x = x^1$). The simplified polynomial is a monomial because it has only one term.

◀◀◀ Now Try Exercises 27 and 33.

OBJECTIVE 3   Evaluate polynomials.   A polynomial usually represents different numbers for different values of the variable, as shown in the next example.

EXAMPLE 3 Evaluating a Polynomial

Evaluate $5x^3 - 4x - 4$ when $x = -2$ and when $x = 3$.

First, replace x with -2.

$$
\begin{aligned}
5x^3 - 4x - 4 &= 5(\mathbf{-2})^3 - 4(\mathbf{-2}) - 4 \\
&= 5(-8) - 4(-2) - 4 \qquad \text{Apply the exponent.} \\
&= -40 + 8 - 4 \qquad \text{Multiply.} \\
&= -36 \qquad \text{Add and subtract.}
\end{aligned}
$$

Next, replace x with 3.

$$
\begin{aligned}
5x^3 - 4x - 4 &= 5(\mathbf{3})^3 - 4(\mathbf{3}) - 4 \\
&= 5(27) - 4(3) - 4 \\
&= 135 - 12 - 4 \\
&= 119
\end{aligned}
$$

Continued on Next Page

CAUTION
Notice the use of parentheses around the numbers that are replacing the variable in Example 3. This is particularly important when a negative number replaces a variable that is raised to a power, so the sign of the product will be correct.

Now Try Exercise 41. ▶▶▶

OBJECTIVE 4 Add polynomials.   Polynomials may be added, subtracted, multiplied, and divided.

Adding Polynomials
To add two polynomials, add like terms.

EXAMPLE 4 Adding Polynomials Vertically

(a) Add $6x^3 - 4x^2 + 3$ and $-2x^3 + 7x^2 - 5$ vertically.

Write like terms in columns. Then add column by column.

$$\begin{array}{r} 6x^3 - 4x^2 + 3 \\ -2x^3 + 7x^2 - 5 \\ \hline 4x^3 + 3x^2 - 2 \end{array}$$

(b) Add $2x^2 - 4x + 3$ and $x^3 + 5x$ vertically.

Write like terms in columns and add column by column.

$$\begin{array}{r} 2x^2 - 4x + 3 \\ x^3 \qquad + 5x \qquad \\ \hline x^3 + 2x^2 + x + 3 \end{array}$$ Leave spaces for missing terms.

Now Try Exercise 45. ▶▶▶

The polynomials in Example 4 above also could be added horizontally by combining like terms, as shown in the next example.

EXAMPLE 5 Adding Polynomials Horizontally

Find each sum by adding horizontally.

(a) $(6x^3 - 4x^2 + 3) + (-2x^3 + 7x^2 - 5)$
Combine like terms.

$$(6x^3 - 4x^2 + 3) + (-2x^3 + 7x^2 - 5)$$

The sum is the same as in Example 4(a) above: $4x^3 + 3x^2 - 2$

(b) $(2x^2 - 4x + 3) + (x^3 + 5x)$

$= 2x^2 - 4x + 3 + x^3 + 5x$

$= x^3 + 2x^2 + x + 3$ Combine like terms.

Continued on Next Page

> **NOTE**
> Notice the order of the terms in the final answer to Example 5(b). Mathematicians generally write polynomials in descending order of power. The answers in this textbook will also be shown that way.

> **Now Try Exercise 51.**

OBJECTIVE 5 Subtract polynomials. Recall from **Section 1.4** that we changed subtraction problems to addition problems. To subtract two numbers, we *added* the first number to the *opposite* of the second number. Here are two examples as a quick review.

$$7 - 2 \qquad\qquad -8 - (-5)$$
$$7 + (-2) = 5 \qquad\qquad -8 + (+5) = -3$$

A similar method is used to subtract polynomials.

Subtracting Polynomials

To subtract two polynomials, change *all* the signs of the second polynomial and add the result to the first polynomial.

EXAMPLE 6 Subtracting Polynomials

(a) Perform this subtraction: $(5x - 2) - (3x - 8)$

$$(5x - 2) - (3x - 8)$$

Change subtraction to addition.
Change *every* sign in the *second* polynomial to its opposite.

$$= (5x - 2) + (-3x + 8)$$
$$= 2x + 6$$

Combine like terms.

(b) Subtract $6x^3 - 4x^2 + 2$ from $11x^3 + 2x^2 - 8$.

Rewrite the problem with the polynomials in the correct order.

$$(11x^3 + 2x^2 - 8) - (6x^3 - 4x^2 + 2)$$

Change every sign in the *second* polynomial; then *add* the two polynomials.

$$(11x^3 + 2x^2 - 8) + (-6x^3 + 4x^2 - 2)$$
$$= 5x^3 + 6x^2 - 10$$

To check a subtraction problem, use the fact that if $a - b = c$, then $a = b + c$. For example, $6 - 2 = 4$, so check by writing $6 = 2 + 4$, which is correct.

Check the polynomial subtraction above by adding $6x^3 - 4x^2 + 2$ and $5x^3 + 6x^2 - 10$. Since the sum is $11x^3 + 2x^2 - 8$, the subtraction was done correctly.

> **Now Try Exercises 49 and 57.**

10.4 Exercises

FOR EXTRA HELP

 Addison-Wesley Math Tutor Center

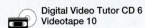 MathXL

Digital Video Tutor CD 6
Videotape 10

Student's Solutions Manual

 MyMathLab

 MathXL Tutorials on CD

Fill in each blank.

1. In the term $7x^5$, the coefficient is _____ and the exponent is _____.

2. The expression $5x^3 - 4x^2$ has _____ term(s).
   (How many?)

3. The degree of the term $-4x^8$ is _____.

4. The polynomial $4x^2 - y^2$ _____ an example of a trinomial.
   (is/is not)

5. Evaluate $x^2 + 10$ when $x = 4$. The result is _____.

6. _____ is an example of a monomial with coefficient 5, in the variable x, having degree 9.

For each polynomial, state the number of terms and identify the coefficient of each term.

7. $6x^4$

8. $-9y^5$

9. t^4

10. s^7

11. $-19r^2 - r$

12. $2y^3 - y$

13. $x + 8x^2 - 8x^3$

14. $v - 2v^3 + 2v^4$

Simplify each polynomial, if possible. Write the result with descending powers. See Example 1.

15. $-3m^5 + 5m^5$

16. $-4y^3 + 3y^3$

17. $2r^5 + (-3r^5)$

18. $-19y^2 + 9y^2$

19. $0.2m^5 - 0.5m^2$

20. $-0.9y + 0.9y^2$

21. $-3x^5 + 2x^5 - 4x^5$

22. $6x^3 - 8x^3 + 9x^3$

23. $-4p^7 + 8p^7 + 5p^9$

24. $-3a^8 + 4a^8 - 3a^2$

25. $-4y^2 + 3y^2 - 2y^2 + y^2$

26. $3r^5 - 8r^5 + r^5 + 2r^5$

Simplify each polynomial, if possible, and write it with descending powers. Then give the degree of the simplified polynomial, and tell whether it is a monomial, binomial, trinomial, *or* none of these. *See Examples 1 and 2.*

27. $6x^4 - 9x$

28. $7t^3 - 3t$

29. $5m^4 - 3m^2 + 6m^5 - 7m^3$

30. $6p^5 + 4p^3 - 8p^4 + 10p^2$

31. $\dfrac{5}{3}x^4 - \dfrac{2}{3}x^4 + \dfrac{1}{3}x^2 - 4$

32. $\dfrac{4}{5}r^6 + \dfrac{1}{5}r^6 - r^4 + \dfrac{2}{5}r$

33. $0.8x^4 - 0.3x^4 - 0.5x^4 + 7x$

34. $1.2t^3 - 0.9t^3 - 0.3t^3 + 9$

Evaluate each polynomial **(a)** *when* $x = 2$ *and* **(b)** *when* $x = -1$. *See Example 3.*

35. $-2x + 3$

36. $5x - 4$

37. $2x^2 + 5x + 1$

38. $-3x^2 + 14x - 2$

39. $2x^5 - 4x^4 + 5x^3 - x^2$

40. $x^4 - 6x^3 + x^2 + 1$

41. $-4x^5 + x^2$

42. $2x^6 - 4x$

Add each pair of polynomials vertically. See Example 4.

43. $3m^2 + 5m$ and $2m^2 - 2m$

44. $4a^3 - 4a^2$ and $6a^3 + 5a^2$

45. $-6x^3 - 4x + 1$
and $5x + 5$

46. $7n^2 + 2n - 4$
and $-6n^2 - 8$

47. $3w^3 - 2w^2 + 8w$
and $2w^2 - 8w + 5$

48. $-5r^3 - 4r^2 - 3r$
and $r^4 - 4r^2 + 3r$

Perform the indicated operations. See Examples 5 and 6.

49. $(12x^4 - x^2) - (8x^4 + 3x^2)$

50. $(13y^5 - y^3) - (7y^5 + 5y^3)$

51. $(2r^2 + 3r - 12) + (6r^2 + 2r)$

52. $(3r^2 + 5r - 6) + (2r - 5r^2)$

53. $(8m^2 - 7m) - (3m^2 + 7m - 6)$

54. $(x^2 + x) - (3x^2 + 2x - 1)$

55. $(16x^3 - x^2 + 3x) + (-12x^3 + 3x^2 + 2x)$

56. $(-2b^6 + 3b^4 - b^2) + (b^6 + 2b^4 + 2b^2)$

57. Subtract $5m^2 - 4$ from $12m^3 - 8m^2 + 6m + 7$

58. Subtract $a^2 + a - 1$ from $-3a^3 + 2a^2 - a + 6$

59. Subtract $9x^2 - 3x + 7$ from $-2x^2 - 6x + 4$.

60. Subtract $-5w^3 + 5w^2 - 7$ from $6w^3 + 8w + 5$.

Find the perimeter of each rectangle or triangle.

61.

$4x^2 + 3x + 1$

$x + 2$

62.

$5y^2 + 3y + 8$

$y + 4$

63.

$6t + 4$

$3t^2 + 2t + 7$

$5t^2 + 2$

64.

$6p^2 + p$

$2p + 5$

$9p^3 + 2p^2 + 1$

RELATING CONCEPTS (EXERCISES 65–68) For Individual or Group Work

At the start of this chapter, we gave a polynomial that models the distance in feet that a car going approximately 68 mph will skid in t seconds. If we let D represent the skidding distance, then we have this formula.

$$D = 100t - 13t^2$$

Use this formula as you **work Exercises 65–68 in order.**

65. To find the distance that the car will skid in 1 second, evaluate the polynomial when *t* is 1. Finish the work that is started below.

$D = 100t - 13t^2$ Replace *t* with 1.

$D = 100(1) - 13(1)^2$

$D =$

66. From Exercise 65, you should have found that when *t* is 1 second, then *D* is 87 ft. Now evaluate the polynomial for the other values of *t* listed in the table at the right. Write the ordered pairs in the table. Round distance (*D*) to the nearest whole number.

t (sec)	*D* (ft)	(*t, D*)
0		
0.5		
1	87	(1, 87)
1.5		
2		
2.5		
3		
3.5		

67. Graph each of the ordered pairs by making a dot at the appropriate point on the grid at the right.

68. Connect the points on the grid with a smooth curve. Describe how this graph is different from the graphs of linear equations in **Section 9.5.**

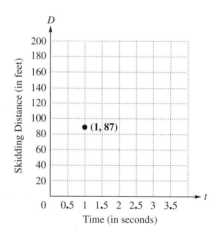

10.5 Multiplying Polynomials: An Introduction

OBJECTIVE 1 Multiply a monomial and a polynomial. As shown in **Section 10.1,** we find the product of two monomials by using the rules for exponents and the commutative and associative properties. Here is an example.

$$(-8m^6)(-9n^6) = (-8)(-9)(m^6)(n^6) = 72m^6n^6$$

OBJECTIVES

1 Multiply a monomial and a polynomial.

2 Multiply two polynomials.

CAUTION
Do not confuse *addition* of terms with *multiplication* of terms. Here is an example.

Adding Terms	Multiplying Terms
$7q^5 + 2q^5 = (7 + 2)q^5 = 9q^5$	$(7q^5)(2q^5) = (7 \cdot 2)q^{5+5} = 14q^{10}$

To find the product of a monomial and a polynomial with more than one term, we use the distributive property and multiplication of monomials.

EXAMPLE 1 **Multiplying a Monomial and a Polynomial**

Use the distributive property to find each product.

(a) $4x^2(3x + 5)$

$$4x^2(3x + 5) = 4x^2(3x) + 4x^2(5) \qquad \text{Distributive property}$$
$$= 12x^3 + 20x^2 \qquad \text{Multiply monomials.}$$

(b) $-8m^3(4m^3 + 3m^2 + 2m)$

$$= (-8m^3)(4m^3) + (-8m^3)(3m^2) + (-8m^3)(2m) \quad \text{Distributive property}$$
$$= -32m^6 + (-24m^5) + (-16m^4) \quad \text{Multiply monomials.}$$
$$= -32m^6 - 24m^5 - 16m^4$$

> **Now Try Exercises 1 and 7.** ▶▶▶

OBJECTIVE 2 Multiply two polynomials. We can use the distributive property repeatedly to find the product of any two polynomials. For example, to find the product of the polynomials $x^2 + 3x + 5$ and $x - 4$, think of $x - 4$ as a single quantity and use the distributive property as follows.

$$(x^2 + 3x + 5)(x - 4) = x^2(x - 4) + 3x(x - 4) + 5(x - 4)$$

Now use the distributive property three times to first find $x^2(x - 4)$, then $3x(x - 4)$, and finally $5(x - 4)$.

$$x^2(x - 4) + 3x(x - 4) + 5(x - 4)$$
$$= x^2(x) + x^2(-4) + 3x(x) + 3x(-4) + 5(x) + 5(-4)$$
$$= (x^3) + (-4x^2) + (3x^2) + (-12x) + (5x) + (-20)$$
$$= x^3 - x^2 - 7x - 20 \qquad \text{Combine like terms.}$$

This example suggests a rule for multiplying polynomials.

Multiplying Polynomials

To multiply two polynomials, multiply each term of the second polynomial by each term of the first polynomial; then add the products.

EXAMPLE 2 Multiplying Two Polynomials Horizontally

(a) Multiply $(x + 1)(x - 4)$.

Multiply each term of the second polynomial by each term of the first. Then combine like terms.

$$(x + 1)(x - 4) = x(x) + x(-4) + 1(x) + 1(-4)$$
$$= x^2 + (-4x) + x + (-4)$$
$$= x^2 - 3x - 4 \qquad \text{Combine like terms.}$$

(b) Multiply $(m^2 + 5)(4m^3 - 2m^2 + 4m)$.

Multiply each term of the second polynomial by each term of the first.

$$(m^2 + 5)(4m^3 - 2m^2 + 4m)$$
$$= m^2(4m^3) + m^2(-2m^2) + m^2(4m) + 5(4m^3) + 5(-2m^2) + 5(4m)$$
$$= (4m^5) + (-2m^4) + (4m^3) + (20m^3) + (-10m^2) + (20m)$$

Now combine like terms.

$$= 4m^5 - 2m^4 + 24m^3 - 10m^2 + 20m$$

◀◀◀ Now Try Exercises 9 and 13.

When multiplying polynomials with three or more terms, simplify the process by writing one polynomial above the other vertically.

EXAMPLE 3 Multiplying Polynomials Vertically

Multiply $(x^3 + 2x^2 + 4x + 1)(3x + 5)$ using the vertical method. Write the polynomials vertically, as shown below.

$$\begin{array}{r} x^3 + 2x^2 + 4x + 1 \\ 3x + 5 \\ \hline \end{array}$$

Here are the steps we will take.

Step 1 Multiply each term in the top row by 5.

Step 2 Multiply each term in the top row by $3x$. Be careful to place like terms in columns.

Step 3 Add like terms.

$$\begin{array}{r} x^3 + 2x^2 + 4x + 1 \\ 3x + 5 \\ \hline 5x^3 + 10x^2 + 20x + 5 \\ 3x^4 + 6x^3 + 12x^2 + 3x \\ \hline 3x^4 + 11x^3 + 22x^2 + 23x + 5 \end{array}$$

$\leftarrow$ *Step 1* $5(x^3 + 2x^2 + 4x + 1)$

$\leftarrow$ *Step 2* $3x(x^3 + 2x^2 + 4x + 1)$

$\leftarrow$ *Step 3* Add like terms.

The product is $3x^4 + 11x^3 + 22x^2 + 23x + 5$.

◀◀◀ Now Try Exercise 15.

10.5 Exercises

Find each product. See Example 1.

1. $5y(8y^2 - 3)$

2. $3a(4a^2 + 6a)$

3. $-2m(3m + 2)$

4. $-6p(3p - 1)$

5. $4x^2(6x^2 - 3x + 2)$

6. $7n^2(5n^2 + n - 4)$

7. $-3k^3(2k^3 - 3k^2 - k + 1)$

8. $-2b^4(-3b^3 - b^2 + 6b - 8)$

Find each product horizontally. See Example 2.

9. $(n - 2)(n + 3)$

10. $(r - 6)(r + 8)$

11. $(4r + 1)(2r - 3)$

12. $(5x + 2)(2x - 7)$

13. $(6x + 1)(2x^2 + 4x + 1)$

14. $(9y - 2)(8y^2 - 6y + 1)$

Find each product vertically. See Example 3.

15. $(4m + 3)(5m^3 - 4m^2 + m - 5)$

16. $(y + 4)(3y^3 - 2y^2 + y + 3)$

17. $(5x^2 + 2x + 1)(x^2 - 3x + 5)$

18. $(2m^2 + m - 3)(m^2 - 4m + 5)$

19. $(-x - 3)(-x - 4)$

20. $(-x + 5)(-x - 2)$

21. $(2x + 1)(x^2 + 3x)$

22. $(x + 4)(3x^2 + 2x)$

Match each product with the correct polynomial.

23. **Product**

$(a + 4)(a + 5)$ ___

$(a - 4)(a + 5)$ ___

$(a + 4)(a - 5)$ ___

$(a - 4)(a - 5)$ ___

Polynomial

A. $a^2 - a - 20$

B. $a^2 + a - 20$

C. $a^2 - 9a + 20$

D. $a^2 + 9a + 20$

24. **Product**

$(x - 5)(x + 3)$ ___

$(x + 5)(x + 3)$ ___

$(x - 5)(x - 3)$ ___

$(x + 5)(x - 3)$ ___

Polynomial

A. $x^2 + 8x + 15$

B. $x^2 - 8x + 15$

C. $x^2 - 2x - 15$

D. $x^2 + 2x - 15$

RELATING CONCEPTS (EXERCISES 25–30) For Individual or Group Work

Work Exercises 25–30 in order. *(All units are in yards.) Refer to the rectangular figure as necessary.*

25. Find a polynomial that represents the area of the rectangle.

26. Suppose you know that the area of the rectangle is 600 yd². Use this information and the polynomial from Exercise 25 to write an equation in x, and solve it.

27. What are the dimensions of the rectangle?

28. Suppose the rectangle represents a lawn and it costs $3.50 per square yard to lay sod on the lawn. How much will it cost to sod the entire lawn?

29. Use the result of Exercise 27 to find the perimeter of the lawn.

30. Again, suppose the rectangle represents a lawn and it costs $9.00 per yard to fence the lawn. How much will it cost to fence the lawn?

31. Perform these multiplications: $(x + 4)(x - 4)$; $(y + 3)(y - 3)$; $(r + 7)(r - 7)$.
Look at the products and explain the pattern that you see.

Chapter **10** Group Activity: EARTHQUAKE INTENSITIES MEASURED BY THE RICHTER SCALE

Objective   Write numbers in scientific notation; perform calculations involving scientific notation.

Charles F. Richter devised a scale in 1935 to compare the intensities, or relative power, of earthquakes. The **intensity** of an earthquake is measured relative to the intensity of a standard **zero-level** earthquake of intensity I_0. The relationship is equivalent to $I = I_0 \times 10^R$, where R is the **Richter scale** measure. For example, if an earthquake has magnitude 5.0 on the Richter scale, then its intensity is calculated as $I = I_0 \times 10^{5.0} = I_0 \times 100,000$, which is 100,000 times as intense as a zero-level earthquake. The following diagram illustrates the intensities of earthquakes and their Richter scale magnitudes.

To compare two earthquakes to each other, a ratio of the intensities is calculated. For example, to compare an earthquake that measures 8.0 on the Richter scale to one that measures 5.0, find the ratio of the intensities.

$$\frac{\text{intensity } 8.0}{\text{intensity } 5.0} = \frac{I_0 \times 10^{8.0}}{I_0 \times 10^{5.0}} = \frac{10^8}{10^5} = 10^{8-5} = 10^3 = 1000$$

Therefore an earthquake that measures 8.0 on the Richter scale is 1000 times as intense as one that measures 5.0.

The table below lists Richter scale measurements for several earthquakes.

Earthquake		Richter Scale Measurement
1960	Concepción, Chile	9.5
1906	San Francisco, California	8.3
1939	Erzincan, Turkey	8.0
1998	Sumatra, Indonesia	7.0
1998	Adana, Turkey	6.3

Source: World Almanac.

1. Compare the intensity of the 1939 Erzincan earthquake to the 1998 Sumatra earthquake.

2. Compare the intensity of the 1998 Adana earthquake to the 1906 San Francisco earthquake.

3. Compare the intensity of the 1939 Erzincan earthquake to the 1998 Adana earthquake.

4. Suppose an earthquake measures 7.2 on the Richter scale. How would the intensity of a second earthquake compare if its Richter scale measure differed by $+3.0$? by -1.0?

KEY TERMS

10.3 **scientific notation**
Scientific notation is used to write very large or very small numbers using exponents. The numbers are written as $a \times 10^n$, where $1 \leq |a| < 10$ and n is an integer.

10.4 **polynomial**
A polynomial is a term, or the sum of terms, with whole number exponents.

descending powers
A polynomial in x is written in descending powers if the exponents on x decrease from left to right.

degree of a term
The degree of a term is the sum of the exponents on the variables in that term.

degree of a polynomial
The degree of a polynomial is the highest degree of any term of the polynomial.

monomial
A monomial is a polynomial with only one term.

binomial
A binomial is a polynomial with exactly two terms.

trinomial
A trinomial is a polynomial with exactly three terms.

NEW SYMBOLS

a^{-n} a to the negative n power a^0 a to the zero power $a \times 10^n$ scientific notation

---

TEST YOUR WORD POWER

See how well you have learned the vocabulary in this chapter. Answers follow the Quick Review.

1. A **polynomial** is an algebraic expression made up of
   - **A.** a term or the product of several terms with whole number exponents
   - **B.** a term or the sum of several terms with whole number exponents
   - **C.** the product of two or more terms with negative exponents
   - **D.** the sum of two or more terms with negative exponents.

2. The **degree of a term** is
   - **A.** the number of variables in the term
   - **B.** the product of the exponents on the variables
   - **C.** the smallest exponent on the variables
   - **D.** the sum of the exponents on the variables.

3. A **trinomial** is a polynomial with
   - **A.** only one term
   - **B.** exactly two terms
   - **C.** exactly three terms
   - **D.** more than three terms.

4. A **binomial** is a polynomial with
   - **A.** only one term
   - **B.** exactly two terms
   - **C.** exactly three terms
   - **D.** more than three terms.

5. A **monomial** is a polynomial with
   - **A.** only one term
   - **B.** exactly two terms
   - **C.** exactly three terms
   - **D.** more than three terms.

6. An example of a polynomial written in **descending powers** is
   - **A.** $7x + 6x^2 - 4x^3 - 2x^4$
   - **B.** $x^6 - 2x^5 + 3x^2 - 6x$
   - **C.** $16 + x^5 - x^3 + x^2$
   - **D.** $3x^4 - 3x^2 - 3 - x$.

7. An example of a number written in **scientific notation** is
   - **A.** 0.25×10^5
   - **B.** 57.6×10^{-3}
   - **C.** 4.42×100^6
   - **D.** 3.61×10^{-4}.

8. The **degree of a polynomial** is
   - **A.** the number of terms in the polynomial
   - **B.** the product of the exponents on the variables
   - **C.** the highest degree of any term in the polynomial
   - **D.** the smallest exponent on the variables.

Concepts	Examples

10.1 The Product Rule and Power Rules for Exponents

For any integers m and n:

Product rule $a^m \cdot a^n = a^{m+n}$

Power rule (a) $(a^m)^n = a^{mn}$

Power rule (b) $(ab)^m = a^m b^m$

Power rule (c) $\left(\dfrac{a}{b}\right)^m = \dfrac{a^m}{b^m}$ when $b \neq 0$

$2^4 \cdot 2^5 = 2^{4+5} = 2^9$

$(3^4)^2 = 3^{4 \cdot 2} = 3^8$

$(6a)^5 = 6^5 a^5$

$\left(\dfrac{2}{3}\right)^4 = \dfrac{2^4}{3^4}$

10.2 Integer Exponents and the Quotient Rule

If $a \neq 0$, for integers m and n:

Zero exponent $a^0 = 1$

Negative exponent $a^{-n} = \dfrac{1}{a^n}$

Quotient rule $\dfrac{a^m}{a^n} = a^{m-n}$

$15^0 = 1$

$5^{-2} = \dfrac{1}{5^2} = \dfrac{1}{25}$

$\dfrac{4^8}{4^3} = 4^{8-3} = 4^5$

10.3 An Application of Exponents: Scientific Notation

To write a number in scientific notation ($a \times 10^n$), move the decimal point to the right of the first nonzero digit. The number of places moved is the absolute value of the exponent n. If the *original* number was "large" (10 or more), n is positive. If the *original* number was "small" (between 0 and 1), n is negative.

$247 = 2.47 \times 10^2$

$0.0051 = 5.1 \times 10^{-3}$

$3.25 \times 10^5 = 325{,}000$

$8.44 \times 10^{-6} = 0.00000844$

10.4 Adding and Subtracting Polynomials

Addition: Add like terms.

Add: $\begin{array}{r} 2x^2 + 5x - 3 \\ 5x^2 - 2x + 7 \\ \hline 7x^2 + 3x + 4 \end{array}$

Subtraction: Change the signs of all the terms in the second polynomial and add to the first polynomial.

Subtract: $(2x^2 + 5x - 3) - (5x^2 - 2x + 7)$
$= (2x^2 + 5x - 3) + (-5x^2 + 2x - 7)$
$= -3x^2 + 7x - 10$

10.5 Multiplying Polynomials

Multiply each term of the first polynomial by each term of the second polynomial. Then add like terms.

Multiply: $\begin{array}{r} 3x^3 - 4x^2 + 2x - 7 \\ 4x + 3 \\ \hline 9x^3 - 12x^2 + 6x - 21 \\ 12x^4 - 16x^3 + 8x^2 - 28x \\ \hline 12x^4 - 7x^3 - 4x^2 - 22x - 21 \end{array}$

ANSWERS TO TEST YOUR WORD POWER

1. B; *Example:* $5x^3 + 2x^2 - 7$
2. D; *Examples:* The term 6 has degree 0, the term $3x$ has degree 1, the term $-2x^8$ has degree 8, and the term $5x^2 y^4$ has degree 6.
3. C; *Example:* $2a^2 - 3ab + b^2$
4. B; *Example:* $3t^3 + 5t$
5. A; *Examples:* -5 and $4xy^5$ are both monomials.
6. B; *Example:* $3n^4 + n^3 - 2n + 8$ is written in descending powers.
7. D; *Example:* 2.5×10^4 is written in scientific notation.
8. C; *Example:* $x^2 y + xy - 5$ is a polynomial of degree 3, because the sum of the exponents in $x^2 y$ is $2 + 1 = 3$.

Chapter **10**

REVIEW EXERCISES

[10.1] *Use the product rule or power rules to simplify each expression. Write the answer in exponential form.*

1. $4^3 \cdot 4^8$

2. $(-5)^6(-5)^5$

3. $(-8x^4)(9x^3)$

4. $(2x^2)(5x^3)(x^9)$

5. $(19x)^5$

6. $(-4y)^7$

7. $5(pt)^4$

8. $\left(\dfrac{7}{5}\right)^6$

9. $(3x^2y^3)^3$

10. Explain why the product rule for exponents does not apply to the expression $7^2 + 7^4$.

[10.2] *Evaluate each expression.*

11. $5^0 + 8^0$

12. 2^{-5}

13. $10w^0$

14. $4^{-2} + 4^{-1}$

Simplify. Write each answer in exponential form, using only positive exponents. Assume that all variables are nonzero.

15. $\dfrac{6^{-3}}{6^{-5}}$

16. $\dfrac{x^{-7}}{x^{-9}}$

17. $\dfrac{p^{-8}}{p^4}$

18. $\dfrac{r^{-2}}{r^{-6}}$

19. $\dfrac{5^4}{5^5}$

20. $10^3(10^{-10})$

21. $n^{-2} \cdot n^3 \cdot n^{-4}$

22. $(10^4)(10^{-1})$

23. $(2^{-4})(2^{-4})$

24. $h^{10} \cdot h^{-4} \cdot h^{-3}$

25. $\dfrac{10^{-4}}{10^9}$

26. $\dfrac{x^2}{x^{-4}}$

[10.3] *Write each number in scientific notation.*

27. 48,000,000

28. 28,988,000,000

29. 0.000065

30. 0.0000000824

Write each number without exponents.

31. 2.4×10^4 **32.** 7.83×10^7 **33.** 8.97×10^{-7} **34.** 9.95×10^{-12}

Perform the indicated operations. Write your answers both in scientific notation and as numbers without exponents.

35. $(2 \times 10^{-3})(4 \times 10^5)$

36. $\dfrac{8 \times 10^4}{2 \times 10^{-2}}$

37. $\dfrac{12 \times 10^{-8}}{4 \times 10^{-3}}$

38. $\dfrac{(2.5 \times 10^5)(4.8 \times 10^{-4})}{(7.5 \times 10^8)(1.6 \times 10^{-5})}$

Solve each application problem. Write your answers both in scientific notation and as numbers without exponents.

39. The population of the United States in 2002 was about 2.88×10^8 people spread over 3.54×10^6 square miles of land. What is the average number of people per square mile? (*Source:* U.S. Bureau of the Census.)

40. Light travels at a speed of 3.0×10^5 kilometers per second. It takes about 36 seconds for light from the sun to reach Venus. How far is Venus from the sun? (*Source: World Almanac.*)

[10.4] *Simplify each polynomial, if possible, and write it with descending powers of the variable. Give the degree of the simplified polynomial, and tell whether it is a* monomial, binomial, trinomial, *or* none of these.

41. $9m^2 + 11m^2 + 2m^2$

42. $-4p + p^3 - p^2 + 8p + 2$

43. $12a^5 - 9a^4 + 8a^3 + 2a^2 - a + 3$

44. $-7y^5 - 8y^4 - y^5 + y^4 + 9y$

Add or subtract as indicated.

45. $(-2a^3 + 5a^2) + (-3a^3 - a^2)$ **46.** $(-4r^3 - 8r^2) + (6r + 4r^3)$ **47.** Subtract $-5y^2 + 2y - 7$
from $6y^2 - 8y + 2$.

48. $(2m^3 - 8m^2 + 4) + (8m^3 + 2m^2 - 7)$ **49.** $(-5y^2 + 3y + 11) + (4y^3 - 7y + 15)$

50. $(6p^2 - p - 8) - (-4p^2 + 2p + 3)$ **51.** $(12r^4 - 7r^3 + 2r^2) - (5r^4 - 3r^3 + 2r^2 + 1)$

[10.5] *Find each product.*

52. $5x(2x + 14)$ **53.** $-3p^3(2p^2 - 5p)$ **54.** $(3r - 2)(2r^2 + 4r - 3)$

55. $(2y + 3)(4y^2 - 6y + 9)$ **56.** $(5p^2 + 3p)(p^3 - p^2 + 5)$ **57.** $(3k - 6)(2k + 1)$

58. $(6p - 3q)(2p - 7q)$

59. $(m^2 + m - 9)(2m^2 + 3m - 1)$

MIXED REVIEW EXERCISES

Perform the indicated operations. Write answers with positive exponents.
Assume that all variables are nonzero.

60. $19^0 - 3^0$

61. $(3p)^4$

62. 7^{-2}

63. $-m^5(8m^2 + 10m + 6)$

64. $2^{-1} + 4^{-1}$

65. $(a + 2)(a^2 - 4a + 1)$

66. $(5y^3 - 8y^2 + 7) - (-3y^3 + y^2 + 2)$

67. Write a polynomial that represents the perimeter and one that represents the area of this rectangle.

Chapter 10

T E S T

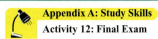
Appendix A: Study Skills
Activity 12: Final Exam

Evaluate each expression.

1. 5^{-4}

2. $(-3)^0 + 4^0$

3. $4^{-1} + 3^{-1}$

Simplify, and write each answer using only positive exponents. Assume that variables represent nonzero numbers.

4. $6^{-3} \cdot 6^4$

5. $12(xy)^3$

6. $\dfrac{10^5}{10^9}$

7. $r^{-4} \cdot r^{-4} \cdot r^3$

8. $(7x^2)(-3x^5)$

9. $\dfrac{m^{-5}}{m^{-8}}$

10. $(3a^4b)^2$

11. $3^4 + 3^2$

Write each number in scientific notation.

12. 300,000,000,000

13. 0.00000557

Write each number without exponents.

14. 2.9×10^7

15. 6.07×10^{-8}

1. _____

2. _____

3. _____

4. _____

5. _____

6. _____

7. _____

8. _____

9. _____

10. _____

11. _____

12. _____

13. _____

14. _____

15. _____

16. _____

16. The mass of the sun is about 330,000 times the mass of Earth. If Earth's mass is about 5.98×10^{24} kg, find the approximate mass of the sun. Write your answer in scientific notation.

Simplify each polynomial, when possible, and write it with descending powers of the variable. Give the degree of the simplified polynomial, and tell whether it is a monomial, binomial, trinomial, *or* none of these.

17. _____

17. $5x^2 + 8x - 12x^2$

18. _____

18. $13n^3 - n^2 + n^4 + 3n^4 - 9n^2$

Perform the indicated operations.

19. _____

19. $(5t^4 - 3t^2 + 3) - (t^4 - t^2 + 3)$

20. _____

20. $(2y^4 - 8y + 8) + (-3y^2 + 2y - 8)$

21. _____

21. Subtract $9t^3 + 8t^2 - 6$ from $9t^3 - 4t^2 + 2$.

22. _____

22. $-4x^3(3x^2 - 5x)$

23. _____

23. $(y + 2)(3y - 1)$

24. _____

24. $(2r - 3)(r^2 + 2r - 5)$

25. _____

25. Write a polynomial that represents the perimeter of this square, and a polynomial that represents the area.

$3x + 9$

Cumulative Review Exercises

CHAPTERS 1–10

1. Write these numbers in words.

   (a) 10.035

   (b) 410,000,351,109

2. Write these numbers in digits.

   (a) Three hundred million, six thousand, eighty

   (b) Fifty-five ten-thousandths

3. Round each number as indicated.

   (a) 0.8029 to the nearest hundredth

   (b) 340,519,000 to the nearest million

   (c) 14.973 to the nearest tenth

4. Find the mean, median, and mode for this set of ticket prices: \$41, \$65, \$37, \$90, \$41, \$65, \$48, \$41, \$59, \$40.

Simplify.

5. $-12 + 7.829$

6. $7 + 5(3 - 8)$

7. $1\dfrac{2}{3} + \dfrac{3}{5}$

8. $\dfrac{(-6)^2 + 9(0 - 4)}{10 \div 5(-7 + 5) - 10}$

9. $\dfrac{8x^2}{9} \cdot \dfrac{12w}{2x^3}$

10. $\dfrac{-0.7}{5.6}$

11. $\dfrac{10^5}{10^6}$

12. $2\dfrac{1}{4} - 2\dfrac{5}{6}$

13. $(-6)^2 + (-2)^3$

14. $(-4y^3)(3y^4)$

15. $\dfrac{5b}{2a^2} \div \dfrac{3b^2}{10a}$

16. $\dfrac{r}{8} + \dfrac{6}{t}$

17. $\dfrac{-4(6)}{3^3 - 27}$

18. $\dfrac{5}{6}$ of 900

19. $(-0.003)(0.04)$

20. $(2a^3b^2)^4$

21. $\dfrac{4}{9} - \dfrac{6}{m}$

22. $t^5 \cdot t^{-2} \cdot t^{-4}$

23. $\dfrac{-\dfrac{15}{16}}{-6}$

24. $\left(-\dfrac{1}{2}\right)^4 + 6\left(\dfrac{2}{3}\right)$

25. $\dfrac{n^{-3}}{n^{-4}}$

26. $-8 - 88$

27. $8^0 + 2^{-1}$

28. $9 - 7\dfrac{4}{5}$

Evaluate each expression when w = −2, x = 4, and y = −1.

29. $-3xy^3$

30. $15y - 6w$

31. $x^2 - 2xy + 6$

Solve each equation. Show your work.

32. $4 + h = 3h - 6$

33. $-1.65 = 0.5x + 2.3$

34. $3(a - 5) = -3 + a$

35. Translate into an equation and solve. If three times a number is subtracted from 20, the result is twice the number. Find the number.

36. Use the six problem-solving steps from **Chapter 3.** A mouse takes 16 times as many breaths as an elephant. If the two animals together take 170 breaths per minute, how many breaths does each take?

37. The Cardinals pitcher gave up 78 runs in 234 innings. At that rate, how many runs will he give up in a 9-inning game?

38. This month Dallas received 2.61 inches of rain. This is what percent of the normal rainfall of 1.8 inches?

39. Calbert bought a digital camera priced at $64.95. He paid $7\frac{1}{2}\%$ sales tax. Find the amount of tax, to the nearest cent, and the total cost of the camera.

40. The distance that light will travel in one year is 5.87×10^{12} miles. How far will light travel in one day? (Assume there are 365 days in a year.) Write your answer both in scientific notation and without exponents.

41. Use the table on toddler shoe sizes at the right.

   (a) What foot length corresponds to size 6 shoes?

   (b) If Kamesha measured her child's foot at $5\frac{7}{16}$ in., what shoe size should she order?

   (c) What is the difference in length between a size 5 and a size 6 shoe?

TODDLER SHOES

Trace bare foot, then measure from heel to longest toe.

Toddler	
Foot Length (in.)	Shoe Size
$4\frac{13}{16}$	5
$5\frac{1}{8}$	6
$5\frac{1}{2}$	7
$5\frac{13}{16}$	8

Source: Lands' End.

42. Find the perimeter and area of this shape.

6 m
3.2 m
4 m
3.8 m
3.2 m
6 m

43. Find the circumference and area of this circle. Use 3.14 for π and round answers to the nearest tenth.

9 cm

Whole Numbers Computation: Pretest

This test will check your skills in doing whole numbers computation, using paper and pencil. Each part of the test is keyed to a section in the Review Chapter, which follows this test. Based on your test results, work the appropriate section(s) in the Review Chapter *before you start Chapter 1.*

Adding Whole Numbers   *(Do not use a calculator.)*

1. 368	**2.** 7093	**3.** 85	**4.** 57,208
+ 22	+ 6073	+ 2968	915
			+ 59,387

5. 714 + 3728 + 9 + 683,775

1. _____

2. _____

3. _____

4. _____

5. _____

Subtracting Whole Numbers   *(Do not use a calculator.)*

1. 426	**2.** 3358	**3.** 30,602
− 76	− 2729	− 5708

4. 4006 − 97   **5.** 679,420 − 88,033

1. _____

2. _____

3. _____

4. _____

5. _____

Multiplying Whole Numbers   *(Do not use a calculator.)*

1. $3 \times 3 \times 0 \times 6$ **2.** 3841   **3.** (520)(3000)
                               $\times$ 7

1. _____

2. _____

(continued)

3. _____

4. _____

5. _____

6. _____

Do not use a calculator; show your work.

4.   71
      $\times$ 26

5. Multiply 359 and 48.

6. 853×609

Dividing Whole Numbers   *(Do not use a calculator; show your work.)*

1. _____

1. $3\overline{)69}$

2. $12 \div 0$

3. $\dfrac{25,036}{4}$

4. $7\overline{)5655}$

2. _____

3. _____

4. _____

5. $52\overline{)1768}$

6. $45,000 \div 900$

7. $38\overline{)2300}$

8. $83\overline{)44,799}$

5. _____

6. _____

7. _____

8. _____

Now check your answers on page A–00 in the Answers section at the back of the book. Record the number of problems you worked correctly in each part of the test.

Adding Whole Numbers: _____ correct out of 5.

   If you got 0, 1, or 2 correct, work **Section R.1** in the Review Chapter.

Subtracting Whole Numbers: _____ correct out of 5.

   If you got 0, 1, or 2 correct, work **Section R.2** in the Review Chapter.

Multiplying Whole Numbers: _____ correct out of 6.

   If you got 0, 1, 2, or 3 correct, work **Section R.3** in the Review Chapter.

Dividing Whole Numbers: _____ correct out of 8.

   If you got 0, 1, or 2 correct, work **Sections R.4** and **R.5** in the Review Chapter.

   If you got 3 or 4 correct, work **Section R.5** in the Review Chapter.

Whole Numbers Review

NOTE

Use the **Whole Numbers Computation: Pretest** on pages 749–750 to determine which sections you need to work in this chapter.

R.1 Adding Whole Numbers

There are 4 triangles at the left and 2 at the right. In all, there are 6 triangles.

The process of finding the total is called *addition*. Here 4 and 2 were added to get 6. Addition is written with a + sign, as shown below.

$$4 + 2 = 6$$

OBJECTIVE 1 Add two or three single-digit numbers.   In addition problems, the numbers being added are called **addends,** and the resulting answer is called the **sum** or **total.**

$$
\begin{array}{r}
4 \quad \leftarrow \text{Addend} \\
+\,2 \quad \leftarrow \text{Addend} \\
\hline
6 \quad \leftarrow \text{Sum (answer)}
\end{array}
$$

Addition problems can also be written horizontally as follows.

$$
\underset{\text{Addend}}{4} \quad + \quad \underset{\text{Addend}}{2} \quad = \quad \underset{\text{Sum}}{6}
$$

Commutative Property of Addition

The **commutative property of addition** states that changing the *order* of the addends in an addition problem does not change the sum.

For example, the sum of $4 + 2$ is the same as the sum of $2 + 4$. Both sums are 6. This allows the addition of the same numbers in a different order.

OBJECTIVES

1. Add two or three single-digit numbers.
2. Add more than two numbers.
3. Add when regrouping is not required.
4. Add with regrouping.
5. Use addition to solve application problems.
6. Check the sum in addition.

Associative Property of Addition

By the **associative property of addition,** changing the *grouping* of addends does not change the sum.

For example, the sum of $3 + 5 + 6$ may be found in several ways.

$(3 + 5) + 6 = 8 + 6 = 14$ Parentheses tell you to add $3 + 5$ first.

$3 + (5 + 6) = 3 + 11 = 14$ Parentheses tell you to add $5 + 6$ first.

Either method gives a sum of 14 because of the associative property of addition.

OBJECTIVE 2 Add more than two numbers.   To add several numbers, first write them in a column. Add the first number to the second. Add this sum to the third number; continue until all the numbers are used.

EXAMPLE 1 Adding More Than Two Numbers

Add 2, 5, 6, 1, and 4.

$$
\begin{array}{r}
2 \\
5 \\
⑥ \\
① \\
+④ \\
\hline
18
\end{array}
$$

$2 + 5 = 7$
$7 + 6 = 13$
$13 + 1 = 14$
$14 + 4 = 18$

Now Try Exercise 1.

NOTE

By the commutative and associative properties of addition, you may also add numbers by starting at the bottom of a column. Adding down from the top or adding up from the bottom will give the same sum.

OBJECTIVE 3 Add when regrouping is not required.   If numbers have two or more digits, first you must arrange the numbers in columns so that the ones digits are in the same column, tens are in the same column, hundreds are in the same column, and so on. Next, you add column by column, starting at the right.

EXAMPLE 2 Adding without Regrouping

Add $511 + 23 + 154 + 10$.

First line up the numbers in columns, with the ones column at the right.

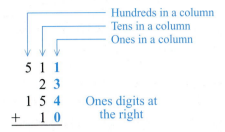

Hundreds in a column
Tens in a column
Ones in a column

$$
\begin{array}{r}
5\ 1\ 1 \\
2\ 3 \\
1\ 5\ 4 \\
+\ \ \ 1\ 0 \\
\end{array}
$$

Ones digits at the right

Now start at the right and add the ones digits. Add the tens digits next, and finally, the hundreds digits.

$$
\begin{array}{r}
5\ 1\ 1 \\
2\ 3 \\
1\ 5\ 4 \\
+\ \ \ 1\ 0 \\
\hline
6\ 9\ 8 \\
\end{array}
$$

Sum of ones
Sum of tens
Sum of hundreds

The sum of the four numbers is 698.

Now Try Exercise 3(a). ▶▶▶

OBJECTIVE 4 Add with regrouping. If the sum of the digits in a column is more than 9, use **regrouping** (sometimes called *carrying*).

EXAMPLE 3 Adding with Regrouping

Add 47 and 29.

Add the digits in the ones column.

$$
\begin{array}{r}
4\mathbf{7} \\
+\ 2\mathbf{9} \\
\end{array}
$$

7 ones + 9 ones = 16 ones

Regroup 16 ones as 1 ten and 6 ones. Write 6 ones in the ones column and write 1 ten in the tens column.

Write 1 ten in the tens column.

$$
\begin{array}{r}
1 \\
47 \\
+\ 29 \\
\hline
6 \\
\end{array}
$$

7 ones + 9 ones = 16 ones

Write 6 ones in the ones column.

Add the digits in the tens column, including the regrouped 1.

$$
\begin{array}{r}
1 \\
4\mathbf{7} \\
+\ 2\mathbf{9} \\
\hline
76 \\
\end{array}
$$

1 ten + 4 tens + 2 tens = 7 tens

Now Try Exercise 5. ▶▶▶

EXAMPLE 4 **Adding with Regrouping**

Add $324 + 7855 + 23 + 7 + 86$.

Step 1 Add the digits in the ones column.

Notice that 25 ones are regrouped as 2 tens and 5 ones.

Step 2 Add the digits in the tens column, including the regrouped 2.

Notice that 19 tens are regrouped as 1 hundred and 9 tens.

Step 3 Add the hundreds column, including the regrouped 1.

Notice that 12 hundreds are regrouped as 1 thousand and 2 hundreds.

Step 4 Add the thousands column, including the regrouped 1.

$$
\begin{array}{r}
1\,1\,2 \\
324 \\
7855 \\
23 \\
7 \\
+\ 86 \\
\hline
8295
\end{array}
$$

Sum of the thousands column is 8 thousands.

Thus, $324 + 7855 + 23 + 7 + 86 = 8295$.

Now Try Exercise 21.

OBJECTIVE 5 Use addition to solve application problems.

EXAMPLE 5 Applying Addition Skills

On this map, the distance in miles from one location to another is written alongside the road. Find the shortest route from Altamonte Springs to Clear Lake.

Approach   Add the mileage along various routes from Altamonte Springs to Clear Lake. Then select the shortest route.

Solution   One way from Altamonte Springs to Clear Lake is through Orlando. Add the mileage numbers along this route.

$$
\begin{array}{rl}
8 & \text{Altamonte Springs to Pine Hills} \\
5 & \text{Pine Hills to Orlando} \\
+\ 8 & \text{Orlando to Clear Lake} \\
\hline
21 & \rightarrow \text{miles from Altamonte Springs to Clear Lake,} \\
& \quad\ \text{going through Orlando}
\end{array}
$$

Another way is through Bertha and Winter Park. Add the mileage numbers along this route.

$$
\begin{array}{rl}
5 & \text{Altamonte Springs to Castleberry} \\
6 & \text{Castleberry to Bertha} \\
7 & \text{Bertha to Winter Park} \\
+\ 7 & \text{Winter Park to Clear Lake} \\
\hline
25 & \rightarrow \text{miles from Altamonte Springs to Clear Lake} \\
& \quad\ \text{through Bertha and Winter Park}
\end{array}
$$

The shortest route from Altamonte Springs to Clear Lake is 21 miles through Orlando.

Now Try Exercise 29.

EXAMPLE 6 Applying Addition Skills

Using the map in Example 5 above, find the total mileage from Shadow Hills to Castleberry to Orlando and back to Shadow Hills through Clear Lake.

Approach   Add the mileage from Shadow Hills to Castleberry to Orlando and back to Shadow Hills to find the total.

Solution   Use the numbers from the map.

$$
\begin{array}{rl}
9 & \text{Shadow Hills to Bertha} \\
6 & \text{Bertha to Castleberry} \\
5 & \text{Castleberry to Altamonte Springs} \\
8 & \text{Altamonte Springs to Pine Hills} \\
5 & \text{Pine Hills to Orlando} \\
8 & \text{Orlando to Clear Lake} \\
+\ 11 & \text{Clear Lake to Shadow Hills} \\
\hline
52 & \rightarrow \text{miles from Shadow Hills to Castleberry to} \\
& \quad\ \text{Orlando and back to Shadow Hills}
\end{array}
$$

Now Try Exercise 41.

OBJECTIVE **6** **Check the sum in addition.** Checking the answer is an important part of problem solving. A common method for checking addition is to re-add from the bottom to the top. This is an application of the commutative and associative properties of addition.

EXAMPLE 7 **Checking Addition**

Check each sum. If the sum is incorrect, find the correct sum.

(a)   Add down.

$$
\begin{array}{r}
\mathbf{1428} \\
738 \\
63 \\
125 \\
17 \\
+\ 485 \\
\hline
\mathbf{1428}
\end{array}
$$

To check, add up.

Adding down and adding up should give the same sum. In this case, the answers agree, so the sum is probably correct.

(b)

$$
\begin{array}{r}
785 \\
63 \\
+\ 185 \\
\hline
1033
\end{array}
\qquad
\begin{array}{r}
\mathbf{1033} \\
785 \\
63 \\
+\ 185 \\
\hline
\mathbf{1033}
\end{array}
$$

Correct, because both answers are the same

To check, add up.

(c)

$$
\begin{array}{r}
635 \\
73 \\
831 \\
+\ 915 \\
\hline
2444
\end{array}
\qquad
\begin{array}{r}
\mathbf{2454} \\
635 \\
73 \\
831 \\
+\ 915 \\
\hline
\mathbf{2444}
\end{array}
$$

Error, because answers are different

To check, add up.

Re-add to find that the correct sum is 2454.

Now Try Exercise 25.

R.1 Exercises

Add. See Examples 1 and 2.

1. (a)
```
   5
   7
   6
 + 5
```
(b)
```
   9
   2
   1
   3
 + 4
```
2. (a)
```
   2
   9
   5
 + 1
```
(b)
```
   3
   8
   6
   4
 + 8
```

3. (a) 3213 + 5715

(b) 38,204 + 21,020

4. (a) 6344 + 1655

(b) 63,251 + 36,305

Add, regrouping as necessary. See Examples 3–4.

5.
```
  67
+ 83
```
6.
```
  78
+ 36
```
7.
```
  746
+ 905
```
8.
```
  621
+ 359
```
9.
```
  798
+ 206
```

10.
```
  172
+ 156
```
11.
```
  7968
+ 1285
```
12.
```
  1768
+ 8275
```
13.
```
  7896
+ 3728
```
14.
```
  9382
+ 7586
```

15.
```
  3705
  3916
+ 9037
```
16.
```
  6629
  6076
+ 8218
```
17.
```
    32
+ 4977
```
18.
```
   402
+ 9938
```
19.
```
  3077
     8
+  421
```

20.
```
    56
  7721
+  172
```
21.
```
  9056
    78
  6089
+  731
```
22.
```
  4022
   709
  8621
+   37
```
23.
```
    18
   708
  9286
+  636
```
24.
```
  1708
   321
    61
+ 8926
```

Check each sum by adding from bottom to top. If an answer is incorrect, find the correct sum. See Example 7.

25.
```
 - - - -
   179
   214
 + 376
   759
```
26.
```
 - - - -
    17
   296
   713
 +  94
  1220
```
27.
```
 - - - -
  4713
    28
   615
 +  64
  5420
```
28.
```
  - - - -
  6 215
    744
     36
+ 4 284
 11,279
```

Using the map below, find the shortest route between each pair of cities. See Examples 5 and 6.

29. Southtown and Rena

30. Elk Hill and Oakton

31. Thomasville and Murphy

32. Austin and Wilson

Solve each application problem.

33. The sale price for a basic auto tune-up is $79, a tire rotation is $24, and an oil change is $19. Find the total cost for all the services.

34. Jane Lim bought an 11-piece set of golf clubs for $120, a dozen golf balls for $9, golf shoes for $45, and a golf glove for $12. How much did she spend in all? (*Source:* Sportmart.)

35. There are 413 women and 286 men on the sales staff. How many people are on the sales staff?

36. One department in an office building has 283 employees while another department has 218 employees. How many employees are in the two departments?

37. According to the latest census, the two states with the highest populations are California with 33,871,648 people and Texas with 20,851,820 people. How many people live in those two states? (*Source:* U.S. Bureau of the Census.)

38. The two states with the smallest populations are Wyoming with 493,782 people and Vermont with 608,827 people. What is the total population of the two states? (*Source:* U.S. Bureau of the Census.)

Find the perimeter of (total distance around) each figure.

39.

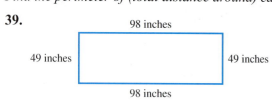

98 inches

49 inches   49 inches

98 inches

40.

65 meters

73 meters   73 meters

98 meters

41.

286 feet

114 feet

308 feet

42.

206 yards   197 yards

109 yards   109 yards

327 yards

R.2 Subtracting Whole Numbers

OBJECTIVE 1 Change addition problems to subtraction problems or the reverse. Suppose you have $18, and you spend $15 for gasoline. You then have $3 left. There are two different ways of looking at these numbers.

As an addition problem:

$$\$15 \;+\; \$3 \;=\; \$18$$

Amount   Amount   Original
spent   left   amount

As a subtraction problem:

$$\$18 \;-\; \$15 \;=\; \$3$$

Original Subtraction Amount Amount
amount   symbol   spent   left

OBJECTIVES

1 Change addition problems to subtraction problems or the reverse.

2 Identify the minuend, subtrahend, and difference.

3 Subtract when no regrouping is needed.

4 Use addition to check subtraction answers.

5 Subtract with regrouping.

6 Use subtraction to solve application problems.

As this example shows, an addition problem can be changed to a subtraction problem and a subtraction problem can be changed to an addition problem. Another example is illustrated below.

$$4 + 1 = 5$$

Two subtraction problems are possible.

$$5 - 1 = 4 \qquad \text{or} \qquad 5 - 4 = 1$$

$$5 - 1 = 4 \qquad\qquad 5 - 4 = 1$$

We can also change a subtraction problem to an addition problem. Here is an example.

$$8 - 3 = 5$$
$$8 = 3 + 5$$

It is also correct to write $8 = 5 + 3$ (using the commutative property).

OBJECTIVE 2 Identify the minuend, subtrahend, and difference. In subtraction, as in addition, the numbers in a problem have names. For example, in the problem $8 - 5 = 3$, the number 8 is the **minuend,** 5 is the **subtrahend,** and 3 is the **difference** or answer.

$$8 \;-\; 5 \;=\; 3 \leftarrow \text{Difference (answer)}$$

Minuend   Subtrahend

$$\begin{array}{r} 8 \leftarrow \text{Minuend} \\ -\,5 \leftarrow \text{Subtrahend} \\ \hline 3 \leftarrow \text{Difference} \end{array}$$

OBJECTIVE 3 Subtract when no regrouping is needed. Subtract two numbers by lining up the numbers in columns so that the digits in the ones place are in the same column. Next, subtract by columns, starting at the right with the ones column. Three subtraction problems are shown below.

Ones digits are lined up in the same column.

(a)
$$\begin{array}{r} 53 \\ -\ 21 \\ \hline 32 \end{array}$$

3 ones − 1 one = 2 ones
5 tens − 2 tens = 3 tens

Ones digits are lined up.

(b)
$$\begin{array}{r} 385 \\ -\ 165 \\ \hline 220 \end{array}$$ ← 5 ones − 5 ones = 0 ones

8 tens − 6 tens = 2 tens
3 hundreds − 1 hundred = 2 hundreds

(c)
$$\begin{array}{r} 9837 \\ -\ 210 \\ \hline 9627 \end{array}$$ ← 7 ones − 0 ones = 7 ones

3 tens − 1 ten = 2 tens
8 hundreds − 2 hundreds = 6 hundreds
9 thousands − 0 thousands = 9 thousands

OBJECTIVE 4 Use addition to check subtraction answers. You can check 8 − 3 = 5 by *adding* 3 and 5.

$$3 + 5 = 8, \quad \text{so} \quad 8 - 3 = 5 \quad \text{is correct.}$$

EXAMPLE 1 Checking Subtraction

Use addition to check each answer. If the answer is incorrect, find the correct answer.

(a)
$$\begin{array}{r} 89 \\ -\ 47 \\ \hline 42 \end{array}$$

Rewrite as an addition problem.

Subtraction problem
$$\begin{array}{r} 89 \\ -\ 47 \\ \hline 42 \\ \hline 89 \end{array}$$
Addition problem
$$\begin{array}{r} 47 \\ +\ 42 \\ \hline 89 \end{array}$$

Because 47 + 42 = 89, the subtraction was done correctly.

Continued on Next Page

(b) $72 - 41 = 21$

Rewrite as an addition problem.

$$72 = 41 + 21$$

But, $41 + 21 = 62$, *not* 72, so the subtraction was done *incorrectly*. Rework the original subtraction to get the correct answer of 31. Then, $41 + 31 = 72$.

(c)

$$\begin{array}{r} 374 \\ -141 \\ \hline 233 \end{array} \quad \begin{array}{l} \longleftarrow \text{Match} \longrightarrow \\ \Big\} \; 141 + 233 = 374 \end{array}$$

The answer checks.

Now Try Exercise 3. ▶▶▶

OBJECTIVE 5 Subtract with regrouping. If a digit in the minuend is *less* than the one directly below it, **regrouping** is necessary (also called *borrowing*).

EXAMPLE 2 Subtracting with Regrouping

Subtract 19 from 57.

Write the problem in vertical format.

$$\begin{array}{r} 5\;7 \\ -\;1\;9 \\ \hline \end{array}$$

In the ones column, 7 is *less* than 9, so, in order to subtract, you must regroup 1 ten as 10 ones.

5 tens $-$ 1 ten $=$ 4 tens $\longrightarrow$

$$\begin{array}{r} {}^{4}\;\;{}^{17} \\ \not5\;\not7 \\ -\;1\;9 \\ \hline \end{array}$$

$\leftarrow$ 1 ten = 10 ones, and 10 ones + 7 ones = 17 ones

Now subtract 17 ones minus 9 ones in the ones column. Then subtract 4 tens minus 1 ten in the tens column.

$$\begin{array}{r} {}^{4}\;\;{}^{17} \\ \not5\;\not7 \\ -\;1\;9 \\ \hline 3\;8 \end{array}$$

$\leftarrow$ Difference

Thus, $57 - 19 = 38$. Check by adding 19 and 38. You should get 57.

Now Try Exercise 7. ▶▶▶

EXAMPLE 3 Subtracting with Regrouping

Subtract, regrouping when necessary.

(a)

$$\begin{array}{r} 7856 \\ -\;137 \\ \hline \end{array}$$

Regroup 1 ten as 10 ones. ⟶ ⟶   10 ones + 6 ones = 16 ones

$$\begin{array}{r} {}^{4}\;\;{}^{16} \\ 7\;8\;\not5\;\not6 \\ -\;\;1\;3\;7 \\ \hline 7\;7\;1\;9 \end{array}$$

$\leftarrow$ Difference

Continued on Next Page

(b)   635
        − 546

Regroup 1 ten
as 10 ones.

10 ones + 5 ones
= 15 ones

Regroup 1 hundred
as 10 tens.

10 tens + 2 tens = 12 tens

```
      2 15
    6 3̶ 5̶        Need to regroup
  − 5 4 6        further because
        9        2 is less than 4
                 in tens column.
```

```
    5 12 15
    6̶ 3̶ 5̶
  − 5 4 6
      8 9   ← Difference
```

(c)   412
        − 225

```
    0 12
  4 1̶ 2̶        Need to regroup
− 2 2 5        further because
      7        0 is less than 2
               in tens column.
```

```
  3 10 12
  4̶ 1̶ 2̶
− 2 2 5
  1 8 7   ← Difference
```

Now Try Exercise 19.

Sometimes a minuend has zeros in some of the positions. In such cases, regrouping may be a little more complicated than what we have shown so far.

EXAMPLE 4   **Regrouping with Zeros**

Subtract.

$$\begin{array}{r} 4607 \\ - 3168 \\ \hline \end{array}$$

There are no tens that can be regrouped into ones. So you must first regroup 1 hundred as 10 tens.

Regroup 1 hundred
as 10 tens.

Write 10 tens.

```
      5 10
    4 6̶ 0̶ 7
  − 3 1 6 8
```

Now regroup 1 ten as 10 ones.

Regroup 1 ten as 10 ones;
10 tens − 1 ten = 9 tens

10 ones + 7 ones = 17 ones

```
          9  ←
    5 1̶0̶ 17  ←
    4 6̶ 0̶ 7̶
  − 3 1 6 8
          9
```

Complete the problem.

```
          9
    5 1̶0̶ 17
    4 6̶ 0̶ 7̶
  − 3 1 6 8
    1 4 3 9   ← Difference
```

Check by adding 1439 and 3168; you should get 4607.

Now Try Exercise 27.

EXAMPLE 5 Regrouping with Zeros

Subtract.

(a)   708
      − 149

Write 10 tens. ────────────╮   ┌──── Regroup 1 ten as 10 ones.
                             ▼
                             9
Regroup 1 hundred ─────→ 6 10 18 ←──── 10 ones + 8 ones = 18 ones
as 10 tens.   7 0̸ 8̸
                        − 1 4 9
                        ─────────
                          5 5 9

(b)   380
      − 276

Regroup 1 ten ──────╮   ╮   ┌──── Write 10 ones.
as 10 ones.   ▼   ▼
                    7 10
                 3 8̸ 0̸
               − 2 7 6
               ───────────
                 1 0 4

(c)   9000
      − 6999

            9 9
         8 10 10 10
         9̸ 0̸ 0̸ 0̸
       − 6 9 9 9
       ────────────
         2 0 0 1

Now Try Exercise 31. ▶▶▶

Recall that an answer to a subtraction problem can be checked by adding.

EXAMPLE 6 Checking Subtraction

Use addition to check each answer.

                                    Check
(a)   **613**   275
      − 275   *Matches*   + 338
      ─────   ─────
        338   **613**   Correct

                                    Check
(b)   **1915**   1635
      − 1635   *Matches*   + 280
      ──────   ──────
        280   **1915**   Correct

                                    Check
(c)   **15,803**   7 325
      − 7 325   *Does not match* + 8 578
      ────────   ───────
        8 578   **15,903** Error

Rework the original problem to get the correct answer, 8478.

Now Try Exercise 35. ▶▶▶

OBJECTIVE 6 **Use subtraction to solve application problems.**

EXAMPLE 7 Applying Subtraction Skills

Diana Lopez drives a United Parcel Service delivery truck. Using the table below, decide how many more deliveries were made by Lopez on Monday than on Thursday.

PACKAGE DELIVERY (LOPEZ)

Day	Number of Deliveries
Monday	137
Tuesday	126
Wednesday	119
Thursday	89
Friday	147
Saturday	0

Lopez made 137 deliveries on Monday, but had only 89 deliveries on Thursday. Find how many more deliveries were made on Monday than on Thursday by subtracting 89 from 137.

$$
\begin{array}{r}
137 \\
-\ \ 89 \\
\hline
48
\end{array}
$$

137 ← Deliveries on Monday
− 89 ← Deliveries on Thursday
48 ← More deliveries on Monday

Lopez made 48 more deliveries on Monday than she made on Thursday.

Now Try Exercise 39.

R.2 Exercises

Use addition to check each subtraction. If an answer is incorrect, find the correct answer.
See Example 1.

1. $\begin{array}{r} 89 \\ -27 \\ \hline 63 \end{array}$

2. $\begin{array}{r} 47 \\ -35 \\ \hline 13 \end{array}$

3. $\begin{array}{r} 382 \\ -261 \\ \hline 131 \end{array}$

4. $\begin{array}{r} 838 \\ -516 \\ \hline 322 \end{array}$

Subtract, regrouping when necessary. See Examples 2–5.

5. $\begin{array}{r} 36 \\ -28 \end{array}$

6. $\begin{array}{r} 97 \\ -39 \end{array}$

7. $\begin{array}{r} 83 \\ -58 \end{array}$

8. $\begin{array}{r} 65 \\ -28 \end{array}$

9. $\begin{array}{r} 45 \\ -29 \end{array}$

10. $\begin{array}{r} 93 \\ -37 \end{array}$

11. $\begin{array}{r} 719 \\ -658 \end{array}$

12. $\begin{array}{r} 916 \\ -618 \end{array}$

13. $\begin{array}{r} 771 \\ -252 \end{array}$

14. $\begin{array}{r} 973 \\ -788 \end{array}$

15. $\begin{array}{r} 9861 \\ -684 \end{array}$

16. $\begin{array}{r} 6171 \\ -1182 \end{array}$

17. $\begin{array}{r} 9988 \\ -2399 \end{array}$

18. $\begin{array}{r} 3576 \\ -1658 \end{array}$

19. $\begin{array}{r} 38,335 \\ -29,476 \end{array}$

20. $\begin{array}{r} 61,278 \\ -3\,559 \end{array}$

21. $\begin{array}{r} 40 \\ -37 \end{array}$

22. $\begin{array}{r} 80 \\ -73 \end{array}$

23. $\begin{array}{r} 60 \\ -37 \end{array}$

24. $\begin{array}{r} 70 \\ -27 \end{array}$

25. $\begin{array}{r} 6020 \\ -4078 \end{array}$

26. $\begin{array}{r} 7050 \\ -6045 \end{array}$

27. $\begin{array}{r} 8503 \\ -2816 \end{array}$

28. $\begin{array}{r} 16,004 \\ -5\,087 \end{array}$

29. $\begin{array}{r} 80,705 \\ -61,667 \end{array}$

30. $\begin{array}{r} 72,000 \\ -44,234 \end{array}$

31. $\begin{array}{r} 66,000 \\ -444 \end{array}$

32. $\begin{array}{r} 77,000 \\ -308 \end{array}$

33. $\begin{array}{r} 20,080 \\ -96 \end{array}$

34. $\begin{array}{r} 80,056 \\ -69 \end{array}$

Use addition to check each subtraction. If an answer is incorrect, find the correct answer.
See Example 6.

35. $\begin{array}{r} 3070 \\ -576 \\ \hline 2596 \end{array}$

36. $\begin{array}{r} 1439 \\ -1169 \\ \hline 270 \end{array}$

37. $\begin{array}{r} 27,600 \\ -807 \\ \hline 26,793 \end{array}$

38. $\begin{array}{r} 34,021 \\ -33,708 \\ \hline 727 \end{array}$

Solve each application problem. See Example 7.

39. A man burns 103 calories during 30 minutes of bowling while a woman burns 88 calories. How many fewer calories does a woman burn than a man?

40. Lynn Couch had $553 in her checking account. She wrote a check for $308 for school fees. How much is left in her account?

41. An airplane was carrying 254 passengers. When it landed in Atlanta, 133 passengers got off. How many passengers were left on the plane?

42. On Tuesday, 5822 people went to a soccer game, and on Friday, 7994 people went to a soccer game. How many more people went to the game on Friday?

This table shows the median yearly earnings for various occupations. Use the table to answer Exercises 43 and 44.

ALL WALKS OF LIFE
Median yearly earnings for selected occupations; starting salaries are lower

Executive secretary / administrative asst.	$31,090
Flight attendant	$38,820
Photographer	$22,350
News reporter	$29,110
Registered nurse	$44,840
State police officer	$44,400
Auto mechanic	$27,400
Accountant	$43,500
Electrical engineer	$64,910

Source: U.S. Department of Labor. *Occupational Outlook Handbook.*

43. (a) Identify the occupations with the highest and lowest earnings.

(b) What is the difference in the earnings for these two occupations?

44. (a) How much less are the earnings of an auto mechanic than an executive secretary?

(b) How much more are the earnings of an accountant than a flight attendant?

45. Downtown Toronto's skyline is dominated by the CN Tower, which rises 1815 feet. The Sears Tower in Chicago is 1450 feet high. Find the difference in height between the two structures. (*Source:* SkyscraperPage.com)

1815 ft

d

1450 ft

CN Tower   Sears Tower

46. The fastest animal in the world, the peregrine falcon, dives at 185 miles per hour. A Boeing 747 cruises at 580 miles per hour. How much faster does the plane fly than the falcon? (*Source: Top 10 of Everything.*)

Diving peregrine
185 mph

Boeing 747
580 mph

R.3   Multiplying Whole Numbers

Suppose we want to know the total number of computers in a computer lab. The computers are arranged in four rows with three computers in each row. Adding the number 3 a total of 4 times gives 12.

$$3 + 3 + 3 + 3 = 12$$

This result can be illustrated by the figure below.

OBJECTIVE **1** **Identify the parts of a multiplication problem.**   Multiplication is a shortcut for repeated addition. The numbers being multiplied are called **factors.** The answer is called the **product.** For example, the product of 3 and 4 can be written with the symbol $\times$, a raised dot, parentheses, or, in computer work, an asterisk.

$$
\begin{array}{r}
3 \leftarrow \text{Factor (also called } \textit{multiplicand}) \\
\times\ 4 \leftarrow \text{Factor (also called } \textit{multiplier}) \\
\hline
12 \leftarrow \text{Product (answer)}
\end{array}
$$

$$3 \times 4 = 12 \qquad 3 \cdot 4 = 12 \qquad (3)(4) = 12 \qquad 3 * 4 = 12$$

Raised dot   In computer work

Commutative Property of Multiplication

By the **commutative property of multiplication,** changing the *order* of two factors does not change the product.

For example: $3 \times 5 = 15$ and $5 \times 3 = 15$

Both products are 15.

> **CAUTION**
> Remember, addition also has a commutative property. For example, $4 + 2$ has the same sum as $2 + 4$. Subtraction, however, is *not* commutative.

OBJECTIVE **2** **Do chain multiplications.**   Some multiplications involve more than two factors.

Associative Property of Multiplication

By the **associative property of multiplication,** changing the *grouping* of factors does not change the product.

> **EXAMPLE 1** **Multiplying Three Numbers**

Multiply $2 \times 3 \times 5$.

$$(2 \times 3) \times 5 \qquad \text{Parentheses tell what to do first.}$$
$$6 \quad \times 5 = 30$$

Also,

$$2 \times (3 \times 5)$$
$$2 \times \quad 15 \quad = 30$$

By the associative property, either grouping results in the same product.

> **Calculator Tip** The calculator approach to Example 1 above uses chain calculations. Notice that you can enter *all* the factors before pressing the ⊜ key.
>
>
> 2 ⓧ 3 ⓧ 5 ⊜ **30**

Now Try Exercises 5 and 9.

OBJECTIVE 3 **Multiply by single-digit numbers.** Regrouping may be needed in multiplication problems with larger factors.

> **EXAMPLE 2** **Regrouping with Multiplication**

Multiply.

(a)
$$\begin{array}{r} 53 \\ \times \ \ 4 \\ \hline \end{array}$$

Start by multiplying in the ones column. Multiply 4 times 3 ones.

$$\begin{array}{r} \mathbf{1} \quad \longleftarrow \qquad \text{Write 1 ten in the tens column.} \\ 53 \\ \times \ \ \underline{4} \qquad 4 \times 3 \text{ ones} = \mathbf{12 \text{ ones}} \\ 2 \quad \longleftarrow \qquad \text{Write 2 ones in the ones column.} \end{array}$$

Next, multiply 4 times 5 tens.

$$\begin{array}{r} \mathbf{1} \\ 53 \\ \times \ \ \underline{4} \qquad 4 \times 5 \text{ tens} = 20 \text{ tens} \\ 2 \end{array}$$

Add the 1 ten that was written at the top of the tens column.

$$\begin{array}{r} \mathbf{1} \\ 53 \\ \times \ \ \underline{4} \\ \mathbf{21}2 \qquad 20 \text{ tens} + 1 \text{ ten} = \mathbf{21} \text{ tens} \end{array}$$

Continued on Next Page

(b)   724
       × 5

Work as shown below.

$$\begin{array}{r} ^{12} \\ 724 \\ \times \quad 5 \\ \hline 3620 \end{array}$$ ← 5 × 4 ones = 20 ones; write 0 ones;
write 2 tens in the tens column.
5 × 2 tens = 10 tens; add the 2 regrouped tens to get 12 tens;
write 2 tens; write 1 hundred in the hundreds column.
5 × 7 hundreds = 35 hundreds; add the
1 regrouped hundred to get 36 hundreds.

Now Try Exercises 11 and 15.

OBJECTIVE 4 Multiply quickly by numbers ending in zeros.   The product of two whole number factors is also called a **multiple** of either factor. For example, since 4 • 2 = 8, the number 8 is a multiple of 4, and 8 is also a multiple of 2. Multiples of 10 are very useful when multiplying. A *multiple of 10* is a whole number that ends in 0, such as 10, 20, or 30; 100, 200, or 300; 1000, 2000, or 3000; and so on. There is a short way to multiply by multiples of 10. Look at the following examples.

$$26 \times 1 = 26$$
$$26 \times 10 = 260$$
$$26 \times 100 = 2600$$
$$26 \times 1000 = 26{,}000$$

Do you see a pattern in the multiplications? These examples suggest the following rule.

> **Multiplying by Multiples of 10**
>
> To multiply a whole number by 10, by 100, or by 1000, attach one, two, or three zeros to the right of the whole number.

EXAMPLE 3 Multiplying by Multiples of 10

Multiply.

(a) $59 \times 10 = 590$
Attach 0

(b) $74 \times 100 = 7400$
Attach 00

(c) $803 \times 1000 = 803{,}000$
Attach 000

Now Try Exercise 19.

You can also find the product of other multiples of ten by attaching zeros.

EXAMPLE 4 **Multiplying with Other Multiples of 10**

Multiply.

(a) 75×3000

Multiply 75 by 3, and then attach three zeros.

$$\begin{array}{r} 75 \\ \times\ \ 3 \\ \hline 225 \end{array} \qquad 75 \times 3000 = 225{,}000$$

(b) 150×70

Multiply 15 by 7, and then attach two zeros.

$$\begin{array}{r} 15 \\ \times\ \ 7 \\ \hline 105 \end{array} \qquad 150 \times 70 = 10{,}500$$

Now Try Exercise 23.

OBJECTIVE 5 **Multiply by numbers having more than one digit.** The next example shows multiplication when both factors have more than one digit.

EXAMPLE 5 **Multiplying with More Than One Digit**

Multiply.

(a) $(46)(23)$

Rewrite the problem in vertical format. Then start by multiplying 46 by 3.

$$\begin{array}{r} 1 \\ 46 \\ \times\ \ 3 \\ \hline 138 \end{array} \quad \leftarrow 46 \times 3 = 138$$

Next, multiply 46 by 20.

$$\begin{array}{r} 1 \\ 46 \\ \times\ 20 \\ \hline 138 \\ 920 \end{array} \quad \leftarrow 46 \times 20 = 920$$

Add the results.

$$\begin{array}{r} 46 \\ \times\ \ 23 \\ \hline 138 \\ +\ \ 92 \\ \hline 1058 \end{array} \begin{array}{l} \leftarrow 46 \times 3 \\ \leftarrow 46 \times 20 \\ \leftarrow \text{Add to find the product.} \end{array}$$

Both 138 and 920 are called *partial products*. To save time, the 0 in 920 is usually not written.

$$\begin{array}{r} 46 \\ \times\ \ 23 \\ \hline 138 \\ 92 \\ \hline 1058 \end{array} \quad \begin{array}{l} \leftarrow \text{0 not written. Be very careful to} \\ \ \ \ \text{place the 2 in the tens column.} \end{array}$$

Continued on Next Page

(b)
```
      2 3 3
 ×    1 3 2
      4 6 6
    6 9 9          Tens lined up
  2 3 3            Hundreds lined up
  3 0,7 5 6   ←  Product
```

(c)
```
    5 3 8
 ×   4 6
```

First multiply by 6.
```
       24
      538
 ×    46          Regrouping is
      3228          needed here.
```

Now multiply by 4, being careful to line up the tens.

```
      1 3
      2 4
      5 3 8
 ×      4 6
      3 2 2 8  ⎤
      2 1 5 2  ⎦  Add the partial products.
      2 4,7 4 8
```

Now Try Exercise 33. ▶▶▶

When 0 appears in the multiplier, be sure to move the partial product to the left to account for the position held by the 0.

EXAMPLE 6 **Multiplication with Zeros**

Multiply.

(a)
```
      1 3 7
 ×    3 0 6
      8 2 2
    0 0 0          Tens lined up
  4 1 1            Hundreds lined up
  4 1,9 2 2
```

(b)
```
      1 4 0 6                      1 4 0 6
 ×    2 0 0 1                 ×    2 0 0 1
      1 4 0 6                      1 4 0 6
    0 0 0 0  ← 0 to line up tens ⎤
    0 0 0 0  ← 0 to line up hundreds ⎦    2 8 1 2 0 0  ← Zeros are written
  2 8 1 2                                                so that the
  2,8 1 3,4 0 6                    2,8 1 3,4 0 6         partial product
                                                        2812 starts in
                                                        the thousands
                                                        column.
```

> **CAUTION**
>
> In Example 6(b) above, in the solution on the right, be careful to insert zeros so that thousands are lined up in the thousands column.

Now Try Exercise 39. ▶▶▶

OBJECTIVE **6** **Use multiplication to solve application problems.**

EXAMPLE 7 **Applying Multiplication Skills**

Find the total cost of 24 cell phones priced at $59 each.

Approach   To find the cost of all the phones, multiply the cost of one phone ($59) by the number of phones (24).

Solution   Multiply $59 by 24.

$$
\begin{array}{r}
59 \\
\times\ \ 24 \\
\hline
236 \\
118\ \ \\
\hline
1416
\end{array}
$$

The total cost of 24 cell phones is $1416.

Calculator Tip   If you are using a calculator for Example 7 above, press the following keys.

59 ⊗ 24 ⊜ **1416**

Now Try Exercise 43.

R.3   Exercises

FOR EXTRA HELP

 Addison-Wesley Math Tutor Center

Math XL MathXL

📖 Student's Solutions Manual

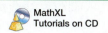 MyMathLab

💿 MathXL Tutorials on CD

Find each product. Try to do the work mentally. See Example 1.

1. $3 \times 1 \times 3$

2. $2 \times 8 \times 2$

3. $9 \times 1 \times 7$

4. $2 \times 4 \times 5$

5. $9 \cdot 5 \cdot 0$

6. $6 \cdot 0 \cdot 8$

7. $4 \cdot 1 \cdot 6$

8. $1 \cdot 5 \cdot 7$

9. $(2)(3)(6)$

10. $(4)(1)(9)$

Multiply. See Examples 2–6.

11.
$$\begin{array}{r} 35 \\ \times\ 7 \\ \hline \end{array}$$

12.
$$\begin{array}{r} 76 \\ \times\ 9 \\ \hline \end{array}$$

13.
$$\begin{array}{r} 28 \\ \times\ 6 \\ \hline \end{array}$$

14.
$$\begin{array}{r} 83 \\ \times\ 5 \\ \hline \end{array}$$

15.
$$\begin{array}{r} 3182 \\ \times\quad 6 \\ \hline \end{array}$$

16.
$$\begin{array}{r} 7326 \\ \times\quad 5 \\ \hline \end{array}$$

17.
$$\begin{array}{r} 36{,}921 \\ \times\qquad 7 \\ \hline \end{array}$$

18.
$$\begin{array}{r} 28{,}116 \\ \times\qquad 4 \\ \hline \end{array}$$

19.
$$\begin{array}{r} 125 \\ \times\ 100 \\ \hline \end{array}$$

20.
$$\begin{array}{r} 246 \\ \times\ 100 \\ \hline \end{array}$$

21.
$$\begin{array}{r} 1485 \\ \times\quad 30 \\ \hline \end{array}$$

22.
$$\begin{array}{r} 8522 \\ \times\quad 50 \\ \hline \end{array}$$

23.
$$\begin{array}{r} 900 \\ \times\ 300 \\ \hline \end{array}$$

24.
$$\begin{array}{r} 400 \\ \times\ 700 \\ \hline \end{array}$$

25.
$$\begin{array}{r} 43{,}000 \\ \times\quad 2000 \\ \hline \end{array}$$

26.
$$\begin{array}{r} 11{,}000 \\ \times\quad 9000 \\ \hline \end{array}$$

27.
$$\begin{array}{r} 68 \\ \times 22 \\ \hline \end{array}$$

28.
$$\begin{array}{r} 82 \\ \times 32 \\ \hline \end{array}$$

29.
$$\begin{array}{r} 83 \\ \times 45 \\ \hline \end{array}$$

30. $(43)(27)$

31. $(32)(475)$

32. $(67)(218)$

33. $(729)(45)$

34. $(681)(47)$

35.
$$\begin{array}{r} 538 \\ \times\ 342 \\ \hline \end{array}$$

36.
$$\begin{array}{r} 3228 \\ \times\quad 751 \\ \hline \end{array}$$

37.
$$\begin{array}{r} 8162 \\ \times\quad 407 \\ \hline \end{array}$$

38.
$$\begin{array}{r} 528 \\ \times\ 106 \\ \hline \end{array}$$

39.
$$\begin{array}{r} 6310 \\ \times\quad 3008 \\ \hline \end{array}$$

40.
$$\begin{array}{r} 3533 \\ \times\quad 5001 \\ \hline \end{array}$$

Solve each application problem. See Example 7.

41. Giant kelp plants in the ocean can grow 18 inches each day. How much could kelp grow in two weeks? How much could it grow in a 30-day month? (*Source: Natural Bridges State Park, CA.*)

42. A hospital has 20 bottles of thyroid medication, with each bottle containing 2500 tablets. How many of these tablets does the hospital have in all?

43. There are 12 tomato plants to a flat. If a garden center has 48 flats, find the total number of tomato plants.

44. A hummingbird's wings beat about 65 times per second, a chickadee's wings about 27 times per second. How many times does each bird's wings beat in one minute? (*Source: Birder's Handbook.*)

45. A Prius hybrid automobile gets 55 miles per gallon on the highway. How many miles can it travel on 11 gallons of gas?

46. Find the total cost of 16 gallons of paint at $18 per gallon.

Ruby-throated hummingbird
65 wing beats per second

Use addition, subtraction, or multiplication to solve each problem.

47. The distance from Reno, Nevada, to the Atlantic Ocean is 2695 miles, while the distance from Reno to the Pacific Ocean is 255 miles. How much farther is it to the Atlantic Ocean than it is to the Pacific Ocean? If you make three round trips from Reno to the Atlantic Ocean, how many frequent flier miles will you earn?

48. The largest living land mammal is the African elephant, and the largest mammal of all time is the blue whale. An African bull elephant may weigh 15,225 pounds and a blue whale may weigh 12 to 25 times that amount. Find the range of weights for the blue whale and the difference between the lightest and heaviest. (*Source: Big Book of Knowledge.*)

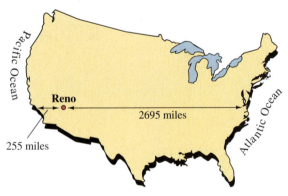

49. A high-fat meal contains 1406 calories, while a low-fat meal contains 348 calories. How many more calories are in seven high-fat meals than in seven low-fat meals?

50. Dannie Sanchez bought four tires at $110 each, two seat covers at $49 each, and six socket wrenches at $3 each. Find the total amount that he spent.

R.4 Dividing Whole Numbers

Suppose $12 is to be divided into 3 equal parts. Each part would be $4, as shown here.

OBJECTIVES

1 Write division problems in three ways.

2 Identify the parts of a division problem.

3 Divide 0 by a number.

4 Recognize that division by 0 is undefined.

5 Divide a number by itself.

6 Use short division.

7 Use multiplication to check quotients.

8 Use tests for divisibility.

$12 total

3 equal parts

OBJECTIVE 1 Write division problems in three ways.   Just as $3 \cdot 4$, 3×4, and $(3)(4)$ are different ways of writing the multiplication of 3 and 4, there are several ways to write 12 divided by 3.

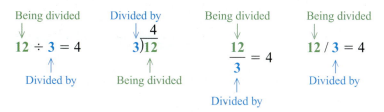

Being divided
↓
$12 \div 3 = 4$
↑
Divided by

Divided by
↓ 4
$3\overline{)12}$
↑
Being divided

Being divided
↓
$\dfrac{12}{3} = 4$
↑
Divided by

Being divided
↓
$12\,/\,3 = 4$
↑
Divided by

We will use three division symbols, $\div$, $\overline{)}\,$, and —. In algebra the bar, —, is frequently used. In computer science the slash, /, is used.

OBJECTIVE 2 Identify the parts of a division problem.   In division, the number being divided is the **dividend,** the number divided by is the **divisor,** and the answer is the **quotient.**

$$\text{dividend} \div \text{divisor} = \text{quotient}$$

$$\text{divisor}\overline{)\;\text{dividend}}^{\text{quotient}} \qquad \dfrac{\text{dividend}}{\text{divisor}} = \text{quotient}$$

Here are examples of three different ways to write division with the dividend, divisor, and quotient identified.

(a) $35 \div 7 = 5$

$35 \div 7 = 5$ ← Quotient
↗   ↖
Dividend   Divisor

(b) $\dfrac{100}{20} = 5$

Dividend
↓
$\dfrac{100}{20} = 5$ ← Quotient
↑
Divisor

(c) $12\overline{)72}^{\,6}$

$\phantom{12\overline{)}}6$ ← Quotient
$12\overline{)72}$ ← Dividend
↑
Divisor

OBJECTIVE 3 Divide 0 by a number. If no money, or $0, is divided equally among five people, each person gets $0. There is a general rule for dividing 0.

> **Dividing 0**
>
> When 0 is divided by any other number (except 0), the quotient is 0.

EXAMPLE 1 Dividing 0 by a Number

Divide. Then rewrite each division using two other symbols.

(a) $0 \div 12 = \textcolor{red}{0}$ can also be written $12\overline{)0}$ or $\dfrac{0}{12} = 0$

(b) $\dfrac{0}{375} = \textcolor{red}{0}$ can also be written $0 \div 375 = 0$ or $375\overline{)0}$

Now Try Exercise 5.

Recall that a subtraction such as $8 - 3 = 5$ can be written as the addition $8 = 3 + 5$. In a similar way, any division can be written as a multiplication. For example, $12 \div 3 = 4$ can be written as $3 \times 4 = 12$ or, by the commutative property, $4 \times 3 = 12$

Here are three more division problems and their corresponding multiplications.

(a) $\dfrac{20}{4} = 5$ becomes $4 \cdot 5 = 20$

(b) $8\overline{)48}$ becomes $8 \cdot 6 = 48$

(c) $72 \div 9 = 8$ becomes $9 \cdot 8 = 72$

OBJECTIVE 4 Recognize that division by 0 is undefined. Division by 0 cannot be done. To see why, try to find the answer to this division.

$$9 \div 0 = ?$$

As we have just seen, any division problem can be changed to a multiplication problem so that

$$\text{divisor} \cdot \text{quotient} = \text{dividend}.$$

If you convert the problem $9 \div 0 = \textcolor{blue}{?}$ to its multiplication counterpart, it reads as follows.

$$0 \cdot ? = 9$$

You already know that 0 times any number must always equal 0. Try any number you like to replace the **?** and you'll always get 0 instead of 9. Therefore, the division problem $9 \div 0$ *cannot* be done. Mathematicians say it is *undefined* and have agreed never to divide by 0. However, $0 \div 9$ *can* be done. Check by rewriting it as a multiplication problem.

$$0 \div 9 = 0 \quad \text{because} \quad 9 \cdot 0 = 0 \quad \text{is true.}$$

Dividing by 0

Dividing by 0 cannot be done. We say that division by 0 is *undefined*. It is impossible to compute an answer.

EXAMPLE 2 **Dividing by 0 Is Undefined**

All of these divisions are *undefined*.

(a) $\dfrac{6}{0}$ is undefined.

(b) $0\overline{)8}$ is undefined.

(c) $18 \div 0$ is undefined.

(d) $\dfrac{0}{0}$ is undefined.

Division Involving 0

$$\frac{0}{\text{nonzero number}} = 0 \qquad \textbf{but} \qquad \frac{\text{number}}{0} \text{ is } \textit{undefined.}$$

CAUTION

When 0 is the divisor in a problem, write *undefined*. You can never divide by 0.

Now Try Exercise 3.

Calculator Tip   Try these two problems on your calculator. Jot down your answers.

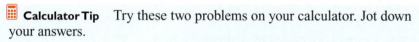

When you try to divide by 0, the calculator cannot do it, so it shows the word "Error" in the display, or the letters "ERR" or "E" (for "error").

OBJECTIVE **5** **Divide a number by itself.**   What happens when a number is divided by itself? For example, $4 \div 4$ or $97 \div 97$?

Dividing a Number by Itself

When a nonzero number is divided by itself, the quotient is 1.

EXAMPLE 3 **Dividing a Nonzero Number by Itself**

Divide.

(a) $16 \div 16 = 1$

(b) $32\overline{)32}$ with 1 above

(c) $\dfrac{57}{57} = 1$

Now Try Exercise 1.

OBJECTIVE 6 Use short division.   **Short division** is a quick method of dividing a number by a one-digit divisor.

EXAMPLE 4 Using Short Division

Divide $2\overline{)86}$.

First, divide 8 by 2.

$$\overset{4}{2\overline{)86}}\qquad \frac{8}{2}=4$$

Next, divide 6 by 2.

$$\overset{43}{2\overline{)86}}\qquad \frac{6}{2}=3$$

The quotient is 43.

Now Try Exercise 11.

When two numbers do *not* divide evenly, the leftover portion is called the **remainder.** The remainder is always *less* than the divisor.

EXAMPLE 5 Using Short Division with a Remainder

Divide 147 by 4.

Rewrite the problem. $4\overline{)147}$

Because 1 cannot be divided by 4, divide 14 by 4.

$$\overset{3}{4\overline{)14^{2}7}}\qquad \frac{14}{4}=3 \text{ with 2 left over}$$

Next, divide 27 by 4. The final number left over is the remainder. Write the remainder to the side. "R" stands for remainder.

$$\overset{3\ 6\ \mathbf{R3}}{4\overline{)14^{2}7}}\qquad \frac{27}{4}=6 \text{ with 3 left over}$$

The quotient is 36 **R**3.

Now Try Exercise 15.

EXAMPLE 6 **Writing Zeroes in the Quotient**

Divide 1439 by 7.

Rewrite the problem. Then, divide 14 by 7.

$$\frac{2}{7)1439} \qquad \frac{14}{7} = 2 \text{ with 0 left over}$$

No need to write 0

Next, divide 3 by 7. But 7 will not go into 3 even one time, so write a 0 in the quotient.

$$\frac{20}{7)143^39} \qquad \frac{3}{7} = 0 \text{ with 3 left over}$$

Finally, divide 39 by 7.

$$\frac{20\ 5\ \mathbf{R4}}{7)143^39} \qquad \frac{39}{7} = 5 \text{ with 4 left over}$$

Now Try Exercise 17.

OBJECTIVE 7 **Use multiplication to check quotients.**   Check the answer to a division problem as follows.

Checking Division

(divisor × quotient) + remainder = dividend

Parentheses tell you what to do first. In this case, multiply the divisor by the quotient first and then add the remainder.

EXAMPLE 7 **Using Multiplication to Check Division**

Check each quotient.

(a) $\frac{91\ \mathbf{R3}}{5)458}$

(divisor × quotient) + remainder = dividend

$$(5 \times 91) + 3$$
$$455 + 3 = \mathbf{458}$$

Matches original dividend so the division was done correctly

Continued on Next Page

$$
\begin{array}{r}
29\ \mathbf{R}4 \\
\textbf{(b)}\ 6\overline{)1258}
\end{array}
$$

(divisor × quotient) + remainder = dividend

$$
(6\ \times\ 29)\ +\ 4
$$

$$
174\ \ +\ 4\ =\ \mathbf{178}
$$

Does **not** match original dividend of 1258

The quotient does *not* check. Rework the original problem to get the correct quotient, 209 **R**4. Then, to check, $(6 \times 209) + 4 = 1254 + 4 = 1258$, the original dividend.

> **CAUTION**
> A common error is forgetting to add the remainder. Be sure to add any remainder when checking a division problem.

Now Try Exercise 31.

OBJECTIVE 8 Use tests for divisibility. It is often important to know whether a number is divisible by another number. You will find this useful in **Chapter 4** when working with fractions.

> **Divisibility**
> A whole number is *divisible* by another whole number if the remainder is 0.

There are some quick tests you can use to decide whether one number is divisible by another.

> **Tests for Divisibility**
> **A number is divisible by**
>
> 2   if it ends in 0, 2, 4, 6, or 8.
>
> 3   if the sum of its digits is divisible by 3.
>
> 4   if the last two digits make a number that is divisible by 4.
>
> 5   if it ends in 0 or 5.
>
> 6   if it is divisible by both 2 and 3.
>
> 8   if the last three digits make a number that is divisible by 8.
>
> 9   if the sum of its digits is divisible by 9.
>
> 10   if it ends in 0.

The most commonly used divisibility tests are those for 2, 3, 5, and 10.

Divisibility by 2

A number is divisible by **2** if the number ends in 0, 2, 4, 6, or 8.

EXAMPLE 8   **Testing for Divisibility by 2**

Which numbers are divisible by 2?

(a) 986

         Ends in 6

Because the number ends in 6, which is in the list (0, 2, 4, 6, or 8), the number 986 is divisible by 2.

(b) 3255 is *not* divisible by 2.

         Ends in 5, and *not* in 0, 2, 4, 6, or 8

Now Try Exercise 47(a).

Divisibility by 3

A number is divisible by **3** if the sum of its digits is divisible by **3.**

EXAMPLE 9   **Testing for Divisibility by 3**

Which numbers are divisible by 3?

(a) 4251

Add the digits.

$$4\ 2\ 5\ 1$$
$$4 + 2 + 5 + 1 = 12$$

Because 12 is divisible by 3, the number 4251 is divisible by 3.

(b) 29,806

Add the digits.

$$2\ 9\ 8\ 0\ 6$$
$$2 + 9 + 8 + 0 + 6 = 25$$

Because 25 is *not* divisible by 3, the number 29,806 is *not* divisible by 3.

> **CAUTION**
> Be careful when testing for divisibility by *adding the digits*. This method works only when testing for divisibility by 3 or by 9.

Now Try Exercise 47(b).

Divisibility by 5 and by 10

A number is divisible by **5** if it ends in 0 or 5.

A number is divisible by **10** if it ends in 0.

EXAMPLE 10 **Determining Divisibility by 5**

Which numbers are divisible by 5?

(a) 12,900   ends in 0, so it is divisible by 5.

(b) 4325   ends in 5, so it is divisible by 5.

(c) 392   ends in 2, so it is *not* divisible by 5.

Now Try Exercise 47(c).

EXAMPLE 11 **Determining Divisibility by 10**

Which numbers are divisible by 10?

(a) 700 and 9140   end in 0, so both numbers are divisible by 10.

(b) 355 and 18,743   do *not* end in 0, so these numbers are *not* divisible by 10.

Now Try Exercise 49 and 55.

R.4 Exercises

Divide. Then rewrite each division using two other division symbols. See Examples 1–3.

1. $\dfrac{12}{12}$

2. $\dfrac{9}{0}$

3. $24 \div 0$

4. $4 \div 4$

5. $\dfrac{0}{4}$

6. $0 \div 8$

7. $0 \div 12$

8. $\dfrac{0}{7}$

9. $0\overline{)21}$

10. $2 \div 0$

Divide by using short division. See Examples 4–6. Also, in Exercises 11–14, identify the dividend, the divisor, and the quotient.

11. $4\overline{)84}$

12. $2\overline{)66}$

13. $3\overline{)639}$

14. $4\overline{)488}$

15. $6\overline{)9137}$

16. $9\overline{)8371}$

17. $6\overline{)1858}$

18. $8\overline{)859}$

19. $4024 \div 4$

20. $16,024 \div 8$

21. $15,019 \div 3$

22. $32,013 \div 8$

23. $\dfrac{26,684}{4}$

24. $\dfrac{16,398}{9}$

25. $\dfrac{74,751}{6}$

26. $\dfrac{72,543}{5}$

27. $\dfrac{71,776}{7}$

28. $\dfrac{77,621}{3}$

29. $\dfrac{128,645}{7}$

30. $\dfrac{172,255}{4}$

Check each quotient. If a quotient is incorrect, find the correct quotient. See Example 7.

31. $7\overline{)4692}$ quotient 67 **R**2

32. $9\overline{)5974}$ quotient 663 **R**5

33. $6\overline{)21,409}$ quotient $3\ 568$ **R**2

34. $4\overline{)103,516}$ quotient $25,879$

35. $6\overline{)18,023}$ quotient $3\ 003$ **R**5

36. $8\overline{)33,664}$ quotient $4\ 208$

37. $6\overline{)69,140}$ quotient $11,523$ **R**2

38. $3\overline{)82,598}$ quotient $27,532$ **R**1

Solve each application problem.

39. Kaci Salmon, a supervisor at Albany Electric, earns $184 for an 8-hour shift. The workers she supervises each earn $112 or $152 for an 8-hour shift. What are the hourly wages for Kaci and for her workers?

40. Ken Griffey, Jr., has a nine-year baseball contract for $117,000,000. Find his pay for each year. (*Source: USA Today.*)

41. Six identical delivery vans for Rosita's Bakery cost a total of $99,600. Find the cost of each van.

42. Ted Slauson, coordinator of Toys for Tots, has collected 2628 toys. If his group gives four toys to each child, how many children will receive the gifts?

A college theater group is looking at two different budgets to cover the cost of props, costumes, and programs for the spring play. They are not sure whether to charge $5, $7, or $9 per ticket. The play's director organized the information in the table below. Use the table for Exercises 43–46.

Budget Amount	Number of $5 Tickets	Number of $7 Tickets	Number of $9 Tickets
$1890			
$2205			

43. For the $1890 budget, find the number of tickets that would need to be sold at $5 each; at $7 each; at $9 each. Write your answers in the table at the left.

44. For the $2205 budget, find the number of tickets that would need to be sold at $5 each; at $7 each; at $9 each. Write your answers in the table at the left.

45. If the director thinks that 300 tickets can be sold, which ticket price will cover the $1890 budget? How much extra money will there be?

46. If the director thinks that 300 tickets can be sold, which ticket price will cover the $2205 budget? How much money will be left over?

47. Circle the numbers that are

   (a) divisible by 2

     358   2047   190   85

   (b) divisible by 3

     736   10,404   5603   78

   (c) divisible by 5.

     53   13,740   985   5506

48. Circle the numbers that are

   (a) divisible by 2

     443   3500   256   74

   (b) divisible by 3

     26,001   9316   92   840

   (c) divisible by 5.

     5051   95   30,652   710

Put a ✓ mark in the blank if the number at the left is divisible by the number at the top of each column. Use the divisibility tests from Examples 8–11.

	2	3	5	10
49. 30	___	___	___	___
51. 184	___	___	___	___
53. 445	___	___	___	___
55. 903	___	___	___	___
57. 5166	___	___	___	___
59. 21,763	___	___	___	___

	2	3	5	10
50. 25	___	___	___	___
52. 192	___	___	___	___
54. 897	___	___	___	___
56. 500	___	___	___	___
58. 8302	___	___	___	___
60. 32,472	___	___	___	___

R.5 Long Division

Long division is used to divide by a number with more than one digit.

OBJECTIVES

1 Do long division.
2 Divide multiples of 10.
3 Use multiplication to check quotients.

OBJECTIVE 1 Do long division. In long division, estimate the various numbers by using a *trial divisor* to get a *trial quotient*.

EXAMPLE 1 Using a Trial Divisor and a Trial Quotient

Divide. $42\overline{)3066}$

Because 42 is closer to 40 than to 50, use 4 as a trial divisor.

$$42$$
↑ —— Trial divisor is 4

Try to divide the first digit of the dividend by 4. Because 3 cannot be divided by 4, use the first *two* digits, 30.

$$\frac{30}{4} = 7 \text{ with remainder 2}$$

$$\begin{array}{r} 7 \leftarrow \text{Trial quotient} \\ 42\overline{)3066} \end{array}$$

↑ —— 7 goes over the 6, because
$\frac{306}{42}$ is about 7

Multiply 7 and 42 to get 294; next, subtract 294 from 306.

$$\begin{array}{r} 7 \\ 42\overline{)3066} \\ \underline{294} \leftarrow 7 \times 42 = 294 \\ 12 \leftarrow 306 - 294 = 12 \end{array}$$

Bring down the 6 at the right.

$$\begin{array}{r} 7 \\ 42\overline{)3066} \\ \underline{294}\downarrow \\ 126 \leftarrow 6 \text{ brought down} \end{array}$$

Use the trial divisor, 4.

$$\begin{array}{r} 73 \\ 42\overline{)3066} \\ \underline{294} \\ 126 \leftarrow \text{First two digits of 126} \rightarrow \frac{12}{4} = 3 \\ \underline{126} \leftarrow 3 \times 42 = 126 \\ 0 \end{array}$$

Check the quotient by multiplying 42 and 73. The product should be 3066, which matches the original dividend.

CAUTION
The first digit in the answer in long division must be placed in the proper position over the dividend.

Now Try Exercises 1 and 9. ▶▶▶

EXAMPLE 2 **Dividing to Find a Trial Quotient**

Divide. $58\overline{)2730}$

Use 6 as a trial divisor, because 58 is closer to 60 than to 50.

First two digits → $\dfrac{27}{6} = 4$ with remainder 3
of dividend

$$
\begin{array}{r}
4 \quad \leftarrow \text{ Trial quotient} \\
58\overline{)2730} \\
\underline{232} \quad \leftarrow 4 \times 58 = 232 \\
41 \quad \leftarrow 273 - 232 = 41 \text{ (smaller than 58,} \\
\text{the divisor)}
\end{array}
$$

Bring down the 0.

$$
\begin{array}{r}
4 \\
58\overline{)2730} \\
\underline{232}\downarrow \\
410 \quad \leftarrow 0 \text{ brought down}
\end{array}
$$

$$
\begin{array}{r}
46 \\
58\overline{)2730} \\
\underline{232} \\
410 \quad \rightarrow \text{First two digits} \\
\underline{348} \quad \leftarrow 6 \times 58 = 348 \\
62 \quad \leftarrow \text{Greater than 58}
\end{array}
$$

First two digits of 410 $\dfrac{41}{6} = 6$ with remainder 5

The remainder, 62, is *greater than the divisor,* 58, so **7** should be used in the quotient instead of **6**.

$$
\begin{array}{r}
47 \ \textbf{R4} \quad \leftarrow \\
58\overline{)2730} \\
\underline{232} \\
410 \\
\underline{406} \quad \leftarrow 7 \times 58 = 406 \\
4 \quad \leftarrow 410 - 406 = 4
\end{array}
$$

Now Try Exercise 13.

Sometimes it is necessary to write a 0 in the quotient.

EXAMPLE 3 **Writing Zeros in the Quotient**

Divide. $42\overline{)8734}$

Start as above in Example 2.

$$
\begin{array}{r}
2 \\
42\overline{)8734} \\
\underline{84} \quad \leftarrow 2 \times 42 = 84 \\
3 \quad \leftarrow 87 - 84 = 3
\end{array}
$$

Bring down the 3.

$$
\begin{array}{r}
2 \\
42\overline{)8734} \\
\underline{84}\downarrow \\
33 \quad \leftarrow 3 \text{ brought down}
\end{array}
$$

Continued on Next Page

Since 33 cannot be divided by 42, write a 0 in the quotient as a placeholder.

$$\begin{array}{r} 2\textbf{0} \quad \leftarrow \text{0 in quotient} \\ 42\overline{)8734} \\ \underline{84} \\ 33 \end{array}$$

Bring down the final digit, the 4.

$$\begin{array}{r} 20 \\ 42\overline{)873\textbf{4}} \\ \underline{84} \downarrow \\ 33\textbf{4} \quad \leftarrow \text{4 brought down} \end{array}$$

Complete the problem.

$$\begin{array}{r} 207 \ \textbf{R}40 \\ 42\overline{)8734} \\ \underline{84} \\ 334 \\ \underline{294} \quad \leftarrow 7 \times 42 = 294 \\ 40 \quad \leftarrow 334 - 294 = 40 \end{array}$$

The quotient is 207 **R**40.

> **CAUTION**
> There must be a digit in the quotient (answer) above *every* digit in the dividend *once the answer has begun*. Notice that in Example 3 above, a 0 was used to assure an answer digit above every digit in the dividend.

Now Try Exercise 17. ⫸

OBJECTIVE 2 Divide multiples of 10.   When the divisor and dividend both contain zeros at the far right, recall that these numbers are multiples of 10. There is a short way to divide multiples of 10. Look at the following examples.

$$26,000 \div 1 = 26,000$$
$$26,000 \div 10 = 2600$$
$$26,000 \div 100 = 260$$
$$26,000 \div 1000 = 26$$

Do you see a pattern in these divisions using multiples of 10? These examples suggest the following rule.

> **Dividing by Multiples of 10**
> Divide a whole number by 10, by 100, or by 1000 by dropping one, two, or three zeros from the whole number.

EXAMPLE 4 Dividing by Multiples of 10

Divide.

(a) $60 \div 10 = 6$

— 0 in divisor
— Drop 0 from dividend.

Continued on Next Page

(b) 3500 ÷ 100 = 35

              ━ 00 in divisor

           ━ Drop 00 from dividend.

(c) 915,000 ÷ 1000 = 915

              ━ 000 in divisor

           ━ Drop 000 from dividend.

You can also find the quotient for other multiples of 10 by dropping zeros.

(d) 40)11,000   Drop 0 from the divisor and the dividend.

$$
\begin{array}{r}
275 \\
4\overline{)1100} \\
\underline{8} \\
30 \\
\underline{28} \\
20 \\
\underline{20} \\
0
\end{array}
$$

(e) 3500)31,500   Drop two zeros from the divisor and the dividend.

$$
\begin{array}{r}
9 \\
35\overline{)315} \\
\underline{315} \\
0
\end{array}
$$

Now Try Exercise 23.

OBJECTIVE 3 Use multiplication to check quotients. Quotients in long division can be checked just as quotients in short division were checked. Multiply the quotient and divisor, then add any remainder. The result should match the original dividend.

EXAMPLE 5 Checking Division

Check the quotient.

$$
\begin{array}{r}
114 \textbf{ R43} \\
48\overline{)5324}
\end{array}
$$

$$
\begin{array}{r}
114 \\
\times\ 48 \\
\hline
912 \\
456 \\
\hline
5472 \\
+\ \ 43 \\
\hline
5515
\end{array}
$$

Multiply the quotient and the divisor.

← Add the remainder.

← Does **not** match original dividend of 5324

The quotient does *not* check. Rework the original problem to get 110 **R44**.

Now Try Exercise 31.

R.5   Exercises

*First, indicate what number to use as the trial divisor. Then use the trial divisor to find the first digit in the quotient and write it in the correct position. Do **not** finish the division.*

1. $24\overline{)768}$

Use _____ as the trial divisor.

2. $35\overline{)805}$

Use _____ as the trial divisor.

3. $18\overline{)4500}$

Use _____ as the trial divisor.

4. $28\overline{)3500}$

Use _____ as the trial divisor.

5. $86\overline{)10,327}$

Use _____ as the trial divisor.

6. $51\overline{)24,026}$

Use _____ as the trial divisor.

7. $52\overline{)38,025}$

Use _____ as the trial divisor.

8. $63\overline{)34,400}$

Use _____ as the trial divisor.

9. $77\overline{)249,826}$

Use _____ as the trial divisor.

10. $92\overline{)247,892}$

Use _____ as the trial divisor.

11. $420\overline{)470,800}$

Use _____ as the trial divisor.

12. $190\overline{)901,050}$

Use _____ as the trial divisor.

Divide by using long division. See Examples 1–3. In Exercises 21–24, drop zeros before dividing. See Example 4.

13. $29\overline{)1859}$

14. $58\overline{)2204}$

15. $47\overline{)11,121}$

16. $83\overline{)39,692}$

17. $26\overline{)62,583}$

18. $28\overline{)84,249}$

19. $63\overline{)78,072}$

20. $55\overline{)43,223}$

21. $150\overline{)499,760}$

22. $720\overline{)52,560}$

23. $400\overline{)340,000}$

24. $900\overline{)153,000}$

Check each division. If a quotient is incorrect, find the correct quotient. See Example 5.

25. 56)5943   106 **R**17

26. 87)3254   37 **R**37

27. 600)394,800   658 **R**9

28. 300)139,100   463 **R**200

29. 410)25,420   62 **R**3

30. 760)132,600   174 **R**360

31. 72)32,465   450 **R**65

32. 47)9570   23 **R**29

Solve each application problem by using addition, subtraction, multiplication, or division.

33. In 1900, the average workweek was 59 hours. Today it is 38 hours. How many more hours were worked each year in 1900 than today? Assume 50 weeks of work per year. (*Source:* Reiman Publications.)

34. The U.S. Government Printing Office uses 255,000 pounds of ink each year. If it does an equal amount of printing on each of 200 workdays in a year, find the weight of the ink used each day.

35. The most expensive standard hotel room in a recent study was the Ritz-Carlton at $375 per night, while the least expensive was Motel 6 at $32 per night. Find the amount saved by staying four nights at Motel 6 instead of the Ritz-Carlton. (*Source: USA Today.*)

36. Two divorced parents share their child's education costs, which amount to $3718 per year. If one parent pays $1880 each year, find the amount paid by the other parent over five years.

37. Judy Martinez owes $11,088 on a loan. Find her monthly payment if the loan is to be paid off in 36 months.

38. A consultant charged $28,160 for studying the environmental impact of a new shopping mall. If the consultant worked 220 hours, find the rate charged per hour.

39. Clarence Hanks can assemble 42 circuits in one hour. How many circuits can he assemble in a 5-day workweek of 8 hours per day?

40. There are two conveyer lines in a factory, each of which packages 240 sacks of salt per hour. If the lines operate for 8 hours, find the total number of sacks of salt packaged by the two lines.

41. A youth soccer association brought in $7588 from fund-raising projects. There were expenses of $838 that had to be paid first, with the balance of the money divided evenly among the 18 teams. How much did each team receive?

42. Feather Farms Egg Ranch collected 3545 eggs in the morning and 2575 eggs in the afternoon. If the eggs are packed in flats containing 30 eggs each, find the number of flats needed for packing.

Chapter R

KEY TERMS

R.1 **addends** — Addends are the numbers being added in an addition problem.

sum (total) — The answer in an addition problem is the sum (total).

commutative property of addition — The commutative property of addition states that changing the *order* of the addends in an addition problem does not change the sum.

associative property of addition — The associative property of addition states that changing the *grouping* of addends does not change the sum.

regrouping — Regrouping is used in an addition problem when the sum of the digits in a column is greater than 9; also called *carrying*.

R.2 **minuend** — In the subtraction problem $8 - 5 = 3$, the 8 is the minuend. It is the number from which another number is subtracted.

subtrahend — In the subtraction problem $8 - 5 = 3$, the 5 is the subtrahend. It is the number being subtracted.

difference — The answer in a subtraction problem is the difference.

regrouping — Regrouping is used in subtraction if a digit is less than the one directly below it; also called *borrowing*.

R.3 **factors** — The numbers being multiplied are factors. For example, in $3 \times 4 = 12$, both 3 and 4 are factors.

product — The answer in a multiplication problem is the product.

commutative property of multiplication — The commutative property of multiplication states that changing the *order* of the factors in a multiplication problem does not change the product.

associative property of multiplication — The associative property of multiplication states that changing the *grouping* of factors does not change the product.

multiple — The product of two whole number factors is a multiple of both those numbers.

R.4 **dividend** — In division, the number being divided is the dividend.

divisor — In division, the number being used to divide another number is the divisor.

quotient — The answer in a division problem is the quotient.

short division — Short division is a quick method of dividing a number by a one-digit divisor.

remainder — The remainder is the number left over when two numbers do not divide evenly. The remainder is always less than the divisor.

R.5 **long division** — The process of long division is used to divide by a number with more than one digit.

QUICK REVIEW

Concepts

R.1 *Addition of Whole Numbers*

Add from top to bottom, starting with the ones column and working left. To check, add from bottom to top.

Examples

```
        1 1 4 0
          6 8 7   ⎫
           2 6    ⎬ Addends
              9   ⎪
Add up   + 4 1 8  ⎭
to check.  1 1 4 0    Sum
```

Concepts	Examples

R.1 *Commutative Property of Addition*

Changing the *order* of the addends in an addition problem does not change the sum.

$$2 + 4 = 6$$
$$4 + 2 = 6$$

By the commutative property, the sum is the same.

R.1 *Associative Property of Addition*

Changing the *grouping* of addends does not change the sum.

$$(2 + 3) + 4 = 5 + 4 = 9$$
$$2 + (3 + 4) = 2 + 7 = 9$$

By the associative property, the sum is the same.

R.2 *Subtraction of Whole Numbers*

Subtract the subtrahend from the minuend to get the difference; use regrouping when necessary. To check, add the difference to the subtrahend to get the minuend.

Problem

```
      6 12 18
    4 7 3 8   ← Minuend      Matches
  -   6 4 9   ← Subtrahend
    4 0 8 9   ← Difference
```

Check

```
      6 4 9
  + 4 0 8 9
    4 7 3 8
```

R.3 *Multiplication of Whole Numbers*

The numbers being multiplied are called *factors*. The multiplicand is being multiplied by the multiplier, giving the product. When the multiplier has more than one digit, partial products must be found and then added to find the product.

```
      7 8   ← Multiplicand  }
  ×   2 4   ← Multiplier    } Factors
      3 1 2 ← Partial product
    1 5 6   ← Partial product (move one position left)
    1 8 7 2 ← Product
```

R.3 *Commutative Property of Multiplication*

Changing the *order* of the factors in a multiplication problem does not change the product.

$$3 \times 4 = 12$$
$$4 \times 3 = 12$$

By the commutative property, the product is the same.

R.3 *Associative Property of Multiplication*

Changing the *grouping* of factors does not change the product.

$$(2 \times 3) \times 4 = 6 \times 4 = 24$$
$$2 \times (3 \times 4) = 2 \times 12 = 24$$

By the associative property, the product is the same.

R.4 *Division of Whole Numbers*

$\div$ and $\overline{)}$ mean divide.

Also a —, as in $\frac{25}{5}$, means to divide the top number (dividend) by the bottom number (divisor).

$$88 \div 4 = 22$$

Dividend ↑ Quotient ↑ Divisor ↑

```
             22   ← Quotient
Divisor → 4 )88   ← Dividend
             88
              0
```

$$\text{Dividend} \longrightarrow \frac{88}{4} = 22 \quad \leftarrow \text{Quotient}$$
$$\text{Divisor} \longrightarrow$$

Chapter **R**
REVIEW EXERCISES

If you need help with any of these review exercises, look in the section indicated in the red brackets.

[R.1] *Add.*

1. $\begin{array}{r} 74 \\ + 18 \\ \hline \end{array}$

2. $\begin{array}{r} 35 \\ + 78 \\ \hline \end{array}$

3. $\begin{array}{r} 807 \\ 4606 \\ + \quad 51 \\ \hline \end{array}$

4. $\begin{array}{r} 8215 \\ 9 \\ + 7433 \\ \hline \end{array}$

[R.2] *Subtract.*

5. $\begin{array}{r} 238 \\ - 199 \\ \hline \end{array}$

6. $\begin{array}{r} 573 \\ - 389 \\ \hline \end{array}$

7. $\begin{array}{r} 2210 \\ - 1986 \\ \hline \end{array}$

8. $\begin{array}{r} 99{,}704 \\ - 73{,}838 \\ \hline \end{array}$

[R.3] *Multiply. Do the work mentally.*

9. $2 \times 4 \times 6$

10. $9 \times 1 \times 5$

11. $(6)\,(1)\,(8)$

12. $7 \cdot 7 \cdot 0$

Multiply.

13. $\begin{array}{r} 43 \\ \times \ 4 \\ \hline \end{array}$

14. $\begin{array}{r} 781 \\ \times \ \ 7 \\ \hline \end{array}$

15. $\begin{array}{r} 5440 \\ \times \quad 6 \\ \hline \end{array}$

16. $\begin{array}{r} 93{,}105 \\ \times \qquad 5 \\ \hline \end{array}$

Multiply by using the shortcuts for multiples of 10.

17. $\begin{array}{r} 320 \\ \times \ \ 60 \\ \hline \end{array}$

18. $\begin{array}{r} 280 \\ \times \ \ 90 \\ \hline \end{array}$

19. $\begin{array}{r} 517 \\ \times \ \ 400 \\ \hline \end{array}$

20. $\begin{array}{r} 16{,}000 \\ \times \quad 8\,000 \\ \hline \end{array}$

Multiply.

21. $\begin{array}{r} 34 \\ \times \ 18 \\ \hline \end{array}$

22. $\begin{array}{r} 52 \\ \times \ 36 \\ \hline \end{array}$

23. $\begin{array}{r} 655 \\ \times \ \ 21 \\ \hline \end{array}$

24. $\begin{array}{r} 392 \\ \times \ \ 77 \\ \hline \end{array}$

[R.4] *Divide.*

25. $42 \div 7$

26. $18 \div 18$

27. $\dfrac{125}{0}$

28. $\dfrac{0}{35}$

[R.4–R.5] *Divide.*

29. $4\overline{)432}$

30. $9\overline{)216}$

31. $76\overline{)26{,}752}$

32. $2704 \div 18$

Solve each application problem.

33. There are 52 playing cards in a deck. How many cards are there in two decks? In six decks?

34. The college bookstore received mathematics textbooks packed 12 books per carton. How many textbooks were in a delivery of 238 cartons?

35. Raoul rented a car for four days at $47 per day. There was also a $24 charge to fill the gas tank when he returned the car. Raoul had budgeted $200 for the rental. Was he over or under his budget? By how much?

36. The Village Middle School wants to raise $315,280 for a new library. If $87,340 has already been raised, how much more is needed to reach the goal?

37. It takes 2000 hours of work to build one home. How many hours of work are needed to build five homes? To build 12 homes?

38. A Japanese bullet train travels 80 miles in one hour. Find the number of miles traveled in three hours, and the number of miles traveled in seven hours.

39. If an acre needs 250 pounds of fertilizer, how many acres can be fertilized with 5750 pounds of fertilizer?

40. Each home in a subdivision requires 180 feet of fencing. Find the number of homes that can be fenced with 5760 feet of fencing material.

41. Susan Hessney had $382 in her checking account. She wrote a check for $135 and deposited her $563 tax refund. What is the new balance in her account?

42. Find the total cost of four T-shirts at $14 each and three sweatshirts at $29 each.

43. The Houston Space Center charges $18 for each adult admission, $14 for each child, and $17 for each senior. Find the total cost to admit 18 adults, 6 children, and 15 seniors.

44. A newspaper carrier has 56 customers who take the paper daily and 23 customers who take the paper on weekends only. A daily customer pays $18 per month and a weekend-only customer pays $11 per month. Find the total monthly collections.

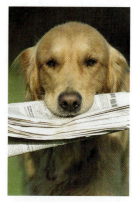

Chapter R

TEST

Fill in the blanks to complete each sentence.

1. In an addition problem, the numbers being added are called
   _____ and the answer is called the _____.

2. In a multiplication problem, the numbers being multiplied are called
   _____ and the answer is called the _____.

3. In a subtraction problem the answer is called the _____
   and in a division problem the answer is called the _____.

*Add, subtract, multiply, or divide, as indicated. Do **not** use a calculator.*

4. $984 + 65 + 7561$

5.
$$\begin{array}{r} 17{,}063 \\ 7 \\ 12 \\ 1\ 505 \\ 93{,}710 \\ +\quad 333 \end{array}$$

6. $17{,}002 - 54$

7.
$$\begin{array}{r} 5062 \\ -\ 1978 \end{array}$$

8. $5 \times 7 \times 4$

9. $57 \cdot 3000$

10. $(85)\,(21)$

11.
$$\begin{array}{r} 7381 \\ \times\quad 603 \end{array}$$

12. $6\overline{)1236}$

13. $\dfrac{791}{0}$

14. $38{,}472 \div 84$

15. $280\overline{)44{,}800}$

1. _____

2. _____

3. _____

4. _____

5. _____

6. _____

7. _____

8. _____

9. _____

10. _____

11. _____

12. _____

13. _____

14. _____

15. _____

Solve each application problem.

16. _____

16. Find the cost of 48 glass beakers for a chemistry lab at $11 per beaker.

17. _____

17. In a consumer survey of cell phones, the most expensive model cost $350 and the least expensive model was $59. Find the difference in price between the most expensive and the least expensive phone.

18. _____

18. A stamping machine produces 900 license plates each hour. How long will it take to produce 31,500 license plates?

19. _____

19. Kenée Shadbourne paid $690 for tuition, $185 for books, and $68 for supplies. These amounts were withdrawn from her checking account, which had a balance of $1108. Find the new balance in her account.

20. _____

20. An appliance manufacturer assembles 118 self-cleaning ovens each hour during the first four hours and 139 standard ovens each hour during the next four hours. Find the total number of ovens assembled in the 8-hour period.

21. _____

21. The monthly rents collected from the four units in an apartment building are $785, $800, $815, and $725. After expenses of $1085 are paid, find the amount that remains.

22. _____

22. Describe the divisibility tests for 2, 5, and 10. Also give two examples for each test: one number that is divisible and one number that is not divisible.

Appendix A: Study Skills

Developed by Linda C. Russell
Minneapolis Community and Technical College

Your Brain-Friendly Guide to Success in Math

	Activity	To Be Used With	Starts on Page
Activity 1:	How Your Brain Learns	Diagnostic Pretest before Chapter 1	A-2
Activity 2:	Using Your Textbook	Section 1.2	A-4
Activity 3:	Homework: How, Why, and When	Section 1.2 Exercises	A-8
Activity 4:	Taking Lecture Notes	Section 1.8	A-11
Activity 5:	Reviewing a Chapter	Chapter 1 Summary	A-12
Activity 6:	Managing Your Time	Chapter 1 Test	A-14
Activity 7:	Making and Using Study Cards	Section 2.1	A-17
Activity 8:	Preparing for Tests: Performance Health Tips	Chapter 2 Test	A-22
Activity 9:	Making and Using a Mind Map	Sections 3.2, 4.4, 4.8, 5.10 Exercises	A-23
Activity 10:	Tips for Taking Tests	Chapter 3 Test	A-28
Activity 11:	Analyzing Your Test Results	Chapter 4 Test	A-30
Activity 12:	Preparing for a Final Exam	Chapter 7 or 9 or 10 Test	A-32

Activity 1: How Your Brain Learns
To Accompany the Diagnostic Pretest in This Textbook

Your brain knows how to learn, just as your lungs know how to breathe. However, there are important things you can do to maximize your brain's ability to do its work. This short introduction will help you choose effective strategies to use while learning mathematics. This is a simplified explanation of a complex process.

Your brain's outer layer is called the **neocortex**, which is where higher level thinking, language, reasoning, and purposeful behavior occur. The neocortex has about 100 billion (100,000,000,000) brain cells called **neurons**.

LEARNING SOMETHING NEW

- As you learn something new, threadlike branches grow out of each neuron. These branches are called **dendrites.**

- When the dendrite from one neuron grows close enough to the dendrite from another neuron, a connection is made. There is a small gap at the connection point called a **synapse.** One dendrite sends an electrical signal across the gap to another dendrite.

- *Learning = growth and connecting of dendrites.*

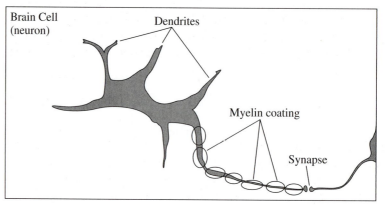

The above drawing shows a neuron with several dendrites. One of the dendrites has developed a myelin coating through repeated practice.

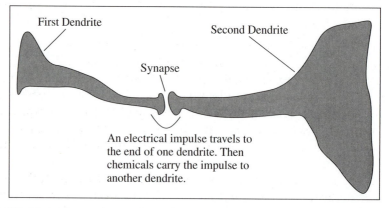

This drawing shows a close up view of the connection point (synapse) between two dendrites.

REMEMBERING NEW SKILLS

- When you practice a skill just once or twice, the connections between neurons are very weak. If you do not practice the skill again, the dendrites at the connection points wither and die back. You have forgotten the new skill!

- If you practice a new skill many times, the dendrites for that skill become coated with a fatty protein called **myelin.** Each time one dendrite sends a signal to another dendrite, the myelin coating becomes thicker and smoother, allowing the signals to move faster and with less interference. Thinking can now occur more quickly and easily, and *you will remember the skill for a long time* because the dendrite connections are very strong.

OTHER IMPORTANT POINTS

- You grow dendrites specifically for the thing you are studying. If you practice adding fractions, you will grow specialized dendrites just for fractions. If you *watch other people* solve fraction problems, *you will grow dendrites for watching, not for solving.* So, be sure you are actively learning and practicing.

- If you practice something the *wrong* way, you will develop strong dendrite connections for doing it the wrong way! So, as you study, check frequently that you are getting correct answers.

- As you study a new topic that is related to things you already know, you will grow new dendrites, but your brain will also send signals throughout the network of dendrites for the related topics. In this way, you build a complex **neural network** that allows you to apply concepts, see differences and similarities between ideas, and understand relationships between concepts.

Periodically in this textbook you will see an icon that looks like this:

This desk light directs you to "brain friendly" activities in Appendix A which are designed to help you grow and develop your own reliable neural networks for mathematics. Since you must grow your own dendrites (no one can grow them for you), these activities show you what you can do to

- develop new dendrites,

- strengthen existing ones, and

- encourage the myelin coating to become thicker so signals are sent with less effort.

When you incorporate the activities into your regular study routine, you will discover that you understand better, remember longer, and forget less.

Also remember that it does take *time* for dendrites to grow. Trying to cram several new concepts and skills in at the last minute is not possible. Your brain simply can't grow new dendrites that quickly. You can't expect to develop huge muscles by lifting weights for just one evening before a body building competition! Practice the techniques *throughout the course* to facilitate good growth.

If you are under stress or feeling anxious, such as during a test, your body secretes **adrenaline** into your system. The presence of adrenaline in the brain blocks connections between neurons: in other words, you can't think! If you've ever experienced "blanking out" on a test, you've seen first hand what adrenaline does. You'll learn several solutions to that problem in Activities 10 and 11.

CAUTION

In order to make the brain-friendly activities work for you, there is a key prerequisite.

You must be in the right math course.

Why?

- If the course is too *advanced*, you won't have the dendrites for the background knowledge.

- You won't be able to connect the new information to your previously learned concepts.

- You'll be lost!

- If you are not sure you are in the right course, talk with your instructor **at once.**

Often, the first chapter of a math book will be review to help you *reactivate* your background knowledge. So even if it seems too simple at the beginning, look in later chapters to see the new material that lies ahead. Also, take the **Diagnostic Pretest** (just before Chapter 1) to see an overview of what the course covers.

START YOUR COURSE RIGHT!

- Attend all class sessions (especially the first one),
- Gather the necessary supplies
- Carefully read the syllabus for the course.

Ask your instructor about class policies if you don't understand them. Then you will put yourself in the best situation for success. You will not only learn mathematics, but also how to become an effective student, using your brain's natural capabilities.

Activity 2: Using Your Textbook

To Accompany Section 1.2 in This Textbook

Your math textbook has a number of features that will be very helpful. Right now, early in the semester, you need to know what your book has to offer. Later, you will want to dig deeper into these parts of the book to help you in different chapters. First, let's look at some general features that will help in all chapters.

Table of Contents

Look on page iii in the very front of the book to see the beginning of the Table of Contents. The three things listed before Chapter 1 that you will find most helpful are "An Introduction to Scientific Calculators," "To the Student" and the "Diagnostic Pretest." Before you start Chapter 1, you'll also want to look at Chapter R, Whole Numbers Review, near the end of the book.

On page x in the Preface is a list of supplementary resources for students. If you are interested in any of these resources, ask your instructor if they are available at your school.

An Introduction to Scientific Calculators

Now turn to "An Introduction to Scientific Calculators" on page xvii. Each calculator model is slightly different, so you must learn to use your own calculator by studying the manual that came with it. However, these pages contain some basic information about using a *scientific* calculator to help you get started.

To the Student

There are ten tips for succeeding in mathematics in "To the Student." **Mark the ones that will be particularly helpful for you.**

Diagnostic Pretest

The *Diagnostic Pretest*, which begins on page xxv, is very important! It covers all the skills and concepts included in the textbook. Hopefully you've already taken it to help you confirm that you are registered for the correct math course. The answers are in the Answers section in the back of this book. There is a Diagnostic Pretest in most math books, and you should *always* take it.

> What did you find out about your skills when you took the Diagnostic Pretest?
> _____

Whole Numbers Computation: Pretest

This textbook assumes that you know how to add, subtract, multiply, and divide whole numbers. It is important that you have dendrites in your brain for these skills, so new dendrites can grow off of the old ones. *Dendrites cannot grow in a void.*

Use the Whole Numbers Computation Pretest, which starts on page 749, to check your dendrite connections. Take the test *without using a calculator.* Check the answers in the Answers section at the back of this text. Then use the scoring guide at the end of the test. Each section of the test matches a section in Chapter R, Whole Numbers Review. If some of your dendrite connections are rusty or have disappeared, work through the corresponding sections in Chapter R *before* you start Chapter 1. That way your brain will be primed and ready to start learning algebra.

> Which whole number skills did you need to review?
> _____

Chapter Summary

Go back to page iii and look at the Table of Contents again. Notice that at the end of each chapter there are three important sections. First, notice the **Chapter Summary.** Turn to page 71 to find the summary for Chapter 1. It lists the chapter's Key Terms (arranged in the order that they appear in the chapter), New Symbols, and a vocabulary quiz called Test Your Word Power. Be sure to try the quiz. Next is a Quick Review section. Notice the red topics, which tell you the concepts you learned in each section of the chapter. Also, in the right column, notice the worked examples, which have tips and vocabulary in blue.

The second feature at the end of the chapter is the **Review Exercises** section. Use the review exercises as a way to check your understanding of each concept in the chapter. You can practice every type of problem. If you get stuck, the red numbers in brackets tell you which section of the chapter to refer to for more explanations. Make sure you do the Mixed Review Exercises to practice for tests.

The third feature at the end of the chapter is the **Chapter Test**. Plan to take the test as a practice exam. That way you can be sure you really know how to work all types of problems without looking back at the chapter.

> How will you make good use of these three features at the end of each chapter?
>
> _____

Section Numbering

Each chapter in your math textbook is divided into several sections. At the beginning of each section you will see a number, such as 1.3 or 3.5. The first number *refers to the chapter* you are in. The number to the right of the decimal point refers to which *section of the chapter* you are in. When you use the Chapter Summaries and Review Exercises, the red numbers direct you back to the appropriate section of the chapter in which that concept is taught.

Your instructor will use these numbers also, to assign readings and homework.

Answers

How do you find out if you've worked the exercises and sample problems correctly? Your textbook provides many of the answers you are looking for. Throughout the chapter you will find suggestions to try some of the odd-numbered exercises at the end of each section. Work these exercises to check your understanding.

You can find the answers to all of the *odd numbered section exercises* in the **Answers to Selected Exercises** section near the end of this textbook. Also, *all* of the answers are given for the Whole Numbers Computation Pretest, Diagnostic Pretest, Chapter Review Exercises, Chapter Tests, and Cumulative Review Exercises. Check your textbook now, and find the page on which the Answers section begins.

> **NOTE**
> Our suggestion is that you flag the Answers section with a sticky note or other device, so that you can turn to it quickly.

Index

At the very end of this textbook you will find an Index. All of the topics, vocabulary, and concepts are listed in alphabetical order in the Index. For example, look up the words below. Go to the page or pages listed and find each word. Write down the page that introduces or defines each one.

> *Commutative property of multiplication* is defined on page _____.
>
> *Factors* are defined on page _____.
>
> *Rounding of mixed numbers* is explained on page _____.

There may be several subheadings listed under the main word, or several page numbers listed. Be aware that the *first* place that word appears in the textbook is probably where that word is introduced and defined. If you are having trouble remembering it, the earliest page number would be a good place to start.

Chapter Features

Much of your time will be spent using the chapters in your textbook. There are four features to pay special attention to as you use your book.

- **OBJECTIVES** Each section of a chapter lists the objectives in the upper corner of the page in the light green box. The objectives are listed again as each one is introduced. An objective tells you *what you will be able to do after you complete the section*. An excellent way to check your learning is to go back to the list of objectives when you are finished with a section and ask yourself if you can do them all.

- **Now Try** exercises. At the end of each worked example, there is a suggestion to try one or two of the odd-numbered exercises at the end of the section. These exercises give you immediate practice. **This is a perfect way to get your dendrites growing right away!** The answers are given in the Answers section at the back of this textbook, so you can check yourself easily.

- **Cautions, notes, and calculator tips.**

    - **CAUTION** A bright yellow box is a comment about a common error that students make, or a common trouble spot you will want to avoid.
    - Look for the specially marked purple **NOTE** boxes. They contain hints and other explanations that can clarify ideas. Also, they may add an interesting side comment about a topic.
    - You might see a small picture of a red calculator 🖩 in two areas of your textbook. In the main part of the chapter, the icon means that there is a **Calculator Tip**. These tips help you become familiar with the way your scientific calculator works and the typical ways in which you should use it. A picture of a calculator in the Exercises sections means that it is recommended that you use your calculator to work that exercise.

> List a page number from Chapter 1 for each of these features:
>
> A **CAUTION** appears on page _____.
>
> A **NOTE** appears on page _____.
>
> A **Calculator Tip** appears on page _____.

- **Color boxes and color print.** You will see blue shaded boxes throughout your textbook. Page through Chapter 1 and figure out what they are used for.

> Blue boxes are used for _____.

Color print is also designed to stand out so you notice it. What do you think the blue print and red print are used for?

Blue print is used for _____.

Red print is used for _____.

Why Are These Textbook Features Brain Friendly?

Your textbook authors have made every effort to provide text features that make it easier for you to understand the mathematics. *Your brain naturally seeks organization and predictability.* When you pay attention to the regular features of your textbook, you are allowing your brain to get familiar with all of the helpful tips, suggestions, and explanations that your book has to offer. Then in later chapters, your brain will continue to notice and take advantage of those features. This way you are making the best possible use of your textbook.

Activity 3: Homework: How, Why, and When
To Accompany Section 1.2 Exercises in This Textbook

It is best for your brain if you keep up with the reading and homework in your math class. Remember that the more times you work with the information, the more dendrites you grow! So, give yourself every opportunity to read, work problems, and review your mathematics.

You have two choices for reading your math textbook. Read the short descriptions below and decide which will be best for you.

- **Preview before class, read carefully after class**
  Maddy learns best by listening to her teacher explain things. She "gets it" when she sees the instructor work problems on the board. She likes to ask questions in class and put the information in her notes. She has learned that it helps if she has *previewed* the section before the lecture, so she knows generally what to expect in class. *But after the class instruction,* when Maddy gets home, she finds that she can understand the math textbook easily. She remembers what her teacher said, and she can double-check her notes if she gets confused. So, Maddy does her **careful** reading of the section in her text **after** hearing the classroom lecture on the topic.

- **Read carefully before class**
  De'Lore, on the other hand, feels he learns well by reading on his own. He prefers to read the section and try working the example problems before coming to class. That way, he already knows what the teacher is going to be talking about. Then, he finds that he can follow the teacher's examples more easily. It is also easier for him to take notes in class. De'Lore likes to have his questions answered right away, which he can do if he has already read the chapter section. So, De'Lore **carefully** reads the section in his text **before** he hears the classroom lecture on the topic.

Notice that there is **no one right way** to work with your textbook. You always must figure out what works best for you. Note also that neither Maddy nor De'Lore tries to read an entire chapter at once. They work with one section at a time. **The key is that you read the textbook regularly!** The rest of this activity will give you some ideas of how to make the most of your reading.

Try the following steps as you **read** your math textbook.

- Read slowly. Read only one section—or even part of a section—at a time.

- Do the "Now Try" odd-numbered exercises **as you go**. Check them right away. The answers are at the back of this textbook.

- If your mind wanders, work problems on separate paper and write explanations in your own words.

- Make study cards as you read each section. Pay special attention to the yellow and blue boxes in the book. Make cards for new vocabulary, rules, procedures, formulas, and sample problems.

- **NOW**, you are ready to do your homework assignment!

Why Are These Reading Techniques Brain Friendly?

- The steps above encourage you to be **actively** working with the material in your text. Your brain grows dendrites when it is **doing** something.

- These methods require you to try several different techniques, not just the same thing over and over. Your brain loves variety!

- Also, the techniques allow you to take small breaks in your learning. Those rest periods are crucial for good dendrite growth.

Which three steps will be most helpful for you when reading the textbook?

1. _____   2. _____   3. _____

_____   _____   _____

Homework

Teachers assign homework so you can grow your own dendrites (learn the material) and then coat the dendrites with myelin through practice (remember the material). Really! In learning, you get good at what you practice. So, completing homework every day will strengthen your neural network and prepare you for exams.

If you have read each section in your textbook according to the steps above, you will probably encounter few difficulties with the exercises in the homework. But, here are some additional suggestions that will help you succeed with the homework.

- If you **have trouble with a problem,** find a similar worked example problem in the section. Pay attention to **every line** of the worked example to see how to get from step to step. Work it yourself too, on separate paper; don't just look at it.

- If it is **hard to remember the steps** to follow for certain procedures, write the steps on a separate card. Then write a short explanation of each step. Keep the card nearby while you do the exercises, but try **not** to look at it.

- If you **aren't sure you are working the assigned exercises correctly**, choose two or three odd numbered problems that are a similar type and work them. Then check the answers in the Answers section near the back of the book and see if you are doing them correctly. If you aren't, go back to the section in the text and review the examples and find out how to correct your errors. Finally, when you are sure you understand, try the assigned problems again.

- **Make sure you do some homework every day**, even if the math class does not meet each day!

Why Are These Homework Suggestions Brain Friendly?

- **Your brain will grow dendrites** as you study the worked examples in the text and try doing them yourself on separate paper. So, when you see similar problems in the homework, you will already have dendrites to work from.

- When you try to remember the steps (without looking at your card), you are **giving yourself a practice test.** This is an excellent way to reinforce what you are learning.

- **Correcting errors right away** is how you learn and reinforce the correct procedures. It is hard to unlearn a mistake, so always check to see that you are on the right track!

What are your biggest homework concerns?

List your two or three main concerns about homework below. Then write a **brain friendly solution** for each one.

1. Concern: _____

   Solution: _____

2. Concern: _____

   Solution: _____

3. Concern: _____

   Solution: _____

Activity 4: Taking Lecture Notes
To Accompany Section 1.8 in This Textbook

Study the set of sample math notes in this activity, and read the comments about them. Then try to incorporate the techniques into your own math note-taking in class.

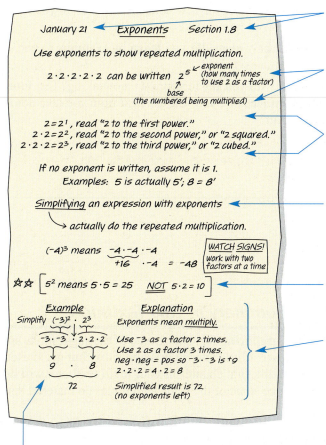

The date and title of the day's lecture topic are always at the top of every page. **Always begin a new day with a new page.**

Note **the definitions** of base and exponent are written in parentheses— don't trust your memory!

Skipping lines makes the notes easier to read.

See how the direction word (*simplify*) is emphasized and explained.

Stars mark an important concept. This is a warning to avoid future mistakes. **Note the underlining too**, which highlights the importance.

Notice the two columns, which allow for the example and its explanation to be close together. **Whenever you're given a series of steps to follow, try the two-column method.**

Note the brackets and arrows in the left column, which clearly show how the problem is set up to be simplified.

Why Are These Lecture Notes Brain Friendly?

The notes are *easy to look at*, and you know that the brain responds to things that are visually pleasing. Other techniques that are visually memorable are the use of spacing (the two columns), stars, underlining, and circling. All of these methods *allow your brain to take note of important concepts and steps*.

The notes are also *systematic*, which means that they use certain techniques regularly. This way, your brain easily recognizes the topic of the day, the signals that show an important point, and steps to follow for procedures. When you develop a system that you always use in your notes, your notes are easy to understand later when you are reviewing for a test.

NOW TRY THIS:

Find one or two people in your math class to work with. Compare each other's lecture notes over a period of a week or so. Ask yourself the following questions as you examine the notes.

1. What are you doing in your notes to show the **main points** or larger concepts? (Such as underlining, boxing, using stars, capital letters, etc.)

2. In what ways do you **set off the explanations** for worked problems, examples, or smaller ideas (subpoints)? (Such as indenting, using arrows, circling or boxing)

3. What does **your instructor do** to show that he/she is moving from one idea to the next? (Such as saying "Next" or "any questions," "now," or erasing the board, etc.)

4. **How do you mark** that in your notes? (Such as skipping lines, using dashes or numbers, etc.)

5. What **explanations (in words) do you give yourself** in your notes, so when those new dendrites you grew in lecture are fading, you can read your notes and still remember the new concepts later when you try to do your homework?

6. What **new ideas did you learn** by examining your classmates' notes?

7. What **will you try** in your own notetaking? List **four** techniques that you will use next time you take notes in math.

Activity 5: Reviewing a Chapter
To Accompany Chapter 1 Summary in This Textbook

This activity is really about **preparing for tests**. Some of the suggestions are ideas that you will learn to use a little later in this appendix, but get started trying them out now. Often, the first chapter in your math textbook will be review, so it is good to practice some of the study techniques on material that is not too challenging.

Use these chapter reviewing techniques:

- **Make a study card for each vocabulary word and concept.** Include a definition, an example, a sketch, and a page reference. Include the symbol or formula if there is one. See Activity 7 for a quick look at some sample study cards.

- **Go back to the section** to find more explanations or information about any new vocabulary, formulas, or symbols.

- **Use the Chapter Summary** to practice each type of problem. Do not expect the Summary to substitute for reading and working through the whole chapter! First, take the "Test Your Word Power" quiz to check your understanding of new vocabulary. The answers are at the end of the Quick Review. Then read the Quick Review. **Pay special attention to the red headings**. Check the explanations for the solutions to problems given. Try to think about the **whole chapter.**

- **Study your lecture notes** to see what your instructor has emphasized in class. Then review that material in your text.

- **Do the Review Exercises.**
  - ✓ Check your answers **after** you're done with each **section of exercises**.
  - ✓ If you get stuck on a problem, **first** check the Chapter Summary. If that doesn't clear up your confusion, then check the section and your lecture notes.
  - ✓ Pay attention to **direction words** for the problems, such as *simplify, evaluate, round, solve,* and *estimate.*
  - ✓ Make **study cards for especially difficult problems**.

- **Do the Mixed Review exercises.** This is a good check to see if you can still do the problems when they are in mixed up order. **Check your answers carefully** in the Answers section in the back of the book. Are your answers **exact** and **complete?** Make sure you are **labeling** answers correctly, using the right **units.** For example, does your answer need to include $\$, cm^2, ft$, and so on?

- **Take the Chapter Test as if it is a real test.** If your instructor has skipped sections in the chapter, figure out which problems to skip on the test before you start.
  - ✓ **Time yourself** just as you would for a real test.
  - ✓ **Use a calculator or notes** just as you would be permitted to (or not) on a real test.
  - ✓ **Take the test in one sitting**, just like a real test is given in one sitting.
  - ✓ **Show all your work.** Practice showing your work just the way your instructor has asked you to show it.
  - ✓ **Practice neatness**. Can someone else follow your steps?
  - ✓ **Check your answers in the back of the book.** Note that section numbers are shown in red. For problems you missed, you can go back to the sections for help.

Notice that reviewing a chapter will take some time. Remember that it takes time for dendrites to grow! You cannot grow a good network of dendrites by rushing through a review in one night. But if you use the suggestions over a few days or evenings, you will notice that you understand the material more thoroughly and remember it longer.

Why Are These Reviewing Activities Brain Friendly?

You have already become familiar with the features of your textbook. This activity requires you to make good use of them! Your *brain needs repetition* in order to strengthen dendrites and the connections between them. By following the steps outlined here, you will be reinforcing the concepts, procedures, and skills you need to use for tests (and for the next chapters).

The combination of techniques provides repetition in different ways. Variety *promotes good branching of dendrites* instead of just relying on one branch, or route, to connect to the other dendrites. A thorough review of each chapter will *solidify your dendrite connections*. It will help you be sure that you understand the concepts *completely and accurately*. Also, taking the Chapter Test will *simulate the testing situation*, which gives you practice in test taking conditions.

NOW TRY THIS:

Try the reviewing techniques listed in this activity before your next test. For each technique, **write a comment** *about how it worked for you.*

1. **Make a study card for each vocabulary word and concept.** This was helpful because _____.

2. **Go back to the section** to find more explanations or information. This was helpful because _____.

3. **Take the Test Your Word Power Quiz and use the Quick Review** to review each concept in the chapter. This was helpful because _____.

4. **Study your lecture notes** to see what your instructor has emphasized in class. This was helpful because

_____ .

5. **Do the Review Exercises,** following the specific suggestions on the previous page. This was helpful because _____.

6. **Do the Mixed Review exercises.** This was helpful because _____.

7. **Take the Chapter Test** as if it is a real test. This was helpful because _____.

Activity 6: Managing Your Time
To Accompany Chapter 1 Test in This Textbook

Many college students find themselves juggling a difficult schedule and multiple responsibilities. Perhaps you are going to school, working part time, and managing family demands. Here are some tips to help you develop good time management skills and habits.

- **Read the syllabus for each class.** Check on class policies, such as attendance, late homework, and make up tests. Find out how you are graded. Keep the syllabus in your notebook.

- **Make a semester or quarter calendar.** Put test dates and major due dates for *all* your classes on the same calendar. That way you will see which weeks are the really busy ones. Try using a different color pen for each class. Your brain responds well to the use of color. A sample semester calendar is at the end of this activity. Make a larger copy so you have room to write.

- **Make a weekly schedule.** After you fill in your classes and other regular responsibilities (such as work, picking up kids from school, etc.), block off some study periods during the day that you can guarantee you will use for studying. Aim for two hours of study for each one hour you are in class. A sample weekly schedule is at the end of this activity. Enlarge it and make a copy for each week.

- **Make "To Do" lists.** Then use them by crossing off the tasks as you complete them. You might even number them in the order they need to be done (most important ones first).

- **Break big assignments into smaller chunks.** They won't seem so big that way. Make deadlines for each small part so you stay on schedule.

- **Give yourself small breaks in your studying.** Do not try to study for hours at a time! Your brain needs rest between periods of learning. Try to give yourself a 10 minute break each hour or so. You will learn more and remember it longer.

- **If you get off schedule, just try to get back on schedule tomorrow.** We all slip from time to time. All is not lost! Make a new "To Do" list and start doing the most important things first.

- **Get help when you need it.** Talk with your instructor during office hours. Also, most colleges have some kind of Learning Center, tutoring center, or counseling office. If you feel lost and overwhelmed, ask for help. Someone can help you decide what to do first and what to spend your time on right away.

Why Are These Time Management Tips Brain Friendly?

- Your brain appreciates some order. It enjoys a little routine, for example, choosing the same study time and place each day. You will find that you quickly settle in to your reading or homework.

- Also, your brain functions better when you are calm. Too much rushing around at the last minute to get your homework and studying done sends hostile chemicals to your brain and makes it more difficult for you to learn and remember. So, a little planning can really pay off.

- Building rest into your schedule is good for your brain. Remember, it takes time for dendrites to grow.

- We've suggested using color on your calendars. This too, is brain friendly. Remember, your brain likes pleasant colors and visual material that are nice to look at. Messy and hard to read calendars will not be helpful, and you probably won't look at them often.

NOW TRY THIS:

Choose two or three of the suggestions above to try this week. Describe how you think each one will help you.

1. _____ 2. _____ 3. _____

_____   _____   _____

SEMESTER CALENDAR

WEEK	MON	TUES	WED	THUR	FRI	SAT	SUN
1							
2							
3							
4							
5							
6							
7							
8							
9							
10							
11							
12							
13							
14							
15							
16							
17							

WEEKLY SCHEDULE

HOUR	MON	TUES	WED	THUR	FRI	SAT	SUN
AM 6:00–7:00							
7:00–8:00							
8:00–9:00							
9:00–10:00							
10:00–11:00							
11:00–12:00							
PM 12:00–1:00							
1:00–2:00							
2:00–3:00							
3:00–4:00							
4:00–5:00							
5:00–6:00							
6:00–7:00							
7:00–8:00							
8:00–9:00							
9:00–10:00							
10:00–11:00							
11:00–12:00							
AM 12:00–1:00							

Activity 7: Making and Using Study Cards
To Accompany Section 2.1 in This Textbook

You may have used "flash cards" in other classes before. In math, study cards can be helpful too. However, they are different because the main things to remember in math are *not* necessarily terms and definitions, they are *sets of steps to follow* to solve problems (and how to know which set of steps to follow), and *concepts about how math works* (principles). So, the cards will look different, but be just as useful.

In this activity, you will find five types of study cards to use in math. Look carefully at what kinds of information to put on them and where to put it. Then use them the way you would any flash card:

- to quickly review when you have a few minutes,

- to do daily reviews,

- to review before a test.

But remember, the most helpful thing about study cards is making them. When you are making them, you have to do the kind of thinking that is most brain friendly and will improve your neural network of dendrites. After each card description you will find an assignment to try.

1. New Vocabulary Cards

Put the word (spelled correctly) and the page number on the front of the card. On the back, write:

- the definition (in your own words if possible),

- an example, an exception (if there are any),

- any related words, and

- a sample problem (if appropriate).

Front of Card

variable   From page ___

<u>Def</u>: A letter representing a number that varies (changes)
      depending on the situation.

<u>Example</u>:   C + 5   where c represents the <u>c</u>lass limit
           ↗ ↑   which varies for different classes.
     variable ↑
         constant
               <u>varies</u> ⟷ <u>variable</u>

– Use any letter you like: a, b, c, d, h, k, m, n, p, x, y, etc.
– Related words: constant, coefficient

Back of Card

*List four new vocabulary words you need to learn right now. Make a study card for
each one.*

_____   _____

_____   _____

2. Procedure ("Steps") cards

Write the name of the procedure at the top on the front of the card. Then
write each step *in words*. If you need to know abbreviations for some words,
include them along with the whole words written out. On the back, put an
example of the procedure, showing each step you need to take. You can review by looking at the front and practicing a new worked example, or by
looking at the back, and remembering what the procedure is called and what
the steps are.

Front of Card

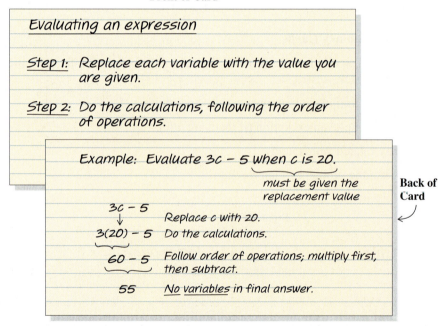

NOW TRY THIS:

*What procedure are you learning right now? Write below the steps that you will put on
your study card.*

Procedure: _____

Step 1. _____

Step 2. _____

Step 3. _____

Step 4. _____

Step 5. _____

3. Typical Mistakes card

This is a warning card to remind you either of typical errors everybody seems to make, or your personal typical errors. Use corrected homework, quizzes, or tests to find your own common errors. On the front, label the procedure or the warning. List the steps to follow, including things you should *not* do or that you might forget (like "remember to bring down the negative signs!!" or "numbers cannot be divided by zero!"). Your card should show the correct way to do the procedure, and also remind you what *not* to do.

Front of Card

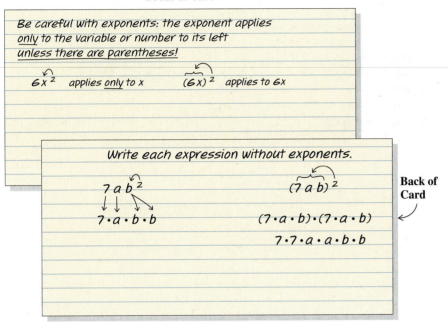

Be careful with exponents: the exponent applies <u>only</u> to the variable or number to its left unless there are parentheses!

$6x^2$ applies only to x $(6x)^2$ applies to 6x

Write each expression without exponents.

$7ab^2$ $(7ab)^2$

$7 \cdot a \cdot b \cdot b$ $(7 \cdot a \cdot b) \cdot (7 \cdot a \cdot b)$

$7 \cdot 7 \cdot a \cdot a \cdot b \cdot b$

Back of Card

NOW TRY THIS:

Identify a concept or procedure which has given you trouble on homework or quizzes: _____

Make a study card which clarifies the concept or procedure and points out your typical mistakes.

4. Tough Problems card

When you are doing your homework and find yourself saying, "This is really hard," make a study card! On the front write out the procedure *in words* for working this type of problem. If there are special notes (like what <u>not</u> to do), include them. On the back, work at least one example; make sure you label what you are doing.

Front of Card

Back of Card

> Evaluating expressions with negative numbers
>
> 1. When replacing the variables, be sure to write all the negative signs.
> 2. When multiplying a series of negative numbers, work with two factors at a time and <u>watch</u> <u>the</u> <u>signs</u>!

> <u>Example</u>
> Evaluate $^-2m^2n$ when <u>m is $^-3$</u> and <u>n is $^-1$</u>.
>
> $$^-2m^2n$$
>
> $^-2 \cdot ^-3 \cdot ^-3 \cdot ^-1$ neg · neg is pos
> $6 \cdot ^-3 \cdot ^-1$ pos · neg is neg
> $^-18 \cdot ^-1$ neg · neg is pos
> 18

NOW TRY THIS:

Choose three types of difficult problems and work them out on study cards. Be sure to put the words for solving the problem on one side and the worked problem on the other side.

---

5. Practice Quiz Cards

Make up a few quiz cards for each type of problem you learn and use them to prepare for a test. Choose two or three problems from the different sections of the chapter. Be sure you don't just choose the easiest problems! Put the problem **with the direction words** (like **solve**, **simplify**, **evaluate**) on the front, and work the problem on the back. If you like, put the page number from the text there too. When you review, you work the problem on a separate paper, and check it by looking at the back.

Front of Card

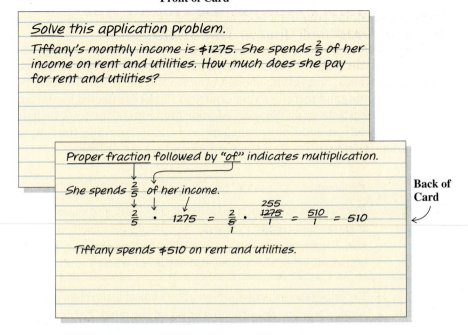

Back of Card

NOW TRY THIS:

*Choose two or three problems from **Section 2.1** and make a study card for each one. Write the direction words on the front and work the problem on the back.*

Why Are Study Cards Brain Friendly?

First, making the study cards is an **active technique** that really gets your dendrites growing. You have to make decisions about what is most important and how to put it on the card. This kind of thinking is more in depth than just memorizing, and as a result, you will understand the concepts better and remember them longer.

Second, the cards are **visually appealing** (if you write neatly and use some color). Your brain responds to pleasant visual images, and again, you will remember longer and may even be able to "picture in your mind" how your cards look. This will help you during tests.

Third, because study cards are small and portable, you can review them easily whenever you have a few minutes. Even while you're waiting for a bus or have a few minutes between classes, you can take out your cards and read them to yourself. Your **brain really benefits from repetition**; each time you review your cards your dendrites are growing thicker and stronger. After a while, the information will become automatic and you will remember it for a long time.

Activity 8: Preparing for Tests: Performance Health Tips
To Accompany Chapter 2 Test in This Textbook

Many things can improve your test scores. But you may not realize that eating the right foods, and getting enough exercise and sleep, can also improve them. Your brain (and therefore your ability to think) is affected by the condition of your whole body. So, part of your preparation for tests includes keeping yourself in good physical shape as well as spending time on the actual course material. Try these suggestions and see the difference.

To Improve Your Test Score:	Explanation
Get **7 to 8 hours of sleep** the night before the exam. (It's helpful to get that much sleep **every** night.)	**Fatigue and exhaustion** reduce efficiency. They also cause **poor memory and recall**. If you didn't sleep much the night before a test, 20 minutes of relaxation or meditation can help. Also see the comments below about eating carbohydrates to help you sleep.
To help you sleep the night before the test, or any time you need to calm down, **eat high carbohydrate foods** such as popcorn, bread, rice, crackers, muffins, bagels, pasta, corn, baked potatoes (not fries or chips), and cereals.	Carbohydrates increase the level of a chemical in the brain called serotonin, which has a **calming effect on the mind**. It reduces feelings of tension and stress and improves your ability to concentrate. You only need to eat a small amount, like half a bagel, to get this effect.
Do 15–20 minutes of **moderate exercise** (like walking) shortly before the test. Daily exercise is even better!	**Exercise reduces stress** and will help prevent a panic attack or "blanking out" on a test. Exercise also **increases your alertness, clear thinking, and energy level**.
Practice slow, deep breathing for 5 minutes each day. Then do a minute or two of deep breathing right before the test. Also, stop during the test if you feel your anxiety building and do a minute of deep breathing.	When test anxiety hits, you breathe more quickly and shallowly, which causes hyperventilation. Symptoms may be confusion, inability to concentrate, shaking, dizziness, and more. Slow, **deep breathing will calm you and prevent panic.**
Eat a **small, high-energy meal** about two hours before the test. Start the meal with a small amount of protein such as fish, chicken, or non-fat yogurt. Include carbohydrates if you like, but no high-fat foods.	Just 3 to 4 ounces of protein increases the amount of a chemical in the brain called tyrosine, which **improves your alertness, accuracy, and motivation**. High-fat foods dull your mind and slow down your brain.
Before the test, **go easy on caffeinated beverages** such as coffee, tea, and soft drinks. Do not eat candy bars or other sugary snacks.	Extra caffeine can **make you jittery**, "hyper", and shaky for the test. It can increase the tendency to panic. Too much sugar causes negative emotional reactions in some people.
Drink plenty of water. Don't wait until you feel thirsty; your body is already dehydrated by the time you feel it.	Research suggests that staying well hydrated improves the electro-chemical communications in your brain.
Give your brain the time it needs to grow dendrites!	Cramming doesn't work; your brain cannot grow dendrites that quickly. **Studying every day using the techniques in this appendix** is the way to give your brain the time it needs.

List below the three or four health tips you think will help you on your next test.

1. _____
2. _____
3. _____
4. _____

What changes will you have to make in order to try the tips you chose?

_____

_____

_____

_____

Activity 9: Making and Using a Mind Map
To Accompany Sections 3.2, 4.4, 4.8, and 5.10 Exercises in This Textbook

Mind mapping is a visual way to show information that you have learned. It is an excellent way to review. Mapping is flexible and can be personalized, which is helpful for your memory. Your brain likes to see things that are **pleasing** to look at, **colorful**, and which **show connections** between ideas. Take advantage of that by creating maps that are

- easy to read,
- use color in a systematic way, and
- clearly show you how different concepts are related (using arrows or dotted lines, for example).

Why is mapping brain friendly?

Remember that your brain grows dendrites when you are **actively thinking** about and working with information. Making a map requires you to think hard about *how to place the information, how to show connections* between parts of the map, and *how color will be useful*. It also takes a lot of thinking to fill in all related details and *show how those details connect* to the larger concept. All that thinking will let your brain grow a complex, many-branched neural network of interconnected dendrites. It is time well spent.

Directions for Making a Mind Map

Here are some general directions for making a map. After you read them, work on completing the map below that has been started for you. It is from Sections 3.1, 3.2, and 4.8. On the following pages are "starter" maps for the details about perimeter (3.1), area (3.2), and volume (4.8). Also, we have included a starter map for fractions (4.4). As you learn about circles and cylinders in 5.10, add the new information to the appropriate maps.

- To begin a mind map, write the concept in the center of a piece of paper and either circle it or draw a box around it.
- Make a line out from the center concept and draw a box large enough to write the definition of the concept.
- Think of the other aspects (subpoints) of the concept that you have learned, such as procedures to follow or formulas. Make a separate line and box connecting each subpoint to the center.
- From each of the new boxes, add the information you've learned. You can continue making new lines and boxes or circles, or you can list items below the new information.
- Use color to highlight the major points. For example, everything related to one subpoint might be the same color. That way you can easily see related ideas.
- You may also use arrows, underlining, or small drawings to help yourself remember.

NOW TRY THIS: "Big Picture" Map of Perimeter, Area, and Volume (Sections 3.1, 3.2, 4.8)

Fill in the missing parts on the map below as you study Sections 3.1, 3.2, and 4.8. It is a "big picture" map, which will help you see how the details you will be learning fit into an overall system. We have filled in the information on Perimeter (from 3.1) so you can see what kinds of information should go in a "big picture" map.

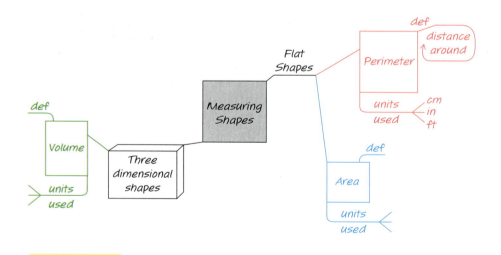

NOW TRY THIS: Detailed Perimeter Map (Section 3.1, 5.10)

*Make a detailed map for **Section 3.1, Perimeter**. Follow the directions below to create a map for finding the perimeter of different shapes. A sample way to* begin *the map is shown.*

- Use the map below as a starting point. Copy it onto a separate paper.

- As you review **Section 3.1** and the study cards you have made, put in the information you learned about finding the perimeter of different shapes. Begin with a Square, drawing a square directly above your box (at the 12 o'clock position).

- Look at the example below. As you continue through **Section 3.1**, work clockwise (to the right), with Rectangle, Parallelogram, Triangle, and Irregular Shapes to complete your map. When you get to **Section 5.10**, add Circles to your map (the perimeter of a circle is called its circumference.)

- For each shape give the *definition,* the *formula for perimeter,* and a *worked example.* Also, draw the shape correctly to provide a visual clue.

- You may wish to use a different color for each shape.

- You may also use arrows, underlining, or small drawings to help yourself remember.

NOW TRY THIS: Detailed Area Map (Sections 3.2, 5.10) and Detailed Volume Map (Sections 3.2, 4.8, 5.10)

*Make a detailed map for **Section 3.2**, Area. Follow the directions below to create a map for finding the area of different shapes. A sample way to begin the map is shown.*

- Use the map below as a starting point. Copy it onto a separate paper.

- As you continue through **Section 3.2**, work clockwise (to the right), with Square and Parallelogram to continue your map. You can add Circles to this map when you get to **Section 5.10**.

- For each shape give the *formula for area,* and a *worked example*.

- Make a separate detailed map for Volume (from **Section 4.8**) following the same pattern. When you get to **Section 5.10**, add Cylinders to your volume map.

By the time you finish Section 5.10, you should have four complete geometry maps.

- "Big Picture" map of Perimeter, Area, and Volume
- Detailed Perimeter map
- Detailed Area map
- Detailed Volume map

NOW TRY THIS: Fractions Computation Map (Sections 4.3, 4.4)

*On separate paper make a map that summarizes Computation with Fractions (**Sections 4.3** and **4.4**). Use the starter map below.*

- The longest rectangles (the one that starts "Rewrite all numbers as fractions . . ." and the one at the bottom of the map) are *instructions* for all four operations.

- Notice the wavy dividing lines that separate the map into two sides.

- Your job is to complete the map by writing the steps used in multiplying fractions (from **Section 4.3**) and the steps used in adding and subtracting fractions (from **Section 4.4**).

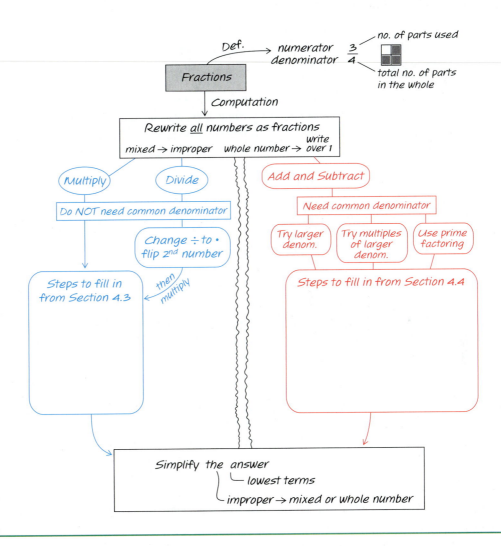

Activity 10: Tips for Taking Tests
To Accompany Chapter 3 Test in This Textbook

To Improve Your Test Score	Explanation
Do not try to review up until the last minute before the test. Instead, go for a walk, do several minutes of deep breathing, and arrive just in time for the test. Ignore what other students are saying just before the test.	Realizing you cannot do a problem 5 minutes before the test **may cause you to panic**. Then you'll do poorly on the whole test instead of just missing one or two questions you forgot to study. Moderate exercise and deep breathing will calm your mind.
Come prepared with a pencil (not a pen) and an eraser, calculator, etc. If you are easily distracted, sit in the corner farthest from the door.	**Working in pencil means you can erase** things instead of crossing them out and making a mess. Be sure your calculator is working properly.
Do a "knowledge dump" as soon as you get the test paper. Write important notes to yourself in a corner of the test paper: formulas you've *memorized,* common errors you want to watch out for, etc.	Writing down warnings and things that you've memorized **lets you relax**; you won't have to worry about forgetting those things and can refer to them as needed.
Scan the entire test, note the point value of different problems, and plan your time accordingly. Allow at least 5 minutes to check your work at the end of the testing time.	If you have 50 minutes to do 20 problems, 50 divided by 20 gives you about 2.5 minutes per problem. **Spend less time on easy ones**, more on problems with higher point values.
Do *not* work the problems in order from number 1 through to the end. **Pick out easy problems** and do those first in order to build confidence. If you feel your anxiety starting to build, *immediately* stop for a minute, close your eyes, and take several slow, deep breaths.	Doing the easier problems first will **improve your confidence**, and you will be sure to get those problems completed correctly. Anxiety causes shallow breathing, which leads to confusion and inability to concentrate. Deep breathing calms and relaxes you.
Read directions carefully and circle any significant words. When you finish a problem, read the directions again to make sure you did what was asked.	**Pay attention to any announcements** written on the board or made by your instructor. If any are unclear, ask your instructor to clarify them. There is nothing worse than getting a bunch of problems wrong because you misread the directions!
For a more difficult problem, **read it twice**. Write down anything you think of that might help solve the problem: a formula, a picture you draw, etc. If you still can't get it, circle the problem and **come back to it later**. Do *not* erase any of the things you wrote down.	If you know even a *little* bit about the problem, write it down. The **answer may come to you** as you work on it, or at least you may get partial credit. But don't spend too long on any one problem. Your subconscious mind will work on the tough problem while you go on with the test.
As you work on more difficult problems, **watch your "inner voice."** You may have negative thoughts going through your mind, such as, "I can't do it," or "I'll flunk, so why try?" or "who cares about this test anyway." In your mind, yell, "STOP" and take several deep, slow breaths. Or, replace the negative thought with a positive one.	Here are **examples of positive statements**. Try writing one of them on the top of your test paper. • I know I can do it. • I can do this one step at a time. • I've studied hard, and I'll do the best I can. • This test is a positive challenge for me to show what I've learned.

(continued)

To Improve Your Test Score	Explanation
Show your work. Most math teachers give partial credit if some of the steps in your work are correct, even when the final answer is wrong. **Write neatly.** If you like to scribble when first working or checking a problem, do it on scratch paper.	**If your teacher can't read your writing, he/she can't give you credit for it.** If you need more space to work, ask if you can use extra pieces of paper that you hand in with your test paper. Put your name and the problem numbers on them.
If you still can't solve a difficult problem when you come back to it the second time, **make a guess and do not change it.** In this situation, your first guess is your best bet. On problems you were able to solve, do not change the answer just because you're a little unsure. **Change it only if you find an obvious mistake.**	If you are thinking about changing an answer, be sure you can explain to yourself exactly why you are changing it. If you cannot find a specific error, leave your first answer alone. **When the tests are returned, check to see if changing answers helped or hurt you.**
Check that the **answer to a word problem is reasonable** and make sense. Read the problem again to make sure you've answered all the questions.	**Use common sense.** Can the father really be 7 years old? Would a month's rent be $32,140? Label your answer: $, years, inches, etc.
To check for careless errors, you need to **rework the problem again, without looking at your previous work.** Cover up your work with a piece of scratch paper and pretend you are doing the problem for the first time. Then compare the two answers.	If you just "look over" your work, your mind can easily make the same mistake again without noticing it. Reworking the problem from the beginning **forces you to rethink it.** If possible, use a different method to solve the problem the second time.
Ignore students who finish early. Use the entire test time. You do not get extra credit for finishing early. Use the extra time to rework problems and correct careless errors.	Students who leave early are often the ones who didn't study or who are too anxious to continue working. If they bother you, **sit as far from the door as possible.**

Why Are These Test-Taking Tips Brain Friendly?

Several of the suggestions address anxiety. *Reducing your anxiety* allows your brain to make the connections between dendrites; in other words, you can *think*.

Other suggestions remind you that *your brain continues to work* on difficult problems even if you skip them and go on to the next one. Your subconscious mind will come through for you if you are open to the idea!

Finally, some of the suggestions ask you to *use your common sense*. Following the directions, being neat, showing your work, and paying attention to whether your answers really make sense are examples of using your common sense to help you.

NOW TRY THIS:

Choose three test-taking strategies that you will try on your next math test. Explain how each one will help you.

1. _____

2. _____

3. _____

Activity 11: Analyzing Your Test Results
To Accompany Chapter 4 Test in This Textbook

After taking a test, many students heave a big sigh of relief and try to forget it ever happened. Don't fall into this trap! An exam is a learning opportunity. It gives you clues about *what your instructor thinks is important*, what *concepts and skills are valued* in mathematics, and *if you are on the right track.*

- **Immediately after the test**, jot down problems that caused you trouble. Find out how to solve them by checking your textbook, notes, or asking your instructor or tutor (if available). You might see those same problems again on a final exam.

- **After the test is returned,** find out what you got wrong or where you had points deducted. Write down the problems so you can learn how to do them correctly. Sometimes you only have a short time in class to review your test. *If you need more time*, ask your instructor if you can look at the test in his/her office.

- **Find out why you made the errors you made.** Here is a list of typical reasons you might make errors on math tests.
    1. You read the directions wrong.
    2. You read the question wrong or skipped over something.
    3. You made a computation error (maybe even an easy one).
    4. Your answer is not accurate.
    5. Your answer is not complete.
    6. You labeled your answer wrong. For example, you labeled it "ft" and it should have been "ft^2."
    7. You didn't show your work.

    *8. You didn't understand the concept.
    *9. You were unable to go from words (in an application problem) to setting up the problem.
    *10. You were unable to apply a procedure to a new situation.

    11. You were so anxious that you made errors even when you knew the material.

The first seven errors are **test taking errors.** They are easy to correct if you decide to carefully read test questions and directions, proofread or rework your problems, show all your work, and double check units and labels every time.

* The three starred errors are **test preparation errors.** Remember that in order to grow a complex neural application network, you need to practice the kinds of problems that you will see on the tests. So, for example, if application problems are difficult for you, you must *do more application problems!* If you have practiced the study skills techniques in this appendix, however, you are less likely to make these kinds of errors on tests because you will have a deeper understanding of the math concepts and you will be able to remember them better.

The last error isn't really an error. **Anxiety** can play a big part in your test results. Go back to Activity 8 in this appendix and read the suggestions about exercise and deep breathing. Recall from Activity 1 that when you are anxious, your body produces adrenaline. The presence of *adrenaline in the brain blocks connections* between dendrites. If you can *reduce the adrenaline* in your system, you will be able to *think more clearly* on your test. Even just five minutes of brisk walking right before your test can help do that.

Also, *practicing a relaxation technique while you do your homework* will make it more likely that you can benefit from using it during a test. *Deep breathing* is helpful because it gets *oxygen into your brain*. When you are anxious you tend to breathe more shallowly, which can make you feel confused and easily distracted.

- Last, you must **make a plan for the next test**, based on your results from this one. You might review the Chapter Summary and work the problems in the Chapter Review Exercises or the Chapter Test. Ask your instructor or a tutor (if available) for more help if you are confused about any of the problems.

NOW TRY THIS:

Below is a record sheet for you to track your progress in test taking. Use it to find out if you tend to make particular kinds of errors on tests. Then you can work specifically on correcting them. Just check in the box when you made one of the errors below. If you take more than five tests, make your own grid on separate paper.

Test Taking Errors

Test #	Read directions wrong	Read question wrong	Computation error	Not exact or accurate	Not complete	Labeled wrong	Didn't show work
1							
2							
3							
4							
5							

What will you do to avoid test taking errors on your next test?

_____

_____

Test Preparation Errors

Test #	Didn't understand concept	Didn't set up problem correctly	Couldn't apply concept to new situation
1			
2			
3			
4			
5			

What will you do to avoid test preparation errors on your next test?

_____

_____

Anxiety

Test #	Felt anxious *before* the test	Felt anxious *during* the test	Blanked out on questions	Got questions wrong that I knew how to do
1				
2				
3				
4				
5				

What will you do to reduce your anxiety before the next test?

_____

_____

Activity 12: Preparing for a Final Exam
To Accompany Chapter 7, 9, or 10 Test in This Textbook

Your math final exam is likely to be a *comprehensive exam*. This means that it will cover material from the *entire course*. The end of the term will be less stressful if you *make a plan* for how you will prepare for each of your exams.

First, figure out the **grade you need to earn on the final exam** in order to get the course grade you are aiming for. Check your course syllabus for grading policies, or ask your instructor if you are not sure of them. This allows you to set a goal for yourself. Also, you will know which of your courses you have to spend more time on in your final exam preparations.

> How many points do you need to get on your mathematics final exam to get the grade you want?
>
> _____

Second, create a **final exam week plan for your work and personal life**. If you need to make an adjustment in your work schedule, do it in advance, so you aren't scrambling at the last minute. If you have family members to care for, you might want to enlist some help from others so you can spend extra time studying. Again, try to plan in advance so you don't create additional stress for yourself. You will have to set some priorities, and studying has to be at the top of the list! Although life doesn't stop for finals, some things can be ignored for a short time. You don't want to "burn out" during final exam week; **get enough sleep and healthy food so you can perform your best**.

> What adjustments do you need to make in your personal life for final exam week?
>
> _____
>
> _____

Third, use the following suggestions to guide your studying and reviewing.

- **Make sure you know exactly which chapters and sections will be on the final exam.**

- **Divide up the chapters** and decide how much you will work on each day.

- Begin your reviewing **several days before the exam.**

- **Use returned quizzes and tests** to review earlier material (if they have been returned to you).

- **Practice all types of problems**, but emphasize the types that are most difficult for you. Use the **Cumulative Reviews** that are at the end of each chapter in this textbook.

- **Rewrite your notes or make mind maps** to create summaries of important information.

- **Make study cards for all types of problems.** Be sure to use the same **direction words** (*simplify, solve, estimate*) that your exam will use. Carry the cards with you and review them whenever you have a few spare minutes.

MANAGING STRESS

Of course, a week of final exams produces stress. **Students who develop skills for reducing and managing stress do better on their final exams and are less likely to "bomb" an exam.** You already know the damaging effect of adrenaline on your ability to think clearly. But several days (or weeks) of elevated stress is also harmful to your brain and your body. You will feel better if you make a conscious effort to reduce your stress level. Even if it takes you away from studying for a little while each day, the time will be well spent.

REDUCING PHYSICAL STRESS

Physical ways to reduce stress are very helpful. Some examples are listed below. Can you add any to the list?

- *Laugh until your eyes water.* Watching your favorite funny movie, exchanging a joke with a friend, or listening to a comedy bit on the radio are all ways to generate a healthy laugh. Laughing raises the level of *calming* chemicals in your brain (endorphins).

- *Exercise for 20–30 minutes.* If you normally exercise regularly, do NOT stop during final exam week! Exercising helps relax muscles, diffuses adrenaline, and raises the level of endorphins in your body. If you don't usually exercise much, do try to get some gentle exercise, such as a daily walk, to help you relax.

- *Practice deep breathing.* Several minutes of deep, smooth breathing will calm you. Close your eyes too.

- *Visualize a relaxing scene.* Choose something that you find peaceful and picture it. Imagine what it feels like and sounds like Try to put yourself in the picture.

- If you feel stress in your muscles, such as your shoulders or back, *slowly squeeze the muscles as much as you can, and then release them.* Sometimes we don't realize we are clenching our teeth or holding tension in our shoulders until we consciously work with them. Try to notice what it feels like when they are relaxed and loose. Squeezing and then releasing muscles is also something you can do during an exam if you feel yourself tightening up.

REDUCING MENTAL STRESS

Mental stress reduction is also a powerful tool that will help you both before and during an exam. In addition to the suggestions below, do you have any of your own techniques?

- *Talk positively to yourself.* Tell yourself you will get through it.

- *Reward yourself.* Give yourself small breaks, a little treat, a nice bath—something that makes you feel cared for—every day of final exam week.

- *Make a list of things to do* and feel the sense of accomplishment when you cross each item off.

- When you take time to relax or exercise, *make sure you are relaxing your mind too*. Use your mind for something *completely* different from the kind of thinking you do when you study. Plan your garden, play your favorite music, walk your dog, read a good book.

- *Visualize.* Picture yourself completing exams and projects successfully. Picture yourself taking the test calmly and confidently.

NOW TRY THIS:

Choose three techniques for reducing stress from the lists above that you plan to try during final exams week. If you have other ideas to add, include them as well.

1. _____
2. _____
3. _____

Other ideas for reducing stress

_____

_____

Appendix B: Inductive and Deductive Reasoning

Appendix B: Inductive and Deductive Reasoning

OBJECTIVE 1 **Use inductive reasoning to analyze patterns.** In many scientific experiments, conclusions are drawn from specific outcomes. After many repetitions and similar outcomes, the findings are generalized into statements that appear to be true. When general conclusions are drawn from specific observations, we are using a type of reasoning called **inductive reasoning.** The next several examples illustrate this type of reasoning.

OBJECTIVES

1 Use inductive reasoning to analyze patterns.
2 Use deductive reasoning to analyze arguments.
3 Use deductive reasoning to solve problems.

EXAMPLE 1 Using Inductive Reasoning

Find the next number in the sequence 3, 7, 11, 15, ...

To discover a pattern, calculate the difference between each pair of successive numbers.

$$7 - 3 = 4$$
$$11 - 7 = 4$$
$$15 - 11 = 4$$

Notice that the difference is always 4. Each number is 4 greater than the previous one. Thus, the next number in the pattern is 15 + 4, or 19.

Now Try Exercise 1. ▷▷▷

EXAMPLE 2 Using Inductive Reasoning

Find the next number in this sequence.

7, 11, 8, 12, 9, 13, . . .

The pattern in this example involves addition and subtraction.

$$7 + 4 = 11$$
$$11 - 3 = 8$$
$$8 + 4 = 12$$
$$12 - 3 = 9$$
$$9 + 4 = 13$$

To get the second number, we add 4 to the first number. To get the third number, we subtract 3 from the second number. To obtain subsequent numbers, we continue the pattern. The next number is 13 − 3 = 10.

Now Try Exercise 3. ▷▷▷

EXAMPLE 3 **Using Inductive Reasoning**

Find the next number in the sequence 1, 5, 25, 125, . . .

Each number after the first is obtained by multiplying the previous number by 5. So the next number would be 125 · **5** = 625.

Now Try Exercise 5.

EXAMPLE 4 **Using Inductive Reasoning**

(a) Find the next geometric shape in this sequence.

The figures alternate between a blue circle and a red triangle. Also, the number of dots increases by 1 in each subsequent figure. Thus, the next figure should be a blue circle with five dots inside it.

(b) Find the next geometric shape in this sequence.

The first two shapes consist of vertical lines with horizontal lines at the bottom extending first *left* and then *right*. The third shape is a vertical line with a horizontal line at the top extending to the *left*. Therefore, the next shape should be a vertical line with a horizontal line at the top extending to the *right*.

Now Try Exercise 11.

OBJECTIVE **2** **Use deductive reasoning to analyze arguments.**   In the previous discussion, specific cases were used to find patterns and predict the next event. There is another type of reasoning called **deductive reasoning,** which moves from general cases to specific conclusions.

EXAMPLE 5 **Using Deductive Reasoning**

Does the conclusion follow from the premises in this argument?

All Buicks are automobiles. ← Premise

All automobiles have horns. ← Premise

∴   All Buicks have horns. ← Conclusion

In this example, the first two statements are called *premises* and the third statement (below the line) is called a *conclusion.* The symbol ∴ is a mathematical symbol meaning "**therefore.**" The entire set of statements is called an *argument.*

Continued on Next Page

The focus of deductive reasoning is to determine whether the conclusion follows (is valid) from the premises. A set of circles called **Euler circles** is used to analyze the argument.

In Example 5, the statement "All Buicks are automobiles" can be represented by two circles, one for Buicks and one for automobiles. Note that the circle representing Buicks is totally inside the circle representing automobiles because the first premise states that *all* Buicks are automobiles.

Now, a circle is added to represent the second statement, vehicles with horns. This circle must completely surround the circle representing automobiles because the second premise states that *all* automobiles have horns.

To analyze the conclusion, notice that the circle representing Buicks is *completely* inside the circle representing vehicles with horns. Therefore, it must follow that all Buicks have horns. ***The conclusion is valid.***

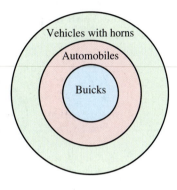

Now Try Exercise 15. ▶▶▶

EXAMPLE 6 **Using Deductive Reasoning**

Does the conclusion follow from the premises? Draw a set of Euler circles to analyze the argument.

All tables are round.

All glasses are round.

∴   All glasses are tables.

Use Euler circles. Draw a circle representing tables *inside* a circle representing round objects, because the first premise states that *all* tables are round.

The second statement requires that a circle representing glasses must now be drawn inside the circle representing round objects, but not necessarily inside the circle representing tables. Therefore, the conclusion does *not* follow from the premises. This means that ***the conclusion is invalid.***

Now Try Exercise 17. ▶▶▶

OBJECTIVE 3 **Use deductive reasoning to solve problems.** Another type of deductive reasoning problem occurs when a set of facts is given in a problem and a conclusion must be drawn using these facts.

EXAMPLE 7 Using Deductive Reasoning

There were 25 students enrolled in a ceramics class. During the class, 10 of the students made a bowl and 8 students made a birdbath. Three students made both a bowl and a birdbath. How many students did not make either a bowl or a birdbath?

This type of problem is best solved by organizing the data using a drawing called a **Venn diagram.** Two overlapping circles are drawn, with each circle representing one item made by the students, as shown below.

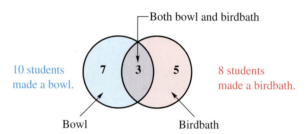

In the region where the circles overlap, write the number of students who made *both* items, namely, 3. In the remaining portion of the birdbath circle, write the number 5, which when added to 3 will give the total number of students who made a birdbath, namely, 8. In a similar manner, write 7 in the remaining portion of the bowl circle, since $7 + 3 = 10$, the total number of students who made a bowl. The total of all three numbers written in the circles is 15. Since there were 25 students in the class, this means $25 - 15$ or 10 students did not make either a birdbath or a bowl.

Now Try Exercise 19.

EXAMPLE 8 Using Deductive Reasoning

Four cars in a race finish first, second, third, and fourth. The following facts are known.

(a) Car A beat Car C.

(b) Car D finished between Cars C and B.

(c) Car C beat Car B.

In which order did the cars finish?

To solve this type of problem, it is helpful to use a line diagram.

1. *Write A before C,* because Car A beat Car C (fact **a**).

    A   C

2. *Write B after C,* because Car C beat Car B (fact **c**).

    A   C   B

3. *Write D between C and B,* because Car D finished between Car C and Car B (fact **b**).

The correct order of finish is shown below.

    A   C   D   B

Now Try Exercise 21.

Appendix B Exercises

Find the next number in each sequence. Describe the pattern in each sequence. See Examples 1–3.

1. 2, 9, 16, 23, 30, . . .

2. 5, 8, 11, 14, 17, . . .

3. 0, 10, 8, 18, 16, . . .

4. 3, 9, 7, 13, 11, . . .

5. 1, 2, 4, 8, . . .

6. 1, 4, 16, 64, . . .

7. 1, 3, 9, 27, 81, . . .

8. 3, 6, 12, 24, 48, . . .

9. 1, 4, 9, 16, 25, . . .

10. 6, 7, 9, 12, 16, . . .

Find the next shape in each sequence. See Example 4.

11.

12.

13.

14.

For each argument, draw Euler circles and then state whether or not the conclusion follows from the premises. See Examples 5 and 6.

15. All animals are wild.
All lions are animals.

∴   All lions are wild.

16. All students are hard workers.
All business majors are students.

∴   All business majors are hard workers.

17. All teachers are serious.
All mathematicians are serious.

∴   All mathematicians are teachers.

18. All boys ride bikes.
All Americans ride bikes.

∴   All Americans are boys.

Solve each application problem. See Examples 7 and 8.

19. In a given 30-day period, a husband watched television 20 days and his wife watched television 25 days. If they watched television together 18 days, how many days did neither watch television? Draw a Venn diagram.

20. In a class of 40 students, 21 students take both calculus and physics. If 30 students take calculus and 25 students take physics, how many do not take either calculus or physics? Draw a Venn diagram.

21. Tom, Dick, Mary, and Joan all work for the same company. One is a secretary, one is a computer operator, one is a receptionist, and one is a mail clerk.

   (a) Tom and Joan eat dinner with the computer operator.

   (b) Dick and Mary carpool with the secretary.

   (c) Mary works on the same floor as the computer operator and the mail clerk.

   Who is the computer operator?

22. Four cars—a Ford, a Buick, a Mercedes, and an Audi—are parked in a garage in four spaces.

   (a) The Ford is in the last space.

   (b) The Buick and Mercedes are next to each other.

   (c) The Audi is next to the Ford but not next to the Buick.

   Which car is in the first space?

Answers to Selected Exercises

In this section we provide the answers that we think most students will obtain when they work the exercises using the methods explained in the text. If your answer does not look exactly like the one given here, it is not necessarily wrong. In many cases there are equivalent forms of the answer that are correct. For example, if the answer section shows $\frac{3}{4}$ and your answer is 0.75, you have obtained the correct answer but written it in a different (yet equivalent) form. Unless the directions specify otherwise, 0.75 is just as valid an answer as $\frac{3}{4}$.

In general, if your answer does not agree with the one given in the text, see whether it can be transformed into the other form. If it can, then it is the correct answer. If you still have doubts, talk with your instructor.

DIAGNOSTIC PRETEST

(page xxv)
1. 25,003,701 **2.** 6 **3.** 4,400,000 **4.** $-\$18$ **5.** -40
6. $4r - 33$ **7.** $t = 5$ **8.** $n = 1$ **9.** 58 cm **10.** 66 in.^{2}
11. $4n + 13 = n - 8; n = -7$ **12.** $x + x + 12 = 238$; 113 ft, 125 ft
13. $\frac{3}{4}$ **14.** $-\frac{1}{4}$ **15.** $x = -20$ **16.** 30 cm^2 **17.** \$18.17 (rounded)
18. $r = 2.3$ **19.** \$10.20 **20.** 16.3 cm (rounded) **21.** $x = 108$
22. 121.6 miles **23.** 117° **24.** $x = 8$ ft; $y = 7.5$ ft **25.** **(a)** 2.5
(b) 30% **26.** **(a)** $\frac{19}{50}$ **(b)** 7.5% **27.** \$249.10 **28.** \$13.80
29. **(a)** 4.32 m **(b)** 80 g **30.** 5 ft **31.** 10.4 L **32.** **(a)** m
(b) mg **33.** \$252 **34.** **(a)** 2003, 2004 **(b)** 2006; 30 sections
35. **(a)** quadrant II **(b)** no quadrant **(c)** quadrant III
(d) quadrant I
36.

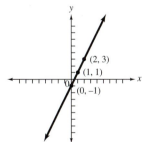

x	y	(x, y)
0	−1	(0, −1)
1	1	(1, 1)
2	3	(2, 3)

37. $\frac{2}{3}$ **38.** $-4a^8b^6$ **39.** **(a)** 5.45×10^7
(b) 0.00407 **40.** $6x^2 - 7x - 5$

CHAPTER 1

Section 1.1 (page 5)
1. 15; 0; 83,001 **3.** 7; 362,049 **5.** hundreds
7. hundred-thousands **9.** ten-millions **11.** hundred-billions
13. ten-trillions, hundred-billions, millions, hundred-thousands, ones
15. eight thousand, four hundred twenty-one **17.** forty-six
thousand, two hundred five **19.** three million, sixty-four thousand,
eight hundred one **21.** eight hundred forty million, one hundred
eleven thousand, three **23.** fifty-one billion, six million, eight
hundred eighty-eight thousand, three hundred twenty-one
25. three trillion, seven hundred twelve million **27.** 46,805
29. 5,600,082 **31.** 271,900,000 **33.** 12,417,625,310

35. 600,000,071,000,400 **37.** six thousand, three hundred
seventy-five **39.** 101,280,000 **41.** ninety-three million, nine
hundred seventy-two thousand, six hundred two dollars
43. 55,000,800 **45.** six million, four hundred thousand each day;
two billion, three hundred thirty-six million in one year
47. largest: 97,651,100; ninety-seven million, six hundred fifty-one
thousand, one hundred; smallest: 10,015,679; ten million, fifteen
thousand, six hundred seventy-nine **48.** Answers will vary.
49. sixty-fours; thirty-twos; sixteens; eights **(a)** 101
(b) 1010 **(c)** 1111 **50.** **(a)** Answers will vary but should
mention that the location or place in which a digit is written gives it
a different value. **(b)** 8 = VIII; 38 = XXXVIII; 275 = CCLXXV;
3322 = MMMCCCXXII **(c)** The Roman system is *not* a place
value system because no matter what place it's in, M = 1000,
C = 100, etc. One disadvantage is that it takes much more space to
write many large numbers; another is that there is no symbol for zero.

Section 1.2 (page 13)
1. $^+$29,035 feet or 29,035 feet **3.** $^-$128.6 degrees **5.** $^-$18 yards
7. $^+$\$100 or \$100 **9.** $^-6\frac{1}{2}$ pounds
11.
13.
15.
17. > **19.** < **21.** < **23.** > **25.** < **27.** > **29.** >
31. < **33.** 15 **35.** 3 **37.** 0 **39.** 200 **41.** 75 **43.** 8042
45.
46. $^-$1.5, $^-$1, 0, 0.5 **47.** A: may be at risk; B: above normal;
C: normal; D: normal **48.** **(a)** The patient would think the
interpretation was "above normal" and wouldn't get treatment.
(b) Patient D's score of 0; zero is neither positive nor negative.

Section 1.3 (page 20)
1. 3

3. $^-$7

5. $^-$1

7. **(a)** $^-$10 **(b)** 10 **9.** **(a)** 12 **(b)** $^-$12 **11.** **(a)** $^-$50
(b) 50 **13.** **(a)** 158 **(b)** $^-$158 **15.** The absolute values are
the same in each pair of answers so the only difference in the sums
is the common sign. **17.** **(a)** 2 **(b)** $^-$2 **19.** **(a)** $^-$7 **(b)** 7

21. (a) −5 **(b)** 5 **23. (a)** 150 **(b)** −150 **25.** Each pair of answers differs only in the sign of the answer. This occurs because the signs of the addends are reversed. **27.** −3 **29.** 7 **31.** −7 **33.** 1 **35.** −8 **37.** −20 **39.** −17 **41.** −22 **43.** −19 **45.** 6 **47.** −5 **49.** 0 **51.** 5 **53.** −32 **55.** 13 + −17 = −4 yards **57.** −62 + 50 = −$12 **59.** −88 + 35 = −$53 **61.** Jeff: −20 + 75 + −55 = 0 points; Terry: 42 + −15 + 20 = 47 points **63.** −14 **65.** 0 **67.** −5 + −18; −23 **69.** 15 + −4; 11 **71.** 6 + (−14 + 14); 6 + 0 = 6 **73.** (−14 + −6) + −7; −20 + −7 = −27 **75.** Some possibilities are: −6 + 0 = −6; 10 + 0 = 10; 0 + 3 = 3 **77.** −4116 **79.** 8686 **81.** −96,077

Section 1.4 (page 26)

1. −6; 6 + −6 = 0 **3.** 13; −13 + 13 = 0 **5.** 0; 0 + 0 = 0 **7.** 14 **9.** −2 **11.** −12 **13.** −25 **15.** −23 **17.** 5 **19.** 20 **21.** 11 **23.** −60 **25.** 0 **27.** 0 **29.** −6 **31. (a)** 8 **(b)** −2 **(c)** 2 **(d)** −8 **33. (a)** −3 **(b)** 11 **(c)** −11 **(d)** 3 **35.** −6 **37.** −5 **39.** 3 **41.** −10 **43.** 12 **45.** −5 **47. (a)** 21 °F; 30 − 21 = 9 degrees difference; **(b)** 0 °F; 15 − 0 = 15 degrees difference **(c)** −17 °F; 5 − −17 = 22 degrees difference; **(d)** −41 °F; −10 − −41 = 31 degrees difference **49.** The student forgot to change 6 to its opposite, −6. It should be −6 + −6 = −12. **51.** −11 **53.** −5 **55.** −10 **57.** Answers on left: −8; 8. On right: −1; 1 Subtraction is *not* commutative; the absolute value of the answer is the same, but the sign changes. **58.** Subtracting 0 from a number does *not* change the number. For example −5 − 0 = −5. But subtracting a number from 0 *does* change the number to its opposite. For example, 0 − −5 = 5.

Section 1.5 (page 34)

1. 42 is closer to 40 **3.** −691 is closer to −700 **5.** 54,402 is closer to 54,000 **7.** 630 **9.** −1080 **11.** 7900 **13.** −86,800 **15.** 42,500 **17.** −6000 **19.** 15,800 **21.** −78,000 **23.** 6000 **25.** 53,000 **27.** 600,000 **29.** 5,400,000 **31.** −9,000,000 **33.** 140,000,000 **35.** 290,000,000 **37.** 20,000,000,000 **39.** 9,000,000,000 **41.** 30,000 miles **43.** −60 degrees **45.** $10,000 **47.** 60,000,000 Americans **49.** −300 feet **51.** 600,000 people in Alaska; 30,000,000 people in California **53.** Answers will vary but should mention looking only at the second digit, rounding first digit up when second digit is 5 or more, leaving first digit unchanged when second digit is 4 or less. Examples will vary. **55.** *Estimate:* −40 + 90 = 50; *Exact:* −42 + 89 = 47 **57.** *Estimate:* 20 + −100 = −80; *Exact:* 16 + −97 = −81 **59.** *Estimate:* −300 + −400 = −700; *Exact:* −273 + −399 = −672 **61.** *Estimate:* 3000 + 7000 = 10,000; *Exact:* 3081 + 6826 = 9907 **63.** *Estimate:* 20 + −80 = −60; *Exact:* 23 − 81 = −58 **65.** *Estimate:* −40 + −40 = −80; *Exact:* −39 − 39 = −78 **67.** *Estimate:* −100 + 30 + +70 = 0; *Exact:* −106 + 34 − −72 = 0 **69.** *Estimate:* 80,000 − 50,000 = $30,000; *Exact:* 78,650 − 52,882 = $25,768 **71.** *Estimate:* 2000 − 700 − 300 − 300 − 200 − 200 = $300; *Exact:* 1920 − 685 − 325 − 320 − 182 − 150 = $258 **73.** *Estimate:* −100 + 40 + 50 = −10 degrees; *Exact:* −102 + 37 + 52 = −13 degrees **75.** *Estimate:* 400 + 100 = 500 doors and windows; *Exact:* 412 + 147 = 559 doors and windows

Section 1.6 (page 45)

1. (a) 63 **(b)** 63 **(c)** −63 **(d)** −63 **3. (a)** −56 **(b)** −56 **(c)** 56 **(d)** 56 **5.** −35 **7.** −45 **9.** −18 **11.** −50 **13.** −40 **15.** −56 **17.** 32 **19.** 77 **21.** 0 **23.** 133 **25.** 13 **27.** 0 **29.** 48 **31.** −56 **33.** −160 **35.** 5 **37.** −3 **39.** −1 **41.** 0 **43.** 5 **45.** −4 **47.** Commutative property: changing the *order* of the factors does not change the product.   Associative property: changing the *grouping* of the factors does not change the product. Examples will vary.   **49.** Examples will vary. Some possibilities are: **(a)** 6 • −1 = −6; 2 • −1 = −2; 15 • −1 = −15

(b) −6 • −1 = 6; −2 • −1 = 2; −15 • −1 = 15   The result of multiplying any nonzero number times −1 is the number with the opposite sign.   **50.** The products are 4, −8, 16, −32. The absolute value doubles each time and the sign changes. The next three products are 64, −128, 256.   **51.** = 9 • −3 + 9 • 5 Both results are 18. **53.** = 8 • 25 Both products are 200.   **55.** = (−3 • 2) • 5 Both products are −30.   **57.** *Estimate:* 300 • 50 = $15,000; *Exact:* 324 • 52 = $16,848   **59.** *Estimate:* −10,000 • 10 = −$100,000; *Exact:* −9950 • 12 = −$119,400   **61.** *Estimate:* 200 • 10 = $2000; *Exact:* 182 • 13 = $2366   **63.** *Estimate:* 20 • 400 = 8000 hours; *Exact:* 24 • 365 = 8760 hours   **65.** −512 **67.** 0 **69.** −355,299 **71.** $247 **73.** −22 degrees **75.** 772 points

Section 1.7 (page 54)

1. (a) 7 **(b)** 7 **(c)** −7 **(d)** −7 **3. (a)** −7 **(b)** 7 **(c)** −7 **(d)** 7 **5. (a)** 1 **(b)** 35 **(c)** −13 **(d)** 1 **7. (a)** 0 **(b)** undefined **(c)** undefined **(d)** 0 **9.** −4 **11.** −3 **13.** 6 **15.** −11 **17.** undefined **19.** −14 **21.** 10 **23.** 4 **25.** −1 **27.** 0 **29.** 191 **31.** −499 **33.** 2 **35.** −4 **37.** 40 **39.** −48 **41.** 5 **43.** 0 **45.** 2 ÷ 1 = 2 but 1 ÷ 2 = 0.5, so division is not commutative.   **46.** (12 ÷ 6) ÷ 2 = 2 ÷ 2 = 1; 12 ÷ (6 ÷ 2) = 12 ÷ 3 = 4; different quotients. Division is not associative.   **47.** Similar: If the signs match, the result is positive. If the signs are different, the result is negative. Different: Multiplication is commutative, division is not. You can multiply by 0, but dividing by 0 is undefined.   **48.** Examples will vary. The properties are: Any nonzero number divided by itself is 1. Any number divided by 1 is the number. Division by 0 is undefined. Zero divided by any other number (except 0) is 0.

49. (a) $\frac{-6}{-1} = 6; \frac{-2}{-1} = 2; \frac{-15}{-1} = 15$

(b) $\frac{6}{-1} = -6; \frac{2}{-1} = -2; \frac{15}{-1} = -15$ When dividing by −1, change the sign of the dividend to its opposite to get the quotient.

50. Division is not commutative. $\frac{0}{-3} = 0$ because 0 • −3 = 0.

But $\frac{-3}{0}$ is undefined because when $\frac{-3}{0} = ?$ is rewritten as ? • 0 = −3, no number can replace ? and make a true statement. **51.** *Estimate:* −40,000 ÷ 20 = −2000 feet; *Exact:* −35,836 ÷ 17 = −2108 feet **53.** *Estimate:* −200 + 500 = $300; *Exact:* −238 + 450 = $212 **55.** *Estimate:* 400 − 100 = 300 days; *Exact:* 365 − 106 = 259 days **57.** *Estimate:* −700 • 40 = −28,000 feet; *Exact:* −730 • 37 = −27,010 feet **59.** *Estimate:* 300 ÷ 5 = 60 miles; *Exact:* 315 ÷ 5 = 63 miles **61.** Average score of 168 **63.** The back shows 520 grams, which is 10 grams more than the front. **65.** −$15 **67.** 16 hours, with 40 minutes left over **69.** 33 rooms, with space for 2 people unused. **71.** −10 **73.** undefined **75.** 31.70979198 rounds to 32 years.

Summary Exercises on Operations with Integers (page 58)

1. −6 **2.** 0 **3.** −7 **4.** −7 **5.** 63 **6.** −1 **7.** −56 **8.** −22 **9.** 12 **10.** 8 **11.** −13 **12.** 0 **13.** 0 **14.** −17 **15.** −48 **16.** −10 **17.** −50 **18.** undefined **19.** −14 **20.** −6 **21.** 0 **22.** 16 **23.** −30 **24.** undefined **25.** 48 **26.** −19 **27.** 2 **28.** −20 **29.** 0 **30.** 16 **31.** −3 **32.** −36 **33.** 6 **34.** −7 **35.** −5 **36.** −31 **37.** −2 **38.** −32 **39.** −5 **40.** −4 **41.** −9732 **42.** 100 **43.** 4 **44.** −343 **45.** 5 **46.** −6 **47.** −10 **48.** −5 **49. (a)** The quotient is 0.   **(b)** The product is 0.   **(c)** The quotient is 1.   **50. (a)** The quotient is −1. *Adding* 15 + −15 would give a *sum* of 0.   **(b)** Dividing by zero cannot be done; the correct answer is "undefined."   **(c)** Work with two numbers at a time; −10 ÷ −2 is 5; then 5 ÷ −5 is −1, not positive 1.

Section 1.8 (page 66)

1. $4 \cdot 4 \cdot 4$; 4 cubed or 4 to the third power   3. 2^7; 128; 2 to the seventh power   5. 5^4; 625; 5 to the fourth power   7. 7^2; $7 \cdot 7$; 49   9. 10^1; 10; 10   11. (a) 10 (b) 100 (c) 1000 (d) 10,000   13. (a) 4 (b) 16 (c) 64 (d) 256   15. 9,765,625   17. 4096   19. 4   21. 25   23. $^-64$ 25. 81   27. $^-1000$ 29. 1   31. 108   33. 200   35. $^-750$ 37. $^-32$ 39. (a) The answers are 4, $^-8$, 16, $^-32$, 64, $^-128$, 256, $^-512$. When a negative number is raised to an even power, the answer is positive; when raised to an odd power, the answer is negative. (b) negative; positive   41. $^-6$ 43. 0   45. $^-39$ 47. 16   49. 23   51. $^-43$ 53. 7   55. $^-3$ 57. 0   59. $^-38$ 61. 41   63. $^-2$ 65. 13   67. 126   69. 8   71. $\dfrac{27}{^-3} = ^-9$ 73. $\dfrac{^-48}{^-4} = 12$ 75. $\dfrac{^-60}{^-1} = 60$ 77. $^-4050$ 79. 7   81. $\dfrac{27}{0}$ is undefined.

Chapter 1 Review Exercises (page 77)

1. 86, 0, 35,600   2. eight hundred six   3. three hundred nineteen thousand, twelve   4. sixty million, three thousand, two hundred   5. fifteen trillion, seven hundred forty-nine billion, six   6. 504,100   7. 620,080,000   8. 99,007,000,356

9.
```
 ◄─●─┼─┼●┼─┼─┼─●─┼─●─┼─┼─┼─►
  ⁻5⁻4⁻3⁻2⁻1  0  1  2  3  4  5
```

10. >   11. <   12. >   13. <   14. 5   15. 9   16. 0   17. 125   18. $^-1$ 19. $^-13$ 20. $^-3$ 21. 0   22. 1   23. $^-24$ 24. $^-7$ 25. 3   26. 0   27. $^-17$ 28. 5; $^-5 + 5 = 0$ 29. $^-18$; $18 + ^-18 = 0$ 30. $^-7$ 31. 17   32. $^-16$ 33. 13   34. 18   35. $^-22$ 36. 0   37. $^-20$ 38. $^-1$ 39. $^-3$ 40. 14   41. 15   42. $^-16$ 43. 3   44. $^-8$ 45. 0   46. 210   47. 59,000   48. 85,000,000   49. $^-3000$ 50. $^-7,060,000$ 51. 400,000   52. $^-200$ pounds   53. $^-1000$ feet   54. 400,000,000 directories   55. 9,000,000,000 people   56. $^-54$ 57. 56   58. $^-100$ 59. 0   60. 24   61. 17   62. $^-48$ 63. 125   64. $^-36$ 65. 50   66. $^-72$ 67. 9   68. $^-7$ 69. undefined   70. 5   71. $^-18$ 72. 0   73. 15   74. $^-1$ 75. $^-5$ 76. 18   77. 0   78. 156 days and 2 extra hours   79. 10,000   80. 32   81. 27   82. 16   83. $^-125$ 84. 8   85. 324   86. $^-200$ 87. $^-25$ 88. $^-2$ 89. 10   90. $^-28$ 91. $\dfrac{8}{^-8} = ^-1$ 92. $\dfrac{11}{0}$ is undefined.   93. associative property of addition   94. commutative property of multiplication   95. addition property of 0   96. multiplication property of 0   97. distributive property   98. associative property of multiplication   99. *Estimate:* $10,000 \cdot 200 = \$2,000,000$; *Exact:* $\$11,900 \cdot 192 = \$2,284,800$ 100. *Estimate:* $\$200 + \$400 - \$700 = ^-\$100$; *Exact:* $\$185 + \$428 - \$706 = ^-\$93$ 101. *Estimate:* $800 \div 20 = 40$ miles; *Exact:* $840 \div 24 = 35$ miles   102. *Estimate:* $(\$40 \cdot 20) + (\$90 \cdot 10) = \$1700$; *Exact:* $(\$39 \cdot 19) + (\$85 \cdot 12) = \$1761$ 103. $^-\$700, ^-\$100, \$700, \$900, \$0, \$700$ 104. January; April   105. $2100   106. $^-\$1850$

Chapter 1 Test (page 81)

[1.1] 1. twenty million, eight thousand, three hundred seven
2. 30,000,700,005
[1.2] 3.
```
 ◄─┼─┼─●┼─┼●●┼─┼─┼─●─►
  ⁻3⁻2⁻1  0  1  2  3
```

4. >; <   5. 10; 14   [1.3, 1.4, 1.6–1.8] 6. $^-6$ 7. $^-5$ 8. 7   9. $^-40$ 10. 10   11. 64   12. $^-50$ 13. undefined   14. $^-60$ 15. $^-5$ 16. $^-45$ 17. 6   18. 25   19. 0   20. 9   21. 128   22. $^-2$ 23. 8   24. $^-16$ [1.8] 25. An exponent shows how many times to use a factor in repeated multiplication. Examples will vary. Some possibilities are $(2)^4 = 2 \cdot 2 \cdot 2 \cdot 2 = 16$ and $(^-3)^2 =$

$(^-3)(^-3) = 9$.   [1.3] 26. Commutative property: changing the *order* of addends does not change the sum. Associative property: changing the *grouping* of addends does not change the sum. Examples will vary.   [1.5] 27. 900   28. 36,420,000,000   29. 350,000   30. *Estimate:* $\$200 + \$300 + ^-\$500 = \$0$; *Exact:* $\$184 + \$293 + ^-\$506 = ^-\$29$ 31. *Estimate:* $^-1000 \div 10 = ^-100$ yards; *Exact:* $^-1104 \div 12 = ^-95$ yards   32. *Estimate:* $30(200 - 100) = 3000$ calories; *Exact:* $31(220 - 110) = 3410$ calories   [1.4] 33. $^-10 - ^-100 = ^-10 + ^+100 = 90$ degrees difference   [1.7] 34. 27 cartons because 26 cartons would leave 28 books unpacked

CHAPTER 2

Section 2.1 (page 91)

1. c is variable; 4 is constant.   3. m is variable; $^-3$ is constant.   5. h is variable; 5 is coefficient.   7. c is variable; 2 is coefficient; 10 is constant.   9. x and y are variables.   11. g is variable; $^-6$ is coefficient; 9 is constant   13. (a) $654 + 10$ is 664 robes. (b) $208 + 10$ is 218 robes. (c) $95 + 10$ is 105 robes.   15. (a) $3 \cdot 11$ inches is 33 inches. (b) $3 \cdot 3$ feet is 9 feet.   17. (a) $3 \cdot 12 - 5$ is 31 brushes. (b) $3 \cdot 16 - 5$ is 43 brushes.   19. (a) $\dfrac{332}{4}$ is 83 points. (b) $\dfrac{637}{7}$ is 91 points.   21. $12 + 12 + 12 + 12$ is 48, $4 \cdot 12$ is 48; $0 + 0 + 0 + 0$ is 0, $4 \cdot 0$ is 0; $^-5 + ^-5 + ^-5 + ^-5 = ^-20$, $4 \cdot ^-5$ is $^-20$ 23. $^-2(^-4) + 5$ is $8 + 5$ is 13; $^-2(^-6) + ^-2$ is $12 + ^-2$ is 10; $^-2(0) + ^-8$ is $0 + ^-8$ is $^-8$ 25. A variable is a letter that represents the part of a rule that varies or changes depending on the situation. An expression expresses, or tells, the rule for doing something. For example, $c + 5$ is an expression, and c is the variable.   27. $b \cdot 1 = b$ or $1 \cdot b = b$ 29. $\dfrac{b}{0}$ is undefined or $b \div 0$ is undefined.   31. $c \cdot c \cdot c \cdot c \cdot c \cdot c$ 33. $x \cdot x \cdot x \cdot x \cdot y \cdot y \cdot y$ 35. $^-3 \cdot a \cdot a \cdot a \cdot b$ 37. $9 \cdot x \cdot y \cdot z$ 39. $^-2 \cdot c \cdot c \cdot c \cdot c \cdot c \cdot d$ 41. $a \cdot a \cdot a \cdot b \cdot c \cdot c$ 43. 16   45. $^-24$ 47. $^-18$ 49. $^-128$ 51. $^-18,432$ 53. 311,040   55. 56   57. $\dfrac{36}{0}$ is undefined.   59. (a) 3 miles (b) 2 miles (c) 1 mile   60. (a) $\dfrac{1}{2}$ mile; take half of the distance for 5 seconds (b) $7\dfrac{1}{2}$ seconds; find the number halfway between 5 seconds and 10 seconds (c) $12\dfrac{1}{2}$ seconds; find the number halfway between 10 seconds and 15 seconds

Section 2.2 (page 102)

1. $2b^2$ and b^2; The coefficients are 2 and 1.   3. ^-xy and $2xy$; The coefficients are $^-1$ and 2.   5. 7, 3, and $^-4$; The like terms are constants.   7. $12r$ 9. $6x^2$ 11. ^-4p 13. $^-3a^3$ 15. 0   17. xy 19. $6t^4$ 21. $4y^2$ 23. ^-8x 25. $12a + 4b$ 27. $7rs + 14$ 29. $a + 2ab^2$ 31. $^-2x + 2y$ 33. $7b^2$ 35. cannot be simplified   37. $^-15r + 5s + t$ 39. $30a$ 41. $^-8x^2$ 43. $^-20y^3$ 45. $18cd$ 47. $21a^2bc$ 49. $12w$ 51. $6b + 36$ 53. $7x - 7$ 55. $21t + 3$ 57. $^-10r - 6$ 59. $^-9k - 36$ 61. $50m - 300$ 63. $8y + 16$ 65. $6a^2 + 3$ 67. $9m - 34$ 69. $^-25$ 71. $24x$ 73. $5n + 13$ 75. $11p - 1$ 77. A simplified expression still has variables, but is written in a simpler way. When *evaluating* an expression, the variables are all replaced by specific numbers and the final result is a numerical answer.   79. Like terms have matching variable parts, that is, matching letters and exponents. The coefficients do not have to match. Examples will vary.   81. Keep the variable part unchanged when combining like terms. The correct answer is $5x + 8$.   83. $^-2y + 9$ 85. 0   87. ^-9x

Summary Exercises on Variables and Expressions (page 106)

1. (a) 32 yards   **(b)** 120 inches   **2. (a)** \$13,080   **(b)** \$22,342
3. 60   **4.** $^{-}8$ **5.** 120   **6.** $^{-}216$ **7.** $^{-}800$ **8.** 120,960   **9.** $24b$
10. $7x + 7$ **11.** $^{-}8c - 32$ **12.** 0   **13.** $12c^2d$ **14.** $^{-}6f$
15. $6w + 8$ **16.** $^{-}2a - 6b$ **17.** $50x^3y^2$ **18.** $5r^3 + 3r^2 + 2r$
19. $7h^2$ **20.** $^{-}9$ **21.** $^{-}32y + 25$ **22.** $36x - 10$ **23.** $19n + 1$
24. $5h^2 - 4h - 4$

Section 2.3 (page 113)

1. 58 is the solution.   **3.** $^{-}16$ is the solution.
5. $^{-}12$ is the solution.
7. $p = 4$ **Check** $\underbrace{4 + 5}_{9} = 9$

$9 = 9$

9. $r = 10$ **Check** $8 = \underbrace{10 - 2}$

$8 = 8$

11. $n = {}^{-}8$ **Check** $^{-}5 = n + 3$

$^{-}5 = \underbrace{{}^{-}8 + 3}$

$^{-}5 = {}^{-}5$

13. $k = 18$ **Check** $^{-}4 + k = 14$

$\underbrace{{}^{-}4 + 18} = 14$

$14 = 14$

15. $y = 6$ **Check** $y - 6 = 0$

$\underbrace{6 - 6} = 0$

$0 = 0$

17. $r = {}^{-}6$ **Check** $7 = r + 13$

$7 = \underbrace{{}^{-}6 + 13}$

$7 = 7$

19. $x = 11$ **Check** $x - 12 = {}^{-}1$

$\underbrace{11 + {}^{-}12} = {}^{-}1$

${}^{-}1 = {}^{-}1$

21. $t = {}^{-}3$ **Check** $^{-}5 = {}^{-}2 + t$

$^{-}5 = \underbrace{{}^{-}2 + {}^{-}3}$

$^{-}5 = {}^{-}5$

23.
$\underbrace{{}^{-}2 + {}^{-}5} = 3$

Does not balance $^{-}7 \neq 3$
The correct solution is 8. $\underbrace{8 - 5} = 3$

Balances $3 = 3$

25.
$7 + x = {}^{-}11$

$7 + {}^{-}18 = {}^{-}11$

Balances $^{-}11 = {}^{-}11$
$^{-}18$ is the correct solution.

27.
$^{-}10 = {}^{-}10 + b$

$^{-}10 = \underbrace{{}^{-}10 + 10}$

Does not balance $^{-}10 \neq 0$
The correct solution is 0. $^{-}10 = \underbrace{{}^{-}10 + 0}$

Balances $^{-}10 = {}^{-}10$

29. $c = 6$; see *Student's Solutions Manual* for **Checks** of odd-numbered Exercises 29–39.   **31.** $y = 5$ **33.** $b = {}^{-}30$ **35.** $t = 0$

37. $z = {}^{-}7$ **39.** $w = 3$ **41.** $x = {}^{-}10$ **43.** $a = 0$
45. $y = {}^{-}25$ **47.** $x = 15$ **49.** $k = 113$ **51.** $b = 18$
53. $r = {}^{-}5$ **55.** $n = {}^{-}105$ **57.** $h = {}^{-}5$ **59.** No, the solution is $^{-}14$, the number used to replace x in the original equation.
61. $g = 295$ graduates   **63.** $c = 55$ chirps   **65.** $p = \$110$ per month in winter   **67.** $m = {}^{-}19$ **69.** $x = 2$ **71.** Equations will vary. Some possibilities are   **(a)** $n - 1 = {}^{-}3$ and $8 = x + 10$.

(b) $y + 6 = 6$ and $^{-}5 = {}^{-}5 + b$.   **72. (a)** $x = \dfrac{1}{2}$

(b) $y = \dfrac{5}{4}$ **(c)** $n = \$0.85$ **(d)** Equations will vary.

Section 2.4 (page 123)

1. $z = 2$ **Check** $\underbrace{6 \cdot 2} = 12$

$12 = 12$

3. $r = 4$ **Check** $48 = 12r$

$48 = 12 \cdot 4$

$48 = \underbrace{48}$

5. $y = 0$ **Check** $3y = 0$

$\underbrace{3 \cdot 0} = 0$

$0 = 0$

7. $k = {}^{-}10$ **Check** $^{-}7k = 70$

$\underbrace{{}^{-}7 \cdot {}^{-}10} = 70$

$70 = 70$

9. $r = 6$ **Check** $^{-}54 = {}^{-}9r$

$^{-}54 = {}^{-}9 \cdot 6$

$^{-}54 = \underbrace{{}^{-}54}$

11. $b = {}^{-}5$ **Check** $^{-}25 = 5b$

$^{-}25 = 5 \cdot {}^{-}5$

$^{-}25 = \underbrace{{}^{-}25}$

13. $r = 3$ **Check** $\underbrace{2 \cdot 3} = 6$

$6 = 6$

15. $p = {}^{-}3$ **Check** $^{-}12 = 5p - p$

$^{-}12 = 5 \cdot {}^{-}3 - {}^{-}3$

$^{-}12 = \underbrace{{}^{-}15 + {}^{+}3}$

$^{-}12 = {}^{-}12$

17. $a = {}^{-}5$ **19.** $x = {}^{-}10$ **21.** $w = 0$ **23.** $t = 3$ **25.** $t = 0$
27. $m = 9$ **29.** $y = {}^{-}1$ **31.** $z = {}^{-}5$ **33.** $p = {}^{-}2$ **35.** $k = 7$
37. $b = {}^{-}3$ **39.** $x = {}^{-}32$ **41.** $w = 2$ **43.** $n = 50$
45. $p = {}^{-}10$ **47.** Each solution is the opposite of the number in the equation. So the rule is: When you change the sign of the variable from negative to positive, then change the number in the equation to its opposite. In $^{-}x = 5$, the opposite of 5 is $^{-}5$, so $x = {}^{-}5$.   **49.** Divide by the coefficient of x, which is 3, *not* by the opposite of 3. The correct solution is 5.   **51.** $s = 15$ ft
53. $s = 24$ meters   **55.** $y = 27$ **57.** $x = 1$

Section 2.5 (page 131)

1. $p = 1$ **Check** $\underbrace{7(1)} + 5 = 12$

$\underbrace{7 + 5} = 12$

$12 = 12$

2. $y = 1$ **Check** $2 = 8y - 6$

$$2 = 8(1) - 6$$
$$2 = 8 - 6$$
$$2 = 2$$

5. $m = 0$ **Check** $-3m + 1 = 1$

$$-3(0) + 1 = 1$$
$$0 + 1 = 1$$
$$1 = 1$$

7. $a = -2$ **Check** $28 = -9a + 10$

$$28 = -9(-2) + 10$$
$$28 = 18 + 10$$
$$28 = 28$$

9. $x = -4$ **Check** $-5x - 4 = 16$

$$-5(-4) - 4 = 16$$
$$20 - 4 = 16$$
$$16 = 16$$

11. $p = 4; 4 = p$ **Check** $6(4) - 2 = 4(4) + 6$

$$24 - 2 = 16 + 6$$
$$22 = 22$$

13. $k = -2; -2 = k$ **Check** $-2k - 6 = 6k + 10$

$$-2(-2) - 6 = 6(-2) + 10$$
$$4 + -6 = -12 + 10$$
$$-2 = -2$$

15. $a = 5; 5 = a$ **Check** $-18 + 7a = 2a + 7$

$$-18 + 7(5) = 2(5) + 7$$
$$-18 + 35 = 10 + 7$$
$$17 = 17$$

17. $w = 6$ **19.** $y = -9$ **21.** $t = -5$ **23.** $x = 0$ **25.** $h = 1$
27. $y = -2$ **29.** $m = -3$ **31.** $w = 2$ **33.** $x = 5$ **35.** $a = 3$
37. $b = -3$ **39.** $k = 4$ **41.** $c = 0$ **43.** $y = -5$ **45.** $n = 21$
47. $c = 30$ **49.** $p = -2$ **51.** $b = -2$
53. The series of steps may vary. One possibility is:

$-2t - 10 = 3t + 5$ Change subtraction to adding the opposite.

$-2t + -10 = 3t + 5$ Add $2t$ to both sides

$\dfrac{2t \qquad\qquad 2t}{0 + -10 = 5t + 5}$ (addition property).

Add -5 to both sides

$\dfrac{-5 \qquad\quad -5}{}$ (addition property).

$\dfrac{-15}{5} = \dfrac{5t}{5}$ Divide both sides by 5 (division property).

$-3 = t$

55. Check $-8 + 4(3) = 2(3) + 2$

$$-8 + 12 = 6 + 2$$
$$4 \neq 8$$

The check does not balance, so 3 is not the correct solution. The student added $-2a$ to -8 on the left side, instead of adding $-2a$ to $4a$. The correct solution is 5.   **57. (a)** It must be negative.
(b) The sum of x and a positive number is negative, so x must be negative.   **58. (a)** It must be positive.   **(b)** The sum of d and a negative number is positive, so d must be positive.   **59. (a)** It must be positive; when the signs are different, the product is negative.

(b) The product of n and a negative number is negative, so n must be positive.   **60. (a)** It must be negative also; when the signs match, the product is positive.   **(b)** The product of y and a negative number is positive, so y must be negative.

Chapter 2 Review Exercises (page 145)

1. (a) Variable is k; coefficient is 4; constant is -3.
(b) $-9y + 20$ **2. (a)** 70 test tubes   **(b)** 106 test tubes
3. (a) $x \cdot x \cdot y \cdot y \cdot y \cdot y$ **(b)** $5 \cdot a \cdot b \cdot b \cdot b$ **4. (a)** 9
(b) -27 **(c)** -128 **(d)** 720   **5.** $ab^2 + 3ab$ **6.** $-4x + 2y - 7$
7. $16g^3$ **8.** $12r^2t$ **9.** $5k + 10$ **10.** $-6b - 8$ **11.** $6y$
12. $20x + 2$ **13.** Expressions will vary. One possibility is
$6a^3 + a^2 + 3a - 6$.
14. $n = -11$ **Check** $16 + -11 = 5$

$$5 = 5$$

15. $a = 4$ **Check** $-4 + 2 = 2(4) - 6 - 4$

$$-2 = 8 + -6 + -4$$
$$-2 = -2$$

16. $m = -8$ **17.** $k = 10$ **18.** $t = 0$ **19.** $p = -6$ **20.** $r = 2$
21. $h = -12$ **22.** $w = 4$ **23.** $c = -2$ **24.** $n = 15$ employees
25. $a = -5$ **26.** $p = 10$ **27.** $y = 5$ **28.** $m = 3$ **29.** $x = 9$
30. $b = 7$ **31.** $z = -3$ **32.** $n = 4$ **33.** $t = 0$ **34.** $d = 5$
35. $b = -2$

Chapter 2 Test (page 147)

[2.1] 1. -7 is the coefficient; w is the variable; 6 is the constant.
2. Buy 177 hot dogs.   **3.** $x \cdot x \cdot x \cdot x \cdot x \cdot y \cdot y \cdot y$
4. $4 \cdot a \cdot b \cdot b \cdot b \cdot b$ **5.** -200 **[2.2] 6.** $-4w^3$ **7.** 0   **8.** c
9. cannot be simplified   **10.** $-40b^2$ **11.** $15k$ **12.** $21t + 28$
13. $-4a - 24$ **14.** $6x - 15$ **15.** $-9b + c + 6$
[2.3, 2.4] 16. $x = 5$ **Check** $-4 = 5 - 9$

$$-4 = -4$$

17. $w = -11$ **Check** $-7(-11) = 77$

$$77 = 77$$

18. $p = -14$ **Check** $-1(-14) = 14$

$$14 = 14$$

19. $a = 3$ **Check** $-15 = -3(3 + 2)$

$$-15 = -3(5)$$
$$-15 = -15$$

[2.5] 20. $n = -8$ **21.** $m = 15$ **22.** $x = -1$ **23.** $m = 2$
24. $b = 54$ **25.** $c = 0$ **26.** Equations will vary. Two
possibilities are $x - 5 = -9$ and $-24 = 6y$.

Cumulative Review Exercises: Chapters 1–2 (page 149)

[1.1, 1.2] 1. three hundred six billion, four thousand, two hundred ten
2. 800,066,000   **3. (a)** $>$ **(b)** $<$ **[1.3, 1.6] 4. (a)** commutative
property of addition   **(b)** multiplication property of 0
(c) distributive property   **[1.5] 5. (a)** 9000   **(b)** 290,000
[1.3–1.7] 6. -8 **7.** 10   **8.** 30   **9.** 25   **10.** 7   **11.** 0   **12.** -64
13. undefined   **14.** -60 **15.** -40 **16.** -9 **17.** -25
18. -28 **19.** $\dfrac{11}{-11} = -1$ **[1.5] 20.** *Estimate:* $600 \div 20 =$
30 miles; *Exact:* $616 \div 22 = 28$ miles   **21.** *Estimate:* $-50 + 20$
$= -30$ degrees; *Exact:* $-48 + 23 = -25$ degrees   **22.** *Estimate:*
$\$2000 - (50 \cdot 10) = \$1500$; *Exact:* $\$2132 - (52 \cdot 8) = \$1716$
23. *Estimate:* $10(600 + 40) = \$6400$; *Exact:* $12(552 + 35) =$
$\$7044$ **[2.1] 24.** $-4 \cdot a \cdot b \cdot b \cdot b \cdot c \cdot c$ **25.** 120   **[2.2] 26.** h
27. 0   **28.** $5n^2 - 4n - 2$ **29.** $-30b^2$ **30.** $28p - 28$
31. $-9w^2 - 12$
[2.3–2.5] 32. $x = -4$ **Check** $3(-4) = -4 - 8$

$$-12 = -12$$

33. $y = {}^-6$ **Check**
$$^-44 = {}^-2 + 7({}^-6)$$
$$^-44 = \underbrace{{}^-2 + {}^-42}$$
$$^-44 = {}^-44$$

34. $k = 7$ **Check** $2(7) - 5(7) = {}^-21$
$$\underbrace{14 - 35} = {}^-21$$
$${}^-21 = {}^-21$$

35. $m = 4$ **Check** $\underbrace{4 - 6} = {}^-2(4) + 6$
$${}^-2 = \underbrace{{}^-8 + 6}$$
$${}^-2 = {}^-2$$

36. $x = {}^-1$ **37.** $r = {}^-18$ **38.** $b = {}^-5$ **39.** $t = 1$
40. $y = {}^-12$

CHAPTER 3

Section 3.1 (page 157)
1. $P = 36$ cm **3.** $P = 100$ in. **5.** $P = 4$ miles **7.** $P = 88$ mm
9. $s = 30$ ft **11.** $s = 1$ mm **13.** $s = 23$ yards **15.** $s = 2$ ft
17. $P = 28$ yd **19.** $P = 70$ cm **21.** $P = 72$ ft **23.** $P = 26$ in.
25. $l = 9$ cm **Check** 9 cm + 9 cm + 6 cm + 6 cm = 30 cm
27. $w = 1$ mile **Check** 4 mi + 4 mi + 1 mi + 1 mi = 10 mi
29. $w = 2$ ft **Check** 6 ft + 6 ft + 2 ft + 2 ft = 16 ft
31. $l = 2$ m **Check** 2 m + 2 m + 1 m + 1 m = 6 m
33. $P = 208$ m **35.** $P = 320$ ft **37.** $P = 54$ mm **39.** $P = 48$ ft
41. $P = 78$ in. **43.** $P = 125$ m **45.** ? = 40 cm **47.** ? = 12 in.
49. (a) Sketches will vary. (b) Formula for perimeter of an equilateral triangle is $P = 3s$, where s is the length of one side. (c) The formula will *not* work for other kinds of triangles because the sides will have different lengths. **51.** (a) 140 miles (b) 350 miles (c) 560 miles **52.** (a) 70 miles (b) 175 miles (c) 280 miles (d) The rate is half of 70 miles per hour, so the distance will be half as far; divide each result in Exercise 51 by 2. **53.** (a) 50 hours (b) 60 hours (c) 150 hours **54.** (a) 61 miles per hour (b) 57 miles per hour (c) 65 miles per hour

Section 3.2 (page 167)
1. $A = 77$ ft^2 **3.** $A = 100$ m^2 **5.** $A = 775$ mm^2 **7.** $A = 36$ in.^{2}
9. $A = 105$ cm^2 **11.** $A = 72$ ft^2 **13.** $A = 625$ mi^2 **15.** $A = 1$ m^2
17. $l = 6$ ft **Check** $A = 6$ ft • 3 ft; $A = 18$ ft^2 **19.** $w = 80$ yd
Check $A = 90$ yd • 80 yd; $A = 7200$ yd^2 **21.** $l = 14$ in.
Check $A = 14$ in. • 11 in.; $A = 154$ in.^{2} **23.** $s = 6$ m **25.** $s = 2$ ft
27. $h = 20$ cm **Check** $A = 25$ cm • 20 cm; $A = 500$ cm^2
29. $b = 17$ in. **Check** $A = 17$ in. • 13 in. $A = 221$ in.^{2}
31. $h = 1$ m **Check** $A = 9$ m • 1 m; $A = 9$ m^2 **33.** Height is not part of perimeter. Square units are used for area, not perimeter. $P = 25$ cm + 25 cm + 25 cm + 25 cm; $P = 100$ cm
35. $P = 48$ m; $A = 144$ m^2 **37.** $108 **39.** $725 **41.** 53 yd
43. Panoramic: $P = 36$ in., $A = 56$ in.^{2}; 4 in. × 6 in.: $P = 20$ in., $A = 24$ in.^{2}; 4 in. × 7 in.: $P = 22$ in., $A = 28$ in.^{2}
45.

46. (a) 5 ft by 1 ft has area of 5 ft^2; 4 ft by 2 ft has area of 8 ft^2; 3 ft by 3 ft has area of 9 ft^2 (b) The square plot 3 ft by 3 ft has the greatest area.
47.

48. (a) 7 ft^2, 12 ft^2, 15 ft^2, 16 ft^2 (b) Square plots have the greatest area.

Summary Exercises on Perimeter and Area (page 171)
1. Rectangle; $P = 32$ m; $A = 39$ m^2
2. Square; $P = 104$ ft; $A = 676$ ft^2
3. Parallelogram; $P = 34$ yd; $A = 56$ yd^2
4. Rectangle; $P = 36$ cm; $A = 80$ cm^2
5. Square; $P = 36$ in.; $A = 81$ in.^{2}
6. Parallelogram; $P = 30$ m; $A = 45$ m^2
7. Rectangle; $P = 26$ ft; $A = 36$ ft^2
8. Parallelogram; $P = 188$ ft; $A = 2100$ ft^2
9. $w = 1$ ft

Check: 6 ft + 6 ft + 1 ft + 1 ft = 14 ft

10. $b = 4$ yd

Check: $A = 4$ yd • 1 yd
$A = 4$ yd^2

11. $s = 9$ in.

Check: $A = 9$ in. • 9 in.
$A = 81$ in.^{2}

12. $w = 3$ m

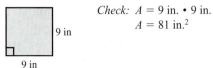

Check: $A = 31$ m • 3 m
$A = 93$ m^2

13. $s = 16$ cm

Check: $P = 16$ cm + 16 cm + 16 cm + 16 cm = 64 cm

14. $\ell = 15$ ft

Check: $P = 15$ ft + 15 ft + 9 ft + 9 ft = 48 ft

15. 504 meters of fencing **16.** 13,000 yd^2 − 5000 yd^2 = 8000 yd^2
17. 9 inches **18.** 176 cm of braid **19.** 7(15 ft^2) = 105 ft^2 of fabric; 7(16 ft) = 112 ft of binding **20.** 48 ft

Section 3.3 (page 178)
1. $14 + x$ or $x + 14$ **3.** $^-5 + x$ or $x + ^-5$ **5.** $20 - x$
7. $x - 9$ **9.** $x - 4$ **11.** ^-6x **13.** $2x$ **15.** $\dfrac{x}{2}$ **17.** $2x + 8$ or
$8 + 2x$ **19.** $7x - 10$ **21.** $2x + x$ or $x + 2x$
23. $4n - 2 = 26$; $n = 7$ **Check** $\underbrace{4 \cdot 7} - 2$ does equal 26.
$$\underbrace{28 - 2}$$
$$26$$

25. $2n + n = {}^-15$ or $n + 2n = {}^-15$; $n = {}^-5$
Check $\underbrace{2 \cdot {}^-5} + {}^-5$ does equal $^-15$.
$$\underbrace{{}^-10 + {}^-5}$$
$$^-15$$

27. $5n + 12 = 7n$; $n = 6$ **Check** $\underbrace{5 \cdot 6} + 12 = \underbrace{7 \cdot 6}$
$$\underbrace{30 + 12} = 42$$
$$42 = 42$$

29. $30 - 3n = 2 + n$; $n = 7$ **Check** $30 - \underbrace{3 \cdot 7} = \underbrace{2 + 7}$
$$\underbrace{30 - 21} = 9$$
$$9 = 9$$

31. Let w be Ricardo's original weight. $w + 15 - 28 + 5 = 177$. He weighed 185 pounds originally.   **33.** Let c be the number of cookies the children ate. $18 - c + 36 = 49$.   Her children ate 5 cookies.   **35.** Let p be the number of pens in each box. $6p - 32 - 35 = 5$. There were 12 pens in each box.   **37.** Let d be each member's dues. $14d + 340 - 575 = {}^-25$.   Each member paid \$15.   **39.** Let a be Tamu's age. $4a - 75 = a$. Tamu is 25 years old.   **41.** Let m be the amount Brenda spent. $2m - 3 = 81$. Brenda spent \$42.   **43.** Let b be the number of pieces in each bag. $5b - 3 \cdot 48 = b$.   There were 36 pieces of candy in each bag. **45.** Let d be the daily amount for an infant. $2d + 3 = 15$.   An infant should receive 6 mg of iron.

Section 3.4 (page 185)

1. a is my age; $a + 9$ is my sister's age. $a + a + 9 = 51$. I am 21; my sister is 30.   **3.** m is husband's earnings; $m + 1500$ is Lien's. $m + m + 1500 = 37{,}500$. Husband earned \$18,000; Lien earned \$19,500.   **5.** m is printer's cost; $5m$ is computer's cost; $5m + m = \$1320$. Printer cost \$220; computer cost \$1100.
7. Shorter piece is x; longer piece is $x + 10$. $x + x + 10 = 78$. Shorter piece is 34 cm; longer piece is 44 cm.   **9.** Longer piece is x; shorter piece is $x - 7$. $x + x - 7 = 31$. Longer piece is 19 ft; shorter piece is 12 ft.   **11.** s is number of Senators; $5s - 65$ is number of Representatives. $s + 5s - 65 = 535$. 100 Senators; 435 Representatives   **13.** First part is x; second part is x; third part is $x + 25$. $x + x + x + 25 = 706$. First part is 227 m; second part is 227 m; third part is 252 m.   **15.** Length is 19 yd. **17.** Length is 12 ft; width is 6 ft.   **19.** Length is 13 in.; width is 5 in. **21.** $P = 52$ in.; $A = 168$ in.^{2}

Chapter 3 Review Exercises (page 195)

1. Square; $P = 112$ cm   **2.** Rectangle; $P = 22$ mi **3.** Parallelogram; $P = 42$ yd   **4.** $P = 141$ m   **5.** 12 ft $= 4s$; $s = 3$ ft   **6.** 128 yd $= 2l + 2(31$ yd$)$; $l = 33$ yd **7.** 72 in. $= 2(21$ in.$) + 2w$; $w = 15$ in.   **8.** $A = 40$ ft^2 **9.** $A = 625$ m^2 **10.** $A = 208$ yd^2 **11.** 126 ft$^2 = 14$ ft $\cdot$ w; $w = 9$ ft   **12.** 88 cm$^2 = 11$ cm $\cdot$ h; $h = 8$ cm   **13.** 100 mi$^2 = s \cdot s$; $s = 10$ mi   **14.** $57 - x$ **15.** $15 + 2x$ or $2x + 15$ **16.** ${}^-9x$ **17.** $4n + 6 = {}^-30$; $n = {}^-9$ **18.** $10 - 2n = 4 + n$; $n = 2$ **19.** m is money originally in account. $m - 600 + 750 + 75 = 309$. \$84 was originally in Grace's account.   **20.** c is number of candles in each box. $4c - 25 = 23$. There were 12 candles in each box. **21.** p is Reggie's prize money; $p + 300$ is Donald's prize money. $p + p + 300 = 1000$. Reggie gets \$350; Donald gets \$650. **22.** w is the width; $2w$ is the length. $84 = 2(2w) + 2(w)$. The width is 14 cm; the length is 28 cm.   **23. (a)** 36 ft $= 4s$; $s = 9$ ft **(b)** $A = 9$ ft $\cdot$ 9 ft; $A = 81$ ft^2 **24.** Rectangles will vary. Two possibilities are: $A = 7$ ft $\cdot$ 3 ft, $A = 21$ ft^2; $A = 6$ ft $\cdot$ 4 ft, $A = 24$ ft^2 **25.** Let f be the fencing for the garden. $36 - f + 20 = 41$. 15 ft of fencing was used on the garden.   **26.** w is the width; $w + 2$ is the length. $36 = 2(w + 2) + 2 \cdot w$. Width is 8 ft; length is 10 ft.

Chapter 3 Test (page 197)

[3.1] **1.** $P = 262$ m   **2.** $P = 40$ in.   **3.** $P = 12$ miles   **4.** $P = 12$ ft **5.** $P = 110$ cm   [3.2] **6.** $A = 486$ mm^2 **7.** $A = 140$ cm^2 **8.** $A = 3740$ mi^2 **9.** $A = 36$ m^2 [3.1, 3.2] **10.** 12 ft $= 4s$; $s = 3$ ft **11.** 34 ft $= 2l + 2(6$ ft$)$; $l = 11$ ft   **12.** 65 in.$^2 = 13$ in. $\cdot$ h; $h = 5$ in. **13.** 12 cm$^2 = 4$ cm $\cdot$ w; $w = 3$ cm   **14.** 16 ft$^2 = s^2$; $s = 4$ ft **15.** Linear units like ft are used to measure length, width, height, and perimeter. Area is measured in square units like ft^2 (squares that measure 1 ft on each side).   [3.3] **16.** $4n + 40 = 0$; $n = {}^-10$ **17.** $7n - 23 = n + 7$; $n = 5$ **18.** Let m represent money spent on groceries. $43 - m + 16 = 44$.   Son spent \$15.   **19.** Let d represent daughter's age. $39 = 5d + 4$.   Daughter is 7 years old.

[3.4] **20.** Let p be the length of one piece; $p + 4$ length of second piece. $p + p + 4 = 118$.   One piece is 57 cm; one piece is 61 cm. **21.** Let w be the width; $4w$ the length. $2(4w) + 2(w) = 420$. Sketches may vary; length is 168 ft; width is 42 ft.   **22.** Let h represent Marcella's hours, $h - 3$ represent Tim's hours. $h + h - 3 = 19$. Marcella worked 11 hours; Tim worked 8 hours.

Cumulative Review Exercises: Chapters 1–3 (page 199)

[1.1, 1.2] **1.** four billion, two hundred six thousand, three hundred **2.** 70,005,489   **3.** $<$ $>$ [1.3, 1.6] **4. (a)** multiplication property of 1   **(b)** addition property of 0   **(c)** associative property of multiplication   [1.5] **5. (a)** 3800   **(b)** 490,000 [1.3–1.7] **6.** ${}^-24$ **7.** 27   **8.** ${}^-3$ **9.** ${}^-100$ **10.** 5   **11.** 0
12. ${}^-13$ **13.** 25   **14.** ${}^-32$ **15.** $\dfrac{27}{0}$ is undefined.

[2.1] **16.** $10 \cdot w \cdot w \cdot x \cdot y \cdot y \cdot y \cdot y$ **17.** 240   [2.2] **18.** $2k$ **19.** $3m^2 + 2m$ **20.** 0   **21.** ${}^-20a$ **22.** ${}^-2x^2 - 3$ **23.** ${}^-12n + 1$ [2.3–2.5] **24.** $x = 2$ Check $\underbrace{6 - 20}_{{}^-14} = \underbrace{2(2) - 9(2)}_{4 \ - \ 18}$
$$ {}^-14 = {}^-14 $$
25. $y = {}^-1$ Check $\underbrace{{}^-5({}^-1)}_{5} = \underbrace{{}^-1 + 6}_{5}$
$$ 5 = 5 $$
26. $b = 4$ **27.** $h = {}^-2$ **28.** $x = 0$ **29.** $a = {}^-3$ [3.1, 3.2] **30.** $P = 60$ in.; $A = 198$ in.^{2} **31.** $P = 60$ m; $A = 225$ m^2 **32.** $P = 24$ ft; $A = 32$ ft^2 [3.3] **33.** $5n + {}^-50 = 0$; $n = 10$ **34.** $10 - 3n = 2n$; $n = 2$ **35.** Let p be number of people originally. $p - 3 + 6 - 2 = 5$. There were 4 people in line. **36.** Let m be amount paid by each player. $12m - 2200 = {}^-40$. Each player paid \$180.   [3.4] **37.** Let g be one group; $3g$ the other. $g + 3g = 192$. 48 in one group; 144 in the other.   **38.** w is the width; $w + 14$ is the length. $2(w + 14) + 2(w) = 92$.   Width is 16 ft; length is 30 ft.

CHAPTER 4

Section 4.1 (page 209)

1. $\dfrac{5}{8}; \dfrac{3}{8}$ **3.** $\dfrac{2}{3}; \dfrac{1}{3}$ **5.** $\dfrac{3}{2}; \dfrac{1}{2}$ **7.** $\dfrac{11}{6}; \dfrac{1}{6}$ **9.** $\dfrac{2}{11}; \dfrac{3}{11}; \dfrac{4}{11}$

11. $\dfrac{8}{25}; \dfrac{17}{25}$ **13.** $\dfrac{13}{71}; \dfrac{58}{71}$ **15.** $\dfrac{6}{20}$ **17.** $\dfrac{19}{20}$ **19.** N: 3; D: 4;
4 parts in the whole   **21.** N: 12; D: 7; 7 parts in the whole

23. Proper: $\dfrac{1}{3}, \dfrac{5}{8}, \dfrac{7}{16}$; Improper: $\dfrac{8}{5}, \dfrac{6}{6}, \dfrac{12}{2}$

25. Proper: $\dfrac{3}{4}, \dfrac{9}{11}, \dfrac{7}{15}$; Improper: $\dfrac{3}{2}, \dfrac{5}{5}, \dfrac{19}{18}$

27. Fractions will vary. The denominator shows the number of equal parts in the whole and the numerator shows how many of the parts are being considered. The fraction bar separates the numerator from the denominator. Show your drawing to your instructor.

29. (number line: points at $-\frac{1}{4}$ and $\frac{1}{4}$)   **31.** (number line: points at $-\frac{3}{5}$ and $\frac{3}{5}$)

33. (number line: points at $-\frac{7}{8}$ and $\frac{7}{8}$)

35. $-\dfrac{3}{4}$ pound   **37.** $-\dfrac{1}{2}$ quart   **39.** $\dfrac{3}{10}$ mile

41. $\frac{2}{5}$ **43.** $\frac{9}{10}$ **45.** $\frac{13}{6}$ **47. (a)** $\frac{12}{24}$ **(b)** $\frac{8}{24}$ **(c)** $\frac{16}{24}$ **(d)** $\frac{6}{24}$
(e) $\frac{18}{24}$ **(f)** $\frac{4}{24}$ **(g)** $\frac{20}{24}$ **(h)** $\frac{3}{24}$ **(i)** $\frac{9}{24}$ **(j)** $\frac{15}{24}$ **49. (a)** $-\frac{1}{3}$
(b) $-\frac{2}{3}$ **(c)** $-\frac{2}{3}$ **(d)** $-\frac{1}{3}$ **(e)** $-\frac{2}{3}$ **(f)** Some possibilities
are: $-\frac{4}{12} = -\frac{1}{3}$; $-\frac{8}{24} = -\frac{1}{3}$; $-\frac{20}{30} = -\frac{2}{3}$; $-\frac{24}{36} = -\frac{2}{3}$.
51. (a) $\frac{1467}{3912}$ **(b)** Divide 3912 by 8 to get 489; multiply 3 by 489
to get 1467.   **52. (a)** $\frac{4256}{5472}$ **(b)** Divide 5472 by 9 to get 608;
multiply 7 by 608 to get 4256.   **53. (a)** $-\frac{1}{5}$ **(b)** Divide 3485 by
2, by 3, and by 5 to see that dividing by 5 gives 697. Or divide 3485
by 697 to get 5.   **54. (a)** $-\frac{1}{6}$ **(b)** Divide 4902 by 4, by 6, and
by 8 to see that dividing by 6 gives 817. Or divide 4902 by 817 to
get 6.   **55.** 7   **56.** 5   **57.** 15   **58.** 16   **59.** Multiply or divide
the numerator and denominator by the same nonzero number. Some
possibilities are: $\frac{2}{3} = \frac{2 \cdot 4}{3 \cdot 4} = \frac{8}{12}$ and $\frac{10}{16} = \frac{10 \div 2}{16 \div 2} = \frac{5}{8}$
61. You cannot do it if you want the numerator to be a whole number,
because 5 does not divide into 18 evenly. You could use multiples of 5
as the denominator, such as 10, 15, 20, etc.   **63.** 10   **65.** -1
67. -6 **69.** 3   **71.** 1   **73.** 2   **75.** -5 **77.** -5 **79.** 1
81. -8 **83.** There are many correct ways to draw the answers for
Exercises 83–100, so ask your instructor to check your drawings.
$\frac{2}{5}$ is unshaded.   **85.** $\frac{5}{8}$ is unshaded.   **87.** $\frac{1}{4}$ is unshaded.
89. $\frac{2}{3}$ is unshaded.   **91.** $\frac{0}{6}$ is unshaded.
93. Divide each rectangle into 5 equal parts, then shade both
rectangles completely.   **95.** Drawings will vary.
97. One possibility is shown. ◯ ⬛ ☐ ☐ ☐ ☐ ☐ ▲ ▲ △
99. One possibility is shown. (!)!!!,...???

Section 4.2 (page 223)
1. (a) yes   **(b)** No, 5 is a common factor.   **(c)** yes   **(d)** yes
(e) No, 3 is a common factor.   **(f)** No, 7 is a common factor.
3. (a) $\frac{2}{3}$ **(b)** $\frac{2}{3}$ **(c)** $-\frac{1}{4}$ **(d)** $-\frac{1}{2}$ **(e)** $\frac{8}{9}$ **(f)** $-\frac{2}{5}$
5. comp. prime comp. neither prime prime comp. comp.
7. $2 \cdot 3$ **9.** $2 \cdot 2 \cdot 5$ **11.** $5 \cdot 5$ **13.** $2 \cdot 2 \cdot 3 \cdot 3$
15. (a) $2 \cdot 2 \cdot 11$ **(b)** $2 \cdot 2 \cdot 2 \cdot 11$ **17. (a)** $3 \cdot 5 \cdot 5$
(b) $2 \cdot 2 \cdot 17$ **(c)** $3 \cdot 3 \cdot 3 \cdot 7$
19. $\dfrac{\cancel{2} \cdot \cancel{2} \cdot \cancel{2}}{\cancel{2} \cdot \cancel{2} \cdot \cancel{2} \cdot 2} = \dfrac{1}{2}$ **21.** $\dfrac{\cancel{2} \cdot \cancel{2} \cdot \cancel{2} \cdot \cancel{2} \cdot 2}{\cancel{2} \cdot \cancel{2} \cdot \cancel{2} \cdot \cancel{2} \cdot 3} = \dfrac{2}{3}$
23. $\dfrac{2 \cdot \cancel{7}}{3 \cdot \cancel{7}} = \dfrac{2}{3}$ **25.** $\dfrac{\cancel{2} \cdot 2 \cdot \cancel{3} \cdot 3}{\cancel{2} \cdot \cancel{3} \cdot 7} = \dfrac{6}{7}$ **27.** Already in lowest terms.
29. $\dfrac{\cancel{3} \cdot \cancel{3} \cdot 3}{\cancel{3} \cdot \cancel{3} \cdot 5} = \dfrac{3}{5}$ **31.** $\dfrac{\cancel{2} \cdot 2 \cdot \cancel{3}}{\cancel{2} \cdot \cancel{3} \cdot 3} = \dfrac{2}{3}$ **33.** $\dfrac{\cancel{3} \cdot 7}{2 \cdot 2 \cdot 2 \cdot \cancel{3}} = \dfrac{7}{8}$
35. $\dfrac{\cancel{2} \cdot \cancel{3} \cdot \cancel{3} \cdot \cancel{3}}{\cancel{2} \cdot 2 \cdot \cancel{3} \cdot \cancel{3} \cdot \cancel{3}} = \dfrac{1}{2}$ **37.** $\dfrac{2 \cdot \cancel{3} \cdot \cancel{3} \cdot \cancel{7}}{3 \cdot \cancel{3} \cdot \cancel{3} \cdot \cancel{7}} = \dfrac{2}{3}$

39. $\dfrac{\cancel{3} \cdot \cancel{M} \cdot 13}{\cancel{3} \cdot 3 \cdot 5 \cdot \cancel{M}} = \dfrac{13}{15}$ **41. (a)** $\frac{1}{4}$ **(b)** $\frac{1}{2}$ **(c)** $\frac{1}{10}$ **(d)** $\frac{60}{60} = 1$
43. (a) $\frac{1}{3}$ **(b)** $\frac{1}{5}$ **(c)** $\frac{7}{15}$ **45. (a)** $\frac{5}{24}$ **(b)** $\frac{2}{3}$ **(c)** $\frac{1}{8}$
47. (a) The result of dividing 3 by 3 is 1, so 1 should be written
above and below all the slashes. The numerator is $1 \cdot 1$, so the correct
answer is $\frac{1}{4}$.   **(b)** You must divide numerator and denominator by
the *same* number. The fraction is already in lowest terms because
9 and 16 have no common factor besides 1.   **49.** $\frac{2c}{5}$ **51.** $\frac{4}{7}$
53. $\frac{6r}{5s}$ **55.** $\frac{1}{7n^2}$ **57.** already in lowest terms   **59.** $\frac{7}{9y}$ **61.** $\frac{7k}{2}$
63. $\frac{1}{3}$ **65.** $\frac{c}{d}$ **67.** $\frac{6ab}{c}$ **69.** already in lowest terms   **71.** $3eg^2$

Section 4.3 (page 235)
1. $-\frac{3}{16}$ **3.** $\frac{9}{10}$ **5.** $\frac{1}{2}$ **7.** -6 **9.** 36   **11.** $\frac{15}{4y}$ **13.** $\frac{1}{2}$ **15.** $\frac{6}{5}$
17. -9 **19.** $-\frac{1}{6}$ **21.** $\frac{11}{15d}$ **23.** b
25. (a) Forgot to write 1s in numerator when dividing out common
factors. Answer is $\frac{1}{6}$.   **(b)** Used reciprocal of $\frac{2}{3}$ in multiplication,
but the reciprocal is used only in division. Correct answer is $\frac{16}{3}$.
27. (a) Forgot to use reciprocal of $\frac{4}{1}$; correct answer is $\frac{1}{6}$.
(b) Used reciprocal of $\frac{5}{6}$ instead of reciprocal of $\frac{10}{9}$; correct answer is $\frac{3}{4}$.
29. Rewrite division as multiplication. Leave the first number
(dividend) the same. Change the second number (divisor) to its
reciprocal by "flipping" it. Then multiply.
31. $\frac{4}{15}$ **33.** $-\frac{9}{32}$ **35.** 21   **37.** 15   **39.** undefined   **41.** $-\frac{55}{12}$
43. $8b$ **45.** $\frac{3}{5d}$ **47.** $\frac{2x^2}{w}$ **49.** $\frac{3}{10}$ yd^2 **51.** 80 dispensers
53. earn \$9300; borrow \$3100   **55.** 9 trips   **57.** 70 infield players
59. 3200 times   **61. (a)** \$58,000   **(b)** \$11,600   **63.** \$25,375
65. 15 million birds   **67.** 129 million dogs and cats

Section 4.4 (page 246)
1. (a) 10   **(b)** 18   **(c)** 20   **(d)** 21
3. (a) LCD $= 2 \cdot 2 \cdot 2 \cdot 2 \cdot 3 = 48$
(b) LCD $= 2 \cdot 2 \cdot 3 \cdot 5 = 60$
(c) LCD $= 2 \cdot 2 \cdot 2 \cdot 3 \cdot 7 = 168$
5. $\frac{1}{2}$ **7.** $\frac{7}{8}$ **9.** $-\frac{9}{40}$ **11.** $-\frac{3}{5}$ **13.** $\frac{7}{10}$ **15.** $\frac{8}{7}$ **17.** $-\frac{3}{8}$
19. $\frac{3 + 5c}{15}$ **21.** $\frac{10 - m}{2m}$ **23.** $\frac{8}{b^2}$ **25.** $\frac{bc + 21}{7b}$ **27.** $\frac{-4 - cd}{c^2}$
29. $-\frac{44}{105}$ **31.** You cannot add or subtract until all the fractional
pieces are the same size.   **33. (a)** Cannot add fractions with unlike
denominators; use 20 as the LCD. The correct answer is
$\frac{15}{20} + \frac{8}{20} = \frac{23}{20}$.   **(b)** When rewriting fractions with 18 as a
denominator, you must multiply denominator and numerator by the
same number. The correct answer is $\frac{15}{18} - \frac{8}{18} = \frac{7}{18}$.
35. (a) $\frac{1}{12}; \frac{1}{12}$ Addition is commutative.   **(b)** $\frac{1}{3}; -\frac{1}{3}$
Subtraction is *not* commutative.   **(c)** $-\frac{3}{5}; -\frac{3}{5}$ Multiplication is
commutative.   **(d)** $6; \frac{1}{6}$ Division is *not* commutative.

36. (a) 0; 0; The sum of a number and its opposite is 0.
(b) 1; 1; When a nonzero number is divided by itself, the
quotient is 1. **(c)** $\frac{5}{6}$; $-\frac{17}{20}$; Multiplying by 1 leaves a number
unchanged. **(d)** 1; 1; A number times its reciprocal is 1.

37. $\frac{47}{60}$ in. **39.** $\frac{23}{24}$ cubic yard **41.** $\frac{3}{4}$ acre **43.** $\frac{23}{50}$ of workers

45. $\frac{4}{25}$ of workers **47.** $\frac{7}{24}$ of the day; 7 hours **49.** $\frac{1}{8}$ of the day

51. $\frac{5}{16}$ inch **53.** $\frac{1}{32}$ inch **55.** $\frac{1}{12}$ mile

Section 4.5 (page 260)

1.
-4 -3 -2 -1 0 1 2 3 4

3.
-4 -3 -2 -1 0 1 2 3 4

5. $\frac{9}{2}$ **7.** $-\frac{8}{5}$ **9.** $\frac{19}{8}$ **11.** $-\frac{57}{10}$ **13.** $\frac{161}{15}$ **15.** $4\frac{1}{3}$

17. $-2\frac{1}{2}$ **19.** $3\frac{2}{3}$ **21.** $-5\frac{2}{3}$ **23. (a)** 10 **(b)** 2 **(c)** 7

(d) 5 **(e)** 4 **(f)** 9 **25.** $7\frac{7}{8}$; $2 \cdot 4 = 8$ **27.** $1\frac{5}{21}$; $3 \div 3 = 1$

29. $5\frac{1}{2}$; $4 + 2 = 6$ **31.** $3\frac{2}{3}$; $4 - 1 = 3$ **33.** $\frac{17}{18}$; $6 \div 6 = 1$

35. $6\frac{1}{5}$; $8 - 2 = 6$ **37.** $P = 7$ in.; $A = 3\frac{1}{16}$ in.^{2}

39. $P = 19\frac{1}{2}$ yd; $A = 21\frac{1}{8}$ yd^2 **41.** $13 + 9 = 22$ ft; $21\frac{1}{6}$ ft

43. $2 \cdot 6 = 12$ ounces; $9\frac{5}{8}$ ounces **45.** $4 - 2 = 2$ miles;

$2\frac{3}{10}$ miles **47.** $4 \cdot 5 = 20$ yd; $18\frac{3}{4}$ yd **49.** $10 - 2 - 3 =$

5 cubic yards; $5\frac{1}{8}$ cubic yards **51.** $19 \div 5$ is about 4 hours; $3\frac{3}{4}$ hours

53. $30 - 6 - 2 = 22$ in.; $21\frac{3}{8}$ in. **55.** $24 + 35 + 24 + 35 =$

118 in.; $116\frac{1}{2}$ in. **57.** $25{,}730 \div 10 = 2573$ anchors; 2480 anchors

59. $(4 \cdot 23) + (5 \cdot 24) + (3 \cdot 25) = 287$ in.; $280\frac{1}{8}$ in.

Summary Exercises on Fractions (page 265)

1. (a) $\frac{3}{8}$; $\frac{5}{8}$ **(b)** $\frac{4}{5}$; $\frac{1}{5}$ **2.**
-1   0   1
$-\frac{2}{3}$ $\frac{2}{3}$

3. (a) 24 **(b)** 4 **4. (a)** 1 **(b)** -4 **(c)** 9

5. (a) $2 \cdot 2 \cdot 2 \cdot 3 \cdot 3$ **(b)** $3 \cdot 5 \cdot 7$ **6. (a)** $\frac{4}{5}$ **(b)** $\frac{7}{8}$ **7.** $\frac{1}{2}$

8. $-\frac{5}{24}$ **9.** $\frac{17}{16}$ **10.** $\frac{5}{6}$ **11.** $-\frac{2}{15}$ **12.** $-\frac{3}{8}$ **13.** 56 **14.** $\frac{11}{24}$

15. $-\frac{7}{6}$ **16.** $-\frac{19}{12}$ **17.** $\frac{25}{12}$ **18.** 35 **19.** $7\frac{7}{12}$; $5 + 3 = 8$

20. $11\frac{3}{7}$; $2 \cdot 5 = 10$ **21.** $3\frac{3}{10}$; $6 - 3 = 3$

22. $\frac{16}{35}$; $2 \div 4 = \frac{2}{4}$ or $\frac{1}{2}$ **23.** 4; $5 \div 1 = 5$ **24.** $2\frac{2}{3}$; $3 - 1 = 2$

25. (a) $2\frac{1}{16}$ in. **(b)** $\frac{5}{16}$ in. **26.** $P = 3\frac{1}{2}$ in.; $A = \frac{49}{64}$ in.^{2}

27. 12 batches **28.** 6¢ **29.** Not sure, 225 adults; Real, 675 adults;

Imaginary, 600 adults **30.** diameter $= \frac{1}{4}$ in.; $P = 4\frac{1}{2}$ in.

31. 23 bottles **32.** 11 lots

Section 4.6 (page 270)

1. $\frac{9}{16}$ **3.** $\frac{8}{125}$ **5.** $-\frac{1}{27}$ **7.** $\frac{1}{32}$ **9.** $\frac{49}{100}$ **11.** $\frac{36}{25}$ or $1\frac{11}{25}$

13. $\frac{12}{25}$ **15.** $\frac{1}{100}$ **17.** $-\frac{3}{2}$ or $-1\frac{1}{2}$ **19. (a)** The answers are

$\frac{1}{4}$, $-\frac{1}{8}$, $\frac{1}{16}$, $-\frac{1}{32}$, $\frac{1}{64}$, $-\frac{1}{128}$, $\frac{1}{256}$, $-\frac{1}{512}$. **(b)** When a negative

number is raised to an even power, the answer is positive.
When a negative number is raised to an odd power, the answer
is negative. **20. (a)** Ask yourself, "What number, times itself,
is 4?" This is the numerator. Then ask, "What number, times itself,
is 9?" This is the denominator. The number under the ketchup is

either $\frac{2}{3}$ or $-\frac{2}{3}$. **(b)** The number under the ketchup is $-\frac{1}{3}$ because

$\left(-\frac{1}{3}\right)\left(-\frac{1}{3}\right)\left(-\frac{1}{3}\right) = -\frac{1}{27}$. **(c)** Either $\frac{1}{2}$ or $-\frac{1}{2}$.

(d) No real number works, because both $\left(\frac{3}{4}\right)^2$ and $\left(-\frac{3}{4}\right)^2$ give a

positive result. **(e)** Either $\frac{1}{3}$ or $-\frac{1}{3}$ inside one set of parentheses

and $\frac{1}{2}$ or $-\frac{1}{2}$ inside the other. **21.** -4 **23.** $\frac{5}{16}$ **25.** $\frac{1}{3}$ **27.** $\frac{1}{6}$

29. $-\frac{17}{24}$ **31.** $-\frac{4}{27}$ **33.** $\frac{1}{36}$ **35.** $\frac{9}{64}$ in.^{2} **37.** $\frac{19}{10}$ or $1\frac{9}{10}$ miles

39. 4 **41.** $-\frac{25}{2}$ or $-12\frac{1}{2}$ **43.** $\frac{1}{14}$ **45.** $\frac{5}{18}$ **47.** -8 **49.** $\frac{9}{100}$

Section 4.7 (page 279)

1. $a = 30$ Check $\frac{1}{3}(30) = 10$
$$10 = 10$$

3. $b = -24$ Check $-20 = \frac{5}{6}(-24)$
$$-20 = -20$$

5. $c = 6$ Check $-\frac{7}{2}(6) = -21$
$$-21 = -21$$

7. $m = \frac{3}{4}$ Check $\frac{9}{16} = \frac{3}{4}\left(\frac{3}{4}\right)$
$$\frac{9}{16} = \frac{9}{16}$$

9. $d = -\frac{6}{5}$ Check $\frac{3}{10} = -\frac{1}{4}\left(-\frac{6}{5}\right)$
$$\frac{3}{10} = \frac{3}{10}$$

11. $n = 12$ Check $\frac{1}{6}(12) + 7 = 9$
$$2 + 7 = 9$$
$$9 = 9$$

13. $r = -9$ Check $-10 = \frac{5}{3}(-9) + 5$
$$-10 = -15 + 5$$
$$-10 = -10$$

15. $x = 24$ Check $\frac{3}{8}(24) - 9 = 0$
$$9 - 9 = 0$$
$$0 = 0$$

17. $y = 45$ **19.** $n = -18$ **21.** $x = \dfrac{1}{12}$ **23.** $b = -\dfrac{1}{8}$

25. (a) $\dfrac{1}{6}(18) + 1 = -2$

$$3 + 1 = -2$$
$$4 \neq -2$$

Does *not* balance; correct solution is -18.

(b) $-\dfrac{3}{2} = \dfrac{9}{4}\left(-\dfrac{2}{3}\right)$

$$-\dfrac{3}{2} = -\dfrac{3}{2}$$

Balances, so $-\dfrac{2}{3}$ is correct solution.

27. Some possibilities are: $\dfrac{1}{2}x = 4$; $-\dfrac{1}{4}a = -2$; $\dfrac{3}{4}b = 6$.

29. Let a be the man's age.

$$109 = 100 + \dfrac{a}{2}$$

The man is 18 years old.

31. Let a be the woman's age.

$$122 = 100 + \dfrac{a}{2}$$

The woman is 44 years old.

33. Let p be the penny size. $\dfrac{p}{4} + \dfrac{1}{2} = 3$. The penny size is 10.

35. Let p be the penny size. $\dfrac{p}{4} + \dfrac{1}{2} = \dfrac{5}{2}$. The penny size is 8.

37. Let h be the man's height. $\dfrac{11}{2}h - 220 = 209$. The man is 78 in. tall.   **39.** Let h be the woman's height. $\dfrac{11}{2}h - 220 = 132$. The woman is 64 in. tall.

Section 4.8 (page 288)

1. $P = 202$ m; $A = 1914$ m^2 **3.** $P = 5$ ft; $A = \dfrac{27}{32}$ ft^2

5. $P = 26\dfrac{1}{4}$ yd; $A = 30\dfrac{3}{4}$ yd^2 **7.** $P = \dfrac{454}{15}$ yd or $30\dfrac{4}{15}$ yd;

$A = \dfrac{115}{3}$ yd^2 or $38\dfrac{1}{3}$ yd^2 **9.** $A = 198$ m^2 **11.** $A = 1716$ m^2

13. Perimeter is the distance around the outside edges of a flat shape and is measured in linear units. Area is the space inside a flat shape and is measured in square units. Volume is the space inside a solid shape and is measured in cubic units.   **15.** $A = \dfrac{63}{8}$ ft^2 or $7\dfrac{7}{8}$ ft^2

17. 132 m of curbing; 726 m^2 of sod   **19. (a)** 175 yd of frontage **(b)** $A = 8750$ yd^2 **21.** Rectangular solid; $V = 528$ cm^3

23. Rectangular solid or cube; $V = 15\dfrac{5}{8}$ in.^{3}

25. Pyramid; $V = 800$ cm^3 **27.** $V = 106\dfrac{2}{3}$ ft^3 **29.** $V = 18$ in.^{3}

31. $V = 651{,}775$ m^3 **33.** $V = 513$ cm^3 **35.** The correct answers are $A = 135$ ft^2 (square feet, not squaring 135) and $5\dfrac{1}{2}$ in. (perimeter is in linear units, not square units).

Chapter 4 Review Exercises (page 301)

1. $\dfrac{2}{5}; \dfrac{1}{5}$ **2.** $\dfrac{3}{10}; \dfrac{7}{10}$

3.

4. (a) -4 **(b)** 8 **(c)** -1 **5.** $\dfrac{7}{8}$ **6.** $\dfrac{3}{5}$ **7.** already in lowest terms   **8.** $\dfrac{3x}{8}$ **9.** $\dfrac{1}{5b}$ **10.** $\dfrac{4n}{7m^2}$ **11.** $\dfrac{1}{16}$ **12.** -12 **13.** $\dfrac{8}{27}$

14. $\dfrac{1}{6x}$ **15.** $2a^2$ **16.** $\dfrac{6}{7k}$ **17.** $\dfrac{5}{24}$ **18.** $-\dfrac{2}{15}$ **19.** $\dfrac{19}{6}$ or $3\dfrac{1}{6}$

20. $\dfrac{3}{2}$ or $1\dfrac{1}{2}$ **21.** $\dfrac{4n + 15}{20}$ **22.** $\dfrac{3y - 70}{10y}$ **23.** $1\dfrac{5}{13}; 2 \div 2 = 1$

24. $2\dfrac{1}{2}; 7 - 5 = 2$ **25.** $4\dfrac{1}{20}; 2 + 2 = 4$ **26.** $-\dfrac{27}{64}$ **27.** $\dfrac{1}{36}$

28. $-\dfrac{4}{5}$ **29.** $\dfrac{7}{6}$ or $1\dfrac{1}{6}$ **30.** 10 **31.** $-\dfrac{4}{27}$ **32.** $w = 20$

33. $r = -15$ **34.** $x = \dfrac{1}{2}$ **35.** $A = 14$ ft^2 **36.** Rectangular solid; $V = 32\dfrac{1}{2}$ in.^{3} **37.** Pyramid; $V = 93\dfrac{1}{3}$ m^3 **38.** $\dfrac{1}{4}$ pound;

$7\dfrac{1}{2}$ pounds   **39.** $\dfrac{5}{6}$ hour; $10\dfrac{11}{12}$ hours   **40.** 12 preschoolers, 40 toddlers, 8 infants   **41.** $P = 2\dfrac{1}{10}$ miles; $A = \dfrac{9}{40}$ mi^2

Chapter 4 Test (page 303)

[4.1] 1. $\dfrac{5}{6}; \dfrac{1}{6}$ **2.**

[4.2] 3. $\dfrac{1}{4}$

4. already in lowest terms   **5.** $\dfrac{2a^2}{3b}$ **[4.3, 4.4] 6.** $\dfrac{13}{15}$ **7.** -2

8. $-\dfrac{7}{40}$ **9.** 14 **10.** $-\dfrac{2}{27}$ **11.** $\dfrac{25}{8}$ or $3\dfrac{1}{8}$ **12.** $\dfrac{4}{9}$ **13.** $\dfrac{9}{16}$

14. $\dfrac{4}{7y}$ **15.** $\dfrac{24 - n}{4n}$ **16.** $\dfrac{10 + 3a}{15}$ **17.** $\dfrac{1}{18b}$ **[4.6] 18.** $-\dfrac{1}{18}$

19. $-\dfrac{31}{30}$ or $-1\dfrac{1}{30}$ **[4.5] 20.** $5 \div 1 = 5; \dfrac{64}{15}$ or $4\dfrac{4}{15}$

21. $3 - 2 = 1; \dfrac{3}{2}$ or $1\dfrac{1}{2}$ **[4.7] 22.** $d = 35$ **23.** $t = -\dfrac{15}{7}$ or $-2\dfrac{1}{7}$

24. $b = 8$ **25.** $x = -15$ **[4.8] 26.** $A = 52$ m^2

27. $A = \dfrac{117}{2}$ yd^2 or $58\dfrac{1}{2}$ yd^2 **28.** Rectangular solid; $V = 6480$ m^3

29. Pyramid; $V = 16$ yd^3 **[4.3–4.5] 30.** $14\dfrac{3}{4}$ hours; $1\dfrac{5}{6}$ hours

31. $\dfrac{7}{2}$ days or $3\dfrac{1}{2}$ days   **32.** 7392 students work

Cumulative Review Exercises: Chapters 1–4 (page 305)

[1.1] 1. five hundred five million, eight thousand, two hundred thirty-eight   **2.** 35,600,000,916   **[1.5] 3. (a)** 60,700   **(b)** 100,000 **(c)** 3210   **[1.3, 1.6] 4. (a)** commutative property of multiplication **(b)** associative property of addition   **(c)** distributive property **[1.3–1.8] 5.** -54 **6.** -20 **7.** undefined   **8.** 81 **9.** 10

10. 4 **11.** -16 **12.** $\frac{27}{27} = 1$ **[4.3–4.6] 13.** -8 **14.** 35

15. $\frac{15a}{2}$ **16.** $\frac{3}{xy^2}$ **17.** $-\frac{8}{15}$ **18.** $\frac{5}{16}$ **19.** $\frac{14 - 3b}{21}$

20. $\frac{8n + 15}{5n}$ **21.** $\frac{13}{9}$ or $1\frac{4}{9}$ **22.** $\frac{13}{20}$ **23.** -1 **24.** $-\frac{5}{8}$

[2.5, 4.7] 25. $a = 1$ **26.** $y = -5$ **27.** $k = -36$ **28.** $x = -4$

29. $m = 2$ **[3.1, 3.2, 4.8] 30.** Square; $P = 18$ in.; $A = 20\frac{1}{4}$ in.^{2}

31. Triangle; $P = 20$ yd; $A = 18\frac{3}{4}$ yd^2 **32.** Parallelogram;

$P = 78$ mm; $A = 288$ mm^2 **[3.3, 3.4] 33.** Let b be the original weight of each bag. $3b - 16 - 25 = 79$; There were 40 pounds in each bag. **34.** Let w be the width; $2w$ is the length. $2(2w) + 2(w) = 102$; Width is 17 yd; length is 34 yd.

CHAPTER 5

Section 5.1 (page 313)

1. (a) $\frac{1}{10}$; 0.1; one tenth **(b)** $\frac{3}{10}$; 0.3; three tenths **(c)** $\frac{9}{10}$; 0.9; nine tenths **3.** 7; 0; 4 **5.** 5; 1; 8 **7.** 4; 7; 0 **9.** 1; 6; 3

11. 1; 8; 9 **13.** 410.25 **15.** 6.5432 **17.** 5406.045 **19.** $\frac{7}{10}$

21. $13\frac{2}{5}$ **23.** $\frac{7}{20}$ **25.** $\frac{33}{50}$ **27.** $10\frac{17}{100}$ **29.** $\frac{3}{50}$ **31.** $\frac{41}{200}$

33. $5\frac{1}{500}$ **35.** $\frac{343}{500}$ **37.** five tenths **39.** seventy-eight hundredths

41. one hundred five thousandths **43.** twelve and four hundredths **45.** one and seventy-five thousandths **47.** 6.7 **49.** 0.32 **51.** 420.008 **53.** 0.0703 **55.** 75.030 **57.** Anne should not say "and" because that denotes a decimal point. **59.** ten thousandths

inch; $\frac{10}{1000} = \frac{1}{100}$ inch **61.** 12 pounds **63.** 3-C **65.** 4-A

67. one and six hundred two thousandths centimeters

69. millionths, ten-millionths, hundred-millionths, billionths; these match the words on the left side of the chart with "ths" attached.

70. First place to the left of the decimal point is ones. If the first place to the right were one*ths*, it would mean a fraction with a denominator of 1, which would equal 1 or more. Anything that is 1 or more is to the *left* of the decimal point. **71.** seventy-two million four hundred thirty-six thousand nine hundred fifty-five hundred-millionths **72.** six hundred seventy-eight thousand five hundred fifty-four billionths **73.** eight thousand six and five hundred thousand one millionths **74.** twenty thousand, sixty and five hundred five millionths **75.** 0.0302040

76. 9,876,543,210.100200300

Section 5.2 (page 322)

1. 16.9 **3.** 0.956 **5.** 0.80 **7.** 3.661 **9.** 794.0 **11.** 0.0980 **13.** 49 **15.** 9.09 **17.** 82.0002 **19.** $0.82 **21.** $1.22 **23.** $0.50 **25.** $48,650 **27.** $310 **29.** $849 **31.** $500 **33.** $1.00 **35.** $1000 **37. (a)** 322 miles per hour **(b)** 107 miles per hour **39. (a)** 186.0 miles per hour **(b)** 763.0 miles per hour **41.** Rounds to $0 (zero dollars) because $0.499 is closer to $0 than to $1. **42.** Round $0.499 to the nearest cent to get $0.50. Guideline: Round amounts less than $1.00 to nearest cent instead of nearest dollar. **43.** Rounds to $0.00 (zero cents) because $0.0015 is closer to $0.00 than to $0.01. **44.** Both round to $0.60. Rounding to nearest thousandth (tenth of a cent) would allow you to identify $0.597 as less than $0.601.

Section 5.3 (page 330)

1. 17.72 **3.** 11.98 **5.** 115.861 **7.** 59.323 **9.** 6 should be written 6.00; sum is 46.22.

11. $0.3000 = \frac{3000 \div 1000}{10,000 \div 1000} = \frac{3}{10} = 0.3$ **13.** 89.7

15. 0.109 **17.** 0.91 **19.** 6.661 **21.** The student subtracted in the wrong order; 15.32 should be on top; correct answer is 7.87 **23. (a)** 24.75 in. **(b)** 3.95 in. **25. (a)** 62.27 in. **(b)** 0.39 in. **27.** 23.013 **29.** -45.75 **31.** -6.69 **33.** -6.99 **35.** -4.279 **37.** -0.0035 **39.** 5.37 **41.** *Estimate:* $20 - 10 = 10$; *Exact:* 6.275 **43.** *Estimate:* $-7 + 1 = -6$; *Exact:* -5.8 **45.** *Estimate:* $-40 + (-200) = -240$; *Exact:* -237.571 **47.** *Estimate:* $8 + (+50) = 58$; *Exact:* 59.23 **49.** *Estimate:* $19 - 15 = 4$ million people; *Exact:* 3.6 million people **51.** *Estimate:* $100 + 30 + 20 + 20 + 20 + 9 = 199$ million people; *Exact:* 234.4 million people **53.** *Estimate:* $2 + 2 + 2 = 6$ m; *Exact:* 6.09 m, which is 0.31 m less than the rhino's height. **55.** *Estimate:* $20 - 9 = 11$; *Exact:* $10.88 **57.** *Estimate:* $5 - 5 = 0$; *Exact:* $0.30 **59.** *Estimate:* $19 + 2 + 2 + 10 + 2 = 35$; *Exact:* $35.25 **61.** $1939.36 **63.** $3.97 **65.** $b = 1.39$ cm **67.** $q = 23.843$ ft

Section 5.4 (page 337)

1. 0.1344 **3.** -159.10 **5.** 15.444 **7.** $34,500.20 **9.** -43.2 **11.** 0.432 **13.** 0.0432 **15.** 0.00432 **17.** 0.0000312 **19.** 0.000009 **21.** 59.6; 4.76; 7226; 32; 803.5; 9. Multiplying by 10, decimal point moves one place to the right; by 100, two places to the right; by 1000, three places to the right. **22.** 5.96; 0.0476; 6.5; 0.32; 8.035; 52.3. Multiplying by 0.1, decimal point moves one place to the left; by 0.01, two places to the left; by 0.001, three places to the left. **23.** *Estimate:* $40 \times 5 = 200$; *Exact:* 190.08 **25.** *Estimate:* $40 \times 40 = 1600$; *Exact:* 1558.2 **27.** *Estimate:* $7 \times 5 = 35$; *Exact:* 30.038 **29.** *Estimate:* $3 \times 7 = 21$; *Exact:* 19.24165 **31.** unreasonable; $289.00 **33.** reasonable **35.** unreasonable; $3.49 **37.** unreasonable; 9.5 pounds **39.** $945.87 (rounded) **41.** $2.45 (rounded) **43.** $53.50 (rounded) **45.** $20,265 **47. (a)** Area before 1929 ≈ 23.2 in.^{2}; Area today ≈ 16.0 in.^{2} **(b)** 7.2 in.^{2} **49. (a)** 0.43 inch **(b)** 4.3 inches **51.** $984.04; $2207.80 **53.** $76.50 **55.** $4.09 (rounded) **57.** $129.25 **59. (a)** $70.05 **(b)** $25.80

Section 5.5 (page 348)

1. (a) -3.9 **(b)** 0.47 **3. (a)** 0.1625 **(b)** -2.77 **5.** 400.2 **7.** 36 **9.** 0.06 **11.** 6000 **13.** 60 **15.** 0.0006 **17.** 25.3 **19.** 516.67 (rounded) **21.** -26.756 (rounded) **23.** 10,082.647 (rounded) **25.** 0.377; 0.91; 0.0886; 3.019; 40.65; 662.57
(a) Dividing by 10, decimal point moves one place to the left; 100, two places to the left; by 1000, three places to the left. **(b)** The decimal point moved to the *right* when multiplying by 10, by 100, or by 1000; here it moves to the *left* when dividing by 10, by 100, or by 1000. **26.** 402; 71; 3.39; 157.7; 460; 8730
(a) Dividing by 0.1, decimal point moves one place to the right; by 0.01, two places to the right; by 0.001, three places to the right. **(b)** The decimal point moved to the *left* when multiplying by 0.1, 0.01, or 0.001; here it moves to the *right* when dividing by 0.1, 0.01, or 0.001. **27.** unreasonable;

Estimate: $40 \div 8 = 5$; $8)\overline{37.8}$ (4.725) **29.** reasonable;
Estimate: $50 \div 50 = 1$ **31.** unreasonable; *Estimate:* $300 \div 5 = 60$;
$5.1)\overline{307.02}$ (60.2) **33.** unreasonable; *Estimate:* $9 \div 1 = 9$; $1.25)\overline{9.3}$ (7.44)
35. $4.00 (rounded) **37.** $67.08 (rounded) **39.** $0.30 **41.** $11.92 per hour **43.** 21.2 miles per gallon (rounded) **45.** 7.37 meters (rounded) **47.** 0.08 meter **49.** 22.49 meters

51. 14.25   **53.** 3.8   **55.** -16.155 **57.** 3.714   **59.** $0.03 (rounded)
61. (a) 1,541,667 pieces (rounded)   (b) 25,694 pieces (rounded)
(c) 428 pieces (rounded)   **63.** 100,000 box tops   **65.** 2632 box
tops (rounded)

Summary Exercises on Decimals (page 353)

1. $\frac{4}{5}$ **2.** $6\frac{1}{250}$ **3.** $\frac{7}{20}$ **4.** ninety-four and five tenths   **5.** two
and three ten-thousandths   **6.** seven hundred six thousandths
7. 0.05   **8.** 0.0309   **9.** 10.7   **10.** 6.19   **11.** 1.0   **12.** 0.420
13. $0.89   **14.** $3.00   **15.** $100   **16.** -0.945 **17.** 49.6199
18. -50 **19.** 15.03   **20.** 0.00488   **21.** -2.15 **22.** 9.055
23. 18.4009   **24.** -6.995 **25.** -808.9 **26.** 2.12   **27.** 0.04
28. $P = 52.1$ in.   **29.** $P = 9.735$ meters   **30.** $1.80   **31.** $2052.50
32. (a) 169 ft^2 (b) $0.75 (rounded)   **33.** 144 ft^2; $0.69 (rounded)
34. (a) 9 days (rounded)   (b) 15.6 pounds (rounded)   **35.** Average
weight of food eaten by bee is 0.32 ounce.   **36.** 0.112 ounce

Section 5.6 (page 359)

1. 0.5   **3.** 0.75   **5.** 0.3   **7.** 0.9   **9.** 0.6   **11.** 0.875   **13.** 2.25
15. 14.7   **17.** 3.625   **19.** 6.333 (rounded)   **21.** 0.833 (rounded)
23. 1.889 (rounded)   **25.** (a) A proper fraction is less than 1.
$\frac{5}{9}$ is a proper fraction so it cannot be equivalent to a decimal number
greater than 1.   (b) $\frac{5}{9}$ means $5 \div 9$ or $9\overline{)5}$ so correct answer is
0.556 (rounded). This makes sense because both the fraction and
decimal are less than 1.   **26.** (a) $2.035 = 2\frac{35}{1000} = 2\frac{7}{200}$, not
$2\frac{7}{20}$ (b) Adding the whole number part gives $2 + 0.35$, which is
2.35, not 2.035. To check, $2.35 = 2\frac{35}{100} = 2\frac{7}{20}$ **27.** Just add the
whole number part to 0.375. So $1\frac{3}{8} = 1.375$; $3\frac{3}{8} = 3.375$;
$295\frac{3}{8} = 295.375$.   **28.** It works only when the fraction part has a
one-digit numerator and a denominator of 10, or a two-digit
numerator and a denominator of 100, and so on.
29. $\frac{2}{5}$ **31.** $\frac{5}{8}$ **33.** $\frac{7}{20}$ **35.** 0.35   **37.** $\frac{1}{25}$ **39.** $\frac{3}{20}$ **41.** 0.2
43. $\frac{9}{100}$ **45.** shorter; 0.72 inch   **47.** too much; 0.005 gram
49. 0.9991 cm, 1.0007 cm   **51.** more; 0.05 inch   **53.** (a) $<$
(b) $=$ (c) $>$ (d) $<$ **55.** 0.5399, 0.54, 0.5455   **57.** 5.0079,
5.79, 5.8, 5.804   **59.** 0.6009, 0.609, 0.628, 0.62812   **61.** 2.8902,
3.88, 4.876, 5.8751   **63.** 0.006, 0.043, $\frac{1}{20}$, 0.051   **65.** 0.37, $\frac{3}{8}$, $\frac{2}{5}$,
0.4001   **67.** (a) red box; green box   (b) 0.01 inch   **69.** 1.4 in.
(rounded)   **71.** 0.3 in. (rounded)   **73.** 0.4 in. (rounded)

Section 5.7 (page 369)

1. 7 months   **3.** 69.8 (rounded)   **5.** $39,622   **7.** $58.24
9. $35,500   **11.** 6.1 (rounded)   **13.** 17.2 hours (rounded)
15. 2.60   **17.** (a) 2.80   (b) 2.93 (rounded)   (c) 3.13 (rounded)
19. 15 messages   **21.** 516 students   **23.** 48.5 pounds
25. 4142 miles (rounded)   **27.** (a) 4050 miles   (b) List the
values from smallest to largest. There are an even number of values,
so find the average of the middle two values:
$(2875 + 5525) \div 2 = 4050$.   **29.** 8 samples
31. 68 and 74 years (bimodal)   **33.** no mode
35. (a) -10 degrees; 4 degrees   (b) 14 degrees warmer
37. Barrow's range is $-2 - (-18) = 16$ degrees; Fairbanks' range is
$31 - (-10) = 41$ degrees; Fairbanks' temperatures have greater
variability.   **39.** (a) mean = 71 degrees; median $\approx$ 71 degrees

(b) They are nearly identical because there is so little variation in the
temperatures.   **41.** (a) Student P: mean = 69.5; median = 70.5;
range = 43. Student Q: mean $\approx$ 70.3; median = 71; range = 17
(b) Student P   (c) Answers will vary.   **42.** (a) Golfer G:
mean = 86.2; median = 87; range = 6; Golfer H: mean = 86.4;
median = 88; range = 18.   (b) Golfer G   (c) Answers will vary.
43. (a) A's range = $74 - 62 = 12$; B's range = $26 - 18 = 8$;
C's range = $69 - 25 = 44$ (b) Building B's ages have the least
variability.   (c) Answers will vary.   **44.** (a) P's range =
$7200 - 4900 = $2300; Q's range = $6000 - 5500 = $500;
R's range = $6400 - 5200 = $1200; Company P's salaries have the
greatest variability.   (b) Answers will vary.

Section 5.8 (page 377)

1. 4   **3.** 8   **5.** 3.317 (rounded)   **7.** 2.236 (rounded)   **9.** 8.544
(rounded)   **11.** 10.050 (rounded)   **13.** 19   **15.** 31.623 (rounded)
17. 30 is about halfway between 25 and 36, so $\sqrt{30}$ should be
about halfway between 5 and 6, or about 5.5. Using a calculator,
$\sqrt{30} \approx 5.477$. Similarly, $\sqrt{26}$ should be a little more than $\sqrt{25}$;
by calculator $\sqrt{26} \approx 5.099$. And $\sqrt{35}$ should be a little less
than $\sqrt{36}$; by calculator $\sqrt{35} \approx 5.916$.   **19.** $\sqrt{1521} = 39$ ft
21. $\sqrt{289} = 17$ in.   **23.** $\sqrt{144} = 12$ mm   **25.** $\sqrt{73} \approx 8.5$ in.
27. $\sqrt{65} \approx 8.1$ yd   **29.** $\sqrt{195} \approx 14.0$ cm
31. $\sqrt{7.94} \approx 2.8$ m   **33.** $\sqrt{65.01} \approx 8.1$ cm
35. $\sqrt{292.32} \approx 17.1$ km   **37.** $\sqrt{65} \approx 8.1$ ft
39. $\sqrt{360,000} = 600$ m   **41.** $\sqrt{32.5} \approx 5.7$ ft
43. $\sqrt{135} \approx 11.6$ ft.   **45.** The student used the formula for
finding the hypotenuse but the unknown side is a leg, so use
$? = \sqrt{(20)^2 - (13)^2}$. Also, the final answer should be m, not m^2.
Correct answer is $\sqrt{231} \approx 15.2$ m.   **47.** $\sqrt{16,200} \approx 127.3$ ft
48. (a)   Second   (b) $\sqrt{7200} \approx 84.9$ ft

     60 ft
Third   First
     60 ft
     Home

49. The distance from third to first is the same as the distance
from home to second because the baseball diamond is a square.
50. (a) The side length is less than 60 ft.   (b) $80^2 = 6400$;
$6400 \div 2 = 3200$; $\sqrt{3200} \approx 56.6$ ft

Section 5.9 (page 385)

Please see *Student's Solutions Manual* for a sample *check* for
odd-numbered Exercises 1–11.   **1.** $h = 4.47$ **3.** $n = -1.4$
5. $b = 0.008$ **7.** $a = 0.29$ **9.** $p = -120$ **11.** $t = 0.7$
13. $x = -0.82$ **15.** $z = 0$ **17.** $c = 0.45$ **19.** $w = 28$
21. $p = -40.5$ **23.** Let d be the adult dose. $0.3\,d = 9$; The adult
dose is 30 milligrams.   **25.** Let d be the number of days.
$65.95d + 12 = 275.80$; The saw was rented for 4 days.
27. $0.7(220 - a) = 140$; The person is 20 years old.
28. $0.7(220 - a) = 126$; The person is 40 years old.
29. $0.7(220 - a) = 134$; $a \approx 28.57$, which rounds to 29. The
person is about 29 years old.   **30.** $0.7(220 - a) = 117$; $a \approx 52.86$,
which rounds to 53. The person is about 53 years old.

Section 5.10 (page 395)

1. $d = 18$ mm   **3.** $r = 0.35$ km   **5.** $C \approx 69.1$ ft; $A \approx 379.9$ ft^2
7. $C \approx 8.2$ m; $A \approx 5.3$ m^2 **9.** $C \approx 47.1$ cm, $A \approx 176.6$ cm^2
11. $C \approx 23.6$ ft; $A \approx 44.2$ ft^2 **13.** $C \approx 27.2$ km; $A \approx 58.7$ km^2
15. $A \approx 150.8$ m^2 **17.** $A \approx 57$ cm^2 **19.** π is the ratio of the
circumference of a circle to its diameter. If you divide the
circumference of any circle by its diameter, the answer is always a

little more than 3. The approximate value is 3.14, which we call π (pi). Your test question could involve finding the circumference or the area of a circle.   **21.** $A \approx 7850 \text{ yd}^2$
23. $C \approx 91.4$ in.; Bonus: 693 revolutions/mile (rounded)
25. $A \approx 70,650 \text{ mi}^2$

27. watch: $C \approx 3.1$ in.; $A \approx 0.8 \text{ in.}^2$; wall clock: $C \approx 18.8$ in.; $A \approx 28.3 \text{ in.}^2$

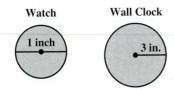

29. $A \approx 78.5 \text{ mi}^2 - 12.6 \text{ mi}^2$; difference $\approx 65.9 \text{ mi}^2$

31. (a) $d \approx 45.9$ cm   **(b)** Divide the circumference by π
(144 cm $\div$ 3.14).   **33.** $C \approx 14.6$ ft   **35.** $1170.33 (rounded)
37. $A \approx 13.8 \text{ ft}^2$ **39.** $A \approx 44.2 \text{ in.}^2$ **40.** $A \approx 132.7 \text{ in.}^2$
41. $A \approx 201.0 \text{ in.}^2$ **42.** small: $0.063 (rounded); medium: $0.049
(rounded); large: $0.046 (rounded) Best Buy   **43.** small: $0.084
(rounded); medium: $0.067 (rounded) Best Buy; large: $0.071
(rounded)   **44.** small: $0.077 (rounded) Best Buy; medium: $0.083
(rounded); large: $0.078 (rounded)   **45.** $V \approx 471 \text{ ft}^3$; $SA \approx 345.4 \text{ ft}^2$
47. $V = 1617 \text{ m}^3$; $SA = 849.4 \text{ m}^2$ **49.** $V \approx 763.0 \text{ in.}^3$;
$SA \approx 678.2 \text{ in.}^2$ **51.** $V = 5550 \text{ mm}^3$; $SA = 2150 \text{ mm}^2$
53. Student should use radius of 3.5 cm instead of diameter of 7 cm
in the formula; units for volume are cm^3 not cm^2. Correct answer is
$V \approx 192.3 \text{ cm}^3$ **55.** $V \approx 3925 \text{ ft}^3$ **57.** $SA = 163.6 \text{ in.}^2$

Chapter 5 Review Exercises (page 411)

1. 0; 5   **2.** 0; 6   **3.** 8; 9   **4.** 5; 9   **5.** 7; 6   **6.** $\frac{1}{2}$ **7.** $\frac{3}{4}$

8. $4\frac{1}{20}$ **9.** $\frac{7}{8}$ **10.** $\frac{27}{1000}$ **11.** $27\frac{4}{5}$ **12.** eight tenths   **13.** four
hundred and twenty-nine hundredths   **14.** twelve and seven
thousandths   **15.** three hundred six ten-thousandths   **16.** 8.3
17. 0.205   **18.** 70.0066   **19.** 0.30   **20.** 275.6   **21.** 72.79
22. 0.160   **23.** 0.091   **24.** 1.0   **25.** $15.83   **26.** $0.70
27. $17,625.79   **28.** $350   **29.** $130   **30.** $100   **31.** $29
32. -5.67 **33.** -0.03 **34.** 5.879   **35.** -6.435
36. *Estimate:* 80 million $-$ 50 million = 30 million; *Exact:*
31.4 million people   **37.** *Estimate:* $300 - $200 - $40 = $60;
Exact: $45.80   **38.** *Estimate:* $2 + $5 + $20 = $27;
$30 - $27 = $3; *Exact:* $4.14   **39.** *Estimate:* 2 + 4 + 5 =
11 kilometers; *Exact:* 11.55 kilometers   **40.** $6 \times 4 = 24$; 22.7106
41. $40 \times 3 = 120$; 141.57   **42.** 0.0112   **43.** -0.000355
44. reasonable; $700 \div 10 = 70$ **45.** unreasonable; $30 \div 3 = 10$;
$2.8\overline{)26.60}^{\,9.5}$ **46.** 14.467 (rounded)   **47.** 1200   **48.** -0.4
49. $708 (rounded)   **50.** $2.99 (rounded)   **51.** 133 shares (rounded)

52. $3.47 (rounded)   **53.** -4.715 **54.** 10.15   **55.** 3.8
56. 0.64   **57.** 1.875   **58.** 0.111 (rounded)   **59.** 3.6008, 3.68,
3.806   **60.** 0.209, 0.2102, 0.215, 0.22   **61.** $\frac{1}{8}, \frac{3}{20}$, 0.159, 0.17

62. mean: 28.5 digital cameras; median: 19.5 digital cameras
63. mean: 44 claims; median: 39 claims   **64.** $51.05
65. (a) Store J: $107 and $160 (bimodal); Store K: $119
(b) Range for Store J = $160 - $69 = $91; range for Store
K = $139 - $95 = $44; Store J's prices have greater variability.
66. $\sqrt{289} = 17$ in.   **67.** $\sqrt{49} = 7$ cm   **68.** $\sqrt{104} \approx 10.2$ cm
69. $\sqrt{52} \approx 7.2$ in.   **70.** $\sqrt{6.53} \approx 2.6$ m   **71.** $\sqrt{71.75} \approx 8.5$ km
72. $b = 0.25$ **73.** $x = 0$ **74.** $n = -13$ **75.** $a = -1.5$

76. $y = -16.2$ **77.** $d = 137.8$ m   **78.** $r = 1\frac{1}{2}$ in. or 1.5 in.

79. $C \approx 6.3$ cm; $A \approx 3.1 \text{ cm}^2$ **80.** $C \approx 109.3$ m; $A \approx 950.7 \text{ m}^2$
81. $C \approx 37.7$ in.; $A \approx 113.0 \text{ in.}^2$ **82.** $V \approx 549.5 \text{ cm}^3$; $SA \approx 376.8 \text{ cm}^2$
83. $V = 1808.6 \text{ m}^3$; $SA \approx 1205.8 \text{ m}^2$ **84.** $V \approx 7.9 \text{ ft}^3$; $SA = 25.5 \text{ ft}^2$
85. 404.865   **86.** -254.8 **87.** 3583.261 (rounded)   **88.** 29.0898
89. 0.03066   **90.** 9.4   **91.** -15.065 **92.** -9.04 **93.** -15.74
94. 8.19   **95.** 0.928   **96.** 35   **97.** -41.859 **98.** 0.3
99. $3.00 (rounded)   **100.** $2.17 (rounded)   **101.** $35.96
102. $199.71   **103.** $78.50   **104.** $y = -0.7$ **105.** $x = 30$
106. rubber strip that is 15.7 ft long; $A \approx 19.6 \text{ ft}^2$ **107.** mean is
67.6; median is 82   **108.** 20 miles   **109.** $V \approx 87.9 \text{ in.}^3$

Chapter 5 Test (page 417)

[5.1] 1. $18\frac{2}{5}$ **2.** $\frac{3}{40}$ **3.** sixty and seven thousandths

4. two hundred eight ten-thousandths   **[5.2–5.5] 5.** *Estimate:*
8 + 80 + 40 = 128; *Exact:* 129.2028   **6.** *Estimate:* $-6(1) = -6$;
Exact: -6.948 **7.** *Estimate:* $-80 - 4 = -84$; *Exact:* -82.702
8. *Estimate:* $-20 \div (-5) = 4$; *Exact:* 4.175   **9.** 669.004
10. 480   **11.** 0.000042   **12.** $4.55 per meter   **13.** Davida, by
0.441 minute   **14.** $5.35 (rounded)   **[5.9] 15.** $y = -6.15$
16. $x = -0.35$ **17.** $a = 10.9$ **18.** $n = 2.85$

[5.6] 19. 0.44, $\frac{9}{20}$, 0.4506, 0.451   **[5.5] 20.** 32.09

[5.7] 21. 75 books   **22.** 103° and 104° (bimodal)
23. $11.25   **24. (a)** $50.55   **(b)** $50.45   **(c)** math books:
$89 - $39.75 = $49.25; biology books: $97.55 - $30.95 = $66.60;
math books have less variability   **[5.8] 25.** $\sqrt{85} \approx 9.2$ cm
26. $\sqrt{279} \approx 16.7$ ft   **[5.10] 27.** $r = 12.5$ in.   **28.** $C \approx 5.7$ km
29. $A \approx 206.0 \text{ cm}^2$ **30.** $V \approx 5086.8 \text{ ft}^3$ **31.** $SA \approx 2599.9 \text{ ft}^2$

Cumulative Review Exercises: Chapters 1–5 (page 419)

[1.1, 5.1] 1. (a) forty-five and two hundred three ten-thousandths
(b) thirty billion, six hundred fifty thousand, eight
2. (a) 160,000,500   **(b)** 0.075   **[1.5, 5.2] 3.** 46,900   **4.** 6.20
5. 0.661   **6.** 10,000   **[1.3–1.8, 4.3–4.6, 5.3–5.5] 7.** -13
8. -0.00006 **9.** undefined   **10.** 44   **11.** 18   **12.** -6
13. 56.621   **14.** $\frac{3}{2}$ or $1\frac{1}{2}$ **15.** $\frac{3}{8b}$ **16.** 11.147   **17.** $-\frac{9}{20}$
18. $-\frac{1}{32}$ **19.** 1.3 (rounded)   **20.** $\frac{5x + 6}{10}$ **21.** -8 **22.** -32
23. $\frac{18}{18} = 1$ **24.** -5.76 **25.** $\frac{55}{24}$ or $2\frac{7}{24}$ **[5.7] 26.** 24.2 years
(rounded)   **27.** 23.5 years   **28.** 20 and 26 years (bimodal)
[2.5, 5.9] 29. $h = -4$ **30.** $x = 5$ **31.** $r = 10.9$ **32.** $y = -2$
33. $n = 4$ **[3.1, 3.2] 34.** $P = 54$ ft; $A = 140 \text{ ft}^2$
[5.10] 35. $C \approx 40.8$ m; $A \approx 132.7 \text{ m}^2$ **[5.8] 36.** $x \approx 12.6$ km;
$A \approx 37.8 \text{ km}^2$ **[4.8, 5.10] 37.** $V \approx 3.4 \text{ in.}^3$; $SA = 13.5 \text{ in.}^2$
[5.3] 38. *Estimate:* $40 - $20 - $1 = $19; *Exact:* $21.17

[4.4] **39.** *Estimate:* 2 + 4 = 6 yards; *Exact:* $6\frac{5}{24}$ yards

[5.5] **40.** *Estimate:* $6 ÷ 3 = $2 per pound; *Exact:* $2.29 per pound (rounded)   [1.7] **41.** *Estimate:* $80,000 ÷ 100 = $800; *Exact:* $729 (rounded)   [3.4] **42.** Let x be first student's money and $3x$ be second student's money. $x + 3x = 30,000$; one student receives $7500; the other receives $22,500.   **43.** Let w be the width and $w + 4$ be the length. $2(w + 4) + 2w = 48$; The width is 10 in. and the length is 14 in.

CHAPTER 6

Section 6.1 (page 428)

1. $\frac{8}{9}$ **3.** $\frac{2}{1}$ **5.** $\frac{1}{3}$ **7.** $\frac{8}{5}$ **9.** $\frac{3}{8}$ **11.** $\frac{9}{7}$ **13.** $\frac{6}{1}$ **15.** $\frac{5}{6}$

17. $\frac{8}{5}$ **19.** $\frac{1}{12}$ **21.** $\frac{5}{16}$ **23.** $\frac{4}{1}$ **25.** $\frac{1}{2}$ **27.** $\frac{36}{1}$

29. Answers will vary. One possibility is stocking cards of various types in the same ratios as those in the table.   **31.** $\frac{6}{5}, \frac{36}{17}$

33. *White Christmas* to *It's Now or Never; White Christmas* to *I Will Always Love You; Candle in the Wind* to *I Want to Hold Your Hand*

35. $\frac{2}{1}$ **37.** (a) $\frac{3}{8}$ (b) $\frac{7}{5}$ **39.** $\frac{7}{5}$ **41.** $\frac{6}{1}$ **43.** $\frac{38}{17}$ **45.** $\frac{1}{4}$

47. $\frac{34}{35}$ **49.** $\frac{1}{1}$; as long as the sides all have the same length, any measurement you choose will maintain the ratio.
50. Answers will vary. Some possibilities are:
$\frac{4}{5} = \frac{8}{10} = \frac{12}{15} = \frac{16}{20} = \frac{20}{25} = \frac{24}{30} = \frac{28}{35}$ **51.** It is not possible. Amelia would have to be older than her mother to have a ratio of 5 to 3.   **52.** Answers will vary, but a ratio of 3 to 1 means your income is 3 times your friend's income.

Section 6.2 (page 436)

1. $\frac{5 \text{ cups}}{3 \text{ people}}$ **3.** $\frac{3 \text{ feet}}{7 \text{ seconds}}$ **5.** $\frac{1 \text{ person}}{2 \text{ dresses}}$ **7.** $\frac{5 \text{ letters}}{1 \text{ minute}}$

9. $\frac{\$21}{2 \text{ visits}}$ **11.** $\frac{18 \text{ miles}}{1 \text{ gallon}}$ **13.** $12 per hour or $12/hour

15. 5 eggs per chicken or 5 eggs/chicken   **17.** 1.25 pounds/person
19. $103.30/day   **21.** 325.9; 21.0 (rounded)   **23.** 338.6; 20.9 (rounded)   **25.** 4 ounces for $3.09   **27.** 15 ounces for $3.15
29. 18 ounces for $1.79   **31.** Answers will vary. For example, you might choose Brand B because you like more chicken, so the cost per chicken chunk may actually be the same or less than Brand A.
33. 1.75 pounds/week   **35.** $12.26/hour   **37.** (a) Radiant $0.44; IDT $0.25; Access $0.235   (b) Radiant $0.088/min, IDT $0.05/min, Access $0.047/min; Access America is the best buy.
39. For a 15-minute call: Radiant $0.036/min; IDT $0.031/min; Access $0.047/min; IDT is the best buy. For a 20-minute call: Radiant $0.030/min; IDT $0.029/min; Access $0.047/min; IDT is the best buy.   **41.** 0.11 second/meter; 9.$\overline{09}$ or 9.1 meters/second (rounded)   **43.** One battery for $1.79; like getting 3 batteries so $1.79 ÷ 3 ≈ $0.597 per battery   **45.** Brand P with the 50¢ coupon is the best buy. ($3.39 − $0.50 = $2.89 and $2.89 ÷ 16.5 ounces ≈ $0.175 per ounce)   **47.** $2.92 (rounded) per month for Verizon, T-Mobile, and Nextel; $3 per month for Sprint.   **48.** Average weekdays per month is about 21.7; Verizon ≈ 18 min/weekday; T-Mobile ≈ 28 min/weekday; Nextel and Sprint ≈ 23 min/weekday.
49. Round to hundredths (nearest cent) to see that T-mobile is the best buy at $0.07 per "anytime minute"; Verizon ≈ $0.16; Sprint ≈ $0.12; Nextel ≈ $0.10.   **50.** Verizon ≈ $1.65 per "anytime minute"; T-mobile ≈ $1.57; Nextel ≈ $1.63; Sprint = $1.48

Section 6.3 (page 447)

1. $\frac{\$9}{12 \text{ cans}} = \frac{\$18}{24 \text{ cans}}$ **3.** $\frac{200 \text{ adults}}{450 \text{ children}} = \frac{4 \text{ adults}}{9 \text{ children}}$

5. $\frac{120}{150} = \frac{8}{10}$ **7.** $\frac{2.2}{3.3} = \frac{3.2}{4.8}$ **9.** $\frac{3}{5} = \frac{3}{5}$; true

11. $\frac{5}{8} = \frac{5}{8}$; true   **13.** $\frac{3}{4} \neq \frac{2}{3}$; false   **15.** 54 = 54; true

17. 336 ≠ 320; false   **19.** 2880 ≠ 2970; false

21. 28 = 28; true   **23.** $42\frac{1}{2} \neq 45$; false   **25.** 66 = 66; true

27. $\frac{16 \text{ hits}}{50 \text{ at bats}} = \frac{128 \text{ hits}}{400 \text{ at bats}}$ $\begin{array}{l}50 \cdot 128 = 6400 \\ 16 \cdot 400 = 6400\end{array}$ Cross products
are equal so the proportion is *true;* they hit equally well.   **29.** $x = 4$
31. $x = 2$ **33.** $x = 88$ **35.** $x = 91$ **37.** $x = 5$ **39.** $x = 10$
41. $x \approx 24.44$ (rounded)   **43.** $x = 50.4$ **45.** $x \approx 17.64$ (rounded)

47. $x = 1$ **49.** $x = 3\frac{1}{2}$ **51.** $x = 0.2$ or $x = \frac{1}{5}$

53. $x = 0.005$ or $x = \frac{1}{200}$ **55.** Find the cross products: 20 ≠ 30, so the proportion is false.

$\frac{6\frac{2}{3}}{4} = \frac{5}{3}$ or $\frac{10}{6} = \frac{5}{3}$ or $\frac{10}{4} = \frac{7.5}{3}$ or $\frac{10}{4} = \frac{5}{2}$
56. Find the cross products: 192 ≠ 180, so the proportion is false.
$\frac{6.4}{8} = \frac{24}{30}$ or $\frac{6}{7.5} = \frac{24}{30}$ or $\frac{6}{8} = \frac{22.5}{30}$ or $\frac{6}{8} = \frac{24}{32}$

Summary Exercises on Ratios, Rates, and Proportions (page 451)

1. $\frac{6}{5}$ **2.** $\frac{2}{7}$ **3.** $\frac{2}{1}$ **4.** $\frac{13}{11}$ **5.** Comparing the violin to piano,
guitar, organ, clarinet and drums gives ratios of $\frac{1}{11}, \frac{1}{10}, \frac{1}{3}, \frac{1}{2},$ and $\frac{2}{3},$
respectively.   **6.** (a) guitar to clarinet   (b) organ to drums, or clarinet to violin   **7.** 2.1 points/min; 0.5 min/point
8. 1.6 points/min; 0.6 min/point   **9.** $16.32/hour; $24.48/hour of overtime   **10.** $0.50/channel; $0.39/channel; $0.32/channel
11. 12 ounces, at about $0.60 per ounce   **12.** Brand P with the $2 coupon is the best buy at $0.57 per pound.

13. $\frac{4}{3} = \frac{4}{3}$ or 924 = 924; true   **14.** 2.0125 ≠ 2.07; false

15. $68\frac{1}{4} = 68\frac{1}{4}$; true   **16.** $x = 28$ **17.** $x = 3.2$ **18.** $x = 182$

19. $x \approx 3.64$ (rounded)   **20.** $x \approx 0.93$ (rounded)   **21.** $x = 1.56$

22. $x \approx 0.05$ (rounded)   **23.** $x = 1$ **24.** $x = \frac{3}{4}$

Section 6.4 (page 456)

1. 22.5 hours   **3.** $7.20   **5.** 42 pounds   **7.** $273.45
9. 10 ounces (rounded)   **11.** 5 quarts   **13.** 14 ft, 10 ft
15. 14 ft, 8 ft   **17.** 96 pieces of chicken; 33.6 pounds lasagna;
10.8 pounds deli meats; $5\frac{3}{5}$ pounds cheese; 7.2 dozen (about 86)
buns; 14.4 pounds potato salad.   **19.** 2065 students (reasonable); about 4214 students with incorrect setup (only 2950 students in the group)   **21.** about 30 people (reasonable); about 1904 people with incorrect setup (only 238 people attended)   **23.** 107,114,000 households (reasonable); about 111,530,612 households with incorrect setup (only 109,300,000 U.S. households)   **25.** 625 stocks
27. 4.06 meters (rounded)   **29.** 311 calories (rounded)
31. 10.53 meters (rounded)   **33.** You cannot solve this problem

using a proportion because the ratio of age to weight is not constant. As Jim's age increases, his weight may decrease, stay the same, or increase.   **35.** 5070 students use cream.   **37.** 120 calories and 12 grams of fiber   **39.** $1\frac{3}{4}$ cups water, 3 tablespoons margarine, $\frac{3}{4}$ cup milk, 2 cups flakes   **40.** $5\frac{1}{4}$ cups water, 9 tablespoons margarine, $2\frac{1}{4}$ cups milk, 6 cups flakes   **41.** $\frac{7}{8}$ cup water, $1\frac{1}{2}$ tablespoons margarine, $\frac{3}{8}$ cup milk, 1 cup flakes   **42.** $2\frac{5}{8}$ cups water, $4\frac{1}{2}$ tablespoons margarine, $1\frac{1}{8}$ cups milk, 3 cups flakes

Section 6.5 (page 470)

1. line named $\overleftrightarrow{CD}$ or $\overleftrightarrow{DC}$ **3.** line segment named $\overline{GF}$ or $\overline{FG}$ **5.** ray named $\overrightarrow{PQ}$ **7.** perpendicular   **9.** parallel   **11.** intersecting   **13.** $\angle AOS$ or $\angle SOA$ **15.** $\angle CRT$ or $\angle TRC$ **17.** $\angle AQC$ or $\angle CQA$ **19.** right (90°)   **21.** acute   **23.** straight (180°)   **25.** $\angle EOD$ and $\angle COD$; $\angle AOB$ and $\angle BOC$ **27.** $\angle HNE$ and $\angle ENF$; $\angle ACB$ and $\angle KOL$; $\angle HNE$ and $\angle ACB$; $\angle ENF$ and $\angle KOL$ **29.** 50°   **31.** 4°   **33.** 50°   **35.** 90°   **37.** $\angle SON \cong \angle TOM$; $\angle TOS \cong \angle MON$ **39.** $\angle GOH$ measures 63°; $\angle EOF$ measures 37°; $\angle AOC$ and $\angle GOF$ both measure 80°.   **41.** True, because $\overleftrightarrow{UQ}$ is perpendicular to $\overleftrightarrow{ST}$.   **42.** True, because they form a 90° angle, as indicated by the small red square.   **43.** False; the angles have the same measure (both are 180°).   **44.** False; $\overleftrightarrow{ST}$ and $\overleftrightarrow{PR}$ are parallel.   **45.** False; $\overleftrightarrow{QU}$ and $\overleftrightarrow{TS}$ are perpendicular.   **46.** True, because both angles are formed by perpendicular lines, so they both measure 90°.   **47.** corresponding angles; $\angle 1$ and $\angle 8$, $\angle 2$ and $\angle 5$, $\angle 3$ and $\angle 6$, $\angle 4$ and $\angle 7$; alternate interior angles; $\angle 4$ and $\angle 5$, $\angle 3$ and $\angle 8$.   **49.** $\angle 2$, $\angle 4$, $\angle 6$, $\angle 8$ all measure 130°; $\angle 1$, $\angle 3$, $\angle 5$, $\angle 7$ all measure 50°.   **51.** $\angle 6$, $\angle 1$, $\angle 3$, $\angle 8$ all measure 47°; $\angle 5$, $\angle 2$, $\angle 7$, $\angle 4$ all measure 133°.   **53.** $\angle 6$, $\angle 8$, $\angle 4$, $\angle 2$ all measure 114°; $\angle 7$, $\angle 5$, $\angle 3$, $\angle 1$ all measure 66°.   **55.** $\angle 1 \cong \angle 3$, both are 138°; $\angle 2 \cong \angle ABC$, both are 42°.

Section 6.6 (page 480)

1. $\angle 1$ and $\angle 4$; $\angle 2$ and $\angle 5$; $\angle 3$ and $\angle 6$; $\overline{AB}$ and $\overline{DE}$, $\overline{BC}$ and $\overline{EF}$, $\overline{AC}$ and $\overline{DF}$ **3.** $\angle 1$ and $\angle 6$; $\angle 2$ and $\angle 4$; $\angle 3$ and $\angle 5$; $\overline{ST}$ and $\overline{YW}$; $\overline{TU}$ and $\overline{WX}$; $\overline{SU}$ and $\overline{YX}$ **5.** $\angle 1$ and $\angle 6$; $\angle 2$ and $\angle 5$; $\angle 3$ and $\angle 4$; $\overline{LM}$ and $\overline{TS}$; $\overline{LN}$ and $\overline{TR}$, $\overline{MN}$ and $\overline{SR}$ **7.** SAS   **9.** SSS   **11.** ASA   **13.** use SAS: $BC = CE$, $\angle ABC \cong \angle DCE$, $BA = CD$ **14.** use SSS: $WP = YP$, $ZP = XP$, $WZ = YX$ **15.** use SAS: $PS = RS$, $m\angle QSP = m\angle QSR = 90°$, $QS = QS$ (common side)   **16.** use SAS: $LM = OM$, $PM = NM$, $\angle LMP \cong \angle OMN$ (vertical angles)   **17.** $a = 6$ cm; $b = 15$ cm   **19.** $a = 5$ mm; $b = 3$ mm   **21.** $a = 24$ in.; $b = 20$ in.   **23.** $x = 24.8$ m; Perimeter = 72.8 m; $y = 15$ m; Perimeter = 54.6 m   **25.** Perimeter = 8 cm + 8 cm + 8 cm = 24 cm; Area = (0.5)(8 cm)(6.9 cm) = 27.6 cm^2 **27.** $h = 24$ ft   **29.** One dictionary definition is "resembling, but not identical." Examples of similar objects are sets of different size pots or measuring cups; small and large size cans of beans; child's tennis shoe and adult's tennis shoe.   **31.** $x = 50$ m   **33.** $n = 110$ m

Chapter 6 Review Exercises (page 493)

1. $\frac{3}{4}$ **2.** $\frac{4}{1}$ **3.** great white shark to whale shark; whale shark to blue whale   **4.** $\frac{2}{1}$ **5.** $\frac{2}{3}$ **6.** $\frac{5}{2}$ **7.** $\frac{1}{6}$ **8.** $\frac{3}{1}$ **9.** $\frac{3}{8}$ **10.** $\frac{4}{3}$

11. $\frac{1}{9}$ **12.** $\frac{10}{7}$ **13.** $\frac{7}{5}$ **14.** $\frac{5}{6}$ **15.** $\frac{\$11}{1 \text{ dozen}}$ **16.** $\frac{12 \text{ children}}{5 \text{ families}}$ **17.** 0.2 page/minute or $\frac{1}{5}$ page/minute; 5 minutes/page   **18.** \$8/hour; 0.125 hour/dollar or $\frac{1}{8}$ hour/dollar   **19.** 13 ounces for \$2.29   **20.** 25 pounds for \$10.40 − \$1 coupon   **21.** $\frac{3}{5} = \frac{3}{5}$ or 90 = 90; true   **22.** $\frac{1}{8} \neq \frac{1}{4}$ or 432 ≠ 216; false   **23.** $\frac{47}{10} \neq \frac{49}{10}$ or 980 ≠ 940; false   **24.** 4.8 = 4.8; true   **25.** 14 = 14; true   **26.** $x = 1575$ **27.** $x = 20$ **28.** $x = 400$ **29.** $x = 12.5$ **30.** $x \approx 14.67$ (rounded)   **31.** $x \approx 8.17$ (rounded)   **32.** $x = 50.4$ **33.** $x \approx 0.57$ (rounded)   **34.** $x \approx 2.47$ (rounded)   **35.** 27 cats   **36.** 46 hits   **37.** \$15.63 (rounded)   **38.** 3299 students (rounded)   **39.** 68 ft   **40.** $27\frac{1}{2}$ hours or 27.5 hours   **41.** 511 calories (rounded)   **42.** 14.7 milligrams   **43.** line segment named $\overline{AB}$ or $\overline{BA}$ **44.** line named $\overleftrightarrow{CD}$ or $\overleftrightarrow{DC}$ **45.** ray named $\overrightarrow{OP}$ **46.** parallel   **47.** perpendicular   **48.** intersecting   **49.** acute   **50.** obtuse   **51.** straight; 180°   **52.** right; 90°   **53. (a)** 10° **(b)** 45° **(c)** 83°   **54. (a)** 25° **(b)** 90° **(c)** 147°   **55.** $\angle 1$ and $\angle 4$ measure 30°; $\angle 3$ and $\angle 6$ measure 90°; $\angle 5$ measures 60°   **56.** $\angle 8$, $\angle 3$, $\angle 6$, $\angle 1$ all measure 160°; $\angle 4$, $\angle 7$, $\angle 2$, $\angle 5$ all measure 20°   **57.** SSS   **58.** SAS   **59.** ASA   **60.** $y = 30$ ft; $x = 34$ ft; $P = 104$ ft   **61.** $y = 7.5$ m; $x = 9$ m; $P = 22.5$ m   **62.** $x = 12$ mm; $y = 7.5$ mm; $P = 38$ mm   **63.** $x = 105$ **64.** $x = 0$ **65.** $x = 128$ **66.** $x \approx 23.08$ (rounded)   **67.** $x = 6.5$ **68.** $x \approx 117.36$ (rounded)   **69.** $\frac{8}{5}$ **70.** $\frac{33}{80}$ **71.** $\frac{15}{4}$ **72.** $\frac{4}{1}$ **73.** $\frac{4}{5}$ **74.** $\frac{37}{7}$ **75.** $\frac{3}{8}$ **76.** $\frac{1}{12}$ **77.** $\frac{45}{13}$ **78.** 24,900 fans (rounded)   **79.** $\frac{8}{3}$ **80.** 75 ft for \$1.99 − \$0.50 coupon   **81.** 21 ft long; 15 ft wide   **82. (a)** 1400 milligrams   **(b)** 100 milligrams   **83.** 21 points (rounded)   **84.** $\frac{1}{2}$ or 0.5 teaspoon   **85.** parallel lines   **86.** line segment   **87.** acute angle   **88.** intersecting lines   **89.** right angle; 90°   **90.** ray   **91.** straight angle; 180°   **92.** obtuse angle   **93.** perpendicular lines   **94. (a)** The car turned around in a complete circle.   **(b)** The governor took the opposite view, for example, having once opposed taxes but now supporting them.   **95. (a)** No; because obtuse angles are > 90°, their sum would be > 180°.   **(b)** Yes; because acute angles are < 90°, their sum could equal 90°.   **96.** $\angle 1$ measures 45°; $\angle 3$ and $\angle 6$ measure 35°; $\angle 4$ measures 55°; $\angle 5$ measures 90°   **97.** $\angle 5$, $\angle 2$, $\angle 7$, $\angle 4$ all measure 75°; $\angle 6$, $\angle 1$, $\angle 3$, $\angle 8$ all measure 105°.

Chapter 6 Test (page 499)

[6.1, 6.2] 1. $\frac{\$1}{5 \text{ minutes}}$ **2.** $\frac{9}{2}$ **3.** $\frac{15}{4}$ **4.** 18 ounces for \$1.89 − \$0.25 coupon ≈ \$0.91 per ounce   **[6.3] 5.** $x = 25$ **6.** $x \approx 2.67$ (rounded)   **7.** $x = 325$ **8.** $x = 10\frac{1}{2}$ **[6.4] 9.** 576 words   **10.** 87 students (rounded)   **11.** 23.8 grams (rounded)   **12.** 60 feet   **[6.5] 13.** (e)   **14.** (a); 90°   **15.** (d)   **16.** (g); 180°   **17.** Parallel lines are lines in the same plane that never intersect. Perpendicular lines intersect to form a right angle. Sketches will vary; show them to your instructor.   **18.** 9°   **19.** 160°   **20.** $m\angle 1 = 50°$, $m\angle 2 = 35°$, $m\angle 3 = 95°$, $m\angle 5 = 35°$

21. measures of $\angle 1$, $\angle 3$, $\angle 5$, and $\angle 7$ are all 65°; measures of $\angle 2$, $\angle 4$, $\angle 6$, and $\angle 8$ are all 115°.   **[6.6] 22.** ASA (Angle–Side–Angle)
23. SAS (Side–Angle–Side)   **24.** $y = 12$ cm; $z = 6$ cm
25. In the larger triangle, $x = 12$ mm, $P = 46.8$ mm; in the smaller triangle, $y = 14$ mm, $P = 39$ mm

Cumulative Review Exercises: Chapters 1–6 (page 501)
[1.1, 5.1] 1. (a) seventy-seven billion, one million, eight hundred five
(b) two hundredths   **2. (a)** 3.040   **(b)** 500,037,000

[1.7] 3. 0   **[1.8] 4.** 14   **[4.5] 5.** $2\dfrac{11}{12}$ **[4.4] 6.** $\dfrac{2h-3}{10}$

[5.3] 7. 99.9905   **[1.8] 8.** $\dfrac{3}{0}$ is undefined.   **9.** -18

[4.6] 10. -14 **[4.3] 11.** $\dfrac{n}{4m}$ **[5.4] 12.** -0.00042 **[4.3] 13.** $\dfrac{1}{2y^2}$

[4.4] 14. $\dfrac{27+2n}{3n}$ **[1.8] 15.** -64 **[5.5] 16.** 5.4   **[4.6] 17.** $-\dfrac{1}{18}$

[2.5] 18. $y = -1$ **[4.7] 19.** $x = 15$ **[6.3] 20.** $x = 10$
[3.3] 21. Let t be the starting temperature. $t + 15 - 23 + 5 = 71$.
The starting temperature was 74 degrees.   **[3.4] 22.** Let l be the length; $l - 14$ is the width. $2(l) + 2(l - 14) = 100$. Length is 32 ft; width is 18 ft.   **[3.1, 3.2] 23.** $P = 7$ km; $A \approx 2.0$ km²
[5.10] 24. $d = 17$ m; $C \approx 53.4$ m; $A \approx 226.9$ m²
25. $V \approx 339.1$ ft³; $SA \approx 282.6$ ft²   **[4.3] 26.** 250 pounds
[6.4] 27. $1\dfrac{1}{4}$ teaspoons   **[4.5] 28.** $7\dfrac{3}{20}$ miles

[6.2] 29. 18.0 miles per gallon (rounded)   **[5.6] 30.** $0.08, 0.10,
0.104, 0.111, 0.14, 0.172, 0.333   **[4.3] 31.** 71 minutes (rounded)
32. 30 minutes (rounded) per $10 card so buy 4 cards for
120 minutes (2 hours)   **[6.1] 33.** $\dfrac{125}{100} = \dfrac{5}{4}$ in lowest terms

---

CHAPTER 7

Section 7.1 (page 512)
1. 0.25   **3.** 0.30 or 0.3   **5.** 0.06   **7.** 1.40 or 1.4   **9.** 0.078
11. 1.00 or 1   **13.** 0.005   **15.** 0.0035   **17.** 50%   **19.** 62%
21. 3%   **23.** 12.5%   **25.** 62.9%   **27.** 200%   **29.** 260%

31. 3.12%   **33.** $\dfrac{1}{5}$ **35.** $\dfrac{1}{2}$ **37.** $\dfrac{11}{20}$ **39.** $\dfrac{3}{8}$ **41.** $\dfrac{1}{16}$ **43.** $\dfrac{1}{6}$

45. $1\dfrac{3}{10}$ **47.** $2\dfrac{1}{2}$ **49.** 25%   **51.** 30%   **53.** 60%   **55.** 37%

57. $37\dfrac{1}{2}$% or 37.5%   **59.** 5%   **61.** exactly $55\dfrac{5}{9}$%, or

55.6% (rounded)   **63.** exactly $14\dfrac{2}{7}$%, or 14.3% (rounded)   **65.** 0.08

67. 0.42   **69.** 3.5%   **71.** 200%   **73.** $\dfrac{95}{100}$ or 95% shaded; $\dfrac{5}{100}$ or

5% unshaded   **75.** $\dfrac{3}{10}$ or 30% shaded; $\dfrac{7}{10}$ or 70% unshaded

77. $\dfrac{3}{4}$ or 75% shaded; $\dfrac{1}{4}$ or 25% unshaded   **79.** 0.01; 1%

81. $\dfrac{1}{5}$; 20%   **83.** $\dfrac{3}{10}$; 0.3   **85.** 0.5; 50%   **87.** $\dfrac{9}{10}$; 0.9

89. $1\dfrac{1}{2}$; 150%   **91. (a)** employee entertainment/company product

discounts   **(b)** 40%; $\dfrac{2}{5}$ **93. (a)** personal development training

(b) telecommuting   **95.** $\dfrac{1}{4}$; 0.25; 25%   **97.** $\dfrac{3}{8}$; 0.375; 37.5%

99. $\dfrac{1}{7}$; exactly $14\dfrac{2}{7}$%, or 14.3% (rounded)   **101. (a)** The student

forgot to move the decimal point in 0.35 two places to the right.

So $\dfrac{7}{20} = 35\%$.   **(b)** The student did the division in the wrong

order. Enter $16 \div 25$ to get 0.64 and then move the decimal point

two places to the right. So $\dfrac{16}{25} = 0.64 = 64\%$.   **103. (a)** $78

(b) $39   **105. (a)** 15 inches   **(b)** $7\dfrac{1}{2}$ inches   **107.** 20 children

109. (a) 2.8 miles   **(b)** 1.4 miles   **111. (a)** $142.50   **(b)** 50%
113. (a) 4100 students   **(b)** 50%   **115. (a)** 35 problems
(b) 0 problems   **117.** 50% means 50 out of 100 parts. That's half
of the number. A shortcut for finding 50% of a number is to divide
the number by 2. Examples will vary.

Section 7.2 (page 524)
1. (a) 10%   **(b)** 3000 runners   **(c)** unknown   **(d)** $\dfrac{10}{100} = \dfrac{n}{3000}$;

300 runners   **3. (a)** 4%   **(b)** 120 ft   **(c)** unknown

(d) $\dfrac{4}{100} = \dfrac{n}{120}$; 4.8 feet   **5. (a)** unknown   **(b)** 32 pizzas

(c) 16 pizzas   **(d)** $\dfrac{p}{100} = \dfrac{16}{32}$; 50%   **7. (a)** unknown

(b) 200 calories   **(c)** 16 calories   **(d)** $\dfrac{p}{100} = \dfrac{16}{200}$; 8%

9. (a) 90%   **(b)** unknown   **(c)** 495 students   **(d)** $\dfrac{90}{100} = \dfrac{495}{n}$;

550 students   **11. (a)** $12\dfrac{1}{2}$%   **(b)** unknown   **(c)** $3.50

(d) $\dfrac{12.5}{100} = \dfrac{3.50}{n}$; $28   **13.** $\dfrac{250}{100} = \dfrac{n}{7}$; 17.5 hours

15. $\dfrac{p}{100} = \dfrac{32}{172}$; 18.6% (rounded)   **17.** $\dfrac{110}{100} = \dfrac{748}{n}$; 680 books

19. $\dfrac{14.7}{100} = \dfrac{n}{274}$; $40.28 (rounded)   **21.** $\dfrac{p}{100} = \dfrac{105}{54}$;

194.4% (rounded)   **23.** $\dfrac{4}{100} = \dfrac{0.33}{n}$; $8.25

25. 150% of $30 cannot be *less* than $30; 25% of $16 cannot be

greater than $16.   **27.** The correct proportion is $\dfrac{p}{100} = \dfrac{14}{8}$. The

answer should be labeled with the % symbol. Correct answer is 175%.

Section 7.3 (page 532)
1. 1500 patients   **3.** $15   **5.** 4.5 pounds   **7.** $7.00
9. 52 students   **11.** 870 phones   **13.** 4.75 hours   **15. (a)** 10%

means $\dfrac{10}{100}$ or $\dfrac{1}{10}$. The denominator tells you to divide the

whole by 10. The shortcut for dividing by 10 is to move the
decimal point one place to the left.   **(b)** Once you find 10%
of a number, multiply the result by 2 for 20% and by 3 for 30%.
17. 231 programs   **19.** 50%   **21.** 680 circuits   **23.** 1080 people
25. 845 species   **27.** $20.80   **29.** 76%   **31.** 125%
33. 4700 employees   **35.** 83.2 quarts   **37.** 2%   **39.** 700 tablets
41. 325 salads   **43.** 1.5%   **45.** 5 gallons   **47.** 1029.2 meters
49. (a) Multiply 0.2 by 100 to change it from a decimal to a percent.
So, $0.2\underline{0} = 20\%$.   **(b)** The correct equation is $50 = p \cdot 20$, so the

solution is 250%.   **51. (a)** $\dfrac{1}{3} \cdot 162 = n$; the solution is $54.

(b) $(0.333333333)(162) = n$; depending upon how your calculator
rounds numbers, the solution is either $54 or $53.99999995.
(c) There is no difference or the difference is insignificant.

52. (a) $22 = \dfrac{2}{3} \cdot n$; the solution is 33 cans.

(b) $22 = (0.666666667)(n)$; depending upon how your calculator
rounds numbers, the solution is either 33 cans or 32.99999998 cans.
(c) There is no difference or the difference is insignificant.

Summary Exercises on Percent (page 536)

1. (a) 0.03; 3%   **(b)** $\frac{3}{10}$; 0.3   **(c)** $\frac{3}{8}$; 37.5%   **(d)** $1\frac{3}{5}$; 1.6

(e) 0.0625; 6.25%   **(f)** $\frac{1}{20}$; 0.05   **(g)** 2; 200%   **(h)** 0.8; 80%

(i) $\frac{9}{125}$; 7.2%   **2. (a)** 3.5 ft   **(b)** 19 miles   **(c)** 105 cows

(d) $0.08   **(e)** 500 women   **(f)** $45   **(g)** $87.50   **(h)** 12 pounds

(i) 95 students   **3.** $12\frac{1}{2}$% or 12.5%   **4.** 75 DVDs   **5.** $0.53
(rounded)   **6.** 175%   **7.** 5.9 pounds (rounded)   **8.** 600 hours
9. 50%   **10.** 500 camp sites   **11.** 98 golf balls   **12.** 2.7%
(rounded)   **13.** 2300 apartments   **14.** 0.63 ounce   **15.** 145%
16. 244 voters (rounded)   **17.** 0.021 inch   **18.** 400%   **19.** 8%
20. $0.68   **21.** flowers; $922.50   **22.** $1230   **23.** $2993
24. $51.25

Section 7.4 (page 544)

1. $37.80   **3. (a)** 10%   **(b)** 5%   **(c)** 2%   **(d)** 1%
5. (a) 101.6 pounds (rounded)   **(b)** 10.1 pounds (rounded)
7. 13.1% female; 86.9% male (both rounded)   **9.** 281 million
people (rounded)   **11.** 138%   **13.** 40 problems
15. 1360 shots (rounded)   **17.** 23.7 miles per gallon (rounded)
19. March; 24.5 million cans, or 24,500,000 cans
21. 52.5 million cans (January); 38.5 million cans (February)
23. 64.8% (rounded)   **25.** 9.1% (rounded)   **27.** 40%
29. 565% (rounded)   **31.** No. 100% is the entire price, so a
decrease of 100% would take the price down to 0. Therefore, 100%
is the maximum possible decrease in the price of something.
33. George ate more than 65 grams, so the percent must be >100%.
Use $p \cdot 65 = 78$ to get 120%.   **34.** The team won more than half the
games, so the percent must be >50%. Correct solution is $0.72 = 72\%$.
35. The brain could not weigh 375 pounds, which is more than the
person weighs. $2\frac{1}{2}$%, = 2.5% = 0.025, so $(0.025)(150) = n$ and
$n = 3.75$ pounds   **36.** If 80% were absent, then only 20% made it
to class. $800 - 640 = 160$ students, or use $(0.20)(800) = n$.

Section 7.5 (page 554)

1. $6; $106   **3.** 3%; $70.04   **5.** $29.28 (rounded); $395.26

7. $0.12 (rounded); $2.22   **9.** $4\frac{1}{2}$%; $13,167

11. $3 + $1.50 = $4.50; $4.83 (rounded); 2($3) = $6;
$6.43 (rounded)   **13.** $8 + $4 = $12; $11.75 (rounded);
2($8) = $16; $15.67 (rounded)   **15.** $1 + $0.50 = $1.50;
$1.43 (rounded); 2($1) = $2; $1.91   **17.** $15; $85   **19.** 30%;
$126   **21.** $4.38 (rounded); $13.12   **23.** $3.79 (rounded); $34.09
25. $42; $342   **27.** $33.30; $773.30   **29.** $213.75; $1713.75
31. $919.67 (rounded); $18,719.67   **33.** $1170
35. $7978.13 (rounded)   **37.** $106.49 (rounded)   **39.** 5%
41. $74.25   **43.** $25.13   **45.** $106.20; $483.80
47. $2016.38 (rounded)   **49.** $21 (rounded)   **51.** $92.38
53. $230.12   **55. (a)** $18.43 (rounded to nearest cent)
(b) When calculating the discount, the *whole* is $18.50. But
when calculating the sales tax, the *whole* is only $17.39 (the
discounted price).   **56. (a)** $396.05 (rounded to nearest cent).
(b) 7.53% sales tax (rounded to the nearest hundredth) would give
a final cost of $398.01.

Chapter 7 Review Exercises (page 567)

1. 0.25   **2.** 1.8   **3.** 0.125   **4.** 0.07   **5.** 265%   **6.** 2%

7. 30%   **8.** 0.2%   **9.** $\frac{3}{25}$ **10.** $\frac{3}{8}$ **11.** $2\frac{1}{2}$ **12.** $\frac{1}{20}$ **13.** 75%

14. 62.5% or $62\frac{1}{2}$%   **15.** 325%   **16.** 6%   **17.** 0.125

18. 12.5%   **19.** $\frac{3}{20}$ **20.** 15%   **21.** $1\frac{4}{5}$ **22.** 1.8   **23.** $46

24. $23   **25.** 9 hours   **26.** $4\frac{1}{2}$ hours   **27.** 242 meters

28. 17,000 cases   **29.** 27 cellular phones   **30.** 870 reference
books   **31.** 9.5% (rounded)   **32.** 225%   **33.** $2.60 (rounded)
34. 80 days   **35.** 4%   **36.** $575   **37.** 20 people   **38.** 175%
39. (a) 3000 patients   **(b)** 31.5% decrease (rounded)
40. (a) $364;   **(b)** 224%   **41. (a)** 80%
(b) 6.3% increase (rounded)   **42.** 10.4% (rounded)

43. $0.11 (rounded); $2.90   **44.** $7\frac{1}{2}$%; $838.50

45. $4 + $2 = $6; $6.41 (rounded); 2($4) = $8; $8.55 (rounded)
46. $0.80 + $0.40 = $1.20; $1.21 (rounded); 2($0.80) = $1.60;
$1.61   **47.** $3.75; $33.75   **48.** 25%; $189   **49.** $68.25; $418.25

50. $183.60; $1713.60   **51.** $\frac{1}{3}$; $33\frac{1}{3}$% (exact) or 33.3% (rounded)

52. cards; 68%; 0.68; $\frac{17}{25}$ **53.** 277 adults (rounded)   **54.** cards,

188 adults (rounded); sci-fi/simulation, 102 adults (rounded)
55. 48.4% (rounded)   **56.** 7161 dogs (rounded)
57. 36.4% (rounded)   **58.** 180.3% increase (rounded)
59. 38.1% (rounded)   **60.** 265 animals (rounded)

Chapter 7 Test (page 571)

[7.1] 1. 0.75   **2.** 60%   **3.** 180%   **4.** 7.5% or $7\frac{1}{2}$%   **5.** 3.00 or 3

6. 0.02   **7.** $\frac{5}{8}$ **8.** $2\frac{2}{5}$ **9.** 5%   **10.** 87.5% or $87\frac{1}{2}$%

11. 175%   **[7.2–7.4] 12.** 320 laptops   **13.** 400%   **14.** $19,500
15. $8466.75   **16.** 34% increase (rounded)   **[7.3] 17.** To find 50%
of a number, divide the number by 2. To find 25% of a number, divide the number by 4. Examples will vary.   **[7.5] 18.** Round $31.94
to $30. Then 10% of $30. is $3 and 5% of $30 is half of $3 or $1.50,
so a 15% tip estimate is $3 + $1.50 = $4.50. A 20% tip estimate is
2($3) = $6.   **19.** Exact tip is $4.79 (rounded). Each person pays
$12.24 (rounded).   **20.** $3.84; $44.16   **21.** $41.39 (rounded);
$188.56   **22.** $815.66 ÷ 6 ≈ $135.94   **23.** $1650; $6650
24. $51.60; $911.60

Cumulative Review Exercises: Chapters 1–7 (page 573)

[1.1, 5.1] 1. (a) ninety and one hundred five thousandths
(b) one hundred twenty-five million, six hundred seventy
2. (a) 30,005,000,000   **(b)** 0.0078   **3. (a)** 50,000   **(b)** 0.70
(c) 8900   **[1.3, 1.6] 4. (a)** commutative property of multiplication
(b) associative property of addition   **[1.2] 5.** <; >

[5.6] 6. 0.7005; 0.705; $\frac{3}{4}$; 0.755   **[5.7] 7.** Mean is $771.25;

median is $725.   **[6.2] 8.** 17 ounces for $2.89   **[5.3] 9.** 48.901

[1.8] 10. 17   **[4.3] 11.** $\frac{9b^2}{8}$ **[4.5] 12.** $\frac{1}{2}$ **[1.8] 13.** −10

[4.4] 14. $\frac{16 + 5x}{20}$ **[1.4] 15.** −40   **[5.4] 16.** −0.001

[4.6] 17. $\frac{12}{11}$ or $1\frac{1}{11}$ **[4.3] 18.** 123   **[1.8] 19.** 0   **[4.4] 20.** $\frac{7m - 16}{8m}$

[5.5] 21. −30   **[4.3] 22.** $\frac{4}{9n}$ **[4.5] 23.** $3\frac{1}{2}$ **24.** $3\frac{2}{9}$ **[1.8] 25.** −44

[5.5] 26. −0.02   **[4.6] 27.** $\frac{3}{10}$ **[2.1] 28.** 19   **29.** −1   **30.** −24

31. 160   **[2.2] 32.** $-9x^2 + 5x$ **33.** 0   **34.** $-40w^3$ **35.** $3h - 10$
[2.4] 36. $n = 5$ **[2.5] 37.** $h = -6$ **[5.9] 38.** $a = 2.2$

[4.7] **39.** $y = -\dfrac{14}{3}$ **40.** $b = 20$ [6.3] **41.** $x = 162.5$

[2.5] **42.** $h = 3$ **43.** $x = -1$ [3.3] **44.** $4n - 5 = -17$; $n = -3$
45. $n + 31 = 3n + 1$; $n = 15$ **46.** Let d be the diapers in each package. $3d - 17 - 19 = 12$. There were 16 diapers in each package.   [3.4] **47.** Let p be Susanna's pay. Let $2p$ be Neoka's pay. $p + 2p = 1620$. Susanna receives \$540; Neoka receives \$1080.
[3.1, 3.2] **48.** $P = 50$ yd; $A = 142$ yd^2 [5.10] **49.** $C \approx 31.4$ ft; $A \approx 78.5$ ft^2 [6.6] **50.** $x \approx 8.1$ ft   [4.8, 5.10] **51.** $V \approx 7.4$ m^3; $SA = 30.1$ m^2 **52.** $V \approx 2255.3$ cm^3 [5.8] **53.** $y \approx 24.2$ mm; $A = 142.5$ mm^2 [4.3] **54.** 2100 students   [3.1, 3.2] **55.** $P = 64$ ft; $A = 252$ ft^2 [4.4] **56.** $\dfrac{1}{6}$ cup more than the amount needed

[7.5] **57.** \$56.70; \$132.30   [6.4] **58.** 93.6 km
[7.4] **59.** 88.6% (rounded)   [5.5] **60.** \$0.23 (rounded)

[1.7] **61.** (a) Britain   (b) Brazil   [6.1] **62.** (a) $\dfrac{2}{1}$ (b) $\dfrac{2}{3}$

[7.4] **63.** (a) 47.6% (rounded)   (b) 180%   **64.** (a) 20% (b) 50%   (c) 100%

CHAPTER 8

Section 8.1 (page 584)

1. (a) 3   (b) 12   **3.** (a) 8   (b) 2   **5.** (a) 5280   (b) 3
7. (a) 2000   (b) 16   **9.** (a) 60   (b) 60   **11.** (a) 2   (b) 240
13. (a) $\dfrac{1}{2}$ (b) 78   **15.** 14,000 to 16,000 lb   **17.** 27   **19.** 112
21. 10   **23.** $1\dfrac{1}{2}$ or 1.5   **25.** $\dfrac{1}{4}$ or 0.25   **27.** $1\dfrac{1}{2}$ or 1.5
29. $2\dfrac{1}{2}$ or 2.5   **31.** snowmobile/ATV; person walking   **33.** 5000
35. 17   **37.** 4 to 8 in.   **39.** 216   **41.** 28   **43.** 518,400
45. 48,000   **47.** (a) pound/ounces   (b) quarts/pints or pints/cups
(c) minutes/hours or seconds/minutes   (d) feet/inches
(e) pounds/tons   (f) days/weeks   **49.** 174,240   **51.** 800
53. 0.75 or $\dfrac{3}{4}$ **55.** \$1.83 (rounded)   **57.** \$140   **59.** (a) 1056 sec
(b) 17.6 min   **61.** (a) $12\dfrac{1}{2}$ qt   (b) 4 containers, because you can't
buy part of a container   **63.** (a) Travel 1391 times around Earth (rounded).   (b) Drive 12,436 times from Los Angeles to New York (rounded).   **64.** (a) $35,000,000 \div 25,000$ becomes $35,000 \div 25 = 1400$ times around Earth   (b) 12,500 times from Los Angeles to New York   (c) They are very close and just as useful for comparing such large numbers.   **65.** (a) 88,176 ft tall   (b) 5.5 mi (rounded)   (c) about 3 times taller   **66.** (a) 65 lb (rounded)
(b) 3 lb (rounded)   (c) Answers will vary but should be about $\dfrac{1}{3}$ of your weight on Earth.

Section 8.2 (page 593)

1. 1000; 1000   **3.** $\dfrac{1}{1000}$ or 0.001; $\dfrac{1}{1000}$ or 0.001
5. $\dfrac{1}{100}$ or 0.01; $\dfrac{1}{100}$ or 0.01   **7.** Answers will vary; about 8 to 10 cm.
9. Answers will vary; about 15 to 25 mm.   **11.** cm   **13.** m
15. km   **17.** mm   **19.** cm   **21.** m   **23.** Some possible answers are: 35 mm film for cameras, track and field events, metric auto parts, and lead refills for mechanical pencils.   **25.** 700 cm
27. 0.040 m or 0.04 m   **29.** 9400 m   **31.** 5.09 m   **33.** 40 cm
35. 910 mm   **37.** less; 18 cm or 0.18 m
39. ⟥5 mm = 0.5 cm; 1 mm = 0.1 cm
41. 0.018 km   **43.** 164 cm; 1640 mm   **45.** 0.0000056 km

Section 8.3 (page 601)

1. mL   **3.** L   **5.** kg   **7.** g   **9.** mL   **11.** mg   **13.** L   **15.** kg
17. unreasonable; too much   **19.** unreasonable; too much
21. reasonable   **23.** reasonable   **25.** Some capacity examples are 2 L bottles of soda and shampoo bottles marked in mL; weight examples are grams of fat listed on cereal boxes and vitamin doses in milligrams.   **27.** Unit for your answer (g) is in numerator; unit being changed (kg) is in denominator so it will divide out. The unit fraction is $\dfrac{1000\ \text{g}}{1\ \text{kg}}$.   **29.** 15,000 mL   **31.** 3 L   **33.** 0.925 L
35. 0.008 L   **37.** 4150 mL   **39.** 8 kg   **41.** 5200 g   **43.** 850 mg
45. 30 g   **47.** 0.598 g   **49.** 0.06 L   **51.** 0.003 kg   **53.** 990 mL
55. mm   **57.** mL   **59.** cm   **61.** mg   **63.** 0.3 L   **65.** 1340 g
67. 0.9 L   **69.** 3 kg to 4 kg   **71.** greater; 5 mg or 0.005 g
73. 200 nickels   **75.** (a) 1,000,000
(b) $\dfrac{3.5\ \text{Mm}}{1} \cdot \dfrac{1,000,000\ \text{m}}{1\ \text{Mm}} = 3,500,000\ \text{m}$
76. (a) 1,000,000,000
(b) $\dfrac{2500\ \text{m}}{1} \cdot \dfrac{1\ \text{Gm}}{1,000,000,000\ \text{m}} = 0.0000025\ \text{Gm}$
77. (a) 1,000,000,000,000   (b) 1000; 1,000,000
78. 1,000,000; 1,000,000,000; $2^{20} = 1,048,576$; $2^{30} = 1,073,741,824$

Section 8.4 (page 607)

1. \$0.83 (rounded)   **3.** 89.5 kg   **5.** 71 beats (rounded)
7. 180 cm; \$0.02/cm (rounded)   **9.** 5.03 m   **11.** 4 bottles;
0.175 L or 175 mL   **13.** 1.89 g   **15.** (a) 10 km   (b) 600 km
(c) 36,000 km   **17.** 215 g; 4.3 g; 4300 mg   **18.** 330 g; 0.33 g;
330 mg   **19.** 1550 g; 1500 g; 3 g   **20.** 55 g; 50 g; 0.5 g
21. 0.45 mg

Section 8.5 (page 613)

1. 21.8 yd   **3.** 262.4 ft   **5.** 4.8 m   **7.** 5.3 oz   **9.** 111.6 kg
11. 30.3 qt   **13.** (a) about 0.2 oz   (b) probably not   **15.** about
31.8 L   **17.** about 1.3 cm   **19.** 18 kg; 56 cm by 36 cm by 23 cm
(all rounded)   **21.** Converting ounces to grams, 0.09 oz $\approx$ 2.55 g, which rounds to 2.6 g. However, converting grams to ounces, 2.5 g $\approx$ 0.0875 oz, which does round to 0.09 oz.   **23.** 3.5 kg $\approx$ 7.7 lb so the baby is heavy enough. But 53 cm $\approx$ 20.7 in. so the baby is not long enough to be in the carrier.   **25.** $-8\,^{\circ}$C   **27.** 40 $^{\circ}$C
29. 150 $^{\circ}$C   **31.** 16 $^{\circ}$C (rounded)   **33.** $-20\,^{\circ}$C   **35.** 46 $^{\circ}$F
(rounded)   **37.** 23 $^{\circ}$F   **39.** 58 $^{\circ}$C (rounded); $-89\,^{\circ}$C (rounded)
41. 10 $^{\circ}$C and 41 $^{\circ}$C (rounded)   **43.** More. There are 180 degrees between freezing and boiling on the Fahrenheit scale, but only 100 degrees on the Celsius scale, so each Celsius degree is a greater change in temperature.   **45.** (a) pleasant weather, above freezing but not hot   (b) 75 $^{\circ}$F to 39 $^{\circ}$F (rounded)   (c) Answers will vary. In Minnesota, it's 0 $^{\circ}$C to $-40\,^{\circ}$C; in California, 24 $^{\circ}$C to 0 $^{\circ}$C.
47. about 1.8 m   **48.** about 5.0 kg   **49.** about 3040.2 km
50. 38 hr 23 min   **51.** about 300 m   **52.** less than 3780 mL
53. about 59.4 mL   **54.** about 1.6%

Chapter 8 Review Exercises (page 625)

1. 16   **2.** 3   **3.** 2000   **4.** 4   **5.** 60   **6.** 8   **7.** 60
8. 5280   **9.** 12   **10.** 48   **11.** 3   **12.** 4   **13.** $\dfrac{3}{4}$ or 0.75
14. $2\dfrac{1}{2}$ or 2.5   **15.** 28   **16.** 78   **17.** 112   **18.** 345,600
19. (a) $4153\dfrac{1}{3}$ yd (exact) or 4153.3 (rounded)
(b) 2.4 mi (rounded)   **20.** \$2465.20   **21.** mm   **22.** cm   **23.** km
24. m   **25.** cm   **26.** mm   **27.** 500 cm   **28.** 8500 m   **29.** 8.5 cm
30. 3.7 m   **31.** 0.07 km   **32.** 930 mm   **33.** mL   **34.** L

35. g **36.** kg **37.** L **38.** mL **39.** mg **40.** g **41.** 5 L
42. 8000 mL **43.** 4580 mg **44.** 700 g **45.** 0.006 g **46.** 0.035 L
47. 31.5 L **48.** 357 g (rounded) **49.** 87.25 kg
50. $1.12 (rounded) **51.** 6.5 yd (rounded)
52. 11.7 in. (rounded) **53.** 67.0 mi (rounded) **54.** 1288 km
55. 21.9 L (rounded) **56.** 44.0 qt (rounded) **57.** 0 °C
58. 100 °C **59.** 37 °C **60.** 20 °C **61.** 25 °C **62.** −15 °C
63. 28 °F (rounded) **64.** 120 °F (rounded) **65.** L **66.** kg
67. cm **68.** mL **69.** mm **70.** km **71.** g **72.** m **73.** mL
74. g **75.** mg **76.** L **77.** 105 mm **78.** $\frac{3}{4}$ hr or 0.75 hr
79. $7\frac{1}{2}$ ft or 7.5 ft **80.** 130 cm **81.** 77 °F **82.** 14 qt
83. 0.7 g **84.** 810 mL **85.** 80 oz **86.** 60,000 g
87. 1800 mL **88.** −1 °C **89.** 36 cm **90.** 0.055 L
91. 1.26 m **92.** 18 tons **93.** 113 g; 177 °C (both rounded)
94. 178.0 lb; 6.0 ft (both rounded) **95.** 13.2 m **96.** 22.0 ft
97. 3.6 m **98.** 39.8 in. **99.** 484,000 lb **100.** 18.9 to 22.7 L
101. (a) 242 tons (b) 144 in. **102.** (a) 1.02 m (b) 670 cm

Chapter 8 Test (page 629)

[8.1] **1.** 36 qt **2.** 15 yd **3.** 2.25 hr or $2\frac{1}{4}$ hr **4.** 0.75 ft or $\frac{3}{4}$ ft
5. 56 oz **6.** 7200 min [8.2, 8.3] **7.** kg **8.** km **9.** mL **10.** g
11. cm **12.** mm **13.** L **14.** cm **15.** 2.5 m **16.** 4600 m
17. 0.5 cm **18.** 0.325 g **19.** 16,000 mL **20.** 400 g **21.** 1055 cm
22. 0.095 L [8.1] **23.** 3.2 ft (rounded) [8.4] **24.** (a) 2310 mg
(b) 500 mg or 0.5 g more [8.5] **25.** 95 °C **26.** 0 °C
27. 1.8 m (rounded) **28.** 56.3 kg (rounded) **29.** 13 gal
30. 5.0 mi (rounded) **31.** 23 °C (rounded) **32.** 10 °F (rounded)
[8.4, 8.5] **33.** 6 m ≈ 6.54 yd; $26.03 (rounded)
[8.1, 8.2] **34.** Possible answers: Use same system as rest of the
world; easier system for children to learn; less use of fractional
numbers; compete internationally.

Cumulative Review Exercises: Chapters 1–8 (page 631)

[1.1, 5.1] **1.** (a) six hundred three billion, five million, forty thousand
(b) nine and forty thousandths **2.** (a) 80.08 (b) 200,065,004
[1.5, 5.2] **3.** (a) 1.0 (b) 500 (c) 306,000,000 [5.7] **4.** mean
is 26.1 hours (rounded); median is 21.5 hours [1.8] **5.** −17
[4.5] **6.** $1\frac{13}{20}$ [5.5] **7.** −25 [4.3] **8.** $\frac{3}{4xy}$ [5.4] **9.** 0.00015
[5.3] **10.** −41.917 [4.4] **11.** $\frac{18 - 5c}{6c}$ [1.4] **12.** −30
[1.8] **13.** 13 **14.** undefined [4.3] **15.** −240 [4.5] **16.** $-\frac{5}{12}$
[4.3] **17.** $\frac{27xy}{4}$ [4.2] **18.** $\frac{2m + 3n}{3m}$ [4.5] **19.** $2\frac{5}{6}$ [4.6] **20.** $-\frac{7}{45}$
[1.8] **21.** 2 [5.5] **22.** −3.2 [4.6] **23.** 0 [2.1] **24.** 22 **25.** −34
26. −6 [2.2] **27.** $-3p^2 - p$ **28.** $-5x - 14$ **29.** $-14r^3$
[2.5] **30.** $b = -1$ [5.9] **31.** $a = 8.05$ [2.5] **32.** $t = 0$ **33.** $n = 2$
[4.7] **34.** $w = 20$ [6.3] **35.** $x = 0.4$ [2.5] **36.** $y = -5$ **37.** $k = 4$
[3.3] **38.** $11n - 8n = -9$; $n = -3$ **39.** $2n - 8 = n + 7$; $n = 15$
40. Let l be the length. $2l + 2(25) = 124$; The length is 37 cm.
[3.4] **41.** Let p be the longer piece; $p - 6$ is the shorter piece.
$p + p - 6 = 90$; The pieces are 48 ft and 42 ft long.
[3.1, 3.2] **42.** $P = 9$ ft; $A = 5\frac{1}{16}$ ft^2 or ≈ 5.1 ft^2
[5.10] **43.** $C ≈ 28.3$ mm; $A ≈ 63.6$ mm^2 [3.1, 3.2] **44.** $P ≈ 6.5$ cm;
$A ≈ 2.2$ cm^2 [4.8, 5.8] **45.** $y = 12$ yd; $A = 96$ yd^2
46. $V = 96$ cm^3; $SA = 128$ cm^2 [6.6] **47.** $x = 14$ in.

[8.1] **48.** 54 in. **49.** 3 days [8.2, 8.3] **50.** 3700 g **51.** 0.6 m
52. 0.007 L [8.5] **53.** −4 °F [7.4] **54.** 58% (rounded)
[6.2] **55.** Brand T at 15.5 ounces for $2.99 − $0.30 coupon
[6.4, 8.5] **56.** (a) 93.6 km (b) 58.0 mi (rounded)
[8.4] **57.** 5.97 kg [4.5] **58.** $1\frac{1}{12}$ yard [8.4] **59.** $9.74 (rounded)
[7.5] **60.** $18.99 (rounded); $170.95 [6.4] **61.** 120 rows
[7.5] **62.** $3631.25 [1.7] **63.** 23 blankets; $9 left over
[5.5] **64.** $0.15 [7.4] **65.** 37.5% decrease

CHAPTER 9

Section 9.1 (page 640)

1. (a) 31,419 points (b) Michael Jordan **3.** (a) Michael Jordan
(b) Allen Iverson **5.** 17,856 points **7.** West, 27.0;
Iverson, 27.0; Pettit, 26.4; round to nearest tenth. **9.** The asterisks
next to O'Neal's and Iverson's names mean that they played during
the 2003–2004 season. **11.** (a) 255 calories (b) moderate jogging **13.** (a) moderate jogging, aerobic dance, racquetball
(b) moderate jogging, moderate bicycling, aerobic dance, racquetball, tennis **15.** 370 calories **17.** (a) 366 calories (rounded)
(b) 239 calories **19.** about 39 calories **21.** (a) 70 million
or 70,000,000 (b) 35 million or 35,000,000 **23.** 205 million or
205,000,000 **25.** 15 million or 15,000,000 **27.** 265 million
or 265,000,000 **29.** Answers will vary. One possibility: choose
Southwest because it has the best on-time performance.
30. Answers will vary. Possibilities include planning more time
between each flight, or doing some or all of your business via
conference calls or e-mail. **31.** Answers will vary. One
possibility: choose Airtran because it has the fewest luggage
problems. **32.** Answers will vary. Possibilities include buying
heavy-duty luggage or shipping the golf clubs via a delivery
service. **33.** Answers will vary. Possibilities include a lot of
bad weather, maintenance problems, new computer system.
34. Answers will vary. Possibilities include availability of
nonstop flights, convenience of departure times, type and size
of aircraft, availability of low-cost fares.

Section 9.2 (page 648)

1. (a) $32,000 (b) carpentry, $12,100 **3.** (a) $\frac{\$9800}{\$32,000} = \frac{49}{160}$
(b) $\frac{\$3000}{\$2000} = \frac{3}{2}$ **5.** $\frac{\$16,000}{\$32,000} = \frac{1}{2}$ **7.** (a) don't know
(b) quicker **9.** $\frac{720}{6000} = \frac{3}{25}$ **11.** $\frac{1020}{1200} = \frac{17}{20}$ **13.** $\frac{1740}{180} = \frac{29}{3}$
15. $522,000 **17.** $174,000 **19.** $261,000 **21.** 160 people
23. mustard; 960 people **25.** 64 people **27.** First find the
percent of the total that is to be represented by each item. Next,
multiply the percent by 360° to find the size of each sector. Finally,
use a protractor to draw each sector. **29.** (a) 90° (b) 20%
(c) 10%; 36° (d) 10%; 36° (e) 15%; 54° (f) 5%; 18°
(g) 15%; 54° (h) see circle graph below.

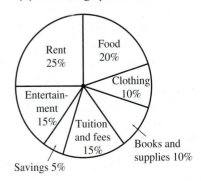

31. (a) $200,000 **(b)** 6.25%; 20%; 30%; 25%; 18.75%
(c) 22.5°; 72°; 108°; 90°; 67.5°
(d) See circle graph below.

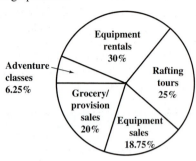

Section 9.3 (page 657)

1. can shop during off hours; 74% **3.** 57%; 342 people **5.** $\frac{1}{2}$,
compare products more easily; nearly $\frac{3}{4}$, shop during off hours

7. May; 10,000 unemployed **9.** 1500 workers **11.** 2500 workers;
45% increase (rounded) **13.** 150,000 gallons **15.** 2001;
250,000 gallons **17.** 550,000 gallons; 367% increase (rounded)
19. 24.1 million or 24,100,000 PCs **21.** 185.6 million or
185,600,000 more PCs **23.** 82.3 million or 82,300,000 PCs;
132% increase (rounded) **25. (a)** 3,000,000 CDs **(b)** 1,500,000
CDs **27. (a)** 2,500,000 CDs **(b)** 3,000,000 CDs **29.** Answers
will vary. Possibilities include: Both stores had decreased sales from
2001 to 2002 and increased sales from 2003 to 2005; Store B had
lower sales than Store A in 2001–02 but higher sales than Store A in
2003–2005. **31.** 2003; $25,000 **33.** 2002, 29% (rounded); 2003,
20%; 2004, 17% (rounded); 2005, 38% (rounded) **35.** Answers
will vary. Possibilities include: The decrease in sales may have
resulted from poor service or greater competition; the increase in
sales may have been a result of more advertising or better service.
37. Shipments have increased at a rapid rate since 1985.
38. Answers will vary. Some possibilities are: lower prices; more uses
and applications for students, home use, and businesses; improved
technology. **39. (a)** 104% (rounded) **(b)** 159% (rounded)
(c) 132% (rounded) **(d)** 37% (rounded) **40.** Since 1990, the
percent of increase for each 5-year period has been dropping.
41. Answers will vary. Some possibilities are: More people will already
own a computer and not want to buy another; some new invention will
replace computers. **42. (a)** no change; 0% **(b)** 167% increase
(rounded) **43. (a)** Answers will vary; perhaps 3,500,000 CDs in
2006. **(b)** Answers will vary; perhaps 4,500,000 CDs in 2006.
44. (a) Most people will probably pick Store B because of its greater
sales and more consistent upward trend. **(b)** Answers will vary.
Some possibilities include: age and physical condition of the store,
annual expenses, sales of other products, annual profit.

Section 9.4 (page 665)

1.

3.

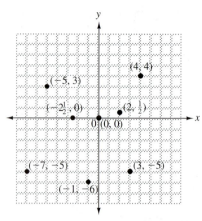

5. A is $(3, 4)$; B is $(5, -5)$; C is $(-4, -2)$; D is approximately
$\left(4, \frac{1}{2}\right)$; E is $(0, -7)$; F is $(-5, 5)$; G is $(-2, 0)$; H is $(0,0)$.

7. III, none, IV, II **9. (a)** any positive number **(b)** any negative
number **(c)** 0 **(d)** any negative number **(e)** any positive number
11. Starting at the origin, move left or right along the x-axis to the
number a; then move up if b is positive or move down if b is negative.

Section 9.5 (page 675)

1. 4; $(0, 4)$; 3; $(1, 3)$; 2; $(2, 2)$;
All points on the line are solutions.

3. -1; $(0, -1)$; -2; $(1, -2)$; -3; $(2, -3)$;
All points on the line are solutions.

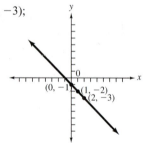

5. 4; -1; -6; 99

7.

x	y	(x, y)
1	−1	(1, −1)
2	0	(2, 0)
3	1	(3, 1)

9.

x	y	(x, y)
0	2	(0, 2)
−1	1	(−1, 1)
−2	0	(−2, 0)

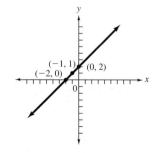

11.

x	y	(x, y)
0	0	(0, 0)
1	−3	(1, −3)
2	−6	(2, −6)

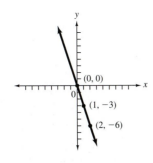

13. (a) positive; increases, or, does the same   **(b)** positive; decreases, or, does the same   **(c)** negative; decreases, or, does the opposite

15.

x	y	(x, y)
0	0	(0, 0)
3	1	(3, 1)
6	2	(6, 2)

17.

x	y	(x, y)
−1	−1	(−1, −1)
−2	−2	(−2, −2)
−3	−3	(−3, −3)

19.

x	y	(x, y)
0	3	(0, 3)
1	1	(1, 1)
2	−1	(2, −1)

21.

negative slope

23.

positive slope

25.

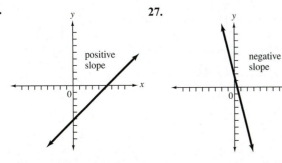

positive slope

27.

negative slope

Chapter 9 Review Exercises (page 689)

1. (a) gymnastics   **(b)** volleyball   **2. (a)** cross country   **(b)** golf
3. (a) 4870 men   **(b)** 12,081 women   **4.** men's, 14.2 (rounded); women's, 13.3 (rounded)   **5. (a)** 100 inches   **(b)** 15 inches
6. (a) 50 inches   **(b)** 55 inches   **7. (a)** 35 inches   **(b)** 45 inches
8. 95 inches difference between Juneau and Memphis
9. (a) lodging; $560   **(b)** food; $400   **(c)** $1700
10. $\frac{560}{400} = \frac{7}{5}$ **11.** $\frac{300}{1700} = \frac{3}{17}$ **12.** $\frac{280}{1700} = \frac{14}{85}$
13. $\frac{300}{160} = \frac{15}{8}$ **14.** painting and wall papering; 63%
15. construction work; 33%   **16.** 43%; 147 homeowners (rounded)
17. 184 homeowners (rounded)   **18. (a)** interior decorating
(b) construction work   **19.** Answers will vary. Possibilities include: painting and wall papering are easier to do, take less time, or cost less than construction work.   **20.** March; 8,000,000 acre-feet
21. June; 2,000,000 acre-feet   **22.** 5,000,000 acre-feet
23. 6,000,000 acre-feet   **24.** 3,000,000 acre-feet; 37.5% decrease
25. 3,000,000 acre-feet; 60% decrease   **26.** $50,000,000
27. $20,000,000   **28.** $20,000,000   **29.** $40,000,000
30. Sales decreased for two years and then moved up slightly. Answers will vary. Perhaps there is less new construction, remodeling, and home improvement in the area near Center A. Or, better product selection and service may have reversed the decline in sales.
31. Sales are increasing. Answers will vary. New construction may have increased in the area near Center B, or greater advertising may attract more attention.   **32.** 36°   **33.** 35%; 126°   **34.** 20% 72°
35. 25% 90°   **36.** $2240; 10%   **37.** See graph below.

38.

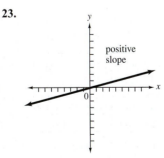

39. *A* is (0, 6); *B* is approximately $\left(-2, 2\frac{1}{2}\right)$; *C* is (0, 0); *D* is (−6, −6); *E* is (4, 3); *F* is approximately $\left(3\frac{1}{2}, 0\right)$; *G* is (2, −4).

40.

x	y	(x, y)
0	−2	(0, −2)
1	−3	(1, −3)
2	−4	(2, −4)

There are many answers because all points on the line are solutions. Some possibilities are (−1, −1) and (−2, 0).

41.

x	y	(x, y)
0	3	(0, 3)
1	4	(1, 4)
2	5	(2, 5)

42.

x	y	(x, y)
−1	4	(−1, 4)
0	0	(0, 0)
1	−4	(1, −4)

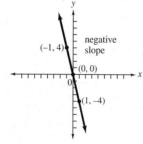

All points on the line are solutions. Some possibilities are (−1, 2) and (−2, 1).

All points on the line are solutions. Some possibilities are $\left(-\frac{1}{2}, 2\right)$ and $\left(\frac{1}{2}, -2\right)$.

Chapter 9 Test (page 695)

[9.1] 1. (a) sardines **(b)** cream cheese **2.** 166 calories (rounded) **3.** 259 mg (rounded) **4. (a)** birds **(b)** 80 bird species **5.** 55 species **6.** 265 species **[9.2] 7.** television; $812,000 **8.** miscellaneous; $84,000 **9.** $308,000 **10.** $644,000 **[9.3] 11.** 2005; $4000 **12.** $5000; 38% increase (rounded) **13.** 2005; explanations will vary. Some possibilities are: laid off from work, changed jobs, was ill, cut down on hours worked. **14.** 5500 students; 3000 students **15.** College B; 1000 students **16.** Explanations will vary. For example, College B may have added new courses or lowered tuition or added child care. **[9.2] 17.** 35%; 126° **18.** 5%; 18° **19.** 20%; 72° **20.** 30%; 108° **21.** $48,000; 10%
22.

[9.4] 23–26.

27. (0, 0); no quadrant **28.** (−5, −4); quadrant III **29.** (3, 3); quadrant I **30.** (−2, 4); quadrant II
[9.5] 31.

x	y	(x, y)
0	−4	(0, −4)
1	−3	(1, −3)
2	−2	(2, −2)

32. (a) Answers will vary; all points on the line are solutions. Some possibilities are (3, −1) and (4, 0). **(b)** positive slope

Cumulative Review Exercises: Chapters 1–9 (page 699)

[1.1, 5.1] 1. (a) six hundred two ten-thousandths **(b)** three hundred million, five hundred sixty **2. (a)** 70,005,000,043 **(b)** 18.09 **[1.5, 5.2] 3. (a)** 3.0 **(b)** 0.80 **(c)** 69,000,000 **[5.7] 4.** Mean is 29 years; median is 25 years; mode is 23 years.

[4.5] 5. $3\frac{2}{15}$ **[5.5] 6.** −0.25 **[1.8] 7.** −39 **[5.4] 8.** −0.00010 or −0.0001 **[4.3] 9.** $\frac{4b}{3a^2}$ **[4.5] 10.** $-2\frac{7}{12}$ **[5.3] 11.** −10.007 **[4.4] 12.** $\frac{2x-56}{7x}$ **[1.4] 13.** −36 **[4.3] 14.** $\frac{25}{3mn}$ **[4.6] 15.** $3\frac{1}{3}$ **[1.8] 16.** undefined **[4.5] 17.** $7\frac{3}{8}$ **[4.4] 18.** $\frac{18+wx}{6w}$ **[1.8] 19.** −4 **[4.3] 20.** 840 **[1.8] 21.** $\frac{0}{-2} = 0$ **[5.5] 22.** −2.06 **[4.6] 23.** $-\frac{7}{20}$ **[2.1] 24.** 0 **25.** 41 **26.** −100 **[2.2] 27.** $x^2 - 5x$ **28.** $5y - 12$ **29.** $-40h^2$ **[2.5] 30.** $x = -7$ **31.** $y = -6$ **32.** $x = 2$ **[4.7] 33.** $m = -15$ **[6.3] 34.** $n = 16.2$ **[2.4] 35.** $w = 20$ **[5.9] 36.** $x = 7$ **[2.5] 37.** $h = -9$ **[3.3] 38.** $12 - 5n = n$; $n = 2$ **39.** $-8 + 2n = -28$; $n = -10$ **[3.4] 40.** Let m be the smaller amount; $m + 500$ the larger amount. $m + m + 500 = 1800$. The two amounts are $650 and $1150. **41.** Let w be the width; $3w$ the length. $2(3w) + 2w = 280$. The width is 35 ft; length is 105 ft. **[3.1, 3.2] 42.** $P = 36$ ft; $A = 70$ ft^2 **43.** $P = 188$ m; $A = 1636$ m^2 **[4.8, 5.8] 44.** $x = 25$ mi; $P = 56$ mi; $A = 84$ mi^2 **[5.10] 45.** $C \approx 18.8$ ft; $A \approx 28.3$ ft^2 **46.** $V \approx 3768$ m^3 **[4.8] 47.** $V = 30$ yd^3 **[8.1] 48.** 135 minutes **49.** 4.5 ft or $4\frac{1}{2}$ ft **[8.2, 8.3] 50.** 1850 mL **51.** 3.5 cm **52.** 0.01 kg **[8.5] 53.** −4 °C (rounded) **[6.4] 54.** 556 nonsmoking employees (rounded) **[7.5] 55.** $116.78 (rounded) **[7.4] 56.** 92.8% (rounded) **[1.7] 57. (a)** 31 computers; $333 left over **(b)** 17 calculators; $10 left over **[4.3] 58.** $1\frac{5}{12}$ pounds

[9.4] 59.

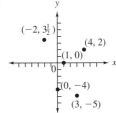

$\left(-2, 3\frac{1}{2}\right)$ is in second quadrant; none in third quadrant.

[9.5] 60.

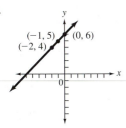

Positive slope; all points on the line are solutions, so there are many possibilities such as $(-3, 3)$ and $(-4, 2)$.

CHAPTER 10

Section 10.1 (page 709)

1. 1 **3.** false; $3^3 = 27$ **5.** false; $(a^2)^3 = a^6$ **7.** $(-2)^5$ **9.** $\left(\frac{1}{2}\right)^6$
11. $(-8p)^2$ **13.** The expression $(-3)^4 = (-3)(-3)(-3)(-3) = 81$, while $-3^4 = -(3 \cdot 3 \cdot 3 \cdot 3) = -81$. **15.** Base is 3; exponent is 5; 243. **17.** Base is -3; exponent is 5; -243. **19.** Base is $-6x$; exponent is 4. **21.** Base is x; exponent is 4. **23.** The product rule does not apply to $5^2 + 5^3$ because it is a *sum*, not a product. $5^2 + 5^3 = 25 + 125 = 150$ **25.** 5^8 **27.** 4^{12} **29.** $(-7)^9$ **31.** t^{17}
33. $-56r^7$ **35.** $42p^{10}$ **37.** $5x^4 + 9x^4 = 14x^4$; $(5x^4)(9x^4) = 45x^8$
39. $-7a^2 + 2a^2 + 10a^2 = 5a^2$; $(-7a^2)(2a^2)(10a^2) = -140a^6$
41. 4^6 **43.** t^{20} **45.** 7^3r^3 **47.** $5^5x^5y^5$ **49.** $8q^3r^3$ **51.** $\frac{1^3}{2^3}$ **53.** $\frac{a^3}{b^3}$
55. $\frac{9^8}{5^8}$ **57.** $(-2)^3x^6y^3$ **59.** $3^2a^6b^4$ **61.** $A = (4x^3)(3x^2) = 12x^5$

Section 10.2 (page 715)

1. negative **3.** negative **5.** positive **7.** 0 **9.** 1 **11.** -1
13. 0 **15.** 0 **17.** 2 **19.** 1 **21.** 15 **23.** $\frac{1}{64}$ **25.** $\frac{8}{15}$ **27.** $\frac{1}{x^4}$
29. 1 **30.** $\frac{5^2}{5^2}$ **31.** 5^0 **32.** $5^0 = 1$; This supports the definition of an exponent of 0. **33.** 6^5 **35.** 10^3 **37.** $\frac{1}{y^4}$ **39.** c^1 or c
41. $\frac{1}{5^4}$ **43.** $\frac{1}{m}$ **45.** 3^4 **47.** a^3 **49.** $\frac{1}{2^8}$ **51.** $\frac{1}{r^4}$ **53.** 10^{10}
55. $\frac{1}{10^5}$ **57.** 10^{11} **59.** $\frac{1}{10^8}$ **61.** 10^4 **63.** $\frac{1}{a^1}$ or $\frac{1}{a}$ **65.** $\frac{1}{2^8}$
67. $\frac{1}{x^{11}}$ **69.** 10^2 **71.** $\frac{1}{y^2}$ **73.** m^9 **75.** $\frac{1}{10^6}$

Summary Exercises on Exponents (page 717)

1. 9 **2.** 150 **3.** 1 **4.** $-\frac{4}{45}$ **5.** 1 **6.** $\frac{1}{64}$ **7.** -9 **8.** 72
9. -8 **10.** -1 **11.** $\frac{5}{12}$ **12.** 9 **13.** $\frac{1}{32}$ **14.** 1 **15.** -8
16. 1 **17.** 6^8 **18.** $\frac{1}{x^7}$ **19.** $\frac{1}{10^8}$ **20.** 2^9 **21.** $y^0 = 1$

22. $192m^{10}n^5$ **23.** $(-5)^{10}$ **24.** w^3 **25.** m^{24} **26.** $\frac{1}{n^2}$ **27.** $\frac{1}{10^6}$
28. d^9 **29.** $16x^8$ **30.** $\frac{1}{c}$ **31.** 10^8 **32.** 4^5 **33.** $\frac{x^6}{y^6}$ **34.** 3^{11}
35. 7^1 or 7 **36.** h^1 or h **37.** $-16a^4$ **38.** $\frac{1^4}{2^4}$ or $\frac{1}{16}$
39. $5^3x^3y^3$ or $125x^3y^3$ **40.** m^5 **41.** $\frac{1}{10^7}$ **42.** $\frac{1}{b^1}$ or $\frac{1}{b}$
43. b^{15} **44.** $\frac{1}{5^7}$ **45.** $\frac{1}{10^3}$ **46.** $21y^8$ **47.** 10^{13} **48.** $\frac{1}{k^8}$
49. $10a^4b^4$ **50.** $\frac{1}{x^9}$

Section 10.3 (page 723)

1. 2.1×10^{-3}; 3.1783×10^2 **3.** 8.312×10^8; 3.19×10^8
5. in scientific notation **7.** not in scientific notation; 5.6×10^6
9. not in scientific notation; 4×10^{-3} **11.** not in scientific notation; 8×10^1 **13.** To write a number in scientific notation $(a \times 10^n)$ move the decimal point to the right of the first nonzero digit. The number of places moved is the absolute value of the exponent n. If the original number is "large," then n is positive. If the original number is "small," then n is negative. **15.** 5.876×10^9
17. 8.235×10^4 **19.** 7×10^{-6} **21.** 2.03×10^{-3}
23. 750,000 **25.** 5,677,000,000,000 **27.** 6.21 **29.** 0.00078
31. 0.000000005134 **33.** 6×10^{11}; 600,000,000,000
35. 1.5×10^7; 15,000,000 **37.** 6.426×10^4; 64,260
39. 3×10^{-4}; 0.0003 **41.** 4×10^1; 40 **43.** 2.6×10^{-3}; 0.0026
45. 3.52×10^2 (rounded) or 352 people per square mile
47. **(a)** 1×10^{-4} or 0.0001 meter **(b)** 1×10^{-1} or 0.1 meter
49. 6.94×10^1 (rounded) or about 69 items

Section 10.4 (page 729)

1. 7; 5 **3.** 8 **5.** 26 **7.** one term; 6 **9.** one term; 1 **11.** two terms; -19, -1 **13.** three terms; 1, 8, -8 **15.** $2m^5$ **17.** $-r^5$
19. already simplified and written in descending powers **21.** $-5x^5$
23. $5p^9 + 4p^7$ **25.** $-2y^2$ **27.** already simplified and written in descending powers; degree 4; binomial **29.** already simplified; write in descending powers as $6m^5 + 5m^4 - 7m^3 - 3m^2$; degree 5; none of these **31.** $x^4 + \frac{1}{3}x^2 - 4$; degree 4; trinomial
33. $7x$; degree 1; monomial **35.** **(a)** -1 **(b)** 5 **37.** **(a)** 19 **(b)** -2 **39.** **(a)** 36 **(b)** -12 **41.** **(a)** -124 **(b)** 5
43. $5m^2 + 3m$ **45.** $-6x^3 + x + 6$ **47.** $3w^3 + 5$
49. $4x^4 - 4x^2$ **51.** $8r^2 + 5r - 12$ **53.** $5m^2 - 14m + 6$
55. $4x^3 + 2x^2 + 5x$ **57.** $12m^3 - 13m^2 + 6m + 11$
59. $-11x^2 - 3x - 3$ **61.** $8x^2 + 8x + 6$ **63.** $8t^2 + 8t + 13$
65. $D = 100(1) - 13(1)^2$; $D = 100 - 13$; $D = 87$ ft **66.** 0, (0, 0); 47, (0.5, 47); 121, (1.5, 121); 148, (2, 148); 169, (2.5, 169); 183, (3, 183); 191, (3.5, 191)
67.

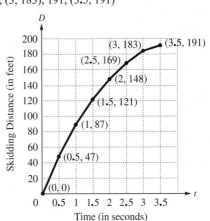

68. Answers will vary. This graph is a curve instead of the straight lines graphed for linear equations. Also, the rate of change in the distance is not consistent. From 0.5 sec to 1 sec, the distance increases 40 ft, but from 3 sec to 3.5 sec, the distance only increases 8 ft.

Section 10.5 (page 735)

1. $40y^3 - 15y$ **3.** $-6m^2 - 4m$ **5.** $24x^4 - 12x^3 + 8x^2$
7. $-6k^6 + 9k^5 + 3k^4 - 3k^3$ **9.** $n^2 + n - 6$ **11.** $8r^2 - 10r - 3$
13. $12x^3 + 26x^2 + 10x + 1$ **15.** $20m^4 - m^3 - 8m^2 - 17m - 15$
17. $5x^4 - 13x^3 + 20x^2 + 7x + 5$ **19.** $x^2 + 7x + 12$
21. $2x^3 + 7x^2 + 3x$ **23.** D; B; A; C **25.** $(30x + 60)$ yd^2
26. $30x + 60 = 600; x = 18$ **27.** 10 yd by 60 yd **28.** $2100
29. 140 yd **30.** $1260 **31.** The answers are $x^2 - 16$, $y^2 - 9$, and $r^2 - 49$. Each product is the difference of the square of the first term and the square of the last term from the binomial.

Chapter 10 Review Exercises (page 741)

1. 4^{11} **2.** $(-5)^{11}$ **3.** $-72x^7$ **4.** $10x^{14}$ **5.** 19^5x^5 **6.** $(-4)^7y^7$
7. $5p^4t^4$ **8.** $\dfrac{7^6}{5^6}$ **9.** $3^3x^6y^9$ **10.** The product rule for exponents

does not apply because it is the *sum* of 7^2 and 7^4, not their *product*.

11. 2 **12.** $\dfrac{1}{32}$ **13.** 10 **14.** $\dfrac{5}{16}$ **15.** 6^2 **16.** x^2 **17.** $\dfrac{1}{p^{12}}$

18. r^4 **19.** $\dfrac{1}{5}$ **20.** $\dfrac{1}{10^7}$ **21.** $\dfrac{1}{n^3}$ **22.** 10^3 **23.** $\dfrac{1}{2^8}$ **24.** h^3

25. $\dfrac{1}{10^{13}}$ **26.** x^6 **27.** 4.8×10^7 **28.** 2.8988×10^{10}

29. 6.5×10^{-5} **30.** 8.24×10^{-8} **31.** 24,000 **32.** 78,300,000
33. 0.000000897 **34.** 0.00000000000995 **35.** 8×10^2; 800
36. 4×10^6; 4,000,000 **37.** 3×10^{-5}; 0.00003 **38.** 1×10^{-2}; 0.01
39. 8.1×10^1 (rounded) or about 81 people per square mile
40. 1.08×10^7 or 10,800,000 kilometers **41.** $22m^2$; degree 2; monomial **42.** $p^3 - p^2 + 4p + 2$; degree 3; none of these
43. already simplified and written in descending powers; degree 5; none of these **44.** $-8y^5 - 7y^4 + 9y$; degree 5; trinomial
45. $-5a^3 + 4a^2$ **46.** $-8r^2 + 6r$ **47.** $11y^2 - 10y + 9$
48. $10m^3 - 6m^2 - 3$ **49.** $4y^3 - 5y^2 - 4y + 26$
50. $10p^2 - 3p - 11$ **51.** $7r^4 - 4r^3 - 1$ **52.** $10x^2 + 70x$
53. $-6p^5 + 15p^4$ **54.** $6r^3 + 8r^2 - 17r + 6$ **55.** $8y^3 + 27$
56. $5p^5 - 2p^4 - 3p^3 + 25p^2 + 15p$
57. $6k^2 - 9k - 6$ **58.** $12p^2 - 48pq + 21q^2$
59. $2m^4 + 5m^3 - 16m^2 - 28m + 9$ **60.** 0 **61.** 3^4p^4 or $81p^4$
62. $\dfrac{1}{7^2}$ or $\dfrac{1}{49}$ **63.** $-8m^7 - 10m^6 - 6m^5$ **64.** $\dfrac{3}{4}$
65. $a^3 - 2a^2 - 7a + 2$ **66.** $8y^3 - 9y^2 + 5$
67. $P = 6x - 2$; $A = 2x^2 + x - 6$

Chapter 10 Test (page 745)

[10.1] 1. $\dfrac{1}{625}$ **[10.2] 2.** 2 **3.** $\dfrac{7}{12}$ **[10.1] 4.** 6^1 or 6 **5.** $12x^3y^3$

[10.2] 6. $\dfrac{1}{10^4}$ **7.** $\dfrac{1}{r^5}$ **8.** $-21x^7$ **9.** m^3 **[10.1] 10.** $3^2a^8b^2$ or

$9a^8b^2$ **11.** Cannot use product rule; $81 + 9 = 90$
[10.3] 12. 3.0×10^{11} **13.** 5.57×10^{-6} **14.** 29,000,000
15. 0.0000000607 **16.** 1.97×10^{30} kg (rounded)
[10.4] 17. $-7x^2 + 8x$; degree 2; binomial **18.** $4n^4 + 13n^3 - 10n^2$; degree 4; trinomial **19.** $4t^4 - 2t^2$ **20.** $2y^4 - 3y^2 - 6y$
21. $-12t^2 + 8$ **[10.5] 22.** $-12x^5 + 20x^4$ **23.** $3y^2 + 5y - 2$
24. $2r^3 + r^2 - 16r + 15$ **[10.4, 10.5] 25.** $P = 12x + 36$;
$A = 9x^2 + 54x + 81$

Cumulative Review Exercises: Chapters 1–10 (page 747)

[1.1, 5.1] 1. (a) ten and thirty-five thousandths **(b)** four hundred ten billion, three hundred fifty-one thousand, one hundred nine
2. (a) 300,006,080 **(b)** 0.0055 **[1.5, 5.2] 3. (a)** 0.80
(b) 341,000,000 **(c)** 15.0 **[5.7] 4.** Mean is $52.70; median is $44.50; mode is $41. **[5.3] 5.** -4.171 **[1.8] 6.** -18

[4.4] 7. $2\dfrac{4}{15}$ **[1.8] 8.** $\dfrac{0}{-14} = 0$ **[4.3] 9.** $\dfrac{16w}{3x}$ **[5.5] 10.** -0.125

[10.2] 11. $\dfrac{1}{10}$ **[4.5] 12.** $-\dfrac{7}{12}$ **[1.8] 13.** 28 **[10.1] 14.** $-12y^7$

[4.3] 15. $\dfrac{25}{3ab}$ **[4.4] 16.** $\dfrac{rt + 48}{8t}$ **[1.8] 17.** undefined

[4.3] 18. 750 **[5.4] 19.** -0.00012 **[10.1] 20.** $2^4a^{12}b^8$ or $16a^{12}b^8$

[4.4] 21. $\dfrac{4m - 54}{9m}$ **[10.2] 22.** $\dfrac{1}{t}$ **[4.6] 23.** $\dfrac{5}{32}$ **24.** $4\dfrac{1}{16}$

[10.2] 25. n^1 or n **[1.4] 26.** -96 **[10.2] 27.** $1\dfrac{1}{2}$ **[4.5] 28.** $1\dfrac{1}{5}$

[2.1] 29. 12 **30.** -3 **31.** 30 **[2.5] 32.** $h = 5$
[5.9] 33. $x = -7.9$ **[2.5] 34.** $a = 6$ **[3.3] 35.** $20 - 3n = 2n$;
$n = 4$ **[3.4] 36.** Let b be number of elephant breaths.
$b + 16b = 170$; $b = 10$; Elephant takes 10 breaths; mouse takes 160.
[6.4] 37. 3 runs **[7.4] 38.** 145% **[7.5] 39.** $4.87 (rounded);
$69.82 **[10.3] 40.** 1.61×10^{10} (rounded) or 16,100,000,000 miles
[4.4, 9.1] 41. (a) $5\dfrac{1}{8}$ in. **(b)** size 7 **(c)** $\dfrac{5}{16}$ in.

[3.1, 3.2, 4.8] 42. $P = 22.4$ m; $A = 31.6$ m^2
[5.10] 43. $C \approx 28.3$ cm; $A \approx 63.6$ cm^2

WHOLE NUMBERS COMPUTATION: PRETEST

(page 749)
Adding Whole Numbers
1. 390 **2.** 13,166 **3.** 3053 **4.** 117,510 **5.** 688,226

Subtracting Whole Numbers
1. 350 **2.** 629 **3.** 24,894 **4.** 3909 **5.** 591,387

Multiplying Whole Numbers
1. 0 **2.** 26,887 **3.** 1,560,000 **4.** 1846 **5.** 17,232
6. 519,477

Dividing Whole Numbers
1. 23 **2.** undefined **3.** 6259 **4.** 807 R6 **5.** 34 **6.** 50
7. 60 R20 **8.** 539 R62

CHAPTER R

Section R.1 (page 757)
1. (a) 23 **(b)** 19 **3. (a)** 8928 **(b)** 59,224 **5.** 150 **7.** 1651
9. 1004 **11.** 9253 **13.** 11,624 **15.** 16,658 **17.** 5009
19. 3506 **21.** 15,954 **23.** 10,648 **25.** incorrect; should be 769
27. correct **29.** 33 miles **31.** 38 miles **33.** $122
35. 699 people **37.** 54,723,468 people **39.** 294 inches
41. 708 feet

Section R.2 (page 765)
1. incorrect; should be 62 **3.** incorrect; should be 121 **5.** 8
7. 25 **9.** 16 **11.** 61 **13.** 519 **15.** 9177 **17.** 7589
19. 8859 **21.** 3 **23.** 23 **25.** 1942 **27.** 5687 **29.** 19,038
31. 65,556 **33.** 19,984 **35.** incorrect; should be 2494
37. correct **39.** 15 calories **41.** 121 passengers

43. **(a)** Highest is electrical engineer; lowest is photographer.
(b) $42,560 **45.** 365 feet

Section R.3 (page 773)

1. 9 **3.** 63 **5.** 0 **7.** 24 **9.** 36 **11.** 245 **13.** 168
15. 19,092 **17.** 258,447 **19.** 12,500 **21.** 44,550 **23.** 270,000
25. 86,000,000 **27.** 1496 **29.** 3735 **31.** 15,200 **33.** 32,805
35. 183,996 **37.** 3,321,934 **39.** 18,980,480
41. 252 inches; 540 inches **43.** 576 plants **45.** 605 miles
47. 2440 miles; 16,170 miles **49.** 7406 calories

Section R.4 (page 783)

1. 1; $12\overline{)12}$; $12 \div 12$ **3.** undefined; $\frac{24}{0}$; $0\overline{)24}$ **5.** 0; $4\overline{)0}$; $0 \div 4$

7. 0; $\frac{0}{12}$; $12\overline{)0}$ **9.** undefined; $\frac{21}{0}$; $21 \div 0$ **11.** 21; dividend: 84;

divisor: 4; quotient: 21. **13.** 213; dividend: 639; divisor: 3;
quotient: 213 **15.** 1522 **R**5 **17.** 309 **R**4 **19.** 1006 **21.** 5006 **R**1
23. 6671 **25.** 12,458 **R**3 **27.** 10,253 **R**5 **29.** 18,377 **R**6
31. incorrect; should be 670 **R**2 **33.** incorrect; should be 3568 **R**1
35. correct **37.** correct **39.** $23; $14; $19 **41.** $16,600
43. 378 $5 tickets; 270 $7 tickets; 210 $9 tickets **45.** $7; $210
47. **(a)** 358; 190 **(b)** 10,404; 78 **(c)** 13,740; 985 **49.** 2, 3, 5, 10
51. 2 **53.** 5 **55.** 3 **57.** 2, 3 **59.** none of the above

Section R.5 (page 789)

1. $24\overline{)768}$ $\quad$ $\frac{3}{}$ use 2 as the trial divisor. **3.** $18\overline{)4500}$ $\frac{2}{}$ use 2 as the trial divisor.
5. $86\overline{)10,327}$ $\frac{1}{}$ use 9 as the trial divisor. **7.** $52\overline{)38,025}$ $\frac{7}{}$ use 5 as the trial divisor.
9. $77\overline{)249,826}$ $\frac{3}{}$ use 8 as the trial divisor. **11.** $420\overline{)470,800}$ $\frac{1}{}$ use 4 as the trial divisor.
13. 64 **R**3 **15.** 236 **R**29 **17.** 2407 **R**1 **19.** 1239 **R**15
21. 3331 **R**110 **23.** 850 **25.** incorrect; should be 106 **R**7
27. incorrect; should be 658 **29.** incorrect; should be 62
31. correct **33.** 1050 hours **35.** $1372 **37.** $308
39. 1680 circuits **41.** $375

Chapter R Review Exercises (page 793)

1. 92 **2.** 113 **3.** 5464 **4.** 15,657 **5.** 39 **6.** 184 **7.** 224
8. 25,866 **9.** 48 **10.** 45 **11.** 48 **12.** 0 **13.** 172 **14.** 5467
15. 32,640 **16.** 465,525 **17.** 19,200 **18.** 25,200 **19.** 206,800
20. 128,000,000 **21.** 612 **22.** 1872 **23.** 13,755 **24.** 30,184
25. 6 **26.** 1 **27.** undefined **28.** 0 **29.** 108 **30.** 24 **31.** 352
32. 150 **R**4 **33.** 104 cards; 312 cards **34.** 2856 textbooks
35. $12 over his budget **36.** $227,940 **37.** 10,000 hours;
24,000 hours **38.** 240 miles; 560 miles **39.** 23 acres
40. 32 homes **41.** $810 **42.** $143 **43.** $663 **44.** $1261

Chapter R Test (page 795)

[R.1] **1.** addends; sum or total [R.3] **2.** factors; product
[R.2, R.4] **3.** difference; quotient [R.1] **4.** 8610 **5.** 112,630
[R.2] **6.** 16,948 **7.** 3084 [R.3] **8.** 140 **9.** 171,000
10. 1785 **11.** 4,450,743 [R.4] **12.** 206 **13.** undefined
[R.5] **14.** 458 **15.** 160 [R.1–R.5] **16.** $528 **17.** $291
18. 35 hours **19.** $165 **20.** 1028 ovens **21.** $2040
[R.4] **22.** A number is divisible by 2 if it ends in 0, 2, 4, 6, or 8.
A number is divisible by 5 if it ends in 0 or 5. A number is divisible
by 10 if it ends in 0. Examples will vary.

Appendix B Exercises (page A-39)

1. 37; add 7 **3.** 26; add 10, subtract 2 **5.** 16; multiply by 2
7. 243; multiply by 3 **9.** 36; add 3, add 5, add 7, add 9, etc.;
or 1^2, 2^2, 3^2, 4^2, etc.

11. ⌐ **13.** ⌐

15. Conclusion follows; it is valid.

17. Conclusion does not follow; it is invalid.

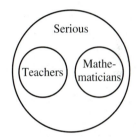

19. **21.** Dick
Together
2 + 18 + 7 = 27
30 − 27 = 3
Neither watched TV on 3 days.

Index